Lecture Notes in Mathematics

Edited by A. Dold and B. Eckmann

Series: Australian National University, Canberra
Advisers: L. G. Kovács, B. H. Neumann and

372

Proceedings of the Second International Conference on The Theory of Groups

Australian National University, August 13–24, 1973

Edited by M. F. Newman

Springer-Verlag
Berlin · Heidelberg · New York 1974

M. F. Newman
Dept. of Mathematics
Australian National University
Institute of Advanced Studies
Canberra, ACT 2600 / Australia

Library of Congress Cataloging in Publication Data

International Conference on the Theory of Groups, 2d,
Australian National University, 1973.
Proceedings.

(Lecture notes in mathematics (Berlin), 372)
1. Groups, Theory of--Congresses. I. Newman, Michael Frederick, 1934- ed. II. Series.
QA3.L28 no. 372 [QA171] 512'.22 74-13872

AMS Subject Classifications (1970): 20xx, 22xx

ISBN 3-540-06845-7 Springer-Verlag Berlin · Heidelberg · New York
ISBN 0-387-06845-7 Springer-Verlag New York · Heidelberg · Berlin

Offsetdruck: Julius Beltz, Hemsbach/Bergstr.

PREFACE

This volume consists of papers presented to the Second International Conference on the Theory of Groups held in Canberra in August 1973 together with a report by the chairman of the Organizing Committee and a collection of problems.

The manuscripts were typed by Mrs Geary, the bulk of the bibliographic work was done by Mrs Pinkerton, and a number of colleagues helped with proof-reading; Professor Neumann, Drs Cossey, Kovács, McDougall, Praeger, Pride, Rangaswamy and Stewart. I here record my thanks to all these people for their lightening of the editorial burden.

M.F. Newman

CONTENTS

INTRODUCTION

Report on the Second International Conference on the Theory of Groups

The Second International Conference on the Theory of Groups at the Australian National University took place 8 years after the first one, from 13 to 24 August 1973. It was again sponsored by the International Mathematical Union, the Australian Academy of Science, and the Australian National University. There were 127 registered participants (38 more than in 1965), of whom 52 (7 more) came from outside Australia, and represented 13 countries (1 more): Canada, Federal Republic of Germany, India, Israel, Japan, Lesotho, New Zealand, Papua New Guinea, Singapore, Switzerland, USSR, United Kingdom, USA.

The scientific programme of the conference followed the pattern of the first conference. There were 20 invited one-hour lectures, given (in chronological order) by: W. Magnus (USA), G. Higman (UK), J. McKay (Canada), D.G. Higman (USA), R. Baer (Switzerland), P.M. Neumann (UK), G. Baumslag (USA), R.C. Lyndon (USA), W.W. Boone (USA), R.A. Bryce (Australia), R. Brauer (USA), O.H. Kegel (UK), P.E. Schupp (USA), P.J. Cossey (Australia), S.I. Adyan (USSR), M. Herzog (Israel), A.I. Kostrikin (USSR), E.E. Shult (USA), M.I. Kargapolov (USSR), B.H. Neumann (Australia). In addition there were numerous informal sessions, at which shorter, contributed papers were presented and discussed. Much of the life of the conference pulsed in the informal gatherings in the various common rooms of Bruce Hall, where participants from outside Canberra were again accommodated: mathematics will owe a debt to the coffee facilities in the library of Bruce Hall and to the adjoining bar, specially kept open late for the convenience and refreshment of participants.

There were again recreational activities of various kinds: a sherry party, a conference dinner, a tour to a sheep station, a climb to the top of Pidgeon House Mountain (some 150 road kilometers east of Canberra - 3 of our cars got stuck trying to ford the Clyde River, and had to be rescued). Private hospitality and entertainment was provided, too, and again the diplomatic missions in Canberra played their part.

The value of such a conference can not be assessed: the consensus of the participants put it very high. It is also nearly impossible to estimate the cost; while many costs had risen steeply since 1965, air fares, which take a large slice of the expenses of overseas visitors, had not, and in many cases had, in fact, been reduced. The Australian National University again contributed a very large amount, which is not easily computed, but can be roughly estimated at $A9000. The International Mathematical Union contributed $US7000, and the Australian Government, through the Australian Academy of Science, $A5000. Industry in Australia was not approached for contributions, but generous contributions were, nevertheless, received, both of money and of computer time on large, fast, expensive computers. IBM World Trade Corporation donated $US1000, IBM Australia Ltd gave $A500 and much valuable computer time at their Systems Development Institute in Canberra, and Control Data Australia Pty Ltd also donated much valuable computer time on their CDC 6600. Numerous contributions were made by overseas governments, academies, universities both in Australia and elsewhere, and other learned institutions in the form of travel grants to some participants or payment of their registration fees. Several Australian universities made visiting positions available to participants.

To make use of the presence of numerous overseas experts, the Second Australian Conference on Combinatorial Mathematics was convened for the weekend immediately after the group theory conference; it took place at the University of Melbourne on 25 and 26 August 1973.

A list of the Organizing and Advisory Committees is appended, but they do not exhaust the crowd of helpers who have earned our gratitude. My warm thanks go to all those who have helped to make the conference possible, to make it real, and to make it memorable.

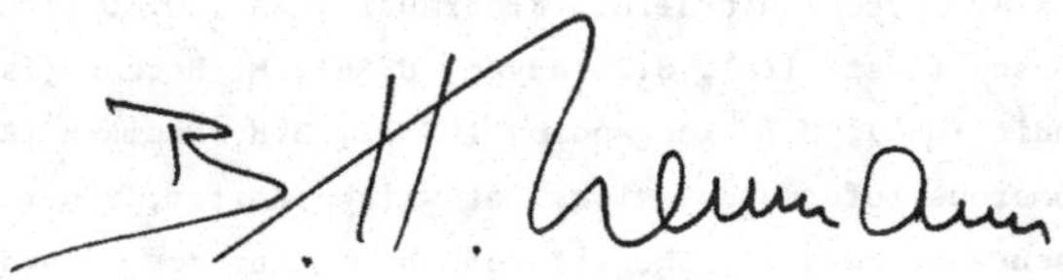

APPENDIX A

The Organising and the Advisory Committees of the Conference

Professor A. Adrian Albert (a) (c) (died 1972-06-06)
Professor A.J. Birch (a) (d) (to April 1973)
Dr R.M. Bryant (b)
Dr R.A. Bryce (b)
Dr C. Christensen (b)
Dr P.J. Cossey (a) (b)
Dr B.J. Gardner (b)
Professor Nathan Jacobson (a) (c)
Dr L.G. Kovács (a) (b)
Dr D. McDougall (b)
Professor B.H. Neumann (a) (b)
Professor Hanna Neumann (a) (b) (died 1971-11-14)
Dr M.F. Newman (a) (b)
Professor G.J.V. Nossal (a) (d) (from April 1973)
Dr K.M. Rangaswamy (b)
Professor John G. Thompson (a) (c)
Professor G.E. Wall (a) (b)
Dr M.A. Ward (b)

(a) Member of Organising Committee
(b) Member of Advisory Committee
(c) Representing the International Mathematical Union
(d) Treasurer of the Australian Academy of Science

Mr P.D. O'Connor, Assistant Secretary of the Australian Academy of Science, acted as Secretary of the Organizing Committee and also assisted the Advisory Committee.

APPENDIX B

List of participants

Mr R.G. Addie, Postmaster General's Department, Melbourne
Professor S.I. Adyan, Akad. Nauk SSSR, Moscow, USSR
Dr Joan Aldous, Open University, UK
Mr W.A. Alford, Australian National University
Professor S. Bachmuth, University of California, Santa Barbara, USA
Professor R. Baer, Eidgenössische Technische Hochschule, Switzerland
Dr D.W. Barnes, University of Sydney
Professor G. Baumslag, Rice University, USA
Professor W.W. Boone, University of Illinois at Urbana-Champaign, USA
Mr J.R. Bowker, University of Queensland
Professor R. Brauer, Harvard University, USA
Professor W. Brisley, University of Newcastle, NSW
Dr M.S. Brooks, Canberra College of Advanced Education
Mr A.M. Brunner, Australian National University
Dr R.M. Bryant, University of Manchester Institute of Science and Technology, UK
Dr R.A. Bryce, Australian National University
Dr R.G. Burns, York University, Canada
Dr K.K.-H. Butler, Pembroke State University, USA
Dr M.C.R. Butler, University of Liverpool, UK
Dr J.J. Cannon, University of Sydney
Dr C. Christensen, Australian National University
Dr S.B. Conlon, University of Sydney
Dr C.D.H. Cooper, Macquarie University
Mr W.A. Coppel, Australian National University
Dr P.J. Cossey, Australian National University
Dr I.M.S. Dey, Open University, UK
Professor J.D. Dixon, Carleton University, Canada
Professor V. Dlab, Carleton University, Canada
Dr G.B. Elkington, University of Sydney
Dr H. Enomoto, Tokyo University, Japan
Mr C.D. Fox, Australian National University
Professor J.S. Frame, Michigan State University, USA
Ms Mary R. Freislich, University of New South Wales
Dr T.M. Gagen, University of Sydney
Dr B.J. Gardner, University of Tasmania
Mr D.J. Glover, Australian National University
Mr L.M. Gordon, University of Sydney
Dr J.R.J. Groves, University of Melbourne
Mrs Susan C. Groves, Australian National University
Mr J. Grunau, Australian National University
Professor N.D. Gupta, University of Manitoba, Canada
Dr V.W.D. Hale, University of York, UK
Mr L.F. Harris, Australian National University
Mr G. Havas, Canberra College of Advanced Education
Professor H. Heineken, Universität Würzburg, Federal Republic of Germany
Mr G.J. Hemion, Australian National University
Professor M. Herzog, Tel-Aviv University, Israel

Mr L.J. Hesterman, Australian National University
Professor D.G. Higman, University of Michigan, USA
Professor G. Higman, University of Oxford, UK
Professor K.A. Hirsch, Queen Mary College, University of London, UK
Miss Kathryn J. Horadam, Australian National University
Mr R.B. Howlett, University of Adelaide
Dr D.C. Hunt, University of New South Wales
Mr R.H. Hunter, Australian National University
Mr M.F. Hutchinson, University of Sydney
Dr S.A. Huq, Australian National University
Dr Y.L. Ilamed, Soreq Research Institute, Yavneh, Israel
Dr G. Ivanov, University of Melbourne
Mrs Masako Izumi, Australian National University
Dr S. Izumi, Australian National University
Mr D.R. Jackett, Australian National University
Dr R.K. James, University of New South Wales
Professor J.N. Kapur, Meerut University, India
Professor M.I. Kargapolov, Akad. Nauk SSSR, Sibirsk. Otd., Novosibirsk, USSR
Dr J. Kautsky, Flinders University of South Australia
Professor O.H. Kegel, Queen Mary College, University of London, UK
Mr B.W. King, Riverina College of Advanced Education
Professor A.I. Kostrikin, Moskovsk. Gos. Universitet, Moscow, USSR
Dr L.G. Kovács, Australian National University
Dr H. Lausch, Monash University
Mr R.H. Levingston, University of Sydney
Professor P.J. Lorimer, University of Auckland, New Zealand
Professor R.C. Lyndon, University of Michigan, USA
Dr D.J. McCaughan, University of Otago, New Zealand
Dr D. McDougall, Australian National University
Professor J. McKay, McGill University, Canada
Dr J.R. McMullen, University of Sydney
Dr J.M. McPherson, Australian National University
Dr Sheila O. Macdonald, University of Queensland
Professor W. Magnus, New York University, USA
Professor E.J. Mayland, York University, Canada
Professor S.A. Meskin, University of Connecticut, USA
Professor H.Y. Mochizuki, University of California, Santa Barbara, USA
Dr I.J. Mohamed, University of Botswana, Lesotho and Swaziland, Lesotho
Mr R.A.R. Monzo, Australian National University
Miss Elizabeth J. Morgan, University of Queensland
Dr S.A. Morris, University of New South Wales
Mr B. Morrison, University of Newcastle, NSW
Mr B.G. Neil, University of Queensland
Professor B.H. Neumann, Australian National University
Dr P.M. Neumann, University of Oxford, UK
Dr B.B. Newman, James Cook University of North Queensland
Dr M.F. Newman, Australian National University
Professor E.T. Ordman, University of Kentucky, USA
Dr M.F. O'Reilly, University of Papua and New Guinea, Papua New Guinea
Sr Elizabeth A. Ormerod, Australian National University
Dr I.B.S. Passi, Kurukshetra University, India
Professor Sophie Piccard, Université de Neuchâtel, Switzerland
Dr Cheryl E. Praeger, Australian National University
Mr S.J. Pride, Australian National University
Mr A.J. Rahilly, University of Sydney
Dr K.M. Rangaswamy, Madurai University, India, and Australian National University
Dr G. Rosenberger, Universität Hamburg, Federal Republic of Germany
Professor R. Schmidt, Universität Kiel, Federal Republic of Germany
Professor P.E. Schupp, University of Illinois at Urbana-Champaign, USA
Mr P.E. Sheridan, Monash University
Mr D.B. Shield, Australian National University
Professor E.E. Shult, University of Florida, USA

Dr H.L. Silcock, Australian National University
Dr N.F. Smythe, Australian National University
Mr J.B. Southcott, University of Queensland
Dr D.E. Taylor, La Trobe University
Professor H.F. Trotter, Princeton University, USA
Dr P.G. Trotter, University of Tasmania
Professor G.E. Wall, University of Sydney
Dr Jennifer R.S. Wallis, Australian National University
Dr W.D. Wallis, University of Newcastle, NSW
Dr J.W. Wamsley, Flinders University of South Australia
Dr J.N. Ward, University of Sydney
Dr M.A. Ward, Australian National University
Professor P.M. Weichsel, University of Illinois at Urbana-Champaign, USA
Dr M.J. Wicks, University of Singapore, Singapore
Dr J. Wiegold, University College of South Wales and Monmouthshire, UK
Dr J.S. Wilson, University of Cambridge, UK
Mr W.H. Wilson, University of Sydney
Dr H. Yamaki, Osaka University, Japan

APPENDIX C

List of Informal Sessions

A Computation in groups
B Background to Lecture "On the structure of blocks of characters of finite groups"
C Sylow subgroups of finite simple groups
D Permutation groups
E One-relator groups
F Fuchsian groups
G Group algebras (and rings)
H Generalized soluble groups
I Topological groups
J Character theory
K Abelian groups
L Varieties of groups
M Representation theory
N Burnside problem
O Free groups and generalizations
P HNN-construction
Q p-groups
R Automorphism groups
S Knot groups
T Geometry
U Finite soluble groups
V Homological methods
W Decision problems

PROC. SECOND INTERNAT. CONF. THEORY OF GROUPS, 20F50
CANBERRA 1973, pp. 8-12.

PERIODIC GROUPS OF ODD EXPONENT

S.I. Adyan

In 1902 Burnside [5] posed the following problem:

Is every group finite when it has a finite number of generators and satisfies the identical relation

$$x^n = 1 \ ?$$

The negative answer to this question was given in [9, 10, 11]. Let $B(m, n)$ denote the group generated by finitely many elements $a_1, a_2, \ldots, a_m$ subject to the identical relation $x^n = 1$. It was proved in [9, 10, 11] that $B(m, n)$ is infinite for $m > 1$ whenever n is odd and at least 4381 . To prove that this group is infinite, it is necessary to be able to establish not only equalities but also inequalities between words in $B(m, n)$. A method for doing this was developed in [9, 10, 11].

As often happens when difficult problems are solved, the method developed in [9, 10, 11] enables us to solve not only the Burnside problem, but also certain other problems in group theory. Of course, it is necessary to refine the method while preserving the basic ideas.

I. I shall say a word first about the possibility of lowering the limit on n . It has now been proved that $B(m, n)$ is infinite for odd $n \geq 665$. The proof of this result, and of all the results discussed below, is contained in the author's book *The Burnside problem and identical relations in groups*, which will be published in 1974 by the Nauka Press in Moscow. The proof given in the book is simpler and shorter than that in [9, 10, 11].

I believe that, using our method, it should be possible to reduce the lower limit on n to $n \geq 101$. To do this it would be necessary only to make a more detailed analysis and to effect certain changes of a technical nature. It is curious that the proposed new limit should satisfy the condition

$$\frac{4381}{665} \simeq \frac{665}{101} \simeq 6.6 \ .$$

Theoretically it is possible to alter our method to deal with the case $n \geq 33$. But for that a fundamentally new idea would be necessary. Finally to investigate

$B(m, n)$ for $n < 33$ we are forced to seek essentially new methods. In particular this is the case for $n = 5$.

Another very important problem that arose after [5] was published and still remains open is the Burnside problem for $n = 2^k$. I believe that $B(m, 2^k)$ is infinite for $m > 1$ and sufficiently large k. But as yet we have made no progress towards a solution.

I shall now state some other results that have been obtained with our method. Many of them have been published for $n \geq 4381$, but they are true for $n \geq 665$.

II. The following property of $B(m, n)$ was established in [3]:

$$\forall x, y\,(xy = yx \Rightarrow \exists z, t, r\,(x = z^t \;\&\; y = z^r))\,,\text{ where } t, r \text{ are integers.}$$

It follows that all abelian subgroups of $B(m, n)$ are cyclic and that the centre is trivial. It was also proved there that $B(3, n)$ is embedded in $B(2, n)$, so that $B(2, n)$ does not satisfy the minimum condition.

It can be shown that $B(2, n)$ does not satisfy either the maximum or the minimum condition for normal subgroups when $n = pk$, $p > 1$, $k \geq 665$ and odd. The question for prime $n > 665$ remains open and is a very interesting problem.

III. It was proved in [1] that for each $n \geq 4381$ the set of identical relations

$$(1) \qquad (x^{rn}y^{rn}x^{-rn}y^{-rn})^n = 1\,,$$

where r runs over all prime numbers, is an independent set of identical relations in groups; that is, each of them fails to follow from all the others.

This result is connected with the well-known finite basis problem for varieties of groups. But it is also of independent interest. The problem of the equivalence of identical relations is a very hard problem in group theory, and it is regrettable that there are very few ways of studying such problems.

It should also be noted that the relations (1) involve the smallest number of variables and have a rather simple structure.

IV. It is well-known that the additive group R of rational numbers has the following properties for subgroups:

(1) $\forall F, G\,(F \subset R \;\&\; G \subset R \;\&\; (F, G \neq \{0\}) \Rightarrow F \cap G \neq \{0\})$,

(2) R is torsion-free.

Some time ago Professor P.G. Kontorovič posed the following problem: *do there exist non-abelian groups with this property?*

Our method enables us to construct examples giving the positive answer to this question [2]. To this end it is necessary to consider a maximal set $\underline{\underline{M}}$ consisting of elementary words of the form A^n with period A and satisfying the condition: for every elementary word E of rank α, $\alpha > 1$, there exists just one word A^n in $\underline{M}$ which is related to E or E^{-1}. It is proved first that the set of relations

$$A^n = 1 \quad (A^n \in \underline{M})$$

is a defining set of relations for $B(m, n)$. Let $A(m, n)$ be the group on generators

$$a_1, a_2, \ldots, a_m, d$$

with defining relations

$$a_i d = d a_i \quad (1 \le i \le m) ,$$

$$A^n = d \quad (A^n \in \underline{M}) .$$

Clearly, $A(m, n)/\langle d\rangle = B(m, n)$, and so $A(m, n)$ is non-abelian. Since $Z(B(m, n)) = \{1\}$, $Z(A(m, n)) = \langle d\rangle$. Using a theory of transformation of words analogous to that considered in [9, 10, 11], it can be shown that $A(m, n)$ has the following properties:

(1) $\forall x \exists t (x^n = d^t)$,

(2) $\forall x, r (x^r = 1 \;\&\; x \ne 1 \Rightarrow r = 0)$,

(3) $\forall x, y \ne 1 \exists z \ne 1 (\langle x\rangle \cap \langle y\rangle = \langle z\rangle)$.

It should be noted that every abelian group with these properties is locally cyclic, and that we have here non-abelian analogues of the additive group of rationals having arbitrary generating number $m > 1$.

We can use $A(m, n)$ to show that every finite subgroup G of $B(m, n)$ is cyclic. Indeed, let G be a finite subgroup of $B(m, n)$ generated by elements $x_1, x_2, \ldots, x_k$, and $\overline{G}$ the subgroup of $A(m, n)$ generated by $G \cup \{d\}$. Since $d \in Z(\overline{G})$, the finiteness of G gives that $\overline{G}/Z(\overline{G})$ is also finite. Thus the commutator subgroup $\overline{G}'$ of $\overline{G}$ is finite and so $\overline{G}' = \{1\}$; thus G is abelian and so is cyclic.

Professor M.I. Kargapolov has drawn my attention to the fact that $A(m, n)$ can be used to solve yet another problem, one described in the Kourov notebook [7]. It is known that if a Sylow subgroup G_p of a locally finite group G is central, then it is a direct factor. The question arises naturally: *is the same true for periodic groups?* Consider the group $\overline{A}(m, n)$ obtained from $A(m, n)$ by adding the extra relation $d^2 = 1$. It is clear that $\overline{A}(m, n)$ has exponent $2n$, that $\langle d\rangle \subset Z(\overline{A}(m, n))$, and $\overline{A}_2(m, n) = \langle d\rangle$ since n is odd. At the same time it follows trivially from the definition of $\underline{M}$ that each of the words a_1^n, a_2^n and $\left(a_1a_2^{-1}\right)^n$ is equal to d or d^{-1} in $A(m, n)$. From this we get easily that

$$\neg\exists N\left(\bar{A}(m, n) = N \times \langle d \rangle\right) .$$

In conclusion I come to one further property of $B(m, n)$ for $m > 1$ and odd $n \geq 665$. This property is connected with a conjecture of Milnor. If G is a group generated by a set

$$S = \{a_1, a_2, \ldots, a_n\}$$

of elements, then its growth function $\gamma_S(m)$ is defined asthe number of elements of G representable as products of not more than m factors in $S \cup S^{-1}$. By definition G has exponential growth if there exist $A > 0$ and $c > 1$ such that $\gamma_S(m) \geq Ac^m$ for all m , and G has polynomial growth if there exist $A, B > 0$ and an integer $d \geq 0$ such that $Am^d \leq \gamma_S(m) \leq Bm^d$ for all m . These conditions do not depend on the choice of S . In 1968 Milnor [7] posed the problem: *does every finitely generated group have exponential or polynomial growth?* It is known that nilpotent-by-finite groups have polynomial growth, while all other soluble-by-finite groups have exponential growth [4, 8, 12].

It is clear that free groups of rank $m > 1$ have exponential growth. On the basis of our method it is easy to prove that for $m > 1$ and odd $n \geq 665$, $B(m, n)$ also has exponential growth.

References

[1] С.И. Адян [S.I. Adyan], "Бесконечные неприводимые системы групповых тождеств" [Infinite irreducible systems of group identities], *Izv. Akad. Nauk SSSR Ser. Mat.* 34 (1970), 715-734; *Math. USSR-Izv.* 4 (1970), 721-739 (1971). MR44#4078.

[2] С.И. Адян [S.I. Adyan], "О некоторых группах без кручения" [On some torsion-free groups], *Izv. Akad. Nauk SSSR Ser. Mat.* 35 (1971), 459-468; *Math. USSR-Izv.* 5 (1971), 475-484. MR44#303.

[3] С.И. Адян [S.I. Adyan], "О подгруппах свободных периодических групп нечетного показателя" [Subgroups of free groups of odd exponent], *Trudy Mat. Inst. Steklov.* 112 (1971), 64-72.

[4] H. Bass, "The degree of polynomial growth of finitely generated nilpotent groups", *Proc. London Math. Soc.* 25 (1972), 603-614.

[5] W. Burnside, "On an unsettled question in the theory of discontinuous groups", *Quart. J. Pure Appl. Math.* 33 (1902), 230-238. FdM33,149.

[6] М.И. Каргаполов, Ю.И. Мерзляков, В.Н. Ремесленников [M.I. Kargapolov, Yu.I. Merzljakov, V.N. Remeslennikov], Коуровская тетрадь (Нерешенные задачи теории групп) [*Kourov notebook. Unsolved problems in the theory of groups*, 3rd edition, supplemented] (Izdat. Sibirsk. Otdel. Akad. Nauk SSSR, Novosibirsk, 1969). MR34#4339 (1st edition); MR37#1448 (2nd edition).

[7] John Milnor, Problem 5603, *Amer. Math. Monthly* 75 (1968), 685-686.

[8] John Milnor, "Growth of finitely generated solvable groups", *J. Differential Geometry* 2 (1968), 447-449. MR39#6212.

[9] П.С. Новиков, С.И. Адян [P.S. Novikov, S.I. Adyan], "О бесконечных периодических группах. I" [Infinite periodic groups. I], *Izv. Akad. Nauk SSSR Ser. Mat.* 32 (1968), 212-244; *Math. USSR-Izv.* 2 (1968), 209-236. MR39#1532a.

[10] П.С. Новиков, С.И. Адян [P.S. Novikov, S.I. Adyan], "О бесконечных периодических группах. II" [Infinite periodic groups. II], *Izv. Akad. Nauk SSSR Ser. Mat.* 32 (1968), 251-524; *Math. USSR-Izv.* 2 (1968), 241-479 (1969). MR39#1532b.

[11] П.С. Новиков, С.И. Адян [P.S. Novikov, S.I. Adyan], "О бесконечных периодических группах. III" [Infinite periodic groups. III], *Izv. Akad. Nauk SSSR Ser. Mat.* 32 (1968), 709-731; *Math. USSR-Izv.* 2 (1968), 665-685 (1969). MR39#1532c.

[12] Joseph A. Wolf, "Growth of finitely generated solvable groups and curvature of Riemannian manifolds", *J. Differential Geometry* 2 (1968), 421-446. MR40#1939.

Steklov Institute,
Ul. Vavilova 42,
Moscow V-333, USSR.

Матиматический ин-тим В.А. Стеклова
Академии Наук СССР
СССР Москва В-333.

PROC. SECOND INTERNAT. CONF. THEORY OF GROUPS, CANBERRA 1973, pp. 13-62.

20F30

EINBETTUNGSEIGENSCHAFTEN VON NORMALTEILERN: DER SCHLUSS VOM ENDLICHEN AUFS UNENDLICHE

Reinhold Baer

Das Problem, mit dem wir uns befassen wollen, kann man kurz folgendermaßen beschreiben: es sei $\underline{\underline{x}}$ eine Einbettungseigenschaft von Normalteilern, so daß für $N \triangleleft G$ allemal feststeht, ob N ein $\underline{\underline{x}}$-Normalteiler von G ist oder nicht. Sei weiter $N \triangleleft G$ und für jeden Homomorphismus σ von G mit endlichem N^σ sei N^σ ein $\underline{\underline{x}}$-eingebetteter Normalteiler von G^σ. Folgt hieraus, daß N ein $\underline{\underline{x}}$-eingebetteter Normalteiler von G ist? Dies kann man natürlich in dieser Allgemeinheit nicht erwarten; man denke nur daran, daß ja $1 = N^\sigma$ aus der Endlichkeit von N^σ für Homomorphismen σ von G folgen kann. Wir wollen deshalb Kriterien dafür gewinnen, daß N ein fast-polyzyklischer, $\underline{\underline{x}}$-eingebetteter Normalteiler von G ist.

Wir werden Lösungen dieses - eingeschränkten - Problems für einige spezielle Eigenschaften $\underline{\underline{x}}$ angeben. Zwei typische Sätze seien erwähnt.

A. Die folgenden Eigenschaften des polyzyklischen Normalteilers N von G sind äquivalent:

(i) Ist σ ein Homomorphismus von G mit $N^\sigma \neq 1$, so gibt es einen Normalteiler J von G^σ mit $1 \subset J \subseteq N^\sigma$, in dem G^σ eine endliche zyklische Gruppe von Automorphismen induziert.

(ii) Es gibt eine positive ganze Zahl e, so daß G^σ für jeden Homomorphismus σ von G mit endlichem N^σ in jedem in N^σ enthaltenen, minimalen Normalteiler von G^σ eine abelsche Automorphismengruppe mit e teilender Ordnung induziert.

B. Der polyzyklische Normalteiler N von G ist dann und nur dann überauflösbar in G eingebettet, wenn N^σ in G^σ für jeden Homomorphismus σ von G mit endlichem N^σ überauflösbar eingebettet ist.

Im §1 werden die Verbindungen unserer Überlegungen mit der algebraischen Zahlentheorie hergestellt; und die Resultate des §2 liefern wesentliche Abschwächungen der Voraussetzung, daß die betrachteten Normalteiler fast polyzyklisch sein sollen. Lemma 1.1 und Lemma 2.1 dürften auch anderwärts sich als nützlich erweisen.

Bezeichnungen

$x \circ y = x^{-1}y^{-1}xy = x^{-1}x^{y}$.

$A \circ b$ = Menge aller $a \circ b$ mit $a \in A$.

$A \circ B$ = von allen $a \circ b$ mit $a \in A$ und $b \in B$ erzeugte Untergruppe.

$A \circ^{(0)} B = A \circ B$, $A \circ^{(i+1)} B = [A \circ^{(i)} B] \circ B$.

$G' = G \circ G$, $G^{(i+1)} = [G^{(i)}]'$.

$o(x)$, $o(X)$ = Ordnung des Elements x bzw. der Gruppe X .

X^n = von allen x^n mit $x \in X$ erzeugte Untergruppe.

$A \triangleleft B$:=: A ist Normalteiler von B .

$\underline{\underline{c}}_G(A/B)$ = Menge aller $x \in G$ mit $A \circ x \subseteq B$.

$\underline{\underline{c}}_U V$ = Zentralisator von V in U .

$\underline{\underline{z}}G$ = Zentrum von G .

$\{\ldots\}$ = von der in der Parenthese eingeschlossenen Elementenmenge erzeugte Untergruppe.

endlich definierbar; siehe Lemma 2.1.

fast polyzyklisch und polyzyklisch; siehe Folgerung 2.3.

residuell minimaler Normalteiler; siehe Definition 2.5.

$\underline{\underline{\text{hek}}}$; siehe Definition 3.1.

$\underline{\underline{\text{khz}}}$; siehe Definition 4.1.

$\underline{\underline{r}}_G N$; siehe Definition 3.3.

rational irreduzibel; siehe §1 und Lemma 3.4.

1. Abelsche Automorphismengruppen freier abelscher Gruppen endlichen Ranges

In diesem Abschnitt wird A stets eine additiv geschriebene abelsche Gruppe sein; und entsprechend werden wir den Effekt einer Abbildung σ von A auf ein Element $a \in A$ als Produkt $a\sigma$ bezeichnen.

Ist Γ eine Gruppe von Automorphismen von A , so spannt Γ einen Ring Θ von Endomorphismen von A auf; und Γ-Zulässigkeit und Θ-Zulässigkeit sind dann äquivalente Eigenschaften einer Untergruppe $U \subseteq A$. Diese Eigenschaften kann man auch kurz durch $U = U\Gamma = U\Theta$ ausdrücken.

Ist Γ eine Gruppe von Automorphismen von A , sind U und V zwei

Γ-zulässige Untergruppen von A mit $U \subset V$, so sagen wir, daß Γ *irreduzibel auf* V/U *operiert* oder auch daß V/U *ein* Γ*-Kompositionsfaktor von* A ist, wenn U und V die einzigen zwischen U und V gelegenen, Γ-zulässigen Untergruppen von A sind.

Ist A eine freie abelsche Gruppe endlichen Ranges, so operiert die Automorphismengruppe Γ *rational irreduzibel* auf A, wenn A/U endlich ist für jede Γ-zulässige Untergruppe $U \neq 0$ von A.

LEMMA 1.1. *Ist* A $[\neq 0]$ *eine freie abelsche Gruppe des endlichen Ranges* n *und* Γ *eine auf* A *rational irreduzibel operierende, abelsche Gruppe von Automorphismen von* A, *so gilt:*

(A) *der von* Γ *aufgespannte Ring* Θ *von Endomorphismen von* A *ist ein Ring ganzer algebraischer Zahlen;*

(B) *ist* K *der Quotientenkörper von* Θ *und* $Q \subseteq K$ *der Primkörper der rationalen Zahlen, so ist der Grad* $[K : Q] = n$ *und* $K = Q(\Gamma)$;

(C) *ist* R *der Ring aller ganzen Zahlen aus* K, *so ist* $\Theta \subseteq R$ *mit endlichem* R/Θ ; *und für fast alle Primzahlen* p *gibt es einen Isomorphismus von* A/pA *auf* R/pR, *der den Verband der* Γ*-zulässigen Untergruppen von* A/pA *auf den Verband der Ideale des Ringes* R/pR *abbildet.*

BEWEIS. Mit Γ ist natürlich auch Θ kommutativ. Ist $\sigma \in \Theta$ und S der Kern des Endomorphismus σ von A, so ist S eine Γ-zulässige Untergruppe von A. Aus der rationalen Irreduzibilität von Γ folgt: entweder ist $S = 0$ oder A/S ist endlich. Im zweiten Falle folgt aber $A = S$ aus der Torsionsfreiheit der abelschen Gruppe A; und dies zieht $\sigma = 0$ nach sich. Also gilt:

(1) Ist $a \in A$ und $\sigma \in \Theta$ mit $a\sigma = 0$, so ist $a = 0$ oder $\sigma = 0$.

Aus (1) ergibt sich sofort die Nullteilerfreiheit von Θ. Mit anderen Worten:

(2) Θ ist ein Integritätsbereich.

Weiter folgt aus (1):

(3) Ist $0 \neq t \in A$, so ist die Abbildung $\sigma \to t\sigma$ ein Isomorphismus von Θ_+ auf die Γ-zulässige Untergruppe $t\Theta \neq 0$ von A, so daß insbesondere $A/t\Theta$ endlich ist.

Da nach Voraussetzung A eine von 0 verschiedene, freie abelsche Gruppe des Ranges n ist, folgt aus (3):

(4) Θ_+ ist eine freie abelsche Gruppe des Ranges n; und die Charakteristik des Integritätsbereichs Θ und seines Quotientenkörpers K ist 0.

Ist $\sigma \in \Theta$, so ist die von $1, \sigma, \ldots, \sigma^i, \sigma^{i+1}, \ldots$ erzeugte Untergruppe von

Θ_+ wegen (4) endlich erzeugbar, wird also von endlich vielen dieser Elemente σ^i erzeugt. Also gibt es eine positive ganze Zahl k, so daß σ^{k+1} in der von $1, \sigma, \ldots, \sigma^k$ erzeugten Untergruppe von Θ_+ enthalten ist. Hieraus folgt:

(5) Die Elemente aus Θ sind ganze algebraische Zahlen.

Aus (2-5) folgt die Gültigkeit von *(A)* und *(B)*.

Es ist wohlbekannt, daß der Ring R aller ganzen algebraischen Zahlen aus K die folgende Eigenschaft hat:

(6) R_+ ist eine freie abelsche Gruppe des Ranges n.

Aus (4) und (6) folgt also:

(7) R_+/Θ_+ ist endlich.

Wir wählen irgendwie $0 \neq t \in A$ aus und setzen $T = t\Theta$. Aus (3) folgt die Endlichkeit von A/T. Das Indexprodukt

(8a) $$j = [A : T][R_+ : \Theta_+]$$

ist also wegen (7) eine wohlbestimmte positive ganze Zahl.

Wir betrachten eine zu j teilerfremde Primzahl p. Dann ist p auch zu $[A : T]$ teilerfremd, so daß

(8b) $$A = pA + T$$

ist. Die Abbildung $\sigma \to t\sigma$ ist wegen (3) ein Isomorphismus von Θ_+ auf T; und sie induziert wegen (8b) offenbar einen Epimorphismus von Θ_+ auf A/pA und also auch einen Epimorphismus von $\Theta_+/p\Theta_+$ auf A/pA. Da [wegen (4)] A und Θ_+ freie abelsche-Gruppen des Ranges n sind, haben $\Theta_+/p\Theta_+$ und A/pA die gleiche Ordnung p^n, so daß unser Epimorphismus ein Isomorphismus ist. Damit haben wir gezeigt:

(8c) Die Abbildung $\sigma \to t\sigma$ induziert einen Isomorphismus β von $\Theta_+/p\Theta_+$ auf A/pA.

Da p nach Voraussetzung zu j teilerfremd ist, ist p auch zu $[R_+ : \Theta_+]$ teilerfremd. Also ist

(8d) $$R = pR + \Theta .$$

Die Abbildung $\sigma \to pR + \sigma$ induziert also einen Epimorphismus dis Ringes Θ auf den Ring R/pR; und aus $p\Theta \subseteq pR$ folgt, daß die Abbildung $p\Theta + \sigma \to pR + \sigma$ ein wohlbestimmter Epimorphismus des Ringes $\Theta/p\Theta$ auf den Ring R/pR ist. Wegen (4) und (6) sind Θ_+ und R_+ freie abelsche Gruppen des Ranges n, so daß die Ringe

$\Theta/p\Theta$ und R/pR beide p^n Elemente enthalten. Unser Epimorphismus ist also ein Isomorphismus. Folglich gilt:

(8e) Es gibt einen und nur einen Isomorphismus des Ringes $\Theta/p\Theta$ auf den Ring R/pR [der $p\Theta + \sigma$ auf $pR + \sigma$ abbildet].

Kombinieren wir (8c+e), so ergibt sich:

(8f) Es gibt einen und nur einen Isomorphismus β von A/pA auf R_+/pR_+ mit folgender Eigenschaft: zu jedem $a \in A$ gibt es wenigstens ein $\sigma \in \Theta$ mit

$$pA + a = pA + t\sigma \quad \text{und} \quad (pA+a)\beta = pR + \sigma .$$

Ist $r \in R$, so folgt aus (8d) die Existenz von $\omega \in \Theta$ mit $pR + r = pR + \omega$; und hieraus folgt die Gültigkeit von

$$pR + r = pR + \omega = (pA+t\omega)\beta$$

wegen (8f). Sind weiter die Elemente $a \in A$ und $\sigma \in \Theta$ gemäß

$$pR + \sigma = (pA+a)\beta = (pA+\sigma)\beta$$

verbunden, so folgt

$$[(pA+a)\beta](pR+r) = (pR+\sigma)(pR+\omega) = pR + \sigma\omega = (pA+t\sigma\omega)\beta = [(pA+a)\omega]\beta ;$$

und hieraus folgt wegen (8f) die Gültigkeit von folgender Tatsache:

(8g) Ist U eine Untergruppe von A/pA , so ist U dann und nur dann Γ-zulässig, wenn $U\beta$ ein Ideal des Ringes R/pR ist.

Da aber fast alle Primzahlen zu j teilerfremd sind, folgt *(C)* aus (8f+g).

BEMERKUNG. Die Aussage *(A)* ist wohlbekannt; siehe etwa Baer [3; p. 143, Lemma 4 und Folgerung 1]. Der Beweis von (1) ist weiter nichts als die Verifizierung eines besonders einfachen Spezialfalls des Schur-schen Lemmas. Wir haben diese Einzelheiten der Bequemlichkeit des Lesers wegen mit ausgeführt.

FOLGERUNG 1.2. *Ist* $A \neq 0$ *eine freie abelsche Gruppe des endlichen Ranges* n *und* Γ *eine auf* A *rational irreduzibel operierende, abelsche Gruppe von Automorphismen von* A *, so gilt:*

(A) die Torsionsuntergruppe von Γ *ist eine endliche, zyklische Gruppe der Ordnung* e *; und* $\varphi(e)$ *ist ein Teiler von* n *;*

(B) die Faktorgruppe von Γ *nach ihrer Torsionsuntergruppe ist eine freie abelsche Gruppe, deren [endlicher] Rang kleiner als* n *ist.*

Beweis. Wegen Lemma 1.1 gibt es einen algebraischen Körper K vom Range n über den rationalen Zahlen, so daß Γ eine Untergruppe der Einheitengruppe des Ringes aller ganzen algebraischen Zahlen aus K ist. Unsere beiden Behauptungen folgen nun direkt aus dem Dirichlet-schen Einheitensatz; vergl. Hecke [p. 124, Satz

100].

FOLGERUNG 1.3. *Ist* A *eine freie abelsche Gruppe des endlichen, positiven Ranges* n *, ist* Γ *eine endliche, zyklische, auf* A *rational irreduzibel operierende Gruppe von Automorphismen der Ordnung* e *, ist* a *die Anzahl der verschiedene Primfaktoren von* e *, ist* b *gleich* 1 *oder* 0 *, je nachdem* 8 *ein Teiler von* e *ist oder nicht, so gilt:*

Ist s *eine positive ganze Zahl, so dass für fast alle Primzahlen* p *die* Γ*-Kompositionsfaktoren von* A/pA *eine* p^s *teilende Ordnung haben, so ist*

$$n \leq s^a 2^b .$$

BEWEIS. Aus Lemma 1.1 folgt, daß der durch Adjunktion der e-ten Einheitswurzeln zum Körper Q der rationalen Zahlen entstehende Körper K die folgenden Eigenschaften hat:

(1a) $$[K : Q] = n .$$

(1b) Ist R der Ring aller ganzen algebraischen Zahlen aus K , so gibt es zu fast allen [rationalen] Primzahlen p einen Isomorphismus von A/pA auf $[R/pR]_+$, der den Verband der Γ-zulässigen Untergruppen von A/pA auf den Verband der Ideale des Ringes R/pR abbildet.

Erinnern wir uns daran, daß K aus Q durch Adjunktion der e-ten Einheitswurzeln entsteht, so folgt

(1c) $$[K : Q] = \varphi(e) ;$$

siehe etwa Hecke [p. 169, Zeile -3].

Ist F eine zu e teilerfremde Restklasse modulo e , deren multiplikative Ordnung genau f ist, dann folgt aus dem Dirichlet-schen Primzahlsatz:

(2a) F enthält unendlich viele [rationale] Primzahlen;

vergl. Hecke [p. 170, Satz 131].

Ist p eine rationale Primzahl, so daß alle Γ-Kompositionsfaktoren von A/pA eine p^s teilende Ordnung haben, ist p eine der Primzahlen, für die gemäß (1b) ein Isomorphismus von A/pA auf $[R/pR]_+$ existiert, der den Verband der Γ-zulässigen Untergruppen von A/pA auf den Verband der Ideale des Ringes R/pR abbildet, so ist der Grad eines jeden p enthaltenden Primideals aus R höchstens s . Aus (1b) und unserer Voraussetzung über s folgt, daß fast alle Primzahlen p diese Eigenschaften haben. Da von 0 verschiedene Zahlen aus R nur durch endlich viele Primideale aus R teilbar sind, so folgt also:

(2b) Es gibt höchstens endlich viele Primideale aus R , deren Grad größer als s ist.

Aus (2a und b) folgt die Existenz unendlich vieler Primzahlen in F, deren Primidealteiler in R einen s nicht überschreitenden Grad haben; und unter diesen sind unendlich viele nicht Diskrimantenteiler. Eine solche Primzahl zerfällt in Primideale des Grades f; vergl. Hecke [p. 112, Satz 92]. Also ist $f \leq s$. Damit haben wir gezeigt:

(2) Ist die positive ganze Zahl f die multiplikative Ordnung einer zu e teilerfremden Restklasse modulo e, so ist $f \leq s$.

Sei

(3a) $$e = \prod_p p^{e(p)}$$

die Zerlegung von e in Primzahlpotenzen.

Ist erstens p ein ungrader Primteiler von e, so ist die Gruppe der zu p teilerfremden Restklassen modulo $p^{e(p)}$ zyklisch der Ordnung $\varphi\left(p^{e(p)}\right) = (p-1)p^{e(p)-1}$; siehe Hecke [p. 48, Satz 44]. Also enthält auch die Gruppe der zu p teilerfremden Restklassen modulo e eine Restklasse der multiplikativen Ordnung $\varphi\left(p^{e(p)}\right)$; und Anwendung von (2) ergibt:

(3b) $\varphi\left(p^{e(p)}\right) \leq s$, falls p ein ungrader Primteiler von e ist.

Ist e grade, so ist die Maximalordnung der ungraden Restklassen modulo $2^{e(2)}$ genau

$$\varphi\left(2^{e(2)}\right)2^{-b} = 2^{e(2)-1-b} \quad ;$$

vergl. Hecke [p. 49, Satz 45]. Also gibt es auch zu e teilerfremde Restklassen der multiplikativen Ordnung $\varphi\left(2^{e(2)}\right)2^{-b}$; und wir folgern aus (2):

(3c) $\left(2^{e(2)}\right)2^{-b} \leq s$, falls e grade ist.

Aus (3a) folgt

$$\varphi(e) = \prod_p \varphi\left(p^{e(p)}\right),$$

wobei das Produkt über alle Primteiler von e zu erstrecken ist. Also folgt aus (3b und c)

(3) $$\varphi(e) \leq s^a 2^b \quad ;$$

und durch Kombination von (1a und c) und (3) ergibt sich unsere Behauptung.

Diskussion der Folgerung 1.3

A. Es sei m eine ganze Zahl mit

(0) $$2 < m .$$

Unter Q verstehen wir weiter den Körper der rationalen Zahlen; und es sei

(1) $$K = Q(\omega)$$

der aus Q durch Adjunktion einer primitiven m-ten Einheitswurzel ω entstehende Körper. Dann ist der Grad

(2) $$[K : Q] = \varphi(m)$$

[Irreduzibilität der Kreisteilungsgleichung]. Weiter sei

(3) J = Integritätsbereich aller ganzen algebraischen Zahlen aus K .

Natürlich gilt dann

(4) J_+ ist eine freie abelsche Gruppe des Ranges $\varphi(m)$.

Aus (1) folgt, daß K eine endliche galois-sche Erweiterung von Q mit abelscher Galoisgruppe ist. Also gilt:

(5) Jede rationale Primzahl p ist das Produkt von e-ten Potenzen der g verschiedenen, p teilenden Primideale aus J , die sämtlich den gleichen Grad f haben; und es ist $efg = \varphi(m)$;

siehe etwa Hasse [I; p. 48, 1)] oder Hecke [p. 107].

Mit diesen Voraussetzungen und Bezeichnungen gilt nun der folgende wohl bekannte

SATZ A*. *Die folgenden Eigenschaften von* m *sind äquivalent:*

(i) pJ *ist für keine rationale Primzahl* p *ein Primideal aus* J ;

(ii) ist p *eine rationale Primzahl und* f *der gemeinsame Grad der* p *enthaltenden Primideale aus* J , *so ist* $f < \varphi(m)$;

(iii) die [multiplikative] Gruppe $\underline{\underline{R}}(m)$ *der zu* m *teilerfremden Restklassen modulo* m *ist nicht zyklisch;*

(iv) m *ist weder eine Primzahlpotenz noch das doppelte einer Primzahlpotenz.*

BEWEIS. Sei p eine rationale Primzahl. Wegen (5) ist pJ dann und nur dann ein Primideal aus J , wenn [$eg = 1$ und also] $f = \varphi(m)$ ist; und damit ist die Äquivalenz von *(i)* und *(ii)* dargetan. Für die Äquivalenz von *(iii)* und *(iv)* vergl. etwa Hecke [§13 insbesondere p. 48/49].

Ist $\underline{\underline{R}}(m)$ zyklisch und E eine $\underline{\underline{R}}(m)$ erzeugende Restklasse, so folgt aus dem Dirichlet-schen Primzahl, daß E unendlich viele Primzahlen enthält; vergl. etwa Hecke [p. 170, Satz 131]. Alle diese Primzahlen sind zu m teilerfremd und unendlich viele von ihnen sind auch zur Diskriminante von K teilerfremd. Auf die unendlich vielen Primzahlen in E , die keine Diskriminantenteiler sind, können wir Hecke [p. 112, Satz 92] anwenden: ihr Grad f ist gleich der Ordnung $\varphi(m)$ der $\underline{\underline{R}}(m)$ erzeugenden Restklasse E . Damit haben wir gezeigt:

(6) Ist $\underline{\underline{R}}(m)$ zyklisch, so gibt es unendlich viele rationale Primzahlen des Grades $\varphi(m)$.

Damit haben wir auch *(iii)* aus *(ii)* abgeleitet.

Um *(ii)* aus *(iii)* abzuleiten, nehmen wir an, daß *(ii)* falsch ist. Dies ist mit folgender Eigenschaft gleichwertig:

(a) Es gibt eine rationale Primzahl p , deren Grad $f = \varphi(m)$ ist, so daß $e = g = 1$ aus (5) folgt.

Hieraus folgt:

(b) p ist kein Diskriminantenteiler;

vergl. etwa Hasse [I; p. 48, 1)]. Angenommen p ist ein Teiler von m . Dann ist $\omega^{mp^{-1}}$ ein Element der multiplikativen Ordnung p , da ω eine primitive m-te Einheitswurzel ist. Aus (a) und (5) folgt, daß pJ ein Primideal und also J/pJ ein Körper der Charakteristik p ist. Weiter ist $pJ + \omega^{mp^{-1}}$ ein Element aus diesem Körper, dessen multiplikative Ordnung p teilt; und hieraus folgt

$$(+) \qquad \omega^{mp^{-1}} \equiv 1 \text{ modulo } pJ .$$

Da p kein Diskriminantenteiler ist, entsteht der Körper J/pJ durch Adjunktion von $\omega + pJ$ zum Primkörper der Charakteristik p . Da die multiplikative Ordnung von $\omega + pJ$ wegen (+) ein Teiler von mp^{-1} ist, so folgt aus (a):

$$\varphi(m) = f \leq \varphi(mp^{-1}) \leq \varphi(m) ,$$

ein Widerspruch aus dem folgt:

(c) p ist kein Teiler von m .

Wegen (b) und (c) können wir Hecke [p. 112, Satz 92] anwenden. Also folgt aus (a):

(d) $f = \varphi(m)$ ist die multiplikative Ordnung der p enthaltenden Restklasse aus $\underline{\underline{R}}(m)$.

Da $\varphi(m)$ die Ordnung von $\underline{\underline{R}}(m)$ ist - vergl. etwa Hecke [p. 48] - so folgt aus (d), daß $\underline{\underline{R}}(m)$ eine zyklische Gruppe ist. Damit haben wir aus der Falschheit von *(ii)* die Falschheit von *(iii)* hergeleitet: also folgt *(ii)* aus *(iii)*, und damit ist die Äquivalenz von *(i)*-*(iv)* dargetan.

Aus dem Satz A* und der im Laufe seines Beweises hergeleiteten Aussage (6) ergibt sich noch der

ZUSATZ. *pJ is entweder für keine oder für unendlich viele rationale Primzahlen p ein Primideal aus J .*

Sei nun m eine ganze Zahl mit $2 < m$, die den äquivalenten Bedingungen *(i)-(iv)* des Satzes A* genügt. Sind weiter Q, K, ω, J bestimmt wie bisher, so ist wegen (4),

$A = J_+$ eine freie abelsche Gruppe des Ranges $n = \varphi(m)$;

und ω induziert durch Multiplikation in A einen Automorphismus der Ordnung m, den wir ebenfalls ω nennen können. Also ist

$\Gamma = \{\omega\}$ eine endliche, zyklische Automorphismengruppe der Ordnung m, die auf A rational irreduzibel operiert.

Ist p eine rationale Primzahl und f [gemäß (5)] der gemeinsame Grad der p enthaltenden Primideale aus J, so sind die Ordnungen der Γ-Kompositionsfaktoren von A/pA genau p^f. Wegen Bedtngung *(ii)* des Satzes A* gilt

$$f < \varphi(m) = n .$$

Setzen wir $s = n - 1$, so folgt also aus dem Bewiesenen, daß

die Γ-Kompositionsfaktoren von A/pA für jede rationale Primzahl p eine p^s teilende Ordnung haben;

und dies zeigt, daß die in der Folgerung 1.3 bewiesene Ungleichung

$n \leq s^a 2^b$ sich gewiß nicht zu $n \leq s$ verbessern läßt, da ja in unserm Beispiel $s = n-1 < n$ ist.

B. Ist n eine positive ganze Zahl, so gibt es unendlich viele Körper K mit folgenden Eigenschaften:

(1a) K entsteht aus dem Körper Q der rationalen Zahlen durch Adjunktion von Quadratwurzeln aus positiven ganzen Zahlen;

(1b) $$[K : Q] = 2^n .$$

Dann ist K eine Kummersche Erweiterung von Q ; und es gilt also:

(2a) K ist eine galois-sche Erweiterung von Q und die Galois-sche Gruppe von K über Q ist elementar abelsch der Ordnung 2^n ;

(2b) ± 1 sind die einzigen Einheitswurzeln in K.

Sei J der Integritätsbereich der ganzen algebraischen Zahlen aus K. Ist $\underline{p}$ ein Primideal aus J, so ist $J/\underline{p}$ ein endlicher Körper mit Primzahlcharakteristik p. Wegen (1a) entsteht $J/\underline{p}$ aus seinem Primkörper durch Adjunktion von Quadratwurzeln. Also ist der Grad von $J/\underline{p}$ über seinem Primkörper höchstens 2 - es gibt ja nur eine einzige Erweiterung des Grades 2. Damit haben wir gezeigt:

(2c) Ist $\underline{p}$ ein Primideal aus J, so ist sein Grad $f \leq 2$.

Setzen wir

(3a) $A = J_+$ und Γ = Einheitengruppe aus J,

so folgt aus (1b):

(3b) A ist eine freie abelsche Gruppe des Ranges 2^n ;

und aus (1a), (1b), (2b) und dem Dirichletschen Einheitensatz folgt:

(3c) Γ ist das direkte Produkt einer zyklischen Gruppe der Ordnung 2 und einer freien abelschen Gruppe des Ranges $2^n - 1$;

siehe Hecke [p. 124, Satz 100]. Da K aus Q auch durch Adjunktion der Einheiten aus J erzeugt werden kann, gilt noch:

(3d) Γ operiert rational irreduzibel auf A .

Aus (2c) folgt, da $K = Q(\Gamma)$ ist:

(4) Ist p eine rationale Primzahl, so haben die Γ-Kompositionsfaktoren von J/pJ [wie auch die von $p^iJ/p^{i+1}J$ für positives i] eine p^2 teilende Ordnung.

Wir können also die in der Folgerung 1.3 auftretende Konstante $s = 2$ wählen, während wegen (3b) der Rang von A gleich 2^n ist. Also ist es unmöglich, eine nur von s und der Ordnung der Torsionsuntergruppe von Γ abhängende Schranke für den Rang von A zu gewinnen, wenn man auf die Voraussetzung verzichtet, daß Γ eine endliche, zyklische Gruppe ist.

FOLGERUNG 1.4. *Ist A eine freie abelsche Gruppe endlichen Ranges und Γ eine auf A rational irreduzibel operierende Automorphismengruppe von A , sind für fast alle Primzahlen p die Γ-Kompositionsfaktoren von A/pA zyklisch der Ordnung p , so ist A zyklisch.*

BEWEIS. Wir bemerken, daß $A \neq 0$ aus unseren Voraussetzungen folgt.

Für fast alle Primzahlen p gibt es eine Γ-zulässige Untergruppe $A(p)$ mit $[A : A(p)] = p$. In $A/A(p)$ induziert dann Γ eine zyklische Gruppe von Automorphismen, deren Ordnung ein Teiler von $p - 1$ ist. Insbesondere gilt

$$A[\Gamma'-1] \subseteq A(p) .$$

Hieraus folgt, daß $A[\Gamma'-1]$ eine Γ-zulässige Untergruppe von A erzeugt, die unendlichen Index in A hat. Da Γ rational irreduzibel operiert, folgt hieraus $A[\Gamma'-1] = 0$; und dies ist mit $\Gamma' = 1$ äquivalent. Folglich gilt:

(1) Γ ist abelsch.

Wir können Lemma 1.1 anwenden; und es folgt:

(2) Es gibt einen Körper K mit folgenden Eigenschaften:

(a) K ist eine endliche Erweiterung des Körpers Q der rationalen

Zahlen und der Grad $[K : Q]$ ist der Rang von A.

(b) Ist R der Ring aller ganzen algebraischen Zahlen aus K, so gibt es es für fast alle Primzahlen p einen Isomorphismus von A/pA auf $[J/pJ]_+$, der den Verband der Γ-zulässigen Untergruppen von A/pA auf den Verband der Ideale des Ringes J/pJ abbildet.

Da nach Voraussetzung die Γ-Kompositionsfaktoren von A/pA für fast alle Primzahlen p zyklisch der Ordnung p sind, folgt aus (2b):

(2c) Fast alle Primideale aus J haben den Grad 1.

Anwendung der Kronecker-Dedekind-schen Dichtigkeitsätze ergibt also:

(2d) $$[K : Q] = 1;$$

vergl. Hasse [II; p. 128, III und p. 129, IV].

Aus (2a und d) folgt, daß A den Rang 1 hat und also zyklisch ist.

Diskussion der Folgerung 1.4

In der Diskussion B der Folgerung 1.3 haben wir ein Beispiel konstruiert, das zeigt, daß man die Voraussetzung

für fast alle Primzahlen p sind die Γ-Kompositionsfaktoren von A/pA zyklisch der Ordnung p,

nicht durch die Voraussetzung

für alle Primzahlen p sind die Ordnungen der Γ-Kompositionsfaktoren von A/pA Teiler von p^2,

ersetzen kann, wenn man eine von A und Γ sonst unabhängige Schranke für den Rang von A - wie etwa 2 - erhalten will.

2. Fast polyzyklische Normalteiler mit fast polyzyklischer Zentralisatorfaktorgruppe

Das folgende ebenso umfassende wie einfache Kriterium wird sich als nützlich für unsere Anwendungen erweisen.

LEMMA 2.1. *Ist N ein endlich erzeugbarer Normalteiler von G, ist $\underline{\underline{M}}$ eine Menge von Normalteilern $X \triangleleft G$ mit $X \subseteq N$ und endlich definierbarem N/X, gibt es in N enthaltene, nicht $\underline{\underline{M}}$ angehörende Normalteiler von G, so gibt es ein M mit folgenden Eigenschaften:*

(a) $M \triangleleft G$, $M \subseteq N$, $M \notin \underline{\underline{M}}$;

(b) aus $X \triangleleft G$ *und* $M \subset X \subseteq N$ *folgt* $X \in \underline{\underline{M}}$.

TERMINOLOGISCHE ERINNERUNG. Die Gruppe E ist *endlich definierbar*, wenn es eine endlich erzeugbare freie Gruppe F und eine endliche Teilmenge T von F mit

$E \cong F/\{T^F\}$ gibt; vergl. Robinson [I; p. 31, Section B].

BEWEIS. Ist $\underline{\underline{M}}'$ die Menge aller X mit

$$X \triangleleft G\ ,\quad X \subseteq N\ ,\quad X \notin \underline{\underline{M}}\ ,$$

so ist $\underline{\underline{M}}'$ nach Voraussetzung nicht leer.

Ist $\underline{\underline{T}}$ eine nicht leere Teilmenge von $\underline{\underline{M}}'$, die durch die Enthaltenseinsbeziehung linear geordnet ist [$X \subseteq Y$ oder $Y \subseteq X$ für jedes Paar X, Y aus $\underline{\underline{T}}$], so ist die Vereinigungsmenge

$$V = \bigcup_{X \in \underline{\underline{T}}} X$$

ein Normalteiler von G mit $V \subseteq N$. Gehörte V nicht zu $\underline{\underline{M}}'$, so gehörte also V zu $\underline{\underline{M}}$, so daß insbesondere N/V endlich definierbar wäre. Aus der vorausgesetzten endlichen Erzeugbarkeit von N folgt dann nach einem Resultat von Ph. Hall die Existenz einer endlichen Teilmenge E mit

$$V = \{E^N\}\ ;$$

vergl. Robinson [I; p. 32, Lemma 1.43, (i)]. Jedes Element aus E gehört zu wenigstens einer Untergruppe aus $\underline{\underline{T}}$. Aus der Ordnungseigenschaft von $\underline{\underline{T}}$ und der Endlichkeit von E folgt also die Existenz von $W \in \underline{\underline{T}}$ mit $E \subseteq W$. Dann gilt

$$V = \{E^N\} \subseteq W \triangleleft G\ ;$$

und aus $W \in \underline{\underline{T}}$ und der Definition von V folgt $V = W \in \underline{\underline{T}} \subseteq \underline{\underline{M}}'$, was unserer Annahme widerspricht, daß V nicht zu $\underline{\underline{M}}'$ gehört. Also gehört V zu $\underline{\underline{M}}'$; und damit haben wir gezeigt, daß auf $\underline{\underline{M}}'$ das Maximumprinzip der Mengenlehre angewandt werden kann. Also gibt es ein maximales M in $\underline{\underline{M}}'$; und man sieht sofort ein, daß dieses M den Bedingungen (a) und (b) genügt.

Diskussion des Lemma 2.1

A. Ähnliche Kriterien sind gar nicht selten. So wird der vorliegende Schluß beim Beweis von Robinson [I; p. 73, Theorem 3.18] angewandt; und Robinson [II; p. 11, Lemma 6.17] ist offenbar ein Spezialfall von Lemma 2.1.

B. Man beachte, daß das M, dessen Existenz wir im Lemma 2.1 erweisen, ein Normalteiler von G ist, während uns die üblichen Kriterien nur Normalteiler von N liefern würden.

C. Im Lemma 2.1 drückt sich eine der wichtigsten Eigenschaften der Klasse der endlich definierbaren Gruppen aus. Doch hat uns Dr. John Wilson [Cambridge] darauf hingewiesen, daß es auch Klassen endlich erzeugbarer Gruppen mit dieser Eigenschaft gibt, die nicht nur aus endlich definierbaren Gruppen bestehen.

HILFSSATZ 2.2. *Ist G eine endlich erzeugbare Gruppe mit endlich definierbarer Zentrumsfaktorgruppe $G/\underline{\underline{z}}G$, so ist das Zentrum $\underline{\underline{z}}G$ endlich erzeugbar.*

Der einfache Beweis dieses wohl bekannten Satzes sei kurz skizziert: aus den Voraussetzungen und einem Satz von Ph. Hall folgt nämlich die Existenz einer endlichen Menge E mit

$$\underline{\underline{z}}G = \{E^G\} = \{E\} ;$$

vergl. Robinson [I; p. 32, Lemma 1.43, (i)].

FOLGERUNG 2.3. *Ist* $N \triangleleft G$, *ist* N *endlich erzeugbar und* $G/\underline{\underline{c}}_G N$ *fast polyzyklisch, so ist* N *fast polyzyklisch.*

TERMINOLOGISCHE ERINNERUNGEN. Eine Gruppe ist *noethersch*, wenn alle ihre Untergruppen endlich erzeugbar sind. Eine Gruppe ist *polyzyklisch*, wenn sie gleichzeitig noethersch und auflösbar ist; und eine Gruppe ist *fast polyzyklisch*, wenn sie eine polyzyklische Untergruppe von endlichem Index besitzt.

BEWEIS. Aus $N/\underline{\underline{z}}N \cong N\underline{\underline{c}}_G N/\underline{\underline{c}}_G N \subseteq G/\underline{\underline{c}}_G N$ folgt:

(1) $N/\underline{\underline{z}}N$ ist fast polyzyklisch.

Fast polyzyklische Gruppen sind endlich definierbar; siehe Robinson [I; p. 33, Corollary]. Also können wir Hilfssatz 2.2 auf die endlich erzeugbare Gruppe N anwenden:

(2) $\underline{\underline{z}}N$ ist endlich erzeugbar.

Aus (1) und (2) folgt, daß N fast polyzyklisch ist.

SATZ 2.4. *Die folgenden Eigenschaften von* $N \triangleleft G$ *sind äquivalent:*

(i) N *und* $G/\underline{\underline{c}}_G N$ *sind fast polyzyklisch.*

(ii) *(a)* N *ist endlich erzeugbar;*

(b) Ist σ *ein Epimorphismus von* G *auf* H *mit* $N^\sigma \neq 1$, *so gibt es* $J \triangleleft H$ *mit* $1 \subset J \subseteq N^\sigma$, *endlich erzeugbarem* $\underline{\underline{z}}J$ *und fast polyzyklischem* $H/\underline{\underline{c}}_H J$.

BEWEIS. Es ist klar, daß *(ii)* aus *(i)* folgt.

Wir nehmen umgekehrt die Gültigkeit von *(ii)* an; und bezeichnen mit $\underline{\underline{M}}$ die Menge aller X mit folgenden Eigenschaften:

$$\underline{\underline{M}} : X \triangleleft G , \quad X \subseteq N , \quad G/\underline{\underline{c}}_G(N/X) \text{ ist fast polyzyklisch.}$$

Sei $\underline{\underline{M}}'$ die Menge aller $Y \triangleleft G$ mit $Y \subseteq N$ und nicht fast polyzyklischem $G/\underline{\underline{c}}_G(N/Y)$. Dann machen wir die zu widerlegende Annahme:

(0) $\underline{\underline{M}}'$ ist nicht leer.

Aus Bedingung *(iia)* und Folgerung 2.3 ergibt sich:

(1) Ist $X \in \underline{\underline{M}}$, so ist N/X fast polyzyklisch.

Fast polyzyklische Gruppen sind endlich definierbar; siehe Robinson [I; p. 33, Corollary]. Wegen (0) und (1) können wir also Lemma 2.1 auf $\underline{\underline{M}}$ anwenden:

(2) Es gibt $M \triangleleft G$ mit $M \subseteq N$, $M \notin \underline{\underline{M}}$, während $X \in \underline{\underline{M}}$ aus $X \triangleleft G$, $M \subset X \subseteq N$ folgt.

Wir setzen

(3) $x^* = xM$ für jedes $x \in G$, so daß insbesondere $G^* = G/M$ und $N^* = N/M$ ist.

Aus $M \notin \underline{\underline{M}}$ und $M \subseteq N$ folgen $M \subset N$ und $N^* \neq 1$. Anwendung von Bedingung *(iib)* ergibt:

(4) Es gibt $J \triangleleft G^*$ mit $1 \subset J \subseteq N^*$, endlich erzeugbarem $\underline{\underline{z}}J$ und fast polyzyklischem $G^*/\underline{\underline{c}}_{G^*}J$.

Sei $X \triangleleft G^*$ mit $1 \subset X \subseteq N^*$. Dann gibt es einen eindeutig bestimmten Normalteiler Y von G mit $M \subseteq Y$ und $Y^* = X$. Aus $1 \subset X \subseteq N^*$ folgt $M \subset Y \subseteq N$, so daß $Y \in \underline{\underline{M}}$ aus (2) folgt. Also ist

$$G^*/\underline{\underline{c}}_{G^*}(N^*/X) = [G/M]/\underline{\underline{c}}_{G/M}[(N/M)/(Y/M)] \cong G/\underline{\underline{c}}_G(N/Y)$$

fast polyzyklisch; und damit haben wir gezeigt:

(5) Ist $X \triangleleft G^*$ mit $1 \subset X \subseteq N^*$, so ist $G^*/\underline{\underline{c}}_{G^*}(N^*/X)$ fast polyzyklisch.

Aus (2) folgt $M \triangleleft G$, $M \subseteq N$, $M \notin \underline{\underline{M}}$, so daß $G/\underline{\underline{c}}_G(N/M)$ nicht fast polyzyklisch ist; und dies ist offenbar mit der folgenden Aussage äquivalent:

(6) $$G^*/\underline{\underline{c}}_{G^*}N^* \text{ ist nicht fast polyzyklisch.}$$

Wir machen nun die Annahme:

(7a) $$\underline{\underline{z}}J = 1 .$$

Aus (4) und (5) folgt, daß $G^*/\underline{\underline{c}}_{G^*}J$ und $G^*/\underline{\underline{c}}_{G^*}(N^*/J)$ fast polyzyklisch sind; und hieraus folgt:

(7b) $$G^*/[\underline{\underline{c}}_{G^*}J \cap \underline{\underline{c}}_{G^*}(N^*/J)] \text{ ist fast polyzyklisch.}$$

Ist $x \in \underline{\underline{c}}_{G^*}J \cap \underline{\underline{c}}_{G^*}(N^*/J)$, so gehört der von x in N^* induzierte Automorphismus der Stabilitätsgruppe von $J \triangleleft N^*$ an. Aus einem Satz von Kalouinine folgt dann

$$N^* \circ x \subseteq \underline{\underline{z}}J ;$$

vergl. Specht [p. 88, Beweis von Satz 19]. Also folgt $x \in \underline{\underline{c}}_{G^*}N^*$ aus (7a); und damit haben wir

(7c) $$\underline{\underline{c}}_{G^*}N^* = \underline{\underline{c}}_{G^*}J \cap \underline{\underline{c}}_{G^*}(N^*/J)$$

bewiesen. Aus (7b und c) ergibt sich also, daß $G^*/\underline{\underline{c}}_{G^*}N^*$ fast polyzyklisch ist; und dies widerspricht (6). Aus diesem Widerspruch folgt:

(7) $$\underline{\underline{z}}J \neq 1 .$$

Charakteristische Untergruppen von Normalteilern sind Normalteiler. Setzen wir jetzt $A = \underline{\underline{z}}J$, so ergibt sich aus (4), (5) und (7):

(8) $1 \subset A \subseteq N^*$, $A \triangleleft G^*$ ist endlich erzeugbar und abelsch, $G^*/\underline{\underline{c}}_{G^*}(N^*/A)$ ist fast polyzyklisch.

Mit N ist auch N^*/A endlich erzeugbar. Da $G^*/\underline{\underline{c}}_{G^*}(N^*/A)$ fast polyzyklisch ist, ergibt sich aus Folgerung 2.3, daß N^*/A fast polyzyklisch ist. Da A wegen (8) eine endlich erzeugbare, abelsche Gruppe ist, ist die Erweiterung N^* von A durch N^*/A ebenfalls fast polyzyklisch. Aus $A \subseteq J$ folgt $\underline{\underline{c}}_{G^*}J \subseteq \underline{\underline{c}}_{G^*}A$, so daß das epimorphe Bild $G^*/\underline{\underline{c}}_{G^*}A$ von $G^*/\underline{\underline{c}}_{G^*}J$ wegen (4) ebenfalls fast polyzyklisch ist. Aus (8) folgt also

(9) $G^*/[\underline{\underline{c}}_{G^*}A \cap \underline{\underline{c}}_{G^*}(N^*/A)]$ ist fast polyzyklisch.

Die von $\underline{\underline{c}}_{G^*}A \cap \underline{\underline{c}}_{G^*}(N^*/A)$ in N^* induzierte Automorphismengruppe Γ stabilisiert $A \triangleleft N^*$. Also folgt aus dem Satz von Kaloujnine, daß Γ abelsch ist; vergl. Specht [p. 88, Satz 19]. Eine abelsche Gruppe von Automorphismen einer fast polyzyklischen Gruppe ist aber endlich erzeugbar; siehe Robinson [I; p. 82, Theorem 3.27]. Wegen

$$\Gamma \cong [\underline{\underline{c}}_{G^*}A \cap \underline{\underline{c}}_{G^*}(N^*/A)]/\underline{\underline{c}}_{G^*}N^*$$

haben wir also gezeigt:

(10) $[\underline{\underline{c}}_{G^*}A \cap \underline{\underline{c}}_{G^*}(N^*/A)]/\underline{\underline{c}}_{G^*}N^*$ ist eine endlich erzeugbare, abelsche Gruppe.

Kombination von (9) und (10) zeigt, daß $G^*/\underline{\underline{c}}_{G^*}N^*$ fast polyzyklisch ist; und dies widerspricht (6). Unsere Annahme (0) hat uns zu einem Widerspruch geführt. Also ist $\underline{\underline{M}}'$ leer, so daß insbesondere 1 zu $\underline{\underline{M}}$ gehört. Folglich ist $G/\underline{\underline{c}}_G N$ fast polyzyklisch; und es ergibt sich aus (1), daß N selbst ebenfalls fast polyzyklisch ist: wir haben *(i)* aus *(ii)* abgeleitet und die Äquivalenz von *(i)* und *(ii)* dargetan.

Diskussion von Satz 2.4

A. Sei F die freie [nicht abelsche] Gruppe des Ranges 2 und $G = F/F''$. Dann ist bekanntlich $G' = F'/F''$ eine freie abelsche Gruppe abzählbar unendlichen Ranges; und G/G' ist eine freie abelsche Gruppe des Ranges 2 . Insbesondere ist also G eine endlich erzeugbare metabelsche Gruppe $[G'' = 1]$, die nicht fast polyzyklisch ist.

Ist σ ein Homomorphismus von G , so ist entweder G^{σ} abelsch; oder aber G'^{σ} ist ein von 1 verschiedener Normalteiler von G^{σ} und die von G^{σ} in G'^{σ} induzierte Gruppe von Automorphismen ist eine von 2 Elementen erzeugte abelsche Gruppe, insbesondere also fast polyzyklisch.

Setzen wir also $N = G$, so ist die Bedingung *(ii)* des Satzes 2.4 mit einer Ausnahme erfüllt: das Zentrum des *ad (iib)* postulierten Normalteilers J wird i.A. nicht endlich erzeugbar sein. Da *(i)* von dem Paare $N = G$, G sicher nicht erfüllt wird, haben wir damit die Unentbehrlichkeit der Forderung gezeigt, daß $\underline{\underline{z}}J$ [*ad (iib)*] endlich erzeugbar ist.

B. Man kann Robinson [I; p. 82, Theorem 3.27] - diesen Satz haben wir auch im Beweis von Satz 2.4 benutzt - anwenden, um Bedingung *(iib)* noch etwas abzuschwächen.

DEFINITION 2.5. *N ist ein residuell minimaler Normalteiler von G* , wenn [$N \triangleleft G$ und] 1 der Durchschnitt aller $X \triangleleft G$ mit $X \subset N$ und N/X ein endlicher, minimalen Normalteiler von G/X ist.

Diese Begriffsbildung haben wir schon früher benutzt; vergl. Baer [1; p. 169, Definition 1].

TERMINOLOGISCHE ERINNERUNG. Die Gruppe G *operiert rational irreduzibel* auf ihrem Normalteiler N , wenn aus $X \triangleleft G$ und $1 \subset X \subseteq N$ folgt, daß N/X eine Torsionsgruppe ist.

Ist $N \triangleleft G$ und N eine freie abelsche Gruppe endlichen Ranges, so operiert G dann und nur dann rational irreduzibel auf N , wenn die von G in N induzierte Automorphismengruppe auf N im Sinne des §1 rational irreduzibel operiert; und man überzeugt sich mühelos davon, daß N in diesem Falle ein residuell minimaler Normalteiler von G ist.

SATZ 2.6. *Die folgenden Eigenschaften von $N \triangleleft G$ sind äquivalent:*

(i) N *ist fast polyzyklisch.*

(ii) *(a)* N *ist endlich erzeugbar;*

(b) *Ist σ ein Homomorphismus von G mit $N^{\sigma} \neq 1$, ist 1 der einzige endliche, in N^{σ} enthaltene Normalteiler von G^{σ} , so existiert eine freie abelsche Gruppe A endlichen Ranges mit $A \triangleleft G^{\sigma}$ und $1 \subset A \subseteq N^{\sigma}$;*

(c) *es gibt eine positive ganze Zahl m mit folgender Eigenschaft: ist σ ein Homomorphismus von G und M ein nicht abelscher, endlicher, minimaler Normalteiler von G^{σ} mit $M \subseteq N^{\sigma}$, so ist $o(M) \leq m$.*

(iii) *(a)* N *ist endlich erzeugbar;*

(b) *ist* σ *ein Homomorphismus von* G *mit* $N^\sigma \neq 1$ *, so gibt es einen endlich erzeugbaren, residuell minimalen Normalteiler* $J \triangleleft G^\sigma$ *mit* $1 \subset J \subseteq N^\sigma$ *;*

(c) *es gibt eine positive ganze Zahl* m *mit folgender Eigenschaft: ist* σ *ein Homomorphismus von* G *mit endlichem* N^σ *, ist* M *der einzige minimale Normalteiler von* G^σ *mit* $M \subseteq N^\sigma$ *, ist* M *nicht abelsch, so ist* $o(M) \leq m$ *.*

BEWEIS. Ist N fast polyzyklisch, so ist N gewiß auch endlich erzeugbar. Weiter ist das Produkt P aller auflösbaren Normalteiler von N ein Produkt endlich vieler dieser Normalteiler, da ja N noethersch ist. Also ist P auflösbar und N/P ist endlich, da N fast polyzyklisch ist. Wir notieren:

(1) es gibt eine auflösbare charakteristische Untergruppe P von N mit endlichem N/P .

Da charakteristische Untergruppen von Normalteilern selbst Normalteiler sind, gilt weiter:

(2) $$P \triangleleft G ;$$

und wir setzen

(3) $[N : P] = m$ eine wegen (1) positive ganze Zahl.

Ist σ ein Homomorphismus von G mit $N^\sigma \neq 1$, ist weiter 1 der einzige, in N^σ enthaltene, endliche Normalteiler von G^σ , so ist P^σ wegen (1) unendlich. Da P^σ mit P auflösbar ist, gibt es eine abelsche charakteristische Untergruppe $A \neq 1$ von P^σ . Dann ist $A \triangleleft G^\sigma$; und da A mit N noethersch ist, ist die Torsionsuntergruppe von A ein endlicher Normalteiler von G^σ , der in N^σ enthalten ist, also gleich 1 . Folglich ist A eine freie abelsche Gruppe endlichen Ranges mit $1 \subset A \subseteq N^\sigma$; und damit haben wir auch die Gültigkeit von *(iib)* bewiesen.

Ist λ ein Homomorphismus von G und M ein nicht-abelscher, endlicher, minimaler Normalteiler von G^σ , so ist M nicht auflösbar, woraus $M \nsubseteq P^\sigma$ wegen (1) folgt. Also folgt $1 = M \cap P^\sigma$ aus der Minimalität von M , so daß

$$o(M) = [M : M \cap P^\sigma] = [MP^\sigma : P^\sigma]$$

ein Teiler von $[N^\sigma : P^\sigma]$ und also wegen (3) ein Teiler von m ist. Damit haben wir auch die Gültigkeit von *(iic)* dargetan: *(ii)* folgt aus *(i)*.

Gilt *(ii)*, so gilt auch *(iiia)* = *(iia)*, und *(iiic)* ist eine wesentlich abgeschwächte Form von *(iic)*. Ist der Normalteiler $X \triangleleft Y$ eine freie abelsche Gruppe endlichen, positiven Ranges, so gibt es unter den in X enthaltenen Normalteilern von Y einen A minimalen positiven Ranges; und man sieht sofort, daß Y auf A rational irreduzibel operiert. Da A frei abelsch endlichen Ranges ist, folgt hieraus, daß A ein endlich erzeugbarer, residuell minimaler Normalteiler von Y ist. - Endliche, von 1 verschiedene Normalteiler enthalten stets minimale Normalteiler; und diese sind natürlich erst recht endlich erzeugbar und residuell minimal. Nun ist es klar, daß *(iiib)* aus *(iib)* folgt. Also ist *(iii)* eine Folge von *(ii)*.

Wir nehmen schließlich die Gültigkeit von *(iii)* an. Sei σ ein Epimorphismus von G auf H mit fast polyzyklischem N^σ; und es sei M ein endlicher, minimaler, nicht abelscher Normalteiler von H mit $M \subseteq N^\sigma$. Da N^σ fast polyzyklisch ist, folgt aus einem Satze von Kurt Hirsch die Existenz einer torsionsfreien charakteristischen Untergruppe C von N^σ mit endlichem N^σ/C; vergl. Robinson [II; p. 139, Corollary]. Aus $N^\sigma \triangleleft H$ folgt $C \triangleleft H$; und da M endlich, C aber torsionsfrei ist, gilt $M \cap C = 1$. Also gibt es unter den in N^σ enthaltenen, M trivial schneidenden Normalteilern von H, die endlichen Index in N^σ haben, einen maximalen W. Dieser hat offenbar die folgenden Eigenschaften:

$W \triangleleft H$, $W \subseteq N^\sigma$, $W \cap M = 1$, N^σ/W ist endlich,

MW/W ist der einzige, in N^σ/W enthaltene, minimale Normalteiler von H/W.

Insbesondere ist $MW/W \cong M$. Folglich können wir *(iiic)* anwenden, um zu zeigen, daß $o(M) = [MW : W] \leq m$ ist. Damit haben wir bewiesen:

(4) Ist σ ein Homomorphismus von G mit fast polyzyklischem N^σ, ist M ein endlicher, minimaler, nicht abelscher Normalteiler von G^σ mit $M \subseteq N^\sigma$, so ist $o(M) \leq m$.

Wir machen jetzt die zu widerlegende Annahme:

(5) N ist nicht fast polyzyklisch.

Sei $\underline{\underline{M}}$ die Menge aller $X \triangleleft G$ mit $X \subseteq N$ und fast polyzyklischem N/X. Wegen (5) gilt dann:

(6a) $N \in \underline{\underline{M}}$ und $1 \notin \underline{\underline{M}}$.

Fast polyzyklische Gruppen sind endlich definierbar; siehe Robinson [I; p. 33, Corollary]. Wegen (6a) und *(iiia)* können wir also Lemma 2.1 auf $\underline{\underline{M}}$ anwenden; und es folgt:

(6b) Es gibt $W \triangleleft G$ mit $W \subseteq N$, so daß

(6b.1) N/W nicht fast polyzyklisch ist, aber

(6b.2) N/X fast polyzyklisch für jedes $X \triangleleft G$ mit $W \subset X \subseteq N$ ist.

Wir setzen

(7) $x^* = xW$ für $x \in G$ und entsprechend $G^* = G/W$, $N^* = N/W$.

Dann folgt aus (6b):

(7a) N^* ist nicht fast polyzyklisch.

(7b) Ist $X \triangleleft G^*$ und $1 \subset X \subseteq N^*$, so ist N^*/X fast polyzyklisch.

Da Erweiterungen fast polyzyklischer Gruppen durch fast polyzyklische Gruppen wieder fast polyzyklisch sind, ergibt sich aus (7a und b) noch:

(7c) 1 ist der einzige, in N^* enthaltene, fast polyzyklische Normalteiler von G^*.

Aus (7a) folgt insbesondere $N^* \neq 1$. Da der kanonische Epimorphismus von G auf G^* den Normalteiler N auf N^* abbildet, zeigt Anwendung von Bedingung *(iiib)*:

(8a) Es gibt einen endlich erzeugbaren, residuell minimalen Normalteiler $J \triangleleft G^*$ mit $1 \subset J \subseteq N^*$.

Aus (7c) folgt:

(8b) J ist nicht fast polyzyzklisch und insbesondere ist J nicht endlich.

Sei $\underline{\underline{S}}$ die Menge aller $X \triangleleft G^*$ mit $X \subset J$, so daß J/X ein endlicher, nicht abelscher, minimaler Normalteiler von G^*/X ist. Ist $X \in \underline{\underline{S}}$, so ist J/X endlich, so daß aus (8b) die Unendlichkeit von X folgt. Aus (7b) folgt also, daß N^*/X fast polyzyklisch ist. Da J/X ein in N^*/X enthaltener, nicht abelscher, endlicher, minimaler Normalteiler von G^*/X ist, ergibt Anwendung von (4), daß $[J : X] \leq m$ ist. Damit haben wir gezeigt:

(9a) $[J : X] \leq m$ für jedes $X \in \underline{\underline{S}}$.

Da J wegen (8a) endlich erzeugbar ist, besitzt J nur endlich viele Normalteiler, deren Index in J höchstens m ist; vergl. etwa Baer [2; p. 331, Lemma 4]. Also folgt aus (9a):

(9b) $\underline{\underline{S}}$ ist endlich.

Ist

$$S = \bigcap_{X \in \underline{\underline{S}}} X,$$

so ist natürlich $S \triangleleft G^*$ und $S \subseteq J$; und aus (9a und b) und dem Satz von Poincaré folgt:

(9c) J/S ist endlich.

Sei $\underline{\underline{A}}$ die Menge aller $X \triangleleft G^*$ mit $X \subset J$, so daß J/X ein abelscher, endlicher, minimaler Normalteiler von G^*/X ist; und es sei

$$A = \bigcap_{X \in \underline{\underline{A}}} X .$$

Dann ist natürlich

(9d) $A \triangleleft G^*$ und $J' \subseteq A \subseteq J$;

und aus der residuellen Minimalität von J - siehe (8a) - folgt:

(9e) $1 = A \cap S$.

Aus (9d und e) ergibt sich

$$J' \cap S \subseteq A \cap S = 1 ;$$

und hieraus folgt:

$$S = S/(J' \cap S) \cong SJ'/J' \subseteq J/J' .$$

Also gilt

(9f) S ist abelsch.

Wegen (8a) ist J endlich erzeugbar; und wegen (9c) ist J/S endlich. Also folgt aus dem Satz von Reidemeister-Schreier:

(9g) S ist endlich erzeugbar;

vergl. Specht [p. 153, Satz 4]. Aus (9f und g) folgt, daß S ein endlich erzeugbarer, abelscher Normalteiler von G^* mit $S \subseteq J \subseteq N^*$ ist; und also ergibt sich

$$S = 1$$

aus (7c). Dann ist aber

$$J = J/S \text{ endlich}$$

wegen (9c), so daß J ein endlicher, in N^* enthaltener Normalteiler von G^* ist. Anwendung von (7c) zeigt $J = 1$; und dies widerspricht (8a). Also hat die Annahme (5) zu einem Widerspruch geführt: N ist fast polyzyklisch. Damit haben wir *(i)* aus *(iii)* hergeleitet und die Äquivalenz von *(i)-(iii)* bewiesen.

3. Hyper-endlich-klassig eingebettete Normalteiler

DEFINITION 3.1. N <u>hek</u> G gilt dann und nur dann, wenn

(a) $N \triangleleft G$ gilt, und

(b) es zu jedem Epimorphismus σ von G auf H mit $N^\sigma \neq 1$ ein Element s mit $1 \neq s \in N^\sigma$ und endlichem s^H gibt.

Bedenken wir, daß für jedes Gruppenelement $x \in X$ die Gruppe der von X in x^X induzierten Permutationen im wesentlichen identisch ist mit der Gruppe der von X in $\{x^X\} \triangleleft X$ induzierten Automorphismen, so sehen wir ein, daß die obige Bedingung (b) [unter der Voraussetzung (a)] mit folgender Eigenschaft identisch ist:

(b*) ist σ ein Epimorphismus von G auf H mit $N^\sigma \neq 1$, so gibt es $S \triangleleft H$ mit $1 \subset S \subseteq N^\sigma$ und endlichem $H/\underline{\underline{c}}_H S$.

Man verifiziert mühelos die folgende wichtige und nützliche Eigneschaft der Beziehung $\underline{\underline{\text{hek}}}$:

Sind A und B Normalteiler der Gruppe G mit $A \subseteq B$, so gilt dann und nur dann $B\ \underline{\underline{\text{hek}}}\ G$, wenn $A\ \underline{\underline{\text{hek}}}\ G$ und $B/A\ \underline{\underline{\text{hek}}}\ G/A$ gelten.

LEMMA 3.2. *Ist $N\ \underline{\underline{\text{hek}}}\ G$ und N endlich erzeugbar, so sind N und $G/\underline{\underline{c}}_G N$ fast polyzyklisch.*

BEWEIS. Ist σ ein Epimorphismus von G auf H mit $N^\sigma \neq 1$, so gibt es ein Element s mit $1 \neq s \in N^\sigma$ und endlichem s^H. Wir setzen $S = \{s^H\}$. Dann ist S ein endlich erzeugbarer Normalteiler von H mit $1 \subset S \subseteq N^\sigma$ und endlichem $H/\underline{\underline{c}}_H S$. Insbesondere ist auch $S/\underline{\underline{z}}S$ endlich; und hieraus folgt nach einem Schur-schen Satz die Endlichkeit von S'; vergl. Robinson [I: p. 102, Theorem 4.12]. Also ist S' endlich und S/S' eine endlich erzeugbare abelsche Gruppe, so daß S fast-polyzyklisch und insbesondere $\underline{\underline{z}}S$ endlich erzeugbar ist. Damit haben wir gezeigt, daß Bedingung *(ii)* des Satzes 2.4 von $N \triangleleft G$ erfüllt wird; und hieraus folgt, daß N und $G/\underline{\underline{c}}_G N$ fast polyzyklisch sind.

DEFINITION 3.3. Ist $N \triangleleft G$, so ist $\underline{\underline{r}}_G N$ das Produkt aller $X \triangleleft G$ mit $N \circ^{(i)} X = 1$ für fast alle i.

$\underline{\underline{r}}_G N$ ist wohlbestimmt, da ja $X = 1$ gewählt werden kann; und natürlich gilt $\underline{\underline{c}}_G N \subseteq \underline{\underline{r}}_G N \triangleleft G$.

LEMMA 3.4. *Ist $N\ \underline{\underline{\text{hek}}}\ G$ und N endlich erzeugbar, so gilt:*

(a) $G/\underline{\underline{r}}_G N$ ist endlich und $N \circ^{(i)} \underline{\underline{r}}_G N = 1$ für fast alle i;

(b) ist σ ein Epimorphismus von G auf H, so ist $X \cap \underline{\underline{c}}_H\left[\underline{\underline{r}}_H(N^\sigma)\right] \neq 1$ für jeden Normalteiler $X \triangleleft H$ mit $1 \subset X \subseteq N^\sigma$;

(c) ist σ ein Epimorphismus von G auf H und M ein minimaler Normalteiler von H mit $M \subseteq N^\sigma$, so ist $M \circ \underline{\underline{r}}_H(N^\sigma) = 1$;

(d) ist σ *ein Epimorphismus von* G *auf* H *, ist* F *ein in* N^{σ} *enthaltener, torsionsfreier abelscher Normalteiler von* H *, in dem* H *rational irreduzibel operiert, so ist* $F \circ \underline{\underline{r}}_H(N^{\sigma}) = 1$ *.*

TERMINOLOGISCHE ERINNERUNG. Die Gruppe X operiert rational irreduzibel auf ihrem torsionsfreien abelschen Normalteiler F , wenn aus $Y \triangleleft X$ und $1 \subset Y \subseteq F$ folgt, daß F/Y eine Torsionsgruppe ist.

BEWEIS. Wegen Lemma 3.2 ist N noethersch. Also folgt aus der definierenden Eigenschaft (b*) von $\underline{\underline{\text{hek}}}$ die Existenz einer endlichen Kette N_i von Normalteilern von G mit folgenden Eigenschaften:

$$1 = N_0 \text{ , } N_i \subseteq N_{i+1} \text{ , } N_n = N \text{ ;}$$

$$G/\underline{\underline{c}}_G(N_{i+1}/N_i) \text{ ist endlich für } 0 \le i < n \text{ .}$$

Wir setzen

$$J = \bigcap_{i=0}^{n-1} \underline{\underline{c}}_G(N_{i+1}/N_i) \text{ .}$$

Dann ist $J \triangleleft G$; und aus dem Satz von Poincaré folgt die Endlichkeit von G/J . Weiter ist $N_{i+1} \circ J \subseteq N_i$ für $0 \le i < n$; und hieraus folgt $N \circ^{(n)} J = 1$. Damit haben wir gezeigt:

(1) Es gibt $J \triangleleft G$ mit endlichem G/J und $N \circ^{(j)} J = 1$ für fast alle j .

Aus den Produktformeln für Kommutatoren folgert man mühelos:

(2) Aus $X \triangleleft G$ und $Y \triangleleft G$ mit $N \circ^{(i)} X = N \circ^{(i)} Y = 1$ für fast alle i folgt $N \circ^{(i)} XY = 1$ für fast alle i .

Ist $\underline{\underline{M}}$ die Menge aller $X \triangleleft G$ mit $N \circ^{(i)} X = 1$ für fast alle i , so folgt aus (1), daß das Produkt aller $X \in \underline{\underline{M}}$ gleich dem Produkt von J und endlich vielen $X \in \underline{\underline{M}}$ ist; und aus (2) ergibt sich, daß das Produkt $\underline{\underline{r}}_G N$ aller $X \in \underline{\underline{M}}$ ebenfalls zu $\underline{\underline{M}}$ gehört. Aus (1) folgt dann noch die Endlichkeit von $G/\underline{\underline{r}}_G N$; und damit haben wir *(a)* bewiesen.

Sei σ ein Epimorphismus von G auf H und $X \triangleleft H$ mit $1 \subset X \subseteq N^{\sigma}$. Mit N ist auch N^{σ} endlich erzeugbar; und aus N $\underline{\underline{\text{hek}}}$ G folgt N^{σ} $\underline{\underline{\text{hek}}}$ G^{σ} . Wegen $G^{\sigma} = H$ können wir also *(a)* auf $N^{\sigma} \triangleleft H$ anwenden. Setzen wir $R = \underline{\underline{r}}_H(N^{\sigma})$, so ergibt sich folglich:

$$X \circ^{(i)} R \subseteq N^\sigma \circ^{(i)} R = 1 \quad \text{für fast alle } i\ .$$

Aus $X \neq 1$ folgt deshalb die Existenz einer nicht negativen ganzen Zahl a mit

$$1 = X \circ^{(a+1)} R \subset X \circ^{(a)} R = Y\ .$$

Dann ist $Y \triangleleft H$ mit $1 \subset Y \subseteq X$ und $Y \circ R = 1$, woraus

$$1 \subset Y \subseteq X \cap \underline{\underline{c}}_H R$$

folgt; und damit haben wir *(b)* bewiesen.

Ist σ ein Epimorphismus von G auf H und M ein minimaler Normalteiler von H mit $M \subseteq N^\sigma$, so folgt $1 \subset M \cap \underline{\underline{c}}_H\left[\underline{\underline{r}}_H(N^\sigma)\right]$ aus *(b)*. Also ergibt sich

$$M = M \cap \underline{\underline{c}}_H\left[\underline{\underline{r}}_H(N^\sigma)\right] \subseteq \underline{\underline{c}}_H\left[\underline{\underline{r}}_H(N^\sigma)\right]$$

aus der Minimalität von M ; und dies ist mit $M \circ \underline{\underline{r}}_H(N^\sigma) = 1$ äquivalent: es gilt auch *(c)*.

Sei σ ein Epimorphismus von G auf H und F ein in N^σ enthaltener, torsionsfreier abelscher Normalteiler von H , auf dem H rational irreduzibel operiert.

Ist $F = 1$, so ist nichts zu beweisen; und wir können also o.B.d.A. $F \neq 1$ annehmen. Dann folgt

$$1 \neq F \cap \underline{\underline{c}}_H\left[\underline{\underline{r}}_H(N^\sigma)\right]$$

aus *(b)*. Da H rational irreduzibel auf F operiert, ist also $F/\left[F \cap \underline{\underline{c}}_H\left[\underline{\underline{r}}_H(N^\sigma)\right]\right]$ eine Torsionsgruppe. Zu jedem $f \in F$ gibt es deshalb eine positive ganze Zahl n mit $f^n \in \underline{\underline{c}}_H\left[\underline{\underline{r}}_H(N^\sigma)\right]$. Da F ein abelscher Normalteiler ist, ist dies mit

$$\left[f \circ \underline{\underline{r}}_H(N^\sigma)\right]^n = f^n \circ \underline{\underline{r}}_H(N^\sigma) = 1$$

äquivalent; und aus der Torsionsfreiheit des abelschen Normalteilers F folgt $f \circ \underline{\underline{r}}_H(N^\sigma) = 1$. Damit haben wir

$$F \circ \underline{\underline{r}}_H(N^\sigma) = 1$$

bewiesen: es gilt auch *(d)*.

FOLGERUNG 3.5. *Ist N hek G und N endlich erzeugbar, so ist $N \cap \underline{\underline{r}}_G N$ ein nilpotenter Normalteiler von G mit endlichem $N/\left[N \cap \underline{\underline{r}}_G N\right]$.*

Dies folgt sofort aus Lemma 3.4 *(a)*.

BEMERKUNG. Lemma 3.2 und Folgerung 3.5 ergeben zusammen eine leichte Verallgemeinerung von McLain [p. 40, Theorem 2].

SATZ 3.6. *Die folgenden Eigenschaften von* $N \triangleleft G$ *sind äquivalent:*

(i) N <u>hek</u> G *und* N *ist endlich erzeugbar.*

(ii) *(a)* N *ist endlich erzeugbar;*

(b) *es gibt* $J \triangleleft G$ *mit endlichem* G/J *und* $N \circ^{(i)} J = 1$ *für fast alle* i ;

(iii) *(a)* N *und* $G/\underline{c}_G N$ *sind noethersch und fast nilpotent;*

(b) *es gibt eine positive ganze Zahl* m *mit folgender Eigenschaft:*

ist σ *ein Epimorphismus von* G *auf* H , *ist* $X \triangleleft H$ *mit* $1 \subset X \subseteq N^\sigma$, *so gibt es* $Y \triangleleft H$ *mit* $1 \subset Y \subseteq X$ *und* m *teilendem* $[H : \underline{c}_H Y]$.

(iv) *(a)* N *ist endlich erzeugbar;*

(b) *ist* σ *ein Epimorphismus von* G *auf* H *mit* $N^\sigma \neq 1$, *so gibt es einen endlich erzeugbaren, residuell minimalen Normalteiler* $X \triangleleft H$ *mit* $1 \subset X \subseteq N^\sigma$;

(c) *es gibt eine positive ganze Zahl* m *mit folgender Eigenschaft:*

ist σ *ein Epimorphismus von* G *auf* H *mit endlichem* $N^\sigma \neq 1$, *so gibt es einen minimalen Normalteiler* $M \triangleleft H$ *mit* $M \subseteq N^\sigma$ *und* m *teilendem* $[H : \underline{c}_H M]$.

(v) *(a)* N *ist endlich erzeugbar;*

(b) *ist* σ *ein Epimorphismus von* G *auf* H *mit* $N^\sigma \neq 1$, *so gibt es einen endlich erzeugbaren, residuell minimalen Normalteiler* $X \triangleleft H$ *mit* $1 \subset X \subseteq N^\sigma$;

(c) *es gibt eine positive ganze Zahl* e *mit folgenden Eigenschaften:*

(c1) $N/[G^e \cap N]$ *ist endlich;*

(c2) *ist* σ *ein Epimorphismus von* G *auf* H *mit endlichem* $N^\sigma \neq 1$, *so gibt es einen minimalen Normalteiler* $M \triangleleft H$ *mit* $M \subseteq N^\sigma$ *und* $H^e \subseteq \underline{c}_H M$.

(vi) *(a)* N *ist fast polyzyklisch;*

(b) es gibt $J \triangleleft G$ *mit folgenden Eigenschaften:*

(b1) G/J *ist eine endlich erzeugbare Torsionsgruppe;*

(b2) ist σ *ein Epimorphismus von* G *auf* H *mit endlichem* $N^\sigma \neq 1$, *so gibt es einen minimalen Normalteiler* $M \triangleleft H$ *mit* $M \subseteq N^\sigma$ *und* $J^\sigma \subseteq \underline{\underline{c}}_H M$.

TERMINOLOGISCHE ERINNERUNG. G^e ist die von allen Elementen g^e mit $g \in G$ erzeugte charakteristische Untergruppe von G.

BEWEIS. Wegen Lemma 3.4 *(a)* ist *(ii)* eine Folge von *(i)*.

Wir nehmen die Gültigkeit von *(ii)* an. Dann ist N endlich erzeugbar; und es gibt einen Normalteiler $J \triangleleft G$ mit endlichem G/J und $N \circ^{(i)} J = 1$ für fast alle i. Also gibt es positive ganze Zahlen m und n mit

(1) $$[G : J] = m \text{ und } N \circ^{(n)} J = 1 .$$

Wir betrachten einen Epimorphismus σ von G auf H und $X \triangleleft H$ mit $1 \subset X \subseteq N^\sigma$. Dann folgt aus (1):

$$X \circ^{(n)} J^\sigma \subseteq N \circ^{(n)} J^\sigma = [N \circ^{(n)} J]^\sigma = 1 .$$

Also gibt es eine nicht negative ganze Zahl a mit

$$X \circ^{(a+1)} J^\sigma = 1 \subset X \circ^{(a)} J^\sigma = Y ;$$

und Y ist ein Normalteiler von H mit $1 \subset Y \subseteq X$ und $Y \circ J^\sigma = 1$. Hieraus folgt $J^\sigma \subseteq \underline{\underline{c}}_H Y$, so daß $[H : \underline{\underline{c}}_H Y]$ ein Teiler von $[H : J^\sigma]$ ist, das wiederum wegen (1) ein Teiler von m ist. Damit haben wir die Gültigkeit von *(iiib)* dargetan.

Aus (1) folgt sofort:

(2a) $N \cap J$ ist nilpotent der Klasse n und $[N : N \cap J]$ ist ein Teiler von m.

Setzen wir $N_i = N \circ^{(i)} J$, so ist N_i mit N ein Normalteiler von G; und es ist

$$N_0 = N , \quad N_i \circ J = N_{i+1} \subseteq N_i , \quad N_n = 1$$

wegen (1). Die von J in N induzierte Automorphismengruppe stabilisiert die Normalteilerreihe N_i, ist also nach einem Satz von Kaloujnine nilpotent der Klasse n; siehe Specht [p. 366, Satz 44]. Dies ist äquivalent mit folgender Aussage:

(2b) $J\underline{\underline{c}}_G N/\underline{\underline{c}}_G N$ ist nilpotent der Klasse n und $[G : J\underline{\underline{c}}_G N]$ ist wegen (1) ein Teiler von m.

Wegen *(iia)* und *(iiib)* wird die Bedingung *(ii)* des Lemma 2.4 von $N \triangleleft G$ erfüllt. Also gilt:

(2c) N und $G/\mathfrak{c}_G N$ sind fast polyzyklisch.

Wegen (2a und b) sind N und $G/\mathfrak{c}_G N$ fast nilpotent, so daß wegen (2c) auch *(iiia)* gilt: wir haben *(iii)* von *(ii)* abgeleitet.

Wir nehmen die Gültigkeit von *(iii)* an. Dann gilt selbstverständlich *(iva)*. Sei σ ein Epimorphismus von G auf H mit $N^\sigma \neq 1$; und wir nehmen weiter an, daß 1 der einzige in N^σ enthaltene, endliche Normalteiler von H ist. Wegen *(iiia)* ist N fast polyzyklisch; und charakteristische Untergruppen von N^σ sind Normalteiler von H. Also gibt es eine freie abelsche charakteristische Untergruppe endlichen Ranges von N^σ, die von 1 verschieden ist; siehe etwa Robinson [I; p. 65/66]. Es folgt die Existenz von $X \triangleleft H$ mit $1 \subset X \subseteq N^\sigma$, so daß X frei abelsch endlichen Ranges ist. Damit haben wir gezeigt:

(3a) N^σ enthält einen von 1 verschiedenen, endlichen Normalteiler von H; oder N^σ enthält einen von 1 verschiedenen Normalteiler von H, der frei abelsch endlichen Ranges ist.

Enthält N^σ einen von 1 verschiedenen, endlichen Normalteiler von H, so enthält N^σ auch einen endlichen, minimalen Normalteiler von H. Enthält N^σ von 1 verschiedene Normalteiler von H, die frei abelsch endlichen Ranges sind, so gibt es unter diesen einen A minimalen Ranges. Ist p irgendeine Primzahl, so ist $A^p \subset A$ und als charakteristische Untergruppe von $A \triangleleft H$ ist $A^p \triangleleft H$. Weiter ist A/A^p endlich, da der Rang der freien abelschen Gruppe A endlich ist. Also gibt es $A(p) \triangleleft H$ mit $A^p \subseteq A(p) \subset A$, so daß $A/A(p)$ ein minimaler Normalteiler von $H/A(p)$ ist. Da $[A : A(p)]$ für jede Primzahl p durch p teilbar ist, ist $A/\bigcap_p A(p)$ unendlich; und aus der Minimalität des Ranges von A folgt $\bigcap_p A(p) = 1$. Damit haben wir gezeigt, daß A ein residuell minimaler Normalteiler von H ist; und folglich gilt:

(3b) N^σ enthält einen endlichen, minimalen Normalteiler von H; oder N^σ enthält einen von 1 verschiedenen, endlich erzeugbaren, abelschen, residuell minimalen Normalteiler von H.

Aus (3b) folgt sofort die Gültigkeit von *(ivb)*. Die Gültigkeit von *(iva)* ist eine Folge von *(iiia)*; und die Gültigkeit von *(ivc)* leitet man mühelos aus *(iiib)* ab. Also folgt *(iv)* aus *(iii)*.

Wir nehmen als nächstes die Gültigkeit von *(iv)* an. Dann gelten selbstverständlich die Bedingungen *(iiia* und *b)* des Satzes 2.6. Ist weiter σ ein Epimorphismus von G auf H mit endlichem N^σ und M der einzige, in N^σ enthaltene, minimale Normalteiler von H, ist weiter N nicht abelsch, so ist

$$1 = \underline{\underline{z}}M = M \cap \underline{\underline{c}}_H M .$$

Also folgt aus unserer Bedingung *(ivc)*, daß

$o(M) = [M\underline{\underline{c}}_H M : \underline{\underline{c}}_H M]$ ein Teiler von $[H : \underline{\underline{c}}_H M]$ und also von m ist;

und damit haben wir auch die Gültigkeit der Bedingung *(iiic)* des Satzes 2.6 verifiziert. Folglich gilt:

(4a) N ist fast polyzyklisch.

Ist σ ein Homomorphismus von G und N^σ eine Torsionsgruppe, so ist N^σ eine fast polyzyklische Torsionsgruppe; und Anwendung eines Satzes von Kurt Hirsch zeigt die Endlichkeit von N^σ; siehe Robinson [II; p. 139, Corollary]. Also gilt:

(4b) Ist σ ein Homomorphismus von G und N^σ eine Torsionsgruppe, so ist N^σ endlich.

Hieraus folgt insbesondere:

(4) $N/(G^n \cap N)$ ist für jede positive ganze Zahl n endlich.

Man sieht mühelos ein, daß *(v)* aus *(iv)* und (4) folgt.

Wir nehmen als nächstes die Gültigkeit von *(v)* an. Es ist klar, daß die Bedingungen *(iiia* und *b)* des Satzes 2.6 gelten. Aus unserer Bedingung *(vc1)* folgt, daß

(5a) $[N : G^e \cap N] = m$ eine positive ganze Zahl ist.

Wir betrachten einen Epimorphismus σ von G auf H mit endlichem N^σ mit folgenden Eigenschaften:

es gibt einen und nur einen minimalen Normalteiler $M \triangleleft H$ mit $M \subseteq N^\sigma$ und M ist nicht abelsch.

Hieraus folgt erstens

(5b) $1 = \underline{\underline{z}}M = M \cap \underline{\underline{c}}_H M$;

und aus Bedingung *(vc2)* folgt also

(5c) $M \cap H^e \subseteq M \cap \underline{\underline{c}}_H M = 1$.

Aus $M \subseteq N^\sigma$ und (5c) folgt dann

$$M = M/[M \cap H^e] = M/[M \cap N^\sigma \cap H^e] \cong M(N^\sigma \cap H^e)/(N^\sigma \cap H^e) \subseteq N^\sigma/(N^\sigma \cap H^e)$$
$$= N^\sigma/(N^\sigma \cap G^{e\sigma}) ,$$

so daß M ein epimorphes Bild von $N/(N \cap G^e)$ ist. Anwendung von (5a) zeigt dann:

(5d) $o(M)$ ist ein Teiler von m .

Damit haben wir aber gezeigt, daß auch die Bedingung *(iiia)* des Satzes 2.6 erfüllt ist; und hieraus folgt:

(5) N ist fast polyzyklisch.

Wir machen nun die folgende zu widerlegende Annahme:

(6+) $N \underline{\underline{\text{hek}}} G$ ist falsch.

Da N wegen (5) noethersch ist, folgt sofort:

(6a) Es gibt einen Epimorphismus von G auf H , der N in K überführt, so daß $K \underline{\underline{\text{hek}}} H$ falsch ist, während aus $X \triangleleft H$ und $1 \subset X \subseteq K$ die Gültigkeit von $K/X \underline{\underline{\text{hek}}} H/X$ folgt.

Aus der Transitivität der Beziehung $\underline{\underline{\text{hek}}}$ und (6a) ergibt sich:

(6b) Aus $X \underline{\underline{\text{hek}}} H$ und $X \subseteq K$ folgt $X = 1$.

Aus (6b) folgt insbesondere, daß 1 der einzige, in K enthaltene endliche Normalteiler von H ist. Aus (5) und (6a) folgt, daß K fast polyzyklisch ist, und daß also K insbesondere unendlich ist. Hieraus ergibt sich die Existenz eines abelschen Normalteilers $A \triangleleft H$ mit $1 \subset A \subseteq K$; und natürlich ist A endlich erzeugbar. Dann ist die Torsionsuntergruppe von A eine endliche charakteristische Untergruppe von A , also ein endlicher, in K enthaltener Normalteiler von H , also gleich 1 . Folglich ist A ein freier abelscher Normalteiler endlichen Ranges; und dieser enthält einen freien abelschen Normalteiler B minimalen positiven Ranges. Jeder in B enthaltene, von 1 verschiedene Normalteiler von H ist frei abelsch von gleichem Range wie B , hat also endlichen Index in B . Damit haben wir gezeigt:

(6c) Es gibt einen freien abelschen Normalteiler $B \triangleleft H$ mit $1 \subset B \subseteq K$, so daß H rational irreduzibel auf B operiert.

Sei $X \triangleleft H$ mit $X \subset B$, so daß B/X ein minimaler Normalteiler von H/X ist. Wegen (5) und (6a) ist K fast polyzyklisch. Also gibt es unter den $Y \triangleleft H$ mit $X \subseteq Y \subset K$ und $B \not\subseteq Y$ ein maximales V . Aus der Minimalität von B/X ergibt sich

$$B \cap V = X ;$$

und aus der Maximalität von V folgt, daß $B \subseteq L$ aus $V \subset L \subseteq K$ und $L \triangleleft H$ folgt. Insbesondere ist also 1 der einzige in K/V enthaltene torsionsfreie Normalteiler von H/V ; und Anwendung des Satzes von Kurt Hirsch zeigt die Endlichkeit von K/V ; siehe Robinson [II; p. 139, Corollary]. Da BV/V der einzige in K/V enthaltene, minimale Normalteiler von H/V ist, und da K/V endlich ist, können wir Bedingung *(vc2)* anwenden:

$$VH^e/V \text{ zentralisiert } BV/V .$$

Also ist insbesondere

$$B \circ H^e \subseteq B \cap V = X ;$$

und damit haben wir gezeigt:

(6d) ist $X \triangleleft H$ mit $X \subset B$, ist B/X ein minimaler Normalteiler von H/X, so ist $B \circ H^e \subseteq X$.

Ist p eine Primzahl, so ist B/B^p eine elementar abelsche, endliche p-Gruppe, da ja B frei abelsch endlichen Ranges ist; und aus $B \neq 1$ folgt $B^p \subset B$. Da B^p eine charakteristische Untergruppe von $B \triangleleft H$ ist, folgt $B^p \triangleleft H$. Also gibt es $B(p) \triangleleft H$ mit $B^p \subseteq B(p) \subset B$, so daß $B/B(p)$ ein endlicher, minimaler Normalteiler von $H/B(p)$ ist; und Anwendung von (6d) zeigt:

$$B \circ H^e \subseteq B(p) \subset B \text{ mit } p \leq [B : B(p)] .$$

Es folgt, daß $B \circ H^e$ ein Normalteiler von H mit unendlichem $B/(B \circ H^e)$ ist, und daß also aus (6c)

(6e) $$B \circ H^e = 1$$

folgt. Ist Γ die von H in B induzierte Automorphismengruppe, so folgt

(6f) $$\Gamma^e = 1$$

aus (6e). Da B eine freie abelsche Gruppe endlichen Ranges ist, ist Γ im wesentlichen dasselbe wie eine Gruppe linearer Transformationen eines Vektorraumes endlichen Ranges über den rationalen Zahlen. Wegen (6f) können wir also einen Satz von Burnside anwenden, um die Endlichkeit von Γ zu bewiesen; vergl. Curtis-Reiner [p. 251, (36.1)]. Anwendung der definierenden Eigenschaft (b*) der Beziehung <u>hek</u> zeigt also

(6g) B <u>hek</u> H.

Wegen (6c) ist $1 \subset B \subseteq K$ und $B \triangleleft H$; und aus (6g) und (6b) folgt der Widerspruch $B = 1$. Also hat unsere Annahme (6+) zu einem Widerspruch geführt: es gilt N <u>hek</u> G, so daß aus (5) die Gültigkeit von *(i)* folgt. Damit haben wir die Äquivalenz von *(i)-(v)* dargetan.

Wir nehmen die Gültigkeit der äquivalenten Bedingungen *(i)-(v)* an. Aus *(iiia)* folgt die Gültigkeit von *(via)*; und aus *(iib)* folgt die Existenz von $J \triangleleft G$ mit endlichem G/J und $N \circ^{(i)} J = 1$ für fast alle i. Ist σ ein Epimorphismus von

G auf H und M ein minimaler Normalteiler von H mit $M \subseteq N^{\sigma}$, so wird

$$M \circ^{(i)} J^{\sigma} \subseteq N^{\sigma} \circ^{(i)} J^{\sigma} = [N \circ^{(i)} J]^{\sigma} = 1 \quad \text{für fast alle } i \text{ ;}$$

und hieraus folgt $M \circ J^{\sigma} = 1$ wegen der Minimalität von M. Also gilt $J^{\sigma} \subseteq \underline{\underline{c}}_H M$; und damit haben wir *(vib)* [und viel mehr] bewiesen: *(vi)* folgt aus *(i)-(v)*.

Wir nehmen umgekehrt die Gültigkeit von *(vi)* an und betrachten einen Epimorphismus σ von G auf H und einen endlichen, minimalen Normalteiler $M \triangleleft H$ mit $M \subseteq N^{\sigma}$. Wegen *(via)* sind N und also auch N^{σ} fast polyzyklisch. Aus einem Satz von Kurt Hirsch folgt deshalb die Existenz einer torsionsfreien, charakteristischen Untergruppe C von N^{σ} mit endlichem N^{σ}/C ; siehe Robinson [2; p. 139, Corollary]. Als charakteristische Untergruppe eines Normalteilers ist C ein Normalteiler von H ; und aus der Torsionsfreiheit von C sowie der Endlichkeit von M folgt $M \cap C = 1$. Damit haben wir die Existenz von $X \triangleleft H$ mit $X \subseteq N^{\sigma}$, endlichem N^{σ}/X und $X \cap M = 1$ gezeigt; und da N^{σ} [wegen *(via)*] fast polyzyklisch ist, gibt es unter diesen X ein maximales D. Wir bemerken, daß

(8a) $\qquad D \triangleleft H$, $D \subseteq N^{\sigma}$, $D \cap M = 1$, N^{σ}/D endlich ist;

und aus der Maximalität von D folgt:

(8b) MD/D ist der einzige in N^{σ}/D enthaltene, minimale Normalteiler von H/D.

Aus (8a und b) folgt die Anwendbarkeit der Bedingung *(vib2)*:

(8c) $$J^{\sigma}D/D \subseteq \underline{\underline{c}}_{H/D}(MD/D) = \underline{\underline{c}}_H(MD/D)/D = \underline{\underline{c}}_H M/D .$$

Also gilt:

(8) Ist σ ein Epimorphismus von G auf H und M ein endlicher, minimaler Normalteiler von H mit $M \subseteq N^{\sigma}$, so ist $J^{\sigma} \subseteq \underline{\underline{c}}_H M$.

Sei σ wieder ein Epimorphismus von G auf H ; und $A \triangleleft H$ mit $1 \subset A \subseteq N^{\sigma}$ sei eine freie abelsche Gruppe endlichen Ranges, auf der H rational irreduzibel operiert. Ist p irgendeine Primzahl, so ist A^p als charakteristische Untergruppe von A ein Normalteiler von H ; und A/A^p ist endlich, da A frei abelsch endlichen Ranges ist. Also gibt es $A(p) \triangleleft H$ mit $A^p \subseteq A(p) \subset A$, so daß $A/A(p)$ ein endlicher, minimaler Normalteiler von $H/A(p)$ ist. Anwendung von (8) zeigt

$$J^{\sigma}A(p)/A(p) \subseteq \underline{\underline{c}}_{H/A(p)}[A/A(p)] \text{ ;}$$

und hieraus folgt

(9a) $J^\sigma \circ A \subseteq A(p)$ für jede Primzahl p .

Natürlich ist $J^\sigma \circ A \triangleleft H$; und aus (9a) folgt die Unendlichkeit von $A/(J^\sigma \circ A)$. Da H rational irreduzibel auf A operiert, folgt hieraus $J^\sigma \circ A = 1$; und dies ist mit

(9b) $$J^\sigma \subseteq \underline{\underline{c}}_H A$$

äquivalent. Aus *(vib1)* folgt, daß G/J und also auch H/J^σ eine endlich erzeugbare Torsionsgruppe ist. Also ergibt sich aus (9b):

(9c) $H/\underline{\underline{c}}_H A$ ist eine endlich erzeugbare Torsionsgruppe.

Die von H in A induzierte Automorphismengruppe Λ ist im wesentlichen mit $H/\underline{\underline{c}}_H A$ identisch; und es folgt aus (9c), daß Λ eine endlich erzeugbare Torsionsgruppe ist. Da A eine freie abelsche Gruppe endlichen Ranges ist, ist Λ eine Gruppe linearer Transformationen eines Vektorraumes endlichen Ranges über den rationalen Zahlen. Folglich können wir einen Satz von I. Schur anwenden, um die Endlichkeit von Λ zu beweisen; siehe Curtis-Reiner [p. 252, (36.2)]. Also ist $H/\underline{\underline{c}}_H A \cong \Lambda$ endlich; und wir haben bewiesen:

(9) Ist σ ein Epimorphismus von G auf H , ist $A \triangleleft H$, ist $1 \subset A \subseteq N^\sigma$, ist A eine freie abelsche Gruppe endlichen Ranges, auf der H rational irreduzibel operiert, so ist $H/\underline{\underline{c}}_H A$ endlich.

Sei σ ein Epimorphismus von G auf H mit $N^\sigma \neq 1$. Gibt es erstens einen endlichen Normalteiler $E \triangleleft H$ mit $1 \subset E \subseteq N^\sigma$, so ist $H/\underline{\underline{c}}_H E$ mit E endlich. Wir nehmen also zweitens an:

(10a) 1 ist der einzige in N^σ enthaltene, endliche Normalteiler von H .

Aus *(via)* folgt, daß N^σ fast polyzyklisch ist. Also folgt aus $N^\sigma \neq 1$ und (10a) die Existenz einer charakteristischen Untergruppe B von A , die frei abelsch endlichen, positiven Ranges ist; vergl. Robinson [1; p. 66, lines 1-3]. Aus $A \triangleleft H$ folgt $B \triangleleft H$: wir haben die Existenz von in A enthaltenen Normalteilern von H gezeigt, die frei abelsch endlichen, positiven Ranges sind. Unter diesen gibt es einen R minimalen Ranges; und es ist klar, daß H auf R rational irreduzibel operiert. Anwendung von (9) zeigt:

(10b) $H/\underline{\underline{c}}_H R$ ist endlich.

Zusammenfassend haben wir also bewiesen:

(10) Ist σ ein Epimorphismus von G auf H mit $N^\sigma \neq 1$, so gibt es

$S \triangleleft H$ mit $1 \subset S \subseteq N^{\sigma}$ und endlichem $H/\underline{\underline{c}}_H S$.

Dies ist genau die definierende Eigenschaft (b*) der hyper-endlich-klassig eingebetteten Normalteiler: es gilt N $\underline{\underline{\text{hek}}}$ G . Aus *(via)* folgt noch die endliche Erzeugbarkeit von N . Damit haben wir *(i)* von *(vi)* abgeleitet und die Äquivalenz von *(i)*-*(vi)* bewiesen.

BEMERKUNG. Wir haben nicht entscheiden können, ob man die Bedingung *(via)* durch die Bedingungen (*va* und *b*) ersetzen kann.

4. Kommutator hyperzentralisierte Normalteiler

DEFINITION 4.1. $N \triangleleft G$ heißt *kommutator hyperzentralisiert*, wenn zu jedem Epimorphismus σ von G auf H mit $N^{\sigma} \neq 1$ ein $X \triangleleft H$ mit $1 \subset X \subseteq N^{\sigma}$ und $X \circ H' = 1$ existiert.

Diese Einbettungsrelation werden wir kurz durch N $\underline{\underline{\text{khz}}}$ G bezeichnen.

Stets gilt 1 $\underline{\underline{\text{khz}}}$ G . - Sind A und B Normalteiler von G mit $A \subseteq B$, so gilt dann und nur dann B $\underline{\underline{\text{khz}}}$ G , wenn A $\underline{\underline{\text{khz}}}$ G und B/A $\underline{\underline{\text{khz}}}$ G/A gelten. - Dann und nur dann gilt N $\underline{\underline{\text{hkz}}}$ G , wenn es zu jedem Epimorphismus σ von G auf H mit $N^{\sigma} \neq 1$ ein X $\underline{\underline{\text{khz}}}$ H mit $1 \subset X \subseteq N^{\sigma}$ gibt. Hieraus folgert man noch, daß das Produkt P aller $\underline{\underline{\text{khz}}}$-eingebetteten Normalteiler von G der Beziehung P $\underline{\underline{\text{khz}}}$ G genügt, während aus $P \subseteq X \triangleleft G$ und X/P $\underline{\underline{\text{khz}}}$ G/P stets $P = X$ folgt.

SATZ 4.2. *Die folgenden Eigenschaften von* $N \triangleleft G$ *sind äquivalent:*

(i) *(a)* N *und* $G/\underline{\underline{c}}_G N$ *sind polyzyklisch;*

(b) N $\underline{\underline{\text{khz}}}$ G .

(ii) N *ist polyzyklisch mit* N $\underline{\underline{\text{khz}}}$ G .

(iii) N *ist polyzyklisch und* $N \circ^{(i)} G' = 1$ *für fast alle* i .

(iv) *(a)* N *ist endlich erzeugbar;*

(b) *ist* σ *ein Epimorphismus von* G *auf* H *mit* $N^{\sigma} \neq 1$, *so gibt es einen endlich erzeugbaren, residuell minimalen Normalteiler* $X \triangleleft H$ *mit* $1 \subset X \subseteq N^{\sigma}$;

(c) *ist* σ *ein Epimorphismus von* G *auf* H *mit endlichem* $N^{\sigma} \neq 1$, *so gibt es einen minimalen Normalteiler* $M \triangleleft H$ *mit* $M \subseteq N^{\sigma}$ *und* $M \circ H' = 1$.

BEWEIS. Es ist klar, daß *(ii)* aus *(i)* folgt. Gilt *(ii)*, so gibt es eine endliche Kette von Normalteilern $N_i \triangleleft G$ mit

$$1 = N_0 , \quad N_{i+1} \circ G' \subseteq N_i \subseteq N_{i+1} , \quad N_n = N ;$$

und hieraus folgt $N \circ^{(n)} G' = 1$: wir haben *(iii)* aus *(ii)* abgeleitet.

Gilt *(iii)*, so ist die polyzyklische Gruppe N insbesondere endlich erzeugbar: es gilt *(iva)*. Ist σ ein Epimorphismus von G auf H mit $N^\sigma \neq 1$, so folgt aus der Polyzyklizität von N die Existenz eines endlich erzeugbaren, abelschen Normalteilers $A \triangleleft H$ mit $1 \subset A \subseteq N^\sigma$; und hieraus folgert man:

Es gibt einen abelschen Normalteiler $B \triangleleft H$ mit $1 \subset B \,[\subseteq A] \subseteq N^\sigma$, so daß B entweder endlich oder frei abelsch endlichen Ranges ist.

Ein endlicher, von 1 verschiedener Normalteiler enthält einen endlichen, minimalen Normalteiler; und ein von 1 verschiedener, freier abelscher Normalteiler endlichen Ranges enthält einen freien abelschen Normalteiler minimalen positiven Ranges. In beiden Fällen haben wir einen von 1 verschiedenen, residuell minimalen Normalteiler erhalten. Also gilt:

Es gibt einen residuell minimalen Normalteiler $R \triangleleft H$ mit $1 \subset R \subseteq B \subset N^\sigma$.

Damit haben wir die Gültigkeit von *(ivb)* bewiesen.

Sei weiter σ ein Epimorphismus von G auf H mit endlichem $N^\sigma \neq 1$. Dann gibt es einen minimalen Normalteiler $M \triangleleft H$ mit $M \subseteq N^\sigma$. Aus *(iii)* folgt

$$M \circ^{(i)} H' \subseteq N^\sigma \circ^{(i)} G' = [N \circ^{(i)} G']^\sigma = 1 \text{ für fast alle } i ;$$

und aus der Minimalität von M folgt $M \circ H' = 1$.

Also gilt auch *(ivc)*: wir haben *(iv)* von *(iii)* abgeleitet.

Wir nehmen schließlich die Gültigkeit der Bedingung *(iv)* an. Dann gelten selbstverständlich auch die Bedingungen (*iiia* und *b*) des Satzes 2.6. Ist weiter σ ein Homomorphismus von G mit endlichem N^σ , ist M der einzige in N^σ enthaltene, minimale Normalteiler von G^σ , so folgt

$$1 = M \circ (G^\sigma)'$$

aus unserer Bedingung *(ivc)*. Insbesondere ist also $M \circ M' = 1$; und hieraus folgt wegen der Minimalität von M , daß M abelsch ist. Also ist die Bedingung *(iiic)* des Satzes 2.6 [leer] erfüllt; und damit haben wir gezeigt:

(1a) N ist fast polyzyklisch.

Das Produkt P aller polyzyklischen Normalteiler von N ist wegen (1a) eine polyzyklische charakteristische Untergruppe von N mit endlichem N/P . Insbesondere ist $P \triangleleft G$. Wäre $P \subset N$, so gäbe es wegen der Endlichkeit von N/P und Bedingung *(ivc)* einen Normalteiler $W \triangleleft G$ mit folgenden Eigenschaften:

$P \subset W \subseteq N$, W/P ist ein minimaler Normalteiler von G/P , $1 = [W/P] \circ [G/P]'$.
Dann ist

$$W'' = W' \circ W' \subseteq W \circ G' \subseteq P ,$$

so daß W mit P polyzyklisch ist. Hieraus folgt der Widerspruch $W \subseteq P \subset W$. Also ist $P = N$. Folglich gilt:

(1b) N ist polyzyklisch.

Sei λ ein Epimorphismus von G auf L und M ein endlicher, minimaler Normalteiler von L mit $M \subseteq N^\lambda$. Da N^λ wegen (1b) polyzyklisch ist, folgt aus einem Satz von Kurt Hirsch die Existenz einer torsionsfreien, charakteristischen Untergruppe C von N^λ mit endlichem N^λ/C ; vergl. Robinson [2; p. 139, Corollary]. Dann ist $C \triangleleft L$ [als charakteristische Untergruppe eines Normalteilers]; und $M \cap C = 1$, da M endlich und C torsionsfrei ist. Da N^λ/C endlich ist, gibt es unter den zwischen C und N^λ gelegenen, M trivial schneidenden Normalteilern von L einen maximalen J . Dieser hat die folgenden Eigenschaften:

(2a) $J \triangleleft L$, $J \subseteq N^\lambda$, $J \cap M = 1$, N^λ/J ist endlich;

(2b) JM/J ist der einzige, in N^λ/J enthaltene, minimale Normalteiler von L/J .

Wir können *(ivc)* anwenden und erhalten:

(2c) $$1 = [JM/J] \circ [L/J]' = J[JM \circ L']/J ;$$

und hieraus folgt wegen (2a) und $M \triangleleft L$:

(2d) $$M \circ L' \subseteq M \cap J = 1 .$$

Damit haben wir gezeigt:

(2) ist λ ein Epimorphismus von G auf L , ist M ein endlicher, minimaler Normalteiler von L mit $M \subseteq N^\lambda$, so ist $M \circ L' = 1$.

Sei wieder λ ein Epimorphismus von G auf L ; und es sei A ein freier abelscher Normalteiler von L mit $A \subseteq N^\lambda$, so daß L auf A rational irreduzibel operiert. Ist dann p eine Primzahl, so ist A^p eine charakteristische Untergruppe von $A \triangleleft L$, woraus $A^p \triangleleft L$ folgt; und da A wegen (1b) frei abelsch endlichen Ranges ist, ist A/A^p endlich. Also gibt es $A(p) \triangleleft L$ mit $A^p \subseteq A(p) \subset A$, so daß $A/A(p)$ ein endlicher, minimaler Normalteiler von $L/A(p)$ ist. Anwendung von (2) ergibt:

$$1 = [A/A(p)] \circ [L/A(p)]' = A(p)[A \circ L']/A(p) ;$$

und dies ist mit

(3a) $$A \circ L' \subseteq A(p) \text{ für jede Primzahl } p$$

äquivalent. Da $[A : A(p)]$ für jede Primzahl p durch p teilbar ist, ist $A/\bigcap_p A(p)$ unendlich; und aus der Tatsache, daß L rational irreduzibel auf A operiert, folgt

(3b) $$1 = \bigcap_p A(p) .$$

Durch Kombination von (3a und b) ergibt sich

$$A \circ L' \subseteq \bigcap_p A(p) = 1 ;$$

und damit haben wir gezeigt:

(3) ist λ ein Epimorphismus von G auf L, ist A ein freier abelscher, in N^λ enthaltener Normalteiler endlichen Ranges von L, auf dem L rational irreduzibel operiert, so ist $A \circ L' = 1$.

Sei λ ein Epimorphismus von G auf L mit $N^\lambda \neq 1$.

FALL 1. Es gibt einen in N^λ enthaltenen, von 1 verschiedenen, endlichen Normalteiler von L.

Dann gibt es auch einen in N^λ enthaltenen, endlichen, minimalen Normalteiler $M \triangleleft L$; und Anwendung von (2) ergibt $M \circ L' = 1$.

FALL 2. 1 ist der einzige in N^λ enthaltene, endliche Normalteiler von L.

Da $N^\lambda \neq 1$ und wegen (1b) polyzyklisch ist, existieren dann von 1 verschiedene, in N^λ enthaltene, freie abelsche Normalteiler endlichen Ranges von L; und unter diesen gibt es einen A kleinsten Ranges. Dann operiert aber L rational irreduzibel auf A; und Anwendung von (3) ergibt $A \circ L' = 1$.

Fassen wir die in den beiden Fällen erzielten Resultate zusammen, so erhalten wir:

(4) ist λ ein Epimorphismus von G auf L mit $N^\lambda \neq 1$, so gibt es $J \triangleleft L$ mit $1 \subset J \subseteq N^\lambda$ und $J \circ L' = 1$.

Kombination von (1b) und (4) ergibt:

(5) N ist polyzyklisch und N khz G.

Ist σ ein Epimorphismus von G auf H mit $N^\sigma \neq 1$, so folgt aus (4) die Existenz von $J \triangleleft H$ mit $1 \subset J \subseteq N^\sigma$ und $J \circ H' = 1$. Die von H in J induzierte Automorphismengruppe ist also abelsch; und da J wegen (5) mit N polyzyklisch ist,

ist bekanntlich diese Automorphismengruppe auch endlich erzeugbar; siehe Robinson [1; p. 82, Theorem 3.27]. Damit haben wir [wegen (5)] gezeigt, daß $N \triangleleft G$ der Bedingung *(ii)* des Satzes 2.4 genügt. Also gilt:

(6) $$G/\underline{\underline{c}}_G N \text{ ist fast polyzyklisch.}$$

Aus (5) folgt die Gültigkeit unserer Bedingung *(ii)*, aus der wir bereits unsere Bedingung *(iii)* abgeleitet haben. Also gilt:

(7) $$N \circ^{(i)} G' = 1 \quad \text{für fast alle } i .$$

Aus (7) und einem Satz von Kaloujnine folgt, daß G' in N eine nilpotente Automorphismengruppe induziert; vergl. Specht [p. 366, Satz 44]. Also induziert G in N eine auflösbare Automorphismengruppe; und es folgt aus (6):

(8) $$G/\underline{\underline{c}}_G N \text{ ist polyzyklisch.}$$

Die Aussagen (5) und (8) sind identisch mit unserer Bedingung *(i)*; und damit ist die Äquivalenz von *(i)*-*(iv)* bewiesen.

Diskussion von Satz 4.2

A. Es gibt viele Gruppen G mit folgenden Eigenschaften:

(a) G ist endlich erzeugbar;

(b) G' ist eine freie abelsche Gruppe abzählbar unendlichen Ranges.

Das einfachste Beispiel einer solchen Gruppe ist $G = F/F''$, wobei F eine endlich erzeugbare, nicht abelsche, freie Gruppe ist, wie sich aus dem Nielsen-Schreier-schen Untergruppensatz ergibt; siehe Specht [p. 155, Satz 5].

Eine solche Gruppe hat die folgenden weiteren Eigenschaften:

(c) $G'' = 1$;

(d) $(G \circ G') \circ G' = 1$.

Hieraus folgt:

(e) $G \,\underline{\underline{\mathrm{khz}}}\, G$ [und $G' \,\underline{\underline{\mathrm{khz}}}\, G$] .

Dar Paar $N = G$, G genügt also den Bedingungen *(ib)*, zweite Hälfte von *(ii)*, zweite Hälfte von *(iii)*, (*iva* und *c*);

aber G ist nicht polyzyklisch.

Also sind die Bedingungen *(ia)*, erste Hälfte von *(ii)*, erste Hälfte von *(iii)*, *(ivb)* unentbehrlich.

B. Wir haben nicht entscheiden können, ob man in *(ivb)* die geforderte endliche Erzeugbarkeit entbehren kann, und ob man in *(ii)*, *(iii)* das Wort "polyzyklisch" durch "endlich erzeugbar" ersetzen kann.

5. Gleichzeitig hek-und khz-eingebettete Normalteiler

SATZ 5.1. *Die folgenden Eigenschaften von* $N \triangleleft G$ *sind äquivalent:*

(i) *(a)* N *ist endlich erzeugbar;*

(b) N hek G *und* N khz G .

(ii) *(a)* N *und* $G/\underline{c}_G N$ *sind polyzyklisch;*

(b) *es gibt eine positive ganze Zahl* e *mit folgender Eigenschaft:*

ist σ *ein Epimorphismus von* G *auf* H *mit* $N^\sigma \neq 1$, *so gibt es* $X \triangleleft H$ *mit* $X' = 1 \subset X \subseteq N^\sigma$ *und zyklischem* $H/\underline{c}_H X$ *mit* e *teilender Ordnung.*

(iii) *(a)* N *ist endlich erzeugbar;*

(b) *ist* σ *ein Epimorphismus von* G *auf* H *mit* $N^\sigma \neq 1$, *so gibt es einen endlich erzeugbaren, residuell minimalen Normalteiler* $X \triangleleft H$ *mit* $1 \subset X \subseteq N^\sigma$;

(c) *es gibt eine positive ganze Zahl* e *mit folgenden Eigenschaften:*

ist σ *ein Epimorphismus von* G *auf* H *mit endlichem* $N^\sigma \neq 1$, *so gibt es einen minimalen Normalteiler* M *von* H *mit* $M \subseteq N^\sigma$ *und abelschem* $H/\underline{c}_H M$ *mit* e *teilender Ordnung.*

BEWEIS. Wir nehmen zunächst die Gültigkeit von *(i)* an. Aus der endlichen Erzeugbarkeit von N hek G und Satz 3.6, *(iiia)* folgt dann:

(1a) N und $G/\underline{c}_G N$ sind noethersch und fast nilpotent.

Aus N khz G und Satz 4.2 folgt, daß N/X auflösbar ist, wenn nur $X \triangleleft G$, $X \subseteq N$ und N/X endlich ist. Wegen (1a) ist N fast nilpotent; und hieraus folgt bekanntlich, daß das Produkt aller nilpotenten Normalteiler von N eine nilpotente charakteristische Untergruppe C von N mit endlichem N/C ist. Da charakteristische Untergruppen von Normalteilern wieder Normalteiler sind, gilt $C \triangleleft G$. Also ist N/C auflösbar, so daß die Erweiterung N der nilpotenten Gruppe C durch die auflösbare Gruppe N/C selbst auflösbar ist. Wegen (1a) ist N noethersch. Also gilt:

(1b) N ist polyzyklisch.

Wegen (1b) und N khz G gilt Bedingung *(ii)* des Satzes 4.2; und daraus folgt:

(1c) $G/\underline{c}_G N$ ist polyzyklisch.

Aus der endlichen Erzeugbarkeit von N hek G und Satz 3.7, *(iiib)* folgt:

(2a) es gibt eine positive ganze Zahl m mit folgender Eigenschaft:

ist σ ein Epimorphismus von G auf H, ist $X \triangleleft H$ mit $1 \subset X \subseteq N^{\sigma}$, so gibt es $Y \triangleleft H$ mit $1 \subset Y \subseteq X$ und m teilendem $[H : \underline{\underline{c}}_H Y]$.

Aus N $\underline{\underline{\text{khz}}}$ G und (1b) folgt die Gültigkeit von Bedingung *(ii)* des Satzes 4.2. Also gilt:

(2b) $$N \circ^{(i)} G' = 1 \text{ für fast alle } i.$$

Ist σ ein Epimorphismus von G auf H, ist $X \triangleleft H$ mit $1 \subset X \subseteq N^{\sigma}$, so folgt aus (2a) die Existenz von $Y \triangleleft H$ mit $1 \subset Y \subseteq X$ und m teilendem $[H : \underline{\underline{c}}_H Y]$. Aus (2b) folgt dann:

$$Y \circ^{(i)} H' \subseteq N^{\sigma} \circ^{(i)} G'^{\sigma} = [N \circ^{(i)} G']^{\sigma} = 1 \text{ für fast alle } i.$$

Wegen $Y \neq 1$ gibt es dann eine positive ganze Zahl k mit

$$Y \circ^{(k)} H' = 1 \subset Y \circ^{(k-1)} H' = U.$$

Dann gilt also $1 \subset U \subseteq Y \subseteq X$ und $U \circ H' = 1$. Wäre $\underline{\underline{z}}U = 1$, so wäre also auch $U \cap H' = 1$, woraus die Kommutativität von U und also $\underline{\underline{z}}U = U \neq 1$ folgten, ein Widerspruch aus dem sich $1 \neq \underline{\underline{z}}U$ ergibt. Wir setzen $V = \underline{\underline{z}}U$. Als charakteristische Untergruppe des Normalteilers $U \triangleleft H$ ist $V \triangleleft H$; und es gilt $1 \subset V \subseteq U \subseteq Y \subseteq X$. Weiter ist $V \circ H' \subseteq U \circ H' = 1$; und es ist $\underline{\underline{c}}_H Y \subseteq \underline{\underline{c}}_H U \subseteq \underline{\underline{c}}_H V$, so daß auch $[H : \underline{\underline{c}}_H V]$ ein Teiler von m ist. Damit haben wir gezeigt:

(2) es gibt eine positive ganze Zahl m mit folgender Eigenschaft:

ist σ ein Epimorphismus von G auf H, ist $X \triangleleft H$ mit $1 \subset X \subseteq N^{\sigma}$, so gibt es $Y \triangleleft H$ mit $Y' = 1 \subset Y \subseteq X$ und abelschem $H/\underline{\underline{c}}_H Y$ und m teilendem $[H : \underline{\underline{c}}_H Y]$.

Wir betrachten einen Epimorphismus σ von G auf H und einen endlichen, minimalen Normalteiler $M \triangleleft H$ mit $M \subseteq N^{\sigma}$. Kombination von (2) mit der Minimalität von M ergibt:

M und $H/\underline{\underline{c}}_H M$ sind abelsch und $[H : \underline{\underline{c}}_H M]$ ist ein Teiler von m.

Hieraus folgt aber wegen der Minimalität von M, daß $H/\underline{\underline{c}}_H M$ eine endliche zyklische Gruppe mit m teilender Ordnung ist; vergl. etwa Baer [5; p. 29, Lemma 5.1, (2)]. Damit haben wir gezeigt:

(3) ist σ ein Epimorphismus von G auf H und M ein endlicher, minimaler Normalteiler von H mit $M \subseteq N^{\sigma}$, so ist M abelsch und

$H/\underline{\underline{c}}_H M$ ist zyklisch mit m teilender Ordnung.

Wir betrachten einen Epimorphismus σ von G auf H und einen freien abelschen Normalteiler $A \triangleleft H$ endlichen Ranges n mit $1 \subset A \subseteq N^{\sigma}$, so daß H rational irreduzibel auf A operiert. Ist p eine Primzahl, so ist A^p eine charakteristische Untergruppe von $A \triangleleft H$ und also $A^p \triangleleft H$. Weiter ist A/A^p endlich der Ordnung $p^n \neq 1$. Also existiert $A(p) \triangleleft H$ mit $A^p \subseteq A(p) \subset A$, so daß $A/A(p)$ ein minimaler Normalteiler von $H/A(p)$ ist. Weiter ist $A/A(p) \subseteq N^{\sigma}/A(p)$. Kombination von (3) und der Minimalität von $A/A(p)$ zeigt, daß H in $A/A(p)$ eine zyklische Gruppe von Automorphismen mit m teilender Ordnung induziert. Hieraus folgt

$$H^m H' \circ A \subseteq A(p) \quad \text{für jede Primzahl } p .$$

Es folgt insbesondere, daß $H^m H' \circ A$ ein Normalteiler von H mit unendlichem $A/[H^m H' \circ A]$ ist. Da H auf A rational irreduzibel operiert, folgt hieraus

$$H^m H' \circ A = 1 .$$

Ist also Γ die von H in A induzierte Automorphismengruppe, so haben wir gezeigt:

$$\Gamma \text{ ist abelsch und } \Gamma^m = 1 ;$$

$$\Gamma \text{ operiert rational irreduzibel auf } A .$$

Wir können also Folgerung 1.2, (A) auf das Paar A, Γ anwenden; und es folgt, daß Γ zyklisch mit m teilender Ordnung ist. Wegen $\Gamma \cong H/\underline{\underline{c}}_H A$ haben wir also gezeigt:

(4) ist σ ein Epimorphismus von G auf H und A ein freier abelscher Normalteiler endlichen Ranges von H mit $1 \subset A \subseteq N^{\sigma}$, auf dem H rational irreduzibel operiert, so ist $H/\underline{\underline{c}}_H A$ zyklisch mit m teilender Ordnung.

Sei σ ein Epimorphismus von G auf H mit $N^{\sigma} \neq 1$. Enthält erstens N^{σ} einen endlichen Normalteiler $E \triangleleft H$ mit $1 \neq E$, so gibt es auch einen minimalen Normalteiler $M \triangleleft H$ mit $M \subseteq E \subseteq N^{\sigma}$; und aus (3) folgt, daß M abelsch und $H/\underline{\underline{c}}_H M$ zyklisch mit m teilender Ordnung ist. Ist zweitens 1 der einzige endliche, in N^{σ} enthaltene Normalteiler von H, so folgern wir aus (1b), daß N^{σ} polyzyklisch ist und es also von 1 verschiedene, in N^{σ} enthaltene, abelsche Normalteiler von H gibt, die einmal endlich erzeugbar sind [da ja N^{σ} polyzyklisch ist] und die weiter torsionsfrei und also frei abelsch endlichen Ranges sind [da ja 1 der einzige, in N^σ enthaltene, endliche Normalteiler von H ist]. Unter diesen Normalteilern gibt es einen A minimalen Ranges: es ist also $A \triangleleft H$, $1 \subset A \subseteq N^\sigma$ und A ist frei abelsch endlichen Ranges; und aus der Minimalität des Ranges von A folgt, daß H auf A rational irreduzibel operiert. Anwendung von (4) zeigt dann, daß $H/\underline{c}_H A$ zyklisch mit m teilender Ordnung ist; und damit haben wir gezeigt:

(5) ist σ ein Epimorphismus von G auf H mit $N^\sigma \neq 1$, so gibt es $A \triangleleft H$ mit $1 \subset A \subseteq N^\sigma$, so daß A abelsch und $H/\underline{c}_H A$ zyklisch mit m teilender Ordnung ist.

Aus (1a und b), (2) und (5) folgt die Gültigkeit von *(ii)* und damit haben wir *(ii)* aus *(i)* abgeleitet.

Man prüft leicht nach, daß *(iii)* nur eine abgeschwächte Form von *(ii)* ist.

Gilt schließlich *(iii)*, so gilt gewiß auch Bedingung *(iv)* des Satzes 3.6, woraus N <u>hek</u> G folgt; und ebenso folgt aus *(iii)* die Gültigkeit der Bedingung *(iv)* des Satzes 4.2, woraus N <u>khz</u> G folgt. Damit haben wir die Gültigkeit von *(i)* dargetan und die Äquivalenz der Bedingungen *(i)*-*(iii)* bewiesen.

BEMERKUNG 5.2. *Ist N endlich erzeugbar mit N <u>hek</u> G und N <u>khz</u> G, sind e und s positive ganze Zahlen, so dass für jeden Homomorphismus σ von G und jeden in N^σ enthaltenen endlichen, minimalen Normalteiler M von G^σ die Ordnung $o(M)$ die s-te Potenz einer Primzahl teilt und die von G^σ in M induzierte Automorphismengruppe zyklisch mit e teilender Ordnung ist, ist weiter a die Anzahl der verschiedenen Primfaktoren von e und b gleich 1 oder 0, je nachdem 8 ein Teiler von e ist oder nicht, so gilt:*

ist σ ein Homomorphismus von G mit $N^\sigma \neq 1$, ist 1 der einzige, in N^σ enthaltene, endliche Normalteiler von G^σ, so gibt es ein A mit folgenden Eigenschaften:

$$A \triangleleft G^\sigma, \quad 1 \subset A \subseteq N^\sigma;$$

A ist frei abelsch mit $s2^{a}{}^{b}$ nicht überschreitendem Rang;

die von G^σ in A induzierte Automorphismengruppe ist zyklisch mit e teilender Ordnung.

Der Beweis sei dem Leser überlassen; er ergibt sich im wesentlichen aus Satz 5.1 [seinem Beweis] und hauptsächlich der Folgerung 1.3.

6. Hyperzentral und überauflösbar eingebettete Normalteiler

SATZ 6.1. *Die folgenden Eigenschaften von* $N \triangleleft G$ *sind äquivalent:*

(i) *(a)* N *und* $G/\underline{\mathfrak{c}}_G N$ *sind noethersch und nilpotent;*

(b) *ist* σ *ein Epimorphismus von* G *auf* H *mit* $N^\sigma \neq 1$, *so ist* $N^\sigma \cap \underline{\underline{z}}H \neq 1$.

(ii) *(a)* N *ist endlich erzeugbar;*

(b) *ist* σ *ein Epimorphismus von* G *auf* H *mit* $N^\sigma \neq 1$, *so gibt es einen residuell minimalen Normalteiler* $X \triangleleft H$ *mit* $1 \subset X \subseteq N^\sigma$;

(c) *ist* σ *ein Epimorphismus von* G *auf* H *mit endlichem* $N^\sigma \neq 1$, *so ist* $N^\sigma \cap \underline{\underline{z}}H \neq 1$.

(iii) *(a)* N *ist endlich erzeugbar;*

(b) *ist* σ *ein Epimorphismus von* G *auf* H *mit* $N^\sigma \neq 1$, *so* $N^\sigma \cap \underline{\underline{z}}H \neq 1$.

BEWEIS. Es ist klar, daß *(iii)* aus *(i)* folgt. Bedenkt man, daß Zentrumselemente Normalteiler erzeugen, so sieht man, daß auch *(ii)* aus *(iii)* folgt.

Wir nehmen schließlich die Gültigkeit von *(ii)* an. Ist σ ein Epimorphismus von G auf H mit noetherschem N^σ, so sieht man sofort ein, daß $N^\sigma \triangleleft H$ der Bedingung *(iii)* des Satzes 5.1 genügt. Hieraus folgt:

(1) ist σ ein Epimorphismus von G auf H mit noetherschem N^σ, so sind N^σ und $H/\underline{\mathfrak{c}}_H N^\sigma$ polyzyklisch.

Sei wieder σ ein Epimorphismus von G auf H mit noetherschem N^σ; und sei M ein endlicher, minimaler Normalteiler von H mit $M \subseteq N^\sigma$. Wegen (1) ist N^σ polyzyklisch. Also folgt aus einem Satz von Kurt Hirsch die Existenz einer torsionsfreien, charakteristischen Untergruppe C von N^σ mit endlichem N^σ/C; vergl. Robinson [2; p. 139, Corollary]. Charakteristische Untergruppen von Normalteilern sind Normalteiler, so daß $C \triangleleft H$. Da M endlich und C torsionsfrei ist, gilt $M \cap C = 1$. Aus der Endlichkeit von N^σ/C folgt, daß es unter den $X \triangleleft H$ mit $C \subseteq X \subseteq N^\sigma$ und $M \cap X = 1$ ein maximales D gibt. Dann ist also $D \triangleleft H$, $D \subseteq N^\sigma$, $D \cap M = 1$ und N^σ/D ist endlich. Aus der Maximalität von D folgt, daß MD/D der einzige, in N^σ/D enthaltene, minimale Normalteiler von H/D ist. Aus der

Endlichkeit von N^σ/D und Bedingung *(iic)* folgt $[N^\sigma/D] \cap \underline{z}[H/D] \neq 1$, da ja $1 \subset M \subseteq N^\sigma/D$ ist. Aus der Einzigkeit von MD/D folgt also

$$MD/D \subseteq [N^\sigma/D] \cap \underline{z}[H/D] \subseteq \underline{z}[H/D] ;$$

und hieraus folgern wir

$$M \circ H \subseteq M \cap D = 1 ,$$

woraus sich $M \subseteq \underline{z}H$ ergibt. Also gilt:

(2) ist σ ein Epimorphismus von G auf H mit noetherschem N^σ, ist M ein minimaler Normalteiler von H mit $M \subseteq N^\sigma$, so ist $M \subseteq \underline{z}H$.

Sei σ ein Epimorphismus von G auf H mit der Eigenschaft:

(3a) ist β ein Homomorphismus von H, der keinen Isomorphismus in N^σ induziert, so ist $N^{\sigma\beta}$ noethersch.

Ist $N^\sigma \neq 1$, so folgt aus Bedingung *(iib)* die Existenz eines residuell minimalen Normalteilers $R \triangleleft H$ mit $1 \subset R \subseteq N^\sigma$. Aus (3a) folgt dann, daß

(3b) N^σ/R noethersch ist.

Sei $\underline{S}$ die Menge aller $X \triangleleft H$ mit $X \subset R$, so daß R/X ein endlicher, minimaler Normalteiler von H/X ist. Dann folgt

(3c) $$1 = \bigcap_{X \in \underline{S}} X$$

aus der residuellen Minimalität von R.

Ist $X \in \underline{S}$, so ist R/X endlich und aus (3b) folgt, daß N^σ/X noethersch ist. Da R/X ein in N^σ/X enthaltener, endlicher, minimaler Normalteiler von H/X ist, ergibt also Anwendung von (2), daß $R/X \subseteq \underline{z}(H/X)$ gilt. Dies ist mit $R \circ H \subseteq X$ gleichwertig, so daß aus (3c)

$$R \circ H \subseteq \bigcap_{X \in \underline{S}} X = 1$$

folgt; und dies ist mit

$$1 \subset R \subseteq N^\sigma \cap \underline{z}H$$

gleichwertig. Ist $1 \neq r \in R$, so ist $1 \subset \{r\} \triangleleft H$, da Zentrumselemente Normalteiler erzeugen. Wegen (3a) ist $N^\sigma/\{r\}$ noethersch, so daß auch N^σ noethersch ist. Folglich gilt:

(3) ist σ ein Epimorphismus von G auf H mit $N^\sigma \neq 1$, ist $N^{\sigma\beta}$

noethersch für jeden Homomorphismus β von H, der keinen Isomorphismus in N^σ induziert, so ist N^σ ebenfalls noethersch und $1 \neq N^\sigma \cap \underline{\underline{z}}H$.

Sei $\underline{\underline{M}}$ die Menge aller $X \triangleleft G$ mit $X \subseteq N$ und noetherschem N/X. Aus (1) folgt, daß N/X für $X \in \underline{\underline{M}}$ polyzyklisch ist. Wir machen die zu widerlegende Annahme:

(+) $\underline{\underline{M}}$ enthält nicht alle in N gelegenen Normalteiler von G.

Dann folgt aus Lemma 2.1 die Existenz von $M \triangleleft G$ mit folgenden Eigenschaften:

(4a) $M \subseteq N$, N/M ist nicht noethersch;

(4b) aus $X \triangleleft G$ und $M \subset X \subseteq N$ folgt, daß N/X noethersch ist.

Aus (4b) und (3) folgt insbesondere, daß M noethersch ist; und dies widerspricht (4a). Die Annahme (+) hat also zu einem Widerspruch geführt, so daß insbesondere $1 \in \underline{\underline{M}}$ und also

(4) N noethersch ist.

Aus (1) und (4) folgt *(ia)*; und aus (3) und (4) folgt *(ib)* da ja epimorphe Bilder noetherscher Gruppen noethersch sind. Damit haben wir *(i)* aus *(ii)* abgeleitet und die Äquivalenz von *(i)*-*(iii)* dargetan.

Diskussion von Satz 6.1

A. Dieser Satz verallgemeinert u.A. Baer [1; p. 203, Theorem].

B. Man beachte, daß in Bedingung *(iib)* nicht die endliche Erzeugbarkeit des residuell minimalen Normalteilers X gefordert wird.

C. Es gibt viele endlich erzeugbare, unendliche, einfache Gruppen G. Dann erfüllt das Paar $N = G \triangleleft G$ trivialerweise die Bedingungen (*iia* und *c*), ohne aber *(iib)* zu genügen, was die Unentbehrlichkeit von *(iib)* zeigt.

Überauflösbar eingebettet ist $N \triangleleft G$, wenn es zu jedem Epimorphismus σ von G auf H mit $N^\sigma \neq 1$ einen zyklischen Normalteiler $Z \triangleleft H$ mit $1 \subset Z \subseteq N^\sigma$ gibt.

HILFSSATZ 6.2. *Ist der überauflösbar eingebettete Normalteiler* $N \triangleleft G$ *endlich erzeugbar, so ist*

(a) N *polyzyklisch und*

(b) $G/\underline{\underline{c}}_G N$ *ist polyzyklisch und überauflösbar.*

BEWEIS. Die Aussage *(a)* ist wohlbekannt und wird leicht aus Lemma 2.1 abgeleitet.

Wir werden als nächstes durch vollständige Induktion bezgl. n die folgende Aussagenfamilie beweisen.

$(b.n)$ Ist $X \triangleleft Y$, gibt es $X_i \triangleleft Y$ mit

$$1 = X_0 \;,\; X_i \subseteq X_{i+1} \;,\; X_n = X \text{ und zyklischem } X_{i+1}/X_i \text{ für } 0 \le i < n \;,$$

so ist $Y/\underline{\underline{c}}_Y X$ eine polyzyklische, überauflösbare Gruppe.

BEWEIS *von* (*b*.1). Dies folgt sofort daraus, daß $X = X_1$ zyklisch und die Automorphismengruppen zyklischer Gruppen endliche, abelsche Gruppen sind.

ABLEITUNG *von* (*b*.n+1) *aus* (*b*.n) *für* $0 < n$. Wir betrachten eine Normalteilerkette $X_i \triangleleft Y$ mit folgenden Eigenschaften:

(0) $\quad 1 = X_0 \,,\; X_i \subseteq X_{i+1}$ mit zyklischem X_{i+1}/X_i für $0 \le i \le n$.

Anwendung von (*b*.n) ergibt sofort:

(1) $\quad Y/\underline{\underline{c}}_Y X_n$ und $[Y/X_1]/\underline{\underline{c}}_{Y/X_1}[X_{n+1}/X_1]$ sind polyzyklisch und überauflösbar.

Aus

$$[Y/X_1]/\underline{\underline{c}}_{Y/X_1}[X_{n+1}/X_1] = [Y/X_1]/[\underline{\underline{c}}_Y(X_{n+1}/X_1)/X_1] \cong Y/\underline{\underline{c}}_Y(X_{n+1}/X_1)$$

und (1) folgt also, daß $Y/\underline{\underline{c}}_Y(X_{n+1}/X_1)$ und $Y/\underline{\underline{c}}_Y X_n$ beide polyzyklisch und überauflösbar sind. Untergruppen und direkte Produkte polyzyklischer, überauflösbarer Gruppen sind polyzyklisch und überauflösbar. Also ist

(2) $\quad Y/[\underline{\underline{c}}_Y X_n \cap \underline{\underline{c}}_Y(X_{n+1}/X_1)]$ polyzyklisch und überauflösbar.

Wegen $0 < n$ und (0) gilt

(3) $\quad X_1 \subseteq X_n$ und $X_{n+1} = \{X_n, a\}$ für geeignetes a .

Ist

$$D = \underline{\underline{c}}_Y X_n \cap \underline{\underline{c}}_Y(X_{n+1}/X_1) \;,$$

so gilt

(3a) $\quad X_n \circ x = 1$ und $X_{n+1} \circ x \subseteq X_1$ für alle $x \in D$.

Sind x, y irgendwelche Elemente aus D , so wird also

$$a \circ xy = (a \circ y)(a \circ x)^y = (a \circ y)(a \circ x) = (a \circ x)(a \circ y) \;,$$

wie sich aus der Kommutativität von $X_1 \subseteq X_n$ und Huppert [p. 253, 1.2 Hilfssatz b)] ergibt. Also gilt:

(3b) die Abbildung $x \to a \circ x$ ist ein Homomorphismus λ von D in die zyklische Gruppe X_1 .

Der Kern von λ besteht aus Elementen, die sowohl X_n als auch a

zentralisieren; und derartige Elemente zentralisieren auch $X_{n+1} = \{X_n, a\}$. Andererseits ist $\underline{\underline{c}}_Y X_{n+1}$ sowohl in D enthalten als auch im Kern von λ. Also gilt:

(3c) $\underline{\underline{c}}_Y X_{n+1}$ ist der Kern von λ.

Aus (3b und c) folgt, daß $D/\underline{\underline{c}}_Y X_{n+1}$ eine zyklische Gruppe ist; und Kombination dieser Tatsache mit (2) zeigt, daß $Y/\underline{\underline{c}}_Y X_{n+1}$ polyzyklisch und überauflösbar ist. Damit haben wir ($b.n$+1) aus ($b.n$) abgeleitet und den Beweis der Aussagenfamilie ($b.n$) induktiv zu Ende geführt.

Aus *(a)* und ($b.n$) folgt die Aussage *(b)*.

BEMERKUNG. Hilfssatz 6.2, *(b)* ist eine Verallgemeinerung von Huppert [p. 719, 9.8 Hilfssatz]. Der induktive Beweis ist nichts weiter als eine Anpassung des Huppertschen Beweises an die gegenwärtige Situation.

SATZ 6.3. *Die folgenden Eigenschaften von $N \triangleleft G$ sind aquivalent:*

(i) *(a)* *N und $G/\underline{\underline{c}}_G N$ sind noethersch und überauflösbar;*

(b) *N ist ein überauflösbar eingebetteter Normalvon G.*

(ii) *(a)* *N ist endlich erzeugbar;*

(b) *N ist ein überauflösbar eingebetteter Normalteiler von G.*

(iii) *(a)* *N ist endlich erzeugbar;*

(b) *ist σ ein Epimorphismus von G auf H mit $N^\sigma \neq 1$, so gibt es einen endlich erzeugbaren, residuell minimalen Normalteiler $X \triangleleft H$ mit $1 \subset X \subseteq N^\sigma$;*

(c) *ist σ ein Epimorphismus von G auf H mit endlichem N^σ, so ist N^σ ein überauflösbar eingebetteter Normalteiler von G. H.*

BEWEIS. Es ist klar, daß *(ii)* aus *(i)* folgt. Bedenkt man, daß jeder endliche, zyklische, von 1 verschiedene Normalteiler einen Normalteiler von Primzahlordnung enthält, der natürlich minimal ist, während unendliche zyklische Normalteiler offenbar residuell minimal sind, so sieht man ein, daß *(iii)* aus *(ii)* folgt.

Wir nehmen schließlich die Gültigkeit von *(iii)* an. Bedenkt man, daß jeder von 1 verschiedene, endliche, zyklische Normalteiler einen minimalen Normalteiler von Primzahlordnung enthält, und daß die Automorphismengruppen zyklischer Gruppen abelsch sind, so sieht man, daß $N \triangleleft G$ der Bedingung *(iv)* des Satzes 4.2 genügt; und hieraus folgt:

(1a) N und $G/\underline{\underline{c}}_G N$ sind polyzyklisch;

(1b) N <u>khz</u> G .

Sei σ ein Epimorphismus von G auf H und M ein endlicher, minimaler Normalteiler von H mit $M \subseteq N^\sigma$. Aus (1a) folgt, daß N^σ polyzyklisch ist. Anwendung des Satzes von Kurt Hirsch ergibt also die Existenz einer torsionsfreien, charakteristischen Untergruppe C von N^σ mit endlichem N^σ/C ; vergl. Robinson [2; p. 139, Corollary]. Aus $N^\sigma \triangleleft H$ folgt $C \triangleleft H$; und da C torsionsfrei und M endlich ist, ist $C \cap M = 1$. Folglich ist CM/C ein minimaler Normalteiler von H/C mit $CM/C \subseteq N^\sigma/C$ und endlichem N^σ/C . Aus Bedingung *(iiia)* folgt, daß N^σ/C ein überauflösbar eingebetteter Normalteiler von H/C ist; und hieraus ergibt sich, daß

$$M = M/(M \cap C) \cong MC/C$$

zyklisch von Primzahlordnung ist. Also gilt:

(2) ist σ ein Epimorphismus von G auf H und M ein endlicher, minimaler Normalteiler von H mit $M \subseteq N^\sigma$, so ist M zyklisch von Primzahlordnung.

Sei σ ein Epimorphismus von G auf H und R ein endlich erzeugbarer, residuell minimaler Normalteiler von H mit $1 \subset R \subseteq N^\sigma$. Weiter sei $\underline{\underline{M}}$ die Menge aller Normalteiler $X \triangleleft H$ mit $X \subset R$ und R/X einem endlichen, minimalen Normalteiler von H/X . Wegen $X \subset R \subseteq N^\sigma$ folgt aus (2) die Kommutativität von R/X [und mehr]; und dies ist mit $R' \subseteq X$ äquivalent. Aus der residuellen Minimalität von R folgt also

$$R' \subseteq \bigcap_{X \in \underline{\underline{M}}} X = 1 ;$$

und damit haben wir gezeigt, daß

(3a) R abelsch ist.

FALL 1. R enthält einen von 1 verschiedenen, endlichen Normalteiler von H .

Dann enthält R auch einen endlichen, minimalen Normalteiler von $R \subseteq N^\sigma$ und dieser ist wegen (2) endlich von Primzahlordnung.

FALL 2. 1 ist der einzige in R enthaltene, endliche Normalteiler von H .

Die Torsionsuntergruppe der endlich erzeugbaren, abelschen [wegen (3a)] Gruppe R ist eine endliche, charakteristische Untergruppe von R , also gleich 1 als endlicher, in R enthaltener Normalteiler von H . Damit haben wir gezeigt:

(3b) R ist eine freie abelsche Gruppe endlichen Ranges mit $R \triangleleft H$ und

$1 \subset R \subseteq N^{\sigma}$.

Alle in R enthaltenen, von 1 verschiedenen Normalteiler von H sind freie abelsche Gruppen endlichen, positiven Ranges; und unter ihnen gibt es also einen A minimalen Ranges. Jeder von 1 verschiedene, in A enthaltene Normalteiler von H hat folglich gleichen Rang wie A und mithin endlichen Index in A, so daß H rational irreduzibel auf A operiert. Dann operiert auch die von H in A induzierte Gruppe Γ von Automorphismen rational irreduzibel auf A. Ist p eine Primzahl, sind U, V Untergruppen von A mit $A^p \subseteq U \subset V \subseteq A$, ist V/U ein Γ-Kompositionsfaktor von A/A^p, so sind U und V Normalteiler von H; und V/U ist ein endlicher, minimaler Normalteiler von H/U. Wegen $A \subseteq R \subseteq N^{\sigma}$ können wir (2) anwenden; und es folgt, daß V/U zyklisch der Ordnung p ist. Damit haben wir gezeigt, daß das Paar A, Γ die Voraussetzungen der Folgerung 1.4 erfüllt. Also ist A eine unendliche zyklische Gruppe.

Wir haben damit in beiden Fällen bewiesen, daß R einen von 1 verschiedenen, zyklischen Normalteiler enthält. Wir formulieren dieses Resultat:

(3) ist σ ein Epimorphimus von G auf H und R ein endlich erzeugbarer, residuell minimaler Normalteiler von H mit $1 \subset R \subseteq N^{\sigma}$, so enthält R einen von 1 verschiedenen, zyklischen Normalteiler von H.

Aus (3) und Bedingung *(iiib)* ergibt sich sofort:

(4) N ist ein überauflösbar eingebetteter Normalteiler von G.

Da N wegen Bedingung *(iiia)* endlich erzeugbar ist, folgt aus (4) und Hilfssatz 6.2:

(5) $G/\underline{\subseteq}_G N$ ist überauflösbar.

Es ist klar, daß überauflösbar eingebettete Normalteiler auch überauflösbare Gruppen sind. Also folgt aus (4), (5) und (1a) die Gültigkeit von *(ia)*, während *(ib)* mit (4) identisch ist. Wir haben *(i)* aus *(iii)* abgeleitet und die Äquivalenz von *(i)*-*(iii)* bewiesen.

Diskussion von Satz 6.3

A. Satz 6.3 ist eine Verallgemeinerung von Baer [4; p. 20, Satz 3.1]. Es sei darauf hingewiesen, daß der dort angegebene Beweis große Ähnlichkeit mit dem Beweis des Satzes 6.3 hat, daß aber der gegenwärtige Beweis einfacher ist.

B. Die Unentbehrlichkeit der in Bedingung *(iiib)* gemachten Endlichen-Erzeugbarkeits-Voraussetzung zeigt das in Baer [4; p. 26/27] konstruierte Gegenbeispiel. Dies ist deshalb bemerkenswert, weil diese Voraussetzung im Satze 6.1 entbehrlich war.

Literaturverzeichnis

Reinhold Baer

[1] "The hypercenter of a group", *Acta Math.* 89 (1953), 165-208. MR15,395.

[2] "Das Hyperzentrum einer Gruppe. III", *Math. Z.* 59 (1953), 299-338. MR15,598.

[3] "Auflösbare Gruppen mit maximalbedingung", *Math. Ann.* 129 (1955), 139-173. MR16,994.

[4] "Überauflösbare Gruppen", *Abh. Math. Sem. Univ. Hamburg* 23 (1959), 11-28. MR21#2687.

[5] "Principal factors, maximal subgroups and conditional identities of finite groups", *Illinois J. Math.* 13 (1969), 1-52. MR38#5933.

Charles W. Curtis, Irving Reiner

Representation theory of finite groups and associative algebras (Pure and Appl. Math., 11. Interscience [John Wiley & Sons], New York, London, 1962). MR26#2519.

Helmut Hasse

[1] "Bericht über neuere Untersuchungen und Probleme aus der Theorie der algebraischen Zahlkörper. Teil I: Klassenkörpentheorie", *Jber. Deutsche Math.-Verein.* 35 (1926), 1-55. FdM52,150.

[2] "Bericht öber neuere Untersuchungen und Probleme aus der Theorie der algebraischen Zahlkörper. Teil Ia: Beweise zu Teil I", *Jber. Deutsche Math.-Verein.* 36 (1927), 233-311. FdM53,143.

[3] "Bericht über neuere Untersuchungen und Probleme aus der Theorie der algebraischen Zahlkörper. Teil II: Reziprozitätsgesetz", *Ergänzungsband VI zum Jber. Deutsche Math.-Verein.* 39 (1930). FdM56,165.

Erich Hecke

Vorlesungen über die Theorie der Algebraischen Zahlen (Akad. Verlagsges., Leipzig, 1923; reprinted, Chelsea, New York, 1948). FdM49,106.

B. Huppert

Endliche Gruppen I (Die Grundlehren der Mathematischen Wissenschaften, Band 134. Springer-Verlag, Berlin, Heidelberg, New York, 1967). MR37#302.

D.H. McLain

"Remarks on the upper central series of a group", *Proc. Glasgow Math. Assoc.* 3 (1956), 38-44. MR18,870.

Derek J.S. Robinson

[1] *Finiteness conditions and generalized soluble groups*, Part 1 (Ergebnisse der Mathematik und ihrer Grenzgebiete, Band 62. Springer-Verlag, Berlin, Heidelberg, New York, 1972). Zbl.243.20032.

[2] *Finiteness conditions and generalized soluble groups*, Part 2 (Ergebnisse der Mathematik und ihrer Grenzgebiete, Band 63. Springer-Verlag, Berlin, Heidelberg, New York, 1972). Zbl.243. 00

Wilhelm Specht

Gruppentheorie (Die Grundlehren der Mathematischen Wissenschaften in einzeldarstellungen, Band 52. Springer-Verlag, Berlin, Gottingen, Heidelberg, 1956). MR18,189.

Forschungsinstitut für Mathematik,
Eidg. Technische Hochschule Zürich,
Zürich, Switzerland.

PROC. SECOND INTERNAT. CONF. THEORY OF GROUPS, 20D99
CANBERRA 1973, pp. 63-64.

CHARACTERISATION OF THE GROUPS WITH THE GASCHÜTZ COHOMOLOGY PROPERTY

D.W. Barnes

In [3], Gaschütz showed that a finite p-group G has the property:

(P_1) *If* A *is a* $\mathbb{Z}G$*-module of* p*-power order and* $H^1(G, A) = 0$, *then* $H^n(U, A) = 0$ *for all* $n \geq 1$ *and all subgroups* U *of* G.

As was pointed out by Hoechsmann, Roquette and Zassenhaus in [5], Gaschütz's argument can be adapted to show that a finite p-group G has the property:

(P_2) *If* A *is a* $\mathbb{Z}G$*-module of* p*-power order and* $\hat{H}^r(G, A) = 0$ *for some* r, *then* $\hat{H}^n(U, A) = 0$ *for all* n *and all subgroups* U *of* G.

In [1], Babakhanian showed that if a Sylow p-subgroup P of G is a Frobenius complement in G, then G has the property (P_2). We show conversely, that these are the only finite groups with the property (P_2). (We must of course allow as Frobenius groups, the usually excluded cases where $P = G$ or $P = 1$.) This follows immediately from

THEOREM. *Let* P *be a Sylow* p*-subgroup of the finite group* G. *Suppose that for every finite-dimensional* $\mathbb{Z}_p G$*-module* A, $H^1(G, A) = 0$ *implies* $H^1(P, A) = 0$. *Then* P *is a Frobenius complement in* G.

PROOF. Let A be a finite-dimensional $\mathbb{Z}_p G$-module such that $H^1(G, A) = 0$. Then $H^1(P, A) = 0$ and, by Gaschütz [3] Satz 1, $H^n(P, A) = 0$ for all $n \geq 1$. But $\text{res}^n : H^n(G, A) \to H^n(P, A)$ is a monomorphism. Thus $H^n(G, A) = 0$. It follows by dimension shifting, that G has the property P^i of Hoechsmann, Roquette and Zassenhaus [5] for all $i > 0$. By their Proposition 1 (b), G is p-nilpotent. Let N be the normal p-complement of G.

Let A be any finite-dimensional $\mathbb{Z}_p G$-module. Since $|N|$ is prime to p, $A = A^N \oplus [A, N]$ and N acts non-trivially on every composition factor of $[A, N]$.

Thus $[A, N]$ has no composition factor in the principal block, and by Gruenberg [4] 8.12 Proposition 22, $H^n(G, [A, N]) = 0$ and $H^n(G, A) = H^n(G, A^N)$ for all $n \geq 0$. It follows that $H^1(P, [A, N]) = 0$. Thus

$$H^1(P, A) = H^1(P, A^N) \oplus H^1(P, [A, N]) = H^1(P, A^N) \simeq H^1(G/N, A^N) .$$

Since $H^1(N, A) = 0$, the exact sequence

$$0 \to H^1(G/N, A^N) \to H^1(G, A) \to H^1(N, A)^G$$

yields $H^1(G/N, A^N) \simeq H^1(G, A)$. We now have $\hat{H}^1(P, A) \simeq \hat{H}^1(G, A)$ for all finite-dimensional Z_pG-modules A. By Barnes [2] Theorems 3.1 and 3.3, P is a Frobenius complement in G.

References

[1] A. Babakhanian, "Cohomological triviality and finite p-nilpotent groups", *Arch. der Math.* 21 (1970), 40-42. MR41#6977.

[2] Donald W. Barnes, "A functorial transfer theorem", *Math. Z.* 118 (1970), 168-174. MR43#357.

[3] Wolfgang Gaschütz, "Kohomologische Trivialitäten und äussere Automorphismen von p-Gruppen", *Math. Z.* 88 (1965), 432-433. MR33#4137

[4] Karl W. Gruenberg, *Cohomological topics in group theory* (Lectures Notes in Mathematics, 143. Springer-Verlag, Berlin, Heidelberg, New York, 1970). MR43#4923.

[5] K. Hoechsmann, P. Roquette and H. Zassenhaus, "A cohomological characterization of finite nilpotent groups", *Arch. der Math.* 19 (1968), 225-244. MR37#2863.

University of Sydney,
Sydney, NSW 2006.

PROC. SECOND INTERNAT. CONF. THEORY OF GROUPS, 20E15, 20F05
CANBERRA 1973, pp. 65-74.

FINITELY PRESENTED METABELIAN GROUPS

Gilbert Baumslag

My objective here today is to describe what is known about the defining relations of finitely generated metabelian groups. This is not a difficult task because very little is known and much of what is known is of very recent origin. Despite this meagre knowledge, the results obtained so far suggest that the theory of finitely presented solvable groups is far richer than one might have suspected. It is for this reason that I have chosen to discuss finitely presented metabelian groups at this time.

There are two fundamental theorems of P. Hall that are of relevance here. The first of these is

A (P. Hall [16], 1954). *Finitely generated metabelian groups satisfy* max-n , *the maximum condition for normal subgroups.*

This result, which is a consequence of the Hilbert basis theorem, has an important corollary.

A1. *Every finitely generated metabelian group is recursively presented, that is, can be presented by a recursively enumerable system of defining relations.*

The second theorem of P. Hall that is needed is

B (P. Hall [17], 1959). *Finitely generated metabelian groups are residually finite (that is, their subgroups of finite index have trivial intersection).*

Corollary A1 makes meaningful the following result which can be deduced from Theorem B and a slightly stronger version of Corollary A1.

B1. *The word problem is solvable for finitely generated metabelian groups.*

These results of Hall show that finitely generated metabelian groups are rather well-behaved. However they shed little light on the nature of finitely presented metabelian groups which Hall made little effort to elucidate. In fact there seems to be almost no information about finitely presented metabelian groups in the literature prior to the 1970's. The only result I have come across is a theorem of A.I. Šmel'kin which is also implicit in Hall's earlier work.

C (A.I. Šmel'kin [26], 1965). *A finitely generated free metabelian group* M *is finitely presented only if it is cyclic.*

As usual a group M is free metabelian if $M \cong F/F''$ for some free group F. Actually Šmel'kin deduces from Theorem C and a well-known result of M. Auslander and R.C. Lyndon [1] the corollary

C1 (A.I. Šmel'kin [26], 1965). *A finitely generated free solvable group of derived length at least two is finitely presented only if it is cyclic.*

Theorem C may be regarded as a special case of the following unpublished result.

D (G. Baumslag [10], 1973). *A finitely generated parafree metabelian group* P *is finitely presented only if it is cyclic.*

Here (following [2]) a group P is termed parafree metabelian if

(a) P is metabelian;

(b) P is residually nilpotent, that is, $\bigcap_{n=1}^{\infty} P_n = 1$ where P_n is the nth term of the lower central series of P;

(c) $P/P_n \cong M/M_n$ for $n = 1, 2, \ldots$, for some free metabelian group M.

According to a theorem of K.W. Gruenberg [15] free metabelian groups are residually nilpotent and hence parafree. Actually parafree metabelian groups abound (see [2]). So Theorem D is a genuine generalisation of Theorem C.

In general there seem to be two ways of proving that a given finitely generated group G is not finitely presented.

The first of these is a two-stage procedure. The first stage is to produce an explicit presentation of G in terms of a finite set of generators and an infinite set of defining relations. The second stage is to prove that no finite subset of these relations suffice to define G. (The fact that G is not finitely presented no matter which finite system of generators one may choose for G is the content of a theorem of B.H. Neumann [22].) Both parts of this procedure can be difficult to effect; the second tends to be the more awkward and often invokes the use of generalised free products.

The second way to prove that a finitely generated group G is not finitely presented is to show that its multiplicator $m(G)$ is not finitely generated, for the multiplicator of a finitely presented group is always finitely generated. Actually it was an open question for a while whether conversely a finitely generated group with a finitely generated multiplicator is finitely presented. This is false; there are even metabelian counter-examples and we shall exhibit further such counter-examples later in this talk. It should be pointed out that it is in general not an easy task to compute the multiplicator of a group. If the group is metabelian a great deal of information about the multiplicator can be extracted from the Lydon-Hochschild-Serre spectral sequence. However it is not always possible to unravel this information.

Šmel'kin used the first of the two methods I have described to prove Theorem C; his proof is quite straightforward. This approach does not seem to be available in dealing with parafree metabelian groups because it is not easy to find presentations for them. In fact Theorem D is proved by demonstrating that the multiplicator of any parafree group is not finitely generated provided only that it is not cyclic. This is accomplished by exploiting the fact that parafree metabelian groups have a matrix representation of a very nice kind.

There is another collection of results about the defining relations of finitely generated metabelian groups that I would like to describe next. These are concerned with what one might term 1-relator metabelian groups. Specifically we call a group O a 1-relator metabelian group if $O \cong K/K''$ where K is a 1-relator group. In other words O can be defined by adding one extra relation to those needed to define a free metabelian group. We shall write

$$O = \langle\langle a_1, \ldots, a_q; r = 1\rangle\rangle$$

to express the fact that O is generated by its elements $a_1, \ldots, a_q$ and defined in terms of these elements by all relations of the form $\big[[w, x], [y, z]\big] = 1$ together with the relation $r = 1$ (here $[u, v] = u^{-1}v^{-1}uv$ and, for later use, $u^v = v^{-1}uv$).

There is a Freiheitssatz for such 1-relator metabelian groups, due to Barbara Long, which mimics Magnus' Freiheitssatz for 1-relator groups, namely,

E (Barbara Long [20], 1970). *Let* O *be the* 1*-relator metabelian group*

$$O = \langle\langle a_1, \ldots, a_q; r = 1\rangle\rangle$$

If r *"involves" all of the generators* $a_1, \ldots, a_q$ *then* $a_1, \ldots, a_{q-1}$ *freely generate a free metabelian group.*

The precise meaning of the word "involves" is somewhat involved and I shall not attempt to explain it here.

Theorem E shows that 1-relator metabelian groups behave a little like free metabelian groups. This is also borne out by

F (G. Baumslag [11], 1973). *Let* $O = \langle\langle a_1, \ldots, a_q; r = 1\rangle\rangle$ *be a* 1*-relator metabelian group that is not cyclic. Then, if either* $q \geq 3$ *or* r *is a proper power,* $m(O)$ *is not finitely generated (and hence* O *is not finitely presented).*

Harking back to Šmel'kin's Theorem we see that Theorem F may also be viewed as a generalisation of that theorem.

One of the reasons for studying 1-relator metabelian groups is that they provide us with a number of finitely and infinitely related groups of a surprisingly complex kind. We observe that, by Theorem F, 1-relator metabelian groups are uniformly well-behaved if they have at least 3 generators. However 2-generator 1-relator metabelian groups can be spectacularly different as the following theorem shows.

G (G. **Baumslag** [11], 1973). *(i) Let*

$$A = \langle\langle a, b;\ a^{-1}b^m a = b^n \rangle\rangle$$

and assume that neither m *nor* n *divides the other. Then* A *is not finitely related but* $m(A) = 1$.

(ii) Let

$$B = \langle\langle a, b;\ [a, b]^a = [a, b]^b [a, b] \rangle\rangle .$$

*Then** B *is a finitely presented group whose derived group is free abelian of infinite rank; in fact*

$$B = \langle a, b;\ [a, b]^a = [a, b]^b [a, b],\ \left[[a, b], [a, b]^b\right] = 1 \rangle$$

is a 2-generator 2-relator group .

(iii) Let

$$C = \langle\langle a, b;\ [a, b]^{2a} = [a, b]^3 \rangle\rangle .$$

Then $m(C)$ *is finitely generated notwithstanding the fact that* C *is an infinitely related group with an abelian normal subgroup of infinite rank.*

I would like to comment on these three examples in turn.

Firstly *(i)*. Suppose for definiteness that $m = 2$ and $n = 3$. Then A' is isomorphic to the subgroup of the additive group of rational numbers comprising rationals of the form $\frac{r}{2^{\alpha}3^{\beta}}$ $(r, \alpha, \beta \in Z)$. If we identify A' with this group then

$$A = A' \rtimes \langle a \rangle$$

where

$$a^{-1}sa = \frac{2s}{3} \quad (s \in A') .$$

So here A is a split extension of a locally cyclic group by an infinite cyclic group and yet it is not finitely related. A is, I think, the simplest example of a finitely generated infinitely related group. It is rather surprising that it has not been noted as such until now. It is also rather surprising that this very simple

* This was first proved by Jim Boler.

group has trivial multiplicator, underlining the fact that there exist finitely generated groups with finitely generated multiplicators which are not finitely related (*cf.* G. Baumslag [3], [7] and [8]). I would like to point out that the group $A = \langle\langle a, b; a^{-1}b^2a = b^3\rangle\rangle$ was brought to my attention by R. Bieri in May of 1973. He asked at that time whether A is finitely presented. He had already proved that $m(A) = 1$. This group arose in connection with his work on the cohomological dimension of certain solvable groups (R. Bieri [12]).

Consider next the groups in *(ii)*. The interesting feature here is that B contains an abelian normal subgroup of infinite rank although it is finitely presented (see also G. Baumslag [4]).

Finally consider *(iii)*. The group C involved here shows how little influence the internal properties of even a 1-relator metabelian group have on the multiplicator. Thus despite the fact that the group is infinitely related and despite the fact that it contains an abelian normal subgroup of infinite rank its multiplicator is finitely generated (*cf.* G. Baumslag [6]).

The preceding discussion serves to indicate that there are a substantial number of quite complex finitely generated but not finitely related metabelian groups. They are, nevertheless, all recursively presented (by Corollary A1). Thus by appealing to Graham Higman's remarkable embedding theorem [18], every finitely generated metabelian group can be embedded in a finitely presented group. However Higman's procedure leads to a group which is not even solvable. The main result in this theory I have been describing is the following theorem.

H (G. Baumslag [5], 1972). *Every finitely generated metabelian group G can be embedded in a finitely presented metabelian group G^*.*

Theorem H shows that Higman's Theorem can be dispensed with as far as finitely generated metabelian groups are concerned.

The one facet of Theorem H that I want to emphasise is that it shows that the abelian normal subgroup structure of finitely presented metabelian groups, is somewhat surprisingly, no less complex than that of finitely generated metabelian groups. This seems to indicate, as I remarked at the outset, that the theory of finitely presented solvable groups may well turn out to be far richer than one might hitherto have suspected.

Professor M.I. Kargarpolov has informed me at this conference that V.N. Remeslennikov has very recently announced in [25] that he too has proved Theorem H, independently. At the same time Remeslennikov has announced the existence of a finitely presented group with an infinitely generated center. This striking fact means, in particular, that for finitely presented groups as a whole the impact on even the central subgroup structure is negligible. Professor Kargarpolov has further informed me that Remeslennikov has proved that Philip Hall's Theorem A does not carry over to finitely presented solvable groups:

I (V.N. Remeslennikov [24], 1972). *There exists a finitely presented solvable group which does not satisfy* max-n .

In this connection it might also be noted that Theorem B does not carry over to finitely presented solvable groups either, for there exists a finitely presented centre-by-metabelian group which is not residually finite (G. Baumslag [9]).

There is one more fact that I would like to mention which is connected to Theorem H and can be proved in a similar way.

J (J. Boler [13], 1973). *Every finitely generated metabelian group whose derived group is of finite rank can be embedded in a finitely presented metabelian group with the same property.*

I shall sketch the proof of Theorem H at the end of this lecture. Before I do so I would like to give one application of the theory I have described and then raise some problems that arise here.

The application that I have in mind is to finitely presented linear groups*. Some years ago Hyman Bass and later Serge Lang [19], p. 162 asked: If L is a finitely presented linear group over a field of characteristic zero, is there a specialisation of the underlying coefficient ring of L into an algebraic number field which maps L monomorphically? Now the abelian subgroups of any finitely generated solvable linear group over an algebraic number field are necessarily of finite rank (A.I. Mal'cev [21]). So in order to provide a negative answer to the Bass-Lang question we need only produce a finitely presented linear group with an abelian subgroup of infinite rank (where of course the underlying field is of characteristic zero). Such groups are now in plentiful supply because of Theorem H - one need only note that a finitely generated torsion-free metabelian group always has a linear representation over a field of characteristic zero (V.N. Remeslennikov [23]). An explicit example is provided by the group L generated by the matrices (over the quotient field of the ring of integral polynomials in x)

$$a = \begin{pmatrix} 1 & 1 \\ 1 & 0 \end{pmatrix}, \quad s = \begin{pmatrix} 1+x & 0 \\ 0 & 1 \end{pmatrix}, \quad t = \begin{pmatrix} x & 0 \\ 0 & 1 \end{pmatrix} .$$

For L' is free abelian of infinite rank but L has the following presentation

$$L = \langle a, s, t; [a, a^t] = [s, t] = 1, a^s = aa^t \rangle .$$

These facts are established in G. Baumslag [4]. They were also discovered independently by V.N. Remeslennikov according to the announcement [25].

The Bass-Lang problem suggests the following:

PROBLEM 1. *Is there any way of discerning finitely presented metabelian groups from the other finitely generated metabelian groups?*

* This was first pointed out to me by R.C. Lyndon in 1971.

Notice that the multiplicator is not sufficiently discriminating to settle Problem 1.

PROBLEM 2. *Let* G *and* H *be finitely generated metabelian groups. If* G *and* H *have the same finite images is* H *finitely presented whenever* G *is?*

Problem 2 leads naturally to the isomorphism and conjugacy problems for finitely presented metabelian groups which are essentially untouched (*cf.* R.C. Brigham [14]).

Theorem H raises the possibility that there may be analogous results for other varieties of groups. In general one may ask

PROBLEM 3. *Let* $\underline{\underline{V}}$ *be a variety of groups such that there are only countably many finitely generated groups in* $\underline{\underline{V}}$. *Can every finitely generated group in* $\underline{\underline{V}}$ *be embedded in a finitely presented group which is also in* $\underline{\underline{V}}$ *?*

In view of Philip Hall's results [17] one may specifically ask whether every finitely generated abelian-by-polycyclic group can be embedded in a finitely presented group of the same kind.

Another possibility raised by Theorem H that I would like to mention is

PROBLEM 4. *Can the free solvable group of rank two and derived length three be embedded in a finitely presented solvable group with the same derived length?*

I would like now to turn my attention to some of the problems suggested by the results about 1-relator metabelian groups. The first will be couched in somewhat more general terms because of the following partial generalisation of Theorem F.

K (G. Baumslag [11], 1973). *Let*

$$G = \langle\langle a_1, \ldots, a_q; r_1 = 1, \ldots, r_p = 1\rangle\rangle .$$

If $q - p \geq 2$ *then* $m(G)$ *is not finitely generated (and hence* G *is not finitely presented).*

PROBLEM 5. *Let*

$$G = \langle a_1, \ldots, a_q; r_1 = 1, \ldots, r_p = 1\rangle .$$

If $q - p \geq 2$ *and* $n \geq 2$ *is* $G/G^{(n)}$ *necessarily infinitely related (where* $G^{(n)}$ *denotes the* n*th term of the derived series of* G *)?*

PROBLEM 6. *Let* G *be a* 1*-relator group. Is* $G/G^{(n+1)}$ *necessarily infinitely related whenever* $G/G^{(n)}$ *is?*

Finally, I would like to end by sketching the proof of Theorem H. To this end let G be a finitely generated metabelian group. The proof that G can be embedded in a finitely presented metabelian group can be divided up into three steps.

Step 1. By appealing to a theorem of W. Magnus G is embedded in W/N where W

is the wreath product of two finitely generated abelian groups. This embedding is arranged so that N is contained in the base group of W.

Step 2. W is embedded in a finitely presented metabelian group W^*. I shall indicate how this is accomplished when

$$W = \langle a \rangle \operatorname{wr} \langle t \rangle$$

is the wreath product of two infinite cyclic groups. Notice that W may be presented as follows

$$W = \left\langle a,\ t;\ \left[a,\ a^{t^i}\right] = 1 \ (i = 0, \pm 1, \ldots)\right\rangle .$$

We now add an extra generator s to this presentation and two extra relations, giving a group

$$(1) \qquad W^* = \left\langle a,\ s,\ t;\ a^s = aa^t,\ [s,\ t] = 1,\ \left[a,\ a^{t^i}\right] = 1 \ (i = 0, \pm 1, \ldots)\right\rangle .$$

Two things need to be checked. First that $W \leq W^*$ and, more importantly, that W^* is finitely presented. Indeed

$$(2) \qquad W^* = \langle a,\ s,\ t;\ a^s = aa^t,\ [s,\ t] = 1,\ [a,\ a^t] = 1\rangle .$$

The difficulty is to prove that the relations $\left[a,\ a^{t^i}\right] = 1 \quad (i = 0, \pm 1, \ldots)$ follow from the ones exhibited. In fact the trick is made clear by the following simple calculations:

$$(3) \qquad 1 = [a,\ a^t] = [a,\ a^t]^s = [a^s,\ a^{ts}] = [a^s,\ a^{st}] = \left[aa^t,\ a^ta^{t^2}\right] .$$

Now $[a^t,\ a] = 1$ and so $\left[a^t,\ a^{t^2}\right] = 1$, that is, a^t centralises both a and a^{t^2}. This implies that

$$(4) \qquad \left[aa^t,\ a^ta^{t^2}\right] = \left[a,\ a^ta^{t^2}\right] = \left[a,\ a^{t^2}\right] .$$

Hence putting (3) and (4) together yields

$$\left[a,\ a^{t^2}\right] = 1 .$$

In this way it follows that (1) and (2) define the same group, as needed. This is how Step 2 is completed.

Step 3. Step 2 has some freedom in it. If this is utilised it is possible to arrange that N^*, the normal closure of N in W^*, meets W in N:

$$N^* \cap W = N .$$

So

$$G \leq W^*/N^* .$$

But W^* is finitely presented and hence, appealing to Hall's Theorem A, so is every quotient of W^*. In particular W^*/N^* is a finitely presented group and so G has been embedded in a finitely presented metabelian group $G^* = W^*/N^*$, as required.

The above proof of Step 2 is taken from [4]. The full proof of Theorem H will appear in [7].

ACKNOWLEDGEMENTS. I would like to express my thanks to the National Science Foundation, the University of Warwick and the Australian National University for their support.

References

[1] Maurice Auslander and R.C. Lyndon, "Commutator subgroups of free groups", *Amer. J. Math.* 77 (1955), 929-931. MR17,709.

[2] Gilbert Baumslag, "Groups with the same lower central sequence as a relatively free group. I The groups", *Trans. Amer. Math. Soc.* 129 (1967), 308-321. MR36#248.

[3] Gilbert Baumslag, "A finitely generated infinitely related group with trivial multiplicator", *Bull. Austral. Math. Soc.* 5 (1971), 131-136. MR45#6897.

[4] Gilbert Baumslag, "A finitely presented metabelian group with a free abelian group of infinite rank", *Proc. Amer. Math. Soc.* 35 (1972), 61-62.MR45#871 .

[5] Gilbert Baumslag, "On finitely presented metabelian groups", *Bull. Amer. Math. Soc.* 78 (1972), 279. MR45#354.

[6] Gilbert Baumslag, "Some remarks about multiplicators and finitely presented groups", *Math. Z.* 126 (1972), 239-242.

[7] Gilbert Baumslag, "Subgroups of finitely presented metabelian groups", *J. Austral. Math. Soc.* 16 (1973), 98-110.

[8] Gilbert Baumslag, "A remark on groups with trivial multiplicator", *Amer. J. Math.* (to appear).

[9] Gilbert Baumslag, "A finitely presented solvable group that is not residually finite", *Math. Z.* 133 (1973), 125-127.

[10] Gilbert Baumslag, "On parafree metabelian groups", unpublished.

[11] Gilbert Baumslag, "One-relator metabelian groups", unpublished.

[12] Robert Bieri, "Über die cohomologische Dimension der auflösbaren Gruppen", *Math. Z.* 128 (1972), 235-242. Zbl.237.20027.

[13] J. Boler, PhD thesis, Rice University, 1974.

[14] Robert C. Brigham, "On the isomorphism problem for just-infinite groups", *Comm. Pure Appl. Math.* 24 (1971), 789-796. MR44#5377.

[15] K.W. Gruenberg, "Residual properties of infinite soluble groups", *Proc. London Math. Soc.* (3) 7 (1957), 29-62. MR19,386.

[16] P. Hall, "Finiteness conditions for soluble groups", *Proc. London Math. Soc.* (3) 4 (1954), 419-436. MR17,344.

[17] P. Hall, "On the finiteness of certain soluble groups", *Proc. London Math. Soc.* (3) 9 (1959), 595-622. MR22#1618.

[18] G. Higman, "Subgroups of finitely presented groups", *Proc. Roy. Soc. London Ser. A* 262 (1961), 455-475. MR24#152.

[19] Serge Lang, *Diophantine geometry* (Interscience Tracts in Pure and Applied Mathematics, 11. Interscience [John Wiley & Sons], New York, London, 1962). MR26#119.

[20] Barbara Long, PhD thesis, City University of New York, 1970.

[21] А.И. Мальцев [A.I. Mal'cev], "О некоторых классах бесконечных разрешимых групп" [On certain classes of infinite solvable groups], *Mat. Sb. (NS)* 28 (70) (1951), 567-588; *Amer. Math. Soc. Transl.* (2) 2 (1956), 1-21. MR13,203.

[22] B.H. Neumann, "Some remarks on infinite groups", *J. London Math. Soc.* 12 (1937), 120-127. FdM63,64.

[23] В.Н. Ремесленнников [V.N. Remeslennikov], "Представление конечно порожденных метабелевых групп матрицами" [Representation of finitely generated metabelian groups by matrices], *Algebra i Logika* 8 (1969), 72-75; *Algebra and Logic* 8 (1969), 39-40. MR44#335.

[24] В.Н. Ремесленников [V.N. Remeslennikov], "Пример конечноопределенной разрешимой группы без условия максимальности для нормальных подгрупп" [A finitely presented soluble group without maximum condition for normal subgroups], *Mat. Zametki* 12 (1972), 287-293.

[25] В.Н. Ремесленников [V.N. Remeslennikov], "О конечно-определенных группах" [On finitely-presented groups], *Proc. Fourth All-union Symposium on the Theory of Groups*, February 1973, pp. 164-169 (Novosibirsk, 1973).

[26] А.Л. Шмелькин [A.L. Šmel'kin], "О разрешимых произведениях групп" [On soluble products of groups], *Sibirsk. Mat. Ž.* 6 (1965), 212-220. MR32#2464.

Graduate Center of the City University of New York,
New York 10036, USA.

PROC. SECOND INTERNAT. CONF. THEORY OF GROUPS, CANBERRA 1973, pp. 75-81.

20F05

SOME PROBLEMS ON ONE-RELATOR GROUPS

Gilbert Baumslag

A. Classification

The isomorphism problem is perhaps the most important as well as the most intractable problem in the theory of 1-relator groups. Almost nothing is known besides Magnus' Theorem [16]:

A 1-relator group $G = \langle a_1, \ldots, a_q;\ r = 1 \rangle$ *which can be generated by* $q - 1$ *elements is free.*

Let us term a 1-relator group G *cyclically pinched* if G is the free product of two free groups A and B with a cyclic subgroup amalgamated:

$$G = \{A * B;\ a = b\} .$$

The cyclically pinched 1-relator groups constitute a particularly important class of 1-relator groups and are very well-behaved (see, for example, B. Baumslag [1], [2], G. Baumslag [3], [4], Burns [11], Karrass and Solitar [15], Meskin [18] and Wehrfritz [20]). In particular we have the following theorem (G. Baumslag [4])

Cyclically pinched 1-relator groups are residually finite.

This theorem raises the possibility of classifying cyclically pinched 1-relator groups by their subgroups of finite index. Thus we raise the

PROBLEM 1. *Prove that the number of cyclically pinched 1-relator groups with the same finite images as a given cyclically pinched 1-relator group* G *is finite (preferably one).*

The special case of Problem 1 where G is free does not seem to be too difficult.

If G is a group let σG be the sequence whose nth term is the number of subgroups of G of index n. Hall [14] has computed this sequence for finitely generated free groups (*cf.* also Dey [12]). Here we propose

PROBLEM 2. *If* G *is a cyclically pinched 1-relator group find conditions on* σG.

Problem 1 may essentially be reformulated as

PROBLEM 3. *Prove that the number of cyclically pinched 1-relator groups having the same* σ*-sequence as the cyclically pinched 1-relator group* G *is finite (preferably one).*

One may also formulate these problems more generally for residually finite 1-relator groups.

Now there exists a parafree group G, which is not free, such that

$$P/V(P) \cong F/V(F)$$

for every variety $\underline{\underline{V}}$ of groups, where here F is a free group of rank two (see G. Baumslag [7] and the review by Dunwoody [13]). Here we ask

PROBLEM 4. *Let* G *be a parafree 1-relator group,* F *a free group of rank two. If*

$$G/V(G) \cong F/V(F)$$

for every variety $\underline{\underline{V}}$ *of groups, is* G *necessarily free?*

B. Subgroups

One-relator groups have lots of locally free subgroups. For example if

$$G = \langle a, b;\ [a, b[a^l, b^m]] = 1 \rangle ,$$

then G is an extension of a locally free group of finite rank by an infinite cyclic group. Indeed a maximal locally free subgroup M of such a group G is always of finite rank (that is, there exists an integer k such that every finitely generated subgroup of M is contained in a k-generator subgroup of M). This (and other examples) suggest the following

PROBLEM 5. *Is a maximal locally free subgroup of a 1-relator group always of finite rank?*

Karrass and Solitar [15] have shown that

The finitely generated subgroups of a cyclically pinched 1-relator group are finitely presented.

The corresponding problem for 1-relator groups is as yet unanswered. A related problem which should be fairly easily dispensed with is

PROBLEM 6. *If* A *and* B *are non-cyclic, can* $A \times B$ *be a subgroup of a 1-relator group?*

The reason that Problem 6 is related to the question as to whether finitely generated subgroups of 1-relator groups are finitely presented is the following. Most non-abelian subgroups of 1-relator groups contain non-abelian free subgroups.

So it is likely that if $A \times B$ were a subgroup of a 1-relator group, then the direct product D of two free groups of rank two would also be a subgroup of a 1-relator group. But D contains a 3-generator subgroup which is not finitely related!

Now it follows from G. Baumslag [8] that

Every perfect subgroup of a positive 1-relator group is trivial.

Infinitely generated perfect subgroups of 1-relator groups abound, for example if

$$G = \langle a, b; a = [a, a^b] \rangle$$

then its derived group is a (non-trivial) perfect group (see G. Baumslag [6]). However the following question remains unanswered

PROBLEM 7. *Is every non-trivial, finitely generated subgroup of a 1-relator group different from its derived group?*

I should point out that because of Karrass and Solitar's incisive work [15] on the subgroup structure of the generalised free product of two groups, the elaboration of the subgroup structure of 1-relator groups has been made a great deal more tractable.

My final problems on the subgroups of 1-relator groups are analogues of well-known theorems about free groups. To this end, for the remainder of this section, let

$$G = \langle a_1, \ldots, a_q; r = 1 \rangle$$

be a 1-relator group where $q \geq 3$.

PROBLEM 8 (Hempel). *Is a finitely generated normal subgroup of G either of finite index or trivial?*

PROBLEM 9. *If $N \triangleleft G$ and G/N is infinite, does G/N' have a trivial center?*

A good deal of work connected to Problems 8 and 9 has already been carried out (*cf.* Karrass and Solitar [15] and the references cited therein).

Finally I should mention the work of Newman [19] on the solvable subgroups of 1-relator groups.

C. Series, residual properties

The lower central series of a free group is of length at most ω. The object of the next few problems is to raise the possibility that the lengths of a number of related series for 1-relator groups are comparable also to ω.

PROBLEM 10. *Is the lower central series of a 1-relator group of length at most* ωn *for some finite ordinal* n *?*

PROBLEM 11. *Is the derived series of a* 1*-relator group of length at most* ωn *for some finite ordinal* n *?*

PROBLEM 12. *If* G *denotes the intersection of the subgroups of finite index in a group* G *and if, inductively* $f^{m+1}G = f(f^m G)$ *, is* $f^n G = f^{n+1} G$ *for some integer* n *whenever* G *is a* 1*-relator group?*

Results of this type have been obtained for positive 1-relator groups (G. Baumslag [8]).

Next I would like to mention some problems involving residual finiteness (*cf.* for example, G. Baumslag [5]). The first of these is well-known.

PROBLEM 13. *Is every* 1*-relator group with torsion residually finite?*

PROBLEM 14. *If* $G = \langle a, b, \ldots; r = 1 \rangle$ *and* $H = \langle a, b, \ldots; s = 1 \rangle$ *are residually finite, is* $I = \langle a, b, \ldots; [r, s] = 1 \rangle$ *also residually finite?*

PROBLEM 15. *If the derived group of a* 1*-relator group* G *is locally free is* G *residually finite?*

Problem 15 should be viewed in the light of the theorem (G. Baumslag [9]):

If G *is a finitely generated group which is free by cyclic then* G *is residually finite.*

Actually Problem 15 should be formulated more generally as

PROBLEM 16. *If the derived group of a finitely generated group* G *is locally free is* G *residually finite?*

The last two problems bring to mind a rather peculiar quite interesting problem.

PROBLEM 17. *Suppose the finitely generated group* G *is free-by-cyclic. Is* G *finitely presented?*

Concluding the problems in this section I want to raise a problem about cyclically pinched 1-relator groups (*cf.* B. Baumslag [1], G. Baumslag [3]).

PROBLEM 18. *Let* $G = \{A * B;\ a = b\}$ *be the free product of two free groups* A *and* B *amalgamating* $a \in A$ *with* $b \in B$ *. If neither* a *nor* b *is a proper power, is* G *residually free?*

One might observe, before closing this section, that Problem 14 has a positive solution if r and s involve disjoint sets of generators. Indeed there are a large number of positive results about the residual finiteness of 1-relator groups where the relator satisfies certain conditions. One further example of this type of result is the following: If $G = \langle a, b;\ a^l b^m a^r b^s = 1 \rangle$ and if either $(l+r)(m+s) \neq 0$ or if $l + r = m + s = 0$ then G is residually finite.

D. Group algebras, homology, representations

If G is a group and $\mathbb{Q}$ the field of rational numbers then $\mathbb{Q}G$ denotes the group algebra of G with rational coefficients. It is not known whether $\mathbb{Q}G$ is an integral domain whenever G is torsion-free. In G. Baumslag [8] it is proved that

A torsion-free positive 1-relator group is locally indicable, that is, every non-trivial finitely generated subgroup has an infinite cyclic factor group.

Now according to a well-known theorem of Higman the rational group algebra of a locally indicable group has no zero divisors. Thus the rational group algebra of a torsion-free positive 1-relator group is an integral domain. In general one may ask

PROBLEM 19. *Is a torsion-free 1-relator group locally indicable?*

If $H_2(G)$ denotes the second homology group with trivial integral coefficients of the group G we ask (prompted by the results described in G. Baumslag [10])

PROBLEM 20. *Suppose* $G = \langle a_1, \ldots, a_q; r = 1\rangle$, *that* $\underline{V}$ *is a variety of groups and that* F *is a free group of rank* $q - 1$. *If* $q \geq 3$ *is* $H_2(G/V(G))$ *finitely generated if and only if* $H_2(F/V(F))$ *is finitely generated?*

Wehrfritz [20] has recently proved the following theorem:

Let $G = \langle A * B; a = b\rangle$ *be a free product of two free groups* A *and* B *with* $a \in A$ *identified with* $b \in B$. *If neither* a *nor* b *is a proper power then* G *is a linear group over a field of characteristic zero.*

Here we ask

PROBLEM 21. *Is every cyclicly pinched 1-relator group linear over a field of characteristic zero?*

Magnus [17] has, also recently, proved a related result:

Let $G = \left\langle a_1, b_1, \ldots, a_k, b_k; \prod_{i=1}^{k} [a_i, b_i] = 1\right\rangle$. *Then* G *has a faithful representation as a group of* 2×2 *matrices over the rationals.*

Prompted by this result we ask more generally (*cf.* Problem 21)

PROBLEM 22. *Is every cyclically pinched 1-relator group linear over the rationals?*

Finally we raise a problem which, if answered affirmatively, would settle Problem 13.

PROBLEM 23. *Is every 1-relator group with torsion linear?*

ACKNOWLEDGEMENT. I would like to express my thanks to the National Science

Foundation and the Australian National University.

References

[1] Benjamin Baumslag, "Residually free groups", *Proc. London Math. Soc.* (3) 17 (1967), 402-418. MR35#6738.

[2] Benjamin Baumslag, "Generalized free products whose two-generator subgroups are free", *J. London Math. Soc.* 43 (1968), 601-606. MR38#2217.

[3] Gilbert Baumslag, "On generalised free products", *Math. Z.* 78 (1962), 423-438. MR25#3980.

[4] Gilbert Baumslag, "On the residual finiteness of generalised free products of nilpotent groups", *Trans. Amer. Math. Soc.* 106 (1963), 192-209. MR26#2489.

[5] Gilbert Baumslag, "Groups with one defining relator", *J. Austral. Math. Soc.* 4 (1964), 385-392. MR30#3124.

[6] Gilbert Baumslag, "A non-cyclic one-relator group all of whose finite quotients are cyclic", *J. Austral. Math. Soc.* 10 (1969), 497-498. MR40#7337.

[7] Gilbert Baumslag, "Groups with the same lower central sequence as a relatively free group. II. Properties", *Trans. Amer. Math. Soc.* 142 (1969), 507-538. MR39#6959.

[8] Gilbert Baumslag, "Positive one-relator groups", *Trans. Amer. Math. Soc.* 156 (1971), 165-183. MR43#325.

[9] Gilbert Baumslag, "Finitely generated cyclic extensions of free groups are residually finite", *Bull. Austral. Math. Soc.* 5 (1971), 87-94. Zbl.216.83.

[10] Gilbert Baumslag, "One-relator metabelian groups", unpublished.

[11] R.G. Burns, "On the finitely generated subgroups of an amalgamated product of two groups", *Trans. Amer. Math. Soc.* 169 (1972), 293-306.

[12] I.M.S. Dey, "Schreier systems in free products", *Proc. Glasgow Math. Assoc.* 7 (1965-66), 61-79. MR32#5718.

[13] M.J. Dunwoody, Review of "More groups that are just about free" by Gilbert Baumslag, (*Bull. Amer. Math. Soc.* 74 (1968), 752-754), in MR37#1449.

[14] Marshall Hall, Jr., "Subgroups of finite index in free groups", *Canad. J. Math.* 1 (1949), 187-190. MR10,506.

[15] A. Karrass and D. Solitar, "The subgroups of a free product of two groups with an amalgamated subgroup", *Trans. Amer. Math. Soc.* 150 (1970), 227-255. MR41#5499.

[16] W. Magnus, "Über freie Faktorgruppen und freie Untergruppen gegebener Gruppen", *Monatsh. Math. Phys.* 47 (1939), 307-313. FdM65,59.

[17] W. Magnus, "Rational representations of Fuchsian groups", these Proc.

[18] Stephen Meskin, "On some groups with a single defining relator", *Math. Ann.* 184 (1969), 193-196 (1970). MR41#1848.

[19] B.B. Newman, "Some results on one-relator groups", *Bull. Amer. Math. Soc.* 74 (1968), 568-571. MR36#5204.

[20] B.A.F. Wehrfritz, "Generalized free products of linear groups", *Proc. London Math. Soc.* (3) 27 (1973), 402-424.

Graduate Center of the City University of New York,
New York 10036, USA.

PROC. SECOND INTERNAT. CONF. THEORY OF GROUPS, 20E15
CANBERRA 1973, pp. 82-89.

COMPUTATION IN NILPOTENT GROUPS (APPLICATION)

A.J. Bayes, J. Kautsky and J.W. Wamsley

1. Introduction

The method described in [2] is essentially comprised of the following construction: Given the class $n - 1$ factor group of G, $G/\gamma_n(G)$, construct the factor group of class n. To implement this on a computer we need first a suitable description of the input, that is, $G/\gamma_n(G)$. This is dealt with in Section 2. Section 3 describes the algorithm of the computer program for extending the class. The extending is performed through a series of extensions by elementary abelian p-groups. Section 4 comments on the particular computer system used to perform the calculations, results of which are presented in Section 5.

For the purpose of the computer representation the generators are identified with the positive integers commencing with 1, m being the m-th generator introduced. The inverse of m is denoted by $-m$.

2. Representation of $G/\gamma_n(G)$

To describe the factor group $G/\gamma_n(G)$ we need to know the prime p, the class $c = n - 1$, the number of generators s_2 (where $|G/\gamma_n(G)| = p^{s_2}$), the list of the p-th powers and the list of commutators. However, for the construction of the factor group of the next class we need as well the list (explicit or implied) of defining relations, the list of weights of the generators (new generators are given weight n if they are introduced at class n) and the list of definitions (that is, list of those p-th powers and commutators which define the extra generators). It is also useful to carry a parameter $e_{\max}$ which limits the number of elementary extensions used to complete the last class and s_1 which is the number of generators before the final elementary extension of the last class. The additional information also proves useful in abbreviating the lists of p-th powers and commutators as p-th powers of all generators larger than s_1 as well as all commutators of weight greater than n

are trivial.

The nature of this information, that is, arrays of strings of variable length, each element of the information to be readily accessible at random rather than sequentially and subject to alterations in both content and volume, calls for implementation using some list processing technique. Unfortunately no efficient language of that type was available and the magnitude of the problems to be attacked indicated that the capacity of any computer was likely to be reached both time- and space-wise. FØRTRAN was thus chosen as the source language. The p-th powers, commutators and defining relations were packed in natural order and separate lists of pointers used to indicate lengths of individual strings. For fast reference, the arrays of weights and special pointers to characterize the trivial commutators were also included in the basic information on the given class factor group.

To ensure maximal utilization of space a single large (to the size of the computer) array was used to store all the information as well as a working field for all temporary results and actions.

3. Basic program

The flow chart in Figure 1 is mostly self explanatory and reflects the theory presented in [2]. By "data" we mean the information concerning the factor group $G/\gamma_n(G)$ as described in Section 2. The equations are generated from required relations. These are (see [2]) of types

$$(ab)c = a(bc) ,$$

$$(a^p)b = a^{p-1}(ab) ,$$

$$a(b^p) = (ab)b^{p-1} ,$$

with a, b, c satisfying certain weight conditions, and the defining relations. The defining relations are either listed as part of the data or generated by a special routine or both. An equation is obtained from a relation by collecting both sides in the presecribed manner. Each such equation is then transformed into the form

$$d = (d+1)^{\alpha_{d+1}}(d+2)^{\alpha_{d+2}} \dots (s_2)^{\alpha_{s_2}} \tag{1}$$

with $s_1 < d \leq s_2$, $0 \leq \alpha_j < p$ for $j = d+1, \dots, s_2$ and eliminated against all equations as far obtained. This means that at any stage there is a set of equations of type (1) such that none of the generators on the left hand sides occurs in any of the expressions on the right hand sides witn non-zero exponent.

A significant part of the computer time was used in the actual collection and the performance of the program was closely linked to the performance of this routine.

START

enter "data" of the factor group (namely p, c, s_1, s_2, e_{max}, and so on)

set s_3 equal to s_2

increase class c by 1

set e equal to 1

if $s_1 = 0$ go to B , otherwise continue

add new generators of weight c into p-th powers (record their number in s_3 , up date lists of weights, p-th powers)

add new generators of weight c into commutators (record their number in s_3 , up date lists of weights, commutators)

A set s_1 equal to s_2

set s_2 equal to s_3

generate and eliminate equations (for details, see text)

substitute for the eliminated generators and renumber the remaining ones (s_2 decreases, up date lists of weights, p-th powers, commutators)

if $s_1 = s_2$ go to C , otherwise continue

B if $e = e_{max}$ go to D , otherwise continue

increase e by 1

add new generators into commutators (see above for comments)

add new generators into p-th powers (see above for comments)

go to A

C if $e = 1$ EXIT (group is completed, no new class), otherwise continue

set e_{max} equal to $e - 1$

D store "data" (class completed)

EXIT

FIGURE 1

4. The particular computer implementation

The FØRTRAN programs were developed and run on the IBM System/360 Model 67 computer at the Systems Development Institute in Canberra using the Cambridge Monitor

System. The interactive environment allowed us to store the intermediate results for every class factor group as an independent file which was disposed of only when the next class or two were safely constructed and checked. This procedure proved very useful, not only as a restarting device but also made it relatively easy to prepare a number of service programs which were used for example in the search for a complete set of defining relations. These included a program for collection of a supplied expression in a given factor group, a program for finding a set of independent relations and others. Again, use of the interactive environment allowed a quick feed-back of the obtained information.

5. Some results

Firstly we give the three generator Burnside group of exponent 4 .

The basic generators are 1, 2 and 3 . Further generators are defined as follows:

$4 = 1^2$,	$5 = 2^2$,	$6 = 3^2$.
7 = [2, 1],	8 = [3, 1],	9 = [3, 2],
10 = [2, 1, 1],	11 = [2, 1, 2],	12 = [3, 1, 1],
13 = [3, 1, 2],	14 = [3, 1, 3],	15 = [3, 2, 1],
16 = [3, 2, 2],	17 = [3, 2, 3],	18 = [2, 1, 1, 1],
19 = [2, 1, 2, 1],	20 = [2, 1, 2, 2],	21 = [3, 1, 1, 1],
22 = [3, 1, 2, 1],	23 = [3, 1, 2, 2],	24 = [3, 1, 3, 1],
24 = [3, 1, 3, 2],	26 = [3, 1, 3, 3],	27 = [3, 2, 1, 1],
28 = [3, 2, 1, 2],	29 = [3, 2, 1, 3],	30 = [3, 2, 2, 1],
31 = [3, 2, 2, 2],	32 = [3, 2, 3, 1],	33 = [3, 2, 3, 2],
34 = [3, 2, 3, 3],	35 = [2, 1, 2, 2, 1],	36 = [2, 1, 2, 2, 2],
37 = [3, 1, 2, 1, 1],	38 = [3, 1, 3, 2, 1],	39 = [3, 1, 3, 3, 1],
40 = [3, 1, 3, 3, 2],	41 = [3, 1, 3, 3, 3],	42 = [3, 2, 1, 2, 1],
43 = [3, 2, 1, 3, 1],	44 = [3, 2, 1, 3, 2],	45 = [3, 2, 2, 1, 1],
46 = [3, 2, 2, 1, 2],	47 = [3, 2, 2, 1, 3],	48 = [3, 2, 2, 2, 1],
49 = [3, 2, 3, 1, 1],	50 = [3, 2, 3, 1, 2],	51 = [3, 2, 3, 1, 3],
52 = [3, 2, 3, 2, 1],	53 = [3, 2, 3, 3, 1],	54 = [3, 2, 3, 3, 2],
55 = [3, 2, 3, 3, 3],	56 = [3, 2, 1, 2, 1, 1],	57 = [3, 2, 1, 3, 1, 1],
58 = [3, 2, 2, 1, 2, 1],	59 = [3, 2, 3, 1, 2, 1],	60 = [3, 2, 3, 1, 3, 1],
61 = [3, 2, 3, 2, 1, 1],	62 = [3, 2, 3, 2, 1, 2],	63 = [3, 2, 3, 2, 1, 3],
64 = [3, 2, 2, 1, 2, 1, 1],	65 = [3, 2, 3, 1, 2, 1, 1],	66 = [3, 2, 3, 1, 3, 1, 1],
67 = [3, 2, 3, 2, 1, 2, 1],	68 = [3, 2, 3, 2, 1, 3, 1],	69 = [3, 2, 3, 2, 1, 3, 2].

It was found that the group was defined by insisting that it be of class n for $n \geq 7$ and assigning period 4 to the following elements:

1, 2, 3, 7, 8, 9, 1.2, 1.3, 2.3, 1.5, 1.6, 1.9, 2.4, 2.6, 2.8, 3.4, 3.5, 3.7, 4.5, 4.6, 4.7, 4.8, 4.9, 5.6, 5.7, 5.8, 5.9, 6.7, 6.8, 6.9, 7.8, 7.9, 8.9, 1.2.3, 1.2.4, 1.2.6, 1.2.8, 1.2.9, 1.3.4, 1.3.5, 1.3.7, 1.3.9, 1.4.9, 1.5.6, 1.5.8, 1.5.9, 1.6.7, 1.6.9, 1.7.8, 1.7.9, 1.8.9, 2.3.4, 2.3.5, 2.3.7, 2.3.8, 2.4.6, 2.4.8, 2.4.9, 2.5.8, 2.6.7, 2.6.8, 2.7.8, 2.7.9, 2.8.9, 3.4.5, 3.4.7, 3.4.9, 3.5.7, 3.5.8, 3.6.7, 3.7.8, 3.7.9, 3.8.9, 1.2.3.4, 1.2.3.5, 1.2.3.6, 1.2.3.7, 1.2.3.8, 1.2.3.9, 1.2.4.6, 1.2.4.9, 1.2.5.6, 1.2.5.8, 1.2.5.9, 1.2.6.7, 1.2.6.8, 1.2.6.9, 1.3.4.5, 1.3.4.7, 1.3.4.9, 1.3.7.9, 1.4.5.6, 1.4.5.8, 1.4.5.9, 1.4.6.7, 1.4.6.9, 1.4.7.9, 1.4.8.9, 1.5.6.7, 1.5.6.9, 1.5.7.8, 1.5.8.9, 1.6.7.8, 2.3.5.8, 2.3.6.7, 2.3.6.8, 2.3.7.8, 2.3.7.9, 2.3.8.9.

The squares of generators which are non-trivial were found to be:

$1^2 = 4$, $2^2 = 5$, $3^2 = 6$, $7^2 = 18.19.20.35.36$, $8^2 = 21.24.26.39.41$, $9^2 = 31.33.34.54.55$, $10^2 = 35.36$, $11^2 = 36$, $12^2 = 39.41$, $13^2 = 40.47.48.50.52.59.60.61.63$, $14^2 = 41$, $15^2 = 47.48.50.52.53.67.68$, $16^2 = 54.55$, $17^2 = 55$.

The non-trivial commutators are:

[2, 1] = 7, [3, 1] = 8, [3, 2] = 9, [4, 2] = 10.18.19.20, [4, 3] = 12.21.24.26, [5, 1] = 11.18.19.20.35, [5, 3] = 16.31.33.34, [5, 4] = 19.36, [6, 1] = 14.21.24.26.39, [6, 2] = 17.31.33.34.54, [6, 4] = 24.41, [6, 5] = 33.55, [7, 1] = 10, [7, 2] = 11, [7, 3] = 13.15.23.25.27.32.37.40.43.45.46.49.51.52.56.61.62.63.64.68.69, [7, 4] = 18.35.36, [7, 5] = 20.36, [7, 6] = 25.32.40.43.47.52.69, [8, 1] = 12, [8, 2] = 13, [8, 3] = 14, [8, 4] = 21.39.41, [8, 5] = 23.40.47.48.50.52.59.60.61.63, [8, 6] = 26.41, [8, 7] = 23.25.27.28.29.30.37.40.42.44.48.53.56.58.60.61.63.65.66.67.68, [9, 1] = 15, [9, 2] = 16, [9, 3] = 17, [9, 4] = 27.47.48.50.52.53.67.68, [9, 5] = 31.54.55, [9, 6] = 34.55, [9, 7] = 28.30.45.46.56.64.67, [9, 8] = 29.32.49.51.57.66.68, [10, 1] = 18, [10, 2] = 19.35, [10, 3] = 22.23.25.28.29.30.38.43.45.49.50.52.56.57.59.61.62.66.68, [10, 4] = 35.36, [10, 5] = 35.36, [10, 6] = 38.40.44.47.57.60.61.62.63.66.67, [10, 7] = 35.36, [10, 8] = 37.38.43.47.48.50.52.53.56.61.66, [10, 9] = 42.48.62.64, [11, 1] = 19, [11, 2] = 20, [11, 3] = 23.30.42.48.50.52.58.59.62.65.68.69, [11, 4] = 36, [11, 5] = 36, [11, 6] = 40.44.47.50.59.60.63.65.67.69, [11, 7] = 36, [11, 8] = 40.44.45.50.52.59.60.61.64.65.67, [11, 9] = 46.48.58.62.69, [12, 1] = 21, [12, 2] = 22.23.25.27.28.29.30.38.43.47.63.66.68, [12, 3] = 24.39, [12, 4] = 39, 41, [12, 5] = 38.40.42.43.44.47.48.53.57.60.61.66, [12, 6] = 39.41, [12, 7] = 37.38.43.47.48.50.52.53.57.61.64, [12, 8] = 39.41, [12, 9] = 43.53.63.66, [12, 11] = 56.65, [13, 1] = 22, [13, 2] = 23, [13, 3] = 25.29.32.44.47.49.50.52.53.57.59.62.66, [13, 4] = 37, [13, 5] = 40.47.48.50.52.60.61.63.65.67.68, [13, 6] = 40,51.53.63.66,

[13, 7] = 40.42.44.50.52.56.58.60.65.67, [13, 8] = 40.43.44.47.49.53.57.59.61.62.68,
[13, 9] = 44.47.50.52.59.62, [13, 10] = 65, [13, 11] = 58.64.67, [13, 12] = 57,
[14, 1] = 24, [14, 2] = 25, [14, 3] = 26, [14, 4] = 41,
[14, 5] = 40.44.47.50.52.59.60.63.68.69, [14, 6] = 41,
[14, 7] = 40.44.47.49.59.60.61.62.63.66.68, [14, 8] = 41, [14, 9] = 51.53.60.63.69,
[14, 10] = 57.65, [14, 11] = 61.65, [14, 13] = 60.68, [15, 1] = 27, [15, 2] = 28,
[15, 3] = 29, [15, 4] = 47.48.50.52.53.61, [15, 5] = 46.62.69, [15, 6] = 51.63.69,
[15, 7] = 45.48.56.58.62.64.67, [15, 8] = 49.53.57.60.63.66.68, [15, 9] = 47.50.67.68,
[15, 10] = 56.64, [15, 11] = 67, [15, 12] = 57.66, [15, 13] = 68, [15, 14] = 68,
[16, 1] = 30, [16, 2] = 31, [16, 3] = 33.54, [16, 4] = 45, [16, 5] = 54.55,
[16, 6] = 54.55, [16, 7] = 46.48.64.67, [16, 8] = 47.52.61.63.65, [16, 9] = 54.55,
[16, 10] = 58, [16, 12] = 59.61.67.68, [16, 13] = 62.69, [16, 14] = 63.68,
[16, 15] = 62, [17, 1] = 32, [17, 2] = 33, [17, 3] = 34, [17, 4] = 49, [17, 5] = 55,
[17, 6] = 55, [17, 7] = 50.52.61.62.65, [17, 8] = 51.53.66.68, [17, 9] = 55,
[17, 10] = 59, [17, 11] = 62.67, [17, 12] = 60, [17, 13] = 68, [17, 15] = 63,
[18, 1] = 35.36, [18, 2] = 35, [18, 3] = 38.43.56.57.66, [18, 6] = 65, [18, 9] = 64,
[19, 1] = 36, [19, 2] = 35.36,
[19, 3] = 38.40.42.43.44.45.47.48.53.57.60.61.65.66.67.68, [19, 6] = 59.61.67,
[19, 8] = 56, [19, 9] = 58.67, [19, 13] = 64, [19, 14] = 65, [19, 15] = 64,
[19, 17] = 67, [20, 1] = 35, [20, 2] = 36,
[20, 3] = 40.47.50.52.58.60.61.62.63.65.67.68, [20, 6] = 67, [20, 8] = 64,
[21, 1] = 39.41, [21, 2] = 38.43.47.48.50.52.53.61.65.66, [21, 3] = 39, [21, 5] = 65,
[21, 9] = 66, [22, 1] = 37, [22, 2] = 38.40.43.44.45.47.48.53.57.59.60.66.67,
[22, 3] = 38.40.44.47.53.57.62.66, [22, 5] = 58.67, [22, 6] = 68, [22, 7] = 56.64.65,
[22, 8] = 65.66, [22, 9] = 59.61, [22, 13] = 65, [22, 14] = 66, [22, 15] = 65,
[22, 17] = 68, [23, 1] = 38.42.43.45.47.48.50.52.53.57.59.66.67,
[23, 2] = 40.47.48.50.52.60.61.63.65.67.68, [23, 3] = 40.44.50.59.60.61.65,
[23, 4] = 56.65, [23, 6] = 63.68, [23, 7] = 58.67, [23, 8] = 59.65.67.68,
[23, 9] = 62, [23, 10] = 64, [23, 12] = 65, [23, 14] = 68, [23, 15] = 67,
[23, 17] = 69, [24, 1] = 41, [24, 2] = 38.40.44.47.49.57.60.61.62.63.65.66.68,
[24, 3] = 39.41, [24, 5] = 59.67, [24, 7] = 57, [24, 9] = 60.68, [24, 11] = 65,
[24, 15] = 66, [24, 16] = 68, [25, 1] = 38, [25, 2] = 40.44.47.50.52.59.60.63.68.69,
[25, 3] = 40.51.53.60.63.68, [25, 4] = 57, [25, 7] = 59.65, [25, 8] = 60.68,
[25, 9] = 68.69, [25, 11] = 67, [25, 15] = 68, [25, 16] = 69, [26, 1] = 39,
[26, 2] = 40, [26, 3] = 41, [26, 5] = 68, [26, 7] = 66, [27, 1] = 47.48.50.52.53.61,
[27, 2] = 42.45.48.62.67, [27, 3] = 43.49.53.63.68, [27, 5] = 58, [27, 6] = 60,
[27, 7] = 56, [27, 8] = 57, [27, 9] = 61.67.68, [27, 11] = 64, [27, 13] = 65,
[27, 14] = 66, [27, 16] = 67, [27, 17] = 68, [28, 1] = 42, [28, 2] = 46.62.69,
[28, 3] = 44.47.50.69, [28, 4] = 56, [28, 6] = 68, [28, 7] = 58. 64,
[28, 8] = 59.61.65.67, [28, 9] = 67.69, [28, 13] = 67, [28, 14] = 68, [28, 15] = 67,
[28, 17] = 69, [29, 1] = 43, [29, 2] = 44, [29, 3] = 51.63.69, [29, 4] = 57,

[29, 5] = 67, [29, 7] = 59.65.67, [29, 8] = 60.66, [29, 9] = 68.69, [29,11] = 67, [29, 15] = 68, [29, 16] = 69, [30, 1] = 45, [30, 2] = 46, [30, 3] = 47, [30, 6] = 63, [30, 7] = 64.67, [30, 8] = 61.65.68, [30, 9] = 62, [30, 10] = 64, [30, 12] = 65, [30, 13] = 67, [31, 1] = 48, [31, 2] = 54.55, [31, 3] = 54, [31, 4] = 67, [31, 8] = 69, [32, 1] = 49, [32, 2] = 50, [32, 3] = 51, [32, 5] = 62, [32, 7] = 61.65.67, [32, 8] = 66.68, [32, 9] = 63, [32, 10] = 65, [32, 12] = 66, [32, 13] = 68, [33, 1] = 52, [33, 2] = 55, [33, 3] = 54.55, [33, 4] = 61, [33, 7] = 62, [33, 8] = 63, [33, 10] = 67, [33, 12] = 68, [33, 13] = 69, [34, 1] = 53, [34, 2] = 54, [34, 3] = 55, [34, 4] = 68, [34, 7] = 69, [37, 2] = 56, [37, 3] = 57.65.66, [37, 5] = 64, [37, 6] = 66, [38, 1] = 57, [38, 2] = 61.65, [38, 3] = 66.68, [38, 5] = 67, [38, 7] = 65, [38, 8] = 66, [38, 9] = 68, [40, 1] = 66, [40, 2] = 68, [42, 1] = 56, [42, 2] = 67, [42, 3] = 61, [42, 6] = 68, [42, 7] = 64, [42, 8] = 65, [42, 9] = 67, [43, 1] = 57, [43, 2] = 61, [43, 3] = 68, [43, 5] = 67, [43, 7] = 65, [43, 8] = 66, [43, 9] = 68, [44, 1] = 59.61, [44, 2] = 67, [44, 3] = 63.69, [44, 4] = 65, [44, 7] = 67, [44, 8] = 68, [44, 9] = 69, [45, 2] = 58.67, [45, 3] = 59.67.68, [45, 6] = 68, [45, 7] = 64, [45, 8] = 65, [45, 9] = 67, [46, 1] = 58, [46, 3] = 62.67, [46, 4] = 64, [46, 6] = 69, [47, 1] = 59.61.67.68, [47, 2] = 67.69, [47, 3] = 63, [47, 4] = 65, [47, 7] = 67, [47, 8] = 68, [47, 9] = 69, [48, 1] = 67, [48, 3] = 69, [49, 2] = 59.61, [49, 3] = 60.68, [49, 5] = 67, [49, 7] = 65, [49, 8] = 66, [49, 9] = 68, [50, 1] = 59, [50, 2] = 62, [50, 3] = 68.69, [50, 4] = 65, [50, 7] = 67, [50, 8] = 68, [50, 9] = 69, [51, 1] = 60, [51, 2] = 63.68, [51, 4] = 66, [51, 5] = 69, [52, 1] = 61, [52, 2] = 62, [52, 3] = 63, [53, 1] = 68, [53, 2] = 69, [56, 2] = 64, [56, 3] = 65, [57, 2] = 65, [57, 3] = 66, [58, 1] = 64, [58, 3] = 67, [59, 1] = 65, [59, 3] = 68, [60, 1] = 66, [60, 2] = 68, [61, 2] = 67, [61, 3] = 68, [62, 1] = 67, [62, 3] = 69, [63, 1] = 68, [63, 2] = 69.

Secondly a group of exponent 4 with multiplicator of exponent 8 was constructed as follows:

Let G be the group, with presentation,

$$G = \{a, b, c, d \mid [a, b] = [c, d], \text{ exponent } 4 \text{ , class } 4\}$$

and let H be an extension of G by its multiplicator.

Then H is a group of order 2^{206} with the desired property. However once H is obtained then we may form a factor group of H, which still has the desired property of order 2^{21}.

We give this group.

The basic generators are 1, 2, 3 and 4 . Further generators are defined as follows:

5 = [3, 1] , 6 = [3, 2] , 7 = [4, 1] , 8 = [4, 2] , 9 = [4, 3] , 10 = [6, 1] ,

11 = [7, 1] , 12 = [8, 2] , 13 = [9, 1] , 14 = [9, 2] , 15 = [14, 1] ,

16 = [14, 2] , 17 = [2, 1][3, 4] .

The order of 1 and 2 are four while the order of 17 is eight and all other generators are of period 2 .

The non-trivial commutators are:

[2, 1] = 9.17, [3, 1] = 5, [3, 2] = 6, [4, 1] = 7, [4, 2] = 8, [4, 3] = 9,
[5, 4] = $13.15.16.17^4$, [6, 1] = 10, [6, 4] = 16.17^4, [7, 1] = 11, [7, 5] = $15.16.17^4$,
[8, 2] = 12, [8, 6] = 16.17^4, [9, 1] = 13, [9, 2] = 14, [10, 4] = 15.17^4,
[10, 7] = 17^4, [11, 3] = 15.16, [11, 5] = 17^4, [11, 6] = 17^4, [12, 3] = 16,
[12, 6] = 17^4, [13, 1] = 15.16, [13, 2] = 15, [14, 1] = 15, [14, 2] = 16,
[15, 1] = 17^4, [16, 2] = 17^4 .

Note, that a group with this property was first constructed by the last author in conjunction with Macdonald [1].

Next, an attempt was made to construct a group of exponent 5 with multiplicator of exponent 25 as follows:

Let G be the group with presentation,

$$G = \{a, b, c, d \mid [a, b] = [c, d], \text{ exponent } 5, \text{ class } 5\}$$

and let H be an extension of G by its multiplicator. Unfortunately H did not have the desired property. However, it did show that reasonably large groups could be handled since the order of H is 5^{515} and no storage problems were encountered.

Other results obtained included the four generator group of exponent four to class 5 which turned out to be of order 2^{188} and the maximal 3-factor of the Fibonacci group, $F(2, 8)$, to class 12 of order 3^{57} .

References

[1] I.D. Macdonald and J.W. Wamsley, "On the multiplicator of finite p-groups", in preparation.

[2] J.W. Wamsley, "Computation in nilpotent groups (theory)", these Proc.

IBM, Canberra, ACT 2600 (Bayes);

The Flinders University of South Australia,
Bedford Park, SA 5042 (Kautsky, Wamsley).

PROC. SECOND INTERNAT. CONF. THEORY OF GROUPS, 20A99
CANBERRA 1973, pp. 90-102.

BETWEEN LOGIC AND GROUP THEORY

William W. Boone

Introduction and general discussion

This article consists of the talk given by the author to the Congress, but together with various extensions. It is certainly in the area claimed by the title, focusing on connections between recursive functions and groups. Some results are stated and a reference given as to where the proofs may be found, some results are stated and the proofs promised as forthcoming, a weakened form of one of the latter is proved in outline here, but otherwise the article consists of a wild array of obiter dicta, questions, and speculations.

In Part I we discuss the question of generalizing the well-known HNN (Higman-Neumann-Neumann) extension looked at from this point of view:* given a group G with subgroup H, regard the HNN extension process as furnishing a group U which is an extension of G, and having a certain generator t such that for any element, X of G,

(+) $\quad tXt^{-1} = X$ *in* U *if and only if* X *belongs to* H.

Moreover, if G is finitely presented and H is finitely generated, then U is finitely presented. Briefly, one can put matters as follows: Given the finitely presented group G together with the finitely generated subgroup H, the HNN construction yields a finitely presented extension U of G such that a certain expression (+) regarding elements of U characterizes those elements of G which are in the subgroup H.

QUESTION. *How far can one relax the requirement that the subset H of elements of G be a subgroup of G and still obtain a similar kind of result?*

Research currently or recently supported by the United Kingdom Science Research Council, US National Science Foundation, US Air Force Office of Scientific Research, and the University of Illinois under a sabbatical leave grant. The author is also grateful to the Mathematical Institute and All Souls' College, both of Oxford, whose hospitality made his joint work with Graham Higman possible.

* See the original paper, our reference [19]. More precisely, we are generalizing that special case of the HNN construction in which the two isomorphic subgroups of the given group are isomorphic under the identity map.

The case in which H is a subsemigroup has a long history. At one time it was the missing link in the proof of the unsolvability of the word problem for groups because of the then known unsolvability of the quasi-Magnus problem.*

For the formulation of this abstract point of view toward HNN extensions in the mid-nineteen fifties, the author owes a great deal to Kurt Gödel. Indeed, the formulation is as much his as the author's. When the author was working with Bernhard Neumann in Manchester in 1958, Neumann solved the problem for the special case in which the subsemigroup H has one generator, but this is unpublished. Recently the author has been joined in this investigation by Graham Higman and a full report on our work is in preparation. In the present Part I all we do is to state some of the results now known to us; and to give, with a fair amount of detail, a proof for the case in which H is a subsemigroup. While this proof thus yields another example of a finitely presented group with unsolvable word problem - somewhat different from the ones known earlier - it does not seem of any particular interest from that standpoint. Certainly the total argument is no shorter than those in the literature for the unsolvability of the word problem.

Now about Part II. In [12], Higman and the author give an algebraic characterization of finitely generated groups with solvable word problem, namely,

> *A necessary and sufficient condition that a finitely generated group G have a solvable word problem is that there exist a simple group H and a finitely presented group K such that G is a subgroup of H and H is a subgroup of K.*

We have just recently realized that this same class of groups has, in effect, been characterized by Kuznecov [20], namely, *being contained in some finitely presented simple algebraic system,* and by Neumann [24] taken together with Macintyre [21], namely, *being contained in every algebraically closed group.* (No proofs are given in [20]. With a little thought, the Macintyre-Neumann characterization follows from a seemingly weaker analogue asserted in [21].)

In [18] Higman gave an infinite collection of distinct finitely presented simple groups. In Part II we list many open questions, mostly about simple groups, that arise naturally out of the algebraic characterizations mentioned above, or out of this latter work of Higman's. Even hints are given, but it is doubtful that these will help.

Part I

Recall the discussion of the HNN construction in the Introduction. Speaking in

---

* The decision problem, about a given finitely presented group, to determine of an arbitrary word W whether or not W can be expressed as a positive word on a given subset of the generators. See our reference [2], page 232, displayed items (1), (2), and (1'), and (2') for more careful statements of this problem.

very formal terms, we can say that this construction is a uniform functional which, when applied to an arbitrary finite presentation

$$G = (g_1, g_2, \ldots, g_N; R_1, R_2, \ldots, R_K)$$

together with a non-negative integer M, $M \le N$, yields a certain presentation G_M which is an extension of (that is, supergroup of) the given G together with an expression $\phi(X)$. Here

$$G_M = \left[g_1, g_2, \ldots, g_N, t; R_1, R_2, \ldots, R_K, tg_it^{-1} = g_i, i = 1, 2, \ldots, M\right]$$

and $\phi(X)$ is made up of the variable X, the generators of G_M, and the notations for multiplication, inverse, and equality in G_M, that is, $\phi(X)$ is simply $tXt^{-1} = X$. But, moreover, $\phi(X)$ characterizes a certain subset of the set of all words (or elements) of G. Let Γ be a variable for words on $g_1, g_2, \ldots, g_M$. Then for any word Σ of G,

$\phi(\Sigma)$ holds, that is, $t\Sigma t^{-1} = \Sigma$ in G_M if and only if $(\exists\Gamma)\Sigma = \Gamma$ in G.

By a trivial modification of the HNN construction as we have presented it, one has for every presentation G and every set S of words of G, where S is closed under equality, multiplication, and taking inverses, a similar procedure, uniform in G, S, to characterize the words of S in some extension of G. And this extension is finitely presented if the set of group elements S, as a subgroup of G, is finitely generated.

Dropping the requirement that S be closed under the taking of inverses is the subsemigroup case mentioned in the Introduction. We outline, but with much detail, the proof for that case below. We shall assume in the argument that the set of elements S, as a subsemigroup of G, is finitely generated so that the extension of G obtained is finitely presented.

But one can do better than this: in the following theorem the requirement that S be closed under multiplication is dropped as well. The set of elements S has no algebraic structure at all. The "if" part of the theorem is trivial.

THEOREM A (Boone-Higman). *Let G be any finite presentation of a group and S be any set of words of G which is closed under equality in G. Then S is recursively enumerable if and only if there exists a finitely presented group G' in which G is embedded and having an expression $\phi(X)$ such that for each word Σ of G,*

$$\phi(\Sigma) \text{ holds if and only if } \Sigma \in S.$$

It is quite natural to define the set of group elements S to be recursively enumerable if the set of words S is recursively enumerable. In this sense, the theorem can be regarded as characterizing algebraically recursively enumerable subsets of elements of the given group G. Various strengthenings of the theorem are known to Higman and the author. In particular in analogy to the state of affairs with HNN extensions (see, for example, Boone [9], the sequel, Lemma 1, page 58), one has control over the word problem for G' in terms of Turing reducibility, but we do not go into the matter here. Also, in analogy to the special cases of the theorem discussed above, one can exhibit a uniform recursive procedure to effect Theorem A.

OUTLINE OF PROOF FOR THE SUBSEMIGROUP CASE

As above, $G = (g_1, g_2, \ldots, g_N; R_1, R_2, \ldots, R_K)$. Analogous to our exposition of the HNN construction above, we take it, without any loss of generality, that the finitely generated subsemigroup H of G to be characterized is generated by $g_1^{+1}, g_2^{+1}, \ldots, g_M^{+1}$ for some $M \leq N$. That is, H consists of those elements of G expressible as a positive word on $g_1, g_2, \ldots, g_M$.

For the arbitrary G and M, $M \leq N$, we define G_M^+ as follows. Here both i and j range from 1 to M throughout.

Generators: $g_1, g_2, \ldots, g_N$;

x, r_i, q, t, k.

Relations: $R_1, R_2, \ldots, R_K$;

$$xg_j = g_jx^2,$$

$$r_ig_j = g_jxr_ix,$$

$$qg_i = r_i^{-1}qr_i,$$

$$tr_i = r_it,$$

$$xk = kx, \quad r_ik = kr_i,$$

$$(q^{-1}tq)k = k(q^{-1}tq).$$

We use the following notation:

- Σ : variable for words of G;
- Γ : variable for words on $g_1, g_2, \ldots, g_M$;
- Γ^+: variable for *positive* words on $g_1, g_2, \ldots, g_M$;
- L : variable for words on $r_1, r_2, \ldots, r_M$;
- R : variable for words on $x, r_1, \ldots, r_M$;
- P : variable for words which are products of the words $g_ir_i^{-1}$, $i = 1, 2, \ldots, M$.

To make sure the notation is understood: with, say, $M = 3$, $r_1 g_1^{-1} g_2 r_2^{-1}$ and $r_3 x^{-1} r_2^{-1}$ are in the range of P and R respectively, but $g_1^{-1} g_2$ is not in that of Γ^+.

In view of Theorems I and II which we now state, G_M^+ is the desired extension of G, and $[X^{-1}q^{-1}tqX, k] = 1$ the desired $\phi(X)$ characterizing, among the elements of G, those which are in the subsemigroup H.

THEOREM I. *For any G and M, the group G is embedded in the group G_M^+ by the identity map.*

THEOREM II. *For any G and M and for any word Σ of G, $[\Sigma^{-1}q^{-1}tq\Sigma, k] = 1$ in G_M^+ if and only if $(\exists\Gamma^+)\Sigma = \Gamma^+$ in G.*

We first explain the "if" part of Theorem II. Suppose $\Sigma = \Gamma^+$ in G. Then, as is clear by direct inspection of the presentation G_M^+, each of the following holds in G_M^+: $(\exists L)(\exists R)q\Gamma^+ = LqR$; $tL = Lt$; $Rk = kR$; $[q^{-1}tq, k] = 1$. Thus $[\Sigma^{-1}q^{-1}tq\Sigma, k] = 1$ in G_M^+.

We now consider Theorem I and the "only if" part of Theorem II. As the given G and its M are to remain fixed throughout, we let G_{kq} be G_M^+ with both the generators k, q as well as the relations involving k, q dropped. And so on, for any subset of $k, t, q, r_1, r_2, \ldots, r_M$, but writing r as subscript instead of $r_1 r_2 \ldots r_M$ in this connection. Then

(A) G_M^+ is an HNN extension of G_k with stable letter k;

(B) G_k, of G_{kt} with stable letter t; and

(C) G_{kt}, of G_{ktq} with stable letter q.

This is almost exactly as verified for the very similar presentations G, G_1, G_2 and G_3 in Britton [14]. In a moment and several times later we need the rather old-fashioned idea of a "proof" in a given group presentation. Where U and W are words of the presentation K, a proof from U to W in K (alternatively, for naturalness, we say a proof that $U = W$ in K) is a finite sequence of words, called *steps*, each resulting from its predecessor by an application of some defining relation of K, the first word of the sequence being U and the last W. Often

"$U = W$ in K" should be read "there exists a proof from U to W in K". Below, in Lemmas 2 and 3, we also require the equally old-fashioned diagram techniques of Boone [2, 3, 4, 5, 6, 7, 8] to show that there exist proofs having a certain aspect; but our explanation here should be self-contained.

Now suppose $\Sigma = 1$ in G_M^+, Σ any word of G. Then $\Sigma = 1$ in G_{ktq} by (A), (B), (C) above and three applications of the basic fact that a group is embedded in its HNN extension, see, for example, Lemma 3 of Britton [14]. Erasing all occurrences of x, r_i, $i = 1, 2, \ldots, M$, from the proof that $\Sigma = 1$ in G_{ktq} yields a valid proof that $\Sigma = 1$ in G. Thus we have Theorem I.

Everything so far has been more or less trivial. The essential point is the "only if" part of Theorem II. We give the argument, but not quite fully. Throughout, Britton's Lemma means, of course, Lemma 4 of Britton [14].

LEMMA 1. *If* $\Sigma^{-1}q^{-1}tq\Sigma k = k\Sigma^{-1}q^{-1}tq\Sigma$ *in* G_M^+, *then* $(\exists L)(\exists R)q\Sigma = LqR$ *in* G_{kt}.

The lemma follows from (A) and (B) above by two applications of Britton's Lemma. Substituting $q\Sigma$ for Σ it can be read virtually word for word from Britton [14], Part (k) and Part (t), page 24.

LEMMA 2. *If* $q\Sigma = LqR$ *in* G_{kt}, *then* $(\exists L)q\Sigma = LqR$ *in* G_{kt} *in which* q^{-1} *does not occur.*

Under the hypothesis of the lemma, $q\Sigma R^{-1}q^{-1}L^{-1} = 1$ in G_{kt}. Then by (C) above and Britton's Lemma, $\Sigma R^{-1} = P$ in G_{ktq} for a certain P. Now $qP = Lq$ in G_{kt} for a certain L by repeated use of the $qg_i = r_i^{-1}qr_i$, $r_ir_i^{-1} = 1$, $i = 1, 2, \ldots, M$.

Trivially, $1 = R^{-1}R$ in G_{ktq}. Thus the desired proof in G_{kt} can be displayed so:

$$q\Sigma$$
$$q\Sigma R^{-1}R$$
$$qPR$$
$$LqR$$

We now define G^* as follows.

Generators: $q, r_1, r_2, \ldots, r_M, x, g_1, g_2, \ldots, g_N$.

Relations: $qg_j = r_j^{-1}qr_j$,

$r_ig_j = g_jxr_ix$,

$xg_j = g_jx^2$.

Here $i, j = 1, 2, \ldots, M$.

We let G_q^* be G^* with the generator q and the defining relations $qg_j = r_j^{-1}qr_j$, $j = 1, 2, \ldots, M$ dropped. And so on, so that $G_{qr}^* = \left(x, g_1, g_2, \ldots, g_N; xg_j = g_jx^2, j = 1, 2, \ldots, M\right)$, but we write $F(x)$ for G_{qrg}^* , the free group on x . Then

(D) G^* is an HNN extension of G_q^* with stable letter q ;

(E) G_q^* , of G_{qr}^* with stable letters $r_1, r_2, \ldots, r_M$; and

(F) G_{qr}^* , of $F(x)$ with stable letters $g_1, g_2, \ldots, g_N$.

That these conditions hold is verified as for the similar groups G_2, G_3, G_4 in Britton [14]. We do not use (D), (E) or (F) until Lemma 5.

LEMMA 3. *Suppose* $q\Sigma = LqR$ *in* G_{kt} *in which* q^{-1} *does not occur. Then there exist words* U *and* H *of* G_q^* *such that*

(3.1) $q\Sigma = UqH$ *in* G_{kt} *in which* q^{-1} *does not occur and using no applications of any* $qg_j = r_j^{-1}qr_j$, $j = 1, 2, \ldots, M$;

(3.2) $UqH = LqR$ *in* G^* .

Let Π be a given proof satisfying the hypothesis of the lemma. Of Z , the sequence of applications of defining relations effecting Π , let O_F be the first application in Z of some non-trivial defining relation of G , that is, of some R_i , $i = 1, 2, \ldots, K$, which is preceded in Z by an application of some $qg_j = r_j^{-1}qr_j$, $j = 1, 2, \ldots, M$. If O_F exists write Z as $Z_1O_EZ_3O_FZ_5$ where O_E is the last application of some $qg_j = r_j^{-1}qr_j$ preceding O_F in Z . Further, let Π' be the proof displayed in the second column below where Π is as displayed in the first. Here we are assuming O_F replaces R_i by 1 - rather than 1 by R_i - and is applied right - rather than left - of the single occurrence of q in each step of Π , but the definition of Π' intended for the other cases should be obvious. For

Π' , I and D are just the suggested sequences of applications of insertions and deletions respectively, that is, of the trivial defining relations.

$q\Sigma$	$q\Sigma$
Z_1	Z_1
$ABqCD$	$ABqCD$
O_E	I
$AMqND$	$AMM^{-1}BqCN^{-1}ND$
Z_3	Z_3
$PqQR_iT$	$PM^{-1}BqCN^{-1}QR_iT$
O_F	O_F
$PqQT$	$PM^{-1}BqCN^{-1}QT$
Z_5	O_E
LqR	$PM^{-1}MqNN^{-1}QT$
	D
	$PqQT$
	Z_5
	LqR

Thus, by an obvious induction, we have the existence of a proof Π satisfying the hypothesis of the lemma and in which O_F does not exist, that is, Z is of form ST where S contains no application of any $qg_j = r_j^{-1}qr_j$, T no application of any R_i . Since S applies only defining relations of G_{kt} but not any $qg_j = r_j^{-1}qr_j$, and T only those G^* , the lemma is verified if we take the step resulting from the last operation of S to be the desired UqH .

For any word W of G_{kt} let $\dot{W}$ be the word obtained from W by erasing all symbol occurrences except those of $g_1^{\pm 1}, g_2^{\pm 1}, \ldots, g_N^{\pm 1}$; and let $\|W\|$ be W freely reduced.

LEMMA 4. *If* U *and* H *are words of* G_q^* *, and* $q\Sigma$ *and* UqH *satisfy* (3.1) *of Lemma* 3, *then* $\Sigma = \|\dot{H}\|$ *in* G .

The totally trivial argument is left to the reader.

In view of Lemma 1 through 4, to show the "only if" part of Theorem II, we need only show the following lemma which is, indeed, the crux of the matter.

LEMMA 5. *If* $UqH = LqR$ *in* G^* , U *and* H *words of* G_q^* *, then* $\|\dot{H}\|$ *is a* *positive word on* $g_1, g_2, \ldots, g_M$.

Recall (D), (E) and (F) above. By (D) and Britton's Lemma, under the hypothesis of the lemma, $PHR^{-1} = 1$ in G^*_q for a certain P . Thus Lemma 5' implies Lemma 5.

LEMMA 5'. *If* $PHR = 1$ *in* G^*_q *, then* $\|\mathring{H}\|$ *is a positive word on* $g_1, g_2, \ldots, g_M$.

The proof is by induction on the number of occurrences of $r_1^{\pm 1}, r_2^{\pm 1}, \ldots, r_M^{\pm 1}$ in PHR . While matters are not obvious, we leave the details to the interested reader. One uses (E) and (F) above and Britton's Lemma. Needed are the notions of a p-reduction and the function $p[W]$, explained in Boone [9], the sequel, §2, page 57 and through the first three lines of page 58. The required argument is parallel to, but more general then, that for (the duals of) Lemmas 19 to 28, pp. 73-75 of the same article.

Aanderaa [1] suggested this argument.

PART II

The following questions arise naturally from the algebraic characterizations of finitely generated groups with solvable word problem.

(1) *Can one algebraically characterize finitely generated groups with solvable conjugacy (or transformation) problem?*

(2) *Can one give an algebraic condition on a class* C *of finitely presented groups equivalent to the isomorphism problem's being solvable between members of* C *?*

There seems no obvious way to parallel Boone-Higman [12] so as to answer (1) affirmatively. For in the word problem situation, a group's being simple means that, viewed as a logical system, the system is absolutely complete in the sense of Post and Church. Then one can jump, like an agile logician, from *complete* to *decidable*. But there seems no workable analogue of this state of affairs for (1). We do have Collins' Lemma, that is, Lemma 3 of [17], to play the same role for the conjugacy problem in HNN extensions that Britton's Lemma does for the word problem; but Collins' Lemma seems to give no help in *discovering* a characterization for the conjugacy problem.

Regarding (2), Higman has remarked that it is difficult even to imagine what form a characterization of the solvability of the isomorphism problem would take. There is a remote possibility that a technique of Miller [23] (Theorems 1, 2, pp. 79, 80) translating the word problem in a finitely presented group to the isomorphism problem in a class of groups could be used as a tool.

(3) *Can every finitely generated group with solvable word problem be embedded in*

a **finitely generated** *recursively presented simple group?*

(4) *Can every finitely generated group with solvable word problem be embedded in a finitely presented simple group?*

By work of Clapham [16] "no" to (3) and (4) yields "no even if the given group is restricted to finitely presented". It seems possible to Dr Pangloss and the author, that, if the answer to (4) is "no" then the methods of Cannonito and Gatterdam on complexity of algorithms in [15] should be applicable to so demonstrate. But Schupp feels this hope is illusory.

(5) *For what kinds of algebraic and logical systems does an analogue of the characterization of the* decidable *systems given in Kuznecov* [20] *or Neumann-Macintyre* [24], [21] *or Boone-Higman* [12] *hold?*

Since the actual meeting in Canberra, Trevor Evans and his colleagues have found various analogues to [12].

The following questions are psychological and philosophical rather than substantive mathematical questions, so that the author raises them with some misgivings: *Can a particular algebraic characterization of the solvability of the word problem help, at least conceptually, in showing that a particular group has a solvable or unsolvable word problem? Can such a characterization throw any light on Church's Thesis, at least in the sense of giving evidence for or against?*

In [18], Higman gives an infinite number of distinct finitely presented infinite simple groups.

QUESTIONS

(6) *Is the set of all finite presentations of infinite simple groups recursively enumerable?*

Since the set of presentations of Higman's groups is recursively enumerable, a negative answer to (6) would mean that he has not collected all finitely presented simple groups. One can make a stab at showing this along the lines of the negative answer to J.H.C. Whitehead's question - *Is the totality of group presentations with solvable word problem recursively enumerable?* - given in Boone-Rogers [13]; but this is by no means as easy as it looks at first glance.

(7) *Is the conjugacy problem solvable for finitely presented simple groups? But, more generally, which of the usually considered decision problems are solvable for finitely presented simple groups?*

In closing, the author should like to point out how unfortunate it is that the paper of Kuznecov [20] has been so little known in the non-Russian speaking world. While for many of us, it was Richard Thompson in his oral presentation of McKenzie-Thompson [22] who first called our attention to the fact that a finitely presented simple group has solvable word problem, it seems to be Kuznecov who first noted this in [20]. There, too, he gave an algebraic, albeit non-group-theoretic, characterization of groups with solvable word problem we noted above in the Introduction. Further, he proved a universal algebra analogue to the negative answer to Church's question - *Does there exist a uniform partial algorithm solving the word problem in all those finitely presented groups having solvable word problem?* - given in Boone-Rogers [13]. All this in 1958!

The author is grateful to Adyan, H.E. Rose, and Schupp for apprising him of Kuznecov's work, and thanks Schupp and R.D. Hurwitz for a preliminary reading of the present article.

POSTSCRIPT, December 14, 1973. The author has been informed by Richard Thompson that he has solved Question 3 affirmatively.

REFERENCES

[1] Stål Aanderaa, "A proof of Higman's embedding theorem using Britton extension of groups", *Word problems. Decision problems and the Burnside problem in group theory* [Eds. W.W. Boone, F.B. Cannonito, R.C. Lyndon] pp. 1-17 (Studies in Logic and the Foundations of Mathematics, 71. North-Holland, Amsterdam, London, 1973).

[2] William W. Boone, "Certain simple, unsolvable problems of group theory. I", *Nederl. Akad. Wetensch. Proc. Ser. A* 57 (1954), 231-237. MR16,564.

[3] William W. Boone, "Certain simple, unsolvable problems of group theory. II", *Nederl. Akad. Wetensch. Proc. Ser. A* 57 (1954), 492-497. MR16,564.

[4] William W. Boone, "Certain simple, unsolvable problems of group theory. III", *Nederl. Akad. Wetensch. Proc. Ser. A* 58 (1955), 252-256. MR16,564.

[5] William W. Boone, "Certain simple, unsolvable problems of group theory. IV", *Nederl. Akad. Wetensch. Proc. Ser. A* 58 (1955), 571-577. MR20#5230.

[6] William W. Boone, "Certain simple, unsolvable problems of group theory. V", *Nederl. Akad. Wetensch. Proc. Ser. A* 60 (1957), 22-27. MR20#5231.

[7] William W. Boone, "Certain simple, unsolvable problems of group theory. VI", *Nederl. Akad. Wetensch. Proc. Ser. A* 60 (1957), 227-232. MR20#5231.

[8] William W. Boone, "The word problem", *Ann. of Math.* (2) 70 (1959), 207-265.

[9] William W. Boone, "Word problems and recursively enumerable degrees of unsolvability. A first paper on Thue systems", *Ann. of Math.* (2) 83 (1966), 520-571. MR34#1381.

[10] William W. Boone, "Word problems and recursively enumerable degrees of unsolvability. A sequel on finitely presented groups", *Ann. of Math.* (2) 84 (1966), 49-84. MR34#1382.

[11] William W. Boone, "Word problems and recursively enumerable degrees of unsolvability. An emendation", *Ann. of Math.* (2) 94 (1971), 389-391. Zbl.234.02031.

[12] William W. Boone and Graham Higman, "An algebraic characterization of groups with soluble word problem", *J. Austral. Math. Soc.* (to appear).

[13] William W. Boone and Hartley Rogers, Jr, "On a problem of J.H.C. Whitehead and a problem of Alonzo Church", *Math. Scand.* 19 (1966), 185-192. MR35#1465.

[14] John L. Britton, "The word problem", *Ann. of Math.* (2) 77 (1963), 16-32. MR29#5891.

[15] F.B. Cannonito and R.W. Gatterdam, "The computability of group constructions, Part I", *Word problems. Decision problems and the Burnside problem in group theory* [Eds. W.W. Boone, F.B. Cannonito, R.C. Lyndon], pp. 365-400 (Studies in Logic and the Foundations of Mathematics, 71. North-Holland, Amsterdam, London, 1973).

[16] C.R.J. Clapham, "An embedding theorem for finitely generated groups", *Proc. London Math. Soc.* (3) 17 (1967), 419-430. MR36#5199.

[17] Donald J. Collins, "Recursviely enumerable degrees and the conjugacy problem", *Acta Math.* 122 (1969), 115-160. MR39#4001.

[18] Graham Higman, *An infinite family of finitely presented infinite simple groups* (Mimeographed Lecture Notes taken by B.M. Hurley, Oxford University, 1973).

[19] Graham Higman, B.H. Neumann and Hanna Neumann, "Embedding theorems for groups", *J. London Math. Soc.* 24 (1949), 247-254. MR11,322.

[20] А.В. Кузнецов [A.V. Kuznecov], "Алгоритмы как операции в алгебраических системах" [Algorithms as operations in algebraic systems], *Uspehi Mat. Nauk* 13, no. 3 (1958), 240-241.

[21] Angus Macintyre, "Omitting quantifier-free types in generic structures", *J. Symbolic Logic* 37 (1972), 512-520.

[22] Ralph McKenzie and Richard J. Thompson, "An elementary construction of unsolvable word problems in group theory", *Word problems. Decision problems and the Burnside problem in group theory* [Eds. W.W. Boone, F.B. Cannonito, R.C. Lyndon], pp. 457-478 (Studies in Logic and the Foundations of Mathematics, 71. North-Holland, Amsterdam, London, 1973).

[23] Charles F. Miller, III, *On group-theoretic decision problems and their classification* (Annals of Mathematics Studies, 68. Princeton University Press and University of Tokyo Press, Princeton, New Jersey, 1971).

[24] B.H. Neumann, "The isomorphism problem for algebraically closed groups", *Word problems. Decision problems and the Burnside problem in group theory* Eds. W.W. Boone, F.B. Cannonito, R.C. Lyndon , pp. 553-562 (Studies in Logic and the Foundations of Mathematics, 71. North-Holland, Amsterdam, London, 1973).

University of Illinois,
Urbana, Illinois 61801, USA.

PROC. SECOND INTERNAT. CONF. THEORY OF GROUPS, CANBERRA 1973, pp. 103-130. 20C15

ON THE STRUCTURE OF BLOCKS OF CHARACTERS OF FINITE GROUPS

Richard Brauer

1. Introduction

The role played by characters in recent work on finite groups can be described as follows. Given rather incomplete information concerning a finite group G, we try to find the table of irreducible characters of G. If we succeed, we use the character table of G to obtain additional information on G. We may also try a similar procedure in order to show that finite groups with certain given properties do not exist.

For the success of this method, it is of decisive importance to know as much as possible about the connections between the properties of the character table of G and properties of G of a more combinatorial nature. Even for its own sake, an investigation in this direction offers many challenging problems.

Our approach centers around a study of the blocks of characters. It forms a continuation of earlier work [3, 4], and it will be extended in a later paper. We choose a prime p and partition the set of elements of G into disjoint subsets, the p-sections of G, and we partition the set of irreducible characters of G into disjoint subsets, the p-blocks of G. We shall then study the values of the characters in a fixed p-block for the elements in a fixed p-section.

Each element x of G can be written uniquely as the product $x = x_p r$ of a p-element x_p and a *p-regular element* r of the centralizer $C(x_p)$ of x_p (that is of an element $r \in C(x_p)$ of an order not divisible by p). A *p-section* S then consists of the elements x of G whose p-factor x_p is conjugate to a fixed p-element u of G. Thus, S is a union of conjugate classes of G, and one of these classes, that containing u, consists of p-elements. Each element x of S is conjugate to elements of the form ur where r is a p-regular element of $C(u)$. Since characters χ take the same value for conjugate elements, we have $\chi(x) = \chi(ur)$. Hence we know the value of χ on elements of S, if we know the values $\chi(ur)$.

This research has been supported by an NSF-grant.

The definition of p-*blocks* B of G will be given below in §2. We explain here what we mean by a *basic set for* B. Let $\mathbb{Z}$ denote the ring of integers. Let G° be the set of p-regular elements of G. If χ is a character, let $\chi|G^\circ$ be the restriction of χ to G°. If B is a p-block, the linear combinations of the restrictions $\chi|G^\circ$ of the characters $\chi \in B$ with integral coefficients form a $\mathbb{Z}$-module M_B. Any $\mathbb{Z}$-basis of M_B then is a basic set for B. Thus, if M_B has the rank $l(B)$, a basic set for B consists of $l(B)$ complex-valued functions ϕ defined on G°.

Let B be a p-block of G and let u be a p-element. Then, for $\chi \in B$ and for p-regular elements r of $C(u)$, we have

$$\chi(ur) = \sum_{b \in B\ell\{C(u), B\}} \sum_{\phi \in [b]} d(\chi, \phi)\phi(r) . \tag{1.1}$$

Here, $B\ell\{C(u), B\}$ is a set of blocks b of $C(u)$ defined in §2. For each $b \in B\ell\{C(u), b\}$, $[b]$ is a basic set for b. The *decomposition numbers* $d(\chi, \phi)$ are algebraic integers which do not depend on r. If u has order p^m, they belong to the field obtained from the rational field by adjunction of a primitive p^m-th root of unity. The existence of such formulas (1.1) is a consequence of the Second Main Theorem on blocks, *cf.* [2].

Suppose we know the centralizer $C(u)$. We then know the p-blocks b of $C(u)$ and, for each b, we can select a basic set $[b]$. In order to carry out our program of finding the values of the characters $\chi \in B$ for the elements in S, we need to know the decomposition numbers $d(\chi, \phi)$ and the set $B\ell\{C(u), B\}$. If we know the defect of B, (§2), there are only finitely many possibilities for each $d(\chi, \phi)$ .. Further results concerning the $d(\chi, \phi)$ can be found in [4].

In our present paper, we shall study the set $B\ell\{C(x), B\}$ (and a generalization). We will be led to connections between the p-block B and certain subgroups of G. The method is a refinement of the method used in [4].

In an effort to make this paper self-contained, we have given in §2 the basic definitions in a form which requires only some familiarity with Frobenius' theory of group characters (as it is given in most books on finite groups). We have also collected in §2 some known results and lemmas which are used later.

The definitions as given in §2 may not show the full significance of the various concepts. We have to refer to [1, 2] and the older papers quoted there and the books [5] and [6] for further information.

2. Notation. Background material

1. In the following, G will always be a finite group. The set of conjugacy classes of G will be denoted by

$$\mathcal{Cl}(G) = \{K_1, K_2, \ldots, K_k\}$$

and the set of irreducible characters of G by

$$\mathit{Char}(G) = \{\chi_1, \chi_2, \ldots, \chi_k\} .$$

The number k , the *class number of* G , is the same in both sets.

For $\chi \in \mathit{Char}(G)$ and $K \in \mathcal{Cl}(G)$, we set

$$\omega_\chi(K) = |K|\chi(x_K)/\chi(1) \tag{2.1}$$

where x_K is an element of K . The number $|K|$ of elements in K is of course equal to the index $|G : C(x_K)|$ of the centralizer $C(x_K)$ of x_K .

2. We denote the ring of integers by $\mathbb{Z}$. In the following, p will be a fixed prime number. If $m \in \mathbb{Z}$ and $m \neq 0$, we denote by m_p the highest power of p dividing m . Thus, $|G|_p$ is the order of the Sylow p-subgroup of G . Most results will be trivial, if p does not divide the order $|G|$ of G .

In §2, Ω will denote the field obtained from the field of rational numbers by adjoining the $|G|$-th roots of unity and $\underline{p}$ will be a fixed prime ideal divisor of p in Ω . As is well known, the expression $\omega_\chi(K)$ is an algebraic integer in Ω .

3. A *p-block* B of G is a set of irreducible characters of G . Here, $\chi_i, \chi_j \in \mathit{Char}(G)$ belong to the same p-block, if and only if

$$\omega_{\chi_i}(K) \equiv \omega_{\chi_j}(K) \pmod{\underline{p}} \tag{2.2}$$

for all $K \in \mathcal{Cl}(G)$. Since p will be kept fixed, we shall usually simply speak of *blocks* B of G .

The set of all blocks of G is denoted by $\mathcal{Bl}(G)$. Thus,

$$\mathit{Char}(G) = \bigsqcup_{B \in \mathcal{Bl}(G)} B$$

where the union is disjoint.

We say that a character $\chi_j \in B$ has *height* 0 if

$$(\chi_j(1))_p \leq (\chi(1))_p$$

for all $\chi \in B$. If then

$$|G|_p/(\chi_j(1))_p = p^d$$

we call d the *defect* of B . In general, if $\chi \in B$, we set

(2.3) $$p^d\chi(1)_p = |G|_p p^{h(\chi)}$$

where $h(\chi)$, the *height of* χ, is non-negative.

4. If $B \in Bl(G)$ has defect d, there exist classes $K \in Cl(G)$ with the following two properties

(i) $$|G|_p/|K|_p = p^d ,$$

(ii) $$\omega_\chi(K) \not\equiv 0 \pmod{\mathfrak{p}}$$

for $\chi \in B$.

If this is so, each Sylow p-subgroup of the centralizer $C(x)$ of any $x \in K$ is called a *defect group* of B. We mention three facts:

(2A). *The defect groups of a block* B *of defect* d *have order* p^d.

This is equivalent with (i).

(2B). *The defect groups of a block* B *are determined uniquely up to conjugacy in* G, *cf.* [1, (8B)].

(2C). *If* U *is a normal* p-*subgroup of* G, *then* U *is contained in all defect groups of each block;* [1, (9F)].

5. Let H be a subgroup of G and assume that b is a block of H with the defect group D_0. We say that b is *admissible in* G, if the centralizer $C(D_0)$ of D_0 in G satisfies the condition $C(D_0) \subseteq H$.

(2D). *If* $b \in Bl(H)$ *is admissible in* G, *there exists a unique block* B *of* G *such that*

(2.4) $$\omega_\chi(K) \equiv \sum_{\substack{L \in Cl(H) \\ L \subseteq K}} \omega_\psi(L) \pmod{\mathfrak{p}}$$

for all $\chi \in B$ *and all* $\psi \in b$, [2, §2]. *We then write*

(2.5) $$b^G = B .$$

If $B \in Bl(G)$ is given, the set of all $b \in Bl(H)$ admissible in G and satisfying (2.5) is denoted by $Bl(H, B)$.

(2E). *If* $b \in Bl(H, B)$ *has the defect group* D_0, *there exist defects group* D *of* B *with* $D \supseteq D_0$.

This is an easy consequence of the definitions. We then obtain easily:

(2F). *If* $b \in Bl(H)$ *is admissible in* G *and if* F *is a subgroup of* G *with* $F \supseteq H$, *then* b *is admissible in* F, $b^F \in Bl(F)$ *is admissible in* G *and*

$$(b^F)^G = b^G .$$

(2G). *If the subgroup H of G has a normal p-subgroup U such that the centralizer $C(U)$ of U in G is contained in H, then every block b of H is admissible in G.*

This follows from (2C) and [3, (4B)].

REMARK. In the following, the operation of forming b^G will only occur for blocks b of subgroups H such that H satisfies the conditions in (2G). It will therefore not be necessary to check especially that b is admissible in G.

6. If H is a subgroup of G, if $\psi \in \mathrm{Char}(H)$, and if $x \in G$, we set

$$\psi^x(u^x) = \psi(u) \tag{2.6}$$

for $u \in H$. Here, as usual, $u^x = x^{-1}ux$. Then ψ^x is an irreducible character of $H^x = x^{-1}Hx$ and every irreducible character of H^x is obtained in this fashion. Thus, the elements x of G act on the set of all irreducible characters of all subgroups of G. It is clear that every $\chi \in \mathrm{Char}(G)$ remains fixed. The following two results are quite obvious.

(2H). *If ψ ranges over the characters of a block b of a subgroup H of G and if $x \in G$, then ψ^x ranges over the elements of a block b^* of H^x. This block b^* is denoted by b^x.*

(2I). *If H and F are subgroups of G with $H \subseteq F \subseteq G$, if $b \in B\ell(H)$ is admissible in F and if $x \in G$, then b^x is admissible in F^x and*

$$(b^F)^x = (b^x)^{F^x} .$$

If H is a subgroup of G and $b \in B\ell(H)$, the *inertial group of b in G* is defined as the group of all x in G with $b^x = b$.

7. The following results are closely related to the first main theorem on blocks; [1, (10B): 3, §5].

(2J). *If B is a block of G with the defect group D, there exist blocks $b \in B\ell(DC(D), B)$.*

Each such b is called a *root of B in $DC(D)$*. By (2) and (2), b has the unique defect group D.

(2K). *If b is a root of B in $DC(D)$ as in (2J), the most general such root has the form b^x where x is an arbitrary element of the normalizer $N(D)$ of D in G.*

(2L). *Let Q be a p-subgroup of G; let b be a block of $QC(Q)$ with the* *defect group* Q , *and let* $T(b)$ *denote the inertial group of* b *in* $N(Q)$. *The block* b^G *has the defect group* Q , *if and only if*

(2.7) $$|T(b) : QC(Q)| \not\equiv 0 \pmod{p} .$$

We shall denote by $\underline{P}$ the set of all pairs (Q, b) where Q is a p-subgroup of G and b a block of $QC(Q)$ with the defect group Q . The pair will be called a *primitive pair*, if the inertial condition (2.7) is satisfied.

Thus, if $B \in Bl(G)$ has the defect group D , the pair (D, b) consisting of D and a root b of B in $DC(D)$ is primitive. Conversely, if (Q, b) is a primitive pair in $\underline{P}$, b^G has the defect group Q and b is a root of b^G in $QC(Q)$. If $(Q, b) \in \underline{P}$ is not primitive, then by (2L) and (2E), the defect of b^G is larger than that of b .

For $(Q, b) \in \underline{P}$, $T(b)$ will *always* denote the inertial group of b in $N(Q)$.

8. We next discuss some results for the case that H is a normal subgroup of G . We then say that a block B of G *covers* a block b of H , if for some $\chi \in B$, the restriction $\chi|H$ of χ to H contains a character $\psi \in b$ as constituent.

(2M). *If* $H \triangleleft G$, *the blocks of* H *covered by a block* B *of* G *form a family* F *of blocks* b *of* H *conjugate under the action of the elements of* G . *If* $\chi \in B$, *there exists a constituent of* $\chi|H$ *in each* $b \in F$. *If* $\psi \in b$ *and* $b \in F$, *there exist characters* $\chi \in B$, *such that* ψ *is a constituent of* $\chi|H$; [*cf.* 3, (4A)].

(2N). *Assume that* $H \triangleleft G$ *and, as in* (2), *that there exists a normal* p-*subgroup* U *of* H *for which* $C(U) \subseteq H$.

(i) If $b \in Bl(H)$, *then* b^G *covers* b . *If* b^G *has the defect group* D , *we have* $C(D) \subseteq H$. *Moreover,* D *can be chosen such that* $D \cap H$ *is a defect group of* b .

(ii) If b *is the block in (i) and if* $b^x = b$ *for* $x \in G$, *then* HD *contains a Sylow* p-*subgroup of* G .

(iii) If $B \in Bl(G)$ *has the defect group* D *and if* $C(D) \subseteq H$, *there exist blocks* $b \in Bl(H, B)$.

For the proof, *cf.* the propositions (4D), (4C) and (4G) of [3]. The proof of (4C) is given in [7]. In (4G), (used in the proof of *(ii)*), the character ψ is to be taken as the canonical character of b as introduced in [3, §2, 6]. The hypothesis concerning the existence of the subgroup U is not required in [3] since there a more general definition of b^G is used.

3. Linked pairs

Let G be a finite group, let p be our fixed prime, and as in §2, 7, let $\underline{P}$ denote the set of all pairs (Q, b) where Q is a p-subgroup of G and b a block of $QC(Q)$ with the defect group Q. This notation will remain fixed for the rest of the paper.

DEFINITION 3.1. Two pairs (P, b^*) and (Q, b^{**}) of $\underline{P}$ are *linked*, if Q is a normal subgroup of P, $Q \neq P$, and if

$$(b^*)^{PC(Q)} = (b^{**})^{PC(Q)} . \tag{3.1}$$

Note that since P normalizes Q, it normalizes $C(Q)$, and hence $PC(Q)$ is a group. Moreover, $QC(Q) \triangleleft PC(Q)$ and

$$PC(Q)/\{QC(Q)\} = PQC(Q)/\{QC(Q)\} \cong P/\{P \cap QC(Q)\} . \tag{3.2}$$

In particular, the group in (3.2) is a p-group.

(3A). *If the pairs* (P, b^*) *and* (Q, b^{**}) *of* $\underline{P}$ *are linked, the block*

$$\tilde{B} = (b^*)^{PC(Q)} = (b^{**})^{PC(Q)}$$

has the defect group P. *We have* $C_P(Q) \subseteq Q$. *Finally, the inertial group* $T(b^{**})$ *of* b^{**} *contains* $PC(Q)$.

PROOF. By (2E), $\tilde{B}$ has a defect group $\tilde{D} \supseteq P$. Then $C(\tilde{D}) \subseteq C(P) \subseteq C(Q)$ and, by ((2N), *(i)*), $\tilde{B}$ covers b^{**}. Choose $\psi \in \tilde{B}$ of height 0. It follows from (2M), that $\psi|QC(Q)$ has an irreducible constituent $\theta \in b^{**}$. If T is the inertial group of θ in $PC(Q)$, (*i.e.* the group of all $t \in PC(Q)$ for which $\theta^t = \theta$), then Clifford's theorem shows that

$$\psi(1) = e|PC(Q) : T|\theta(1) \tag{3.3}$$

where e is a positive integer. By (2.3),

$$|\tilde{D}|\psi(1)_p = |PC(Q)|_p ; \quad |Q|\theta(1)_p = |QC(Q)|_p p^{h(\theta)} . \tag{3.4}$$

On combining (3.2) and (3.4), we find

$$|P : \{P \cap QC(Q)\}| = |PC(Q)|_p/|QC(Q)|_p = \psi(1)_p p^{h(\theta)}|\tilde{D} : Q|/\theta(1)_p .$$

Now, (3.3) yields

$$|P : \{P \cap QC(Q)\}| = e_p|PC(Q) : T|_p p^{h(\theta)}|\tilde{D} : Q| .$$

Since $\tilde{D} \supseteq P$ and $Q \subseteq P \cap QC(Q)$, this is only possible if $\tilde{D} = P$ and $Q = P \cap QC(Q)$. This proves the first two statements in (3A). Moreover, $|PC(Q) : T|_p = 1$. Since $T \supseteq QC(Q)$, $|PC(Q) : T|$ divides $|PC(Q) : QC(Q)|$, a power of p. Hence $PC(Q) = T$. Thus, $\theta^t = \theta$ for all $t \in PC(Q)$. Since $\theta^t \in (b^{**})^t$, we have $(b^{**})^t = b^{**}$ and this yields the last part of (3A).

The following is an immediate consequence of $C_P(Q) \subseteq Q$.

(3B). *If the pairs* (P, b^*) *and* (Q, b^{**}) *are linked, then* $C_P(Q)$ *is the center* $Z(Q)$ *of* Q, *and* $Z(P) \subseteq Z(Q)$.

We can give a different characterization of linked pairs.

(3C). *Let* P *be a p-subgroup of* G *and let* $Q \neq P$ *be a normal subgroup of* P. *The following conditions are necessary and sufficient in order that two pairs* (P, b^*) *with* $b^* \in B\ell(PC(P))$ *and* (Q, b^{**}) *with* $b^{**} \in B\ell(QC(Q))$ *be in* $\underline{P}$ *and linked:*

(i) $C_P(Q) \subseteq Q$;

(ii) there exists a block $\tilde{B}$ *of* $PC(Q)$ *with the defect group* P *such that* b^* *is a root of* $\tilde{B}$ *in* $PC(P)$ *and that* b^{**} *is a block of* $QC(Q)$ *covered by* $\tilde{B}$.

PROOF. If (P, b^*) and (Q, b^{**}) are in $\underline{P}$ and linked, it follows from (3A) that $C_P(Q) \subseteq Q$ and that $\tilde{B} = (b^*)^{PC(Q)}$ has the defect group P. Then b^* is a root of $\tilde{B}$. As we have already used, (2N) shows that $\tilde{B}$ covers b^{**}. Thus, *(i)* and *(ii)* hold.

Conversely, assume that *(i)* and *(ii)* are satisfied. Since $C(P) \subseteq C(Q)$, it follows from (2N, *(iii)*) that there exist blocks $\tilde{b}$ of $QC(Q)$ for which

$$\tilde{B} = \tilde{b}^{PC(Q)}$$

and, by (2N, *(i)*), $\tilde{B}$ covers $\tilde{b}$. Since $\tilde{B}$ also covers $b^{**} \in B\ell(QC(Q))$, (2M) shows that b^{**} and $\tilde{b}$ are conjugate in $PC(Q)$, say $b^{**} = \tilde{b}^x$ with $x \in PC(Q)$. Then

$$(b^{**})^{PC(Q)} = (\tilde{b}^x)^{PC(Q)} = (\tilde{b}^{PC(Q)})^x = \tilde{B}^x = \tilde{B},$$

cf. (2I). It is now clear that (3.1) holds. Since $(P, b^*) \in \underline{P}$, it remains to show that $(Q, b^{**}) \in \underline{P}$, *i.e.* that b^{**} has the defect group Q. On account of $\{(2N), (i)\}$, a defect group of b^{**} can be taken in the form $Q_1 = P^t \cap QC(Q)$ where P^t is a suitable defect group of $\tilde{B} \in B\ell(PC(Q))$, *i.e.* where t is an element of $PC(Q)$. It is clear that we may choose $t \in C(Q)$. Then $Q_1^{t^{-1}} = P \cap QC(Q)$ too is a defect group of $b^{**} \in B\ell(QC(Q))$. Since *(i)* implies $P \cap QC(Q) = Q$, we are finished.

(3D). *Let* $(P, b^*) \in \underline{P}$ *and let* $Q \neq P$ *be a normal subgroup of* P *such that* $C_P(Q) \subseteq Q$. *The following condition is necessary and sufficient for the existence of blocks* $b^{**} \in Bl(QC(Q))$ *for which* $(Q, b^{**}) \in \underline{P}$ *and for which* (P, b^*) *and* (Q, b^{**}) *are linked:*

$$|T(b^*) \cap PC(Q) : PC(P)| \not\equiv 0 \pmod p . \tag{3.5}$$

If this is so, b^{**} *is uniquely determined by* P, Q *and* b^*.

Indeed, by (2L) applied to $PC(Q)$ instead of G, (3.5) is necessary and sufficient in order that $(b^*)^{PC(Q)} = \tilde{B}$ has the defect group P. The first part of (3D) now follows from (3C). If the pairs are linked, then (3A) in conjunction with (2M) shows that b^{**} is the only block of $QC(Q)$ covered by $\tilde{B}$.

(3E). *If* $(Q, b^{**}) \in \underline{P}$, *there exist pairs* $(P, b^*) \in \underline{P}$ *for which* (P, b^*) *and* (Q, b^{**}) *are linked, if and only if* (Q, b^{**}) *is not primitive, (cf. §2, 7). If* (Q, b^{**}) *is not primitive, we can choose the pair* (P, b^*) *such that* $PC(Q)$ *contains a Sylow* p*-subgroup of* $T(b^{**})$.

PROOF. Note that by definition of $T(b^{**})$, (§2, 7), we have $T(b^{**}) \subseteq N(Q)$ and then $Q \triangleleft T(b^{**})$, $QC(Q) \triangleleft T(b^{**})$.

If the pairs (P, b^*) and (Q, b^{**}) are linked, (3A) shows that P is contained in $T(b^{**})$, but not in $QC(Q)$. It follows that

$$|T(b^{**}) : QC(Q)| \equiv 0 \pmod p , \tag{3.6}$$

and (Q, b^{**}) is not primitive.

Conversely, assume that the pair (Q, b^{**}) is not primitive. Then (3.6) holds. Choose a Sylow p-subgroup P_0 of $T(b^{**})$ and set $M = P_0C(Q)$. It is clear that

$$T(b^{**}) \supseteq M = P_0C(Q) \supset QC(Q) .$$

Set $\tilde{B} = (b^{**})^M$ and let P be a defect group of $\tilde{B}$. Since $Q \triangleleft M$, then $Q \triangleleft P$, (*cf.* (2C)). Now (2L) applied to M instead of G shows that $Q \subset P$. Moreover, by (2N, *(ii)*), $PC(Q)$ contains a Sylow p-subgroup of $T(b^{**})$. Since $PC(Q) \subseteq P_0C(Q) = M$, we must have $M = PC(Q)$, $\tilde{B} = (b^{**})^{PC(Q)}$. If b^* is a root of $\tilde{B}$ in $PC(P)$, then $(P, b^*) \in \underline{P}$ and (P, b^*) and (Q, b^{**}) are linked.

The next proposition will be important later.

(3F). *Suppose that the pairs* (P, b^*) *and* (Q, b^{**}) *of* $\underline{P}$ *are linked. Assume that* U *is a normal subgroup of* P *with* $P \supset U \supset Q$. *Then there exist pairs* $(U, \tilde{b}) \in \underline{P}$ *such that* (P, b^*) *and* $(U, \tilde{b})$ *are linked and that* $(U, \tilde{b})$ *and* (Q, b^{**}) *are linked.*

PROOF. It follows from the assumption that

$$C(P) \subseteq C(U) \subseteq C(Q) .$$

Since $U \triangleleft P$, $PC(U)$ is a group and

$$PC(P) \subseteq PC(U) \subseteq PC(Q) .$$

Set $\tilde{B} = (b^*)^{PC(Q)}$ and $b^\# = (b^*)^{PC(U)}$. By (2F), we have

$$\tilde{B} = (b^\#)^{PC(U)} .$$

Apply now (3A). We see that $\tilde{B}$ has the defect group P and that $C_P(Q) \subseteq Q$ and hence that $C_P(U) \subseteq Q \subseteq U$. By (2E), $b^\#$ has a defect group $D^\# \supseteq P$ and the defect of $b^\#$ is at most equal to that of $\tilde{B}$, that is, $|D^\#| \leq |P|$, *cf.* (2A). Hence $D^\# = P$. Now (3C) shows the existence of a pair $(U, \tilde{b}) \in \underline{P}$ such that (P, b^*) and $(U, \tilde{b})$ are linked. Then $\tilde{b}^{PC(U)} = (b^*)^{PC(U)}$ and, by (2F),

$$\tilde{b}^{PC(Q)} = (b^*)^{PC(Q)} = \tilde{B} .$$

It is seen easily that $UC(Q)$ is a group and that

$$QC(Q) \triangleleft UC(Q) \triangleleft PC(Q) .$$

If we set $\hat{b} = \tilde{b}^{UC(Q)}$, by (2F),

$$\hat{b}^{PC(Q)} = \tilde{b}^{PC(Q)} = \tilde{B} .$$

Now (2N, *(i)*) shows that $\tilde{B}$ covers $\hat{b} \in Bl(UC(Q))$. Since according to (2M), b^{**} is the only block of $QC(Q)$ covered by $\tilde{B}$, it follows that $\hat{b}$ must cover b^{**} . By (2N), (2M) then $\hat{b} = (b^{**})^{UC(Q)}$. Hence $\tilde{b}^{UC(Q)} = (b^{**})^{UC(Q)}$ and Definition 2.1 shows that $(U, \tilde{b})$ and (Q, b^{**}) are linked.

REMARK. Theorem (6D) of [3] shows that linked pairs can be characterized by a relation between the canonical characters of the blocks. We shall not need this here.

4. Double chains

DEFINITION 4.1. A *double chain* C is a sequence

$$(4.1) \qquad (D_0, b_0), (D_1, b_1), \ldots, (D_r, b_r)$$

of elements of $\underline{P}$ such that (D_{j-1}, b_{j-1}) and (D_j, b_j) are linked for $j = 1, 2, \ldots, r$.

We note some consequences of the definition.

(4A). *If* (4.1) *is a double chain, then*

$$Z(D_r) \supseteq Z(D_{r-1}) \supseteq \ldots \supseteq Z(D_0) .$$

This is seen from (3B).

(4B). *If* (4.1) *is a double chain, we have*

$$b_0^G = b_1^G = \dots = b_r^G .$$

Indeed, if we set $M = D_{j-1}C(D_j)$, we have $b_{j-1}^M = b_j^M$ and then $b_{j-1}^G = b_j^G$, *cf.* (2F). This holds for $j = 1, 2, \dots, r$.

(4C). *If* (4.1) *is a double chain, there exist double chains*

$$(D_0^*, b_0^*), (D_1^*, b_1^*), \dots, (D_s^*, b_s^*)$$

with $D_0^* = D_0$, $b_0^* = b_0$, $D_s^* = D_r$, $b_s^* = b_r$ *obtained by inserting elements of* $\underline{P}$ *in* (4.1) *such that* $|D_{j-1}^* : D_j^*| = p$ *for* $j = 1, 2, \dots, s$.

This follows by repeated application of (3F).

(4D). *If* (4.1) *is a double chain,* $r \geq 1$, *and if* $D_r \triangleleft D_0$, *then the pairs* (D_0, b_0) *and* (D_r, b_r) *are linked.*

PROOF. We use induction on r . The case $r = 1$ is trivial. Suppose next that $r = 2$. Since D_0 normalizes D_2 , D_0 normalizes $C(D_2)$ and $M = D_0C(D_2)$. Set $V = D_0C(D_1)$ and $W = D_1C(D_2)$. These are likewise groups. It then follows from the definition of linked pairs and (2F) that

$$b_0^M = \left(b_0^V\right)^M = \left(b_1^V\right)^M = b_1^M = \left(b_1^W\right)^M = \left(b_2^W\right)^M = b_2^M .$$

Now, Definition 3.1 shows that (D_0, b_0) and (D_2, b_2) are linked.

Finally, if $r \geq 3$, then as $D_r \triangleleft D_1$, by induction (D_1, b_1) and (D_r, b_r) are linked. Since (D_0, b_0) and (D_1, b_1) are linked, the case already treated shows that (D_0, b_0) and (D_r, b_r) are linked.

(4E). *Assume that* $\underline{C}$ *is a double chain* (4.1). *Suppose that* U *is a subgroup with* $D_0 \supseteq U \supseteq D_r$. *Then there exist double chains* $\underline{C}^*$ *joining* (D_0, b_0) *and* (D_r, b_r) *in which a term* $(U, b^*) \in \underline{P}$ *occurs.*

PROOF. On account of (4C), we may assume that $|D_{j-1} : D_j| = p$ for $j = 1, 2, \dots, r$. Then $|D : D_0| = p^r$. We use again induction on r .

The cases where $U = D_0$ or $U = D_r$ are trivial. In particular, we may assume that $r \geq 2$. If $D_1 \supseteq U$, we can use (4E) for the part of $\underline{C}$ joining (D_1, b_1)

and (D_r, b_r) and see that (4E) holds for $\underline{C}$ too.

Suppose then that D_1 does not contain U. Then $D_0 = D_1U$. Let Q be a subgroup of index p of D_0 with $Q \supseteq U$. Thus, $Q \neq D_1$ and $R = D_1 \cap Q$ is a normal subgroup of index p^2 of D_0. Since (4E) holds for the part of $\underline{C}$ joining (D_1, b_1) and (D_r, b_r) and since $D_1 \supset R \supseteq D_r$, we see that we may assume without loss of generality that R is one of the groups D_j. Necessarily, $R = D_2$. As $R \triangleleft D_0$, by (4), (D_0, b_0) and (D_2, b_2) are linked. On account of (3F), there exist double chains

$$(D_0, b_0), \ (Q, \tilde{b}), \ (D_2, b_2) \ .$$

If we replace the first three terms of $\underline{C}$ by these three pairs, then in the new double chain $\underline{C}$, we have $D_1 = Q \supseteq U$ and as we noted above, (4E) holds.

DEFINITION 4.2. Let (D, b) be an element of $\underline{P}$. The *net* $\underline{A}(D, b)$, associated with (D, b) in G, is the set of all subgroups Q of D such that (D, b) can be joined with a suitable $(Q, \tilde{b}) \in \underline{P}$ by a double chain. In particular, $D \in \underline{A}(D, b)$.

As a consequence of (4E), we have

(4F). *If* $Q \in \underline{A}(D, b)$ *and if* P *is a subgroup of* D *with* $P \supseteq Q$, *then* $P \in \underline{A}(D, b)$.

The following result also follows from (4E) in conjunction with (4A).

(4G). *If* $Q \in \underline{A}(D, b)$ *is abelian, then* Q *is a maximal abelian subgroup of* D. *In particular, if* D *is abelian,* $\underline{A}(D, b)$ *consists only of* D.

We now come to the main result of this section which is proved by a Jordan-Hölder type argument.

(4H). *Suppose that* $(D, b) \in \underline{P}$. *If* Q *belongs to the net* $\underline{A}(D, b)$, *there exists a unique pair* $(Q, \tilde{b}) \in \underline{P}$ *such that* (D, b) *and* $(Q, \tilde{b})$ *can be linked by a double chain. We then use the notation* b_Q *for* $\tilde{b}$.

PROOF. Suppose we have two double chains

$$\underline{C} = \{(D_i, b_i);\ i = 0, 1, \ldots, r\}, \quad \underline{C}^* = \{(D^*_j, b^*_j);\ j = 0, 1, \ldots, s\}$$

with $D_0 = D^*_0 = D$, $b_0 = b^*_0 = b$ and $D_r = D^*_s = Q$. We have to show that $b_r = b^*_s$. It can be assumed without loss of generality that $|D_{i-1} : D_i| = p$ for all i and $|D^*_{j-1} : D^*_j| = p$ for all j. Then $r = s$.

We use induction on r. If $r = 1$, the statement follows from (3D). If

$r \geq 2$ and $D_1 = D_1^*$, then $b_1 = b_1^*$. Since by induction (4H) holds for $\underline{A}(D_1, b_1)$ and the subgroup Q of D_1, we have $b_r = b_r^*$.

Assume then that $r \geq 2$ and that $D_1 \neq D_1^*$. Then $P = D_1 \cap D_1^*$ is a normal subgroup of D of index p^2; $P \supseteq Q$. It follows from (4E) that there exist double chains joining (D_1, b_1) and (Q, b_r) in which a term $(P, \hat{b})$ appears. Replacing the part of $\underline{C}$ starting with (D_1, b_1) by this double chain, we see that we may assume that $D_2 = P$. Likewise, we may assume that $D_2^* = P$. By (4D), the pairs (D, b) and (P, b_2) are linked and so are (D, b) and (P, b_2^*). Hence $b_2 = b_2^*$, *cf.* (3D). Since by induction (4H) holds for $\underline{A}(P, b_2)$ and the subgroup Q of D_2, we conclude that, indeed, $b_r = b_r^*$.

We mention some corollaries.

(4I). *If* $Q \in \underline{A}(D, b)$, $Q \neq D$ *and if we set* $P = N_D(Q)$, *then* (P, b_P) *and* (Q, b_Q) *are linked.*

Indeed, by (4H) and (4E), there exist double chains joining (P, b_P) and (Q, b_Q) and since $Q \triangleleft P$, we can apply (4D).

(4J). *If* P *and* Q *are as in* (4I), *then*

$$T(b_Q) \supseteq PC(Q) .$$

This is seen by combining (4I) and (3A).

The next result has been first obtained by Jørn Olsson.

(4K). *If* $Q \in \underline{A}(D, b)$, *then* $C_D(Q) = Z(Q)$.

PROOF. The statement is clear, if $Q = D$. If $Q \subset D$, we have $P = N_D(Q) \supset Q$ and (P, b_P) and (Q, b_Q) are linked by (4I). Then (3B) shows that $C_P(Q) = Z(Q) \subseteq C_D(Q)$. Since $C_D(Q) \subseteq N_D(Q) = P$, we find $C_D(Q) = Z(Q)$.

DEFINITION 4.3. A subgroup Q of a p-group D is of *maximal type*, if $C_D(Q) = Z(Q)$.

Then (4K) can be stated by saying that the members Q of the net $\underline{A}(D, b)$ are of *maximal type* in D.

An abelian subgroup Q of the p-group D is of maximal type in D, if and only if Q is maximal abelian in D.

DEFINITION 4.4. Let $(D, b) \in \underline{P}$. Two elements Q and R of the net

$\underline{A}(D, b)$ are *strongly conjugate* (with regard to (D, b)) if there exist elements $x \in G$ such that

$$Q^x = R \; , \quad b_Q^x = b_R \; .$$

We can give a sufficient condition for strong conjugacy.

(4L). *Assume that* $(D, b) \in \underline{P}$ *and that* $Q \in \underline{A}(D, b)$. *Let* $\{(D_j, b_j);\ j = 0, 1, \ldots, r\}$ *be a double chain with* $D_0 = D$, $D_r = Q$. *If*

$$x \in T(b_r)T(b_{r-1}) \cdots T(b_0) \tag{4.2}$$

then $Q^x = R$ *belongs to* $\underline{A}(D, b)$ *and is strongly conjugate to* Q ; *we have* $b_Q^x = b_R$.

PROOF. Note that the right side of (4.2) need not be a subgroup of G . If (4.2) holds, we can write

$$x = t_r t_{r-1} \cdots t_0$$

with $t_j \in T(b_j)$, $0 \leq j \leq r$. Set $x_{-1} = 1$ and

$$x_j = t_j t_{j-1} \cdots t_0 \; ; \quad (1 \leq j \leq r) \; .$$

Then $x_j = t_j x_{j-1}$ for $0 \leq j \leq r$. Set

$$D_j^* = x_j^{-1} D_j x_j \; , \quad b_j^* = (b_j)^{x_j} \; . \tag{4.3}$$

Since $t_j \in T(b_j) \subseteq N(D_j)$, we have

$$D_j^* = x_{j-1}^{-1} D_j x_{j-1} \; , \quad b_j^* = (b_j)^{x_{j-1}} \; ; \quad (1 \leq j \leq r) \; . \tag{4.4}$$

It follows from (4.3) and (4.4) that x_{j-1} maps $D_{j-1}C(D_j)$ onto $D_{j-1}^*C(D_j^*)$ and, since (D_{j-1}, b_{j-1}) and (D_j, b_j) are linked, that (D_{j-1}^*, b_{j-1}^*) and (D_j^*, b_j^*) are linked; *cf.* Definition 3.1 and (2I). Hence $\{(D_j^*, b_j^*);\ 0 \leq j \leq r\}$ is a double chain. Since $D_0^* = D_0 = D$ and $D_r^* = D_r^x$, (4L) is evident.

In our discussion of $\underline{A}(D, b)$, the inertial groups $T(b_Q)$ for $Q \in \underline{A}(D, b)$ are important. If $Q \triangleleft P \subseteq D$ and $P \neq Q$, (P, b_P) and (Q, b_Q) are linked. There exists some connection between $T(b_P)$ and $T(b_Q)$.

(4M). *Assume that* $P, Q \in \underline{A}(D, b)$ *where* $P \neq Q$ *and* $Q \triangleleft P$. *Set* $W = PC(Q)$ *and* $M = T(b_P) \cap N(Q)$, $L = T(b_P) \cap W$, $X = T(b_Q) \cap N(W)$. *Then*

$M = T(b_P) \cap T(b_Q)$, $L = W \cap M$, $X = WM$ *and the subquotients* M/L *of* $T(b_P)$ *and* X/W *of* $T(b_Q)$ *are isomorphic.*

PROOF. Since $T(b_P) \subseteq N(P)$, $M = T(b_P) \cap N(Q)$ normalizes P and Q and then $PC(P)$, $W = PC(Q)$, and $QC(Q)$. Since M fixes b_P , it fixes b_P^W and hence the unique block b_Q covered by b_P^W . Hence $M \subseteq T(b_Q)$. This implies $M = T(b_P) \cap T(b_Q)$.

By (4J), $W \subseteq T(b_Q)$ and hence

$$W \triangleleft X = T(b_Q) \cap N(W) \subseteq T(b_Q) .$$

It is now clear that $L = T(b_P) \cap W$ is contained in M and that, in fact, $L = M \cap W$. Thus, the first two statements of (4M) hold. Of course, $PC(P) \subseteq L$.

Again, X normalizes Q and $QC(Q)$ and X fixes b_Q . Also, X normalizes W . Then, for $x \in X$, we have (*cf.* (2I))

$$\left(b_Q^W\right)^x = \left(b_Q^x\right)^W = b_Q^W .$$

By (3A), $\tilde{B} = b_Q^W = b_P^W$ has the defect group P . Then $\tilde{B}^x$ has the defect group P^x . Since $\tilde{B}^x = \tilde{B}$, P^x and P are both defect groups of $\tilde{B} \in Bl(W)$. On account of (2B), we can set $P^x = P^y$ with $y \in W$. Then $u = xy^{-1}$ belongs to $N(P) \cap X$ and u fixes $\tilde{B}$. Hence

$$b_P^W = \tilde{B} = \tilde{B}^u = \left(b_P^W\right)^u = \left(b_P^u\right)^W , \tag{4.5}$$

cf. (2I). The block b_P is a root of $\tilde{B}$ and, as $u \in N(P)$ normalizes $PC(P)$, (4.5) shows that b_P^u too is a root of $\tilde{B}$ in $PC(P)$. On account of (2K), we can set $b_P^u = b_P^z$ with $z \in N(P) \cap W$. Then $m = uz^{-1} \in T(b_P)$. Since u and z belong to $X \subseteq T(b_Q)$, m belongs to $T(b_P) \cap T(b_Q) = M$. Now,

$$x = uy = mzy \in MW .$$

Thus, $X \subseteq MW$. On the other hand, since M normalizes W and since M and W are contained in $T(b_Q)$ we have $MW \subseteq N(W) \cap T(b_Q) = X$. Hence $MW = X$. As $W \triangleleft X$, this implies $M/L \simeq X/W$ as we wished to show.

DEFINITION 4.5. A double chain $\{(D_j, b_j);\ j = 0, 1, \ldots, r\}$, is *special*, if every $D_{j-1}C(D_j)$ for $1 \le j \le r$ contains a Sylow p-subgroup of $T(b_j)$. If $(D, b) \in \underline{P}$, the set of subgroups of D occurring in special chains with the initial term (D, b) is denoted by $\underline{A}_0(D, b)$. By definition, $D \in \underline{A}_0(D, b)$.

Thus, $\underline{A}_0(D, b) \subseteq \underline{A}(D, b)$.

(4N). *Assume that* $(D, b) \in \underline{P}$. *If* $Q \in \underline{A}_0(D, b)$, *the special double chain linking* (D, b) *and* (Q, b_Q) *is unique. If* $P = N_D(Q)$, *for* $Q \neq D$, *the term preceding* (Q, b_Q) *in this special double chain is* (P, b_P).

PROOF. It suffices to prove the last part of (4N). Let $\{(D_j, b_j);\ j = 0, 1, \ldots, r\}$ be a special double chain with $D_0 = D$, $b_0 = b$ and $D_r = Q$, $r > 0$. Then $D_{r-1} \subseteq P$. We have to show that we have equality.

By (4J), $PC(Q) \subseteq T(b_Q)$ and hence

$$QC(Q) \subseteq D_{r-1}C(Q) \subseteq PC(Q) \subseteq T(b_Q) .$$

The definition of special chain implies that

$$|T(b_Q) : D_{r-1}C(Q)| \not\equiv 0 \pmod{p} .$$

Since $PC(Q)/(QC(Q))$ is a p-group, it follows that

$$D_{r-1}C(Q) = PC(Q) . \tag{4.6}$$

On account of (4K), we have $C_P(Q) \subseteq Q$ and then $P \cap QC(Q) = Q$ and

$$PC(Q)/(QC(Q)) \simeq P/(P \cap QC(Q)) = P/Q .$$

Likewise,

$$D_{r-1}C(Q)/(QC(Q)) \simeq D_{r-1}/Q .$$

Since $D_{r-1} \subseteq P$, (4.6) implies $D_{r-1} = P$.

We state a corollary.

(4O). *Let* $(D, b) \in \underline{P}$. *An element* $Q \neq D$ *of* $\underline{A}(D, b)$ *belongs to* $\underline{A}_0(D, b)$, *if and only if* $N_D(Q) \in \underline{A}_0(D, b)$ *and* $N_D(Q)C(Q)$ *contains a Sylow* p-*subgroup of* $T(b_Q)$.

Since $N_D(Q) \supset Q$ and since $D \in \underline{A}(D, b)$, (4O) gives a characterization of the elements of $\underline{A}_0(D, b)$ on an inductive basis.

For elements of $\underline{A}_0(D, b)$, we can prove the converse of (4L).

(4R). *Let* $(D, b) \in \underline{P}$. *Suppose that* Q *and* V *belong to* $\underline{A}_0(D, b)$ *and are strongly conjugate, say*

$$V = Q^x , \quad b_V = b_Q^x \tag{4.7}$$

with $x \in G$. *If* $\{(D_j, b_j); j = 0, 1, \ldots, r\}$ *is the special double chain with* $D_0 = D$, $b_0 = b$ *and* $D_r = Q$, *we have*

$$x \in T(b_r)T(b_{r-1}) \ldots T(b_0) . \tag{4.8}$$

PROOF. (4P) is trivially true if $r = 0$, *i.e.* if $Q = D_0$. We use induction on r .

If $Q \neq D$ and if we set $P = N_D(Q)$, by (4N), $D_{r-1} = P$. Likewise, if $U = N_D(V)$, the term preceding (V, b_V) in the special double chain joining (D, b) and (V, b_V) is (U, b_U) .

It follows from (4.7) that x maps $T(b_Q)$ onto $T(b_V)$. Since $PC(Q)$ contains a Sylow p-subgroup P_0 of $T(b_Q)$ and $UC(V)$ contains a Sylow p-subgroup U_0 of $T(b_V)$, we can find an element t of $T(b_Q)$ such that $P_0^{tx} = U_0$. If we replace x by $tx \in T(b_Q)x \subseteq N(Q)x$, (4.7) remains valid and $P_0^x = U_0$. Since $PC(Q) = P_0C(Q)$ and $UC(V) = U_0C(V)$, this implies

$$(PC(Q))^x = UC(V) . \tag{4.9}$$

Then, by (4.7) and (2I),

$$\left(b_Q^{PC(Q)}\right)^x = b_V^{UC(V)} . \tag{4.10}$$

Since $b_V^{UC(V)}$ by (3A) has the defect group U , it follows from (4.10) that $b_Q^{PC(Q)}$ has the defect group $U^{x^{-1}}$. Since P is also a defect group, $U^{x^{-1}}$ and P are conjugate in $PC(Q)$, *cf.* (2B). We can then set $P^{yx} = U$ with $y \in PC(Q) \subseteq T(b_Q)$. If we replace x by yx , the relations (4.7), (4.9) and (4.10) remain valid and we have $P^x = U$.

The block $b_V^{UC(V)}$ has the root b_U in $UC(U)$. It now follows from (4.10) that $b_Q^{PC(Q)}$ has the root $(b_U)^{x^{-1}}$ in $PC(P)$. Since b_P also is a root of

$b_Q^{PC(Q)}$ and since these two roots are conjugate in $PC(Q) \cap N(P)$, *cf.* (2K), we can set $b_P^{zx} = b_U$ with $z \in PC(Q) \cap N(P)$. Now $P^{zx} = P^x = U$, the elements P and U of $\underline{A}_0(D, b)$ are strongly conjugate.

Induction yields

$$zx \in T(b_{r-1}) \ldots T(b_1)T(b_0) .$$

Since $z \in PC(Q) \subseteq T(b_r)$, it is clear that (4.8) holds. This is then also true for the original element x .

5. The nets associated with a block of G

DEFINITION 5.1. Let B be a block of G with the defect group D and let b be a root of B in $DC(D)$. We then call $\underline{A}(D, b)$ a *net associated* with B .

If D^* is another defect group of B and b^* a root of B in $D^*C(D^*)$, it follows from (2B) and (2K) that we can set $D^* = D^x$, $b^* = b^x$ with $x \in G$. Clearly, $\underline{A}(D^x, b^x)$ consists of the groups Q^x with Q ranging over $\underline{A}(D, b)$.

(5A). *If* $B \in B\ell(G)$ *and if* $\underline{A}(D, b)$ *is a net associated with* B *, the most general such net is* $\underline{A}(D^x, b^x)$ *where* x *is an arbitrary element of* G .

We note another preliminary result.

(5B). *If* $\underline{A}(D, b)$ *is a net associated with a block* B *of* G *, then* (D, b) *is a primitive element of* $\underline{P}$ *. Conversely, if* (D, b) *is a primitive element of* $\underline{P}$ *, then* $\underline{A}(D, b)$ *is a net associated with* $B = b^G$.

This is clear from (2J) and (2L).

In the following, B will be a block of G , $\underline{A}(D, b)$ will be a fixed associated net, and the notation b_Q for $Q \in \underline{A}(D, b)$ will refer to this net.

(5C). *Let* B *be a block of* G *and let* $\underline{A}(D, b)$ *be an associated net. If* (U, b^*) *is an element of* $\underline{P}$ *such that* $b^{*G} = B$ *, there exist elements* Q *of* $\underline{A}_0(D, b)$ *and* $x \in G$ *for which*

$$Q^x = U , \quad b_Q^x = b^* . \tag{5.1}$$

Conversely, if (5.1) *holds with* $Q \in \underline{A}_0(D, b)$ *or only with* $Q \in \underline{A}(D, b)$ *and* $x \in G$ *, then* $(U, b^*) \in \underline{P}$ *and* $b^{*G} = B$.

PROOF. Consider a special double chain $\underline{C}^* = \{(D_j^*, b_j^*);\ j = 0, 1, 2, \ldots, r\}$

such that $D_r^* = U$, $b_r^* = b^*$ for which the number r has the largest possible value. Then (3E) shows that the initial term (D_0^*, b_0^*) is primitive. By (4B), $b_0^{*G} = b^{*G}$ and hence $b_0^{*G} = B$. It follows that $\underline{A}(D_0^*, b_0^*)$ is a net associated with B , *cf.* (5B). According to (5A), we can set $D_0^* = D^x$, $b_0^* = b^x$ with $x \in G$. If we define D_j, b_j by $D_j^x = D_j^*$, $b_j^x = b_j^*$, $\underline{C} = \{(D_j, b_j);\ j = 0, 1, \ldots, r\}$ is a special double chain and $D_0 = D$, $b_0 = b$. Set $Q = D_r$. Then $Q \in \underline{A}_0(D, b)$ and (5.1) holds.

The last part of (5C) is obvious.

We note

(5D). *Let* (D, b) *be a primitive element of* $\underline{P}$. *If* $U \in \underline{A}(D, b)$, *then* U *is conjugate to elements* $Q \in \underline{A}_0(D, b)$. *In fact, there exist elements* $x \in G$ *and* $Q \in \underline{A}_0(D, b)$ *such that*

$$U = Q^x , \quad b_U = b_Q^x .$$

Indeed, we can apply (5C) with $B = b^G$ since $(U, b_U) \in \underline{P}$ and $b_U^G = b^G = B$.

(5E). *Let* (D, b) *be a primitive pair in* $\underline{P}$. *There exist systems* $\underline{R}$ *of representatives for the strong conjugacy classes of elements of* $\underline{A}(D, b)$ *such that* $\underline{R}$ *consists of elements of* $\underline{A}_0(D, b)$.

This is clear from (5D).

REMARK. For elements of $\underline{A}_0(D, b)$, we have necessary and sufficient conditions for strong conjugacy, *cf.* (4L), (4P). This is important, if we wish to construct systems $\underline{R}$.

(5F). *Let* B *and* $\underline{A}(D, b)$ *be as in* (5C). *Choose* $\underline{R}$ *as in* (5E). *For each* $R \in \underline{R}$, *choose a set* $\underline{Y}_R$ *of coset representatives of* G mod $T(b_R)$,

$$G = \bigsqcup_{y \in \underline{Y}_R} T(b_R)y \quad \text{(disjoint)}.$$

Then in (5.1), Q *can be chosen in* $\underline{R}$ *and* $x \in \underline{Y}_Q$. *If this is done,* Q *and* x *are determined uniquely by* U *and* b^* .

PROOF. The first statement is clear by the definition of strong conjugacy. The second statement is also obvious. If we have

$$Q^x = R^z = U , \quad b_Q^x = b_R^z = b^*$$

with $x, z \in G$ and $Q, R \in \underline{R}$, then

$$Q^{xz^{-1}} = R , \quad b_Q^{xz^{-1}} = b_R .$$

Hence Q and R are strongly conjugate. Since $Q, R \in \underline{R}$, we have $Q = R$. Then $xz^{-1} \in T(b_Q)$. If both x and z belong to $\underline{Y}_Q$, necessarily $x = z$.

In constructing $\underline{A}(D, b)$ for a given pair $(D, b) \in \underline{P}$, it is important to know, given some $Q \in \underline{A}(D, b)$, whether or not a normal subgroup R of Q belongs to $\underline{A}(D, b)$. We prove a result which can be of help.

(5G). *Suppose that* (D, b) *is a primitive element of* $\underline{P}$. *Assume that* $Q \in \underline{A}(D, b)$ *and that* $R \neq Q$ *is a normal subgroup of* Q. *Assume further that whenever* $Q \in \underline{A}(D^x, b^x)$ *for some* $x \in G$, *we have* $D^x \cap C(R) \subseteq R$. *Then* $R \in \underline{A}(D, b)$.

PROOF. Our assumption shows for $x = 1$ that $C_Q(R) \subseteq R$. If $Q = D$ then as (D, b) is primitive, we have (*cf.* §2, 7),

$$|T(b_Q) : QC(Q)| \not\equiv 0 \pmod p .$$

Then (3D) implies that $R \in \underline{A}(D, b)$.

Assume then that $Q \neq D$. If $R \notin \underline{A}(D, b)$, it follows from (3D) that

$$|T(b_Q) \cap QC(R) : QC(Q)| \equiv 0 \pmod p .$$

Here, $QC(Q) \triangleleft T(b_Q)$. We now see that there exists a subgroup Y of $T(b_Q) \cap QC(R)$ containing $QC(Q)$ as a normal subgroup of index p. If P_0 is a Sylow p-subgroup of Y, we have $Y = P_0 QC(Q)$. Clearly, P_0 is contained in $T(b_Q) \cap QC(R)$, but not in $QC(Q)$.

Let $P_1 \supseteq P_0$ be a Sylow p-subgroup of $T(b_Q)$. Since P_1 is not contained in $QC(Q)$, (3E) implies that there exist pairs $(P, b^*) \in \underline{P}$ for which (P, b^*), (Q, b_Q) are linked and for which $PC(Q)$ contains a Sylow p-subgroup of $T(b_Q)$,

$$(5.2) \qquad (P, b^*) , \quad (Q, b_Q)$$

is a special double chain. After transforming P and b^* with a suitable element of $T(b_Q)$, we may assume that $PC(Q) \supseteq P_1 \supseteq P_0$.

Let $\underline{C}^* = \{(D_j^*, b_j^*);\ j = 0, 1, \ldots, r\}$ be a special double chain with the last two terms (5.2), such that r is maximal. As in the proof of (5C), we see that there exist $x \in G$ for which

$$D_0^* = D^x , \quad b_0^* = b^x .$$

Then, $Q = D_r^* \in \underline{A}(D^x, b^x)$ and by hypothesis $D^x \cap C(R) \subseteq R$. This implies $C_P(R) \subseteq R$. As is seen easily, we then have

$$PC(Q) \cap QC(R) \subseteq QC(Q) .$$

This is a contradiction, because P_0 is contained in the group on the left, but not in that on the right.

We give an application.

DEFINITION 5.2. Let G be a finite group and D a subgroup. A subgroup U of D is *extreme in* D with regard to fusion in G, if $C_D(U)$ is not contained properly in $C_{D^*}(U)$ for any conjugate $D^* \supseteq U$ of D in G.

(5H). *Let* (D, b) *be a primitive pair in* $\underline{P}$. *If* U *is a subgroup of* D *which is extreme in* D *with regard to fusion in* G, *then* $V = UC_D(U)$ *belongs to* $\underline{A}(D, b)$.

PROOF. If U is a subgroup of D and $V = UC_D(U)$, we have $C_D(V) \subseteq C_D(U) \subseteq V$. This implies

$$C_D(V) = Z(V) . \tag{5.5}$$

In the terminology of Definition 4.3, V is of maximal type in D. It is also clear that $C_D(U) \subseteq V$ and hence

$$C_D(U) = C_V(U) . \tag{5.6}$$

Consider a normal series

$$D = D_0 \triangleright D_1 \triangleright \dots \triangleright D_{r-1} \triangleright D_r = V .$$

Then, for $j = 0, 1, \dots, r$,

$$C_D(D_j) \subseteq C_D(V) = Z(V) \subseteq D_j .$$

If U is extreme in D with regard to fusion in G, we shall show by induction on j that $D_j \in \underline{A}(D, b)$. This is clear for $j = 0$. Suppose then that we know for some j with $0 \le j < r$ that $Q = D_j$ belongs to $\underline{A}(D, b)$. We set $R = D_{j+1}$ and try to apply (5G). Suppose that, for some $x \in G$, we have $D^x \supseteq Q$. If we set $D^x = D^*$, then since $Q = D_j$ contains V, by (5.6),

$$C_{D^*}(U) \supseteq C_V(U) = C_D(U) .$$

Since U was extreme in D, we must have equality here. It follows that

$$C_{D^*}(R) \subseteq C_{D^*}(V) \subseteq C_{D^*}(U) = C_D(U) \subseteq V \subseteq R .$$

Thus, (5G) applies and yields $R = D_{j+1} \in \underline{A}(D, b)$. It is now clear that all D_j and, in particular, that V belongs to $\underline{A}(D, b)$. This proves (5H).

If Y is a subgroup of D, there exist conjugates U of Y in G which are extreme in D with regard to fusion in G. We only have to choose $U = Y^x$ with $x \in G$ such that $U \subseteq D$ and that $|C_D(U)|$ is maximal. If we then had $C_D(U) \subset C_{D^*}(U)$ for some $D^* = D^y \supseteq U$, $y \in G$, we would have $U^{y^{-1}} \subseteq D$ and

$$\left|C_D\left(U^{y^{-1}}\right)\right| = |C_{D^*}(U)| > |C_D(U)| ,$$

contrary to the choice of U.

We can use (5H) to obtain a generalization of Theorem (3A) of [4].

(5I). *Let B be a block of G and $\underline{A}(D, b)$ an associated net. If the subgroup U of D is extreme in D with regard to fusion in G, there exist blocks $\tilde{B} \in Bl(UC(U), B)$ with the defect group $V = UC_D(U)$. In particular, if U is abelian, then $\tilde{B} \in Bl(C(U), B)$ has the defect group $C_D(U)$.*

PROOF. By (5H), $V = UC_D(U)$ belongs to $\underline{A}(D, b)$. Since $VC(V) \subseteq UC(U)$, $\tilde{B} = b_V^{UC(U)}$ is defined and then $\tilde{B} \in Bl(UC(U), B)$. Let $\tilde{D} \supseteq V$ be a defect group of $\tilde{B}$. Then $\tilde{D}$ belongs to a conjugate D^* of D in G. As in (5.6), $C_D(U) = C_V(U)$ and hence $C_D(U) \subseteq C_{D^*}(U)$. Since $U \subseteq D^*$ and since U was extreme in D, it follows that $C_D(U) = C_{D^*}(U)$. Clearly

$$V \subseteq \tilde{D} \subseteq UC(U) = VC(U) .$$

We now find

$$\tilde{D} = VC_{\tilde{D}}(U) \subseteq VC_{D^*}(U) = VC_D(U) = V .$$

Thus, $\tilde{D} = V$ and this completes the proof. If U is abelian, we have $UC(U) = C(U)$ and $V = C_D(U)$.

The next result is obtained by combining (5H) with (5D).

(5J). *Let (D, b) be a primitive element of $\underline{P}$. If Y is a subgroup of D, there exist conjugates $U \subseteq D$ of Y in G, for which $UC_D(U) \in \underline{A}_0(D, b)$.*

If Y is such that all conjugates $U \subseteq D$ of Y in G are of maximal type in

D, then (5J) shows that some conjugate of the group Y belongs to $\underline{A}_0(D, b)$. It is clear, for instance, that there exist abelian such Y (that is, abelian subgroups Y of D such that any conjugate $U \subseteq D$ of Y in G is maximal abelian). We then obtain

(5K). *If* $(D, b) \in \underline{P}$ *is primitive, then* $\underline{A}_0(D, b)$ *contains abelian groups.*

We conclude this section with another result which is sometimes useful in constructing $\underline{A}(D, b)$.

(5L). *Assume that* $(D, b) \in \underline{P}$ *and that* $Q \neq D$ *belongs to* $\underline{A}_0(D, b)$. *Assume further that the automorphism group of* Q *is a p-group. If* $R \neq Q$ *is a normal subgroup of* Q, *the condition* $C_{N_D(Q)}(R) \subseteq R$ *is necessary and sufficient in order that* $R \in \underline{A}(D, B)$.

PROOF. The necessity of the condition is clear by (3A). Conversely, assume that $C_P(R) \subseteq R$ with $P = N_D(Q)$. Then, $PC(Q)$ contains a Sylow p-subgroup of $T(b_Q)$, *cf.* (4N) and Definition 4.5. If we let $T(b_Q)$ act on Q by conjugation, we obtain a homomorphism of $T(b_Q)$ into the automorphism group of Q; the kernel is $C(Q)$. It follows that $T(b_Q)/Q$ is a p-group. This implies that $T(b_Q) = PC(Q)$. Since $PC(Q) \cap QC(R) = QC(Q)$, (3D) applies and yields $R \in \underline{A}(D, b)$.

6. The set $B\ell(H, B)$. The subsections

(6A). *Let* B *be a block of* G *and let* $\underline{A}(D, b)$ *be an associated net. Let* H *be a subgroup of* G. *If* $\tilde{b} \in B\ell(H, B)$, *(§2, 5), then* $\tilde{b}$ *has the form*

$$\tilde{b} = \left(b_Q^x\right)^H \tag{6.1}$$

where Q *is an element of* $\underline{A}_0(D, b)$, *where* b_Q *is the corresponding block of* $QC(Q)$, *cf.* (4H), *and where* x *is an element of* G *for which* $Q^x \subseteq H$. *Finally, we have*

$$\left|T(b_Q) \cap H^{x^{-1}} : QC(Q)\right| \not\equiv 0 \pmod p . \tag{6.2}$$

Conversely, if these conditions are satisfied, then $\tilde{b}$ *in* (6.1) *belongs to* $B\ell(H, B)$.

PROOF. If $\tilde{b} \in B\ell(H, B)$ has the defect group U, by definition of the set $B\ell(H, B)$, we have $C(U) \subseteq H$. Let b^* be a root of $\tilde{b}$ in $UC(U) \subseteq H$. Then

$$\tilde{b} = (b^*)^H . \tag{6.3}$$

Since b^* is a block of $UC(U)$ with the defect group U, we have $(U, b^*) \in \underline{P}$.

Also,

(6.4) $$|T(b^*) \cap H : UC(U)| \not\equiv 0 \pmod p ,$$

cf. (2L).

It follows from (6.3) and (2F) that

$$(b^*)^G = \tilde{b}^G = B .$$

Now, (5C) applies. We see that there exists an element Q of $\underline{A}_0(D, b)$ and an element x of G such that $Q^x = U \subseteq H$ and $b_Q^x = b^*$. On substituting the latter equation in (6.3), we obtain (6.1). The congruence (6.2) is obtained from (6.4) by transformation with x^{-1}.

It is also clear that if conversely Q and x satisfy the conditions given in (6A), then (6.1) defines a block $\tilde{b} \in B\ell(H, B)$ with the defect group U. If we drop the condition (6.2), we still have $\tilde{b} \in B\ell(H, B)$, but then the defect group of $\tilde{b}$ will contain U as a subgroup.

We study the question of uniqueness in connection with the representation (6.1) of $\tilde{b}$. The method is similar to that used in the proof of (4P).

It is clear that it will suffice to take Q in (6.1) in a fixed set $\underline{R}$ of representatives for the strong conjugacy classes of elements of $\underline{A}(D, b)$. Suppose then that we have

(6.5) $$\tilde{b} = \left(b_Q^x\right)^H = \left(b_R^y\right)^H$$

where $Q, R \in \underline{R}$, where $x, y \in G$ are such that $Q^x \subseteq H$, $R^y \in H$, and where (6.2) and the analogous condition for R and y hold. Then $\tilde{b}$ has the two defect groups $U = Q^x$ and R^y. These groups must be conjugate in H. There exist then elements $h \in H$ such that $R^{yh} = U$. It now follows that b_R^{yh} and b_Q^x are both roots of $\tilde{b}$. This implies that we can choose h such that, in addition to $R^{yh} = U$, we also have $b_R^{yh} = b_Q^x$. We see that Q and R are strongly conjugate and as Q and R belong to $\underline{R}$, we have $Q = R$.

If $Q = R$, our method shows that $xh^{-1}y^{-1} \in T(b_Q)$. Then y belongs to the double coset $T(b_Q)xH$. Conversely, if this is so and if $Q = R$, (6.5) holds.

(6B). *Let B, D, b, H and $\tilde{b}$ be as in* (6A). *We can choose Q in* (6.1) *in a fixed set $\underline{R}$ of representatives for the strong conjugacy classes of the elements of $\underline{A}(D, b)$. For each Q we can choose x in a set of double coset representatives mod $(T(b_Q), H)$. If this is done, the representation* (6.1) *of $\tilde{b}$ is unique.*

Note that if one element of $T(b_Q)xH$ transforms Q into a subgroup of H, the same is true for all elements of the double coset.

In connection with the formula (1.1), we are mainly interested in the case $H = C(u)$ where u is a p-element of G.

DEFINITION 6.1. A *subsection* of G is a pair $(u, \tilde{b})$ consisting of a p-element u of G and a block $\tilde{b}$ of $C(u)$. We speak of $(u, \tilde{b})$ as a *subsection of the p-section S* of G containing u, (§1). We say that $(u, \tilde{b})$ *belongs to the block* $\tilde{b}^G$ of G. If $t \in G$, we set

$$(u, \tilde{b})^t = (u^t, \tilde{b}^t).$$

In this sense the elements of G act on the subsections of a p-section S, which belong to a block B. We call the defect groups of $\tilde{b}$ also the *defect groups of the subsection* $(u, \tilde{b})$.

If the subsection $(u, \tilde{b})$ has the defect group $\tilde{D}$, then as $\langle u \rangle \triangleleft C(u)$, we have $u \in \tilde{D}$. Since $\tilde{D} \subseteq C(u)$, it follows that $u \in Z(\tilde{D})$.

For $u = 1$, $C(u) = G$. The only subsection $(1, \tilde{b})$ belonging to a block B of G is $(1, B)$.

We apply (6A), (6B) to the case $H = C(u)$.

(6C). *Let $B \in B\ell(G)$ and let $\underline{A}(D, b)$ be an associated net. Each subsection belonging to B is conjugate to a subsection*

$$\left(u, (b_Q)^{C(u)}\right) \tag{6.6}$$

where $Q \in \underline{A}(D, b)$, where $u \in Z(Q)$, and where

$$|T(b_Q) \cap C(u) : QC(Q)| \not\equiv 0 \pmod{p}. \tag{6.7}$$

We may choose here Q in a fixed set $\underline{R}$ of representatives for the strong conjugacy classes of elements of $\underline{A}(D, b)$. Moreover, for each Q it suffices to take u in a set of representatives for the conjugacy classes of $T(b_Q)$ consisting of elements u of $Z(Q)$ for which (6.7) *holds. If this is done, the subsections* (6.6) *form a system $\underline{M}_B$ of representatives for the classes of conjugate subsections belonging to B.*

PROOF. It is clear from (6A) that every subsection belonging to B is conjugate to a subsection (6.6) such that $Q \in \underline{R}$, $u \in Z(Q)$ and that (6.7) holds. If two subsections of this kind are conjugate, (6B) shows that the group Q must be the same in both. Suppose that, for a subsection (6.6), we have

$$\left(u, (b_Q)^{C(u)}\right) = \left(v, (b_Q)^{C(v)}\right)^t$$

with $v \in Z(Q)$ and $t \in G$. Then $v^t = u$ and

$$b_Q^{C(u)} = \left[b_Q^{C(v)}\right]^t = \left(b_Q^t\right)^{C(u)} .$$

The method used in the proof of (6B) shows that we may assume without loss of generality first that $t \in N(Q)$ and then that even $t \in T(b_Q)$. Hence u and v are conjugate in $T(b_Q)$. Conversely, if this is so, $\left(v, (b_Q)^{C(v)}\right)$ is conjugate to the subsection (6.6).

We recall that, for a block B, $k(B)$ denotes the number of irreducible characters and $l(B)$ the number of elements in a basic set for B (which is equal to the number of modular irreducible characters in B).

(6D). *Let* $B \in Bl(G)$ *and let* $\underline{M}_B$ *denote a set of representatives for the conjugacy classes of subsections belonging to* B. *Then*

$$k(B) = \sum_{(u,\tilde{b})\in\underline{M}_B} l(\tilde{b}) . \tag{6.8}$$

PROOF. As shown in [2], it follows from (1.1) that

$$k(B) = \sum_{v\in\underline{V}} \sum_{\hat{b}\in Bl(C(v),B)} l(\hat{b}) , \tag{6.9}$$

where $\underline{V}$ is a set of representatives for the conjugacy classes of G consisting of p-elements. Each pair $(v, \hat{b})$ occurring in (6.9) is a subsection belonging to B. It is clear that two such subsections $(v, \hat{b})$ and $(v_1, \hat{b}_1)$ can be conjugate only if $v = v_1$. If now $(v, \hat{b})^t = (v, \hat{b}_1)$ with $t \in G$, then $t \in C(v)$ and $\hat{b}^t = \hat{b}_1$. Since $\hat{b}$ is a block of $C(v)$, then $\hat{b} = \hat{b}_1$, *i.e.* the two subsections are equal.

Every subsection belonging to B is conjugate to some subsection $(v, \hat{b})$ in (6.9). It is now clear that the subsections $(v, \hat{b})$ in (6.9) form a set $\underline{M}_B^*$ of representatives for the conjugacy classes of subsections belonging to B. Since $l(b_1) = l(b_2)$ for conjugate blocks, the formulas (6.8) and (6.9) are equivalent.

We add some remarks.

(6E) (Jørn Olsson). *The condition* (6.7) *in* (6C) *can only be satisfied when* $C_D(u) = Q$.

PROOF. Since $u \in Z(Q)$, we have $C_D(u) \supseteq Q$. If $C_D(u) \supset Q$, there exists a subgroup $P \supset Q$ of $C_D(u)$ with $Q \triangleleft P$. Since $Q \in \underline{A}(D, b)$, the results of §4 show that $P \in \underline{A}(D, b)$, that (P, b_P) and (Q, b_Q) are linked, and that

$P \subseteq T(b_Q)$. Then $P \subseteq T(b_Q) \cap C_D(u)$ while P is not contained in $QC(Q)$. This contradicts (6.7).

(6F). *If $B \in Bl(G)$ and if (u, b) is a subsection of B with the defect group Q there exist defect groups D^* of B with $Q = C(u) \cap D^*$.*

PROOF. It suffices to prove this for the subsection in (6.6). Here, Q is a defect group. By (6E), $C(u) \cap D = Q$.

Concluding remarks.

Let B be a block of G with a given defect group D. One can ask to describe the possibilities for the various invariants of B such as $k(B)$, the number $l(B)$ of modular irreducible characters in B *etc.* in group theoretical terms.

Even if we only know D, information is available. As is shown for instance by (6C) and (6D), much more can be said if we know the groups Q of a net $\underline{A}(D, b)$ associated with B and the inertial groups $T(b_Q)$.

Many of the questions are trivial for $D = \{1\}$. Assume then that $D \neq \{1\}$. Then all $Q \in \underline{A}(D, b)$ are either non-abelian or maximal abelian in D and hence are different from $\{1\}$. Their normalizers $N(Q)$ are then local subgroups of G for the prime p. If D and $b \in Bl(DC(D), B)$ are given and if we know the local subgroups $N(R)$ for all subgroups R of D which are either non-abelian or maximal abelian in D, we can find $\underline{A}(D, b)$, the groups $Q \in \underline{A}(D, b)$, the inertial groups $T(b_Q)$, the subset $\underline{A}_0(D, b)$ of $A(D, b)$ and we can construct a system $\underline{R}$ of representatives for the strong conjugacy classes of elements of $\underline{A}(D, b)$. This is shown by the preceding results. As a matter of fact, much less information is needed. It will be shown in a subsequent paper how the results can be applied for a further study of the blocks of G.

References

[1] Richard Brauer, "Zur Darstellungstheorie der Gruppen endlicher Ordnung", *Math. Z.* 63 (1955/56), 406-444. MR17,824.

[2] Richard Brauer, "Zur Darstellungstheorie der Gruppen endlicher Ordnung. II", *Math. Z.* 72 (1959), 25-46. MR21#7258.

[3] Richard Brauer, "On blocks and sections in finite groups. I", *Amer. J. Math.* 89 (1967), 1115-1136. MR36#2716.

[4] Richard Brauer, "On blocks and sections in finite groups, II", *Amer. J. Math.* 90 (1968), 895-925. MR39#5713.

[5] Charles W. Curtis, Irving Reiner, *Representation theory of finite groups and associative algebras* (Pure and Appl. Math., 11. Interscience [John Wiley & Sons], New York, London, 1962). MR26#2519.

[6] Larry Dornhoff, *Group representation theory, Part B: Modular representation theory* (Pure and Appl. Math., 7. Marcel Dekker, New York, 1972). Zbl.236.20004.

[7] P. Fong, "On the characters of p-solvable groups", *Trans. Amer. Math. Soc.* 98 (1961), 263-284. MR22#11052.

Harvard University,
Cambridge, USA!

PROC. SECOND INTERNAT. CONF. THEORY OF GROUPS, 20F05
CANBERRA 1973, pp. 131-140.

TRANSITIVITY-SYSTEMS OF CERTAIN ONE-RELATOR GROUPS

A.M. Brunner

For different non-negative integers n the pairs $\{a^{2^n}, b\}$, which are generating pairs of $G = \mathrm{gp}(a, b;\ b^{-1}a^2b = a^3)$, are shown to lie in different T-systems of G. The presentation $\{x, y;\ x^{-1}[x, y]^2, [x, x^{y^n}]\}$ is associated with the generating pair $\{a^{2^n}, b\}$ of G for each positive integer n. It is shown that a two generator group which is an extension of an abelian group by an infinite cyclic group has only one T-system of generating pairs. This implies that the quotient group of G by its second derived group has only one T-system of generating pairs.

1. Introduction

Let G be an n-generator group. A generating n-tuple of G,

$$\mathrm{g} = (g_1, g_2, \ldots, g_n) ,$$

is an ordered set of n elements which generate G; let Γ denote the set of all generating n-tuples of G.

Let F denote the free group of rank n with generating n-tuple $\mathrm{x} = (x_1, x_2, \ldots, x_n)$, and let A denote its group of automorphisms.

We define an action by the elements of A on the elements of Γ: if α is an element of A given by

$$x_i\alpha = w_i(x_1, x_2, \ldots, x_n) , \quad (i = 1, 2, \ldots, n) ,$$

and $\mathrm{g} = (g_1, g_2, \ldots, g_n)$ belongs to Γ, then

$$\mathrm{g}\alpha = (g_1', g_2', \ldots, g_n') ,$$

where

$$g_i' = w_i(g_1, g_2, \ldots, g_n) , \quad (i = 1, 2, \ldots, n) .$$

The elements of A acting in this way are clearly permutations of Γ. Moreover, if γ, δ are elements of A then $\mathrm{g}(\gamma\delta) = (\mathrm{g}\delta)\gamma$, so that A has an anti-representation as a group of permutations of Γ.

Incidently, when $G = F$, the elements of A induce permutations of Γ which act as Nielsen transformations of rank n (see p. 130 and p. 160 of [5]).

Now let B denote the automorphism group of G. We define an action by the elements of B on the elements of Γ as follows: if β belongs to Γ, then

$$\mathrm{g}\beta = (g_1\beta, g_2\beta, \ldots, g_n\beta) .$$

Here B has a representation as a group of permutations of Γ.

LEMMA 1.1. *For each* g *in* Γ, α *in* A, *and* β *in* B,

$$(\mathrm{g}\alpha)\beta = (\mathrm{g}\beta)\alpha .$$

PROOF. Suppose that $x_i\alpha = w_i(x_1, x_2, \ldots, x_n)$, $(i = 1, 2, \ldots, n)$, and let $\mathrm{g} = (g_1, g_2, \ldots, g_n)$. We write $w_i(\mathrm{g})$ for $w_i(g_1, g_2, \ldots, g_n)$. Then,

$$\begin{aligned}(\mathrm{g}\beta)\alpha &= (w_1(\mathrm{g}\beta), w_2(\mathrm{g}\beta), \ldots, w_n(\mathrm{g}\beta))\\ &= (w_1(\mathrm{g})\beta, w_2(\mathrm{g})\beta, \ldots, w_n(\mathrm{g})\beta)\\ &= (w_1(\mathrm{g}), w_2(\mathrm{g}), \ldots, w_n(\mathrm{g}))\beta\\ &= (\mathrm{g}\alpha)\beta .\end{aligned}$$

It follows that P, the group of permutations of Γ induced by the elements of A and B, is a central product of the group of permutations induced by the elements of A with the group of permutations induced by the elements of B. The transitivity sets $\mathrm{g}P$, of Γ under the action of elements of P, are called T-systems.

For a detailed account on T-systems we refer to Neumann and Neumann [7], where T-systems were studied in connection with certain characteristic subgroups of free groups.

In this paper we are interested in the information about the presentations of a group which is provided by a knowledge of the T-systems of the group. If g is a generating n-tuple of a group G, and is associated with a presentation $(x_1, x_2, \ldots, x_n; r, s, t, \ldots)$, then any generating n-tuple of G in the same T-system as g is associated with a presentation $(x_1, x_2, \ldots, x_n; r\alpha, s\alpha, t\alpha, \ldots)$, for some α belonging to A. In the special case that G is a one-relator group, and g is associated with a presentation $(x_1, x_2, \ldots, x_n; r)$, every presentation associated with a generating n-tuple in the same T-system as g is of the form $\{x_1, x_2, \ldots, x_n; (r\alpha)^{\pm 1}\}$, where α belongs to A.

In [6], McCool and Pietrowski found that certain of the free products with amalgamation of two infinite cyclic groups have more than one T-system of generating pairs associated with one-relator presentations. Subsequently, Dunwoody and Pietrowski [3] showed that the trefoil knot group has an infinite number of T-systems; however, only one of these contains generating pairs associated with one-relator presentations (see [8]).

The purpose of this paper is to study T-systems of the group $G = \text{gp}\{a, b; b^{-1}a^2b = a^3\}$, and presentations associated with generating pairs in them. We find that G has an infinite number of distinct T-systems of generating pairs. The T-systems containing the pairs (a, b) and $\{a^2, b\}$, respectively, are distinct, and are associated with the one-relator presentations $\{x, y; y^{-1}x^2yx^{-3}\}$ and $\{x, y; x^{-2}y^{-1}xyx^{-1}y^{-1}xy\}$ respectively.

The fact that the generating pairs (a, b) and $\{a^4, b\}$ do not lie in the same T-system can be deduced directly from a result of G. Higman who, according to Baumslag and Solitar [1], has shown that the generating pair $\{a^4, b\}$ is not associated with a one-relator presentation.

This work is part of the author's PhD thesis submitted to the Australian National University. The author gratefully acknowledges the help given him by his supervisors, Professor B.H. Neumann, FAA, FRS, and Dr R.M. Bryant.

2. Preliminaries

In this section we state some results about the automorphism group of the free group of rank 2 which will be used in this paper. Also we prove a theorem about the T-systems of certain metabelian groups.

Throughout the rest of this paper F shall denote the free group of rank 2 with generating pair $\text{x} = \{x_1, x_2\}$, and A shall denote its group of automorphisms.

If K denotes the free abelian group of rank 2, with generating pair $\text{k} = \{k_1, k_2\}$, then its automorphism group, M, can be described as the group of all μ with $k_i\mu = k_1^{s_i}k_2^{t_i}$, $(i = 1, 2)$, where s_i, t_i are integers satisfying $s_1t_2 - s_2t_1 = \pm 1$.

There is an obvious mapping ψ of A onto M. Let α be an element of A given by $x_i\alpha = w_i\{x_1, x_2\}$, $(i = 1, 2)$, and let σ_{ij} denote the exponent sum of

w_j on x_i, $(i, j = 1, 2)$. It can be shown that there is a τ in M with

$k_i\tau = k_1^{\sigma_{1i}}k_2^{\sigma_{2i}}$, $(i = 1, 2)$. We put $\alpha\psi = \tau$. According to [5] (p. 168) the mapping ψ defines an epimorphism of A onto M.

LEMMA 2.1 (Nielsen; Corollary N.4, p. 168 of [5]). *The kernel of the mapping ψ consists of all inner automorphisms of F.*

The following lemma can be obtained from Satz 5.2 of [7].

LEMMA 2.2. *Let G be a two-generator group, and let Γ be the set of all generating pairs of G. Then, the group of permutations of Γ induced by the inner automorphisms of F is contained in the group of permutations of Γ induced by the inner automorphisms of G.*

LEMMA 2.3. *Let H be a cyclic group generated by h, and let α be an element of A with $(1, h)\alpha = (1, h)$. Then $\alpha = \bar{\alpha}\gamma$, where γ is an inner automorphism of F, and $\bar{\alpha}$ is an element of A with $(x_1, x_2)\bar{\alpha} = (x_1^{\varepsilon}, x_1^{s}x_2)$ for some integer s and $\varepsilon = \pm 1$.*

PROOF. Suppose α is defined by $x_i\alpha = w_i(x_1, x_2)$, $(i = 1, 2)$. Clearly

$(1, h)\alpha = (w_1(1, h), w_2(1, h)) = (h^{\sigma_{21}}, h^{\sigma_{22}})$, and as $(1, h)\alpha = (1, h)$ we have $\sigma_{21} = 0$ and $\sigma_{22} = 1$. Moreover, since $\sigma_{11}\sigma_{22} - \sigma_{12}\sigma_{21} = \pm 1$, we conclude that $\sigma_{11} = \pm 1$. However, there is an element $\bar{\alpha}$ in A with $(x_1, x_2)\bar{\alpha} = (x_1^{\varepsilon}, x_1^{\sigma_{12}}x_2)$, and $\bar{\alpha}\psi = \alpha\psi$. The result now follows from Lemma 2.1.

The next theorem ought to be considered together with the following result of Dunwoody (Theorem 4.10 of [2]), which states: a two-generator metabelian group whose commutator quotient group is free abelian of rank 2 has only one T-system of generating pairs.

THEOREM 2.4. *Let G be a two-generator group containing an abelian normal subgroup N such that G/N is infinite cyclic. Then G is a metabelian group with only one T-system of generating pairs.*

PROOF. It is clear that G has a generating pair (a, b) with a in N and G/N generated by bN, and in this case N is the normal closure in G of a. Also, any other generating pair lies in the same T-system as a pair (c, bd), where c and d belong to N.

Now, since G/N is infinite and generated by bN, any relation of G has to be of the form $w(a, b)$; a word with zero exponent sum on b. Thus $w(a, b) = \prod_j a^{l(j)b^j}$ for some integers $l(j)$. Further c, which is an element of

N , can be written in the form $c = \prod_i a^{k(i)b^i}$ for some integers $k(i)$.

We note that, if n is any element of N , and s is an integer, then $n^{(bd)^s} = n^{b^s d^{b^{s-1}} d^{b^{s-2}} \cdots d} = n^{b^s}$ since n^{b^s} and $d^{b^{s-1}}$, $d^{b^{s-2}}$, ..., d are elements of the abelian group N .

Then,

$$\begin{aligned} w(c, bd) &= \prod_j c^{l(j)(bd)^j} \\ &= \prod_j c^{l(j)b^j} \quad \text{by our comment above,} \\ &= \prod_j \left(\prod_i a^{k(i)b^i}\right)^{l(j)b^j} \\ &= \prod_i \left(\prod_j a^{l(j)b^j}\right)^{k(i)b^i} \quad \text{since } N \text{ is abelian,} \\ &= \prod_i (w(a, b))^{k(i)b^i} \\ &= 1 . \end{aligned}$$

Consequently, if we define a mapping V of G into G by $aV = c$, and $bV = bd$ then we see that V extends to an epimorphism of G . However, it is well known that G , which is a metabelian group, is residually finite (see Theorem 1 of [4]), and hence a Hopf group. We conclude that V is an automorphism of G , and thus that G has only the one T-system.

3. The group $G = \mathrm{gp}(a, b; b^{-1}a^2b = a^3)$

The following lemma will be crucial to the proof of Theorem 3.1 below. The proof of the lemma is modelled on the style of proof used to establish the non-Hopficity of G on p. 261 of [5].

We will use the following notations: $g^h = h^{-1}gh$; $[g, h] = g^{-1}h^{-1}gh$; and if r is a positive integer then $[g, rh]$ is specified by $[g, rh] = [[g, (r-1)h], h]$.

LEMMA 3.1. *Let G be the group $\mathrm{gp}(a, b; b^{-1}a^2b = a^3)$. If m and n are non-negative integers with $m > n$ then there is no automorphism θ of G with* $\left[a^{2^m}, b\right]\theta = \left[a^{2^n}, b\right]$.

PROOF. Firstly, $\left[a^{2^r}, b\right] = a^{-2^r}b^{-1}a^{2^r}b = a^{-2^r}a^{3.2^r} = a^{2^{r-1}}$, for any non-negative integer r. Then, as $m > n$, the relations $\left[a^{2^m}, (n+1)b\right] = a^{2^{m-n+1}}$ and $\left[a^{2^m}, (n+1)b\right]^{2^{n+1}} = a^{2^m}$ hold.

Suppose now, by way of contradiction, that there is an automorphism θ of G with $\left(a^{2^m}, b\right)\theta = \left(a^{2^n}, b\right)$. As $\left[a^{2^m}, (n+1)b\right] = a^{2^m}$ we see that $\left[a^{2^n}, (n+1)b\right]^{2^{n+1}} = a^{2^n}$, and, since $a = \left[a^{2^n}, nb\right]$, we have $[a, b]^{2^{n+1}} = a^{2^n}$.

Let N be the normal closure in G of a. Then N is generated by the elements a^{b^i} where i is any integer. Using a_i to denote a^{b^i} we have (according to [5], p. 261),

$$N = \mathrm{gp}\left(\ldots, a_{-1}, a_0, a_1, \ldots; \ldots, a_{-1}^3 = a_0^2, a_0^3 = a_1^2, \ldots\right).$$

Since there is an integer t with $2^n = 3t \pm 1$ we can write $a^{2^n}[a, b]^{-2^{n+1}}$ in the form $a_0^{3t}a_0^{\varepsilon}\left(a_1^{-1}a_0\right)^{2^{n+1}}$ where $\varepsilon = \pm 1$. Written as such, this is an element of N_0; the subgroup of N generated by a_0 and a_1, which (see p. 261 of [5]) is defined by $\mathrm{gp}\left(a_0, a_1; a_0^3 = a_1^2\right)$. Now, appealing to the solution to the word problem for free products with amalgamation (Theorem 4.4 of [5]), we see that $a_0^{3t}a_0^{\varepsilon}\left(a_1^{-1}a_0\right)^{2^{n+1}}$ is not the identity in N_0, and hence not in G, for any non-negative integer n. This is the desired contradiction.

THEOREM 3.2. *Let* G *be the group* $\mathrm{gp}\left(a, b; b^{-1}a^2b = a^3\right)$. *Then, for different non-negative integers* n, *the generating pairs* $\left(a^{2^n}, b\right)$ *lie in different* T-*systems of* G.

PROOF. Let m and n be non-negative integers with $m > n$, and suppose that $\left(a^{2^n}, b\right)$ and $\left(a^{2^m}, b\right)$ are in the same T-system of G.

Thus there is an α in A and a β in B which, by Lemma 1.1, we can choose such that $\left(a^{2^n}, b\right)\alpha = \left(a^{2^m}, b\right)\beta$.

The relation $a = [a^2, b]$ implies that the normal closure in G of a is the derived group G' ; so that, since G' is characteristic in G , we have $\{a^{2^n}, b\}\beta = \{c, b^{\varepsilon}d\}$ where c and d belong to G' . Then, since there is an element δ of A with $(x_1, x_2)\delta = \{x_1, x_2^{-1}\}$, we can assume that $\{a^{2^n}, b\}\beta = (c, bd)$ where c and d belong to G' .

Let ϕ denote the epimorphism of G onto G/G' . Then,

$$\left(\{a^{2^n}, b\}\alpha\right)\phi = \left(\{a^{2^n}, b\}\phi\right)\alpha \quad \text{by Lemma 2.1,}$$
$$= \{b^0N, b^1N\} ;$$

whereas $(c, bd) = \{b^0N, b^1N\}$. Thus $\{b^0N, bN\}\alpha = \{b^0N, bN\}$, so that by applying Lemma 2.3 we see that $\alpha = \bar{\alpha}\gamma$, where $\bar{\alpha}$ is an element of A given by $\{x_1, x_2\}\bar{\alpha} = \{x_1^{\varepsilon}, x_1^{s}x_2\}$, for some integer s and $\varepsilon = \pm 1$, and γ is an inner automorphism of F . In view of Lemma 2.2 we have that $\{a^{2^n}, b\}\bar{\alpha} = \{a^{2^m}, b\}\beta'$ for some element β' of B .

However, $\{a^{2^n}, b\}\bar{\alpha} = \{a^{\varepsilon . 2^n}, a^{2^n . s}b\}$, and, since there are clearly η_1 and η_2 in B with $(a, b)\eta_1 = \{a^{-1}, b\}$ and $(a, b)\eta_2 = \{a, a^{2^n . s}b\}$, we conclude that $\{a^{2^n}, b\} = \{a^{2^m}, b\}\bar{\beta}$ for some $\bar{\beta}$ in B . Lemma 3.1 now reveals the required contradiction.

4. Presentations of $G = \mathrm{gp}(a, b; b^{-1}a^2b = a^3)$

In this section we find particular presentations associated with the generating pairs $\{a^{2^n}, b\}$ of G .

We say that two sets of relations between elements of a group are equivalent if each relation in one set is implied by the relations in the other set.

LEMMA 4.1. *The sets of relations* $\{c = [c, b]^2\}$ *and* $\{b^{-1}c^2b = c^3, [c, c^b] = 1\}$ *are equivalent.*

PROOF. If $c = [c, b]^2$ then $1 = [c, [c, b]] = [c, c^{-1}c^b] = [c, c^b]$. Further using $[c, [c, b]] = 1$, we have

$$1 = c^{-1}[c, b][c, b] = [c, b]c^{-1}[c, b] = c^{-2}[c, b]b^{-1}cb = c^{-3}b^{-1}c^2b .$$

Also, as this argument is clearly reversible, we have established our claim.

LEMMA 4.2. *The sets of relations* $\left\{c = [c, b]^2, \left[c, c^{b^n}\right] = 1\right\}$ *and* $\left\{c = [c, b]^2, \left[c, c^{b^j}\right] = 1;\ j = 1, 2, \ldots, n\right\}$ *are equivalent.*

PROOF. We write $c_i = c^{b^i}$ for each integer i.

Suppose that the relations $c = [c, b]^2$ and $\left[c, c^{b^j}\right] = 1$ for $j = 1, 2, \ldots, n$ hold. By Lemma 4.1 the relation $c = [c, b]^2$ implies that $c_1^2 = c_0^3$ and $[c_0, c_1] = 1$. Also, $c_1^2 = c_0^3$ implies that $c_r^{2^r} = \left(c_0^{2^r}\right)^{d^r} = c_0^{3^r}$ for any integer r, so that $\left[c_0, c_r^{2^r}\right] = 1$.

Conjugating both sides of the relation $c_r^{2^r} = c_0^{3^r}$ by d^{n-r} we see that $c_n^{2^r} = c_{n-r}^{3^r}$, so that $1 = [c_0, c_n] = \left[c_0, c_n^{2^r}\right] = \left[c_0, c_{n-r}^{3^r}\right]$.

Suppose now that j is an integer with $j = 1, 2, \ldots, n$. Then, by the above, $1 = \left[c_0, c_j^{2^j}\right] = \left[c_0, c_j^{3^{n-j}}\right]$. But $\left(2^j, 3^{n-j}\right) = 1$, so that $[c_0, c_j] = 1$. This establishes the assertion of the lemma.

THEOREM 4.3. *Let* G *be the group* $\text{gp}\left(a, b;\ b^{-1}a^2b = a^3\right)$. *Then, for each positive integer* n, *the generating pair* $\left\{a^{2^n}, b\right\}$ *of* G *is associated with a presentation* $\left(x, y;\ x^{-1}[x, y]^2, \left[x, x^{y^n}\right]\right)$.

PROOF. We prove the theorem by induction on n.

Applying Tietze transformations (see §1.5 of [5]) we see that G is defined by $\text{gp}\left(a, b, c;\ c = a^2, a = \left[a^2, b\right]\right)$ and $\text{gp}\left(c, b;\ c = [c, b]^2\right)$. Thus $\left\{a^2, b\right\}$ is associated with a presentation $\left(x, y;\ x^{-1}[x, y]^2\right)$.

Assume now that the assertion is true for some positive integer n. Thus G is the group defined by $\text{gp}\left(d, b;\ d = [d, b]^2, \left[d, d^{b^n}\right] = 1\right)$, where $d = a^{2^n}$. Applying Tietze transformations, the following groups define G;

$$\mathrm{gp}\left(d, b, e; e = d^2, d = [d^2, b], [d, d^b] = 1, \left[d, d^{b^n}\right] = 1\right)$$

(using Lemma 4.3 here), and

$$\mathrm{gp}\left(e, b; e = [e, b]^2, [[e, b], [e, b]^b] = 1, \left[[e, b], [e, b]^{b^n}\right] = 1\right) .$$

By Lemma 4.2 the relations $e = [e, b]^2$ and $\left[e, e^{b^n}\right] = 1$ imply the relations $e = [e, b]^2$ and $\left[e, e^{b^j}\right] = 1$ for $j = 1, 2, \ldots, n$. These last relations clearly imply the relations $e = [e, b]^2$, $\left[[e, b], [e, b]^{b^n}\right] = 1$ and $\left[[e, b], [e, b]^{b^n}\right] = 1.$

Conversely, from $e = [e, b]^2$ and $\left[[e, b], [e, b]^{b^n}\right] = 1$ we conclude that $\left[e, e^{b^n}\right] = 1$. By Lemma 4.2 we have $\left[e, e^{b^j}\right] = 1$ for $j = 1, 2, \ldots, n$. Then, $1 = \left[[e, b], [e, b]^{b^n}\right] = \left[e^{-1}e^b, e^{-b^n}e^{b^{n+1}}\right]$, and, expanding this, we see that $\left[e, e^{b^{n+1}}\right] = 1$.

Thus G is defined by $\mathrm{gp}\left(e, b; e = [e, b]^2, \left[e, e^{b^{n+1}}\right] = 1\right)$, where $e = a^{2^{n+1}}$, which completes the induction.

The group $G = \mathrm{gp}(a, b; b^{-1}a^2b = a^3)$ was shown to be non-Hopfian in [1]. Our investigations (Lemma 3.1) show that G has an infinite ascending chain of non-Hopf kernels N_n : the normal closure in G of $[a, a^b]$ and $\left[a, a^{b^n}\right]$, where n is a non-negative integer. Moreover, the union of the N_n is equal to the second derived group G''. However, by Lemma 2.4, we see that G/G'' has only one T-system of generating pairs, whereas each G/N_n has an infinite number of T-systems of generating pairs.

As a last remark we mention that this paper has not shown that the generating pairs $\left(a^{2^n}, b\right)$, for each non-negative integer n, are representative of every T-system of generating pairs of G.

References

[1] Gilbert Baumslag and Donald Solitar, "Some two-generator one-relator non-Hopfian groups", *Bull. Amer. Math. Soc.* 68 (1962), 199-201. MR26#204.

[2] Martin John Dunwoody, "Some problems on free groups", (PhD thesis, Australian National University, Canberra, 1964).

[3] M.J. Dunwoody and A. Pietrowski, "Presentations of the trefoil group", *Canad. Math. Bull.* (to appear).

[4] P. Hall, "On the finiteness of certain soluble groups", *Proc. London Math. Soc.* (3) 9 (1959), 595-622. MR22#1618.

[5] Wilhelm Magnus, Abraham Karrass, Donald Solitar, *Combinatorial group theory* (Pure and Appl. Math. 13. Interscience [John Wiley & Sons], New York, London, Sydney, 1966). MR34#7617.

[6] James McCool and Alfred Pietrowski, "On free products with amalgamation of two infinite cyclic groups", *J. Algebra* 18 (1971), 377-383. MR43#6296.

[7] Bernhard H. Neumann und Hanna Neumann, "Zwei Klassen charakteristischer Untergruppen und ihrer Faktorgruppen", *Math. Nachr.* 4 (1951), 106-125. MR12,671.

[8] Elvira Strasser Rapaport, "Note on Nielsen transformations", *Proc. Amer. Math. Soc.* 10 (1959), 228-235. MR21#3477.

Institute of Advanced Studies,
Australian National University,
Canberra, ACT.

PROC. SECOND INTERNAT. CONF. THEORY OF GROUPS, CANBERRA 1973, pp. 141-149.

20E05 (20E10)

CHARACTERISTIC SUBGROUPS OF FREE GROUPS

Roger M. Bryant

1.

In [4], B.H. Neumann asked whether it is true that every characteristic subgroup W of a free group F of infinite rank is fully invariant, and in [5] he conjectured that this is so. Cohen [1] provided support for the conjecture by proving that W is always fully invariant when F/W is abelian-by-nilpotent. However two examples will be described here of characteristic subgroups of the free group F of countable rank which are not fully invariant. Also, a proof will be given of the fact that there are continuously many characteristic subgroups of F which are not fully invariant.

Instead of considering absolutely free groups we can consider relatively free groups and ask which varieties have the property that the characteristic subgroups of their free groups of infinite rank are fully invariant. Cohen's result implies that varieties of abelian-by-nilpotent groups have the property. In general it is easy to see that if $\underline{\underline{V}}$ is any variety with the property then every subvariety of $\underline{\underline{V}}$ has the property. The second (more complicated) example described here is designed to show that the property does not hold in the variety of all groups which are both nilpotent of class 2-by-abelian and centre-by-centre-by-metabelian. But I would expect that the property fails already for the variety of all centre-by-metabelian groups. If this is so then the question of which varieties have the property is answered for a large class of varieties because Hall [2] has pointed out that every variety defined by a "commutator-subgroup function" either contains the variety of all centre-by-metabelian groups or is abelian-by-nilpotent.

The construction employed has some features in common with one which has been used several times to produce examples of non finitely-based varieties of groups. (See, for example, Newman [7].) I have made the presentation here self-contained, but some simple facts concerning varieties of groups are used for which I refer to Hanna Neumann [6].

Professor Kostrikin informed me during the conference that A. Ju. Ol'šanskiĭ has found an example similar to the ones described here and that details of this will be published shortly [8]. I understand that this example is essentially in a variety which is nilpotent of class 3-by-abelian.

2.

Let F be the (absolutely) free group on a countable generating set $\{x_1, x_2, \ldots\}$. For any variety $\underline{\underline{V}}$ let $\underline{\underline{V}}(F)$ denote the set of elements of F which are laws of $\underline{\underline{V}}$. (This notation differs from that of [6].) Thus $\underline{\underline{V}}(F)$ is a fully invariant subgroup of F and $F/\underline{\underline{V}}(F)$ is a free group of countable rank in $\underline{\underline{V}}$. For any positive integer n let $\underline{\underline{B}}_n$ denote the variety of all groups of exponent dividing n and $\underline{\underline{A}}_n$ the variety of all abelian groups of exponent dividing n. Let $\underline{\underline{T}}_2$ be the variety generated by the dihedral group of order 8. Thus $\underline{\underline{T}}_2$ is nilpotent of class 2 and has exponent 4. Let A be the free group of $\underline{\underline{A}}_2$ on a countable generating set $\{\alpha_1, \alpha_2, \ldots\}$.

Each of the two examples is obtained in a similar way: the first from a group G_1 constructed in §3 and the second from a group G_2 constructed in §4. Each G_i is a semi-direct product AK_i (K_i normal in G_i) of a group K_i by the group A. For each i let π_i be the projection of G_i onto A, that is the homomorphism π_i from G_i to A satisfying $(ak)\pi_i = a$ for all $a \in A$, $k \in K_i$, and let Λ_i be the set of all homomorphisms λ from F to G_i which have the property that the kernel of $\lambda\pi_i$ is $\underline{\underline{A}}_2(F)$. For example Λ_i contains the homomorphism λ defined by $x_j\lambda = \alpha_j$ for all j. Let W_i be the intersection of the kernels of the elements of Λ_i. If $\lambda \in \Lambda_i$ and ϕ is an automorphism of F then $\phi\lambda\pi_i$ has kernel $\underline{\underline{A}}_2(F)\phi^{-1} = \underline{\underline{A}}_2(F)$, so $\phi\lambda \in \Lambda_i$. Thus if $w \in W_i$ we have $w\phi\lambda = 1$ for all $\lambda \in \Lambda_i$, so that $w\phi \in W_i$. Therefore W_i is a characteristic subgroup of F. In each of the two cases we shall define elements w_i, w_i' of F such that there is an endomorphism of F mapping w_i to w_i', and we shall prove that $w_i \in W_i$ but $w_i' \notin W_i$. This will prove that W_i is not fully invariant.

The group G_2 will be in the variety $\underline{\underline{V}}$ of all groups which are both nilpotent of class 2-by-abelian and centre-by-centre-by-metabelian, and it follows that $W_2 \supseteq \underline{\underline{V}}(F)$. Also, it is easily seen from the definition of W_2 that $W_2/\underline{\underline{V}}(F)$ is a characteristic subgroup of $F/\underline{\underline{V}}(F)$. But, since W_2 is not fully invariant in F, $W_2/\underline{\underline{V}}(F)$ is not fully invariant in $F/\underline{\underline{V}}(F)$. Thus the relatively free group of countable rank in $\underline{\underline{V}}$ has a characteristic subgroup which is not fully invariant.

To see that F has $2^{\aleph_0}$ characteristic subgroups which are not fully invariant we can make use of either example. What is needed is that in each case the group G_i has non-zero exponent: G_1 has exponent dividing 24 and G_2 has exponent dividing 8. So, for a fixed i, write $G = G_i$, $W = W_i$, $w = w_i$ and $w' = w_i'$. Let m be the exponent of G and let $n = p^3$ where p is any prime not dividing m. Then $w^n \in \underline{\underline{B}}_n(F) \cap W$, but $(w')^n \notin W$ because F/W has exponent dividing m. Consequently, if X is any characteristic subgroup of F satisfying $\underline{\underline{B}}_n(F) \cap W \subseteq X \subseteq W$ we have $w^n \in X$ but $(w')^n \notin X$, and therefore X is not fully invariant. Since $\underline{\underline{B}}_n(F)W = F$ there is a one-to-one correspondence between the set of these X and the set of characteristic subgroups of F which contain $\underline{\underline{B}}_n(F)$. Thus it suffices to prove that the latter set has cardinality $2^{\aleph_0}$. In fact we shall see that $\underline{\underline{B}}_n$ has $2^{\aleph_0}$ subvarieties, so there are $2^{\aleph_0}$ fully invariant subgroups of F containing $\underline{\underline{B}}_n(F)$. It is proved in [ ] that there is a finitely based variety $\underline{\underline{D}}$ of soluble groups of exponent dividing n having a subvariety which is not finitely based. Therefore, by a result of Kovács [3], $\underline{\underline{D}}$ has $2^{\aleph_0}$ subvarieties. Consequently $\underline{\underline{B}}_n$ has $2^{\aleph_0}$ subvarieties.

3.

As in §2, let A be the free group of $\underline{\underline{A}}_2$ on a countable generating set $\{\alpha_1, \alpha_2, \ldots\}$. Write $U = \langle \alpha_1 \rangle$ and $V = \langle \alpha_2, \alpha_3, \ldots \rangle$. Let B be the free group of $\underline{\underline{T}}_2$ on a generating set $\{\beta(a) : a \in A\}$ indexed by the elements of A and let $B^2 = \langle b^2 : b \in B \rangle$. Since $\underline{\underline{T}}_2$ has the laws x_1^4 and $\left[x_1^2, x_2\right]$, B^2 is central in B and has exponent 2. Since B/B^2 is abelian B^2 contains the derived group of B. Thus B^2 contains the set Q of all elements of the forms $\beta(a)^2$, $a \in A$, and $[\beta(a_1), \beta(a_2)]$, $a_1, a_2 \in A$, $a_1 \neq a_2$. Now $B/\langle Q \rangle$ is evidently abelian of exponent 2, so Q generates B^2. Since the $\beta(a)$ are free generators of B any relation between the elements of Q leads to a law of $\underline{\underline{T}}_2$. It follows easily that Q is a basis for B^2. Furthermore the elements of Q are distinct as written above except that $[\beta(a_1), \beta(a_2)] = [\beta(a_2), \beta(a_1)]$. The elements of Q are of the following three types:

(1) $$\beta(a)^2 : a \in A ;$$

(2) $$[\beta(a_1),\ \beta(a_2)] \ :\ a_1a_2 \notin U\ ;$$

(3) $$[\beta(v),\ \beta(\alpha_1 v)] \ :\ v \in V\ .$$

We can produce another basis for B^2 by replacing the elements (3) of Q by the element $[\beta(1),\ \beta(\alpha_1)]$ together with the elements

(4) $$[\beta(1),\ \beta(\alpha_1)][\beta(v),\ \beta(\alpha_1 v)] \ :\ v \in V\backslash\{1\}\ .$$

Let M be the subgroup of B^2 generated by the elements (1), (2) and (4). Thus $[\beta(1),\ \beta(\alpha_1)] \notin M$. We now prove that M contains all elements

(5) $$[\beta(a_1),\ \beta(a_2)][\beta(a_1a),\ \beta(a_2a)] \ :\ a,\ a_1,\ a_2 \in A\ .$$

This is obvious for $a_1 = a_2$. If $a_1a_2 \notin U$ then $[\beta(a_1),\ \beta(a_2)]$ and $[\beta(a_1a),\ \beta(a_2a)]$ both have type (2) and so are in M. Finally if $a_1a_2 = \alpha_1$, the element (5) has the form

$$[\beta(v),\ \beta(\alpha_1 v)][\beta(v'),\ \beta(\alpha_1 v')] \ :\ v,\ v' \in V\ ,$$

which may be written as the product

$$[\beta(1),\ \beta(\alpha_1)][\beta(v),\ \beta(\alpha_1 v)]\cdot[\beta(1),\ \beta(\alpha_1)][\beta(v'),\ \beta(\alpha_1 v')]$$

of two elements each of which is either trivial or of type (4). Thus all elements of type (5) lie in M.

The group A has a natural action on B described by $\beta(a)^{a'} = \beta(aa')$ for all $a,\ a' \in A$. The images under this action of the elements (1), (2) and (4) have types (1), (2) and (5), respectively. Hence M is invariant under the action of A. Consequently we may form semi-direct products AB and $A(B/M)$. (These may be regarded as examples of generalised wreath products.) Let C be the free group of $\underline{\underline{A}}_3$ on a generating set $\{\gamma(h) : h \in A(B/M)\}$ indexed by the elements of $A(B/M)$. Let $G_1 = A(B/M)C$ be the semi-direct product of C by $A(B/M)$ in which $\gamma(h)^{h'} = \gamma(hh')$ for all $h,\ h' \in A(B/M)$. (In other words G_1 is the wreath product of a cyclic group of order 3 by $A(B/M)$.) Thus G_1 is evidently a group of exponent dividing 24.

We use the notation for G_1 given in §2, and let

$$w_1 = [x_1^4,\ x_2^4,\ x_3^4]\ ,\quad w_1' = [x_1^4,\ x_2^4,\ x_1^4]\ .$$

It is clear how to define an endomorphism of F mapping w_1 to w_1'. To prove that

$w_1 \in W_1$ we must prove that $\left[(x_1\lambda)^4, (x_2\lambda)^4, (x_3\lambda)^4\right] = 1$ for all $\lambda \in \Lambda_1$. Since C is an abelian normal subgroup of G_1 it is sufficient to show that at least two of $(x_1\lambda)^4$, $(x_2\lambda)^4$ and $(x_3\lambda)^4$ lie in C. Now the elements $x_1^{-1}x_2$, $x_1^{-1}x_3$ and $x_2^{-1}x_3$ are not in $\underline{\underline{A}}_2(F)$, the kernel of $\lambda\pi_1$, so the elements $x_1\lambda\pi_1$, $x_2\lambda\pi_1$ and $x_3\lambda\pi_1$ are distinct. Therefore it suffices to show that if $g \in G_1$ and $g\pi_1 \neq \alpha_1$ then $g^4 \in C$. Equivalently we show that if $a \in A$, $b \in B$, and $a \neq \alpha_1$, then the element ab of AB satisfies $(ab)^4 \in M$. Now b has the form $\beta(a_1) \ldots \beta(a_n)$ where $a_i \in A$ $(1 \leq i \leq n)$. Thus

$$(ab)^4 = \left(\beta(a_1a) \ldots \beta(a_na)\beta(a_1) \ldots \beta(a_n)\right)^2$$

which expands to

$$\prod_{1\leq i\leq n} \beta(a_ia)^2\beta(a_i)^2[\beta(a_i), \beta(a_ia)]$$

$$\times \prod_{1\leq j<i\leq n} [\beta(a_ia), \beta(a_ja)][\beta(a_i), \beta(a_j)]\cdot[\beta(a_i), \beta(a_ja)][\beta(a_j), \beta(a_ia)] .$$

This lies in M because it is a product of elements of types (1), (2) and (5).

To prove that $w_1' \notin W_1$ we must prove that $w_1'\lambda \neq 1$ for some $\lambda \in \Lambda_1$. Let λ be the homomorphism from F to G_1 defined by $x_1\lambda = \alpha_1\beta(1)M$, $x_2\lambda = \alpha_2\gamma(1)$ and $x_n\lambda = \alpha_n$ $(n \geq 3)$. Since $x_n\lambda\pi_1 = \alpha_n$ for all n it is clear that $\lambda \in \Lambda_1$. Now

$$\begin{aligned}(x_1\lambda)^4 = (\alpha_1\beta(1)M)^4 &= (\beta(\alpha_1)\beta(1)M)^2 \\ &= \beta(\alpha_1)^2\beta(1)^2[\beta(1), \beta(\alpha_1)]M \\ &= [\beta(1), \beta(\alpha_1)]M\end{aligned}$$

which is a non-trivial element b, say, of B/M. Also

$$(x_2\lambda)^4 = (\alpha_2\gamma(1))^4 = \gamma(\alpha_2)^2\gamma(1)^2 .$$

Therefore, by an easy calculation,

$$\left[(x_1\lambda)^4, (x_2\lambda)^4, (x_1\lambda)^4\right] = \gamma(1)^2\gamma(\alpha_2)^2\gamma(\alpha_2b)\gamma(b)$$

which is non-trivial because the elements 1, α_2, α_2b and b are distinct elements of $A(B/M)$. This completes the proof that W_1 is not fully invariant.

4.

We continue with the notation of the preceding sections. The derived group B' of B has a basis consisting of the commutators $[\beta(a_1), \beta(a_2)]$, $a_1 \neq a_2$. Let $\sigma = [\beta(1), \beta(\alpha_2)]$ and $\tau = [\beta(\alpha_1), \beta(\alpha_1\alpha_2)]$. Then B' has another basis consisting of σ and τ together with the following elements:

(i) $[\beta(a_1), \beta(a_2)] : a_1a_2 \notin V$;

(ii) $[\beta(a_1), \beta(a_2)]\sigma : a_1, a_2 \in V, a_1 \neq a_2$, $[\beta(a_1), \beta(a_2)] \neq \sigma$;

(iii) $[\beta(a_1), \beta(a_2)]\tau : a_1, a_2 \notin V, a_1 \neq a_2$, $[\beta(a_1), \beta(a_2)] \neq \tau$.

Let N be the subgroup of B' generated by the elements of types (i), (ii) and (iii). We now prove that N contains all elements

(iv) $[\beta(a_1), \beta(a_2)][\beta(a_3), \beta(a_4)] : a_1 \neq a_2, a_3 \neq a_4$; $a_1a_3, a_2a_4 \in V$.

If $a_1a_2 \notin V$ then $a_3a_4 \notin V$ and the element (iv) is a product of two elements of type (i). So we assume that $a_1a_2 \in V$ and hence $a_3a_4 \in V$. If $a_1 \in V$ we can write (iv) as the product $[\beta(a_1), \beta(a_2)]\sigma\cdot[\beta(a_3), \beta(a_4)]\sigma$ of two elements each of which is either trivial or of type (ii). Similarly if $a_1 \notin V$ we can write (iv) as the product of two elements each of which is either trivial or of type (iii). Thus all elements of type (iv) lie in N .

The images under the action of A of elements of types (i), (ii) and (iii) have types (i), (iv) and (iv), respectively. It follows that N is invariant under the action of A and we may form the semi-direct product $G_2 = A(B/N)$. By its construction G_2 is in the product variety $\underline{\underline{T}}_2\underline{\underline{A}}_2$, so G_2 is nilpotent of class 2-by-abelian and has exponent dividing 8 . To see that G_2 is centre-by-centre-by-metabelian we note first that the second derived group of G_2 is contained in the derived group of B/N and that this is equal to $\langle \sigma N, \tau N\rangle$. For all $a \in A$ the element $\sigma\tau(\sigma\tau)^a$ lies in N because $\sigma\sigma^a$ and $\tau\tau^a$ have type (iv) if $a \in V$ while $\sigma\tau^a$ and $\tau\sigma^a$ have type (iv) if $a \notin V$. Thus $\sigma\tau N$ is central in G_2 . The second derived group of $G_2/\langle\sigma\tau N\rangle$ has order at most 2 and so $G_2/\langle\sigma\tau N\rangle$ is centre-by-metabelian. Thus G_2 is centre-by-centre-by-metabelian.

We use the notation for G_2 given in §2 and let

$$w_2 = \left[x_1^2, \left(x_1^2\right)^{x_2}\right]\left[x_1^2, \left(x_1^2\right)^{x_3}\right]^{-1} , \quad w_2' = \left[x_1^2, \left(x_1^2\right)^{x_2}\right] .$$

It is clear how to define an endomorphism of F mapping w_2 to w_2'. But in order to prove that $w_2 \in W_2$ and $w_2' \notin W_2$ we shall need some preliminary facts.

(I) *Every element of* V *centralizes* B'/N.

PROOF. If $a \in V$ and $a_1 \neq a_2$ then $[\beta(a_1), \beta(a_2)][\beta(a_1a), \beta(a_2a)]$ lies in N because it has type (iv). But $[\beta(a_1a), \beta(a_2a)] = [\beta(a_1), \beta(a_2)]^a$ and so a centralizes $[\beta(a_1), \beta(a_2)]N$. Therefore a centralizes B'/N.

(II) *If* $b_1, b_2 \in B$ *and* $a \in V$ *then*

$$\left[b_1b_2, (b_1b_2)^a\right]N = \left[b_1, b_1^a\right]\left[b_2, b_2^a\right]N .$$

PROOF. We can write

$$\begin{aligned}\left[b_1b_2, (b_1b_2)^a\right] &= \left[b_1, b_1^a\right]\left[b_2, b_2^a\right]\left[b_1, b_2^a\right]\left[b_2, b_1^a\right]\\ &= \left[b_1, b_1^a\right]\left[b_2, b_2^a\right]\cdot\left[b_1, b_2^a\right]\left[b_1, b_2^a\right]^a\end{aligned}$$

and the result follows since $\left[b_1, b_2^a\right]\left[b_1, b_2^a\right]^a$ lies in N by (I).

(III) *If* $a_1, a_2 \in A$, $a_1a_2 \in V\backslash\{1\}$ *and* $a \in A\backslash V$ *then*

$$[\beta(a_1), \beta(a_2)][\beta(a_1a), \beta(a_2a)]N = \sigma\tau N .$$

PROOF. Since $a \notin V$ one of a_1 and a_1a lies in V. Since $a_1 = (a_1a)a$ and $a_2 = (a_2a)a$ we can assume without loss of generality that $a_1 \in V$. Thus $[\beta(a_1), \beta(a_2)]\sigma$ and $[\beta(a_1a), \beta(a_2a)]\tau$ lie in N because they have type (iv). This gives the result.

(IV) *Let* $a \in V$, $a' \in A$ *and* $b \in B$. *Then* $[b^ab, (b^ab)^{a'}] \in N$.

PROOF. We can write

$$\begin{aligned}[b^ab, (b^ab)^{a'}] &= [b, b^{a'}][b^a, b^{aa'}][b^a, b^{a'}][b, b^{aa'}]\\ &= [b, b^{a'}][b, b^{a'}]^a\cdot[b^a, b^{a'}][b^a, b^{a'}]^a\end{aligned}$$

which lies in N by (I).

(V) *Let* $a \in A\backslash V$, $a' \in A\backslash\langle a\rangle$, $b \in B$, *and write* $b = \beta(a_1) \ldots \beta(a_n)$, $a_i \in A$ $(1 \leq i \leq n)$. *Then*

$$[b^ab, (b^ab)^{a'}]N = (\sigma\tau N)^n .$$

PROOF. Since $a \notin V$ one of a' and aa' lies in V. Since $aa' \in A\backslash\langle a\rangle$ and

$[b^a b, (b^a b)^{a'}] = [b^a b, (b^a b)^{aa'}]$ we may assume without loss of generality that $a' \in V$. Now

$$b^a b = \beta(a_1 a) \ldots \beta(a_n a)\beta(a_1) \ldots \beta(a_n) .$$

Hence, by (II),

$$[b^a b, (b^a b)^{a'}]N = \prod_{1 \le i \le n} [\beta(a_i a), \beta(a_i aa')][\beta(a_i), \beta(a_i a')]N .$$

Since $a' \in V \backslash \{1\}$ each product on the right hand side is equal to $\sigma\tau N$ by (III). This gives the result.

To prove that $w_2 \in W_2$ we need to show that $\left[x_1^2, \left(x_1^2\right)^{x_2}\right]\lambda$ is equal to $\left[x_1^2, \left(x_1^2\right)^{x_3}\right]\lambda$ for all $\lambda \in \Lambda_2$. Let $\lambda \in \Lambda_2$ and write $x_i\lambda = a_i b_i N$ where $a_i \in A$, $b_i \in B$, for all i. Thus

$$\begin{aligned}\left[x_1^2, \left(x_1^2\right)^{x_2}\right]\lambda &= \left[(a_1 b_1)^2, \left((a_1 b_1)^2\right)^{a_2 b_2}\right]N \\ &= \left[b_1^{a_1} b_1, \left(b_1^{a_1} b_1\right)^{a_2 b_2}\right]N \\ &= \left[b_1^{a_1} b_1, \left(b_1^{a_1} b_1\right)^{a_2}\right]N ,\end{aligned}$$

since B is nilpotent of class 2. Similarly,

$$\left[x_1^2, \left(x_1^2\right)^{x_3}\right]\lambda = \left[b_1^{a_1} b_1, \left(b_1^{a_1} b_1\right)^{a_3}\right]N .$$

By the definition of Λ_2 we have $a_2^{-1} a_1 \ne 1$ and $a_2 \ne 1$. Thus $a_2 \in A \backslash \langle a_1 \rangle$ and, similarly, $a_3 \in A \backslash \langle a_1 \rangle$. The equality

$$\left[b_1^{a_1} b_1, \left(b_1^{a_1} b_1\right)^{a_2}\right]N = \left[b_1^{a_1} b_1, \left(b_1^{a_1} b_1\right)^{a_3}\right]N$$

now follows from either (IV) or (V) according as $a_1 \in V$ or $a_1 \notin V$.

To prove that $w_2' \notin W_2$ let λ be the homomorphism from F to G_2 defined by $x_1\lambda = \alpha_1\beta(1)N$ and $x_i\lambda = \alpha_i$ $(n \ge 2)$. Clearly $\lambda \in \Lambda_2$. But

$$w_2'\lambda = \left[(\alpha_1\beta(1))^2,\ \left((\alpha_1\beta(1))^2\right)^{\alpha_2}\right]N = \left[\beta(1)^{\alpha_1}\beta(1),\ \left(\beta(1)^{\alpha_1}\beta(1)\right)^{\alpha_2}\right]N .$$

By (V) we have $w_2'\lambda = \sigma\tau N \neq 1$. This completes the proof that W_2 is not fully invariant.

References

[1] D.E. Cohen, "Characteristic subgroups of some relatively free groups", *J. London Math. Soc.* 43 (1968), 445-451. MR37#1450.

[2] P. Hall, "Finiteness conditions for soluble groups", *Proc. London Math. Soc.* (3) 4 (1954), 419-436. MR17,344.

[3] L.G. Kovács, "On the number of varieties of groups", *J. Austral. Math. Soc.* 8 (1968), 444-446. MR37#5277.

[4] B.H. Neumann, "Ascending verbal and Frattini series", *Math. Z.* 69 (1958), 164-172. MR20#3218.

[5] B.H. Neumann, "On characteristic subgroups of free groups", *Math. Z.* 94 (1966), 143-151. MR35#240.

[6] Hanna Neumann, *Varieties of groups* (Ergebnisse der Mathematik und ihrer Grenzgebiete, Band 37. Springer-Verlag, Berlin, Heidelberg, New York, 1967). MR35#6734.

[7] M.F. Newman, "Just non-finitely-based varieties of groups", *Bull. Austral. Math. Soc.* 4 (1971), 343-348. MR43#4891.

[8] А.Ю. Ольшанский [A.Ju. Ol'šanskiĭ], "О характеристических подгруппах свободных групп" [On characteristic subgroups of free groups], *Uspehi Mat. Nauk* (to appear).

UMIST,
Manchester M60 1QD, England.

PROC. SECOND INTERNAT. CONF. THEORY OF GROUPS, CANBERRA 1973, pp. 150-157. 20E10

METABELIAN VARIETIES OF GROUPS

R.A. Bryce

1. Introduction

The reader is referred to Hanna Neumann [12], [13] for notation, terminology and basic facts relating to varieties of groups. Recall that a variety of universal algebras is a class of universal algebras closed under the operations of forming subalgebras, cartesian products and quotient algebras. Equivalently a variety is the class of universal algebras satisfying a given set of identical relations (Birkoff [1]; see also Neumann [11] or Cohn [9] for varieties of universal algebras).

Since the intersection of an arbitrary collection of varieties (of Ω-algebras, say), is again a variety it follows that the set of subvarieties of a given variety $\underline{\underline{V}}$ is a complete lattice under the inclusion order, denoted $\mathrm{lat}(\underline{\underline{V}})$.

If $\underline{\underline{U}}$, $\underline{\underline{V}}$ are varieties of groups $\underline{\underline{UV}}$ is the variety of groups G such that $\underline{\underline{V}}(G) \in \underline{\underline{U}}$. Thus, it is easy to see, the class of all metabelian groups is $\underline{\underline{AA}}$ $(= \underline{\underline{A}}^2)$ where $\underline{\underline{A}}$ is the variety of abelian groups.

Varieties of groups will be denoted $\underline{\underline{U}}$, $\underline{\underline{V}}$ etc.

2. Lat$(\underline{\underline{A}}^2)$

In 1967 Cohen [8] proved that $\mathrm{lat}(\underline{\underline{A}}^2)$ has descending chain condition. Thus every element of $\mathrm{lat}(\underline{\underline{A}}^2)$ can be expressed as a finite join of (finitely) join irreducible elements; and the problem of classifying metabelian varieties can be stated as:

(2.1) *determine the join irreducible elements in* $\mathrm{lat}(\underline{\underline{A}}^2)$, *and*

(2.2) *determine when two irredundant joins of join irreducible elements of* $\mathrm{lat}(\underline{\underline{A}}^2)$ *are distinct.*

An account of the present state of play in this classification problem follows. Part of the answer requires the concepts of bigroup and variety of bigroups, and we introduce these now; see [6] for a fuller discussion.

3. Varieties of bigroups

A bigroup is a universal algebra which is a group G together with a unary operation θ which is to be an idempotent endomorphism of G. It is well known that an idempotent endomorphism of a group G determines, and is determined by, a splitting of G : $G\theta$ is a complement in G for $\ker\theta$. Thus bigroups are as well described by triples $\mathsf{G} = (G, A, B)$ where G is a group, A a normal subgroup of G and B a complement for A in G. In these terms: a sub-bigroup of G is a bigroup $\mathsf{H} = (H, A_0, B_0)$ where H is a subgroup of G, $A_0 = H \cap A$, $B_0 = H \cap B$ (the reason being that this is necessary and sufficient for a subgroup H to admit θ); a normal sub-bigroup N of G is a sub-bigroup normal as a subgroup, and the quotient G/N is $(G/N, AN/N, BN/N)$; the cartesian product of bigroups G_i $(i \in I)$ is $\left(\prod_i G_i, \prod_i A_i, \prod_i B_i\right)$. A homomorphism $\alpha : (G, A, B) \to (H, C, D)$ is a group homomorphism $\alpha : G \to H$ with $A\alpha \le C$, $B\alpha \le D$.

If V is a variety of bigroups we denote by $V(\mathsf{G})$ the smallest normal sub-bigroup of G whose quotient is in V; and if U is a variety of bigroups, UV is the variety of all bigroups G such that $V(\mathsf{G}) \in U$.

Note that if $\underline{\mathrm{U}}$, $\underline{\mathrm{V}}$ are varieties of groups then the class of all bigroups (G, A, B) such that $A \in \underline{\mathrm{U}}$, $B \in \underline{\mathrm{V}}$ is a variety, denoted $\underline{\mathrm{U}} \circ \underline{\mathrm{V}}$. Indeed

$$\underline{\mathrm{U}} \circ \underline{\mathrm{V}} = (\underline{\mathrm{U}} \circ \underline{\mathrm{E}})(\underline{\mathrm{E}} \circ \underline{\mathrm{V}})$$

where $\underline{\mathrm{E}}$ denotes the variety of trivial groups. The free bigroups of $\underline{\mathrm{U}} \circ \underline{\mathrm{V}}$ are easily described; the underlying group of the one of rank r is the verbal wreath product $F_r(\underline{\mathrm{U}} \circ \underline{\mathrm{V}})$ obtained by split extending the verbal product $\underline{\mathrm{U}}\Pi\left\{F_r(\underline{\mathrm{U}})^b : b \in F_r(\underline{\mathrm{V}})\right\}$ by $F_r(\underline{\mathrm{V}})$ with wreathing action. If $F_r(\underline{\mathrm{V}})$ is free on $\{z_1, \ldots, z_r\}$ and $F_r(\underline{\mathrm{U}})$ free on $\{y_1, \ldots, y_r\}$ then as *bigroup* $F_r(\underline{\mathrm{U}} \circ \underline{\mathrm{V}})$ is freely generated by $\{z_1y_1, \ldots, z_ry_r\}$. It is a result of Šmelkin [15] that the *group* generated by $\{z_1y_1, \ldots, z_ry_r\}$ is isomorphic to $F_r(\underline{\mathrm{UV}})$.

It is worth noting that the generators $\{y_1, \ldots, y_r, z_1, \ldots, z_r\}$ have the property that for every $\mathsf{G} = (G, A, B) \in \underline{\mathrm{U}} \circ \underline{\mathrm{V}}$, and pair of mappings

$$\{y_1, \ldots, y_r\} \to A, \quad \{z_1, \ldots, z_r\} \to B$$

there is a unique bigroup homomorphism $\mathsf{F}_r(\underline{\mathrm{U}} \circ \underline{\mathrm{V}}) \to \mathsf{G}$ which extends them both.

There are two obvious connexions between $\mathrm{lat}(\underline{\mathrm{UV}})$ and $\mathrm{lat}(\underline{\mathrm{U}} \circ \underline{\mathrm{V}})$. Define

$$\sigma : \mathrm{lat}(\underline{\underline{UV}}) \to \mathrm{lat}(\underline{\underline{U}} \circ \underline{\underline{V}})$$

by

$$\underline{\underline{W}}\sigma = \{(G, A, B) : A \in \underline{\underline{U}}, B \in \underline{\underline{V}}, G \in \underline{\underline{W}}\}$$

and

$$\tau : \mathrm{lat}(\underline{\underline{U}} \circ \underline{\underline{V}}) \to \mathrm{lat}(\underline{\underline{UV}})$$

by

$$\mathcal{W}\tau = \mathrm{var}\{G : (G, A, B) \in \mathcal{W}\} .$$

In general the mappings σ, τ are not well behaved: σ is a meet-, but not a join-, homomorphism and τ is a join-, but not a meet-, homomorphism. Also

$$\underline{\underline{W}}\sigma\tau \le \underline{\underline{W}} \text{ and } \mathcal{W}\tau\sigma \ge \mathcal{W}$$

and the inclusions may be proper. Of special interest are the *closed* subvarieties $\underline{\underline{W}}$ of $\underline{\underline{U}} \circ \underline{\underline{V}}$; $\underline{\underline{W}}\tau\sigma = \underline{\underline{W}}$. (These are called open in [6]; but closed seems more appropriate since $(\tau\sigma)^2 = \tau\sigma$ and so $\tau\sigma$ is a closure operation in the usual sense.)

Šmel'kin's result mentioned above compares with:

Let $F_r(\mathcal{W})$ be free (as bigroup) on $\{x_1, \ldots, x_r\}$, where $\mathcal{W} \in \mathrm{lat}(\underline{\underline{U}} \circ \underline{\underline{V}})$. The *group* generated by $\{x_1, \ldots, x_r\}$ is free on $\{x_1, \ldots, x_r\}$ in the variety of groups $\mathcal{W}\sigma$.

4. Finite exponent: reduction to the case of prime-power exponent

Let $\mathsf{F} = (F, M, H)$ and $\mathsf{D} = (D, N, K)$ be bigroups in which M, N are both abelian. The $\mathbb{Z}$-module $M \otimes_{\mathbb{Z}} N$ may be regarded as a $\mathbb{Z}(H\times K)$-module - denoted $M \# N$ - in the usual outer tensor product fashion:

$$(m \otimes n)^{hk} = m^h \otimes n^k .$$

The splitting extension $(M \# N)(H \times K)$ is denoted $\mathsf{F} \# \mathsf{D}$ and regarded as group or bigroup as suits the context. It is easy to verify the following sequence of results concerning bigroups in $\underline{\underline{A}} \circ \underline{\underline{O}}$.

(4.1) *If* $\mathsf{F}_1 \le \mathsf{F}_2$ *then* $\mathsf{F}_1 \# \mathsf{D} \le \mathsf{F}_2 \# \mathsf{D}$ *(as bigroups).*

(4.2) *If* $\zeta : \mathsf{F} \to \mathsf{F}\zeta$ *is a homomorphism then there exists an onto (bigroup) homomorphism* $\mathsf{F} \# \mathsf{D} \to \mathsf{F}\zeta \# \mathsf{D}$.

(4.3) $\left(\prod_i \mathsf{F}_i\right) \# \mathsf{D}$ *is subdirectly embedded in* $\prod_i (\mathsf{F}_i \# \mathsf{D})$ *(as bigroups).*

(4.4) *If* $\mathsf{F} \in \mathrm{var}\{\mathsf{F}_i : i \in I\}$ *then* $\mathsf{F} \# \mathsf{D} \in \mathrm{var}\{\mathsf{F}_i \# \mathsf{D} : i \in I\}$.

The usefulness of this construction in metabelian varieties is the following result (§4 in [7]; Chapter 3 in [6]).

(4.5) THEOREM. *Let* G *be a non-nilpotent metabelian critical group. There exist bigroups* $\mathsf{F} = (F, M, H)$ *and* $\mathsf{D} = (D, N, K)$ *satisfying*

(i) F *is a* p*-group for some prime* p *and* $\exp M = p^\alpha$, $\alpha \geq 1$ *; and*

(ii) K *is a non-trivial cyclic group of order* n *;* $p \nmid n$ *;* N *is homocyclic of exponent* p^α *and* K *acts faithfully and irreducibly on* N/pN

such that

$$\text{var}\, G = \text{var}\, \mathsf{F} \# \mathsf{D} .$$

Now let $\underline{\mathbf{V}}$ be a join irreducible metabelian variety of finite exponent, which is not of prime power exponent. Since $\underline{\mathbf{V}}$ is generated by critical groups it follows that it is generated by non-nilpotent critical groups, and hence by (4.5) by a set of groups $\{\mathsf{F}_i \# \mathsf{D} : i \in I\}$ where F_i are all p-groups for the same prime p. Let S be the variety of bigroups $\text{var}\{\mathsf{F}_i : i \in I\}$ and let $|K| = n$. Then it is easy to see that

$$\underline{\mathbf{V}} = [S(\underline{\mathbf{E}} \circ \underline{\mathbf{A}}_n) \wedge \underline{\mathbf{A}} \circ \underline{\mathbf{A}}]\tau .$$

Indeed

(4.6) THEOREM. *Let* S *be a subvariety of* $\underline{\mathbf{A}}_{p^\alpha} \circ \underline{\mathbf{A}}_{p^\beta}$ *not contained in* $\underline{\mathbf{E}} \circ \underline{\mathbf{A}}_{p^\beta}$. *Then if* $p \nmid n$

$$[S(\underline{\mathbf{E}} \circ \underline{\mathbf{A}}_n) \wedge \underline{\mathbf{A}} \circ \underline{\mathbf{A}}]\tau$$

is join irreducible if and only if S *is join-irreducible.*

The plausible conjecture that $\underline{\mathbf{V}}$ should be $(S\tau)\underline{\mathbf{A}}_n \wedge \underline{\mathbf{A}}_{p^\alpha}\underline{\mathbf{A}}_{p^\beta n}$ is not true. One can show ((6.3.3) in [6]):

(4.7) THEOREM. $\underline{\mathbf{V}} = (S\tau)\underline{\mathbf{A}}_n \wedge \underline{\mathbf{A}}_{p^\alpha}\underline{\mathbf{A}}_{p^\beta n}$ *if and only if* S *is closed.*

There exist examples where S is not closed - p. 353 in [6]. However we can give sufficient conditions to make S closed ((6.3.6), (6.3.7) in [6]).

(4.8) *Every subvariety of* $\underline{\mathbf{A}}_{p^\alpha} \circ \mathsf{A}_p$ *is closed as is every subvariety of* $\underline{\mathbf{A}}_{p^\alpha} \circ \underline{\mathbf{A}}_{p^\beta} \wedge N_p$.

To sum up this section: The problem of classifying join irreducible metabelian varieties of finite exponent has been reduced to that of classifying metabelian varieties of groups and varieties of bigroups of prime power exponent.

5. Prime-power exponent

The join-irreducible subvarieties of $\underline{\underline{A}} \circ \underline{\underline{A}}$ of prime-power exponent are describable in terms of nilpotent ones $\big((4.2.30)$ in $[6]\big)$.

(5.1) THEOREM. *Let* U *be a non-nilpotent variety of bigroups in* $\underline{\underline{A}} \circ \underline{\underline{A}}$ *of p-power exponent. Then* U *is join irreducible if and only if*

$$U = L\left(\underline{\underline{E}} \circ \underline{\underline{A}}_{p^r}\right) \wedge \underline{\underline{A}} \circ \underline{\underline{A}}$$

for some nilpotent join irreducible $L \in \operatorname{lat} \underline{\underline{A}} \circ \underline{\underline{A}}$ *of p-power exponent.*

In general the nilpotent join-irreducible varieties are unknown. However two special cases are worth recording $\big((4.3.15)$ and $(4.4.6)$ in $[6]\big)$.

(5.2) THEOREM. *The join irreducibles in* $\operatorname{lat}\left(\underline{\underline{A}}_{p^\alpha} \circ \underline{\underline{A}}_p\right)$ *are:*

(i) non-nilpotent: $\underline{\underline{A}}_{p^\sigma} \circ \underline{\underline{A}}_p$, $1 \leq \sigma \leq \alpha$;

(ii) nilpotent: $\underline{\underline{E}} \circ \underline{\underline{A}}_p$, $\underline{\underline{A}}_{p^\sigma} \circ \underline{\underline{E}}$, $0 \leq \sigma \leq \alpha$ *and*

$$\underline{\underline{A}}_{p^\sigma} \circ \underline{\underline{A}}_p \wedge N_c, \quad 1 \leq \sigma \leq \alpha, \quad c \geq (\sigma-1)(p-1) + 2 .$$

The join irreducibles in $\operatorname{lat}\left(\underline{\underline{A}}_{p^\alpha} \circ \underline{\underline{A}}_{p^\alpha} \wedge N_p\right)$ *are*

$$\underline{\underline{A}}_{p^\sigma} \circ \underline{\underline{E}}, \ \underline{\underline{E}} \circ \underline{\underline{A}}_{p^\sigma}, \ \underline{\underline{A}}_{p^\tau} \circ \underline{\underline{A}}_{p^\tau} \wedge N_c, \quad 0 \leq \sigma \leq \alpha, \ 1 \leq \tau \leq \alpha, \ 2 \leq c \leq p .$$

Note that $\operatorname{lat}\left(\underline{\underline{A}}_{p^\alpha} \circ \underline{\underline{A}}_p\right)$ and $\operatorname{lat}\left(\underline{\underline{A}}_{p^\alpha} \circ \underline{\underline{A}}_{p^\alpha} \wedge N_c\right)$ are both distributive (see §7) so every element in them is uniquely expressibly as an irredundant join of join irreducible elements. In these cases, therefore, (5.2) is a classification theorem. Also $\operatorname{lat} \underline{\underline{A}}_m \circ \underline{\underline{A}}_n$ (where $p \mid m$ implies $p^2 \nmid n$) is distributive $\big((6.2.4)$ in $[6]\big)$ so (4.7), (4.8) and (5.2) provide a classification theorem for $\operatorname{lat} \underline{\underline{A}}_m \underline{\underline{A}}_n$.

As regards varieties of *groups* of prime-power exponent not as much is known. There is a complete analogue of (5.2) but not of (5.1).

(5.3) THEOREM (Brooks [5]). *Let* $\underline{U}$ *be a join irreducible non-nilpotent subvariety of* $\underline{\underline{A}}_p \underline{\underline{A}}_{p^2}$. *Then*

$$\underline{\underline{U}} = \begin{cases} \underline{\underline{N}}_c\underline{\underline{A}}_p \wedge \underline{\underline{A}}_p\underline{\underline{A}}_{p^2} \wedge \underline{\underline{B}}_{p^2} , & 1 \le p < c , \\ \underline{\underline{N}}_c\underline{\underline{A}}_p \wedge \underline{\underline{A}}_p\underline{\underline{A}}_{p^2} , & c \le p . \end{cases}$$

The reader is referred to p. 132 in [10] for a description of the join irreducible elements in lat $\underline{\underline{A}}_{p^\alpha}\underline{\underline{A}}_p$.

(5.4) THEOREM (Weichsel [16]; Brisley [2]). *Let* $\underline{\underline{U}}$ *be a join irreducible variety of p-power exponent in* lat $\underline{\underline{A}}^2$ *of class less than* p . *Then*

$$\underline{\underline{U}} = \underline{\underline{B}}_{p^\alpha} \wedge \underline{\underline{N}}_c \wedge \underline{\underline{A}}^2 .$$

In [3] Brisley also gives a description of lat$\left(\underline{\underline{A}}^2 \wedge \underline{\underline{B}}_{p^\alpha} \wedge \underline{\underline{N}}_{p+1}\right)$.

6. Infinite exponent

The main results here are unpublished ones of L.G. Kovács and M.F. Newman (see (6.1.1) and (6.1.2) in [6]; a proof is sketched on p. 354 of [6]).

(6.1) THEOREM. *If* $\underline{\underline{U}}$ *is a proper subvariety of* $\underline{\underline{A}}^2$ *not of finite exponent then there exists a unique torsion free variety* $\underline{\underline{T}}$, *a unique positive integer* m *and a finite exponent variety* $\underline{\underline{P}}$ *such that*

$$\underline{\underline{U}} = \underline{\underline{T}} \vee \underline{\underline{A}}\underline{\underline{A}}_m \vee \underline{\underline{P}} .$$

(6.2) THEOREM. *The proper torsion free join irreducible subvarieties of* $\underline{\underline{A}}^2$ *are precisely those of the form*

$$\underline{\underline{N}}_c\underline{\underline{A}}_s \wedge \underline{\underline{A}}^2 , \quad c \ge 1 , \quad s \ge 1 .$$

Every torsion free variety $\underline{\underline{T}}$ *in* lat $\underline{\underline{A}}^2$ *is uniquely expressible as an irredundant join of torsion free join irreducibles and this is the unique decomposition of* $\underline{\underline{T}}$ *as an irredundant join of join irreducibles.*

It follows that the join irreducibles in lat $\underline{\underline{A}}^2$ are of the form: $\underline{\underline{N}}_c\underline{\underline{A}}_s \wedge \underline{\underline{A}}^2$ $(c, s \ge 1)$, $\underline{\underline{A}}_{p^\alpha}\underline{\underline{A}}$ (p a prime, $\alpha \ge 1$) or of finite exponent. The finite exponent case has been discussed in §§4, 5.

7. Non-distributivity in lat $\underline{\underline{A}}^2$

In a lattice with descending chain condition every element can be written uniquely as an irredundant join of join irreducible elements if and only if the lattice is distributive. In this case in fact it suffices to know the partially ordered set of join irreducible elements to construct the lattice. Unfortunately lat $\mathbf{A}^2$ is not distributive. The known positive results in this direction are covered by:

(7.1) THEOREM ((6.2.4) in [6]). *If* $p \mid m$ *implies* $p^2 \nmid n$ *then* lat$(\mathbf{A}_m\mathbf{A}_n)$ *is distributive.*

(7.2) THEOREM ((6.2.5) in [6]). *If* $\mathbf{W} \in$ lat $\mathbf{A}^2$ *is of bounded exponent and if* p *groups in* $\mathbf{W}$ *have class at most* p *then* lat $\mathbf{W}$ *is distributive.*

(7.3) THEOREM (Brisley [3]). *If* W *is a metabelian variety of* p*-groups of class at most* $p + 1$ *then* lat $\mathbf{W}$ *is distributive.*

The known non-distributive examples are covered by:

(7.4) THEOREM (Brooks [4]). Lat$(\mathbf{A}_3\mathbf{A}_9 \wedge \mathbf{N}_{11})$ *is not distributive.*

(7.5) THEOREM (Ormerod [14]). Lat$(\mathbf{A}_2\mathbf{A}_4 \wedge \mathbf{N}_6)$ *is not distributive.*

(7.6) THEOREM (Footnote to p. 325 in [6]). Lat$\left(\mathbf{A}_{p^2}\mathbf{A}_{p^2} \wedge \mathbf{N}_{p+2}\right)$ *is not distributive.*

The next example shows that, by contrast with (7.3) above, the bigroups get bad before the groups do.

(7.7) THEOREM ((6.2.5) in [6]). *If* $\mathbf{W}$ *is the subvariety of* $\mathbf{A}_{p^2}\mathbf{A}_{p^2n}$ $(p \nmid n \neq 1)$ *which consists of groups whose Sylow subgroups have class at most* $p + 1$, *then* lat $\mathbf{W}$ *is not distributive.*

Using (6.6), Brooks [4] shows that the variety $\mathbf{P}$ in (6.1) need not be uniquely determined by $\mathbf{V}$ even when it is chosen minimally. These results indicate that an answer to (2.2) may not be easy.

References

[1] Garrett Birkhoff, "On the structure of abstract algebras", *Proc. Cambridge Philos. Soc.* 31 (1935), 433-454. FdM61,1026.

[2] Warren Brisley, "On varieties of metabelian p-groups, and their laws", *J. Austral. Math. Soc.* 7 (1967), 64-80. MR34#7646.

[3] Warren Brisley, "Varieties of metabelian p-groups of class p, $p + 1$", *J. Austral. Math. Soc.* 12 (1971), 53-62. MR43#4890.

[4] M.S. Brooks, "On lattices of varieties of metabelian groups", *J. Austral. Math. Soc.* 12 (1971), 161-166. MR45#3526.

[5] M.S. Brooks, "On varieties of metabelian groups of prime-power exponent", *J. Austral. Math. Soc.* 14 (1972), 129-154.

[6] R.A. Bryce, "Metabelian groups and varieties", *Philos. Trans. Roy. Soc. London Ser. A* 266 (1970), 281-355. MR42#349.

[7] R.A. Bryce and John Cossey, "Some product varieties of groups", *Bull. Austral. Math. Soc.* 3 (1970), 231-264. MR42#4618.

[8] D.E. Cohen, "On the laws of a metabelian variety", *J. Algebra* 5 (1967), 267-273. MR34#5929.

[9] P.M. Cohn, *Universal algebra* (Harper's series in modern mathematics. Harper and Row, New York, Evanston, London, 1965). MR31#224.

[10] L.G. Kovács and M.F. Newman, "On non-Cross varieties of groups", *J. Austral. Math. Soc.* 12 (1971), 129-144. MR45#1966.

[11] B.H. Neumann, *Special Topics in Algebra: Universal Algebra* (Courant Institute of Mathematical Sciences, New York University, 1962).

[12] Hanna Neumann, "Varieties of groups", *Proc. Internat. Conf. Theory of Groups* (Canberra, 1965), pp. 251-259 (Gordon and Breach, New York, London, Paris, 1967). MR35#6733.

[13] Hanna Neumann, *Varieties of groups* (Ergebnisse der Mathematik und ihrer Grenzgebiete, Band 37. Springer-Verlag, Berlin, Heidelberg, New York, 1967). MR35#6734.

[14] Elizabeth A. Ormerod, "A non-distributive metabelian variety lattice", these Proc.

[15] А.Л. Шмелькин [A.L. Šmel'kin], "Сплетения и многообразия групп" [Wreath products and varieties of groups], *Izv. Akad. Nauk SSSR Ser. Mat.* 29 (1965), 149-170. MR33#1352.

[16] Paul M. Weichsel, "On metabelian p-groups", *J. Austral. Math. Soc.* 7 (1967), 55-63. MR34#7645.

The Australian National University,
Canberra.

PROC. SECOND INTERNAT. CONF. THEORY OF GROUPS, 20D10
CANBERRA 1973, pp. 158-164.

SUBDIRECT PRODUCT CLOSED FITTING CLASSES

R.A. Bryce and John Cossey

1. Introduction

In [2] we pointed out that the class of finite soluble groups whose socle is central is an R_0-closed Fitting class. It follows that if p, q are primes, the class S_pS_q contains a proper, non-nilpotent, R_0-closed Fitting class. This contrasts with the closure operations S, E_ϕ and, when $q|p-1$, Q - see [2] for details and notation. Here we prove

THEOREM 1.1. *If p, q are primes with $q|p-1$ then S_pS_q contains a unique maximal R_0-closed Fitting class namely the class of those groups whose socle is central.*

2. Preliminaries

We shall freely adopt the convention that if G is a finite group and M a finite dimensional $GF(p)G$-module then MG denotes the splitting extension of M, regarded as elementary abelian p-group, by G according to the implied homomorphism of G into the automorphism group of M. In this spirit also an elementary abelian p-group on which G acts as a group of operators will be regarded as a module or a group as seems to suit the occassion.

In these terms we associate with a class F of groups, a prime p and given group G the class of $GF(p)G$-modules

$$M(G) = \{M : MG \in \mathrm{F}\} .$$

The dependence of $M(G)$ on F and p will not cause confusion.

The lemma following is Lemma 1.1 in [2].

LEMMA 2.1. *Let G be a group with normal subgroups N_1, N_2 such that*

$$N_1 \cap N_2 = 1 \text{ and } G/N_1N_2 \text{ is nilpotent.}$$

If $\mathfrak{F}$ *is a Fitting class containing* G/N_2 *then* $G \in \mathfrak{F}$ *if and only if* $G/N_1 \in \mathfrak{F}$.

COROLLARY 2.2. *Let* G *be a group with a normal subgroup* N *and* C *a supplement for* N *in* G *such that* $C/C \cap N$ *is nilpotent and* $N \cap C$ *is central in* N . *Then if* $\mathfrak{F}$ *is an* R_0*-closed Fitting class containing* G , C *belongs to* $\mathfrak{F}$ *also.*

PROOF. Since $C \cap N$ is central in N , $C/C \cap N = D$ acts as a group of operators on N . We show that the splitting extension ND is in $\mathfrak{F}$. To do this we temporarily replace C by a minimal supplement C_1 of $C \cap N$ in C . Then $NC_1 = G$ and $N \cap C_1 = (N \cap C) \cap C_1 \leq \Phi(C_1)$, the Frattini subgroup of C_1 . Since $C_1/N \cap C_1 \cong C/N \cap C$ is nilpotent, therefore, C_1 is nilpotent.

Take a copy C_2 of C_1 and form the splitting extension NC_2 . There is a natural epimorphism $NC_2 \twoheadrightarrow G$ whose kernel intersects N trivially. Hence $NC_2 \in R_0\{G, C_2\}$. But if $G \in \mathfrak{F}$ then $C_2 \in \mathfrak{F}$ since a Fitting class contains all relevant nilpotent groups (see, for example, Hartley [3, p. 204]). It follows that $NC_2 \in \mathfrak{F}$. If now L is the subgroup of C_2 corresponding to $N \cap C_1$, then L centralizes N so $L \trianglelefteq NC_2$. Since $L \cap N = 1$ and NC_2/LN is nilpotent it follows from Lemma 2.1 that $NC_2/L \in \mathfrak{F}$. But $NC_2/L \cong ND$ and our claim that $ND \in \mathfrak{F}$ is proved.

Finally consider the splitting extension NC_0 where C_0 is a copy of C . As above there is a natural epimorphism $NC_0 \twoheadrightarrow G$ whose kernel intersects C trivially. If K is the subgroup of C corresponding to $N \cap C$ then K is normal in NC_0 . Also $NC_0/K \cong ND$ so

$$NC_0 \in R_0\{G, ND\} \subseteq \mathfrak{F} .$$

Using Lemma 2.1 again we conclude that $C_0 \cong NC_0/N \in \mathfrak{F}$. In other words $C \in \mathfrak{F}$.

We need one other result.

LEMMA 2.3. *Let* p, q *be primes with* $q|p-1$, *and* C *a cyclic group of* q*-power order. If* N *is a non-trivial irreducible* $GF(p)C$*-module then there exists an irreducible* $GF(p)C$*-module* N^* *such that*

$$\ker N \otimes N^* > \ker N .$$

PROOF. Let θ be the representation of C that N affords. Since all factor groups of C are q-power cycles we may suppose N to be faithful. If $|C| = q^\beta$ put $C = \langle c \rangle$ and

$$d = c^{q^{\beta-1}} .$$

Now by Clifford's Theorem $N_{\langle d \rangle}$ is a direct sum of conjugate (whence isomorphic) irreducibles. But $q | p-1$ so these are one dimensional and d is represented on N by a scalar, λ say. Write α for the automorphism of C given by $c \to c^{-1}$ and N^* for the module affording $\alpha\theta$. On N^* d is represented by λ^{-1}, so on $N \otimes N^*$ by 1.

3. Modules

Let $\mathfrak{F}$ be an R_0-closed Fitting class and $G \in \mathfrak{S}_p\mathfrak{S}_q$ be a group.

(3.1). *If* M *is a* GF$(p)G$-*module and* $M_1 \le M$ *with* M/M_1 *a trivial module then* $M \in \mathrm{M}(G)$ *if and only if* $M_1 \in \mathrm{M}(G)$.

For, $M_1G \trianglelefteq MG$ and MG is a normal product $(M_1G)M$.

(3.2). *Suppose* N *is a completely reducible* GF$(p)G$-*module and* $N \in \mathrm{M}\big(G/O_p(G)\big)$. *Then* $M \otimes N \in \mathrm{M}(G)$ *whenever* $M \in \mathrm{M}(G)$; *and if* $M \otimes N \in \mathrm{M}(G)$ *then a copower of* M (*namely* $(\dim N)M$) *is in* $\mathrm{M}(G)$.

For, $O_p(G) \le \ker N$ so N may be regarded as a $G/O_p(G)$-module, N_0 say. Consider $M \# N_0$ as a module for $G \times G/O_p(G)$. Since $(M \# N_0)_G$ is a direct sum of copies of M the R_0-closure of $\mathfrak{F}$ means that

$$(M \# N_0)_G \in \mathrm{M}(G) .$$

Similarly $(M \# N)_{G/O_p(G)} \in \mathrm{M}\big(G/O_p(G)\big)$ and it follows that

$$M \# N_0 \in \mathrm{M}\big(G \times G/O_p(G)\big) .$$

But $G_0 = \{(g, gO_p(G)) : g \in G\}$ is isomorphic to G and subnormal in $G \times G/O_p(G)$; and the isomorphism $g \mapsto (g, gO_p(G))$ turns $(M \# N_0)_{G_0}$ into $M \otimes N$. Hence

$$M \otimes N \in \mathrm{M}(G) .$$

Conversely, since G and G_0 are both subnormal in, and their product is equal to, $G \times G/O_p(G)$, we conclude that if $M \otimes N \in \mathrm{M}(G)$ then

$$M \# N_0 \in \mathrm{M}\big(G \times G/O_p(G)\big) ,$$

whence

$$(M \# N_0)_G \in \mathrm{M}(G) .$$

But $(M \# N_0)_G \cong (\dim N)M$.

LEMMA 3.3. *Let* $G \in S_pS_q$, *where* $q|p-1$; $G/O_p(G)$ *be cyclic; and* $\mathfrak{M}$ *be a class of* GF(p)*G-modules satisfying:*

(a) whenever M/M_1 *is trivial then* $M \in \mathfrak{M}$ *if and only if* $M_1 \in \mathfrak{M}$; *and*

(b) if N *is an irreducible* GF(p)*G-module then* $M \otimes N \in \mathfrak{M}$ *if* $M \in \mathfrak{M}$; *and if* $M \otimes N \in \mathfrak{M}$ *then some copower of* M *is in* $\mathfrak{M}$.

The following conditions are then equivalent:

(i) $\mathfrak{M}$ *is not empty,*

(ii) the zero module is in $\mathfrak{M}$,

(iii) for all GF(p)*G-modules* M *some copower of* M *is in* $\mathfrak{M}$.

PROOF. What we prove in fact is: if $M_0 \leq M$ then $M_0 \in \mathfrak{M}$ only if some copower of M is in $\mathfrak{M}$ and conversely $M \in \mathfrak{M}$ only if some copower of M_0 is in $\mathfrak{M}$.

The proof of this will be a succession of inductions. First we prove by induction on r :

(3.4). *If* $M_0 < M$, M/M_0 *is completely reducible and homogeneous and* $|G : \ker M/M_0| = q^r$ *then* $M_0 \in \mathfrak{M}$ *only if some copower of* M *is in* $\mathfrak{M}$.

First if $r = 0$ then, by *(a)*, $M_0 \in \mathfrak{M}$ means $M \in \mathfrak{M}$. Next, if $r > 0$ write N for an irreducible component of M/M_0 , and let N^* be the module assured by Lemma 2.3. Now $M \otimes N^*/M_0 \otimes N^* = M/M_0 \otimes N^*$ is completely reducible. Hence there exists a chain of modules

$$M_0 \otimes N^* = L_0 < L_1 < \dots < L_t = M \otimes N^*$$

where L_i/L_{i-1} $(1 \leq i \leq t)$ are all homogeneous. Note that

$$\ker M/M_0 = \ker N < \ker M/M_0 \otimes N^* \leq \ker L_i/L_{i-1} \quad (1 \leq i \leq t) .$$

For every positive integers s , $sL_i/sL_{i-1} \cong s\, L_i/L_{i-1}$ is homogeneous and $|G : \ker sL_i/sL_{i-1}| < q^r$. By induction, therefore, $sL_{i-1} \in \mathfrak{M}$ means some copower of sL_i (and therefore of L_i) is in $\mathfrak{M}$. But $M_0 \in \mathfrak{M}$ implies $M_0 \otimes N^* \in \mathfrak{M}$ by *(b)*, so by induction on i , some copower of $M \otimes N^* \in \mathfrak{M}$: say $v(M \otimes N^*) \in \mathfrak{M}$. However

$$v(M \otimes N^*) \cong vM \otimes N^* .$$

Finally some copower of vM is in M by *(b)* so a copower of M is in M. This completes the inductions on r and with it the proof of (3.4).

(3.5). *If* $M_0 < M$ *and there is a chain of submodules*

$$M_0 < M_1 < \ldots < M_t = M$$

where M_i/M_{i-1} $(1 \le i \le t)$ *is completely reducible and homogeneous then* $M_0 \in \mathsf{M}$ *only if some copower of* M *is in* M.

This has been proved for $t = 1$ in (3.4). If $t > 1$ we conclude by induction that if $M_0 \in \mathsf{M}$, a copower of M_{t-1}, say sM_{t-1}, is in M. Then (as above) sM_t/sM_{t-1} is homogeneous so, by (3.4), a copower of sM_t (and therefore of M) is in M. This completes the inductive step and (3.5) is proved.

The proof that $M \in \mathsf{M}$ only if a copower of M_0 is in M is exactly parallel and we omit it.

4. Proof of Theorem 1.1

First we note a corollary of Corollary 2.2.

LEMMA 4.1. *If* F *is an* R_0*-closed Fitting class and a group in* $\mathsf{F} \cap \mathsf{S}_p\mathsf{S}_q$ *(p, q primes) has non-central socle then* F *contains* $C(p, q)$.

PROOF. Let $G \in \mathsf{F} \cap \mathsf{S}_p\mathsf{S}_q$ have a minimal normal subgroup M not in the centre of G. Then $M \le O_p(G)$ and in fact M is in the centre of $O_p(G)$. Hence if C is a Sylow q-subgroup of G, $MC \in \mathsf{F}$ by Corollary 2.2. But C does not centralize M so $C_C(M)$ is proper in C; it is moreover normal in MC so, by Lemma 2.1, $MC/C_C(M) \in \mathsf{F}$. We may therefore suppose M is faithful for C. Choose C_0 a cycle of order q in C. Then $MC_0 \in \mathsf{S}_n\mathsf{F} \subseteq \mathsf{F}$. If M_0 is an irreducible component of M_{C_0} then, by Corollary 2.2, $M_0C_0 \in \mathsf{F}$. But $M_0C_0 \cong C(p, q)$.

LEMMA 4.2. *Let* F *be an* R_0*-closed Fitting class;* p *and* q *primes with* $q \mid p-1$; $G \in \mathsf{S}_p\mathsf{S}_q$ *a group for which* $G/O_p(G)$ *is cyclic; and* $\mathsf{M}(G/O_p(G))$ *contains all irreducible* $GF(p)(G/O_p(G))$*-modules. The following conditions are equivalent:*

(a) $G \in \mathsf{F}$;

(b) $\mathsf{M}(G)$ *is not empty;*

(c) $\mathsf{M}(G)$ *contains every* $GF(p)G$*-module.*

PROOF. First, *(a)* implies *(b)* since if G is in F then $0 \in \mathsf{M}(G)$.

Second, if $M(G)$ is not empty it contains a copower of the module M : this by Lemma 3.3 which applies on account of (3.1) and (3.2). Say $sM \in M(G)$, $s > 1$. However $sM \cong M \oplus (s-1)M$ so MG is isomorphic to a complement of $(s-1)M$ in $(sM)G$. That is $M(MG)$ is not empty which means - Lemma 3.3 again - that $O \in M(MG)$. In other words, $MG \in F$ so $M \in M(G)$. We have shown that *(b)* implies *(c)*. That *(c)* implies *(a)* is easy: $O \in M(G)$ means $G \in F$.

Suppose now that F is an R_0-closed Fitting class properly contained in $S_pS_q(q|p-1)$, and that F is not nilpotent. Then $S_p, S_q \subseteq F$. Choose a group $G \in S_pS_q \backslash F$ as follows: its Sylow q-subgroup C is to have smallest possible exponent and, subject only to that, $|G|$ must be as small as possible. We show that $G \cong C(p, q)$.

First, G is monolithic and non-nilpotent; and it is also co-monolithic, which is to say that C is cyclic. Also C acts faithfully on $O_p(G)$ and hence faithfully on $O_p(G)/\Phi(O_p(G))$. If $\Phi(O_p(G)) \neq 1$ and M_0 is a faithful irreducible component of $O_p(G)/\Phi(O_p(G))$ we conclude from the minimality of G and Corollary 2.2 that

$$M_0C \in F .$$

It follows that $M(C)$ contains all irreducible GF$(p)C$-modules: if M_1 is one it may be obtained up to linear isomorphism as a component of M_0 restricted to a suitable subgroup of C and inflated back to C . In other words

$$M_1C \in \{S_n, R_0\}F \subseteq F .$$

But now if σG is the socle of G , $H = G/\sigma G \in F$ so, using Lemma 4.2,

$$W = \sigma G \text{ wr } H \in F .$$

If B is the base group of W then, by the well-known Krasner-Kaloujnine Theorem G may be embedded in W as a supplement for B . Using a technique in the proof of Lemma 1.5 in [1], we find $(B \oplus B)G$ subdirectly embedded in $W \times W \times W$. Finally we conclude that $B \oplus B \in M(G)$ and therefore $G \in F$ by Lemma 4.2.

Hence it must follows that $\Phi(O_p(G)) = 1$ and then that for some positive integer r ,

$$G \cong C(p, q^r) .$$

We show that $r = 1$. For, if $r > 1$ then all extensions of p-groups by groups of exponent dividing q^{r-1} are in F . Since $V = C_q \text{ wr } C_{q^{r-1}}$ is a normal product of groups of exponent at most q^{r-1} ,

$$MV \in \mathfrak{F}$$

where M is a faithful and irreducible $GF(p)V$-module. But C is isomorphic to a (subnormal) subgroup of V, so

$$MC \in \mathfrak{F} ,$$

and, using Clifford's Theorem, M_C is a direct sum of faithful irreducible $GF(p)C$-modules. If one such is M_0 then $M_0C \in \mathfrak{F}$ by Corollary 2.2. However $M_0C \cong C(p, q^r)$, a contradiction.

It follows that $G \cong C(p, q)$. Now $\mathfrak{F}$ cannot contain a group whose socle is not central by Lemma 4.1 since $G \notin \mathfrak{F}$. The proof of Theorem 1.1 is therefore complete.

References

[1] R.M. Bryant, R.A. Bryce and B. Hartley, "The formation generated by a finite group", *Bull. Austral. Math. Soc.* 2 (1970), 347-357. MR43#4901.

[2] R.A. Bryce and John Cossey, "Metanilpotent Fitting classes", *J. Austral. Math. Soc.* (to appear).

[3] B. Hartley, "On Fischer's dualization of formation theory", *Proc. London Math. Soc.* (3) 19 (1969), 193-207. MR39#5696.

Australian National University,
Canberra.

PROC. SECOND INTERNAT. CONF. THEORY OF GROUPS, 20E40, 20F05
CANBERRA 1973, pp. 165-187.

ON THE RANK OF THE INTERSECTION OF SUBGROUPS OF A FUCHSIAN GROUP

R.G. Burns

1. Introduction

The Fuchsian groups are the discrete subgroups of $LF(2, R)$, the group of all 2×2 matrices over the reals with determinant $+1$. We are interested here in the following group-theoretical property in particular in connection with Fuchsian groups: a group is said to have the *finitely generated intersection property* if the intersection of every pair of finitely generated subgroups is again finitely generated. In his paper [4], Greenberg proved, using geometrical methods, that Fuchsian groups have the finitely generated intersection property thereby extending the result of Howson [7] that free groups have the finitely generated intersection property. In [3] the present author, using purely algebraic methods, extended Greenberg's result by giving a fairly general criterion for an amalgamated product of two groups to have the finitely generated intersection property. Here we show that the methods of [3] can be made to yield an explicit bound for the rank of the intersection of two subgroups of a Fuchsian group in terms of their ranks. (By the *rank* $r(G)$ of a finitely generated group G we mean the smallest number of generators for G .)

It is a result of Poincaré [13] (see also [10]) that a finitely generated Fuchsian group is either a free product of cyclic groups or has a presentation

$$(1) \quad G = \left\langle a_1, b_1, \ldots, a_n, b_n, c_1, \ldots, c_t \mid c_1^{\alpha_1} = \ldots = c_t^{\alpha_t} = c_1^{-1} \ldots c_t^{-1}[a_1, b_1] \ldots [a_n, b_n] = 1 \right\rangle$$

where n , $t \geq 0$, $\alpha_i > 1$ $(1 \leq i \leq t)$, and $[a_i, b_i] = a_i b_i a_i^{-1} b_i^{-1}$ $(1 \leq i \leq n)$. However not all such presentations correspond to Fuchsian groups: if the *measure* $\mu(G)$ of the group G in (1) is defined by

$$\mu(G) = 2n - 2 + \sum_{i=1}^{t} \left(1-1/\alpha_i\right) ,$$

then for G to be Fuchsian it is necessary and sufficient that $\mu(G) > 0$.

Our main result is the following.

THEOREM 1.1. *Let* H_1, H_2 *be nontrivial subgroups of finite ranks* m, n *of a Fuchsian group, and write* $\langle H_1, H_2\rangle$ *for the smallest subgroup containing* H_1 *and* H_2 . *Then the following statements are true:*

(i) *if* $\langle H_1, H_2\rangle$ *has a presentation* (1) *with* $t = 0$, *and at least one of* H_1, H_2 *has finite index in* $\langle H_1, H_2\rangle$ *then* $r(H_1 \cap H_2) \le mn - m - n + 2$;

(ii) *if* $\langle H_1, H_2\rangle$ *is a free product of cyclic groups then* $r(H_1 \cap H_2) \le 2mn - 2m - 2n + 3$;

(iii) *if* $\langle H_1, H_2\rangle$ *has a presentation* (1) *with either* $n > 0$ *or* $t \ge 4$ *(or both), and both* H_1 *and* H_2 *have infinite index in* $\langle H_1, H_2\rangle$ *then*

$$r(H_1 \cap H_2) \le \tfrac{1}{2}\big[\big(9(8m-6)(8n-6)-3(8m-6)-3(8n-6)+4\big)(8m-6)(8n-6)mn \times \{(8m-6)(8n-6)8mn-2\}+2\big] \times (2mn-2m-2n+3) \ . \tag{2}$$

We shall shortly show how bounds for $r(H_1 \cap H_2)$ in the cases not covered by *(i)*, *(ii)*, or *(iii)* follow relatively easily; but first we make a few remarks concerning the theorem.

Statement *(i)* follows as in [1, Remark 2] once it is proven that the rank of a subgroup of index j in $\langle H_1, H_2\rangle$ is at least $j + 1$. This follows from [5, Theorem 1] and the fact that the rank of the group G of (1) in the case $t = 0$, is (by abelianizing) $2n$. It seems likely that a bound close to that of *(i)* holds without the restriction $t = 0$, for all but a few of the Fuchsian groups presented as in (1). This would follow as above from a lower bound in terms of j for the rank of a subgroup H of index j in a Fuchsian group G given by (1). Such a bound, in turn, would follow in all but a few cases from the fact that $\mu(H) = j\mu(G)$ (see, for example, [5, Theorem 3]) and an expression for the rank of a group presented as in (1). Such an expression is given in [16]; however it is incorrect in some instances.

The bound in *(ii)* follows from Corollary 4.6. Note that *(ii)* was established by Howson [7] and Hanna Neumann [11], [12] in the case that $\langle H_1, H_2\rangle$ is free, but that the bound has been improved in this case to $2mn - 3m - 2n + 4$ for $m \le n$ ([2]).

The proof of *(iii)* occupies the rest of the paper. Crucial to the proof are the facts that for either $n > 0$ or $t \ge 4$ the group (1) is, obviously, an amalgamated product $(A * B; U)$ where A and B are free products of cyclic groups and U is infinite cyclic, and that by [6] every subgroup of infinite index in a Fuchsian group is a free product of cyclic groups. Note that the bound in *(iii)* is asymptotically equivalent to m^6n^6. It seems unlikely that this is anything near best possible.

We now indicate how a bound for $r(H_1 \cap H_2)$ can be derived in the case (which includes all cases not covered by Theorem 1.1) that $\langle H_1, H_2\rangle$ is a Fuchsian group of the form (1) with $t > 0$. By the proof of Satz IV.17 of [15] an infinite Fuchsian group (the finite ones are cyclic) with a presentation (1) contains a normal subgroup having a presentation (1) with $t = 0$, of index equal to the order of $\mathrm{LF}(2, p^k)$, where p is any prime such that $(p, 2, \alpha_1, \ldots, \alpha_t) = 1$, and k satisfies $(p^k-1) \equiv 0 \bmod 2\alpha_1 \ldots \alpha_t$. It is then not difficult to derive from Theorem 1.1 an upper bound for $r(H_1 \cap H_2)$ in terms of $m, n, \alpha_1, \ldots, \alpha_t$.

We note that similar bounds can be obtained for groups

$$G = \left\langle a_1, a_2, \ldots, a_n, c_1, \ldots, c_t \mid c_1^{\alpha_1} = \ldots = c_t^{\alpha_t} = c_1^{-1} \ldots c_t^{-1}a_1^2a_2^2 \ldots a_n^2 = 1\right\rangle,$$

(which include the fundamental groups of non-orientable compact two-dimensional manifolds).

Finally we mention the following problem. Does there exist a finitely generated group with the finitely generated intersection property in which for some pairs m, n there is no general bound for the rank of the intersection of a rank m subgroup with a rank n subgroup? It is not difficult to produce a group of infinite rank without such a bound.

I thank N. Purzitsky and A. Karrass for helpful comments.

This research was partially supported by a grant from the National Research Council of Canada.

2. Preliminaries

To prove Theorem 1.1 *(iii)* we shall need a few of the details of the theorem of Karrass and Solitar [8] giving the structure of the subgroups of an amalgamated product of two groups, and some of the lemmas from [3]. To formulate these we first sketch, as in [3], the relevant definitions.

It is well known that the following definition of an amalgamated product of two groups is equivalent to the more usual ones. A group G is called a *free product of two subgroups* A *and* B *amalgamating their intersection* $A \cap B = U$ say, if for each pair T_A, T_B of left transversals for U in A, B respectively, every element

$g \in G$ can be uniquely expressed in the form

(3) $$g = t_1 \ldots t_n u ,$$

where $n \geq 0$, $u \in U$, $t_i \in (T_A \cup T_B)\backslash U$ $(i = 1, \ldots, n)$, and t_i, t_{i+1} do not both belong to T_A nor to T_B $(i = 1, \ldots, n-1)$. We write $G = (A * B; U)$. We shall call the right hand side of (3) the *normal form* for g (relative to T_A, T_B). The element g will be said to have *length* n , to *begin* with t_1 and to *end* with $t_n u$. Finally, the elements $t_1 \ldots t_i$ $(i = 0, \ldots, n)$ will be called *initial segments* of g , and u the *U-syllable* of g .

The following definition is needed for the subgroup theorem. A *cress* for a subgroup H of $G = (A * B; U)$ relative to left transversals T_A, T_B for U in A, B respectively, both containing the identity 1 , is a pair $\{C_A, C_B\}$ of right transversals for H in G , with the following properties:

(a) for all $g \in C_A \cup C_B$, where $g = t_1 \ldots t_n u$ in normal form,

(i) if $g \in C_A$, then $gu^{-1} \in C_A$ (and similarly for C_B);

(ii) if $u = e$ and $t_n \in T_A$, then $g, gt_n^{-1} \in C_A$ (and similarly, if $t_n \in T_B$, then $g, gt_n^{-1} \in C_B$);

(iii) if $gu^{-1} \in C_A \cap C_B$, then $g \in C_A \cap C_B$;

(b) if S_A is the set consisting of 1 and all nontrivial $g \in C_A$ which have $u = 1$ and $t_n \in T_B$, then S_A is a complete double coset representative system for G modulo (H, A) (and similarly for S_B);

(c) if R_A is the set of all $g \in C_A$ which have $u = 1$, then R_A is a complete double coset representative system for G modulo (H, U) (and similarly for R_B).

We can now choose a generating set for the arbitrary subgroup H of G in terms of a given cress $\{C_A, C_B\}$ for H in G . (The existence of a cress for every $H \leq G$ is proved in [8].) For each $g \in G$ let $g\varphi_A$ denote the representative in C_A of Hg . Write

$$R = \{r, r\varphi_A \mid r \in R_B, r \neq r\varphi_A\} ,$$

$$Q_A = \left\{d \mid d \in S_A,\ dAd^{-1} \cap H \nsubseteq dUd^{-1}\right\},$$

$$Q_B = \left\{d \mid d \in S_B,\ dBd^{-1} \cap H \nsubseteq dUd^{-1}\right\}.$$

We shall also need the following concept. A set S of subgroups of a group is said to generate a *tree product* of its members if there is a tree whose vertices are the elements of S such that for every pair G_1, G_2 of adjacent vertices the subgroup K_1 generated by all $G \in S$ whose distance from G_1 is less than its distance from G_2, and K_2, analogously defined, generate their free product amalgamating $G_1 \cap G_2$. (Alternatively, see [8].) A vertex of a tree incident with only one edge is called *extremal*.

We are now able to state the results from [8] that we require.

LEMMA 2.1. *(i) Let* $H \leq G = (A * B; U)$ *and let* $\{C_A, C_B\}$ *be a cress for* H *relative to* T_A, T_B. *Then* H *is generated by the set*

$$R_1 = \left\{r(r\varphi_A)^{-1} \mid r \in R_B,\ r \neq r\varphi_A\right\}$$

together with $H \cap U$ *and all subgroups* $dAd^{-1} \cap H$ $(d \in Q_A)$ *and* $dBd^{-1} \cap H$ $(d \in Q_B)$. *More precisely the subgroup* K *of* H *generated by all* $dAd^{-1} \cap H$ $(d \in Q_A)$ *and* $dBd^{-1} \cap H$ $(d \in Q_B)$ *and* $H \cap U$ *is a tree product of these groups where the amalgamated subgroups are conjugate in* G *to subgroups of* U, *and* H *has a presentation (as a so-called Higman-Neumann-Neumann or HNN group)* $\left\langle R_1, K \mid r^{-1}L(r)r = M(r),\ r \in R_1\right\rangle$, *where for each* r, $L(r)$, $M(r)$ *are isomorphic subgroups of conjugates of* U *(and the precise way in which* r *transforms* $L(r)$ *to* $M(r)$ *is given by some isomorphism).*

(ii) If every subgroup of U *(including* U *itself) is finitely generated, then* H *is finitely generated if and only if* R, Q_A *and* Q_B *are finite, and for all* $g \in G$, $gAg^{-1} \cap H$ *is finitely generated.*

Part *(i)* is just a restatement of part of Theorem 5 of [8]. Part *(ii)* follows from Lemma 3 and Theorems 4, 5 of [8].

The next (and final) lemma is essentially Lemma 3.5 of [3].

LEMMA 2.2. *Let* $H \leq G = (A * B; U)$ *and let* $\{C_A, C_B\}$ *be a cress for* H *relative to* T_A, T_B. *Let* $g \in R_A$ *and write* $g = dq$, *where* $q \in T_A$ *and* $d \in S_A$. *If* $a \in A$ *is the end of any element of* $Hg\backslash U$ *then* $a = uq_1^{-1}a_1q$, *where* $u \in U$,

$a_1 \in A \cap d^{-1}Hd$, $q_1 \in T_A$ *and either* $q_1 = 1$ *or* dq_1 *is an initial segment of some element of* $R \cup Q_A \cup Q_B$. *(The analogous result is of course true with* B *replacing* A *throughout.)*

For the proof see that of [3, Lemma 3.5].

3. An analogue of the Howson-Hanna Neumann formula

Let F be a free group freely generated by a set X . If $w \in F$ is a non-trivial word with reduced form $x_1 \ldots x_n$ where $x_i \in X \cup X^{-1}$ $(i = 1, \ldots, n)$ then x_n is called the *ending* of w . For every subset S of F we denote by $e(S)$ the set of all endings of elements of S . Let $|Y|$ denote the number of elements of any set Y . The Howson-Hanna Neumann formula (see [7], [11]) states that, if H is a finitely generated subgroup of F , then

$$2r(H) = |e(H)| + \sum_W (|e(W)|-2) , \tag{4}$$

where the summation is over those right cosets $W = Hg$, distinct from H , which have at least 3 endings.

In this section we established an analogous formula for $G = (A * B; U)$. The following definition provides the appropriate analogue of "ending".

Choose, as usual, left transversals T_A, T_B , both containing 1 , for U in A, B . Let $H \le G$. We define an *A-ending* of a double coset HgU , to be a double coset $(g^{-1}Hg \cap A)a^{-1}U$, where $a \in T_A \backslash \{1\}$, and a ends some element of HgU . (Clearly the number of A-endings is independent of the representative g of HgU .) The *B-endings* are defined in a similar way. Let $e(HgU)$ denote the set of endings of HgU ; that is, the set of A-endings and B-endings of HgU . We shall say that HgU is *multiple ended* if it has at least one A-ending and at least one B-ending.

We are now able to state our analogue of (4).

THEOREM 3.1. *Let* H *be a finitely generated subgroup of* $G = (A * B; U)$ *and let* $\{C_A, C_B\}$ *be a cress for* H *relative to* T_A, T_B . *Then*

$$|R| + 2|Q_A| + 2|Q_B| = |e(HU)| + \sum_W (|e(W)|-2) , \tag{5}$$

where the summation is over those (H, U) *double cosets* W *, different from* HU *, that are multiple ended.*

PROOF. In essence the proof is the same as the proof by Howson and Hanna Neumann of (4). We may assume $H \not\le U$ since (5) is trivial if $H \le U$. Since H is finitely generated, by the proof of Lemma 8 of [8], R, Q_A and Q_B are finite. For each $d \in Q_A$

choose an element $a_d \in (A \cap d^{-1}Hd)\backslash U$, and write $\hat{Q}_A = \{da_d d^{-1} \mid d \in Q_A\}$. Define $\hat{Q}_B$ similarly. It then follows from the definition of a cress that R, $\hat{Q}_A$, $\hat{Q}_B$ are pairwise disjoint, whence $|R| + |Q_A| + |Q_B| = |R \cup \hat{Q}_A \cup \hat{Q}_B|$. Denote by Γ the (unoriented) graph whose set V of vertices is the set of all initial segments of $R \cup \hat{Q}_A \cup \hat{Q}_B$ (including 1) where two vertices are joined by an edge if their lengths differ by 1 , and one is an initial segment of the other; that is, if one has normal form $t_1 \ldots t_n u$ and the other $t_1 \ldots t_{n-1}$ (with respect to T_A, T_B). It follows from the uniqueness of the normal form that Γ is a tree, and thence that the set of extremal vertices is precisely $R \cup \hat{Q}_A \cup \hat{Q}_B$. Denote by $v(x)$ the valency of (that is, the number of edges incident with) $x \in V$. Then since Γ is a tree the number of terminal vertices is also $v(1) + \sum_x (v(x)-2)$ where the summation is over all vertices different from 1 of valency ≥ 2 , or equivalently (by the definition of a cress) over all nontrivial $x \in S_A \cup S_B$ which are also initial segments of elements of $R \cup Q_A \cup Q_B$. Hence

$$(6) \qquad |R| + |Q_A| + |Q_B| = v(1) + \sum_x (v(x)-2) ,$$

where the summation is as above.

We shall now establish a relationship between the right hand sides of (5) and (6). Let $x \in R_A$ have the form dq where $q \in T_A$ and $d \in S_A$. By Lemma 2.2 if an element of $HxU\backslash U$ ends in $a \in A$, then a has the form $uq_1^{-1}a_1q$ where $u \in U$, $a_1 \in d^{-1}Hd \cap A$, and either $q_1 = 1$ or dq_1 is an initial segment of an element of $R \cup Q_A \cup Q_B$. Hence every A-ending of HxU has the form $(x^{-1}Hx \cap A)q^{-1}a_1^{-1}q_1u^{-1}U$, which is the same as $(x^{-1}Hx \cap A)q^{-1}q_1U$ since $x = dq$ and $a_1 \in d^{-1}Hd \cap A$. Conversely, if dq, dq_1 are distinct initial segments of elements of $R \cup Q_A \cup Q_B$, where $d \in S_A$, $q, q_1 \in T_A$, then for some $u \in U$, $uq_1^{-1}q$ ends some element of $HdqU$. For, suppose for instance that dq_1 is an initial segment of some element g say, of Q_A . Then ga_gg^{-1} $(\in \hat{Q}_A)$ belongs to H , whence $ga_gg^{-1}dq \in HdqU$. It is easy to see that (when written in normal form) $ga_gg^{-1}dq$ ends in $uq_1^{-1}q$ for some $u \in U$. The other possibilities (for example, that dq_1 is an initial segment of an element of R) are treated similarly.

Since the elements dq with $q \in T_A \setminus \{1\}$, $d \in S_A$, which are initial segments of elements of $R \cup Q_A \cup Q_B$ are precisely the vertices of Γ adjacent to the vertex d , and longer than d , it follows that if $d \neq 1$ then HdU and $HdqU$ have $(v(d)-1)$ A-endings each, while if $d = 1$ then HU and HqU each have n_A A-endings, where n_A is the number of beginnings of elements of $R \cup Q_A \cup Q_B$ lying in T_A . Hence the total number of A-endings of all HxU determined by initial segments x of elements of $R \cup Q_A \cup Q_B$ is $2n_A + 2\Sigma(v(d)-1)$, where the summation is over all $d \in S_A$ that are also initial segments of elements of $R \cup Q_A \cup Q_B$.

Since this argument holds with B replacing A throughout, and since $n_A + n_B = v(1)$, we deduce that

$$(7) \qquad \sum_W |e(W)| = 2v(1) + 2\sum_x (v(x)-1) ,$$

where the left hand summation is over all multiple ended (H, U) cosets W and the summation in the right hand side is over all nontrivial $x \in S_A \cup S_B$ that are also initial segments of elements of $R \cup Q_A \cup Q_B$. (We have used here the fact that the multiple ended (H, U) cosets W are precisely the cosets HxU where x is an initial segment of an element of $R \cup Q_A \cup Q_B$: this follows from the proof of Lemma 8 of [8].)

Now the only initial segments of elements of $R \cup Q_A \cup Q_B$ not in $S_A \cup S_B$ are the elements of R with their U-syllables deleted (that is, the longest initial segments of elements of R). Thus since by definition the elements of R determine only $\frac{1}{2}|R|$ (H, U) cosets, it follows that

$$|e(HU)| + \sum_W (|e(W)|-2) = 2v(1) + 2\sum_x (v(x)-2) - |R| ,$$

where the summations are as in (7). The desired equation (5) is now immediate from this and (6).

COROLLARY 3.2. *Let* H *be a finitely generated subgroup of* $G = (A * B; U)$ *and let* T_A, T_B *be left transversals for* U *in* A, B *, both containing* 1 *. Then if* H *is not contained in a conjugate of* U *,*

$$(8) \qquad 2r(H) \leq \left\{|e(HU)| + \sum_W (|e(W)|-2)\right\} \times \max_{g\in G} \{1, r(g^{-1}Hg \cap A), r(g^{-1}Hg \cap B)\} ,$$

where the summation is over all multiple ended (H, U) *cosets* W *.*

PROOF. Let R, Q_A, Q_B be defined as in Section 2 in terms of a cress for H

in G relative to T_A, T_B. Lemma 2.1 gives that

$$r(H) \le \tfrac{1}{2}|R| + \left(|Q_A|+|Q_B|\right) \times \max_{g\in G} \left(r\left(g^{-1}Hg \cap A\right),\ r\left(g^{-1}Hg \cap B\right)\right) .$$

The corollary follows from this and Theorem 3.1.

4. Intersections of double cosets

It is clear that if H_1 and H_2 are finitely generated subgroups of $G = (A * B;\ U)$ then in order to use (5) and (8) to get a bound for $r\left(H_1 \cap H_2\right)$ in terms of $r\left(H_1\right)$ and $r\left(H_2\right)$ we shall need to know, among other things, such a bound when $H_1, H_2 \le A$ or $H_1, H_2 \le B$, and a bound for the number of multiple ended $\left(H_1 \cap H_2,\ U\right)$ cosets contained in the intersection of an $\left(H_1,\ U\right)$ and an $\left(H_2,\ U\right)$ coset. In this section we are concerned chiefly with the latter bound: our result is the following

THEOREM 4.1 (*cf.* [3, Lemma 4.1]). *Let H_1 and H_2 be finitely generated subgroups of $G = (A * B;\ U)$ where A is a free product of cyclic groups and U is infinite and maximal cyclic in A. If neither H_1 nor H_2 is contained in a conjugate of U, then there exist left transversals T_A, T_B for U in A, B such that the number of $\left(H_1 \cap H_2,\ U\right)$ cosets with at least one A-ending contained in each intersection of an $\left(H_1,\ U\right)$ coset with an $\left(H_2,\ U\right)$ coset is at most*

$$\left|R\left(H_1\right) \cup Q_A\left(H_1\right) \cup Q_B\left(H_1\right)\right| \times \left|R\left(H_2\right) \cup Q_A\left(H_2\right) \cup Q_B\left(H_2\right)\right| \times \max_{x\in G}\left(1,\ 2r\left(x^{-1}Ax \cap H_1\right)\right) \times \max_{x\in G}\left(1,\ 2r\left(x^{-1}Ax \cap H_2\right)\right) ,$$

where $R\left(H_i\right)$, $Q_A\left(H_i\right)$, $Q_B\left(H_i\right)$ $(i = 1, 2)$ are defined as in Section 2 in terms of cresses for H_1 and H_2 in G relative to T_A, T_B. It follows that this provides an upper bound also for the number of multiple ended $\left(H_1 \cap H_2,\ U\right)$ cosets contained in the intersection of a multiple ended $\left(H_1,\ U\right)$ coset with a multiple ended $\left(H_2,\ U\right)$ coset.

For the proof we restate, in greater detail, or modified form, various results of [3], indicating where necessary how the proofs need to be modified.

LEMMA 4.2 (*cf.* [3, Theorem 6.1]). *Let F be a free product of cyclic groups C_i, $i \in I$ (some index set) and let U be any infinite cyclic subgroup of F. Then the following hold:*

(i) *there is a left transversal* T *for* U *in* F *such that, for each coset* Hg *of each finitely generated nontrivial subgroup* H *of* F *there is a subset* $V \subseteq U$ *containing at most* $2r(H)$ *elements such that*

$$Hg \subseteq TV\left(g^{-1}Hg \cap U\right) \; ; \tag{9}$$

(ii) *if* U *is maximal infinite cyclic in* F *then there is a transversal* T *which in addition to* (9) *also satisfies*

$$U(T\backslash\{1\}) = T\backslash\{1\} \; . \tag{10}$$

PROOF. Each element of F has a unique normal form $c_{i_1} \ldots c_{i_n}$ where $n \geq 0$, $c_{i_j} \in C_j\backslash\{1\}$ and $i_j \neq i_{j+1}$ $(j = 1, \ldots, n-1)$; n defines the length of the element. We may assume that a generator u of U has normal form $c_{i_1} \ldots c_{i_n}$ with the additional property that $i_k \neq i_n$; for if u does not have this property then some conjugate $f^{-1}uf$ will, and then we may consider fC_if^{-1} in place of C_i $(i \in I)$. As in the proof of Theorem 6.1 of [3] we then define T_1 to be a complete set of representatives of double cosets UaU $(a \in A)$ such that each representative is an element of smallest length in its double coset, and choose a left transversal T for U in F from the set $U\left(T_1\backslash\{1\}\right) \cup \{1\}$.

If we can prove (9) with $gu^n = g_1$ replacing g, where u^n is an element of U, then it will follow for g ; for then

$$Hg = Hg_1u^{-n} \subseteq TV\left(g_1^{-1}Hg_1 \cap U\right)u^{-n} = TVu^{-n}\left(g^{-1}Hg \cap U\right) ,$$

and the set Vu^{-n} has the same number of elements as V. Thus we may suppose g has u^n as terminal segment for any convenient value of n.

Since every element $t \in T_1$ was chosen as a shortest element of UtU, it follows that the longest initial segments that t and t^{-1} have in common with u have lengths at most half the length of u. Thus if l, k are nonzero integers of signs ε, δ $(= \pm 1)$ respectively and v is the normal form of $u^{\varepsilon}tu^{\delta}$, then $u^{l-\varepsilon}vu^{k-\delta}$ is in normal form as written, that is, no cancellation occurs. Now let V_1 be a maximal subset of elements u^s, with $s > 0$, from distinct left cosets of $g^{-1}Hg \cap U$ in U, such that $tu^s \in Hg$ for some $t \in T$. Suppose that $|V_1| > 2r(H)$. Then there is a positive integer m say, with the property that there are $> 2r(H)$ elements of V_1 of the form u^s with $s \le m$. By the above we may assume g has u^m as terminal segment; say $g = g_1u^m$. Then by the preceding and by the definition of T each tu^s $\left(u^s \in V_1\right)$ can be written in normal form (that is, with no cancellation occuring) as $u^l vu^{s-1}$. Since this element is in Hg, we have that $gu^{-(s-1)}v^{-1}u^{-l} \in H$, and since $g = g_1u^m$, the latter element can be written (in non-cancelling form) as $g_1u^{m-s+1}v^{-1}u^{-l}$. Hence for all $u^s \in V_1$, g_1u^{m-s+1} is an initial segment of an element of H.

Write $r = r(H)$ and let $\{a_1, \ldots, a_r\}$ be a set of generators for H given by the Kuroš subgroup theorem as set forth in [9, p. 243] (so that H is the free product of the cyclic subgroups generated by each of $a_1, \ldots, a_r$). Since each element g_1u^{m-s+1} is an initial segment of an element of H, it follows from the Kuroš rewriting process (see [9, p. 230]) that some element of each coset Hg_1u^{m-s+1} is an initial segment of some a_i or its inverse. However the cosets Hg_1u^{m-s+1} are all distinct; for if $u^{s_1}, u^{s_2} \in V_1$ with $s_1 \ne s_2$, then $u^{-s_1}u^{s_2} \notin g^{-1}Hg \cap U$ by definition of V_1, whence $g_1u^{m-s_1+1}u^{-(m-s_2+1)}g_1 \notin H$. Hence V_1 cannot contain $> 2r(H) = 2r$ elements. Part *(i)* now follows since the restriction that V_1 contain only positive powers of u is inessential: we can replace T_1 in the above argument by T_1u^{-n} for any suitable positive integer n.

We now prove *(ii)*. If U is maximal cyclic in F it is not difficult to show that $g^{-1}Ug \cap U = \{1\}$ for all $g \in F\backslash U$. Property (10) then follows for $T = U(T_1 \{1\}) \cup \{1\}$ (where T_1 is as in the proof of part *(i)*) as in the proof of Lemma 2.2 of [3].

LEMMA 4.3 (*cf.* [3, Lemma 3.1]). *Let* $G = (A * B; U)$ *where* A *is a free product of cyclic groups and* U *is infinite and maximal cyclic in* A. *Let* T_A *be a left transversal for* U *in* A *satisfying* (9) *and* (10), *and let* T_B *be any left transversal for* U *in* B, *containing* 1. *Let* H *be any finitely generated subgroup of* G *and* Hg *any right coset of* H *in* G, *and let* R, Q_A, Q_B *be defined as in Section* 2 *in terms of a cress for* H *in* G. *Let* D *denote the set of all elements of* Hg *with ends in* $A\backslash U$, *and* D_1 *the set obtained from* D *by deleting* *the U-syllables from the ends of the normal forms of the elements of* D *. Then, if* H *is contained in no conjugate of* U *, there is a subset* $V \subseteq U$ *containing at most*

$$2|R \cup Q_A \cup Q_B| \max_{x \in G} \left(1, r\left(x^{-1}Ax \cap H\right)\right)$$

elements, such that

$$\dot{D} \subseteq D_1 V\left(g^{-1}Hg \cap U\right) .$$

PROOF. Suppose that $a \in A \backslash U$ ends some element of $Hg \backslash U$. Let $g_1 \in R_A$ be the representative of the double coset HgU , and write $g_1 = dq$ where $q \in T_A$ and $d \in S_A$. Then $g = hg_1u$ where $h \in H$ and $u \in U$. By Lemma 2.2,

$$a = u_1 q_1^{-1} a_1 q u = u_1 q_1^{-1} a_1 q_1 \left(q_1^{-1} q u\right) , \tag{11}$$

where $u_1 \in U$, $q_1 \in T_A$ is either 1 or is such that dq_1 is an initial segment of some element of $R \cup Q_A \cup Q_B$, and $a_1 \in A \cap d^{-1}Hd$. Using (11) and Lemma 4.2 it is not difficult to show (see the proof of Lemma 3.1 of [3] for details) that for each such a ,

$$u_1^{-1} a \in \left\{A \cap \left(dq_1\right)^{-1} H dq_1\right\} q_1^{-1} q u \subseteq T_A V\left(g^{-1}Hg \cap U\right) ,$$

where V is a subset of U containing at most $\max\left\{1, 2r\left\{A \cap \left(dq_1\right)^{-1} a dq_1\right\}\right\}$ elements, and depending on Hg , q and q_1 only. Since T_A satisfies (10) and $a \in A \backslash U$ it follows that $a \in T_A V\left(g^{-1}Hg \cap U\right)$.

Since Hg is fixed and therefore also q we may write $V\left(q_1\right)$ for V . If we write V_1 for the union of all $V\left(q_1\right)$ such that dq_1 is an initial segment of an element of $R \cup Q_A \cup Q_B$, then

$$|V_1| \leq |R \cup Q_A \cup Q_B| \times \max_{x \in G} \left(1, 2r\left(x^{-1}Ax \cap H\right)\right) ,$$

and for all elements $a \in A \backslash U$ that end elements of Hg , $a \in T_A V_1 \left(g^{-1}Hg \cap U\right)$. This completes the proof of the lemma.

PROOF OF THEOREM 4.1. Choose left transversals T_A, T_B as in Lemma 4.3. For arbitrary $g \in G$ let Y be a set of representatives ending in elements of $A \backslash U$, of distinct $\left(H_1 \cap H_2, U\right)$ cosets contained in $H_1gU \cap H_2gU$. We may clearly assume that every $y \in Y$ has U-syllable 1 . By Lemma 4.3, for $i = 1, 2$ there is a subset

$V(H_i)$ of U with at most $2|R(H_i) \cup Q_A(H_i) \cup Q_B(H_i)| \times \max_{x \in G} \left(1, r\left(x^{-1}Ax \cap H_i\right)\right)$ elements such that the endings of all elements of H_igU which end in elements of $A \backslash U$, lie in $T_AV(H_i)\left[g^{-1}H_ig \cap U\right]$. Clearly we may assume that distinct elements of $V(H_i)$ lie in distinct left cosets of $g^{-1}H_ig \cap U$ in U. For each $y \in Y$, $y = h_1gu = h_2gv$ for some $h_1 \in H_1$, $h_2 \in H_2$, and $u, v \in U$. For $i = 1, 2$ choose left transversals W_{H_i} in U for $g^{-1}H_ig \cap U$ such that $W_{H_i} \supseteq V(H_i)$. We may then assume that for all $y \in Y$ the elements u, v defined above are such that $u^{-1} \in W_{H_1}$ and $v^{-1} \in W_{H_2}$. For suppose for instance that $u^{-1} = wu_1$ where $u_1 \in g^{-1}Hg \cap U$ and $w \in W_{H_1}$; then $u = u_1^{-1}w^{-1}$, and $h_1gu = h_1gu_1^{-1}g^{-1}gw^{-1} = \hat{h}_1gw^{-1}$, where $\hat{h}_1 \in H_1$.

Thus for each $y \in Y$ we have that $yu^{-1} = h_1g$ and $yv^{-1} = h_2g$ where $u^{-1} \in W_{H_1}$ and $v^{-1} \in W_{H_2}$. Since every y has u-syllable 1 and $V(H_i) \subseteq W_{H_i}$, it follows that $u^{-1} \in V(H_1)$ and $v^{-1} \in V(H_2)$. Hence Y contains at most $|V(H_1)| \times |V(H_2)|$ elements, since if $y_1, y_2 \in Y$ are such that $y_1 = \hat{h}_1gu^{-1} = \hat{h}_2gv^{-1}$, $y_2 = \check{h}_1gu^{-1} = \check{h}_2gv^{-1}$, then $y_1y_2^{-1} \in H_1 \cap H_2$ and therefore y_1, y_2 cannot represent distinct $(H_1 \cap H_2, U)$ double cosets. The proof is thus complete.

COROLLARY 4.4. *Let F be a free product of cyclic groups, let U be an infinite cyclic subgroup of F and let X, Y be subgroups of F of finite nonzero ranks m, n. Then for each $f \in F$, $XfU \cap YfU$ contains at most $4mn$ $(X \cap Y, U)$ cosets. (If one of X or Y is trivial, then clearly the intersection contains precisely one $(X \cap Y, U)$ coset.)*

The proof is identical with that of Theorem 4.1, except that part *(i)* of Lemma 4.2 is used in place of Lemma 4.3.

COROLLARY 4.5. *Let H_1, H_2 be finitely generated subgroups of $G = (A * B; U)$ where A and B are free products of cyclic groups, and U is infinite and maximal cyclic in A. Let T_A be a left transversal for U in A satisfying* (9) *and* (10) *and T_B be any left transversal containing* 1 *for U in B. Let $R(H_1)$, $R(H_2)$, $R(H_1 \cap H_2)$, and so on, be defined as in Section* 2 *in terms of cresses for H_1, H_2*

and $H_1 \cap H_2$ *relative to* T_A, T_B . *Write briefly*

$$s(H_i) = |R(H_i)| + 2|Q_A(H_i)| + 2|Q_B(H_i)| \quad (i = 1, 2) .$$

Then if $s(H_1 \cap H_2) \geq 4$,

$$(12) \quad s(H_1 \cap H_2) - 2 \leq \left\{\frac{9}{4} s(H_1)s(H_2)-3s(H_1)-3s(H_2)+4\right\} \times s(H_1)s(H_2)\mu(A, H_1)\mu(A, H_2)$$
$$\times \{s(H_1)s(H_2)(\mu(A, H_1)\mu(A, H_2)+\mu(B, H_1)\mu(B, H_2))-2\},$$

where $\mu(I, J) = \max_{x \in G} \{1, 2r(x^{-1}Ix \cap J)\}$.

PROOF. It is not difficult to show (by an argument similar to one in [11]) that if $s(H_1 \cap H_2) \geq 4$ then there is an element $g \in G$ such that $g^{-1}(H_1 \cap H_2)g$ is multiple ended. We may therefore assume that $H_1 \cap H_2$ is multiple ended by replacing, if necessary, H_1, H_2 and $H_1 \cap H_2$ by their respective conjugates by g . (It is also readily seen that for any subgroup L of G and any $g \in G$, $s(L) = s(g^{-1}Lg)$ since, for example, $|Q_A(L)|$ is the number of (L, A) double coset representatives x such that $x^{-1}Lx \cap A \nsubseteq U$, and this number is clearly the same for L as for $g^{-1}Lg$. Alternatively one may use Theorem 3.1.)

Since H_i $(i = 1, 2)$ is multiple ended, formula (5) simplifies to

$$(13) \qquad s(H_i) - 2 = \sum_W (|e(W)|-2) ,$$

where the summation is over all multiple ended (H_i, U) cosets W .

We shall now show that if H_1gU and H_2gU are multiple ended, one of the following three possibilities arises:

(i) every multiple ended $(H_1 \cap H_2, U)$ coset contained in $H_1gU \cap H_2gU$ has precisely 2 endings;

(ii) for $i = 1, 2$ either $H_igU = H_id_iU$ where $d_i \in Q_A(H_i)$, or H_igU has at least 3 endings;

(iii) for $i = 1, 2$ either $H_igU = H_if_iU$ where $f_i \in Q_B(H_i)$, or H_igU has at least 3 endings.

Suppose for instance $|e(H_1gU)| = 2$. Then H_1gU has exactly one A-ending and exactly one B-ending. We may assume $g \in R_A(H_1)$, say $g = dq$ where $d \in S_A(H_1)$ and $q \in T_A$. By the proof of Lemma 8 of [8], g is an initial segment of some element of $R(H_1) \cup Q_A(H_1) \cup Q_B(H_1)$. It follows that we cannot have both $d \in Q_A(H_1)$ and $q \neq 1$, since if these hold then by the proof of Theorem 3.1, $|e(H_1gU)| \geq 3$. Hence either $d \notin Q_A(H_1)$ or $q = 1$ (or both). We examine the three possibilities in turn. If $q = 1$ and $d \notin Q_A(H_1)$, then by the definition of $Q_A(H_1)$, for all $x \in H_1gU$ we have $x^{-1}H_1x \cap A \leq U$, whence $x^{-1}(H_1 \cap H_2)x \cap A \leq U$, and since H_1gU has exactly one A-ending it easily follows that $(H_1 \cap H_2)xU$ has at most one A-ending. If $q \neq 1$ and $d \notin Q_A(H_1)$ then $g^{-1}H_1g \cap A \leq q^{-1}Uq$. The double coset $\left(g^{-1}H_1g \cap A\right)q^{-1}U$ is an A-ending of H_1gU and, by assumption, the only one. Hence if an element of H_1gU ends in $a \in T_A \backslash \{1\}$, then a must have the form $\left(q^{-1}uqq^{-1}u_1\right)^{-1} = u_1^{-1}u^{-1}q$, for some $u, u_1 \in U$. Hence for any $x \in H_1gU$, $\left(x^{-1}(H_1 \cap H_2)x \cap A\right)a^{-1}U$ is the only possible A-ending of $(H_1 \cap H_2)xU$; that is, every $(H_1 \cap H_2, U)$ coset contained in H_1gU has at most one A-ending. The remaining possibility is that $q = 1$ and $d \in Q_A(H_1)$, that is, $g \in Q_A(H_1)$. Since the above argument holds with H_2 replacing H_1 throughout, or B replacing A , or both, we infer that (i), (ii) and (iii) above are the only possibilities.

It follows that if H_1gU, H_2gU have intersection containing a multiple ended $(H_1 \cap H_2, U)$ coset with at least 3 endings, then (ii) or (iii) (or both) hold. If we write $\nu(H_i, U)$ $(i = 1, 2)$ for the number of (H_i, U) cosets with at least 3 endings, then the number of pairs H_1gU, H_2gU satisfying either (ii) or (iii) is at most

$$\nu(H_1, U)\nu(H_2, U) + \nu(H_2, U)\left(|Q_A(H_1)|+|Q_B(H_1)|\right) + \nu(H_1, U)\left(|Q_A(H_2)|+|Q_B(H_2)|\right) + |Q_A(H_1)||Q_A(H_2)| + |Q_B(H_1)||Q_B(H_2)| ,$$

and, in view of the definition of $s(H_i)$ $(i = 1, 2)$ this is at most

$$\nu(H_1, U)\nu(H_2, U) + \tfrac{1}{2}s(H_1)\nu(H_2, U) + \tfrac{1}{2}s(H_2)\nu(H_1, U) + \tfrac{1}{4}s(H_1)s(H_2) . \tag{14}$$

By Theorem 4.1 each intersection $H_1gU \cap H_2gU$ contains at most

$$s(H_1)s(H_2)\mu(A, H_1)\mu(A, H_2) \tag{15}$$

multiple ended $(H_1 \cap H_2, U)$ cosets. Also clearly $\nu(H_i, U) \leq \sum \left(|e(W)|-2\right)$, where the summation is over all (H_i, U) cosets W with at least 3 endings, whence by

(13), $\nu(H_i, U) \le s(H_i) - 2$ $(i = 1, 2)$. This together with (14) and (15) implies that the number of $(H_1 \cap H_2, U)$ cosets with at least 3 endings is at most

$$(16) \quad \left[\left(s(H_1)-2\right)\left(s(H_2)-2\right)+\tfrac{1}{2}s(H_1)\left(s(H_2)-2\right)+\tfrac{1}{2}s(H_2)\left(s(H_1)-2\right)+\tfrac{1}{4}s(H_1)s(H_2)\right] \times s(H_1)s(H_2)\mu(A, H_1)\mu(A, H_2)$$
$$= \left(\tfrac{9}{4}s(H_1)s(H_2)-3s(H_1)-3s(H_2)+4\right) \times s(H_1)s(H_2)\mu(A, H_1)\mu(A, H_2) .$$

To complete the proof we find a bound for $|e((H_1 \cap H_2)gU)|$ for any coset $(H_1 \cap H_2)gU$. By Corollary 4.4 each pair $\left(g^{-1}H_1g \cap A\right)a_1U$, $\left(g^{-1}H_2g \cap A\right)a_2U$ of A-endings of H_1gU, H_2gU respectively, contributes at most $\mu(A, H_1)\mu(A, H_2)$ A-endings to $e((H_1 \cap H_2)gU)$, and the analogous statement is true for the B-endings in $e((H_1 \cap H_2)gU)$. Hence

$$|e((H_1 \cap H_2)gU)| \le |e(H_1gU)||e(H_2gU)| \times \left(\mu(A, H_1)\mu(A, H_2)+\mu(B, H_1)\mu(B, H_2)\right) .$$

By (13) certainly $|e(H_igU)| \le s(H_i)$ for all multiple ended H_igU . Hence for multiple ended $(H_1 \cap H_2)gU$,

$$|e((H_1 \cap H_2)gU)| \le s(H_1)s(H_2)\left(\mu(A, H_1)\mu(A, H_2)+\mu(B, H_1)\mu(B, H_2)\right) .$$

The desired result (12) now follows from this, (13) and (16).

The next result generalizes that of Howson [7] and Hanna Neumann [11], [12].

COROLLARY 4.6. *Let F be a free product of cyclic groups and let H_1, H_2 be nontrivial subgroups of finite ranks m, n respectively. Then*

$$(17) \qquad r(H_1 \cap H_2) \le 2mn - 2m - 2n + 3 .$$

PROOF. By the Kuroš subgroup theorem every subgroup of F is also a free product of cyclic groups, and by Gruško's Theorem the rank of such a free product is the number of its cyclic free factors. Hence we may suppose F is a free product of finitely many cycles; for, if it is not, we may consider instead the subgroup generated by H_1 and H_2 . It is a further easy consequence of the Kuroš subgroup theorem (as stated, for example, in [9]) that then F can be embedded in the free product of an infinite cycle A by a finite cycle B . Hence we may assume without loss of generality that $F = A * B$ with A and B as just described. In this case by Lemma 2.1 *(i)* with $U = \{1\}$, Theorem 3.1 takes the simple form

$$2r(H) = |e(H)| + \sum_W \left(|e(W)|-2\right)$$

for all finitely generated subgroups H of F , where the summation is over all right cosets of H in F , different from H and with at least 3 endings.

If $r(H_1 \cap H_2) \le 1$, (17) is immediate. Thus suppose $r(H_1 \cap H_2) \ge 2$. It follows (as in [11]) that we may assume without loss of generality that $|e(H_1)|$, $|e(H_2)|$ and $|e(H_1 \cap H_2)|$ are all at least 2. Hence

$$2r(H_1) - 2 = \sum_W (|e(W)|-2),$$

where the summation is over all right cosets of H in F with at least 3 endings, and the analogous formulae hold for H_2 and $H_1 \cap H_2$. An argument similar to that given in the proof of Corllary 4.5 (but much simpler since all double cosets reduce now to right cosets) shows that

$$2r(H_1 \cap H_2) - 2 \le (2r(H_1)-2)(2r(H_2)-2),$$

whence (17).

5. Proof of Theorem 1.1 *(iii)*

Let H_1, H_2 be finitely generated subgroups of $G = (A * B;\ U)$ where A and B are free products of cyclic groups and U is infinite and maximal cyclic in A. Let T_A, T_B, $s(H_i)$ $(i = 1, 2)$, $s(H_1 \cap H_2)$, $\mu(A, H_i)$, $\mu(B, H_i)$ $(i = 1, 2)$ be as in Corollary 4.5. By Lemma 2.1 *(ii)*, for all $g \in G$, $g^{-1}H_i g \cap A$ and $g^{-1}H_i g \cap B$ $(i = 1, 2)$ are finitely generated. Thus by Theorem 3.1 and Corollary 3.2, if $H_1 \cap H_2$ is not cyclic then

$$(18)\quad 2r(H_1 \cap H_2) \le s(H_1 \cap H_2) \max_{g \in G} \left\{1,\ r\left[g^{-1}(H_1 \cap H_2)g \cap A\right],\ r\left[g^{-1}(H_1 \cap H_2)g \cap B\right]\right\}.$$

Since $g^{-1}(H_1 \cap H_2)g \cap A = \left[g^{-1}H_1 g \cap A\right] \cap \left[g^{-1}H_2 g \cap A\right]$, we have by Corollary 4.6 that

$$(19)\quad \max_{g \in G} \left\{r\left[g^{-1}(H_1 \cap H_2)g \cap A\right]\right\} \le \tfrac{1}{4}\mu(A, H_1)\mu(A, H_2) - \mu(A, H_1) - \mu(A, H_2) + 3$$

and that the analogous bound holds with B replacing A. Write briefly $\rho(A, H_1, H_2)$ for the right side of (19) and define $\rho(B, H_1, H_2)$ similarly. Then by (18), (19) and Corollary 4.5, provided $s(H_1 \cap H_2) \ge 4$,

$$(20)\quad 2r(H_1 \cap H_2) \le \left[\left(\tfrac{9}{4}s(H_1)s(H_2) - 3s(H_1) - 3s(H_2) + 4\right) \times s(H_1)s(H_2)\mu(A, H_1)\mu(A, H_2) \times \{s(H_1)s(H_2)(\mu(A, H_1)\mu(A, H_2) + \mu(B, H_1)\mu(B, H_2)) - 2\} + 2\right] \times \max(\rho(A, H_1, H_2), \rho(B, H_1, H_2)).$$

In order to deduce from this an upper bound for $r(H_1 \cap H_2)$ in terms of $r(H_1)$ and $r(H_2)$, and thus prove (2), we need to know for any finitely generated subgroup H

of G upper bounds for $s(H)$, $\mu(A, H)$ and $\mu(B, H)$, in terms of $r(H)$. Such bounds are in general difficult to obtain (since an analogue of Gruško's Theorem would be needed). However in the rather special circumstance that H is a free product of cyclic groups, which, by [6], obtains when G is Fuchsian and H has infinite index, such bounds can be found. The precise result is as follows.

THEOREM 5.1. *Let* $G = (A * B; U)$ *where* U *is infinite cyclic and let* T_A, T_B *be left transversals containing* 1 *for* U *in* A, B. *Let* H *be a finitely generated subgroup of* G *which is a free product of cyclic groups and let* R, Q_A, Q_B *be defined as in Section* 2 *in terms of a cress for* H *in* G *relative to* T_A, T_B. *Then, if* H *is nontrivial,*

$$s(H) = |R| + 2|Q_A| + 2|Q_B| \leq 8r(H) - 6 , \tag{21}$$

and

$$\max_{g \in G} \{r(g^{-1}Hg \cap A), r(g^{-1}Hg \cap B)\} \leq r(H) . \tag{22}$$

The bound given in part *(iii)* of Theorem 1.1 is immediate from this theorem, the inequality (20), the result of [6] that subgroups of infinite index in Fuchsian groups are free products of cyclic groups, and the obvious fact that the groups (1) are amalgamated products of the required sort for either $n > 0$ or $t \geq 4$.

For the proof of Theorem 5.1 we require some preliminary results. I thank A. Karrass for pointing out to me the following lemma.

LEMMA 5.2. *Let* F *be a free product of* n *cyclic groups,* n *finite, and let* N *be an isolated normal subgroup of* F $\{$*that is, if* $x^n \in N$ *for* $n \neq 0$, *then* $x \in N\}$ *such that* F/N *is either free or a free product of cyclic groups with the same number of finite free factors as* F *has. Then there is a set* $\{a_1, \ldots, a_n\}$ *of generators for* F *such that* $F = \langle a_1 \rangle * \langle a_2 \rangle * \ldots * \langle a_n \rangle$, *and* N *is the normal closure in* F *of* $\{a_1, \ldots, a_k\}$ *for some* $k \leq n$.

For the case that F is free this is Theorem 3.3 of 9, p. 132. The proof of that theorem is easily adapted to yield our more general lemma.

COROLLARY 5.3. *(i) Let* $G = (A * B; U)$ *where* U *is infinite cyclic and* A, B *and* G *are free products of* α, β *and* γ *cyclic groups respectively. Then* $\gamma \leq \alpha + \beta - 1$.

(ii) Let G *be the HNN group* $\langle x, A \mid x^{-1}ux = v \rangle$, *where* A *and* G *are free products of* α *and* γ *cyclic groups respectively, and* u, v *are elements of* A *of infinite order. Then* $\gamma \geq \alpha$.

PROOF. In *(i)* G is isomorphic to a quotient of $A * B$ by the normal closure of a single element which is not a proper power in $A * B$, while in *(ii)* G is isomorphic to the quotient of $\langle x \rangle * A$ by the normal closure of $x^{-1}uxv^{-1}$. Thus in either case G is a quotient of a free product of cyclic groups (of ranks $\alpha + \beta$ and $\alpha + 1$ respectively) by the normal closure of a single element. The conclusions are therefore immediate from Lemma 5.2.

LEMMA 5.4. *Let F be a free product of n cyclic groups, n finite, and set*

$$H = \left\langle x_1, \ldots, x_m, F \mid x_i^{k_i} = u_i \quad (i = 1, \ldots, m) \right\rangle ,$$

where $k_i > 1$ and u_i lies in F and has infinite order $(i = 1, \ldots, m)$. If H is a free product of cyclic groups then $r(H) = n \geq m$.

PROOF. Write $F = X * Y$ where X is free and Y is a free product of finite cycles. It is clear that H is a tree product with the infinite cycles $\langle x_1 \rangle, \langle x_2 \rangle, \ldots, \langle x_m \rangle$ as extremal vertices all adjacent to the vertex F. Hence all elements of finite order in H are contained in conjugates of F, and therefore in conjugates of Y. Since H is a free product of cyclic groups and no two cyclic factors of Y are conjugate in H, it follows (from the Kuroš subgroup theorem: see [9, p. 243]) that Y is a free factor of H, and that if N is the normal closure of Y in H then H/N is free. We now prove that $u_i \notin N$ $(i = 1, \ldots, m)$. Suppose for instance that $u_1 \in N$; then since $u_1 \in F$, it follows (by elementary properties of tree products) that u_1 is in the normal closure N_1 say, of Y in F. Since N_1 is a free product of (possibly infinitely many) finite cycles it follows that there is a free factor Z say, of N_1 which is a free product of finitely many finite cycles and which contains u_1. Then $\langle x_1 \rangle$ and Z generate their amalgamated product $P = \left\langle x_1, Z \mid x_1^{k_1} = u_1 \right\rangle$, which, as a subgroup of H, is a free product of cycles. As above it follows that $P = \langle x \rangle * Z$ for some $x \in P$ of infinite order. Thus P is isomorphic to a proper factor of itself, an impossibility since it is well known that a free product of finitely many cycles is Hopfian. We have thus reached a contradiction: hence $u_1 \notin N$. In the same way we deduce that $u_i \notin N$ $(i = 1, \ldots, m)$. It follows that H/N, which we have already seen to be free, is isomorphic to

$$H_1 = \left\langle x_1, \ldots, x_m, X \mid x_i^{k_i} = v_i \right\rangle ,$$

where k_i is as before and $1 \neq v_i \in X$. Also $r(H_1) \geq m$ since H_1 has a quotient isomorphic to $C_{k_1} * C_{k_2} * \ldots * C_{k_m}$, where C_{k_i} is cyclic of order k_i

$(i = 1, \ldots, m)$. Since the subgroup of H_1 generated by x_1 and X is one-relator, it follows by Theorem 1 of Shenitzer [14] that v_1 is primitive in X ; that is, there exists a free basis $v_1, a_2, \ldots, a_k$ for X (where $k = r(X)$). Thus the subgroup X_1 generated by $x_1, a_2, \ldots, a_k$ is free on these k generators, whence

$$H_1 = \left\langle x_2, \ldots, x_m, X_1 \mid x_i^{k_i} = v_i \quad (i = 2, \ldots, m) \right\rangle .$$

By repeated applications of Shenitzer's Theorem we deduce that H_1 has rank k . Therefore $k \geq m$ whence $n \geq m$.

PROOF OF THEOREM 5.1. By Lemma 2.1 *(i)*, H is an HNN group

$$\left\langle R_1, K \mid r^{-1}L(r)r = M(r),\ r \in R_1 \right\rangle , \tag{23}$$

where $R_1 = \left\{ r(rQ_A)^{-1} \mid r \in R_B \cap R \right\}$ and K is the tree product of the set $S = \left\{ d_1Ad_1^{-1} \cap H,\ d_2Bd_2^{-1} \cap H,\ H \cap U \mid d_1 \in Q_A,\ d_2 \in Q_B \right\}$, where the amalgamated subgroups and the $L(r), M(r)$ are conjugates of subgroups of U .

We first show that if K is nontrivial then

$$|Q_A| + |Q_B| \leq 3r(K) - 3 , \tag{24}$$

Case 1. Suppose that no pair of adjacent vertices in S has trivial intersection. Then no two cyclic vertices of $S\backslash\{H \cap U\}$ can be adjacent since a pair of adjacent cyclic vertices would generate a non-cyclic subgroup of H with nontrivial centre, contradicting the hypothesis that H is a free product of cyclic groups. For the same reason, if $H \cap U$ were adjacent to a cyclic vertex P_1 it would have to be contained in it; if this happens we contract these two vertices to the single vertex P_1 and consider instead the tree product with vertices $S\backslash\{H \cap U\}$. Thus we have a tree product K (with corresponding tree Γ , say) with infinite cyclic amalgamations and with no adjacent cyclic vertices. This and the fact that, by Lemma 5.4, every vertex has rank at least the number of cyclic vertices adjacent to it, together imply that if C is the set of cyclic vertices of Γ , then

$$|C| \leq \max\left\{1, \sum r(P)\right\} , \tag{25}$$

where the summation is over all vertices P of rank > 1 .

Suppose next that there is at least one vertex P_2 of rank > 1 . (The contrary case is easy.) In the tree Γ contract the subtree with vertices P_2 and all the cyclic vertices adjacent to P_2, to a single vertex $\hat{P}_2$, and call the resulting tree Γ_1. Then in Γ_1 the vertex $\hat{P}_2$ has no adjacent cyclic vertices, and by Lemma 5.4, $r(\hat{P}_2) \geq r(P_2)$. For each vertex P_3 adjacent to $\hat{P}_2$ contract the subtree of Γ_1 whose set of vertices consists of P_3 and the cyclic vertices adjacent to P_3, to a single vertex $\hat{P}_3$ say. Again $r(\hat{P}_3) \geq r(P_3)$. Let Γ_2 be the tree resulting after this has been carried out for each P_3 adjacent to $\hat{P}_2$ in Γ_1. Continue contracting subtrees in this way until the process terminates (as it must, since Γ_1 is finite by Lemma 2.1 *(ii)*) at Γ_k say. Then by repeated application of Corollary 5.3 *(i)* to the tree product K determined by Γ_k we obtain (by counting outwards from P_2) that

$$r(K) \geq 1 + \sum \left(r(\hat{P})-1\right) , \tag{26}$$

where the summation is over all vertices $\hat{P}$ of Γ_k. Since $r(\hat{P}) \geq r(P)$ where here P is the vertex in S of rank > 1 giving rise to $\hat{P}$, we deduce from (26) that

$$r(K) \geq 1 + \sum \left(r(P)-1\right) , \tag{27}$$

where the summation is over all vertices P of Γ of rank > 1. Let $\mathcal{D}$ be the set of non-cyclic vertices of Γ. Then from (25) and (27) we infer that

$$r(K) \geq 1 + |C| - |\mathcal{D}| . \tag{28}$$

Clearly $|\mathcal{D}| \leq \sum \left(r(P)-1\right)$, where the summation is over all vertices P of rank > 1. Hence also, by (27),

$$r(K) \geq 1 + |\mathcal{D}| . \tag{29}$$

Adding (28) and (29) we obtain that $|C| \leq 2r(K) - 2$ and thence by (29), we have $|C| + |\mathcal{D}| \leq 3r(K) - 3$, from which the desired inequality (24) follows (for this case).

Case 2. We now drop the assumption of Case 1. If K is nontrivial then by the definition of tree product, Γ (the tree whose vertices are the elements of S) contains pairwise disjoint non-empty subtrees $\Gamma_1, \Gamma_2, \ldots, \Gamma_l$ whose vertices generate subgroups $K_1, K_2, \ldots, K_l$ of K, such that: every vertex of Γ belongs to some Γ_i; each pair of adjacent vertices of Γ_i has nontrivial intersection ($i = 1, \ldots, l$); and $K = K_1 * K_2 * \ldots * K_l$. By Case 1 above, if C is the set of cyclic, and $\mathcal{D}$ the set of non-cyclic vertices of Γ (possibly with $H \cap U$

omitted), then $|C| + |\mathcal{D}| \leq \sum_{i=1}^{l} (3r(K_i)-3)$, whence, by Gruško's Theorem, (24) certainly follows.

We now turn to the presentation (23) of H as an HNN group. By applying Corollary 5.3 *(ii)*, $|R_1|$ times we obtain that $r(H) \geq r(K)$. Since H has a quotient which is free of rank $|R_1|$, we have also that $r(H) \geq |R_1|$. These two facts together with (24) imply that

$$|R_1| + |Q_A| + |Q_B| \leq r(H) + 3r(H) - 3 = 4r(H) - 3 ,$$

whence (21).

It remains to prove (22). This is immediate from the fact that $r(H) \geq r(K)$ and (27). This completes the proof of Theorem 5.1.

References

[1] R.G. Burns, "A note on free groups", *Proc. Amer. Math. Soc.* 23 (1969), 14-17. MR40#5708.

[2] Robert G. Burns, "On the intersection of finitely generated subgroups of a free group", *Math. Z.* 119 (1971), 121-130. MR43#4892.

[3] R.G. Burns, "On the finitely generated subgroups of an amalgamated product of two groups", *Trans. Amer. Math. Soc.* 169 (1972), 293-306.

[4] Leon Greenberg, "Discrete groups of motions", *Canad. J. Math.* 12 (1960), 415-426. MR22#5932.

[5] A. Howard M. Hoare, Abraham Karrass and Donald Solitar, "Subgroups of finite index of Fuchsian groups", *Math. Z.* 120 (1971), 289-298. MR44#2837.

[6] A. Howard M. Hoare, Abraham Karrass and Donald Solitar, "Subgroups of infinite index of Fuchsian groups", *Math. Z.* 125 (1972), 59-69. Zbl.223.20054.

[7] A.G. Howson, "On the intersection of finitely generated free groups", *J. London Math. Soc.* 29 (1954), 428-434. MR16,444.

[8] A. Karrass and D. Solitar, "The subgroups of a free product of two groups with an amalgamated subgroup", *Trans. Amer. Math. Soc.* 150 (1970), 227-255. MR41#5499.

[9] Wilhelm Magnus, Abraham Karrass, Donald Solitar, *Combinatorial group theory:* (Pure and Appl. Math. 13. Interscience [John Wiley & Sons], New York, London, Sydney, 1966). MR34#7617.

[10] Bernard Maskit, "On Poincaré's theorem for fundamental polygons", *Advances in Math.* 7 (1971), 219-230. MR45#7049.

[11] Hanna Neumann, "On the intersection of finitely generated free groups", *Publ. Math. Debrecen* 4 (1956), 186-189. MR18,11.

[12] Hanna Neumann, "On the intersection of finitely generated free groups", (addendum), *Publ. Math. Debrecen* 5 (1957), 128. MR20#61.

[13] H. Poincaré, "Théorie des groupes Fuchsiens", *Acta Math.* 1 (1882), 1-62. FdM14,338.

[14] Abe Shenitzer, "Decomposition of a group with a single defining relation into a free product", *Proc. Amer. Math. Soc.* 6 (1955), 273-279. MR16,995.

[15] H. Zieschang, E. Vogt and H.-D. Coldewey, *Flächen und ebene diskontinuierliche Gruppen* (Lecture Notes in Mathematics, 122. Springer-Verlag, Berlin, Heidelberg, New York, 1970). Zbl.204,240.

[16] Heiner Zieschang, "Über die Nielsensche Kürzungsmethode in freien Produkten mit Amalgam", *Invent. Math.* 10 (197), 4-37. Zbl.185,52.

York University,
Ontario M3J IP3, Canada.

PROC. SECOND INTERNAT. CONF. THEORY OF GROUPS, 05B25, 20F15, 20M10
CANBERRA 1973, pp. 188-196.

SUBGROUPS OF BINARY RELATIONS

Kim Ki-Hang Butler

A binary relation defined on a set containing n elements can be interpreted as an $n \times n$ incidence matrix. Such matrix may be taken either over the two element boolean algebra or over the field Z_2. The main purpose of this paper is to study the incidence subgroups and the collineation subgroups of semigroups of binary relations. In doing so we shall not only obtain a partial solution to Ryser's Conjecture (see [13], p. 123) but we shall also obtain some combinatorial results concerning the incidence subsemigroups of semigroups of binary relations.

1. Introduction

Various subsemigroups and subgroups of semigroups of binary relations have been studied by many authors (see [1]-[4], [6]- [7], [9]-[11], and [14]-[16]). The main purpose of this paper is to study the incidence subgroups and the collineation subgroups of semigroups of binary relations. In doing so we obtained a partial solution to Ryser's Conjecture [13].

Let $\bar{N} = \{1, \ldots, n\}$. A binary relation, or more briefly relation α on $\bar{N}$ is a collection of ordered pairs of elements of $\bar{N}$. For our purpose, it is convenient to consider relation α as an incidence matrix (that is, $(0, 1)$-matrix), where $a_{ij} = 1$ if $(i, j) \in \alpha$ and $a_{ij} = 0$ otherwise. Such matrix A may be taken either over the two element boolean algebra, $\mathcal{B} = \{0, 1\}$, or over the field Z_2. Let B_n (M_n) denote the set of all $n \times n$ matrices over $\mathcal{B}$ (Z_2). Then B_n (M_n) is a semigroup under matrix multiplication. For simplicity of notation, R_n will represent either B_n or M_n in this paper, where the results and proofs are the same over both structures. Now R_n contains the subgroup S_n of permutation matrices of order n.

In this paper, we will be concerned with the following "general problem".

Suppose A is an element of R_n such that

$$A = PAQ$$

for some permutation matrices P and Q in S_n. Then A is called an *incidence matrix* and P and Q are called *collineation matrices*. It should be noted that if A is an incidence matrix for a finite plane then P and Q determine a collineation of that finite plane. Therefore, the above general problem is equivalent to the Ryser problem ([13], p. 123). We are interested in those subsets of R_n whose incidence matrices obey the semigroup (group) postulates which can be called the *incidence subsemigroups (subgroups)*. Similarly, we are also interested in those subsets of S_n whose collineation matrices obey the semigroup (group) postulates which can be called the *collineation subsemigroups (subgroups)*.

2. Incidence subgroups

It is convenient to have a criterion for determining whether a subset of R_n is an incidence subgroup of R_n. To facilitate this discussion, we first introduce some special notation. For fixed permutation matrix P in S_n, we shall let

$$\langle M|P\rangle = \left\{M \in R_n : M = PMP^{-1}\right\} .$$

For brevity, we shall let XsY (XgY) denote that X is a subsemigroup (subgroup) of Y.

THEOREM 1. $\langle M|P\rangle sR_n$.

PROOF. Assume A and B are elements of $\langle M|P\rangle$. Then $PA = AP$ and $PB = BP$. Therefore, $P(AB) = (PA)B = A(PB) = (AB)P$, in consequence of which AB is an element of $\langle M|P\rangle$.

Let I denote the $n \times n$ identity matrix.

REMARK 1. $\langle M|I\rangle = R_n$.

Let Z denote the $n \times n$ zero matrix, and let J denote the $n \times n$ universal matrix (that is, all entries 1).

REMARK 2. Every $\langle M|P\rangle$ contains Z, I, and J.

The elements of $\langle M|P\rangle$ can be found by considering the action of P on the matrix $C = (c_{ij})_{n\times n}$, where the c_{ij}'s are just variables. Let

$f_P(i, j) = \{(r, s) \in \bar{N} \times \bar{N}$: there exists λ such that $P^\lambda(i) = r$ and $P^\lambda(j) = s\}$

where $1 \le i$, $j \le n$. If $B = (b_{ij})$ is an element of R_n, then B is an element of $\langle M|P\rangle$ iff (r, s) is an element of $f_p(i, j)$. This implies $b_{rs} = b_{ij}$.

We next investigate the nonsingularity in $\langle M|P\rangle$. A matrix A in B_n is said to be *nonsingular* if A is regular (that is, $A \in AB_nA$ [3]) and no row of A is a sum of other rows of A and no column of A is a sum of other columns of A and A has neither zero rows nor zero columns. A matrix A of M_n is said to be nonsingular if $\det A \neq 0$. It is not hard to show that the product of two nonsingular matrices of B_n is not necessarily nonsingular, and therefore, I am very sorry to report that the nonsingular matrices of $\langle M|P\rangle$ in B_n do not form a subgroup. In fact, they do not even form a subsemigroup. However, the nonsingular matrices of $\langle M|P\rangle$ in M_n do form a subgroup.

THEOREM 2. *If* $n\langle M|P\rangle$ *denotes the nonsingular matrices in* $\langle M|P\rangle$, *then*

$$n\langle M|P\rangle g M_n .$$

PROOF. Trivial.

To simplify the matters, we let $S_n^1 = S_n - \{I\}$. For $P, Q \in S_n^1$, we let

$$\langle M|P, Q\rangle = \{M \in R_n : M = PMQ\} .$$

THEOREM 3. $A \in \langle M|P\rangle \Rightarrow A \in \langle M|P, Q\rangle$.

PROOF. Let $R \in S_n^1$, $P = R$, and $Q = R^{-1}$. Then

$$PAQ = RAR^{-1} = A .$$

For $A \in \langle M|P\rangle$, we shall let

$$\langle M^*|P, Q\rangle = \{B \in R_n : B = PAQ\} ,$$

where $P, Q \in S_n^1$.

THEOREM 4. $A \in \langle M^*|P, Q\rangle \Rightarrow A \in \langle M|P, Q\rangle$.

PROOF. Assume A is an element of $\langle M^*|P, Q\rangle$. Then there exists B in R_n such that $A = PBQ$ for $P, Q \in S_n^1$. Let $R \in S_n^1$, $S = PRP^{-1}$, and $T = Q^{-1}R^{-1}Q$. Then

$$\begin{aligned} SAT &= PRP^{-1}AQ^{-1}R^{-1}Q \\ &= PRP^{-1}PBQQ^{-1}R^{-1}Q \\ &= PRBR^{-1}Q \\ &= PBQ \\ &= A\ . \end{aligned}$$

Let R_A (L_A, H_A, D_A) denote the Green's equivalence class of A with respect to R (L, H, D). More details can be found in ([3], Chapter 2).

PROPOSITION 5. *$A \in R_J$ (L_J) $\Longleftrightarrow$ for $P \in S_n^1$ $\left(Q \in S_n^1\right)$ there exist $Q \in S_n$ $(P \in S_n)$ such that*

$$A \in \langle M|P, Q\rangle .$$

PROOF. In the first case let $Q = I$ and in the second case let $P = I$.

A natural question is whether a semigroup R_n can possess the commutative subgroups and, more pointedly, what conditions (if any at all) would insure their existence.

An element P of S_n is called *cyclic* if $P = C^m$, where

$$C = \begin{bmatrix} 0 & 1 & 0 & \dots & 0 \\ 0 & 0 & 1 & \dots & 0 \\ . & . & . & . & . \\ 1 & 0 & 0 & \dots & 0 \end{bmatrix}_{n\times n},$$

and $m \in \bar{N}$. Let σ_n denote the set of all cyclic matrices in S_n. Let C_n denote the set of all circulants in R_n [1].

REMARK 3. $\sigma_n \subset C_n$.

For $\Gamma \subset \sigma_n$, we shall let

$$c\langle C|P\rangle = \left\{C \in C_n : C = PCP^{-1},\ P \in \Gamma\right\} .$$

For brevity, we shall let X as Y (X ag Y) denote that X is an abelian subsemigroup (subgroup) of Y.

The following theorem is derived from ([1], Theorem 1).

THEOREM 6. $c\langle C|P\rangle$ as R_n.

THEOREM 7. *If $C_n\langle C/P\rangle$ denotes the set of all nonsingular matrices in $c\langle C/P\rangle$, then*

$$C_n\langle C/P\rangle \text{ ag } M_n .$$

PROOF. Trivial.

3. Collineation subgroups

We shall determine the conditions that a subset of an arbitrary symmetric group determines a collineation group. In addition, we shall give a partial solution to Ryser's Conjecture; (see [13], p. 123).

RYSER'S CONJECTURE. *Let* A *be the incidence matrix of a projective plane. Do there exist permutation matrices* $P \neq I$ *and* Q *such that*

(**) $$A = PAQ \ ?$$

This is equivalent to the existence of a nontrivial collineation of the geometry. All known finite planes have very large collineation groups [8]. But the general conjecture is very much unsolved so far as the author knows.

We now show that this conjecture could be settled if we can establish the existence of permutation matrices $P \neq I$ and Q satisfies (**).

For $\Omega \subset C_n$, we shall let

$$\mathcal{c}\langle P|C\rangle = \left\{P \in \sigma_n : C = PCP^{-1},\ C \in \Omega\right\} .$$

The next theorem is an immediate consequence of a combination of Theorems 6 and 7 with the definition of cyclic matrix.

THEOREM 8. $\mathcal{c}\langle P|C\rangle$ ag R_n .

Let E_n denote the set of all idempotent matrices $\left(\text{that is, } A^2 = A\right)$ in R_n . For M in E_n we shall let

$$\iota\langle P|M\rangle = \left\{P \in S_n : M = PMP^{-1}\right\} .$$

THEOREM 9. $\iota\langle P|M\rangle$ g R_n .

PROOF. Follows from Theorem 3.5 [9].

Let E_n^n denote the set of all idempotent and nonsingular matrices in R_n . For M in E_n^n, we shall let

$$\iota_n\langle P|M\rangle = \{P \in S_n : \text{there exists } Q \in S_n \text{ such that } M = PMQ\} .$$

THEOREM 10. $\iota_n\langle P|M\rangle$ g R_n .

PROOF. Follows readily from Lemma 3.4 [9].

REMARK 4. The collineation groups $c\langle F|C\rangle$, $\iota\langle P|M\rangle$, $\iota_n\langle P|M\rangle$ are isomorphic to H-classes contained in R_n .

REMARK 5. Let A be a reduced idempotent matrix [9] with rank r $(1 \le r \le n)$. Then there exists a lower (upper) triangular matrix T ,

$$T = \left[\begin{array}{c|c} \begin{bmatrix} 1 & 0 & 0 & \cdots & 0 \\ * & 1 & 0 & \cdots & 0 \\ ** & * & 1 & \cdots & 0 \\ \cdot & \cdot & \cdot & \cdots & \cdot \\ & & & & 1 \end{bmatrix}_{r\times r} & 0 \\ \hline 0 & 0 \end{array}\right]_{n\times n}$$

such that $T = PAQ$ for $P, Q \in S_n$.

Two matrices A and B of R_n are said to be *Vagner inverses* of each other if $A = ABA$ and $B = BAB$ [17].

REMARK 6. If A is an element of E_n , then there exists U in R_A and V in L_A such that $A = UV$, where U and V are Vagner inverses of each other.

The following results are due to Montague and Plemmons [9]. It is included for completeness. A matrix A of R_n is called *Plemmons matrix* if

(i) each column and each row contains a 1 , and

(ii) no two columns or rows of A are equal.

A matrix A of R_n is *Montague matrix* if

$$A = \left[\begin{array}{c|c} [I]_{r\times r} & [Z]_{r\times s} \\ \hline [B]_{s\times r} & [I]_{s\times s} \end{array}\right]_{n\times n} ,$$

where B is a Pelmmons matrix. For any Montague matrix A , we shall let

$$m\langle P|A\rangle = \left\{P \in S_n : A = PAP^{-1}\right\} .$$

For any Plemmons matrix A , we shall let

$$p\langle P|A\rangle = \{P \in S_n : \text{there exists } Q \in S_n \text{ such that } A = PAQ\} .$$

THEOREM 11. *(i)* $m\langle P|A\rangle \text{ g } R_n$,

(ii) $p\langle P|A\rangle \text{ g } R_n$,

(iii) $m\langle P|A\rangle \simeq p\langle P|A\rangle$.

4. Some combinatorial results

In this section, we shall give some counting results concerning the subsemigroup $\langle M|P\rangle$. Let $o(S)$ denote the order of a subsemigroup S . Derivation of our counting formulae rests on the theory of elementary group theory and various simple combinatorial arguments. Following the pattern that Davis has established in [5], and so we shall not present the details of arguments of various assertions. The process is summarized in the following remarks. In the following we assume $M \in B_n$.

(1) If P and Q are conjugate elements of S_n , then $o\langle M|P\rangle = o\langle M|Q\rangle$. Thus it is enough, for our purposes, to evaluate $o\langle M|P\rangle$ on one representative of each conjugate class, $[k(n)]$.

(2) Since every permutation of S_n is associated with a partitions of n into positive integers, namely, the degrees of the cycles into which it is decomposed, therefore, the number of conjugate classes in S_n is equal to the number of partitions of n [12].

(3) Let $[k(n)]$ or $[k_1, \ldots, k_n]$ denote the conjugate class containing permutations of S_n with k_1 1-cycle, ..., k_n n-cycles, so that

$$k_1 + \ldots + nk_n = n .$$

It is well known that the number of elements in the conjugate class $[k(n)]$ is

$$\frac{n!}{1^{k_1} k_1! \ldots n^{k_n} k_n!} .$$

This result is due to Cauchy.

In order to simplify our treatement we shall need the following definition. An element P of S_n is called *involutory* if $P = P^{-1}$; otherwise P is called *noninvolutory*. A conjugate class is called involutory (noninvolutory) if each element of conjugate class is involutory (noninvolutory). If P is an element of involutory (noninvolutory) conjugate class, then we say that an incidence subgroup $\langle M|P\rangle$ is involutory (noninvolutory).

A cycle of the form (ij) is called a transposition; it interchanges the letter i and j and leaves all the others unaltered. Hence $(ij) = (ij)^{-1} = (ji)$. This leads to the following remark.

(4) Every involutory conjugate class $[k(n)]$ can be expressed as

$$[k(n)] = [n-2i, i, 0, \ldots, 0] ,$$

where $0 \le i \le [n/2]$.

Putting the pieces together we have the following results.

THEOREM 12 *(i)* R_n *contains exactly* $1 + [n/2]$ *nonisomorphic involutory incidence subsemigroups.*

(ii) R_n *contains exactly* $P(n) - (1 + [n/2])$ *nonisomorphic noninvolutory incidence subsemigroups, where* $P(n)$ *denotes the partitions of* n .

THEOREM 13. *(i)* R_n *contains exactly*

$$\sum_{i=0}^{[n/2]} \frac{n!}{2^i i!(n-2i)!}$$

distinct involutory incidence subsemigroups.

(ii) R_n *contains exactly*

$$n! - \sum_{i=0}^{[n/2]} \frac{n!}{2^i i!(n-2i)!}$$

distinct noninvolutory incidence subsemigroups.

THEOREM 14. *Let* $P \in [k(n)] = [k_1, \ldots, k_n]$. *Then*

$$o(\langle M|P\rangle) = 2^{\sum_{r,s}^{n} k_r k_s (r,s)} ,$$

where (r, s) *is the g.c.d. of* r *and* s .

ACKNOWLEDGEMENT. The author would like to thank George Markowsky and James Krabill for enjoyable discussions on the subject of this paper.

References

[1] Kim Ki-Hang Butler and J.R. Krabill, "Circulant boolean relation matrices", submitted.

[2] Kim Ki-Hang Butler and George Markowsky, "The number of maximal subgroups of the semigroup of binary relations", *Kyungpook Math. J.* 12 (1972), 1-8. MR46#3649.

[3] A.H. Clifford and G.B. Preston, *The algebraic theory of semigroups*, Volume I (Math. Surveys 7. Amer. Math. Soc., Providence, Rhode Island, 1961). MR46#A2627.

[4] A.H. Clifford and G.B. Preston, *The algebraic theory of semigroups*, Volume II (Math. Surveys, 7. Amer. Math. Soc., Providence, Rhode Island, 1967). MR36#1558.

[5] Robert L. Davis, "The number of structures of finite relations", *Proc. Amer. Math. Soc.* 4 (1953), 486-495. MR14,1053.

[6] József Dénes, "The representation of a permutation as the product of a minimal number of transpositions, and its connection with the theory of graphs", *Magyar Tud. Akad. Mat. Fiz. Oszt. Közl.* 4 (1959), 63-71. MR22#6733.

[7] J. Dénes, "On transformations, transformation-semigroups and graphs", *Theory of graphs* [edited by P. Erdös and G. Katona]. Proc. Colloq., Tihany, Hungary, 1966, pp. 65-75 (Academic Press, New York, 1968). MR38#3367.

[8] D.R. Hughes, "Collineations and generalized incidence matrices", *Trans. Amer. Math. Soc.* 86 (1957), 284-296. MR20#253.

[9] J.S. Montague and R.J. Plemmons, "Maximal subgroups of the semigroup of relations", *J. Algebra* 13 (1969), 575-587. MR40#5759.

[10] M. Petrich, *Topics in semigroups* (Lecture Notes, Pennsylvannia State University, University Park, Pennsylvannia, 1967).

[11] R.J. Pelmmons and B.M. Schein, "Groups of binary relations", *Semigroup Forum* 1 (1970), 267-271. MR43#6351.

[12] John Riordan, *An introduction to combinatorial analysis* (A Wiley publication in mathematical statistics. John Wiley & Sons, New York; Chapman & Hall, London, 1958). MR20#3077.

[13] Herbert John Ryser, *Combinatorial mathematics* (The Carus Mathematical Monographs, 14. The Mathematical Association of America; John Wiley & Sons, New York, 1963). MR27#51.

[14] Boris M. Schein, "Relation algebras and function semigroups", *Semigroup Forum* 1 (1970), 1-62. MR44#2856.

[15] Štefan Schwarz, "On the semigroup of binary relations on a finite set", *Czechoslovak Math. J.* 20 (1970), 632-679. MR45#5251.

[16] Štefan Schwarz, "Circulant boolean relation matrices", submitted.

[17] В.В. Вагнер [V.V. Vagner], "Обобщенные группы" [Generalized groups], *Dokl. Akad. Nauk SSSR* 84 (1952), 1119-1122. MR14,12.

Pembroke State University,
North Carolina 28372, USA.

PROC. SECOND INTERNAT. CONF. THEORY OF GROUPS, CANBERRA 1973, pp. 197-203. 15A21, 16A64, 20C20

THE 2-ADIC REPRESENTATIONS OF KLEIN'S FOUR GROUP

M.C.R. Butler

The indecomposable integral representations of Klein's Four Group over a discrete valuation ring in which 2 is a non-zero prime element were first determined by Nazarova, as a consequence of her study [4] of modules over tetrad rings. It was shown in [2] that her results may be deduced from a functor-induced correspondence of representations with 4-subspace diagrams; this gives a little extra information about endomorphism rings, isomorphisms and direct decompositions. However, the correspondence used in [2] is rather awkward for computational purposes. This note contains a description of very much simpler and more explicit constructions for relating representations and diagrams. These are described in §1, which also contains formal statements of their properties. Their proofs occupy §2. As a result of recent work by Gelfand and Ponomarev [3] (working over an algebraically closed field) and Sheila Brenner [1] (over an arbitrary field), the structure of 4-subspace diagrams may be said to be completely understood. §3 contains a list of parameters (derived from [1]) sufficient to characterise all the indecomposable diagrams; normal forms for each of these indecomposables are given in [1].

1. Statement of results

Throughout this note, G denotes the non-cyclic group of order 4, R a discrete valuation ring in which 2 is a non-zero prime element, K the quotient field of R, and $k = R/2R$ the residue field of R. Finitely generated R-torsionfree RG-modules will be called *RG-lattices*. Such a lattice M is, of course, R-free on a finite basis, with cardinality called the *rank of* M and denoted $rk(M)$. A lattice M will invariably be identified with its canonical image in the KG-module $KM = K \otimes_R M$, and an RG-homomorphism $\phi : M \to N$ between lattices will be identified with its unique extension to a KG-homomorphism of KM into KN.

DEFINITION. An RG-lattice will be called *a reduced lattice* if it is non-trivial and contains neither non-zero RG-free nor rank-1 direct summands. The category of all reduced RG-lattices and RG-homomorphisms will be denoted by R.

DEFINITION. A *reduced* 4-*subspace diagram* over k is defined to be an indexed system $V_* = (V; V_1, V_2, V_3, V_4) = (V; V_i)_{1\le i\le 4}$ in which V is a non-zero finite-dimensional vector space over k, and V_1, V_2, V_3, V_4 are subspaces such that any three of them span V. The symbol $\mathcal{V}$ denotes the category of all reduced 4-subspace diagrams and morphisms, where a morphism from V_* to V'_* is just a k-linear map $\theta : V \to V'$ such that $V_i\theta \subset V'_i$ for $i \in \{1, 2, 3, 4\}$.

Now select arbitrarily an enumeration

$$\chi_1, \chi_2, \chi_3, \chi_4$$

of the four irreducible characters of G (they have values ± 1 in R). Two constructions will be described. The first produces a functor,

$$\Delta : \mathcal{R} \to \mathcal{V},$$

such that the induced map

$$\Delta_{M,N} : \hom_{RG}(M, N) \to \hom_{\mathcal{V}}(\Delta(M), \Delta(N))$$

has the following properties:

(1.1) $\Delta_{M,N}$ *is surjective, preserves and reflects isomorphisms, and (when* $M = N$*) preserves and reflects decompositions of the identity into sums of proper or orthogonal idempotents;*

(1.2) *when* $M = N$, $\ker(\Delta_{M,N})$ *is contained in the Jacobson radical of* $\mathrm{end}_{RG}(M)$.

The second construction simply shows that

(1.3) *for each reduced* 4-*subspace diagram* V_*, *there exists a reduced* RG-*lattice* M *such that* $\Delta(M) \cong V_*$.

It follows that Δ establishes a one-to-one direct sum preserving correspondence between (isomorphisms classes of) reduced RG-lattices and reduced 4-subspace diagrams. The explicit formula (1.7) below for $\ker(\Delta_{M,N})$ makes the relationship between $\mathrm{end}_{RG}(M)$ and $\mathrm{end}(\Delta(M))$ easy to study.

Some technical preliminaries are needed. First of all, the characters $\chi_1, \chi_2, \chi_3, \chi_4$ will be viewed as restrictions to G of K-algebra homomorphisms of KG onto K, also denoted $\chi_1, \chi_2, \chi_3, \chi_4$. These determine the primitive idempotents of KG,

$$e_i = \frac{1}{4}\sum_{g\in G}\chi_i(g^{-1})g \quad (i = 1, 2, 3, 4),$$

such that $e_i e_j = \delta_{ij} e_i$, $\sum_{k=1}^{4} e_k = 1$, $ge_i = \chi_i(g)e_i$, and $\chi_i(e_j) = \delta_{ij}$ (for $g \in G$ and $i, j \in \{1, 2, 3, 4\}$).

There exist, up to isomorphisms, just 4 distinct rank-1 RG-lattices, for which $R_i = Re_i$ $(i = 1, 2, 3, 4)$ are convenient representatives. A finite direct sum of copies of R_i will be called a *lattice of type* χ_i . An RG-lattice is of type χ_i if and only if e_i acts as the identity on it (and the other e_j , $j \neq i$, act as zero on it).

Let M be any RG-lattice, $M \subset KM$. For fixed i , its translate $e_i M$ in KM is an RG-lattice of type χ_i . Let

$$e_* M = e_1 M + e_2 M + e_3 M + e_4 M .$$

This too is an RG-lattice contained in KM . Since $e_i e_j = \delta_{ij} e_i$, the sum is direct, and since $\sum e_i = 1$,

(1.4) $$M \subset e_* M = M + e_i M + e_j M + e_k M$$

whenever i, j, k are different. Notice also that the rank of $e_i M$, denoted $rk_i(M)$, is just the multiplicity of the character χ_i in the character χ_M of the KG-module KM .

The proof of the following characterisation of reduced lattices will be given in §2.

PROPOSITION 1.5. *The non-zero* RG-*lattice is reduced if and only if*

$$M \cap e_i M = 2e_i M \quad (i = 1, 2, 3 \text{ and } 4) .$$

It is easy now to define $\Delta : \mathrm{R} \to V$. Let M be a reduced lattice; then $2e_* M \subset M \subset e_* M$, and the inclusions are strict since $e_* M$ is a direct sum of rank-1 summands and M has no rank-1 summands. Therefore $V(M) = e_* M/M$ is a non-zero finite-dimensional vector space over $R/2R = k$; the action of G on it is trivial since $e_* M$ has an R-basis on which the group elements act as multiplication by 1 or -1 . Let $\pi_M : e_* M \to V(M)$ be the natural map. For $i = 1, 2, 3, 4$, define $V_i(M) = (e_i M)\pi_M = (e_i M + M)/M$. Then, by (1.4),

$$\Delta(M) = \big(V(M);\ V_i(M)\big)_{1 \le i \le 4}$$

is a reduced 4-subspace diagram. The Proposition shows that, for $i = 1, 2, 3, 4$,

(1.6) $$0 \to 2e_i M \to e_i M \xrightarrow{\pi_M} V_i(M) \to 0$$

is an exact sequence, hence that $\dim(V_i(M)) = rk_i(M)$. The definition of $\Delta(\phi)$, for an RG-homomorphism $\phi : M \to N$ between reduced RG-lattices M, N , is evident, for ϕ - as a KG-homomorphism of KM into KN - maps e_iM into e_iN , e_*M into e_*N , and M into N . Observe that $\Delta(\phi) = 0 \iff (e_*M)\phi \subset N$, and this latter condition may be transformed (using Proposition 1.5) into $(e_*M)\phi \subset 2e_*N$. This gives the explicit formula mentioned previously for $\ker(\Delta_{M,N})$, namely,

$$\ker(\Delta_{M,N}) = 2\mathrm{hom}_{RG}(e_*M, e_*N) \tag{1.7}$$

$$\cong 2 \bigoplus_{i=1}^{4} \mathrm{hom}_R(e_iM, e_iN)$$

(the last formula depends on the fact that each R-homomorphism between lattices of the same type χ_i is an RG-homomorphism).

Finally, the second construction, leading to (1.3), is as follows. Let V_* be a reduced 4-subspace diagram. For $i = \{1, 2, 3, 4\}$, choose a lattice H_i of type χ_i and rank $\dim(V_i)$ and a short exact sequence

$$0 \to 2H_i \to H_i \xrightarrow{\sigma_i} V_i \to 0 . \tag{1.8}$$

Define $H = \bigoplus_{i=1}^{4} H_i$ and let $\sigma : H \to V$ be the homomorphism with restriction σ_i to H_i ($i = 1, 2, 3$ and 4) . It will be shown in §2 that the RG-lattice $M = \ker(\sigma)$ satisfies $\Delta(M) \cong V_*$, so proving (1.3).

2. Proofs

PROOF OF PROPOSITION 1.5 'if' part. An easy calculation shows that $RG \cap e_iRG = 4e_iRG \neq 2e_iRG$, and $R_i \cap e_iR_i = R_i \neq 2e_iR_i$. Therefore, if $M \neq 0$ and M is not reduced, then $M \cap e_iM \neq 2e_iM$.

'only if' part. Suppose M is reduced but, for some i , $M \cap e_iM \neq 2e_iM$.

Case 1. Assume that $2e_iM \not\subset M$, and choose $x_0 \in M$ such that $2e_ix_0 \notin M$. Let $f = \sum_{g\in G} g$, so that $f - 4e_i \in 2RG$ and $y_0 = fx_0 \in M\backslash 2M$. Since M is a free R-module and R is a discrete valuation ring and 2 is prime in R , Ry_0 is an R-module summand of M . Choose $\pi \in \mathrm{hom}_R(M, R)$ such that $y_0\pi = 1$, and define $\tau \in \mathrm{hom}_{RG}(M, RG)$ by $x\tau = \sum_{g\in G} (g^{-1}x)\pi g$ for $x \in M$. Then $f(x_0\tau) = y_0\tau = f$, so $x_0\tau$ is a unit in RG and τ must be surjective - which is impossible since M has no free summands. Thus Case 1 cannot occur.

Case 2. Suppose that $2e_iM \subset M \cap e_iM$ is a strict inclusion. Choose $x_i = e_ix_i \in (M \cap e_iM)\backslash 2e_iM$. Then Rx_i is a rank-1 summand of e_iM, hence of e_*M, and is contained in M - which is also impossible since M has no rank-1 summands. So M reduced implies that $M \cap e_iM = 2e_iM$, as required.

PROOF OF (1.1). Let $\theta : \Delta(M) \to \Delta(N)$ be a morphism of diagrams, that is, a vector space homomorphism $\theta : V(M) \to V(N)$ mapping $V_i(M)$ into $V_i(N)$ for $i = 1, 2, 3, 4$. Since, in (1.6), e_iM is a free R-module and π_N is surjective, there is an R-homomorphism $\phi_i : e_iM \to e_iN$ such that, on e_iM, $\phi_i\pi_N = \pi_M\theta$. But e_iM, e_iN both have type χ_i, so ϕ_i is an RG-homomorphism. Let $\phi : e_*M \to e_*N$ be the RG-homomorphism with restriction ϕ_i to e_iM. Then $\phi\pi_N = \pi_M\theta$, so ϕ maps $M = \ker(\pi_M)$ into $N = \ker(\pi_N)$ and $\Delta(\phi) = \theta$. This proves the surjectivity of Δ. If θ is an isomorphism (so in particular maps $V_i(M)$ isomorphically to $V_i(N)$), each ϕ_i is an isomorphism, hence so is ϕ; thus Δ reflects isomorphisms, and it is of course obvious that it preserves isomorphisms. Finally, suppose that $M = N$ and that $1 = \theta + \theta' + \ldots$ is a decomposition of 1 (the identity of $\Delta(M)$) into an orthogonal sum of proper idempotents. Then, for each i, since orthogonal idempotent decompositions may be lifted across the sequences (1.6), one can choose $\phi_i, \phi_i', \ldots$ so that $1 = \phi_i + \phi_i' + \ldots$ is a decomposition of the identity of e_iM into orthogonal idempotents; then - setting $\phi = \bigoplus_{i=1}^{4} \phi_i$, $\phi' = \bigoplus_{i=1}^{4} \phi_i'$, ... - it follows that $\pi_M\theta = \phi\pi_N$, $\pi_M\theta' = \phi'\pi_N$, ... Thus Δ *reflects* orthogonal idempotent decompositions. That Δ preserves them is an easy consequence of (1.2), for no sum or difference of orthogonal idempotents can belong to the Jacobson radical.

PROOF OF (1.2). Let $\phi \in \ker(\Delta_{M,M})$, so by (1.7), $(e_*M)\phi \subset 2e_*M$. Choosing a basis of e_*M, one sees that $\det(1-\phi)$ is a unit in the local ring R, and hence $(1-\phi)^{-1}$ exists and $1 - (1-\phi)^{-1}$ maps e_*M into $2e_*M$. Therefore ϕ has a quasi-inverse in $\ker(\Delta_{M,M})$. Since $\ker(\Delta_{M,M})$ is clearly an ideal in $\operatorname{end}_{RG}(M)$, this proves (1.2).

PROOF OF (1.3). Let $M = \ker(\sigma)$ be the RG-lattice defined by V_* via the maps σ_i in the exact sequences (1.8). Since $M \subset H = \bigoplus_{i=1}^{4} H_i$, $e_iM \subset e_iH = H_i$. Now, let $x_i \in H_i$ so that $x_i\sigma_i \in V_i$. Since V_* is reduced, there exist

$x_j \in H_j$, $j \neq i$, such that $x_i\sigma_i = \sum_{j\neq i} x_j\sigma_j$. Therefore $x = x_i - \sum_{j\neq i} x_j \in M$ and $x_i = e_i x \in e_i M$. This shows that $e_i M = H_i$. Since $M \cap H_i = 2H_i$, this and Proposition 1.5 suffice to show that M is reduced. Also $e_* M = H$, so it is clear that the isomorphism $\alpha : V(M) \to V$ such that $\sigma = \pi_M \alpha$ is the required isomorphism of $\Delta(M)$ with V_* .

3. The indecomposable 4-subspace diagrams

Let $V_* = (V; V_1, V_2, V_3, V_4)$ be an *indecomposable* diagram consisting of a non-zero finite-dimensional vector space V over k and 4 subspaces V_1, V_2, V_3, V_4 . Gelfand and Ponomarev show that the *defect of the diagram*,

$$\rho(V_*) = \sum_{i=1}^{4} \dim(V_i) - 2\dim(V) ,$$

can have only the values 0, ±1 , or ±2 . They obtain normal forms for all the indecomposables over an algebraically closed field, and similar lists are given by Brenner for an arbitrary field. The following classification is taken from [1]: it ignores permutations of subspaces (which correspond in the representation theory to permutations of the characters of G).

$\rho(V_*) = 0$ and $\dim(V) = 2n \geq 2$; for each indecomposable $\alpha \in \mathrm{end}_k(k^n)$, there is one diagram: each V_i has dimension n , and $\mathrm{end}(V_*) \cong k[\alpha]$.

$\rho(V_*) = 0$ and $\dim(V) = 2n + 1 \geq 1$; one diagram, subspace dimensions $n+1, n+1, n, n$, and $\mathrm{end}(V_*) \cong k[t]/(t^{n+1})$.

$\rho(V_*) = 1$ and $\dim(V) = 2n \geq 2$; one diagram, subspace dimensions $n, n, n, n+1$, and $\mathrm{end}(V_*) \cong k$.

$\rho(V_*) = 1$ and $\dim(V) = 2n + 1 \geq 1$; one diagram, subspace dimensions $n, n+1, n+1, n+1$, and $\mathrm{end}(V_*) \cong k$.

$\rho(V_*) = 2$ and $\dim(V) = 2n + 1 \geq 1$; one diagram, subspace dimensions all $n+1$, and $\mathrm{end}(V_*) \cong k$.

The indecomposables of defects −1, −2 are the duals of those of defects 1, 2 , respectively. The only listed diagrams which are not reduced are those V_* of defects −1, −2 with $\dim(V) = 1$.

For example, suppose that M is a reduced indecomposable RG-lattice and also an ideal in RG . Let $\Delta(M) = V_*$. Then $\dim(V_i) \leq 1$ for $i = 1, 2, 3, 4$. Such diagrams occur for $\rho(V_*) = 0$, $\dim(V) = 2$, each $\alpha \in k$, and each $\dim(V_i) = 1$, so that $rk(M) = 4$; for $\rho(V_*) = 0$, $\dim(V) = 1$, $rk(M) = 2$, 2 subspaces equal to V, the other 2 zero; for $\rho(V_*) = 1$, $\dim(V) = 1$, $rk(M) = 3$, 3 subspaces equal to V, the other zero; for $\rho(V_*) = -1$, $\dim(V) = 2$, $rk(M) = 3$, 3 subspaces of dimension 1, the other zero; for $\rho(V_*) = 2$, $\dim(V) = 1$, $rk(M) = 4$, all subspaces of dimension 1; for $\rho(V_*) = -2$, $\dim(V) = 3$, $rk(M) = 4$, all subspaces of dimension 1.

References

[1] Sheila Brenner, "On four subspaces of a vector space", *J. Algebra* (to appear).

[2] M.C.R. Butler, "Relations between diagrams of modules", *J. London Math. Soc.* (2) 3 (1971), 577-587. Zbl.214,57.

[3] I.M. Gelfand and V.A. Ponomarev, "Problems of linear algebra and classification of quadruples of subspaces in a finite-dimensional vector space", *Hilbert space operators and operator algebras*, pp. 163-237 (Colloquia Mathematica Societatis János Bolyai, 5. Tihany, Hungary, 1970; North-Holland, Amsterdam, London, 1972). Zbl.238.00011.

[4] Л.А. Назарова [L.A. Nazarova], "Представления четвериады" [Representations of tetrads], *Izv. Akad. Nauk SSSR Ser. Mat.* 31 (1967), 1361-1378; *Math. USSR Izv.* 1 (1967), 1305-1321 (1969). MR36#6400.

The University of Liverpool,
Liverpool L69 38X, England.

PROC. SECOND INTERNAT. CONF. THEORY OF GROUPS, 20F99, 68A15
CANBERRA 1973, pp. 204-217.

A GENERAL PURPOSE GROUP THEORY PROGRAM

John Cannon

1. Introduction

Questions concerning particular finite groups, which cannot be easily answered by hand methods, can sometimes be resolved with the aid of a computer. Among the more spectacular problems which have been solved using machines, we mention only the construction of character tables for large finite simple groups by Conway, Livingstone, Hunt and others, the construction of the Lyons simple group by Sims and the determination of the four-dimensional crystal groups by Neubüser and colleagues.

Although a wide range of effective computer-orientated algorithms have been developed for group theory their application has been restricted to a very small number of group theorists because of the considerable programming effort required to implement many of these algorithms. For example, the implementation of a current subgroup lattice algorithm involves at least two man-years of programming effort.

In order to make the computer a more practical tool for group theorists we have undertaken to develop a general-purpose group theory program (called GROUP) which is to contain those constructions we believe to be most generally useful. While it is not possible to make such a program completely machine-independent because of efficiency considerations, great care is being taken to ensure that the machine-dependent part of the program is kept to a minimum. Hopefully, the final program, representing 12-15 man-years work, will be transferable to a new machine with about 3 months effort.

In this paper we give an outline of the proposed scope of the program, together with an indication of the potential range of application of the various algorithms. We also give a sampling of elementary applications of those parts of the program already completed.

While most of GROUP is orientated towards finite groups it has some capabilities with respect to infinite groups. Currently implemented or planned algorithms

This research was supported by a grant from the Australian Research Grants Committee.

roughly fall into one of the following categories:

(i) combinatorial group theory;

(ii) calculation with group elements;

(iii) generation and representation of the elements of finite groups;

(iv) permutation groups;

(v) determination of partial information about the structure of a group, for example, classes, Sylow subgroups, normalizers, centralizers, series, and so on;

(vi) subgroup lattices;

(vii) automorphism groups;

(viii) representation theory.

We now briefly consider some details of the algorithms included under each of the above headings. We also indicate the planned implementation date, whenever known.

2. Combinatorial group theory

Under this heading we group algorithms for studying finitely presented groups. The three basic processes currently in use are coset enumeration, rewriting (for example, Reidemeister rewriting), and commutator collection.

(a) Todd-Coxeter algorithm (now available)

The form of the Todd-Coxeter algorithm implemented in GROUP is a packed version of the lookahead algorithm described in Cannon, Dimino, Havas and Watson [4]. Thus on a 128K CDC6000 series machine approximately 40 000 cosets can be stored for a two- or three- generator presentation, allowing the determination of indices up to 30 000 provided that the given presentation is not perverse.

EXAMPLES

(1) The group

$$G(-2, -11) = \langle a, b \mid a^{[a,b]} = a^{-2}, b^{[b,a]} = b^{-11} \rangle$$

belongs to a family of two-generator finite groups having zero deficiency discovered by Macdonald [19]. Enumerating the cosets of the identity, we established that the group has order 26 244 in 50 seconds.

(2) The group generated by a, b and c where $a^2 = b^2 = c^4 = 1$ and order 4 is assigned to the words

ab, ac, bc, ac^2, bc^2, abc, bac, abc^2, $abcb$, $abac$, $acbc$, $ababc$, $abcbc$,
$abcba$, $ababc^2$, $bacabc$, $ac^{-1}bc$, $abcbc^2$, $abc^{-1}abc$,

was found to have order 2^{21} by enumerating the 32 768 cosets of the subgroup $\langle b, c\rangle$ (this presentation is due to Sinkov [27]).

Apart from determining subgroup indices, the Todd-Coxeter algorithm can be put to numerous other uses, some of which are described in Cannon and Havas [6]. These include calculating double cosets HxK given the single coset table for one of the subgroups H and K ; determination of a set of coset representatives for a subgroup; and the calculation of the Schur multipliers for small groups. Thus, using the Todd-Coxeter algorithm it was found that the groups QD(3) and QD(5) (Glauberman [13]) have Schur multipliers of orders 3 and 5 , respectively. Given this result, Ward, and independently, Schult and Gagen, were able to prove that the Schur multiplier of QD(p) , p a prime, is the cyclic group of order p .

(b) Reidemeister-Schreier algorithm (now available)

An experimental Reidemeister-Schreier program is included as part of GROUP. This program enables the user to induce defining relations for subgroups H of index up to 100 in favourable circumstances, given defining relations for a group G containing H . The Todd-Coxeter algorithm is first applied to obtain a coset table for H in G . The Reidemeister rewriting process (Magnus, Karras and Solitar [21]) now uses this table to rewrite the G-relators as H-relators. At this stage we have a presentation for H involving a large number of generators and relations so a number of heuristics are applied so as to reduce the number of generators and simplify the resulting relators wherever possible (Havas [15]). While the number of generators can just about always be reduced to a reasonable number, one often ends up with some very long relators. If the given presentation for G is badly chosen, this process results in very complicated presentations for H , even in the case of very small indices.

EXAMPLE

If this program is applied to the subgroup $\langle a, b^2\rangle$ of index 64 of the two-generator group having exponent 4 , $B_{2,4}$, where

$$B_{2,4} = \langle a, b \mid a^4 = b^4 = (ab)^4 = (a^{-1}b)^4 = (a^2b)^4 = (ab^2)^4 = (a^2b^2)^4 = (a^{-1}b^{-1}ab)^4 = (a^{-1}bab)^4 = 1\rangle ,$$

then the first four relations obtained for the subgroup are

$$x^2 = y^4 = (xy)^4 = (x^2y)^4 = 1 .$$

These four relations are the standard presentation for this group which is

(4, 4 | 2, 4) in the notation of Coxeter. Unfortunately, apart from the above relations, a number of others are obtained including one of length 144.

(c) Low index subgroups

Given a finite presentation for a group G, Lepique [18] has recently programmed an algorithm, based on some ideas of Sims, for constructing all the subgroups of index less than a specified number, in a group G. In its present form, this algorithm is applicable for indices up to 50 and perhaps in favourable situations, for indices up to 100. This algorithm is applicable both to finite and infinite groups. It is a particularly useful tool when attempting to determine structural properties of certain finite groups since it enables one to get at subgroups which are inaccessible to methods which build up from the bottom.

(d) Rewriting words

In [17], Leech describes a modification of the Todd-Coxeter algorithm which rewrites words in the generators of a group G as words in arbitrary generators of a subgroup H of G. Apart from being of use in its own right, this process is needed for certain other computational algorithms. The process can also be used to deduce a large number of further relations from a given presentation of G. Leech has used this to produce a hand proof of the triviality of the group $G^{3,7,17}$.

The above four processes are centred around the Todd-Coxeter algorithm. The next two we describe are rather different.

(e) Nilpotent groups (1974)

Macdonald [20] has developed an algorithm which, given a presentation for a nilpotent group G, constructs $G/Z_i(G)$ for $i = 1, 2, \ldots$ where

$$G = Z_0(G) > Z_1(G) > Z_2(G) > \ldots$$

is a descending central series for G. In particular, this provides a method for determining the order of p-groups which does not depend very heavily on the value of p. Wamsley [1, 28] has further developed this algorithm and used it to show that the three generator Burnside group of exponent 4 has order 2^{69}. He has also successfully applied the method to determine that a certain group had order 5^{515} (see these proceedings).

(f) Construction of defining relations (currently available)

Given a faithful representation for a group G, it is possible to construct concise sets of defining relations for groups of order up to 10^8 (Cannon [3]). The relations are obtained as subsets of sets of fundamental circuits in certain Schreier coset graphs associated with G. The existence of this technique means that we may assume the availability of a presentation for certain subsequent algorithms.

EXAMPLES

(1) The Mathieu simple group M_{11} of order 7 920 has been presented by this algorithm as

$$a^2 = b^4 = (ab)^{11} = (ab^2)^6 = (ab^{-1}abab^2)^3 = ababab^{-1}ab^{-1}abab^{-1}ab^{-1}b^{-1}abab^{-1} = 1 .$$

(2) The unitary simple group $U_3(3)$ of order 6 048 has been presented as

$$a^7 = b^8 = (a^{-1}b)^3 = ba^2baba^{-1}b^{-1}ab^2a^{-1} = bab^{-3}a^{-1}b^{-4}a^{-2} = 1 .$$

(3) The unitary simple group $U_3(4)$ or order 62 400 has been presented as

$$a^2 = b^3 = [a, b]^5 = (ab)^{13} = \left[b^a, a^{b^{-1}}\right] = \left(b^a a^{b^{-1}}\right)^{10} = 1 .$$

3. Calculation with group elements

The algorithms of the previous section are characterized by the fact that one does not need to compute products explicitly. If, however, one wishes to perform calculations with group elements in a systematic manner then one requires an algorithmic method for assigning a unique product to any pair of elements. Although there are many situations where this problem can be solved we restrict our attention to the most commonly occurring cases. If a user wishes to exploit some other method then it is a straightforward matter for him to add the appropriate subroutines. The GROUP system is organized so that in general, subroutines needing to use a subroutine of this section do not need to know how the elements of a particular group are actually represented. Thus if a new type of representation is needed, the appropriate routines are added to this module, and then the subroutines for arbitrary groups described in the following sections will work without change for groups given in terms of the new representation.

The basic operations on group elements included in GROUP are:

product,
inverse,
conjugate,
order,
form specified power of an element,
commutator,
test for identity,
test for a pair of elements commuting.

While some of these operations are logically unnecessary in the sense they can be obtained using the product and inverse operations, it is faster to have separate code for each.

It is planned to include the following types of representation in GROUP:

- permutation (currently available),
- matrices over GF(p) (currently available),
- matrices over GF(p^n) (currently available),
- matrices over Z (1974),
- matrices over quadratic algebraic number fields (1974),
- matrices over arbitrary algebraic number fields (1975),
- solvable groups given by defining relations,
- arbitrary groups given by defining relations (currently available),
- affine transformations,
- root systems for Chevalley groups.

Elements of arbitrary groups given by defining relations are represented by words in the given generators and are multiplied using the appropriate Schreier coset graph for the group over the identity. The Todd-Coxeter algorithm is used to construct the Schreier coset graph and the method is restricted to groups of order about 20,000 because of the necessity to store the Schreier coset graph. In the case of solvable groups, the method described by Jürgensen [16] will be used. This is much less dependent upon the group order than the above method and does not require the storage of any information.

No limitation, apart from storage and execution time constraints, is placed on the degree of permutation or matrix representations. Certainly in some situations it is reasonable to work with permutation representations having degrees greater than a thousand. In the case of matrix representations the time required to form products and inverses rises rapidly with increasing degree so that the practical limit for performing extensive calculations in matrix groups is probably less than degree 20 .

Considerable effort has been put into designing efficient algorithms for working with these representations. Thus, for example, the inverse of a matrix over GF(p^n) is calculated using the Gauss-Jordan method while it is planned to use a so-called 'two-step fraction-free' version of the Gauss-Jordan method to invert matrices over arbitrary algebraic number fields.

In addition to the above operations which are applicable to all types of representation, sometimes operations are provided which are peculiar to a given type of representation. Thus in the case of permutations a routine is provided for determining the cycle structure of a permutation, while in the case of matrices, routines are provided for computing such things as trace, determinant and characteristic polynomial.

4. Generation and representation of the elements of finite groups (1973)

For a number of the algorithms it is necessary that the group be generated and represented in some form. The most convenient and economical method of representing the elements of a group in the machine is to store the set of transversals $\{U_1, \ldots, U_m\}$ for the subgroup chain

$$G = H_m > H_{m-1} > \ldots > H_1 > H_0 = I ;$$

where U_i is the transversal for subgroup H_{i-1} in subgroup H_i .

If G is an arbitrary group a suitable subgroup chain may either be supplied by the user or constructed by the machine. The number of proper subgroups, $m - 1$, in the chain typically ranges from 1 to 4 . Transversals at the top end of the chain can be constructed using the Todd-Coxeter algorithm, while those at the bottom end must be constructed directly.

In the case of permutation groups we use an algorithm due to Sims [24]. This algorithm has the advantage that it generates a chain of subgroups having the property that the transversals are directly obtainable. Thus it is by far the most powerful group generation algorithm currently known. For instance, Sims has used it to generate the Suzuki group of order 448 345 497 600 given permutation generators of degree 1 782 . The algorithm as implemented in GROUP and running on the CDC6600 takes 2.5 seconds to generate M_{24} of order 244 823 040 .

EXAMPLE

Can the simple group $U_3(5)$ of order 126,000 be presented as a factor group of the infinite group

$$a^2 = b^3 = (ab)^7 = 1 ?$$

Starting with two permutations of degree 50 generating $U_3(5)$, we took a fixed element a of order 2 and proceeded to generate the groups $\langle a, b \rangle$ for each element b of $U_3(5)$ satisfying the above relations. As none of these groups was found to have order 126 000 we conclude that $U_3(5)$ cannot be presented as a factor group of (2, 3, 7) .

If all the elements of a group G , which is sufficiently small to fit into core, are required then the 'coset table' method of Dimino [8] is used to generate G . Given r generators for a group G this algorithm takes at most $r|G|$ multiplications and, providing that the list of elements is hash-addressed, about $2r|G|$ element comparisons.

The availability of such a set of transversals for a group G allows us to write each element in a unique form and hence enables us to associate a unique integer in the range 1 to $|G|$ with each element of the group. This number (known as the *element code*) is very useful when it is necessary to store lists of elements because of its compactness.

Since elements are sometimes required as words in the generators, as for instance when the Todd-Coxeter algorithm is used, we provide an option which, concurrently with the construction of the transversals, sets up tables which enable a word corresponding to any element to be quickly generated. Thus during a computation a particular group element can appear as an element of the representation, as code number and as an abstract word.

5. Computation of partial information about group structure (1973/74)

For the analysis of larger groups, the system is to contain routines for constructing

conjugacy classes of elements,
centralizer of a subgroup,
normalizer of a subgroup,
normal closure of a subgroup,
Sylow p-subgroups,
upper and lower central series; derived series.

Having the above routines available enables such things as centre, Sylow intersection, Fitting subgroup, and so on, to be easily obtained. It is also hoped to include some facilities for constructing the maximal subgroups of a group.

The algorithms to be used here will be along the lines of those outlined in Cannon [2], for arbitrary groups, and based on those described in Sims [25, 26] for permutation groups. We expect to be able to perform this kind of analysis for arbitrary groups of orders up to $10^6 - 10^8$, and for permutation groups of considerably larger order.

EXAMPLE

Let $\text{aut}(F_4)$ denote the Weyl group F_4 of order 1 152 extended by the automorphism of order 2 obtained from its Dynkin diagram. We computed the conjugacy classes, listed below, for $\text{aut}(F_4)$.

Number of class	Order of elements	Size of class
1	1	1
2	2	24
3	3	64
4	2	72
5	4	36
6	6	192
7	2	18
8	4	144
9	2	24
10	6	192
11	12	96
12	8	144
13	6	16
14	6	64
15	4	12
16	4	36
17	3	16
18	2	1
19	2	72
20	4	288
21	8	144
22	8	144
23	24	96
24	24	96
25	12	192
26	8	12
27	8	12
28	8	72
29	4	24

6. Subgroup lattices

The central tool provided for the detailed analysis of smaller groups will be an implementation of a subgroup lattice algorithm. This algorithm will consist of three parts:

(a) determination of solvable subgroups of order less than 20 000 using the cyclic extension method (Neubüser [22], Felsch [11], Dryer [10]) (1974);

(b) determination of non-solvable subgroups of order less than 20 000 ;

(c) determination of subgroups of order greater than 20 000 but having index less than 100 using the algorithm of section 2 (c).

The range of application of this algorithm is not known at present as it depends not only on the group order but also on the number of subgroups belonging to the group. Thus the algorithm will be inapplicable to some groups smaller than 20 000 having a very large number of subgroups.

Since every normal subgroup is a union of conjugacy classes it is possible to obtain the lattice of normal subgroups rather easily once one knows the classes and something about how they multiply. Such an algorithm is currently being implemented for GROUP and it is expected that it will be able to find normal subgroup lattices for groups of order up to 10^5 having up to 200 classes.

7. Automorphism groups

Sometime in the future it is planned to incorporate into GROUP an algorithm for determining the automorphism group of a solvable group. Probably the algorithm used will be similar to that described in Gerhards and Altmann [12].

8. Representation theory

The following representation theory routines are planned:

(a) construction of the ordinary character table of a group given a faithful representation for the group, (1974);

(b) construction of further characters given a number of ordinary irreducible characters, (1974);

(c) search for permutation characters given the table of ordinary characters;

(d) construction of all the complex irreducible representations of a small group given a faithful representation for the group.

For (a) it is planned to use Dixon's algorithm [9] which is expected to be applicable to groups of order up to about 10^5 having no more than 100 - 200 classes. In the case of (b) we will provide routines for forming the symmetrized product of characters, for the generation and analysis of "triangles", and for the induction and restriction of characters from subgroups and supergroups. These techniques are independent of the group order and are commonly used in hand/machine methods for determining the character tables of very large simple groups.

9. Implementation and portability

It is our desire that the system be sufficiently portable so that it can be implemented on a range of machines without excessive effort. At the same time useful group theoretic computations are often at the limit of current machine capabilities so that we have a conflict between the desire for machine independence and the need for efficient use of machine resources.

FORTRAN IV was chosen as the language for implementing the system in spite of its great handicaps, simply because no really suitable alternative exists. Since FORTRAN data types are unsuitable for algebraic computations, our first step was to provide a set of routines for manipulating the central data type met in the context of this work, namely, sets of vectors whose components are usually small integers and whose dimension is known at time of first reference but whose cardinality is unknown until all the elements of the set have been created. The fact that the components of the vectors are small means that these components should be packed several to a machine word on word type machines in order to avoid wasting large amounts of space. The collection of routines for storing and fetching elements of such sets, known collectively as the *Stack Handler*, is described in Cannon, Gallagher and McAllister [5].

An effort has been made to restrict most of the machine dependent parts of the GROUP to the Stack Handler. The two major exceptions to this are parts of the multiplication and Todd-Coxeter routines. So the major part of the effort involved in implementing GROUP on a new machine is in implementing the Stack Handler. Hopefully, the entire system when complete can be moved to a new machine with about three man-months of effort.

Once GROUP has been established on a new machine, it can be significantly speeded up by recoding sensitive parts in the host machine's assembly code.

10. Programming GROUP jobs

The current version of GROUP is controlled by a series of commands, interspersed with data, on cards. These commands allow for the definition of groups, the execution of the currently available algorithms and the printing of results.

A full scale programming language (tentatively called GALOIS) specifically designed for programming calculations in algebraic structures is currently under developement. Eventually GROUP jobs will be presented as GALOIS programs, thereby enormously enhancing the problem-solving power of the system to the user.

11. History

GROUP is the direct descendant of the group structure programmes developed at Kiel by the team headed by Neubüser over the period 1958-1969 and at Sydney by Cannon over the period 1965-1971. GROUP will also include the facilities of the GRAPPA system developed at Bell Telephone Laboratories by Dimino and Cannon during the period 1969-1971 (Dimino [7]).

GROUP is being developed cooperatively by the Department of Pure Mathematics, University of Sydney and Lehrstuhl D für Mathematik, RWTH Aachen.

References

[1] A.J. Bayes, J. Kautsky and J.W. Wamsley, "Computation in nilpotent groups (application)", these Proc.

[2] John J. Cannon, "Computing local structure of large finite groups", *Computers in algebra and number theory*, pp. 161-176 (SIAM-AMS Proceedings, 4. Amer. Math. Soc., Providence, Rhode Island, 1971).

[3] John J. Cannon, "Construction of defining relators for finite groups", *Discrete Math.* 5 (1973), 105-129.

[4] John J. Cannon, Lucien A. Dimino, George Havas and Jane M. Watson, "Implementation and analysis of the Todd-Coxeter algorithm", *Math. Comp.* 27 (1973), 463-490.

[5] John J. Cannon, Robyn Gallagher and Kim McAllister, "Stackhander; A scheme for processing packed dynamic arrays (Technical Report, 5, Computer-Aided Mathematics Project, Department of Pure Mathematics, University of Sydney, Sydney, November, 1972).

[6] John J. Cannon and George Havas, "Applications of the Todd-Coxeter algorithm", (Technical Report, 9, Computer-Aided Mathematics Project, Department of Pure Mathematics, University of Sydney, Sydney, June, 1973).

[7] Lucien A. Dimino, *GRAPPA: Group theoretic applications aid* (Bell Telephone Laboratories, Inc., Murray Hill, New Jersey, 1972).

[8] Lucien A. Dimino, Unpublished manuscript, 1970.

[9] John D. Dixon, "High speed computation of group characters", *Numer. Math.* 10 (1967), 446-450. MR37#325.

[10] Peter Dreyer, "Ein Program zur Berechnung der Auflösbaren Untergruppen von Permutationsgruppen", Diplomarbeit, Keil, 1970.

[11] Volkmar Felsch und Joachim Neubüser, "Ein Program zur Berechnung des Untergruppenverbandes einer endlichen Gruppe", *Mitt. Rh.-W. Inst. Math., Bonn* 2 (1963), 39-74.

[12] L. Gerhards and E. Altmann, "A computational method for determining the automorphism group of a finite solvable group", *Computational problems in abstract algebra* (Oxford, 1967), pp. 61-74 (Pergamon Press, Oxford, 1970). MR40#5374.

[13] G. Glauberman, "Global and local properties of finite groups", *Finite simple groups*, pp. 1-64 (Academic Press, London, New York, 1971).

[14] Marshall Hall, Jr, *The theory of groups* (The Macmillan Co., New York, 1959). MR21#1996.

[15] George Havas, "A Reidemeister-Schreier program", these Proc.

[16] H. Jürgensen, "Calculation with the elements of a finite group given by generators and defining relations", *Computational problems in abstract algebra* (Oxford, 1967), pp. 47-57 (Pergamon Press, Oxford, 1970). MR40#5374.

[17] John Leech, "Computer proof of relations in groups", *Proc. Conf. on group theory and computation* (Galway, 1973).

[18] Evelyne Lepique, "Ein Program zur Berechnung von Untergruppen von gegebenem Index in endlich präsentierten Gruppen", Diplomarbeit, Aachen, 1972.

[19] I.D. Macdonald, "On a class of finitely presented groups", *Canad. J. Math.* 14 (1962), 602-613. MR25#3992.

[20] I.D. Macdonald, "A computer application to finite p-groups", *J. Austral. Math. Soc.* 17 (1974), 102-112.

[21] Wilhelm Magnus, Abraham Karrass, Donald Solitar, *Combinatorial group theory* (Pure and Appl. Math. 13. Interscience [John Wiley & Sons], New York, London, Sydney, 1966). MR34#7617.

[22] J. Neubüser, "Untersuchungen des Untergruppenverbandes endlicher Gruppen auf einer programmgesteurten elektronischen Dualmaschine", *Numer. Math.* 2 (1960), 280-292. MR22#8713.

[23] J. Neubüser, "Computing moderately large groups: some methods and applications", *Computers in algebra and number theory*, pp. 183-190(SIAM-AMS Proceedings, 4. Amer. Math. Soc., Providence, Rhode Island, 1971).

[24] Charles C. Sims, "Computational methods in the study of permutation groups", *Computational problems in abstract algebra* (Oxford, 1967), pp. 169-183 (Pergamon Press, Oxford, 1970). MR41#1856.

[25] Charles C. Sims, "Determining the conjugacy classes of a permutation group", *Computers in algebra and number theory*, pp. 191-195 (SIAM-AMS Proceedings, 4. Amer. Math. Soc., Providence, Rhode Island, 1971).

[26] Charles C. Sims, "Computation with permutation groups", *Proc. Second Symposium on Symbolic and algebraic manipulation*, pp. 23-28 (Assoc. for Computing Machinery, New York, 1971).

[27] A. Sinkov, Personal communication, 1972.

[28] J.W. Wamsley, "Computation in nilpotent groups (theory)", these Proc.

University of Sydney,
Sydney, NSW 2006.

PROC. SECOND INTERNAT CONF. THEORY OF GROUPS, CANBERRA 1973, pp. 218-220.

20C20

ON THE MODULAR REPRESENTATIONS OF THE GENERAL LINEAR AND SYMMETRIC GROUPS

R.W. Carter and G. Lusztig

Let V be an n-dimensional vector space over $\mathbb{C}$ with basis $X_1, X_2, \ldots, X_n$ and let $T_r = V \otimes V \otimes \ldots \otimes V$ (r factors). T_r is a module both for $GL_n(\mathbb{C})$, via its action on V, and for the symmetric group S_r by place permutations.

Let $\lambda = (\lambda_1 \geq \lambda_2 \geq \ldots)$ be a partition of r with at most n parts and $\mu = (\mu_1 \geq \mu_2 \geq \ldots \geq \mu_k)$ be the dual partition of λ. Let $\phi_\lambda \in T_r$ be the product of 'non-commutative determinants' given by

$$\phi_\lambda = \begin{vmatrix} X_1 & X_1 & \ldots & X_1 \\ X_2 & X_2 & \ldots & X_2 \\ X_{\mu_1} & X_{\mu_1} & \ldots & X_{\mu_1} \end{vmatrix} \begin{vmatrix} X_1 & X_1 & \ldots & X_1 \\ X_2 & X_2 & \ldots & X_2 \\ X_{\mu_2} & X_{\mu_2} & \ldots & X_{\mu_2} \end{vmatrix} \ldots \begin{vmatrix} X_1 & X_1 & \ldots & X_1 \\ X_2 & X_2 & \ldots & X_2 \\ X_{\mu_k} & X_{\mu_k} & \ldots & X_{\mu_k} \end{vmatrix}$$

where

$$\begin{vmatrix} X_1 & X_1 & \ldots & X_1 \\ X_2 & X_2 & \ldots & X_2 \\ X_m & X_m & \ldots & X_m \end{vmatrix} = \sum_{\sigma \in S_m} \operatorname{sign} \sigma \cdot X_{\sigma(1)} \otimes X_{\sigma(2)} \otimes \ldots \otimes X_{\sigma(m)} .$$

Let $V_{\lambda,\mathbb{Z}}$ be the $\mathbb{Z}[GL_n(\mathbb{Z})]$-submodule of T_r generated by ϕ_λ and $\mathcal{V}_{\lambda,\mathbb{Z}}$ be the $\mathbb{Z}|S_r|$-submodule of T_r generated by ϕ_λ. Thus both $V_{\lambda,\mathbb{Z}}$ and $\mathcal{V}_{\lambda,\mathbb{Z}}$ are free abelian groups of finite rank. Let $V_\lambda = V_{\lambda,\mathbb{Z}} \otimes \mathbb{C}$ and $\mathcal{V}_\lambda = \mathcal{V}_{\lambda,\mathbb{Z}} \otimes \mathbb{C}$. Then V_λ is an irreducible $GL_n(\mathbb{C})$-module giving rise to a polynomial representation, and $\mathcal{V}_\lambda$ is an irreducible S_r-module. In fact as r varies the V_λ form a complete set of irreducible $GL_n(\mathbb{C})$-modules giving polynomial representations, and as n varies the $\mathcal{V}_\lambda$ form a complete set of irreducible S_r-modules.

We now perform a reduction modulo p of the modules V_λ, V_λ. Let K be an algebraically closed field of characteristic p and define $\bar{V}_\lambda$, $\bar{V}_\lambda$ by

$$\bar{V}_\lambda = V_{\lambda,Z} \otimes K , \quad \bar{V}_\lambda = V_{\lambda,Z} \otimes K .$$

Then $\bar{V}_\lambda$ is a $GL_n(K)$-module and $\bar{V}_\lambda$ is an S_r-module, but these modules are not in general irreducible. We consider under what circumstances there exist non-trivial $GL_n(K)$-homomorphisms between modules $\bar{V}_\lambda$ for distinct λ, and non-trivial S_r-homomorphisms between modules $\bar{V}_\lambda$ for distinct λ.

Let $\hat{T}$ be the set of n-tuples $m = (m_1, m_2, \ldots, m_n)$ with $m_i \in Z$ and $\hat{T}_+$ be the subset of $\hat{T}$ satisfying $m_1 \geq m_2 \geq \ldots \geq m_n \geq 0$. Partitions with at most n parts may be regarded as elements of $\hat{T}_+$. $\hat{T} \otimes R$ has a natural Euclidean structure defined by $(m, m') = \sum_{i=1}^{n} m_i m_i'$. We define certain affine hyperplanes $L_j^i(k)$ in $\hat{T} \otimes R$ where $1 \leq i < j \leq n$ and $k \in Z$. Let

$$L_j^i(k) = \{m;\ (m_i - i) - (m_j - j) = kp\}$$

and let $S_j^i(k)$ denote the reflection in the affine hyperplane $L_j^i(k)$. The group W_a of affine orthogonal transformations of $\hat{T} \otimes R$ generated by the elements $S_j^i(k)$ for all i, j, k is called the affine Weyl group. If $\lambda, \lambda' \in \hat{T}_+$ are related by $\lambda = S_j^i(k)\lambda'$ and if $(\lambda_i - i) - (\lambda_j - j) = kp + d$ where $d \geq 0$, then we have

$$\lambda_i' = \lambda_i - d , \quad \lambda_j' = \lambda_j + d , \quad \lambda_h' = \lambda_h \quad \text{for } h \neq i, j .$$

We define a partial ordering on partitions in $\hat{T}_+$ saying that $\lambda \geq \lambda'$ if $\sum_{i=1}^{r} \lambda_i \geq \sum_{i=1}^{r} \lambda_i'$ for all $r = 1, 2, 3, \ldots$. Let $\lambda, \lambda' \in \hat{T}_+$. We write $\lambda' \uparrow\uparrow \lambda$ if $\lambda' \leq \lambda$ and $\lambda = S_j^i(k)\lambda'$ for some i, j, k. We write $\lambda' \uparrow \lambda$ if $\lambda' \uparrow\uparrow \lambda$ and there do not exist $\lambda^{(1)}, \lambda^{(2)}, \ldots, \lambda^{(h)} \in \hat{T}_+$, $h \geq 1$, such that

$$\lambda' \uparrow\uparrow \lambda^{(1)} \uparrow\uparrow \lambda^{(2)} \ldots \uparrow\uparrow \lambda^{(h)} \uparrow\uparrow \lambda .$$

$\uparrow$ is called an indecomposable p-admissible raising operator for partitions. Our results on the existence of non-trivial homomorphisms are expressed in terms of these raising operators. $\lambda \in \hat{T}_+$ is called p-regular if λ does not lie in $L_j^i(k)$ for any i, j, k.

THEOREM 1. *(i) Suppose* $\lambda, \lambda' \in \hat{T}_+$ *and* $\hom_{GL_n(K)}(\bar{V}_{\lambda'}, \bar{V}_\lambda) \neq 0$. *Then* $\lambda' \leq \lambda$ *and* $\lambda' = \omega(\lambda)$ *for some* $\omega \in W_a$.

(ii) Suppose $\lambda, \lambda' \in \hat{T}_+$ *are p-regular and* $\lambda' \uparrow \lambda$. *Then* $\hom_{GL_n(K)}(\bar{V}_{\lambda'}, \bar{V}_\lambda) \neq 0$.

THEOREM 2. *(i) Suppose* λ, λ' *are partitions of* r *and* $\hom_{S_r}(\bar{V}_\lambda, \bar{V}_{\lambda'}) \neq 0$. *Then, provided* $p \neq 2$, *we have* $\lambda' \leq \lambda$ *and* $\lambda' = \omega(\lambda)$ *for some* $\omega \in W_a$.

(ii) Suppose λ, λ' *are partitions of* r *which are p-regular and satisfy* $\lambda' \uparrow \lambda$. *Then* $\hom_{S_r}(\bar{V}_\lambda, \bar{V}_{\lambda'}) \neq 0$.

Part *(i)* of Theorem 1 was conjectured by D.N. Verma and proved by J.E. Humphreys for $p > n$. We conjecture that in fact $\dim \hom_{GL_n(K)}(\bar{V}_{\lambda'}, \bar{V}_\lambda) = 1$ under the hypothesis of Theorem 1 *(ii)*; and that $\dim \hom_{S_r}(\bar{V}_\lambda, \bar{V}_{\lambda'}) = 1$ under the hypothesis of Theorem 2 *(ii)* provided $p \neq 2$.

The proof of the above results involves Kostant's $\mathbb{Z}$-form $U_{\mathbb{Z}}$ of the universal enveloping algebra of the Lie algebra of $GL_n(\mathbb{C})$. It is necessary to construct certain elements of $U_{\mathbb{Z}}$, one for each positive root, which possess favourable commutation properties. Details of the proof will appear in a forthcoming paper.

University of Warwick,
Coventry CV4 7AL, England.

PROC. SECOND INTERNAT. CONF. THEORY OF GROUPS, CANBERRA 1973, pp. 221-225. 20F99

WORDS WHICH GIVE RISE TO ANOTHER GROUP OPERATION FOR A GIVEN GROUP

C.D.H. Cooper

1. Introduction

For any group G, the binary operation $x * y = yx$ is a group operation, expressed in terms of the word yx, which gives a different (unless G is abelian) but isomorphic group structure on the set G. It is natural to ask what other group structures may be defined on G by an operation of the form

$$x * y = W(x, y)$$

where $W(x, y)$ is a word in x, y (we call such words *group words* for G) and whether it is possible for the new group (denoted by G_W) and G to be non-isomorphic. If so, it might be possible to discover facts about one group by considering the other - especially if one group is abelian and the other is not. Thoughts such as these have prompted this paper.

Most of the work that has been done previously on group words has been restricted to what we may call *non-singular* group words - those for which $G = (G_W)_Y$ for some group word Y (evaluated in G_W). Higman and Neumann [8] asked whether $G \cong G_W$ for all non-singular group words W and Hulanicki and Swierczkowski [9] showed that this is true for periodic nilpotent groups of class 2. More recently, Street [11]* not only extended this to "most" nilpotent groups of class 3 and 4 (including, for example, all finite p-groups of class at most 4 for $p \geq 5$) but also gave an infinite family of counter-examples, the smallest of which has order 602.

In this paper we investigate non-abelian groups G for which there is a group word W (necessarily *singular*) for which G_W is abelian. We call such groups *verbally abelian*. Baer [2] pointed out a long time ago that if G is a nilpotent group of class 2 such that G' has finite odd exponent n and

* I am grateful to N.D. Gupta and P.M. Weichsel for drawing my attention to this paper.

$W(x, y) = xy[x, y]^{(n-1)/2}$, then G_W is abelian. This fact proved useful in a paper of Bender [3]. We extend this here to nilpotent groups of class 3 . In addition we discuss, for a general group G , the relationship between the properties of G and G_W .

2. Nilpotent groups of class at most 3

Similar calculations to those of Street [11] yield the following.

THEOREM 1. *The group words for a nilpotent group* G *of class at most* 3 *are those which are equivalent to (that is, are identical to when evaluated in* G *)*

(1) $$xy[x, y]^r[x, y, x]^{-s}[x, y, y]^s$$

for some integers r, s *satisfying the congruence*

(2) $$3s \equiv r^2 + r \pmod{\operatorname{expt}[G', G]} .$$

If W *is such a group word for* G , G_W *is abelian if and only if*

(3) $$2r + 1 \equiv 0 \pmod{\operatorname{expt} G'} .$$

COROLLARY. *If* G *is a nilpotent group of class at most* 3 *then* G *is verbally abelian if and only if* $2 \nmid \operatorname{expt} G'$ *and* $3 \nmid \operatorname{expt}[G', G]$.

From this corollary we see that if G is a finite nilpotent group of class at most 3 then it is verbally abelian if and only if its Sylow subgroups are regular p-groups. There are, however, regular p-groups of class 4 which are not verbally abelian. For example, if $p \geq 7$ and G is the free product of two cyclic groups of order p then it can be shown that $G/[G', G, G, G]$ is not verbally abelian, yet it is a finite regular p-group of class 4 .

If G_W is abelian and the operation is written additively, one asks what words may be used to define a multiplication making G into a ring. A related question was answered in the case of associative rings coming from class 2 groups in [1], but, as the next theorem shows, it is more natural for this question to ask for a Lie ring rather than an associative one.

THEOREM 2. *Suppose* G *is a nilpotent group of class at most* 3 *for which an abelian group operation,* + *, is given by* (1), (2), (3). *The words which produce an operation which is distributive over* + *are those equivalent to*

$$[x, y]^t[x, y, x]^{rt}[x, y, y]^{rt}$$

for some integer t . *Moreover, for any such word,* G *becomes a Lie ring.*

3. Relationship between G and G_W

The following theorem suggests that G_W is at least as "nice" a group as G and leads one to make the following conjecture:

CONJECTURE G_W *is contained in the variety generated by* G .

THEOREM 3. *Suppose* G *is a group and* W *is a group word for* G . *Then*

(i) *(normal) subgroups of* G *are (normal) subgroups of* G_W ,

(ii) $\mathrm{aut}(G) \leq \mathrm{aut}(G_W)$,

(iii) *elements have the same order in* G *as in* G_W ,

(iv) *the descending central series and derived series descend at least as quickly in* G_W *as in* G *and the ascending central series and Fitting series ascend at least as quickly in* G_W *as in* G ,

(v) *if* G *is abelian, nilpotent or soluble so is* G_W .

These facts are not surprising and are easily proved (*cf.* Street [11], Theorem 1). A little less obvious are the following which proceed in the opposite direction.

THEOREM 4. *Suppose* G *is a group and* W *is a group word for* G . *Then*

(i) *if* G *is locally nilpotent, characteristic subgroups of* G_W *are characteristic subgroups of* G ,

(ii) *if* G *is finite,* G *and* G_W *have the same Sylow subgroups and the same Fitting series,*

(iii) *if* G *is finite, and* G_W *is nilpotent or soluble then so is* G .

4. Finite verbally abelian groups

Suppose that G is a finite verbally abelian group. Theorem 4 *(iii)* shows that G must be nilpotent and Theorem 1 (Corollary) gives examples of verbally abelian groups up to class 3 . Theorem 3 *(iii)* shows that the abelian group G_W must be unique up to isomorphism. The following facts are obvious for finite abelian groups and may be extended to finite verbally abelian groups using Theorems 3 and 4.

THEOREM 5. *Suppose* G *is a finite verbally abelian group. Then*

(i) *for all integers* n , $\{g \mid g^n = 1\}$ *and* $\{g^n \mid g \in G\}$ *are characteristic subgroups of* G , *and all elements which have* n*'th roots have the same number of them,*

(ii) *if* θ *is an automorphism of* G *such that* $H^\theta = H$ *for all subgroups* H , *then* θ *has the form* $x \to x^n$ *for some integer* n ,

(iii) G *is conformal with (that is, has the same number of elements of each order as) some abelian group.*

We now return to the original motive for studying group words - knowing something about G_W we might be able to translate it back to the "not-so-nice" group G . We have done this to a small extent in being able to show by Theorem 1 (Corollary) that *(i)* - *(iii)* of Theorem 5 hold for finite p-groups of class less than $\min(p, 4)$. However such groups are precisely the finite regular p-groups of class at most 3 and *(i)* - *(iii)* of Theorem 5 are already known for all finite regular p-groups (see [7] for *(i)*, [5] for *(ii)*, [4] for *(iii)*). So purely as a tool, group words have not (as yet) justified their existence. However they raise a number of interesting questions about themselves. For example, the fact that finite verbally abelian p-groups have very similar arithmetic properties to finite regular p-groups suggests that perhaps they are in fact always regular (as we pointed out above, the converse is false).

References

[1] J.C. Ault and J.F. Watters, "Circle groups of nilpotent rings", *Amer. Math. Monthly* 80 (1973), 48-52.

[2] Reinhold Baer, "Groups with abelian central quotient group", *Trans. Amer. Math. Soc.* 44 (1938), 357-386. FdM64,68.

[3] Helmut Bender, "Über den grössten p'-Normalteiler in p-auflösbaren Gruppen", *Arch. der Math.* 18 (1967), 15-16. MR35#4303.

[4] C.D.H. Cooper, "Conformality and p-isomorphism in finite nilpotent groups", *J. Austral. Math. Soc.* 7 (1967), 165-171. MR35#4304.

[5] Christopher D.H. Cooper, "Power automorphisms of a group", *Math. Z.* 107 (1968), 335-356. MR38#4550.

[6] S. Fajtlowicz, "On fundamental operations in groups", *J. Austral. Math. Soc.* 14 (1972), 445-447.

[7] Marshall Hall, Jr., *The theory of groups* (The Macmillan Co., New York, 1959). MR21#1996.

[8] Graham Higman and B.H. Neumann, "Groups as groupoids with one law", *Publ. Math. Debrecen* 2 (1952), 215-221. MR15,284.

[9] A. Hulanicki and S. Świerczkowski, "On group operations other than xy or yx ", *Publ. Math. Debrecen* 9 (1962), 142-148. MR25#5101.

[10] Hanna Neumann, "On a question of Kertész", *Publ. Math. Debrecen* 8 (1961), 75-78. MR24#A1303.

[11] Anne Penfold Street, "Subgroup-determining functions on groups", *Illinois J. Math.* 12 (1968), 99-120. MR36#3885.

Macquarie University,
North Ryde, NSW 2113.

PROC. SECOND INTERNAT. CONF. THEORY OF GROUPS, 20D10
CANBERRA 1973, pp. 226-237.

CLASSES OF FINITE SOLUBLE GROUPS

John Cossey

1. Closed classes

I want to give here a rather biased account of recent work in the theory of classes of finite soluble groups. I will be concentrating on results which have something to say about the classes themselves, rather than results which use the classes to obtain a picture of the internal structure of finite soluble groups. My main excuse for doing so is that this part of the theory is at a very interesting stage: the classes are proving to be more exotic than might have been expected, and though we know little about them, some results and techniques are appearing, and it seems likely we will not remain so ignorant for long.

The theory started with saturated formations in 1963, though a number of earlier papers gave hints of what was coming: see for example the papers of Hall [29], Baer [1], and Carter [12]. Gaschütz introduced saturated formations in [24] to give a unified account of the theories of Hall [28] and Carter, [13] and [14]. Hall and Carter established the existence of certain characteristic conjugacy classes of subgroups in finite soluble groups (the Hall and Carter subgroups).

We start with formations: a formation $\underline{\underline{X}}$ is a class of groups (from now on, all groups will be finite and soluble) such that $Q\underline{\underline{X}} \subseteq \underline{\underline{X}}$ and $R_0\underline{\underline{X}} \subseteq \underline{\underline{X}}$, where

$$Q\underline{\underline{X}} = \{G : G \text{ is an epimorphic image of some } H \in \underline{\underline{X}}\} ,$$

$$R_0\underline{\underline{X}} = \{G : G \trianglerighteq N_i,\ i = 1, \ldots, n,\ G/N_i \in \underline{\underline{X}},\ \cap N_i = 1\} .$$

A formation is said to be saturated if in addition $E_\phi\underline{\underline{X}} \subseteq \underline{\underline{X}}$, where

$$E_\phi\underline{\underline{X}} = \{G : G \trianglerighteq N,\ N \subseteq \phi G,\ G/N \in \underline{\underline{X}}\} .$$

Two easy examples of saturated formations are the class $\underline{\underline{S}}_\pi$ of all π-groups, for some set π of primes, and the class of all nilpotent groups $\underline{\underline{N}}$. The fundamental theorem of Gaschütz and Lubeseder ([27] and [41]) that a formation is saturated if and only if it has a local definition is the most important single tool for proving results about and constructing examples of saturated formations: for an account of this theorem, see Huppert [35] VI, 7.

It was the following property that gave saturated formations their importance. Let G be a group, and $\underline{\underline{X}}$ a class of groups: we call a subgroup H of G an $\underline{\underline{X}}$-projector of G if H is a maximal $\underline{\underline{X}}$-subgroup, and for every normal subgroup N of G, NH/N is a maximal $\underline{\underline{X}}$-subgroup of G/N. Gaschütz showed in [24] that if $\underline{\underline{X}}$ is a saturated formation, $\underline{\underline{X}}$-projectors always exist, and they are all conjugate. With $\underline{\underline{X}} = \underline{\underline{S}}_\pi$ we get the Hall π-subgroups, and with $\underline{\underline{X}} = \underline{\underline{N}}$ we get the Carter subgroups.

At this stage, a natural question arises. Call a class $\underline{\underline{X}}$ for which $\underline{\underline{X}}$-projectors always exist and are all conjugate a projective class: if $\underline{\underline{X}}$ is a projective class is it a saturated formation? Schunck answered this question negatively [45], and gave a characterisation of projective classes: they are now called Schunck classes. Denote by $\underline{\underline{P}}$ the class of all primitive groups (recall that a primitive group is one which has a self-centralising minimal normal subgroup): then $\underline{\underline{X}}$ is a Schunck class if and only if $Q\underline{\underline{X}} \subseteq \underline{\underline{X}}$, and $G \in \underline{\underline{X}}$ whenever $Q\{G\} \cap \underline{\underline{P}} \subseteq \underline{\underline{X}}$. An example of a Schunck class that is not a formation is given by $\underline{\underline{Q}}_\pi$, the class of those groups which have no π-group as quotient.

Fischer attempted to dualize Gaschütz's theory of formations ([21], [22]). Though new characteristic conjugacy classes of subgroups were obtained (Fischer, Gaschütz, Hartley [23]), the dualization did not work too well. Hartley [30] showed that though a local definition could be used to provide examples, there was no hope of finding a theorem like Gaschütz and Lubeseder's, and these classes remained little understood until very recently. Fischer introduced two types of closed classes. A Fitting class is a class $\underline{\underline{X}}$ such that $S_n\underline{\underline{X}} \subseteq \underline{\underline{X}}$ and $N_0\underline{\underline{X}} \subseteq \underline{\underline{X}}$, where

$$S_n\underline{\underline{X}} = \{G : G \lhd\lhd H \in \underline{\underline{X}}\},$$

$$N_0\underline{\underline{X}} = \{G : G = \langle N_1, \ldots, N_n\rangle, N_i \lhd\lhd G, N_i \in \underline{\underline{X}}\}.$$

A Fischer class is a class $\underline{\underline{X}}$ such that $N_0\underline{\underline{X}} \subseteq \underline{\underline{X}}$, and $F\underline{\underline{X}} \subseteq \underline{\underline{X}}$, when

$$F\underline{\underline{X}} = \{G : G \leq H \in \underline{\underline{X}}, G/\text{core } G \in \underline{\underline{N}}\}.$$

Again, $\underline{\underline{S}}_\pi$ and $\underline{\underline{N}}$ provide examples of Fitting classes: they are also Fischer classes. Let $\underline{\underline{C}}_\pi$ denote the class of groups in which every minimal normal π-subgroup is central; then $\underline{\underline{C}}_\pi$ is a Fitting class which is not a Fischer class, nor a formation (though $R_0\underline{\underline{C}}_\pi \subseteq \underline{\underline{C}}_\pi$).

We have used closure operations to define most of the classes above. A closure operation A is a map from classes of groups to classes of groups satisfying $A^2\underline{\underline{X}} = A\underline{\underline{X}} \supseteq \underline{\underline{X}}$, and if $\underline{\underline{X}} \subseteq \underline{\underline{Y}}$, $A\underline{\underline{X}} \subseteq A\underline{\underline{Y}}$. Schunck classes have not been defined in terms of closure operations, and Gaschütz [25] asked if a class $\underline{\underline{X}}$ satisfying $Q\underline{\underline{X}} \subseteq \underline{\underline{X}}$, $E_\phi\underline{\underline{X}} \subseteq \underline{\underline{X}}$, $D_0\underline{\underline{X}} \subseteq \underline{\underline{X}}$, where

$$D_0\underline{\underline{X}} = \{G : G \cong D_1 \times \ldots \times D_n, D_i \in \underline{\underline{X}}\}$$

is a Schunck class. Hawkes [33] recently gave a negative answer to this question, and produced a new closure operation $\bar{R}_0$ such that $\underline{\underline{X}}$ is a Schunck class if and only if $Q\underline{\underline{X}} \subseteq \underline{\underline{X}}$, $E_\phi\underline{\underline{X}} \subseteq \underline{\underline{X}}$, $\bar{R}_0\underline{\underline{X}} \subseteq \underline{\underline{X}}$.

In investigating the classes above, it is interesting to look at relations between them obtained by adding extra closure properties or by restricting the class of groups considered. The earliest results in this direction are that a formation is a Schunck class if and only if it is saturated, and the result of P.M. Neumann [44] that a formation of nilpotent groups is subgroup closed (see also [5] for a more formation-theoretic proof).

Fischer [22] investigated Fitting formations, that is, classes that are both formations and Fitting classes, and deduced a number of their elementary properties. The next significant contribution came from Hawkes [31]. He showed that a metanilpotent Fitting formation is saturated and subgroup closed, and so has a local definition of a particularly nice form. He also produced a Fitting formation of nilpotent length three which is not saturated. Bryce and Cossey [6] then proved that a subgroup closed Fitting formation is always saturated, and a saturated Fitting formation of nilpotent length three is always subgroup closed. We used Hawkes' example to give an example of a saturated formation of nilpotent length four which is not subgroup closed. Subgroup closed Fitting formations turn out to be more than just saturated. Hawkes [32], in another context, introduced primitive saturated formations: they are defined as follows. Let F_0 denote the family consisting of the empty set, the formation of groups of order one, and the formation of all soluble groups, and for $i > 0$, define F_i inductively by $\underline{\underline{X}} \in F_i$ if $\underline{\underline{X}} \in F_{i-1}$, or $\underline{\underline{X}}$ has a local definition using formations from F_{i-1} . Finally let F be the family comprising all formations $\underline{\underline{X}}$ such that $\underline{\underline{X}} = \bigcup_j \underline{\underline{X}}_j$, with each $\underline{\underline{X}}_j \in \bigcup_i F_i$, and $\underline{\underline{X}}_j \subseteq \underline{\underline{X}}_{j+1}$. Then we call a formation $\underline{\underline{F}}$ primitive if $\underline{\underline{F}} \in F$. It turns out that the primitive saturated formations are just the subgroup closed Fitting formations. It seems likely that subgroup closed Fitting formations can be characterised in another way.

QUESTION 1. *Can every subgroup closed saturated formation be obtained from the family of* $\underline{\underline{S}}_\pi$*'s by the processes of forming products (of classes) and (possibly infinite) intersections?*

This is a variant of [25] Problem 8.5 (which has a negative answer as it stands).

In proving that a subgroup closed Fitting formation is saturated, Q-closure plays a minor role, and one is lead to ask

QUESTION 2. *Is a subgroup closed Fitting class a formation?*

If so, it is also saturated: a positive answer seems most unlikely. However, metanilpotent subgroup closed Fitting classes are formations ([7]: the result can also be deduced without much trouble from the arguments in [31]).

There are a number of troublesome questions of this nature.

QUESTION 3. *Is a Q-closed Fitting class a formation?*

QUESTION 4. *Is a $\{Q, E_\phi\}$-closed Fitting class a formation?*

Question 4 is rather interesting: such a Fitting class is almost a Schunck class, and there is a nice result of Doerk [20] that says: if τ is a closure operation such that $\tau\underline{X} = \underline{X}$ implies $\tau\underline{\underline{N}}\underline{X} = \underline{\underline{N}}\underline{X}$, and if for all such $\underline{X}$, $\tau\underline{X}$ is a Schunck class, $\tau\underline{X}$ is a formation. Thus, Question 4 is equivalent to

QUESTION 5. *Is a $\{Q, E_\phi\}$-closed Fitting class a Schunck class?*

For most of these questions there is some positive evidence: see [7] and [20].

Rather surprisingly, it turns out to be easy to produce an E_ϕ-closed Fitting class that is not a formation, as we shall see later. However, E_ϕ-closed metanilpotent Fitting classes are formations.

Makan [42] has shown that metanilpotent Fischer classes are subgroup closed: this is not true in general, but the answer to the following question is still not known.

QUESTION 6. *Are Fischer classes R_0-closed?*

2. Fitting classes

The structure of Fitting classes is not very well understood: this is probably because we have little knowledge of or expertise in handling normal products. One of the oldest questions in the theory of Fitting classes is to decide if $\mathrm{Fit}(S_3)$, the smallest Fitting class containing S_3, is $\underline{\underline{S}}_3\underline{\underline{S}}_2$: this has only recently been settled negatively by Camina [10].

Camina's proof uses a variation of an example of Blessenohl and Gaschütz [4]. Let $O(G)$ be the largest odd order normal subgroup of G, and I a non-empty set of positive integers. Denote by $O_I(G)$ the set of all elements in $O(G)$ whose orders come from the set I. If $O_I(G)$ is non-empty, G acts on $O_I(G)$ by conjugation as a permutation group. Now define

$\underline{\underline{X}}_I = \{G : O_I(G)$ is empty, or G acts on $O_I(G)$ as an even group of permutations$\}$.

Then $\underline{\underline{X}}_I$ is a Fitting class. The example of Blessenohl and Gaschütz is obtained by taking I to be the set of all positive integers.

Now, for a fixed $n > 1$, set $I = \{3^n\}$. Then $S_3 \in \underline{\underline{X}}_I$. On the other hand $D_{2.3^n} \notin \underline{\underline{X}}_I$, for there are (3^n-3^{n-1}) elements of order precisely 3^n in this group, and any element of order two acts by conjugation as a product of 3^{n-1} involutions, that is, an odd permutation. Thus $D_{2.3^n} \notin \mathrm{Fit}(S_3)$ for any $n > 1$.

On the other hand, using techniques developed by Bryce and Cossey in [7], it can be shown that $C_3 \text{ wr } S_3 \in \mathrm{Fit}(S_3)$. A description of the groups in $\mathrm{Fit}(S_3)$ remains an intriguing problem.

Camina's examples are examples of normal Fitting classes. A Fitting class $\underline{\underline{X}}$ will be called normal if $\underline{\underline{X}} \neq \{1\}$, and for every group G, $G_{\underline{\underline{X}}}$, the largest normal $\underline{\underline{X}}$-subgroup or $\underline{\underline{X}}$-radical of G, is a maximal $\underline{\underline{X}}$-subgroup. We know that $\underline{\underline{N}} \subseteq \underline{\underline{X}}$ for every normal Fitting class $\underline{\underline{X}}$, and that the intersection of normal Fitting classes is normal [4]. Thus there is a unique minimal normal Fitting class, which we will call $\underline{\underline{H}}$. We know very little about $\underline{\underline{H}}$, especially about the groups it contains. It follows from Camina's examples that $D^p_{q^n} \notin \underline{\underline{H}}$, for any prime $p|q-1$, and prime $q \equiv 3 \bmod 4$. ($D^p_{q^n}$ is the group with a unique minimal normal subgroup, a normal abelian Sylow q subgroup of exponent q^n, and a cyclic Hall q'-subgroup of order p). From other examples in [4], $D^p_{q^n} \notin \underline{\underline{H}}$ for p, q primes with $p|q-1$ and $n \not\equiv 0 \bmod p$. It seems unlikely that all these restrictions on p and q are necessary.

CONJECTURE 1. *$D^p_{q^n} \notin \underline{\underline{H}}$ for p, q primes with $p|q-1$.*

In contrast, from [4] we have $G' \in \underline{\underline{H}}$ for all G, and from this we can deduce $D^p_q \in \underline{\underline{H}}$ for p, q primes with $p \nmid q-1$. For we may regard D^p_q as the additive group of a finite field extended by a subgroup of its multiplicative group. The Galois automorphism of the field may be extended to an automorphism σ of the group, and then D^p_q is the derived group of $D^p_q\langle\sigma\rangle$. This can be easily extended to show that $(\underline{\underline{A}} \cap \underline{\underline{S}}_q)\underline{\underline{S}}_p \subseteq \underline{\underline{H}}$ where $\underline{\underline{A}}$ is the class of abelian groups. Again, this seems unlikely to be the full story.

CONJECTURE 2. $\underline{\underline{S}}_q\underline{\underline{S}}_p \subseteq \underline{\underline{H}}$ *for* p, q *primes with* $p \nmid q-1$.

Normal Fitting classes are plentiful, and to demonstrate this I will use products of Fitting classes. We need to be careful without definition, as the usual product of classes does not always give a Fitting class. If $\underline{\underline{X}}$ and $\underline{\underline{Y}}$ are Fitting classes, let $\underline{\underline{XY}}_{\text{Fit}} = \{G : G/G_{\underline{\underline{X}}} \in \underline{\underline{Y}}\}$. It is easy to check that $\underline{\underline{XY}}_{\text{Fit}}$ is a Fitting class. Its usefulness here comes from the fact that $\underline{\underline{XY}}_{\text{Fit}}$ is normal if and only if $\underline{\underline{X}}$ or $\underline{\underline{Y}}$ is normal. From [15] we can deduce that if π_1 and π_2 are non empty sets of primes with $\pi_1 \neq \pi_2$, then $\underline{\underline{N}}_{\pi_1}\underline{\underline{H}}_{\text{Fit}} \neq \underline{\underline{N}}_{\pi_2}\underline{\underline{H}}_{\text{Fit}}$, and $\underline{\underline{N}}_{\pi_i}\underline{\underline{H}}_{\text{Fit}} \neq \underline{\underline{S}}$ the class of all soluble groups, $i = 1, 2$. As an immediate corollary, we get that there are uncoutably many normal Fitting classes. From [15] also, we have $\underline{\underline{NH}}_{\text{Fit}}$ is a Frattini closed normal Fitting class not equal to $\underline{\underline{S}}$. We will see soon that a normal Fitting class is a formation if and only if it is $\underline{\underline{S}}$, and so we have the example promised in §1.

Suppose now that $\underline{\underline{X}}$ is maximal in the partial order given by inclusion. Then we have $\underline{\underline{X}} \subset \underline{\underline{XN}}$, and hence $\underline{\underline{XN}} = \underline{\underline{S}}$ but then $\underline{\underline{X}}$ is normal. It may be deduced from Lausch [36] that the examples given earlier are either maximal or equal to $\underline{\underline{S}}$. However, we can have normal Fitting classes that are not contained in any maximal Fitting class (examples are given in [9]).

As a fairly easy application of our knowledge of normal Fitting classes, consider the following result of Dark [17]: if $S_3 \lhd\lhd G$, then $G' \cap S_3 = S_3'$. This comes immediately from the fact that $S_3 \notin \underline{\underline{H}}$, and a special case of Satz 1 of Gaschütz [26], that if $N \lhd\lhd G$, $N \cap G' \leq N_{\underline{\underline{H}}}$.

One of the peculiar features of normal Fitting classes is that the radical of a direct product need not be the direct product of the radicals. Let $\underline{\underline{Y}}$ be the normal Fitting class of Blessenohl and Gaschütz given earlier. Then $S_3 \notin \underline{\underline{Y}}$, and hence $(S_3)_{\underline{\underline{Y}}} \times (S_3)_{\underline{\underline{Y}}}$ has index 4 in $S_3 \times S_3$: but $(S_3 \times S_3)_{\underline{\underline{Y}}}$ has index at most 2 . In a search for conditions which would ensure that the radical of a direct product was the direct product of the radicals, Lockett, [37] and [40], introduced a new operation which has significance far beyond this question.

To define Lockett's operation, I need some notation. Let G^n be the direct product of n copies of G , and let θ_G denote the projection of G^n onto its first factor. For a Fitting class $\underline{\underline{X}}$, we define $\underline{\underline{X}}^*$ by

$$\underline{\underline{X}}^* = \left\{G \in \underline{\underline{S}} : \text{there is an } n \text{ with } (G^n)_{\underline{\underline{X}}}\theta_G \cong G\right\} .$$

Clearly $\underline{\underline{X}} \subseteq \underline{\underline{X}}^*$: Lockett proves that $\underline{\underline{X}}^*$ is also a Fitting class, and that $(G \times H)_{\underline{\underline{X}}} = G_{\underline{\underline{X}}} \times H_{\underline{\underline{X}}}$ for all groups G and H if and only if $\underline{\underline{X}} = \underline{\underline{X}}^*$. We will say a Fitting class $\underline{\underline{X}}$ is a Lockett class if $\underline{\underline{X}}^* = \underline{\underline{X}}$. Lockett [40] establishes that the Fitting class $\underline{\underline{X}}$ is a Lockett class if $\underline{\underline{X}}$ is any of the following: a Fischer class, R_0-closed, S-closed, Q-closed (and hence if $\underline{\underline{X}}$ is a formation).

In view of the questions we have already asked, the following seems a natural one to ask.

QUESTION 7. *Is a Lockett class R_0-closed?*

Lockett classes are better behaved than Fitting classes in general: the fact that the radical of a direct product is the direct product of the radicals is a very useful fact: also useful is the fact that $\underline{\underline{X}}$ is normal if and only if $\underline{\underline{X}}^* = \underline{\underline{S}}$. Experimental evidence lead Lockett to make the following conjecture.

CONJECTURE 3. *Let $\underline{\underline{X}}_0$ be the smallest normal Fitting class containing $\underline{\underline{X}}$. Then $\underline{\underline{X}} = \underline{\underline{X}}^* \cap \underline{\underline{X}}_0$.*

I want to finish this section with two remarks: one philosophical, one technical.

Firstly, if we restrict the questions asked above (including §1) about Fitting classes to metanilpotent Fitting classes, we get a program for classifying metanilpotent Fitting classes, which I feel has a good chance of success.

Secondly, Gaschütz ([25] Problem 8.6) asked the following question: given a group G , with a normal subgroup $Z \leq Z(G) \cap \phi G$, and a Fitting class $\underline{\underline{X}}$: if $G/Z \in \underline{\underline{X}}$ is $G \in \underline{\underline{X}}$? Lockett [39], modifying examples of Dark [18], gave a negative answer: another, essentially different negative answer is given in Cossey and Gaschütz [15]. However, both examples are of nilpotent length three, and this fact is essential to both. Hence

QUESTION 8. *Let G be metanilpotent, $Z \leq Z(G) \cap \phi G$, $\underline{\underline{X}}$ a Fitting class. If $G/Z \in \underline{\underline{X}}$, is $G \in \underline{\underline{X}}$?*

3. Schunck classes

The natural partial order by inclusion has a major defect for Schunck classes. If $\underline{\underline{X}} \subseteq \underline{\underline{Y}}$, it is not necessarily true that an $\underline{\underline{X}}$-projector of a group G is contained in a $\underline{\underline{Y}}$-projector of G . Consider $\underline{\underline{X}} = \underline{\underline{S}}_2$, $\underline{\underline{Y}} = \underline{\underline{S}}_3\underline{\underline{S}}_2$, and $G = S_4$. Then all $\underline{\underline{X}}$-projectors of G are isomorphic to D_8 , while the $\underline{\underline{Y}}$-projectors are isomorphic to S_3 . We can define a different partial order on the family of Schunck classes by setting $\underline{\underline{X}} \ll \underline{\underline{Y}}$ if for every group G , an $\underline{\underline{X}}$-projector of G is contained in a $\underline{\underline{Y}}$-projector of G . The properties of this partial order were investigated first by

Cline [14] for saturated formations, and by Doerk [19] for Schunck classes. Doerk's methods are very interesting, and I want to say more about them later.

Doerk shows that if $\underline{\underline{X}} \ll \underline{\underline{Y}}_i$, $i \in I$, then $\underline{\underline{X}} \ll \bigcap_i \underline{\underline{X}}_i$, and hence we may define a least upper bound $\underline{\underline{U}}$ for a family of Schunck classes $\underline{\underline{Y}}_i$, $i \in I$ by setting

$$\underline{\underline{U}} = \bigcap\{\underline{\underline{V}} : \underline{\underline{Y}}_i \ll \underline{\underline{V}}, \forall i \in I\} .$$

Blessenohl has defined a "composite" of Schunck classes $\underline{\underline{Y}}_i$ by

$$\langle \underline{\underline{Y}}_i : i \in I \rangle = \{G : S_i \text{ is a } \underline{\underline{Y}}_i\text{-projector of } G, i \in I, \text{ implies } G = \langle S_i : i \in I \rangle\} .$$

Blessenohl has shown that this composite is a Schunck class, and hence that a $\langle \underline{\underline{Y}}_i : i \in I \rangle$-projector of G is generated by a set of $\underline{\underline{Y}}_i$-projectors of G , for any group G (see Gaschütz [25]). Hence $\langle \underline{\underline{Y}}_i : i \in I \rangle$ is the least upper bound of the $\underline{\underline{Y}}_i : i \in I$.

If we now define $\underline{\underline{X}} \wedge \underline{\underline{Y}} = \langle \underline{\underline{U}}_i : \underline{\underline{U}}_i \ll \underline{\underline{X}}, \underline{\underline{U}}_i \ll \underline{\underline{Y}} \rangle$, and $\underline{\underline{X}} \vee \underline{\underline{Y}} = \langle \underline{\underline{X}}, \underline{\underline{Y}} \rangle$, the family of Schunck classes with these operations is a lattice L . The lattice is complete, but not modular (Doerk [20]), and it is complemented (Hawkes [34]).

A number of sublattices have been identified and investigated: saturated formations (Cline [14]), Schunck classes of the form Q_π (these are also the "normal" Schunck classes) (Blessenohl and Gaschütz [4]), D-classes, that is, those Schunck classes $\underline{\underline{X}}$ for which every maximal $\underline{\underline{X}}$-subgroup is an $\underline{\underline{X}}$-projector (Wood [47], [48], and Doerk [20]).

Note that D-classes have some interesting properties: for example, the D-classes that are also formations are just the S_π's (Barnes and Kegel [2]), and they have the property that $\underline{\underline{X}}^2 = \underline{\underline{X}}$.

QUESTION 9. *Are D-classes characterised by the property* $\underline{\underline{X}}^2 = \underline{\underline{X}}$ *? If not, do those Schunck classes* $\underline{\underline{X}}$ *with* $\underline{\underline{X}}^2 = \underline{\underline{X}}$ *form a sublattice of* L *?*

In [20], Doerk introduces antihomomorphs, rechristened Schunck boundaries by Hawkes. The basic idea is to take a class of groups $\underline{\underline{B}}$, and look at the class of $\underline{\underline{B}}$-perfect groups: that is, groups with no quotient in $\underline{\underline{B}}$. Naturally this will not always lead to a Schunck class, so we look for conditions to ensure this. We shall need some notation. For a class $\underline{\underline{X}}$, $\underline{\underline{X}}'$ will denote the class for which $\underline{\underline{X}} \cup \underline{\underline{X}}' = \underline{\underline{S}}$ and $\underline{\underline{X}} \cap \underline{\underline{X}}' = \{1\}$, and

$$(Q\text{-}1)\underline{\underline{B}} = \{G : G \text{ is a } proper \text{ epimorphic image of some group in } \underline{\underline{B}}\} ,$$

if $\underline{\underline{B}} \neq \{1\}$, and $(Q\text{-}1)\{1\} = \{1\}$: also, recall that $\underline{\underline{P}}$ is the class of primitive groups. Then a Schunck boundary will be a class $\underline{\underline{B}}$ satisfying $\underline{\underline{B}} \subseteq \underline{\underline{P}}$ and $(Q\text{-}1)\underline{\underline{B}} \subseteq \underline{\underline{B}}'$. Note that if $\underline{\underline{B}}_0 \subseteq \underline{\underline{B}}$, then $\underline{\underline{B}}_0$ is a Schunck boundary if $\underline{\underline{B}}$ is a Schunck boundary.

If $\underline{\underline{X}}$ is a Schunck class, define $b(\underline{\underline{X}}) = \{G \in \underline{\underline{P}} \cap \underline{\underline{X}}' : (Q\text{-}1)\{G\} \subseteq \underline{\underline{X}}\}$: then $b(\underline{\underline{X}})$ is a Schunck boundary. On the other hand, if $\underline{\underline{B}}$ is a Schunck boundary, define $h(\underline{\underline{B}}) = \{G \in \underline{\underline{B}}' :$ every epimorphic image of G is in $\underline{\underline{B}}'\}$: then $h(\underline{\underline{B}})$ is a Schunck class, the class of "$\underline{\underline{B}}$-perfect" groups. Doerk shows that the maps b, h defined above are mutually inverse bijections between the family of Schunck classes and Schunck boundaries.

If $\underline{\underline{B}} = \{C_p : p \in \pi\}$: then $h(\underline{\underline{B}}) = Q_\pi$: if $\underline{\underline{B}} = \{G \in \underline{\underline{P}} : G/\sigma G \in \underline{\underline{N}}, G \notin \underline{\underline{N}}\}$, then $h(\underline{\underline{B}}) = \underline{\underline{N}}$.

As an application of these ideas, Doerk proves $\underline{\underline{X}}$ is maximal in L if and only if $b(\underline{\underline{X}}) = \{G\}$, for some $G \neq 1$. It follows that L is dually atomic, and so we are lead to the next question.

QUESTION 10. *Is* L *atomic? Find conditions on* $b(\underline{\underline{X}})$ *such that* $\underline{\underline{X}}$ *is minimal in* L .

Examples of minimal elements of L are given by $\underline{\underline{N}}^i$, the Schunck class generated by D_q^p , p and q distinct primes.

Finally, a question posed by Hawkes.

QUESTION 11. *If* $\underline{\underline{X}}$ *is a saturated formation containing* $\underline{\underline{N}}$, *is* $\underline{\underline{X}}$ *minimal in* L ?

4.

I would like to thank the many people who made available to me preprints and unpublished work.

The list of references given below contains some papers not mentioned in the text. I have tried to make a reasonably complete list of papers with results bearing on the subject of this article.

References

[1] Reinhold Baer, "Classes of finite groups and their properties", *Illinois J. Math.* 1 (1957), 115-187. MR19,386.

[2] Donald W. Barnes and Otto H. Kegel, "Gaschütz functors on finite soluble groups", *Math. Z.* 94 (1966), 134-142. MR34#4350.

[3] J.C. Beidleman and A.R. Makan, "On normal Fischer classes", *Bull. London Math. Soc.* 5 (1973), 100-102.

[4] Dieter Blessenohl und Wolfgang Gaschütz, "Über normale Schunck- und Fittingklassen", *Math. Z.* 118 (1970), 1-8. MR43#3344.

[5] R.M. Bryant, R.A. Bryce and B. Hartley, "The formation generated by a finite group", *Bull. Austral. Math. Soc.* 2 (1970), 347-357. MR43#4901.

[6] R.A. Bryce and John Cossey, "Fitting formations of finite soluble groups", *Math. Z.* 127 (1972), 217-223. Zbl.229.20011.

[7] R.A. Bryce and John Cossey, "Metanilpotent Fitting classes", *J. Austral. Math. Soc.* (to appear).

[8] R.A. Bryce and John Cossey, "Subdirect product closed Fitting classes", these Proc.

[9] R.A. Bryce and John Cossey, "Maximal Fitting classes of finite soluble groups *Bull. Austral. Math. Soc.* 10 (1974),

[10] A.R. Camina, "A note on Fitting classes", submitted.

[11] R.W. Carter, "On a class of finite soluble groups", *Proc. London Math. Soc.* (3) 9 (1959), 623-640. MR22#5677.

[12] Roger W. Carter, "Nilpotent self-normalizing subgroups of soluble groups", *Math. Z.* 75 (1961), 136-139. MR23#A928.

[13] R.W. Carter, "Nilpotent self-normalizing subgroups and system normalizers", *Proc. London Math. Soc.* (3) 12 (1962), 535-563. MR25#3988.

[14] Edward Cline, "On an embedding property of generalized Carter subgroups", *Pacific J. Math.* 29 (1969), 491-519. MR42#7773.

[15] John Cossey, "Products of Fitting classes", submitted.

[16] John Cossey and Sheila Oates Macdonald, "On the definition of saturated formations of groups", *Bull. Austral. Math. Soc.* 4 (1971), 9-15. MR43#2084.

[17] R.S. Dark, "On subnormal embedding theorems for groups", *J. London Math. Soc.* 43 (1968), 387-390. MR37#1446.

[18] Rex S. Dark, "Some examples in the theory of injectors of finite soluble groups", *Math. Z.* 127 (1972), 145-156. Zbl.226.20013.

[19] K. Doerk, "Über Homomorphe auflösbarer Gruppen", submitted.

[20] K. Doerk, "Metanilpotente Fittingschunckklassen", submitted.

[21] B. Fischer, *Habilitationsschrift* (Universität Frankfurt).

[22] Bernd Fischer, *Classes of conjugate subgroups in finite solvable groups* (Lecture Notes, Yale University, 1966).

[23] B. Fischer, W. Gaschütz und B. Hartley, "Injektoren endlicher auflösbarer Gruppen", *Math. Z.* 102 (1967), 337-339. MR36#6504.

[24] Wolfgang Gaschütz, "Zur Theorie der endlichen auflösbaren Gruppen", *Math. Z.* 80 (1963), 300-305. MR31#3505.

[25] W. Gaschütz, *Selected topics in the theory of soluble groups* (Lectures given at the 9th Summer Research Institute of the Australian Mathematical Society, Canberra, 1969. Notes by J. Looker).

[26] W. Gaschütz, "Zwei Bemerkungen über normale Fittingklassen", *J. Algebra* (to appear).

[27] Wolfgang Gaschütz und Ursula Lubeseder, "Kennzeichnung gesättigter Formationen", *Math. Z.* 82 (1963), 198-199. MR27#5817.

[28] P. Hall, "A characteristic property of soluble groups", *J. London Math. Soc.* 12 (1937), 198-200. FdM63,69.

[29] P. Hall, "The splitting properties of relatively free groups", *Proc. London Math. Soc.* (3) 4 (1954), 343-356. MR16,217.

[30] B. Hartley, "On Fischer's dualization of formation theory", *Proc. London Math. Soc.* (3) 19 (1969), 193-207. MR39#5696.

[31] Trevor O. Hawkes, "On Fitting formations", *Math. Z.* 117 (1970), 177-182. MR43#3346.

[32] Trevor Hawkes, "Skeletal classes of soluble groups", *Arch. der Math.* 22 (1971), 577-589. MR46#248.

[33] T.O. Hawkes, "Closure operations for Schunck classes", *J. Austral. Math. Soc.* 16 (1973), 316-318.

[34] T.O. Hawkes, in preparation.

[35] B. Huppert, *Endliche Gruppen I* (Die Grundlehren der Mathematischen Wissenschaften, Band 134. Springer-Verlag, Berlin, Heidelberg, New York, 1967). MR37#302.

[36] Hans Lausch, "On normal Fitting classes", *Math. Z.* 130 (1973), 67-72. Zbl.238.20039.

[37] F. Peter Lockett, "On the theory of Fitting classes of finite soluble groups", (PhD thesis, University of Warwick, Coventry, 1971).

[38] F. Peter Lockett, "On the theory of Fitting classes of finite soluble groups", *Math. Z.* 131 (1973), 103-115.

[39] F.P. Lockett, "An example in the theory of Fitting classes", *Bull. London Math. Soc.* 5 (1973), 271-274.

[40] F. Peter Lockett, "The Fitting class $\underline{F}^*$ ", submitted.

[41] Ursula Lubeseder, "Formationsbildungen in endlichen auflösbaren Gruppen", (Dissertation, Kiel, 1963).

[42] A.R. Makan, "On some aspects of finite soluble groups", PhD thesis, Australian National University, Canberra, 1971. Abstract in *Bull. Austral. Math. Soc.* 6 (1972), 157-158.

[43] A.R. Makan, "Fitting classes with the wreath product property are normal", submitted.

[44] Peter M. Neumann, "A note on formations of finite nilpotent groups", *Bull. London Math. Soc.* 2 (1970), 91. MR41#5470.

[45] Hermann Schunck, "$\underline{H}$-Untergruppen in endlichen auflösbaren Gruppen", *Math. Z.* 97 (1967), 326-330. MR35#254.

[46] James Wiegold, "Schunk classes are nilpotent product closed", *Bull. Austral. Math. Soc.* 1 (1969), 27-28. MR40#1478.

[47] Graham John Wood, "On generalizations of Sylow theory in finite soluble groups", (PhD thesis, Queen Mary College, University of London, 1973).

[48] Graham J. Wood, "A lattice of homomorphs", *Math. Z.* 130 (1973), 31-37. Zbl.239.20016.

Australian National University,
Canberra.

PROC. SECOND INTERNAT. CONF. THEORY OF GROUPS, 20C20
CANBERRA 1973, pp. 238-240.

A NOTE ON BLOCKS

John Cossey and Wolfgang Gaschütz

1.

In this note we give elementary proofs of the following two results.

THEOREM 1. *Let G be a p-constrained finite group, k a field of characteristic p. Then kG has precisely one block if and only if G has no normal p'-subgroups.*

Let N be the largest normal p'-subgroup of G, and M/N the largest normal p-subgroup of G/N. Then G is said to be p-constrained if M contains its centraliser.

THEOREM 2. *Let G be a finite group, H/K a p-chief factor of G. Then H/K (as Z_pG-module) lies in the first block of Z_pG.*

Theorems 1 and 2 for the case of soluble groups first appear in Fong and Gaschütz [1]. The proof of Theorem 1 in [1] also applies to p-constrained groups. For another variation on Theorem 2, see Green and Hill [2].

We thank the Mathematics Institute, University of Warwick, for their hospitality while this work was carried out.

2.

Let G be a finite group, k a field, and let

$$kG = B_1 \oplus \ldots \oplus B_s$$

be the decomposition of kG into a direct sum of indecomposable two sided ideals. We assume that the trivial kG-module is an epimorphic image of B_1.

$$Z(kG) = C_1 \oplus \ldots \oplus C_s$$

is the corresponding decomposition of the centre $Z(kG)$ of kG. Each C_i contains a unique maximal ideal R_i, and the radical of $Z(kG)$ is

$$R_1 \oplus \ldots \oplus R_s .$$

The map α defined by $\left(\sum_{g\in G} gk_g\right)\alpha = \sum k_g$ is an epimorphism of kG onto k. Its restriction $\alpha|_{Z(kG)}$ we call β. Clearly, $\ker\alpha$ is generated as vector space by the elements $1-g$, $g \in G$: $\ker\beta$ is generated as vector space by the elements $\sigma_X = \sum_{x\in X} x - |X|$, where X is a conjugacy class of G. In the previous notation, we also have

$$\ker\beta = R_1 \oplus C_2 \oplus \ldots \oplus C_s .$$

If M is an irreducible kG-module, there is an i such that $MB_i \neq 0$, or equivalently, M is isomorphic to a composition factor of B_i.

LEMMA 1. *Let M be an irreducible kG-module. Then $MB_1 \neq 0$ if and only if $M\ker\beta = 0$.*

PROOF. Suppose $MB_1 \neq 0$. Then $MC_i = 0$, $i = 2, \ldots, s$. Since $C_i \leq B_i$, and $MR_1 = 0$ since R_1 is in the radical of kG, it follows that $M\ker\beta = 0$.

If $MB_i \neq 0$ for some $i \neq 1$, then $0 \neq MZ(kG) = MC_i$. Hence $M\ker\beta \neq 0$, and the proof is complete.

The next lemma will be useful in applying Lemma 1. Suppose now that k has characteristic $p \neq 0$, M is an irreducible kG-module, K a normal subgroup of G which acts trivially on M, and H/K a normal p-subgroup of G/K. Observe that H acts trivially on M.

LEMMA 2. *Let X be a conjugacy class in G and C the centralizer of an element x of X. If CK does not contain H, then $M_{\sigma X} = 0$.*

PROOF. If T is a right transversal of $C \cap H$ in H, we have $p \mid |T|$. Let S be a right transversal for CH in G. Then $ST = \{st;\ s \in S,\ t \in T\}$ is a right transversal for C in G. Now

$$\begin{aligned} M\sigma_X &= M\left(\sum_{\substack{s\in S\\ t\in T}} t^{-1}s^{-1}xst - |X|\right) \\ &= M \cdot |T|\left(\sum_{s\in S} s^{-1}xs - |S|\right) \\ &= 0 . \end{aligned}$$

PROOF OF THEOREM 1. Suppose that G has a non-trivial normal p'-subgroup K. Then kG may be written as a sum of two sided ideals

$$kG = B \oplus B^* ,$$

where

$$B = \left\{\sum gk_g : k_g = k_h \text{ if } Kg = Kh\right\} \neq 0$$

and

$$B^* = \left\{\sum gk_g : \sum_{h \in Kh_0} k_h = 0, \text{ for all } h_0 \in G\right\} \neq 0 .$$

Suppose now that G has no normal p' subgroups. From the previous discussion and Lemma 1, we must show that if M is an irreducible kG-module and X is a conjugacy class, $M\sigma_X = 0$.

Let $x_0 \in X$ and P be the largest normal p-subgroup of G . First suppose that P is contained in the centraliser C of x_0 . The definition of p-constraint then gives us that $x_0 \in P$, and hence $X \subseteq P$. But then

$$\begin{aligned} M\sigma_X &= M\left(\sum_{x \in X} x - |X|\right) \\ &= M(|X|-|X|) \\ &= 0 . \end{aligned}$$

Hence suppose that $C \not\geq P$: it follows from Lemma 2 that $M\sigma_X = 0$.

PROOF OF THEOREM 2. By M in the first block we will mean M isomorphic to a composition factor of B_1 . Thus we have to prove that $(H/K)\sigma_X = 0$ for all conjugacy classes X . If $x_0 \in X$, and C is the centraliser of x_0 , as in Theorem 1 there are two cases. If $CK \not\geq H$, $(H/K)\sigma_X = 0$ by Lemma 2. If $CK \geq H$, then

$$\begin{aligned} (H/K)\sigma_X &= (H/K)\left(\sum_{x \in X} x - |X|\right) \\ &= (H/K)(|X|-|X|) \\ &= 0 . \end{aligned}$$

References

[1] P. Fong and W. Gaschütz, "A note on the modular representations of solvable groups", *J. Reine Angew Math.* 208 (1961), 73-78. MR25#2133.

[2] J.A. Green and R. Hill, "On a theorem of Fong and Gaschütz", *J. London Math. Soc.* (2) 1 (1969), 573-576. MR41#6993.

Australian National University, Canberra.

Mathematisches Seminar der Universität, 23 Kiel, Germany.

PROC. SECOND INTERNAT. CONF. THEORY OF GROUPS, 20E30, 20L05, 20L10, 20L15
CANBERRA 1973, pp. 241-280.

THE SUBGROUP THEOREM FOR AMALGAMATED FREE PRODUCTS, HNN-CONSTRUCTIONS AND COLIMITS

R.H. Crowell and N. Smythe

0. Introduction

The purpose of this work is to demonstrate how a subgroup theorem, including Karrass' and Solitar's results [3, 4] for tree products and HNN constructions, may be deduced from the theory of groupoids (here called "groupnets"). It is well-known that many topological proofs of group theoretic results, such as the Nielsen theorem on subgroups of free groups, the Reidemeister-Schreier Theorem, Kurosh' and Grushko's Theorems, may be formalized in purely algebraic terms using groupoids (see for example Ordman [7], Stallings [8], Higgins [2], Crowell and Smythe [1]), however it seems to have been the general feeling, until Higgin's work appeared, that the benefits of such a formalization were negligible, particularly in view of the large amount of preliminary machinery which must be assembled in order to make the theory work. We feel that this effort is in fact justified, in that the theory provides a "theorem-proving" machine for combinatorial group theory. Thus, once the basic machinery has been set up, it is completely obvious that a theorem of the Karrass and Solitar type "A subgroup of an HNN group is again an HNN group" must exist, and it merely remains to find the exact statement of the theorem. Proofs in the theory are usually (not always) entirely straightforward, often amounting to checking that some obvious construction actually satisfies all the requirements made upon it. Of course the details and notation in this present work become complex, because of the nature of the result to be proved, but we trust that the reader will find the basic idea of the argument (which is summarized in Section 8) acceptably simple.

The basic machinery required is briefly developed in Sections 1 to 5. Proofs are omitted, as they will appear elsewhere [1]; in many cases they can also be found in Higgins [2], sometimes in slightly different terminology. In any case, the proofs of these results are in the main straightforward; where this is not so, an indication of the method of proof may be given here.

The subgroup theorem itself is given (with proofs) in Sections 6 to 9.

The subgroup theorem which we obtain (Theorem 9.1) applies to any group G having a presentation of the form

$$\left\| \begin{matrix} \cup X_v\,,\ v \text{ a vertex of } D & \cup R_v \cup \{t_e \mid e \in T\} \cup \\ \cup\{t_e \mid e \text{ an edge of } D\} & \left\{t_e^{-1} x t_e A_e(x^{-1}) \mid x \in X_{\lambda e}\right\} \end{matrix} \right\|$$

where

D is a directed graph,

$\|X_v : R_v\|$ is a presentation of a group A_v for each vertex v of D,

$A_e : A_{\lambda e} \to A_{\rho e}$ is a homomorphism for each edge e of D joining λe to ρe,

T is a maximal tree in D.

Then any subgroup H of G is again of this form, with vertex groups being subgroups of the A_v. If each A_e is one-to-one, then each A_v is embedded in G, and the vertex groups appearing in the graph for H are of the form $u_d^{-1} H u_d \cap A_v$, where u_d is a double coset representative for $d \in G/H : A_v$.

This result covers the work of Karrass and Solitar ([3, 4]). Significantly, however, this method also yields a subgroup theorem for colimits, including for example generalized free products (see Hanna Neumann [5, 6]). (The colimit of the above "diagram (D, A)" is obtained by adding relations $t_e = 1$ for $e \in D$ to the above presentation.) According to Theorem 9.2, any subgroup of such a group is again a colimit of a similar diagram (but now the objects appearing at the vertices are no longer groups, but "covering groupnets").

1. The category of groupnets

DEFINITION 1.1. A *partial product net* consists of two sets A and $E(A)$, two maps $\lambda, \rho : A \to E(A)$, and a partial product $\mu : P \to A$ where $P \subset A \times A$, such that

(i) if $(a, b) \in P$ then $\lambda(b) = \rho(a)$,

(ii) if $(a, b) \in P$ then $\lambda\mu(a, b) = \lambda(a)$ and $\rho\mu(a, b) = \rho(b)$.

We shall denote $\mu(a, b)$ by ab; the phrase "the product ab is defined" shall mean "$(a, b) \in P$". Thus (ii) may be rephrased:

$$\text{if } ab \text{ is defined, then } \lambda(ab) = \lambda(a) \text{ and } \rho(ab) = \rho(b).$$

The set $E(A)$ is called the set of ends of A, and $\lambda(a)$ and $\rho(a)$ are called the left-hand end and the right-hand end of a respectively.

A is a *product net* if in addition

(iii) if $\lambda(b) = \rho(a)$ then ab is defined.

An example of a product net (although not one of concern to us here) is provided by the set of all matrices over a ring, with ends the set of positive integers; $\lambda(a)$ is the number of rows of a, and $\rho(a)$ is the number of columns; the operation is matrix multiplication.

DEFINITION 1.2. A *morphism* $f : \{A, E(A), \lambda, \rho, \mu\} \to \{A', E(A'), \lambda', \rho', \mu'\}$ of partial product nets consists of two maps $f : A \to A'$ and $E(f) : E(A) \to E(A')$ such that

(i) f preserves ends, that is,

$$\begin{array}{ccc} A & \xrightarrow{f} & A' \\ \lambda\downarrow & & \downarrow\lambda' \\ E(A) & \xrightarrow[E(f)]{} & E(A') \end{array} \qquad \begin{array}{ccc} A & \xrightarrow{f} & A' \\ \rho\downarrow & & \downarrow\rho' \\ E(A) & \xrightarrow[E(f)]{} & E(A') \end{array}$$

commute;

(ii) if ab is defined in A, then $f(a)f(b)$ is defined in A', and $f(ab) = f(a)f(b)$.

We shall usually denote a partial product net by its underlying set A, and a morphism by $f : A \to A'$. With this convention, the left-hand and right-hand maps will always be denoted by λ and ρ, regardless of the net under consideration.

An element $i \in A$ is an *identity* in A if $ai = a$ and $ib = b$ whenever ai and ib are defined. Let $\mathrm{Id}(A)$ denote the set of identities in A.

DEFINITION 1.3. A partial product net A is a *partial product net with identities* if there is a one-to-one correspondence $\sigma : E(A) \to \mathrm{Id}(A)$ such that for all $a \in A$, $\sigma\lambda(a).a$ and $a.\sigma\rho(a)$ are defined.

{In this case there is precisely one such correspondence, and $E(A)$ may be identified with $\mathrm{Id}(A)$; λ and ρ are then retractions.}

A *morphism* $f : A \to B$ of partial product nets with identities is a morphism of the underlying partial product nets such that

(i) $f\{\mathrm{Id}(A)\} \subset \mathrm{Id}(B)$,

(ii) $E(f)$ is induced by f (via the identification σ).

DEFINITION 1.4. *A groupnet* G is a product net with identities which is associative and has inverses {that is, for each $a \in G$ there is an element $a^{-1} \in G$ such that $a.a^{-1} = \lambda(a)$ and $a^{-1}.a = \rho(a)$, if we identify $E(G)$ with $\mathrm{Id}(G)$}.

This is essentially the same concept as that of a "Brandt groupoid". However, as this is often called a "groupoid" by topologists, a term which is used in a quite different sense by algebraists, and because we shall have need of the more general notion of a partial product net, we prefer the term "groupnet".

Next we collect a few miscellaneous definitions which will be needed throughout the theory.

DEFINITION 1.5. A groupnet G is *connected* if for every pair $i, j \in \mathrm{Id}(G)$ there exists an element $a \in G$ such that $\lambda(a) = i$ and $\rho(a) = j$ (a is said to connect i to j).

A groupnet G is *acyclic* if the only loops in G are the identities of G: that is, $\lambda(a) = \rho(a)$ implies $a \in \mathrm{Id}(G)$; the groupnet is said to be a *tree* if it is both connected and acyclic.

A morphism $f : G \to H$ is *proper* if f is one-to-one on $\mathrm{Id}(G)$. A morphism f is *monotone* if $f^{-1}(i)$ is connected for every $i \in \mathrm{Id}(H)$.

In general, the image $f(G)$ of a morphism need not be a subgroupnet of H (and epimorphisms need not be surjective). However if f is monotone, $f(G)$ is necessarily a subgroupnet of H. In the case that G is a group, all morphisms $f : G \to H$ are proper and therefore monotone.

2. Homotopy and homotopy type

DEFINITION 2.1. The *unit interval* groupnet I is the 4 element groupnet $\{0, 1, *, *^{-1}\}$ with $\lambda(0) = 0 = \rho(0)$, $\lambda(1) = 1 = \rho(1)$, $\lambda(*) = 0 = \rho(*^{-1})$, $\rho(*) = 1 = \lambda(*^{-1})$, and products

$$0.0 = 0, \quad 0.* = *, \quad *^{-1}.0 = *^{-1}, \quad *.*^{-1} = 0,$$
$$1.1 = 1, \quad *.1 = *, \quad 1.*^{-1} = *^{-1}, \quad *^{-1}.* = 1.$$

For any two groupnets A and B, the cartesian product $A \times B$ has a natural groupnet structure given by co-ordinate-wise multiplication, with identities $\mathrm{Id}(A) \times \mathrm{Id}(B)$. This of course is the categorical product (the categorical coproduct is the disjoint union - *cf.* groups!).

DEFINITION 2.2. Two morphisms $f, g : G \to H$ are said to be *homotopic* (written $f \simeq g$) if there is a morphism $F : G \times I \to H$ such that $F(a, 0) = f(a)$ and $F(a, 1) = g(a)$ for all $a \in G$.

The morphism F is said to be a homotopy between f and g, written $F : f \simeq g$. The homotopy F is completely determined by f and $\{F(i, *) \mid i \in \mathrm{Id}(G)\}$; conversely, if $f : G \to H$ is a morphism of groupnets, and for each $i \in \mathrm{Id}(G)$ an element $h_i \in H$ is given such that $\lambda(h_i) = f(i)$, then there is a homotopy $F : G \times I \to H$ such that $F(a, 0) = f(a)$ and $F(i, *) = h_i$. Thus if G is a group then $f \simeq g$ if and only if there is an element $h \in H$ such that

$g(a) = h^{-1}.f(a).h$ for all $a \in G$.

Two homotopies $F : f \simeq g$ and $G : g \simeq h$ may be "composed" to give a homotopy $H : f \simeq h$ by the rule

$$\begin{cases} H(a, 0) = f(a) , \\ H(i, *) = F(i, *).G(i, *) \quad \text{for } i \text{ an identity.} \end{cases}$$

Hence $\simeq$ is an equivalence relation.

DEFINITION 2.3. A morphism $f : G \to H$ is said to be a *homotopy equivalence* with *homotopy inverse* $g : H \to G$ if $gf \simeq 1_G$ and $fg \simeq 1_H$.

G is said to be of the same *homotopy type* as H , written $G \simeq H$, if there is a homotopy equivalence $f : G \to H$.

THEOREM 2.1. *If G and H are connected then $f : G \to H$ is a homotopy equivalence if and only if for some $i \in \mathrm{Id}(G)$, f is an isomorphism from the loops in G at i onto the loops in H at $f(i)$.*

COROLLARY. *$G \simeq H$ if and only if there is a one-to-one correspondence between the connected components of G and those of H , and for each such pair G', H' of corresponding components an isomorphism of the group of loops in G' at some identity onto the group of loops in H' at some identity.*

Thus a groupnet has the homotopy type of a disjoint union of groups, and these groups uniquely determine the homotopy type.

DEFINITION 2.4. A subgroupnet H of a groupnet G is a strong deformation retract of G if there is a retraction $r : G \to H$ (so that $rj = 1_H : H \to H$, if $j : H \to G$ denotes the inclusion morphism) and a homotopy $D : jr \simeq 1_G$ such that $D(i, *) = i$ for all $i \in \mathrm{Id}(H)$.

If G is a connected groupnet, and H is the group of loops in G at some identity, then H is a strong deformation retract of G .

3. Pregroupnets, the universal groupnet

DEFINITION 3.1. A *congruence* $\equiv$ on a partial product net with identities G is an equivalence relation such that

(i) if $a \equiv b$ then $\lambda(a) \equiv \lambda(b)$ and $\rho(a) = \rho(b)$,

(ii) if $a \equiv a'$ and $b \equiv b'$ and both ab and $a'b'$ are defined, then $ab \equiv a'b'$.

The congruence classes $\{[a] \mid a \in G\}$ form a partial product net with identities, denoted by $G/\equiv$, under the product "$\alpha\beta$ is defined if there exists $a \in \alpha$ and $b \in \beta$ such that ab is defined in G , and then $\alpha\beta = [ab]$ ".

If G is a groupnet, $G/\equiv$ need not be a groupnet. In fact it need not be associative, nor even a product net.

DEFINITION 3.2. A *pregroupnet* is a partial product net with identities such that each element has a two-sided inverse (which need not be unique).

Then if A is a pregroupnet, so is $A/\equiv$.

DEFINITION 3.3. If A is a pregroupnet, a *universal groupnet* for A is a groupnet $G(A)$ and a morphism $\varphi : A \to G(A)$ which is universal for morphisms of A to groupnets, that is, if $\psi : A \to H$ is any morphism to a groupnet H , then ψ factors uniquely through φ .

A universal groupnet for a pregroupnet A (which is of course unique up to a unique isomorphism) may be constructed in the usual manner, as follows. Let $S(A)$ denote the set of non-empty sequences $a_1, a_2, \ldots, a_n$ of elements of A such that $\rho(a_i) = \lambda(a_{i+1})$ for $1 \leq i < n$. Given two sequences u and v , we say v is obtained from u by an elementary contraction, written $u \backslash v$, if $u = a_1, a_2, \ldots, a_n$ and $v = a_1, \ldots, a_i a_{i+1}, \ldots, a_n$ for some i $(a_i a_{i+1}$ must be defined in $A)$. This relation generates an equivalence relation $\sim$, and $G(A) = S(A)/\sim$ with $\varphi : A \to G(A)$ being the composite $A \subset S(A) \to S(A)/\sim$. Note that $\mathrm{Id}(G(A))$ may be identified with $\mathrm{Id}(A)$.

As an example of the use of this construction, suppose we are given two groups A and B , with a third group U which is isomorphic to subgroups of A and B via embeddings $f : U \to A$ and $g : U \to B$. We may define a congruence $\equiv$ on $A \& B$ (the disjoint union of A and B) as the congruence generated by "$a = b$ if $a = f(u)$ and $b = g(u)$ for some $u \in U$ ". This amounts to "pasting A and B together along U ", a procedure familiar in topology. The universal groupnet $G(A\&B/\equiv)$ is just the free product of A and B with $f(U)$, $g(U)$ amalgamated, $A *_U B$.

It is important to know which elements of a pregroupnet A are identified when passing from A to $G(A/\equiv)$. It is usually easy to check which elements are identified under the congruence $\equiv$, but the map $\varphi : A/\equiv \to G(A/\equiv)$ may also kill some elements. In order to discuss this we need the concept of "full associativity".

DEFINITION 3.4. Let A be a pregroupnet. The element $a \in A$ is said to be an *association* of a sequence $u_1 = a_1, a_2, \ldots, a_n \in S(A)$ if there is a chain of elementary contractions $u_1 \backslash u_2 \backslash \ldots \backslash u_m = a$. The pregroupnet A is *fully associative* if whenever a and b are associations of a sequence u , we have $a = b$.

In a groupnet G there is no difficulty with associativity, for given any sequence $a_1, a_2, \ldots, a_n \in S(G)$ all formally possible associations are in fact defined in G, and then the associative law "$a(bc) = (ab)c$ whenever both sides are defined" implies that all associations of a sequence $a_1, a_2, \ldots, a_n$ are equal. However in a pregroupnet, this associative law for 3 elements may hold, but fail for 4 elements, for example.

THEOREM 3.1. *$\varphi : A \to G(A)$ is one-to-one if and only if A is fully associative.*

To prove this result one shows that if two sequences u and $v \in S(A)$ are equivalent, then there is a chain of elementary equivalences $u \,/\, u_1 \,/\, \ldots \,/\, u_n \setminus u_{n+1} \setminus \ldots \setminus v$ in which all the elementary expansions precede all the contractions. If u and v are one-term sequences, that is, belong to A, then they are associations of u_n.

Thus we need a criterion to decide when a pregroupnet is fully associative.

THEOREM 3.2. *Suppose G is a pregroupnet, $G = A \cup B$, where A and B are groupnets such that if a product xy is defined in G then x and y both belong to A or to B. Then G is fully associative.*

Thus for example this shows that groups A and B are embedded in the free product with amalgamation $A *_U B$.

The proof of Theorem 3.2 appears to be highly nontrivial (see [1]). Given this however, it is easy to prove the following generalization.

THEOREM 3.3. *Let G be a pregroupnet, $G = A \cup B$ such that*

(i) A and B are fully associative pregroupnets,

(ii) A and B are full subnets of G (that is, if $x, y \in A$ and xy is defined in G then $xy \in A$; similarly for B),

(iii) if xy is defined in G, then both x and y are in A, or in B,

(iv) $A \cap B$ is a groupnet.

Then G is fully associative.

(For G can be embedded in $G(A) \cup G(B)$.)

Finally we can derive a very useful criterion.

THEOREM 3.4. *Suppose G is a pregroupnet which can be written as a union $\bigcup_{i \in J} G_i$ of subsets such that*

(i) each G_i is a groupnet,

(ii) *if* xy *is defined in* G *then there is an index* $i \in J$ *such that* x *and* y *belong to* G_i ,

(iii) *there is an index* $0 \in J$ *such that* $G_i \cap G_j \subset G_0$ *whenever* $i \neq j$.

Then G *is fully associative.*

(Proof by transfinite induction on J).

For example, suppose D is a directed graph (a partial product net in which no products are defined!). $D \cup E(D)$ may be embedded in a pregroupnet $P(D)$ as follows. For each edge e of D such that $\lambda(e) \neq \rho(e)$, adjoin a symbol e^{-1} ; let G_e denote the groupnet $\{\lambda(e), \rho(e), e, e^{-1}\}$ with the obvious product structure, isomorphic to I . For each edge e such that $\lambda(e) = \rho(e)$, adjoin symbols e^n for n an integer $\neq 0$ or 1 ; let G_e denote the groupnet $\{\lambda(e)\} \cup \{e^n \mid n \text{ an integer}\}$ with the obvious product structure, isomorphic to the infinite cyclic group. Then $P(D) = \cup G_e$, with identities $E(D) = G_0$, is fully associative. Thus $\varphi : P(D) \to G(P(D))$ is one-to-one. The groupnet $G(P(D))$ is the *free groupnet* on D , and will be denoted by $G(D)$.

Identifying D with $\varphi(D)$, it can be seen that $D \cup E(D)$ is a *basis* for $G(D)$, that is, each element of $G(D)$ can be written uniquely as a reduced word (sequence) in $S(P(D))$. Conversely if a groupnet G has a subset D which is a basis for G , then G is free.

At a later point we shall need the following result.

THEOREM 3.5. *Let* $f : A \to A$ *be a monomorphism on a groupnet* A . *Let* $\equiv$ *denote the congruence relation on* $A \times I$ *generated by* $(a, 1) \equiv (f(a), 0)$, *for* $a \in A$. *Then* $A \times I/\equiv$ *is fully associative.*

PROOF. Consider first the case f an isomorphism. Define a net A_f whose underlying set is $A \times Z$ (Z denotes the set of integers), with ends $E(A_f) = \mathrm{Id}(A) \times \{0\}$, $\lambda(a, n) = (\lambda a, 0)$ and $\rho(a, n) = (f^n \rho a, 0)$, and product $(a, m)(b, n) = (a.f^{-m}(b), m+n)$ if $\rho(a, m) = \lambda(b, n)$. Then A_f is a groupnet. There is an embedding of $A \times I/\equiv$ into A_f , defined by $(a, 0) \to (a, 0)$, $(a, 1) \to (f(a), 0)$, $(a, *) \to (a, 1)$, $(a, *^{-1}) \to (f(a), -1)$; hence $A \times I/\equiv$ is fully associative.

Now suppose f is a monomorphism. Define $\overline{A}$ to be the equivalence classes of $A \times Z$ under the equivalence relation generated by $(a, n) \sim (f(a), n+1)$. Let $E(\overline{A})$ be $\mathrm{Id}(A) \times Z/\sim$, with $\lambda[(a, n)] = [(\lambda a, n)]$ and $\rho[(a, n)] = [(\rho a, n)]$, and product $[(a, m)][(b, n)] = [(f^{n+k}(a).f^{m+k}(b), m+n+k)]$ for large k (this is well-defined since f is one-to-one). $\bar{A}$ is a groupnet, and A may be embedded in $\bar{A}$ via $a \to [(a, 0)]$. The map f extends to an isomorphism $\bar{f} : \bar{A} \to \bar{A}$ by $\bar{f}[(a, n)] = [(f(a), n)] = [(a, n-1)]$. Hence $A \times I/\equiv$ may be embedded in $\bar{A} \times I/\equiv$, which by the above is fully associative. //

4. Covering morphisms

DEFINITION 4.1. A morphism $\pi : C \to B$ of groupnets is a *covering map* if it is a surjection and has the unique path lifting property (that is, given any $b \in B$ and any identity $i \in C$ such that $\pi(i) = \lambda(b)$, there is a unique $b^* \in C$ such that $\pi(b^*) = b$ and $\lambda(b^*) = i$).

Suppose H is a *wide* subgroupnet of a groupnet B (that is, $H \supset \mathrm{Id}(B)$). We shall construct a covering map $\pi_H : B_H \to B$ corresponding to H such that B_H has the homotopy type of H (*cf.* the topological notion of covering space).

Define the set of *right cosets* B/H of H to be the collection of subsets $Hx = \{hx \in B \mid h \in H\}$ for $x \in B$. Hx is non-empty since H is wide. There is a map $\rho : B/H \to \mathrm{Id}(B)$ defined by $\rho(Hx) = \rho(x)$; further, for $\sigma \in B/H$ and $b \in B$ with $\rho(\sigma) = \lambda(b)$, there is a well-defined coset σb.

Define $B_H = \{(\sigma, b) \in B/H \times B \mid \rho(\sigma) = \lambda(b)\}$ with ends $E(B_H) = B/H$ and $\lambda, \rho : B_H \to B/H$ given by $\lambda(\sigma, b) = \sigma$ and $\rho(\sigma, b) = \sigma b$. Then B_H is a groupnet under the product $(\sigma, b).(\sigma', b') = (\sigma, bb')$ if $\sigma' = \sigma b$. As usual we shall identify the identity $(\sigma, \rho(\sigma))$ with the end σ. Note that $(\sigma, b)^{-1} = (\sigma b, b^{-1})$.

Finally, the map $\pi_H : B_H \to B$ is defined by $\pi_H(\sigma, b) = b$.

DEFINITION 4.2. Let $\pi : C \to B$ be a covering morphism. A subset $Y \subset \mathrm{Id}(C)$ is said to be a *set of base points* for π if

(i) Y meets every component of C,

(ii) π maps Y isomorphically onto $\mathrm{Id}(B)$.

For such a set, define $H^*(Y) = \{z \in C \mid \lambda(z), \rho(z) \in Y\}$ and $H(Y) = \pi H^*(Y)$. Then π maps $H^*(Y)$ isomorphically onto $H(Y)$.

If π has a set of base points, it is said to be a *regular* covering. Clearly any covering may be expressed as a disjoint union of regular coverings, although not necessarily in a unique way.

For the covering $\pi_H : B_H \to B$ corresponding to a wide subgroupnet H, define $Y_H = \{(Hi, i) \mid i \in \mathrm{Id}(B)\}$. Given $\sigma \in B/H$, choose a coset representative $x_\sigma \in \sigma$. Then $(H\lambda x_\sigma, x_\sigma)$ joins $(H\lambda x_\sigma, \lambda x_\sigma) \in Y_H$ to $(\sigma, \rho(\sigma))$. So Y_H meets every component of B_H . Clearly Y_H is a set of base points for π_H , and $H(Y_H) = H$. There is a strong deformation retraction of B_H onto $H^*(Y)$ (by retracting $(H\lambda x_\sigma, x_\sigma)$), so $B_H \simeq H$.

Moreover, we have

THEOREM 4.1. *Let* $\pi : C \to B$ *be a regular covering map with set of base points* Y *, and let* $H = H(Y)$ *. Then there is a unique isomorphism* $f : C \to B_H$ *such that* $\pi_H f = \pi$ *and* $f(Y) = Y_H$.

Thus each covering of B determines, and is determined by, a collection of subgroupnets of B (a given subgroupnet may occur more than once in the collection). The collection is not unique, depending on both the representation of the covering as a disjoint union of regular coverings and the choice of base set of each regular covering (the latter dependence can be expressed in terms of conjugate subgroups - *cf.* the choice of base point in a topological covering space).

The covering groupnet construction can be used to prove the Nielsen Theorem. For if B is a free groupnet with basis D , then B_H is free with basis $D_H = \{(\sigma, d) \mid d \in D\}$. From $B_H \simeq H$ it is easily seen that H is a free groupnet.

5. Pull-backs and coverings

Let $f : X \to B$ and $g : Y \to B$ be morphisms of groupnets. There is a pull-back construction

$$\begin{array}{ccc} f^*(Y) & \xrightarrow{f'} & Y \\ \downarrow{\scriptstyle g^*} & & \downarrow{\scriptstyle g} \\ X & \xrightarrow{f} & B \end{array}$$

given by $f^*(Y) = \{(x, y) \in X \times Y \mid f(x) = g(y)\}$ with the obvious product, $f'(x, y) = y$ and $g^*(x, y) = x$.

THEOREM 5.1. *If* g *is injective/proper/monotone/a covering morphism/surjective, then so is* g^* *(similarly for* f *and* f' *). If* g *is a homotopy equivalence, and* $f(X')$ *meets* $g(Y)$ *for every component* X' *of* X *, then* g^* *is a homotopy equivalence.*

This construction can be applied to give the Reidemeister-Schreier Theorem. If $f : F \to B$ is a surjection of a free groupnet F onto B , and H is a wide subgroupnet of B , form the pullback

$$\begin{array}{ccc} f^*(B_H) & \xrightarrow{f'} & B_H \\ \downarrow & & \downarrow \pi_H \\ F & \xrightarrow{f} & B \end{array} .$$

Thus $f^*(B_H)$ is free, and the relators of the presentation of B_H associated with f' can be obtained by lifting the relators of f (the Reidemeister rewriting process). The homotopy equivalence $B_H \simeq H$ can then be used to find a presentation for H , introducing relators corresponding to the cosets B/H . The details can be found in [1].

Consider a pull-back square of a covering map π_H ,

$$\begin{array}{ccc} f^*(B_H) & \xrightarrow{f'} & B_H \\ \downarrow \pi_H^* & & \downarrow \pi_H \\ A & \xrightarrow{f} & B \end{array}$$

where f is monotone and A is connected. π_H^* is a covering map - we wish to determine the family of subgroupnets of A to which π_H^* corresponds.

DEFINITION 5.1. For each $b \in B$ with $\rho(b) \in f(A)$, define the *double coset* $H.b.f(A) = \{h.b.f(a) \mid h \in H, a \in A\}$. Let $B/H : f(A)$ denote the set of all such cosets.

Since f is monotone, the double cosets form a partition of $B' = \{b \in B \mid \rho(b) \in f(A)\}$.

For each double coset d choose a map $u_d : \mathrm{Id}(A) \to B$ such that

(i) $\rho u_d(i) = f(i)$, and

(ii) $u_d(i) \in d$, for each $i \in \mathrm{Id}(A)$.

This is possible since A is connected.

Finally, define a subgroupnet K_d of A for each d by

$$K_d = \left\{a \in A \mid f(a) \in u_d(\lambda(a))^{-1}.H.u_d(\rho(a))\right\} .$$

THEOREM 5.2. *π_H^* is the covering map corresponding to the collection $\{K_d \mid d \in B/H : f(A)\}$ of subgroupnets of A , with set of base points $\cup Y_d$, where*

$$Y_d = \{(i, (Hu_d(i), f(i))) \in \mathrm{Id}(f^*(B_H)) \mid i \in \mathrm{Id}(A)\} .$$

In fact, Y_D meets precisely those components of $f^*(B_H)$ containing identities of the form $(i, (\sigma, f(i)))$ with $i \in \mathrm{Id}(A)$, $\sigma \subset d$; and $H(Y_d) = K_d$.

Under the identification of $f^*(B_H)$ with this covering groupnet, we have $\pi_H^*(\sigma, a) = a$ and $f'(\sigma, a) = (Hu_d(\lambda fc(\sigma))fc(\sigma), f(a))$ for $(\sigma, a) \in A_{K_d}$ where $c(\sigma)$ is a coset representative of σ in A .

6. The mapping cylinder

DEFINITION 6.1. A *groupnet diagram* (C, A) consists of a small category C and a functor A from C to the category of groupnets.

Although C can be any small category, for the application we have in mind C will usually be the free category $C(D)$ on a directed graph D . Thus the objects of C will be called vertices, denoted by symbols such as v_1, v_2 , and so forth, and the arrows of C will be called edges, with symbols e_1, e_2 . In particular we shall write e_1e_2 for the product (composite) of two edges e_1 and e_2 such that $\lambda e_2 = \rho e_1$ (that is, C is an associative product net with identities); then $A_{e_1e_2} = A_{e_2} \circ A_{e_1}$. In the case $C = C(D)$ for D a directed graph, we may write (D, A) for $(C(D), A)$; A is then essentially just a collection of groupnets A_v , v a vertex of D , and morphisms $A_e : A_{\lambda e} \to A_{\rho e}$, e an edge of D .

DEFINITION 6.2. A morphism $(t, f) : (C, A) \to (C', A')$ of groupnet diagrams consists of

(i) a functor $t : C \to C'$,

(ii) for each vertex v , a morphism $f_v : A_v \to A'_{tv}$,

(iii) for each edge e , a homotopy $f_e' : A'_{te}f_{\lambda e} \simeq f_{\rho e}A_e$,

such that

(a) if $e = 1_v$ = the identity arrow at the vertex v , then f_e is the constant homotopy $f_v \simeq f_v$,

(b) if $e = e'e''$, then $f_e = f_{e'} \underline{\circ} f_{e''}$, that is,

$$A_{\lambda e'} \xrightarrow{f_{\lambda e'}} A'_{\lambda te'}$$

$$\left.\begin{array}{c} A_{e'} \\ \big\downarrow \end{array}\right. \left(\!f_{e'}\!\right) \quad \left.\begin{array}{c} A'_{te'} \\ \big\downarrow \end{array}\right.$$

$$A_{\lambda e''} \xrightarrow{f_{\lambda e''}} A'_{\lambda te''}$$

$$\left.\begin{array}{c} A_{e''} \\ \big\downarrow \end{array}\right. \left(\!f_{e''}\!\right) \quad \left.\begin{array}{c} A'_{te''} \\ \big\downarrow \end{array}\right.$$

$$A_{\rho e''} \xrightarrow{f_{\rho e''}} A'_{\rho te''}$$

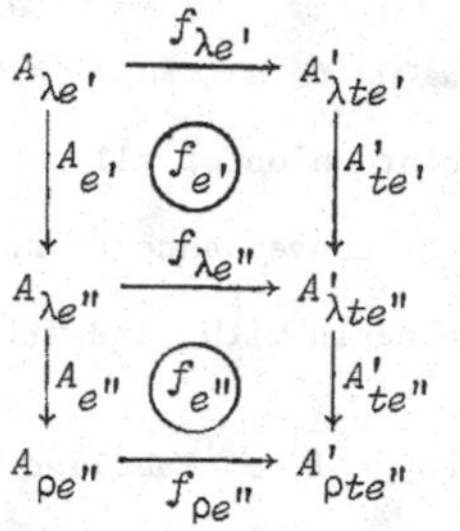

$$\begin{cases} f_e(a,\,0) = A'_{te''}A'_{te'}\,f_{\lambda e}(a) \quad \text{for} \quad a \in A_{\lambda e}\,, \\[2mm] f_e(i,\,*) = A'_{te''}f_{e'}(i,\,*)\cdot\big(f_{e''}\big(A_{e'}(i),\,*\big)\big) \quad \text{for} \quad i \in \mathrm{Id}\big(A_{\lambda e}\big) \end{cases}$$

(then $\quad f_e : A'_{te''}A'_{te'}\,f_{\lambda e'} \simeq f_{\rho e''}A_{e''}A_{e'}$).

Morphisms may be composed by the rule $\quad (s,\,f)(t,\,g) = (u,\,h)\quad$ where

    (i) $u = ts$,

    (ii) $h_v = g_{sv}f_v$,

    (iii) $h_e = f_e \,\phi\, g_{se}\quad$ where this means

$$A_{\lambda e} \xrightarrow{f_{\lambda e}} A'_{\lambda se} \xrightarrow{g_{\lambda se}} A''_{\lambda tse}$$

$$\left.\begin{array}{c} A_e \\ \big\downarrow \end{array}\right. \left(\!f_e\!\right) \quad \left.\begin{array}{c} A'_{se} \\ \big\downarrow \end{array}\right. \left(\!g_{se}\!\right) \quad \left.\begin{array}{c} A''_{tse} \\ \big\downarrow \end{array}\right.$$

$$A_{\rho e} \xrightarrow{f_{\rho e}} A'_{\rho se} \xrightarrow{g_{\rho se}} A''_{\rho tse}$$

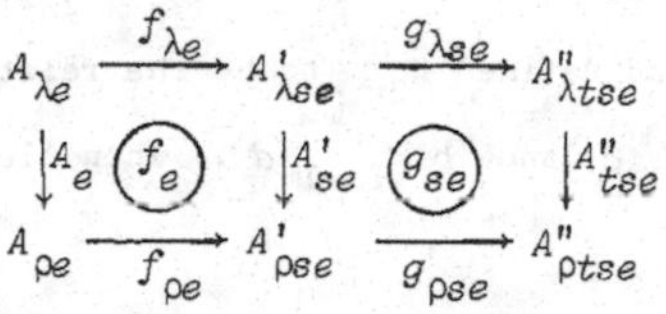

$$\begin{cases} h_e(a,\,0) = A''_{tse}g_{\lambda se}f_{\lambda e}(a) \quad \text{for} \quad a \in A_{\lambda e}\,, \\[2mm] h_e(i,\,*) = g_{se}\big(f_{\lambda e}(i),\,*\big)\cdot g_{\rho se}f_e(i,\,*) \quad \text{for} \quad i \in \mathrm{Id}\big(A_{\lambda e}\big)\,. \end{cases}$$

Let $\cdot$ denote the category consisting of a single object $\cdot$ and a single arrow, the identity arrow on $\cdot$. To each groupnet G corresponds a groupnet diagram $(\cdot,\,G)$ where $G_\cdot = G$; this defines an embedding of the category of groupnets in the category of groupnet diagrams. A morphism $f : (C,\,A) \to G$ from a groupnet diagram to a groupnet is then a morphism $(t,\,f) : (C,\,A) \to (\cdot,\,G)$ in which t is the unique functor from C to $\cdot$; that is, f consists of morphisms $f_v : A_v \to G$ and homotopies $f_e : f_{\lambda e} \simeq f_{\rho e}A_e$.

DEFINITION 6.3. A *mapping cylinder* for a groupnet diagram $(C,\,A)$ is a morphism $m : (C,\,A) \to m(C,\,A)$ to a groupnet $m(C,\,A)$ which is universal for morphisms from $(C,\,A)$ to groupnets (that is, every morphism from $(C,\,A)$ to a groupnet factors uniquely through m).

The name "mapping cylinder" is justified by the construction which follows.

THEOREM 6.1. *For any groupnet diagram* (C, A) *there is a mapping cylinder* $m : (C, A) \to m(C, A)$ *, which is unique up to a unique isomorphism.*

PROOF. Let P denote the disjoint union of all A_v , v a vertex of C , and all $A_{\lambda e} \times I$, e an edge of C . For convenience of notation, we shall assume the vertex groupnets A_v are disjoint to begin with, and write $P = \bigcup_v A_v \bigcup_e A_{\lambda e} \times I_e$, where $I_e = \{0_e, 1_e, *_e, *_e^{-1}\}$ is a copy of I labelled by the edge e .

Define $\equiv$ to be the congruence on P generated by

($) $\qquad a \equiv (a, 0_e)$ and $A_e(a) \equiv (a, 1_e)$ for $a \in A_{\lambda e}$.

LEMMA 1. *Suppose* $i \in \mathrm{Id}(A_v)$, $i' \in \mathrm{Id}(A_{v'})$ *and* $i \equiv i'$. *Then* $v = v'$ *and* $i = i'$.

To prove this, note that the congruence $\equiv$ can be constructed in stages, as follows. Let $\equiv_1$ denote the *equivalence* relation generated by the elementary equivalences of ($), that is, $x \equiv_1 y$ if there is a finite chain $x = x_1, x_2, \ldots, x_k = y$ such that $x_j \equiv x_{j+1}$ or $x_{j+1} \equiv x_j$ is one of the equivalences of ($), for $1 \le j \le k-1$. Inductively, define $\equiv_{2n+1}$ to be the equivalence relation generated by $\equiv_{2n}$, and define $\equiv_{2n}$ to be the relation on $P : x \equiv_{2n} y$ if $x = ab$, $y = cd$ with $a \equiv_{2n-1} c$ and $b \equiv_{2n-1} d$. Then clearly $x \equiv y$ if and only if $x \equiv_n y$ for some n .

Note that if $x \equiv_{2n} y$, then $\lambda x \equiv_{2n-1} \lambda y$ and $\rho x \equiv_{2n-1} \rho y$. And if $x \equiv_{2n+1} y$, then there is a chain $x = x_1, \ldots, x_k = y$ such that $x_j \equiv_{2n} x_{j+1}$, hence $\lambda x_j \equiv_{2n-1} \lambda x_{j+1}$. Since $\equiv_{2n-1}$ is transitive, $\lambda x \equiv_{2n-1} \lambda y$ and $\rho x \equiv_{2n-1} \rho y$. Thus if $x \equiv y$, we have $\lambda x \equiv_1 \lambda y$ and $\rho x \equiv_1 \rho y$. But clearly $\equiv_1$ is the identity relation when restricted to $\bigcup \mathrm{Id}(A_v)$, proving the lemma.

Thus the identities of the pregroupnet $P/\equiv$, and therefore also of the universal groupnet $G(P/\equiv)$, may be identified with the disjoint union $\bigcup \mathrm{Id}(A_v)$.

So far we have $P \xrightarrow{\eta} P/\equiv \xrightarrow{\varphi} G(P/\equiv)$. If C is the free category $C(D)$ on a directed graph D , and the above construction is carried out using only the edges of D , then the groupnet $G(P/\equiv)$ is the required mapping cylinder. In the general case however we must introduce relations corresponding to the relations in C .

Let J be the subgroupnet of $G(P/\equiv)$ generated by

$$\left\{\varphi\eta(i, *_{e'}) \cdot \varphi\eta(A_{e'}(i), *_{e''}) \cdot \varphi\eta\left(i, *_e^{-1}\right) \mid i \in \mathrm{Id}(A_{\lambda e}),\ e = e'e''\right\}.$$

Note that J contains such elements as $\varphi\eta(i, *_e)$, $e = 1_v$.

LEMMA 2. *J consists entirely of loops, and is a normal subgroupnet of $G(P/\equiv)$, that is, if $x \in J$ and axa^{-1} is defined then $axa^{-1} \in J$.*

That J consists of loops is trivial.

To show that J is normal, it is sufficient to consider conjugates of the generating elements by elements of the form $\varphi\eta(a)$, $\varphi\eta(i, *_e)$, $\varphi\eta\left(i, *_e^{-1}\right)$ since such elements generate $G(P/\equiv)$.

Consider $y = \varphi\eta(a)\varphi\eta(i, *_{e'})\varphi\eta(A_{e'}(i), *_{e''})\varphi\eta\left(i, *_e^{-1}\right)\varphi\eta(a^{-1})$ where $e = e'e''$ and $a \in A_v$. For this to be defined, $\varphi\eta\rho(a) = \varphi\eta(i)$; but $\varphi\eta$ is one-to-one on the identities of $\cup A_v$, so $v = \lambda e$ and $a = i$. Further, $\varphi\eta(a) = \varphi\eta(a, 0_e) = \varphi\eta(a, 0_{e'})$.

Then

$$\begin{aligned} y &= \varphi\eta(\lambda a, *_{e'})\varphi\eta(a, 1_{e'})\varphi\eta(A_{e'}(i), *_{e''})\varphi\eta\left(i, *_e^{-1}\right)\varphi\eta(a^{-1}) \\ &= \varphi\eta(\lambda a, *_{e'})\varphi\eta A_{e'}(a)\varphi\eta(A_{e'}(i), *_{e''})\varphi\eta\left(i, *_e^{-1}\right)\varphi\eta(a^{-1}) \\ &= \varphi\eta(\lambda a, *_{e'})\varphi\eta(A_{e'}(\lambda a), *_{e''})\varphi\eta(A_{e'}(a), 1_{e''})\varphi\eta\left(i, *_e^{-1}\right)\varphi\eta(a^{-1}) \\ &= \varphi\eta(\lambda a, *_{e'})\varphi\eta(A_{e'}(\lambda a), *_{e''})\varphi\eta A_e(a)\varphi\eta\left(i, *_e^{-1}\right)\varphi\eta(a^{-1}) \\ &= \varphi\eta(\lambda a, *_{e'})\varphi\eta(A_{e'}(\lambda a), *_{e''})\varphi\eta\left(\lambda a, *_e^{-1}\right) \\ &\in J. \end{aligned}$$

Now consider $y = \varphi\eta(\underline{i}, *_{\underline{e}})\varphi\eta(i, *_{e'})\varphi\eta(A_{e'}(i), *_{e''})\varphi\eta\left(i, *_e^{-1}\right)\varphi\eta\left(\underline{i}, *_{\underline{e}}^{-1}\right)$. For this to be defined, $\rho\underline{e} = \lambda e' = \lambda e$ and $A_{\underline{e}}(\underline{i}) = i$. Then $y = y_1 \cdot y_2 \cdot y_3^{-1}$ where

$$y_1 = \varphi\eta(\underline{i}, *_{\underline{e}})\varphi\eta(i, *_{e'})\varphi\eta\left(\underline{i}, *_{\underline{e}e'}^{-1}\right) \in J,$$

$$y_2 = \varphi\eta(\underline{i}, *_{\underline{e}e'})\varphi\eta(A_{e'}(i), *_{e''})\varphi\eta\left(\underline{i}, *_{\underline{e}e}^{-1}\right) \in J \quad (\underline{e}e'e'' = \underline{e}e),$$

$$y_3 = \varphi\eta(\underline{i}, *_{\underline{e}})\varphi\eta(i, *_e)\varphi\eta\left(\underline{i}, *_{\underline{e}e}^{-1}\right) \in J.$$

Finally to take care of the case of conjugation by $\varphi\eta\left(\underline{i}, *_{\underline{e}}^{-1}\right)$, we need only remark that $\varphi\eta\left(\underline{i}, *_{\underline{e}}^{-1}\right) = y_4 \cdot \varphi\eta(\underline{i}, *_{\underline{e}^{-1}})$ for an appropriate $y_4 \in J$.

Thus the set of cosets $m(C, A) = G(P/\equiv)/J$ forms a groupnet with identities which again we identify with the disjoint union $\cup \mathrm{Id}(A_v)$.

Define $m_v : A_v \to m(C, A)$ and $m_e : A_{\lambda e} \times I \to m(C, A)$ to be the composites $A_v \subset P \xrightarrow{\eta} P/\equiv \xrightarrow{\varphi} G(P/\equiv) \xrightarrow{\nu} m(C,A)$ and

$$A_{\lambda e} \times I \approx A_{\lambda e} \times I_e \subset P \xrightarrow{\eta} P/\equiv \xrightarrow{\varphi} G(P/\equiv) \xrightarrow{\nu} m(C, A) .$$

It follows immediately that $m : (C, A) \to m(C, A)$ is a morphism.

Finally, suppose $f : (C, A) \to G$ is a morphism of (C, A) to a groupnet G . f induces a morphism $\cup f : P \to G$ which is the disjoint union of the maps f_v and f_e . Since $f_e(a, 0) = f_{\lambda e}(a)$ and $f_e(a, 1) = f_{\rho e}(A_e(a))$, the map $\cup f$ is compatible with the congruence $\equiv$, and induces a morphism from $P/\equiv$ to G , therefore also a morphism $f' : G(P/\equiv) \to G$.

Now if $e = e'e''$ and $i \in \mathrm{Id}(A_{\lambda e})$, then

$$\begin{aligned} f'\left[\varphi\eta(i, *_{e'})\varphi\eta(A_{e'}(i), *_{e''})\varphi\eta\left(i, *_e^{-1}\right)\right] &= f_{e'}(i, *)f_{e''}(A_{e'}(i), *)f_e(i, *^{-1}) \\ &= \lambda f_{e'}(i, *) \quad \{\text{see Definition 6.2 (iii) (b)}\} \\ &= f_{\lambda e}(i) . \end{aligned}$$

Thus $f'(J) \subset \mathrm{Id}(G)$, and f' induces $f^* : m(C, A) \to G$ such that $f = f^*m$. Since $m(C, A)$ is generated by $\cup imf_v \cup imf_e$, f^* is the unique such map. Hence $m : (C, A) \to m(C, A)$ is the mapping cylinder for (C, A) . //

We should note that m is the object map of a functor from the category of groupnet diagrams to the category of groupnets. For suppose $(t, f) : (C, A) \to (C', A')$ is a morphism of diagrams, with mapping cylinders m^A and $m^{A'}$ respectively. Then $m^{A'}f : (C, A) \to m(C', A')$ is a morphism, so can be factored uniquely through m^A , giving rise to an induced morphism $mf : m(C, A) \to m(C', A')$. It is straightforward to check that $m(fg) = mf.mg$ and $m1 = 1$.

THEOREM 6.2 (The Monomorphism Theorem). *If each morphism* A_e *is a monomorphism in a groupnet diagram* (C, A) , *then so is each* $m_v : A_v \to m(C, A)$.

PROOF. We shall use the notation of the preceding theorem. We prove first that $P/\equiv$ is fully associative, so that φ is a monomorphism.

Define a net Q with underlying set

$$\cup A_v \cup \{(a, *_e) \mid a \in A_{\lambda e},\ e \text{ an edge}\} \cup \left\{\left(a, *_e^{-1}\right) \mid a \in A_{\lambda e},\ e \text{ an edge}\right\}$$

and with ends $E(Q) = \cup \mathrm{Id}(A_v)$, $\lambda(a)$ and $\rho(a)$ defined as in A_v , $\lambda(a, *_e) = \lambda(a)$ and $\rho(a, *_e) = \rho A_e(a)$, and so forth, and products

(i) $a.b$ defined as in A_v if $a, b \in A_v$,

(ii) $a.(b, *_e) = (ab, *_e)$ if $ab \in A_{\lambda e}$,

(iii) $(a, *_e).b = (aa', *_e)$ if $A_e(a') = b$ and aa' defined

(iv) $(a, *_e).(b, *_e^{-1}) = ab$ if ab defined,

(v) $(a, *_e^{-1}).(b, *_e) = A_e(ab)$ if ab defined,

(vi) $(a, *_e^{-1}).b = (ab, *_e^{-1})$ if ab defined,

(vii) $a.(b, *_e^{-1}) = (a'b, *_e^{-1})$ if $A_e(a') = a$ and $a'b$ defined.

Note that (iii) and (vii) are well-defined since A_e is one-to-one. Then Q is in fact a pregroupnet. Define $\theta : P \to Q$ by

$$\begin{cases} \theta(a) = a \text{ for } a \in Q , \\ \theta(a, 0_e) = a , \\ \theta(a, 1_e) = A_e(a) . \end{cases}$$

Then θ is a morphism, and $\theta(x) = \theta(y)$ if and only if $x \equiv y$, so that θ induces an isomorphism $\theta^* : P/{\equiv} \to Q$.

Hence $P/{\equiv}$ is the disjoint union of a groupnet $\cup\eta(A_v) = P_0$ isomorphic to $\cup A_v$, and pregroupnets $P_e = \eta(A_{\lambda e} \times I_e)$, e an edge of C . If $\lambda e \neq \rho e$, then P_e is a groupnet; if $\lambda e = \rho e$, we can apply Theorem 3.5 to show that P_e is fully associative. Furthermore P_e is a full subnet of $P/{\equiv}$, $P_e \cap P_{e'} \subset P_0$ if $e \neq e'$, and if a product xy is defined in $P/{\equiv}$ then both x and y lie in some P_e or P_0 . Hence by Theorem 3.3, $P/{\equiv}$ is fully associative.

Let K^* be the subgroupnet of $G(P/{\equiv})$ generated by $\{m_e(i, *) \mid i \in \mathrm{Id}(A_{\lambda e}), e \text{ an edge of } C\}$; note $J \subset K^*$. We shall show that $K^* \cap \varphi\eta(A_v) = \mathrm{Id}(A_v)$; this implies that $\nu|\varphi\eta(A_v)$ is a monomorophism, and hence that m_v is a monomorphism.

Let $F(C)$ be the free groupnet with basis the arrows of C , and identities the vertices of C. Define $f : P \to F(C)$ by $f(a) = v$ for $a \in A_v$, $f(a, 0_e) = \lambda(e)$, $f(a, 1_e) = \rho(e)$, $f(a, *_e) = e$, $f\left(a, *_e^{-1}\right) = e^{-1}$. Then f is a morphism, and $f(x) = f(y)$ if $x \equiv y$, so f induces a morphism $f' : P/\equiv \to F(C)$ and hence a morphism $f^* : G(P/\equiv) \to F(C)$. Note $f^*\varphi\eta(A_v) \subset \mathrm{Id}(F(C))$.

Now suppose $k = m_{e_1}\left(i_1, *^{\tau_1}\right).m_{e_2}\left(i_2, *^{\tau_2}\right) \dots m_{e_n}\left(i_n, *^{\tau_n}\right)$, $\tau_j = \pm 1$, is an element of $K^* \cap \varphi\eta(A_v)$; we may suppose n is the smallest integer such that k may be written in this form, unless k is an identity. Then $f^*(k) = e_1^{\tau_1} e_2^{\tau_2} \dots e_n^{\tau_n} \in \mathrm{Id}(F(C))$. So there is an index j such that $e_{j+1} = e_j$ and $\tau_{j+1} = -\tau_j$. If $\tau_j = -1$, then since $m_{e_j}\left(i_j, *^{-1}\right).m_{e_j}(i_{j+1}, *)$ is defined, we have $m_{\lambda e_j}(i_j) = m_{\lambda e_j}(i_{j+1})$, that is, $i_j = i_{j+1}$. If $\tau_j = 1$, then since $m_{e_j}(i_j, *).m_{e_j}\left(i_{j+1}, *^{-1}\right)$ is defined, we have $m_{\rho e_j}A_{e_j}(i_j) = m_{\rho e_j}A_{e_j}(i_{j+1})$; since A_e is proper, again $i_j = i_{j+1}$. But then k can be reduced, contradicting the choice of n. Hence k is an identity. //

DEFINITION 6.4. If each morphism A_e in a groupnet diagram (C, A) is a monomorphism, the mapping cylinder $m(C, A)$ will be called the *categorical product* (or *graph product* if C is a directed graph) of the diagram.

As we shall see in the next section, this definition is motivated by Karrass' and Solitar's "tree product".

DEFINITION 6.5. The *core* of a mapping cylinder $m : (C, A) \to m(C, A)$ is the subgroupnet K of $m(C, A)$ generated by $\{m_e(i, *) \mid i \in \mathrm{Id}(A_{\lambda e}),\ e$ an edge of $C\}$.

Note that $K = K^*/J$, using the notation of the preceding proof, so that $K \cap m_v(A_v) = \mathrm{Id}(A_v)$ if each A_e is proper.

Let $\overline{K}$ denote the normal subgroupnet of $m(C, A)$ generated by K. The double cosets $m(C, A)/\overline{K} : \overline{K}$ form a groupnet under the partial product $\overline{K}a\overline{K}.\overline{K}b\overline{K} = \overline{K}akb\overline{K}$ if there is an element $k \in \overline{K}$ such that akb is defined in $m(C, A)$; the identities $\overline{K}i\overline{K}$ of this groupnet are in one-to-one correspondence with the components of $\overline{K}$. Let $P(C, A)$ denote $m(C, A)/\overline{K} : \overline{K}$ with this groupnet structure, and let $p_v : A_v \to P(C, A)$ and $p_e : A_{\lambda e} \times I \to P(C, A)$ denote composites of m_v and m_e with the quotient morphism $m(C, A) \to P(C, A)$. Observe that $p_e(i, *) = m_{\lambda e}(i)$, so that p_e is the constant homotopy between $p_{\lambda e}$ and $p_{\rho e}A_e$.

THEOREM 6.3. *$(P(C, A), \{p_v\})$ is the colimit of (C, A), that is, given any groupnet G and morphisms $g_v : A_v \to G$ such that $g_{\lambda e} = g_{\rho e}A_e$, there is a unique morphism $h : P(C, A) \to G$ such that $g_v = hp_v$.*

PROOF. Let $g_e : A_{\lambda e} \times I \to G$ be the constant homotopy from $g_{\lambda e}$ to $g_{\rho e}A_e$ (that is, g_e is defined by $g_e(a, 0) = g_{\lambda e}(a)$, $g_e(i, *) = g_{\lambda e}(i)$). Then $g : (C, A) \to G$ is a morphism so factors uniquely through m, that is, $g = h'm$. Since $h'm_e(i, *) = g_{\lambda e}(i)$, we have $h'(\overline{K}) \subset \mathrm{Id}(G)$, and h' factors uniquely through the quotient $m(C, A) \to P(C, A)$. //

7. Examples of categorical products

EXAMPLE 7.1. Let C be a group, with identity (vertex) 1, and let A_1 be a group on which C acts as a group of automorphisms $\{A_e \mid e \in C\}$. Let S be the split extension of A_1 by C, and let $s_1 : A_1 \to S$ be the inclusion $s_1(a) = (a, 1)$. Define $s_e : A_1 \times I \to S$ by $s_e(a, 0) = (a, 1)$ and $s_e(1, *) = (1, e)$. Since $s_e(a, 1) = (1, e)^{-1}.(a, 1).(1, e) = (A_e(a), 1)$, $s : (C, A) \to S$ is a morphism. It is easily checked that this is indeed the mapping cylinder of (C, A). The core of this categorical product is the copy of C in S.

It should be noted that if A_1 is a groupnet of the homotopy type of a group G, then $m(C, A)$ may have the homotopy type of a non-split extension of G by C.

EXAMPLE 7.2. Let (D, A) be a groupnet diagram in which D is an acyclic directed graph, and each A_e is proper.

THEOREM 7.1. *Under these conditions, K is acyclic, and therefore normal in $m(D, A)$: hence the colimit $P(D, A)$ has the same homotopy type as $m(D, A)$.*

PROOF. Suppose $m_{e_1}\left(i_1, *^{\tau_1}\right) \dots m_{e_n}\left(i_n, *^{\tau_n}\right)$ is a loop in K, $\tau_j = \pm 1$. Then $e_1^{\tau_1} e_2^{\tau_2} \dots e_n^{\tau_n}$ is a loop in $F(D)$. Since D is acyclic, there is an index j such that $e_{j+1} = e_j$ and $\tau_{j+1} = -\tau_j$. Then, as in the last part of the proof of Theorem 6.2, the fact that each A_e is proper implies that the expression for this loop may be reduced, and hence is an identity.

In particular then for the diagram in which A, B and C are groups, and f g are one-to-one,

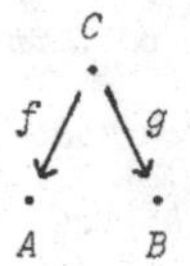

the graph product $m(D, A)$ is simply the free product with amalgamation $A *_C B$, together with two "spines" corresponding to the edges of the graph (note that this groupnet can also be obtained as the mapping cylinder of the graph below, where $A \& B$ denotes the disjoint union of A and B).

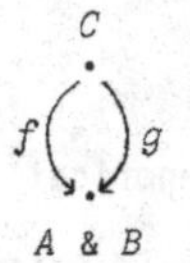

More generally, if D is a tree, the graph product $m(D, A)$ is essentially the "tree product" of Karrass and Solitar. (Their tree is however not quite our D : in their work, an edge of a graph corresponds not to a map, but to a pair of amalgamating subgroups. Thus where they have $A \overset{C}{\text{———}} B$, we use $A \swarrow \overset{C}{} \searrow B$; conversely an edge $C \xrightarrow{f} A$ where the corresponding morphism f is one-to-one, may be written $C \overset{C}{\text{———}} A$ by Karrass and Solitar.) The fact that the vertex groups A_v are embedded in the graph product, and in the colimit, follows immediately from the monomorphism Theorem 6.2 and the above Theorem 7.1.

EXAMPLE 7.3. If D is not acyclic, then the graph product $m(D, A)$ may not have the homotopy type of $P(D, A)$ (which is the "partial generalized free product" in the terminology of Karrass and Solitar). Indeed A_v is in general *not* embedded in $P(D, A)$. The graph product $m(D, A)$ does however have the homotopy type of a kind of HNN construction. Thus, consider the diagram

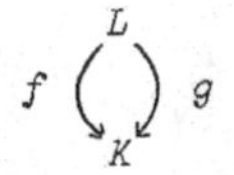

(with f and g one-to-one, and K and L groups).

THEOREM 7.2. *The graph product for this diagram has the homotopy type of the HNN construction with base* K *and associated subgroups* $f(L)$ *and* $g(L)$; *that is, if* K *has a presentation* $\|X : R\|$, *then* $m(D, A)$ *has the homotopy type of the group* $\|X \cup \{t\} : R \cup \{t^{-1}f(x)t = g(x) \mid x \in L\}\|$. *(In fact this result does not depend on* f *and* g *being monomorphisms.)*

PROOF. Let H denote the HNN extension whose presentation is given in the statement, and denote its identity by 1. Adjoin to H a copy of I, identifying

$1 \in H$ with $1 \in I$. This defines a fully associative pregroupnet; let M denote the universal groupnet of this pregroupnet. $M \simeq H$ of course (with "one spine" $*$ which is to correspond to the edge f).

Define $m_K : K \to M$ by $m_K(x) = x$ for $x \in X$ and $m_L : L \to M$ by

$m_L(x) = *.f(x).*^{-1}$ for $x \in L$. Further define $m_f : L \times I \to M$ by

$$\begin{cases} m_f(x,\ 0) = m_L(x)\ , \\ m_f(1,\ *) = * \end{cases}$$

and $m_g : L \times I \to M$ by

$$\begin{cases} m_g(x,\ 0) = m_L(x)\ , \\ m_g(1,\ *) = *.t\ . \end{cases}$$

Clearly $m_f : m_L \simeq m_K f$ and $m_g : m_L \simeq m_K g$.

Now let h be a morphism from the groupnet diagram to a groupnet G. Define $h^* : M \to G$ by

$$\begin{cases} h^*(x) = h_K(x) \quad \text{for} \quad x \in X\ , \\ h^*(t) = h_f(1,\ *^{-1}).h_g(1,\ *)\ , \\ h^*(*) = h_f(1,\ *)\ . \end{cases}$$

It is easily checked that h^* is a morphism, and that $h = h^*m$. Furthermore h^* is the unique morphism with this property since M is generated by $X \cup \{t,\ *\}$. Thus M is the mapping cylinder.

The fact that the base K is embedded in the HNN extension H now follows immediately from Theorem 6.2.

8. Pullbacks and mapping cylinders

We are now in a position to formulate our plan of attack on the subgroup theorem. We have seen that a free product with amalgamation or an HNN construction can be viewed (up to homotopy type) as a mapping cylinder of an appropriate groupnet diagram (and the colimit of the diagram is a quotient of the mapping cylinder). Suppose H is a (wide) subgroupnet of $M = m(C,\ A)$. There is an associated covering map $\pi_H : M_H \to M$, such that $M_H \simeq H$. Now suppose we can form a pullback

$$\begin{array}{ccc} (C,\ B) & \xrightarrow{\pi^*} & (C,\ A) \\ \downarrow{\scriptstyle m^*} & & \downarrow{\scriptstyle m} \\ M_H & \xrightarrow[\pi_H]{} & M \end{array}$$

for an appropriate diagram (C, B) such that each diagram

$$\begin{array}{ccc} B_v & \xrightarrow{\pi_v^*} & A_v \\ \downarrow{\scriptstyle m_v^*} & & \downarrow{\scriptstyle m_v} \\ M_H & \xrightarrow[\pi_H]{} & M \end{array}$$

is a pullback.

Then according to Theorem 5.2, π_v^* is a covering map corresponding to a family of subgroupnets $m_v^{-1}\left(u_d^{-1}.H.u_d\right) \subset A_v$, $d \in M/H : m_v(A_v)$, for a certain double coset representative u_d.

If further we show

(1) that M_H is a mapping cylinder for (B, A), and

(2) the vertex groupnets B_v may be replaced by strong deformation retracts, without changing the homotopy type of M_H,

then we shall have a subgroup theorem. In the case that each A_e is one-to-one, we shall prove that each B_e is one-to-one; thus in this case the subgroup theorem can be loosely expressed as follows - if H is a subgroupnet of a categorical product $M = m(C, A)$, then H is of the homotopy type of a categorical product $m(C, B)$ in which each vertex groupnet B_v is a disjoint union of groupnets $u_d^{-1}.H.u_d \cap A_v$, $d \in M/H : A_v$.

We now construct the pullback of a groupnet diagram. Suppose we are given morphisms

$$\begin{array}{ccc} & & (C, A) \\ & & \downarrow{\scriptstyle g} \\ Y & \xrightarrow{f} & X \end{array}$$

such that either

I: f is a covering map, or

II: f is a fibration and C is (the free category on) a directed graph.

(A morphism f is a *fibration* if it is surjective and has the path-lifting property - hence any surjection of groups is a fibration.)

For each vertex v define B_v, f_v^* and g_v^* via the pullback square

$$\begin{array}{ccc} B_v & \xrightarrow{f_v^*} & A_v \\ \downarrow{\scriptstyle g_v^*} & & \downarrow{\scriptstyle g_v} \\ Y & \longrightarrow & X \end{array} .$$

For each $i \in \mathrm{Id}(Y)$, $j \in \mathrm{Id}(A_{\lambda e})$ with $g_{\lambda e}(j) = f(i)$, choose an element $W(i, j, e) \in Y$ such that

(i) $\lambda W(i, j, e) = i$,

(ii) $fW(i, j, e) = g_e(j, *)$,

(iii) $W(i, j, e) = i$ if e is an identity arrow $e = 1_v$,

(iv) $W(i, j, e'e'') = W(i, j, e').W(i', A_{e'}(j), e'')$ where $i' = \rho W(i, j, e')$.

Case I. Since f is a covering map, there is a unique choice of W satisfying conditions (i) and (ii). Conditions (iii) and (iv) follow from the definition of morphisms of groupnet diagrams.

Case II. Since f is a fibration, we may choose W satisfying conditions (i) and (ii) for each edge of the directed graph. Conditions (iii) and (iv) then serve to define W for arrows e in the category generated by the directed graph. These elements automatically satisfy (i) and (ii).

Define $B_e : B_{\lambda e} \to B_{\rho e}$ by $B_e(y, a) = \left(W(\lambda y, \lambda a, e)^{-1}.y.W(\rho y, \rho a, e), A_e(a)\right)$. It is then straightforward to check that B_e is a well-defined morphism, and B is a functor from C to the category of groupnets. Thus (C, B) is a groupnet diagram.

Next define $f_e^* : B_{\lambda e} \times I \to A_{\rho e}$ to be the constant homotopy between $A_e f_{\lambda e}^*$ and $f_{\rho e}^* B_e$. $f^* : (C, B) \to (C, A)$ is then a morphism of groupnet diagrams.

Finally define $g_e^* : B_{\lambda e} \times I \to Y$ by

$$\begin{cases} g_e^*((y, a), 0) = g_{\lambda e}^*(y, a) = y , \\ g_e^*((i, j), *) = W(i, j, e) . \end{cases}$$

The four conditions above ensure that g_e^* is a homotopy between $g_{\lambda e}^*$ and $g_{\rho e}^* B_e$, and $g^* : (C, B) \to Y$ is a morphism. Furthermore, we have $fg^* = gf^*$, that is, the diagram

$$\begin{array}{ccc} (C, B) & \xrightarrow{f^*} & (C, A) \\ \downarrow{\scriptstyle g^*} & & \downarrow{\scriptstyle g} \\ Y & \xrightarrow{f} & X \end{array}$$

commutes.

THEOREM 8.1. *The diagram constructed above is a pullback square.*

PROOF. Suppose there is a diagram (C', B') with

$$\begin{array}{ccc} (C', B') & \xrightarrow{(t,\ h)} & (C, A) \\ \downarrow k & & \downarrow g \\ Y & \xrightarrow{\ f\ } & X \end{array}$$

commutative, that is, $fk_v = g_{tv}h_v$ and $fk_e = h_e \underline{\circ}\, g_{te}$ (see Definition 6.2).

We wish to define a morphism $(s, \sigma) : (C', B') \to (C, B)$ so that $(t, h) = (1, f^*)(s, \sigma)$ and $k = g^*(s, \sigma)$. Thus we must have $s = t$. The first co-ordinate of $\sigma_v(z)$ is $g^*_{tv}\sigma_v(z) = k_v(z)$, and the second co-ordinate is $f^*_{tv}\sigma_v(z) = h_v(z)$. Thus $\sigma_v(z) = \left(k_v(z), h_v(z)\right)$ is uniquely determined.

The first co-ordinate of $\sigma_e(i, *)$ must be

$$\begin{aligned} g^*_{\rho te}\sigma_e(i, *) &= g^*_{te}\left[\sigma_{\lambda e}(i), *^{-1}\right].\left(\sigma_e \underline{\circ}\, g^*_{te}\right)(i, *) \\ &= W\left(k_{\lambda e}(i), h_{\lambda e}(i), te\right)^{-1}.k_e(i, *) ; \end{aligned}$$

and the second co-ordinate of $\sigma_e(i, *)$ must be

$$\begin{aligned} f^*_{\rho te}\sigma_e(i, *) &= \sigma_e \underline{\circ}\, f^*_{te}(i, *) \quad \left(\text{since } f^*_{te} \text{ is the constant homotopy}\right) \\ &= h_e(i, *) . \end{aligned}$$

It is straightforward to check that with these definitions

(i) σ_e is a well-defined homotopy between $B_{te}\sigma_{\lambda e}$ and $\sigma_{\rho e}B'_e$,

(ii) $\sigma_{1_v}(i, *) = \sigma_v(i)$,

(iii) $\sigma_{e'e''} = \sigma_{e'} \phi\, \sigma_{e''}$ so that (t, σ) is a morphism of groupnet diagrams, and

(iv) $k_e = \sigma_e \underline{\circ}\, g^*_{te}$ and $h_e = \sigma_e \underline{\circ}\, f^*_{te}$, so that k and (t, h) factor through (t, σ). //

DEFINITION 8.1. A morphism $f : G \to H$ of nets is said to be *conservative* if $f(a) = f(b)$, $\lambda(a) = \lambda(b)$ implies $\rho(a) = \rho(b)$ (equivalently, if $f^{-1}\left(\text{Id}(G)\right)$ consists entirely of loops).

In particular, covering maps and proper maps are conservative.

THEOREM 8.2. *In the pullback construction above, if each* A_e *is a monomorphism and* f *is conservative, then each* B_e *is a monomorphism.*

PROOF. Suppose $B_e(y, a) = B_e(y', a')$. Then $A_e(a) = A_e(a')$ so $a = a'$; and $W(\lambda y, \lambda a, e)^{-1}.y.W(\rho y, \rho a, e) = W(\lambda y', \lambda a, e)^{-1}.y.W(\rho y', \rho a, e)$. Now $W(\lambda y, \lambda a, e)^{-1}$ and $W(\lambda y, \lambda a, e)^{-1}$ both have the same left hand endpoint, and both have image $g_e(\lambda a, *)^{-1}$ under f , so they have the same right hand end, that is, $\lambda y = \lambda y'$; hence $W(\lambda y, \lambda a, e) = W(\lambda y', \lambda a, e)$. Similarly $W(\rho y, \rho a, e) = W(\rho y', \rho a, e)$ and hence $y = y'$. //

We come now to the major result of this section, which states that the mapping cylinder of a pullback is the pullback of the mapping cylinder.

Let (C, B), f^*, g^* as constructed above be the pullback of

$$\begin{array}{ccc} & & (C, A) \\ & & \big\downarrow g \\ Y & \xrightarrow{f} & X \end{array}$$

where either f is a covering map or C is the free category on a directed graph and f is a conservative fibration. Let $m(C, A)$ and $m(C, B)$ be the respective mapping cylinders. The universal property of mapping cylinders provides the commutative diagram

$$\begin{array}{ccc} (C, B) & \xrightarrow{f^*} & (C, A) \\ \big\downarrow m^B & & \big\downarrow m^A \\ m(C, B) & \xrightarrow{mf^*} & m(C, A) \\ \big\downarrow g'' & & \big\downarrow g' \\ Y & \xrightarrow{f} & X \end{array}$$

with $g^* = g''m^B$ and $g = g'm^A$.

THEOREM 8.3. *The lower square in the above diagram is a pullback square. In particular, if* X *is the mapping cylinder of* (C, A) *via* g *, then* Y *is the mapping cylinder of* (C, B) *via* g^* .

PROOF. Let Z be the standard pullback, that is, $Z = \{(y, u) \in Y \times m(C, A) \mid f(y) = g'(u)\}$, with projection maps $\alpha : Z \to Y$ and $\beta : Z \to m(C, A)$. The maps g'' and mf^* factor through a morphism $\gamma : m(C, B) \to Z$; in fact γ is defined by

$$\begin{cases} \gamma m^B_v(y, a) = \left(y, m^A_v(a)\right) , \\ \gamma m^B_e\left((i, j), *\right) = \left(W(i, j, e), m^A_e(j, *)\right) . \end{cases}$$

We shall show that γ is an isomorphism.

(i) γ <u>is proper</u>

Each identity of $m(C, B)$ can be written uniquely as $m_v^B(i, j)$, for some vertex v and identities i, j. Then $\gamma m_v^B(i, j) = \left(i, m_v^A(j)\right)$. Thus if $\gamma m_v^B(i, j) = \gamma m_{v'}^B(i', j')$ then $i = i'$, $v = v'$ and $j = j'$.

(ii) γ <u>is onto</u>

Let $(y, u) \in Z$. Now u can be written as a product $u = u_1 u_2 \ldots u_k$ where $u_r = m_{v_r}^A(a_r)$ for some $a_r \in A_{v_r}$ or $u_r = m_{e_r}^A\left(j_r, *^{\pm 1}\right)$ for some $j_r \in \mathrm{Id}\left(A_{\lambda e_r}\right)$. Since f is a fibration, we may choose elements $y_r \in Y$ so that $\lambda y_1 = \lambda y$, $f(y_r) = g'(u_r)$ and $y_1 y_2 \ldots y_k$ is defined. Let $y_{k+1} = y_k^{-1} y_{k-1}^{-1} \ldots y_1^{-1} y$. Then $f(y_{k+1}) = g'(\rho u)$ and $(y, u) = (y_1, u_1)(y_2, u_2) \ldots (y_k, u_k)(y_{k+1}, u_{k+1})$ where $u_{k+1} = \rho u$.

If $u_r = m_{v_r}^A(a_r)$, then $(y_r, u_r) = \gamma m_{v_r}^B(y_r, a_r) \in \mathrm{im}\gamma$.

If $u_r = m_{e_r}^A(j_r, *)$, then $f(y_r) = fW(\lambda y_r, j_r, e_r)$. Let $y_r' = W(\lambda y_r, j_r, e_r)^{-1} y_r$; then $f(y_r') = f(\rho y_r) = g' m_{\rho e_r}^A(j_r')$, where $j_r' = A_{e_r}(j_r)$. Thus

$$
\begin{aligned}
(y_r, u_r) &= \left(W(\lambda y_r, j_r, e_r), u_r\right) \cdot \left(y_r', m_{\rho e_r}^A(j_r')\right) \\
&= \gamma m_{e_r}^B\left((\lambda y_r, j_r), *\right) \cdot \gamma m_{\rho e_r}^B(y_r', j_r') \\
&\in \mathrm{im}\gamma .
\end{aligned}
$$

Finally, if $u_r = m_{e_r}^A\left(j_r, *^{-1}\right)$, then let $y_r' = y_r \cdot W(\rho y_r, j_r, e_r)$, and we have $(y_r, u_r) = \gamma m_{e_r}^B\left((\rho y_r', j_r), *^{-1}\right) \cdot \gamma m_{\lambda e_r}^B(y_r', j_r) \in \mathrm{im}\gamma$.

(iii) γ <u>is one-to-one</u>

We need some preliminary results.

LEMMA 1. *For* $i \in \mathrm{Id}(Y)$ *and* $j \in \mathrm{Id}(A_{\lambda e})$ *such that* $f(i) = g_{\rho e}(A_e(j))$, *there exists* $\underline{i} \in \mathrm{Id}(Y)$ *such that* $f(\underline{i}) = g_{\lambda e}(j)$ *and* $\rho W(\underline{i}, j.e) = i$.

PROOF. Since f is a fibration, there is a lift z of $g_e(j, *^{-1})$ with $\lambda z = i$. Then $W(\rho z, j, e)$ and z^{-1} both cover $g_e(j, *)$ and both start at ρz, so $W(\rho z, j, e) = i$. Hence $\rho z = \underline{i}$ is the required identity.

Next recall the construction of the mapping cylinder (see Theorem 6.1 and Definition 3.3). The three quotient operations involved, the identification $\equiv$, the contraction $\setminus$ and killing the loops J, can be combined into a single operation as follows. Let P be the disjoint union $\bigcup_v A_v \bigcup_e A_{\lambda e} \times I_e$, with ends $E(P) = \bigcup \mathrm{Id}(A_v)$, and with λ and ρ defined by $\lambda(a, *_e) = \lambda(a)$, $\rho(a, *_e) = \rho A_e(a)$, and so forth. A product ab is defined in P if and only if it is defined in some A_v or $A_{\lambda e} \times I_e$. Let $S(P)$ be the product net of non-empty sequences $x_1, x_2, \ldots, x_k$ such that $x_r \in P$ and $\rho(x_r) = \lambda(x_{r+1})$, the product defined by concatenation.

A sequence $V \in S(P)$ is said to arise from a sequence $U = x_1, x_2, \ldots, x_k$ by an elementary expansion of the r-th term if one of the following holds:

(i) $x_r \in A_{\lambda e}$, $V = x_1, x_2, \ldots, x_{r-1}, (x_r, 0_e), x_{r+1}, \ldots, x_k$;

(ii) $x_r = A_e(y)$, $V = x_1, \ldots, x_{r-1}, (y, 1_e), x_{r+1}, \ldots, x_k$;

(iii) $x_r = ab$ for some a, b in P, $V = x_1, \ldots, x_{r-1}, a, b, x_{r+1}, \ldots, x_k$;

(iv) $x_r = (i, *_e)$ where $i \in \mathrm{Id}(A_{\lambda e})$ and $e = e'e''$,
$V = x_1, \ldots, x_{r-1}, (i, *_{e'}), (A_{e'}(i), *_{e''}), x_{r+1}, \ldots, x_k$.

Two sequences are equivalent, $U \sim V$, if there is a finite chain of elementary expansions or contractions taking U to V.

Then it is easily seen that $m(C, A)$ may be identified with $S(P)/\sim$, with m_v^A and m_e^A being the composites $A_v \subset P \subset S(P) \to S(P)/\sim$ and $A_{\lambda e} \times I_e \subset P \subset S(P) \to S(P)/\sim$. The quotient map $S(P) \to m(C, A)$ shall be denoted by ν^A.

Similarly, let Q be the corresponding construction for (C, B), with quotient map $\nu^B : S(Q)/\sim \to m(C, B)$.

The morphisms f^*, g and g^* induce morphisms

$$Sf^* : S(Q) \to S(P) \quad \text{where} \quad \begin{cases} Sf^*(y,\, a) = f^*_v(y,\, a) = a \quad \text{for} \quad (y,\, a) \in B_v \,, \\ Sf^*\big((y,\, a),\, t_e\big) = \big(a,\, t_e\big),\ \big((y,\, a),\, t_e\big) \in B_{\lambda e} \times I_e \,, \\ Sf^*\big(x_1,\, x_2,\, \ldots,\, x_k\big) = Sf^*\big(x_1\big),\, \ldots,\, Sf^*\big(x_k\big) \,; \end{cases}$$

$$Sg = g'\nu^A : S(P) \to X \qquad \begin{cases} Sg(a) = g'm^A_v(a) = g_v(a) \quad \text{for} \quad a \in A_v \,, \\ Sg\big(a,\, t_e\big) = g_e(a,\, t) \quad \text{for} \quad \big(a,\, t_e\big) \in A_{\lambda e} \times I_e \,, \\ Sg\big(x_1,\, \ldots,\, x_k\big) = Sg\big(x_1\big) \ldots Sg\big(x_k\big) \,; \end{cases}$$

$$Sg^* = g''\nu^B : S(Q) \to Y \qquad \begin{cases} Sg^*(y,\, a) = g^*_v(y,\, a) = y \,, \quad \text{for} \quad (y,\, a) \in B_v \,, \\ Sg^*\big((y,\, a),\, t_e\big) = g^*_e\big((y,\, a),\, t_e\big) \,, \\ Sg^*\big(x_1,\, \ldots,\, x_k\big) = Sg^*\big(x_1\big) \ldots Sg^*\big(x_k\big) \,; \end{cases}$$

and we have a commutative diagram

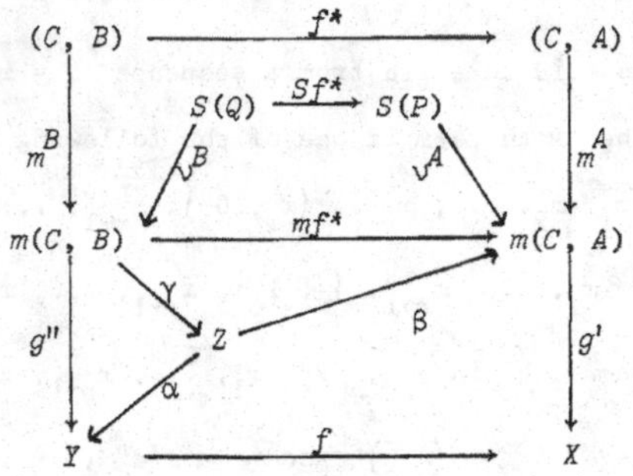

.

LEMMA 2. *If* U *and* V *are sequences in* $S(Q)$ *such that* $Sf^*(U) \sim Sf^*(V)$, *then there exists* $V' \in S(Q)$ *such that* $U \sim V'$ *and* $Sf^*(V') = Sf^*(V)$.

PROOF. By induction on the length of the chain of elementary expansions and contractions taking $Sf^*(U)$ to $Sf^*(V)$, we may assume that $Sf^*(V)$ arises from $Sf^*(U)$ by a single elementary expansion or contraction. There are then eight cases.

Let $U = x_1,\, x_2,\, \ldots,\, x_k$.

Expansion of type (i):

$$x_r = (y,\, a) \in B_{\lambda e} \,,$$

$$Sf^*(V) = Sf^*\big(x_1\big),\, \ldots,\, \big(a,\, 0_e\big),\, \ldots,\, Sf^*\big(x_k\big) \,,$$

Then let $V' = x_1,\, \ldots,\, \big((y,\, a),\, 0_e\big),\, \ldots,\, x_k$.

Contraction of type (i):

$$x_r = \big((y,\, a),\, 0_e\big) \,,$$

$$Sf^*(V) = Sf^*(x_1), \ldots, a, \ldots, Sf^*(x_k) ,$$

Then let $V' = x_1, \ldots, (y, a), \ldots, x_k$.

Expansion of type (ii):

$$x_r = (y, a) \in B_{\rho e} , \text{ with } a = A_e(b) ,$$

$$Sf^*(V) = Sf^*(x_1), \ldots, (b, 1_e), \ldots, Sf^*(x_k) .$$

By Lemma 1, there is an identity $\underline{i} \in Y$ such that $f(\underline{i}) = g_{\lambda e}(\lambda b)$ and $\rho W(\underline{i}, \lambda b, e) = \lambda y$; and there is an identity $\underline{i}'$ such that $f(\underline{i}') = g_{\lambda e}(\rho b)$ and $\rho W(\underline{i}', \rho b, e) = \rho y$. Let $z = W(\underline{i}, \lambda b, e).y.W(\underline{i}', \rho b, e)^{-1}$. Then $B_e(z, b) = (y, a)$, and $V' = x_1, \ldots, ((z, b), 1_e), \ldots, x_k$ is a well-defined element of $S(Q)$.

Contraction of type (ii):

$$x_r = ((y, b), 1_e) \text{ with } a = A_e(b) ,$$

$$Sf^*(V) = Sf^*(x_1), \ldots, a, \ldots, Sf^*(x_k) ,$$

Let $V' = x_1, \ldots, B_e(y, b), \ldots, x_k$.

Expansion of type (iii):

$Sf^*(x_r)$ is a product in P .

There are two possibilities, either $x_r = (y, ab)$ where a and b lie in some A_v , with $Sf^*(V) = Sf^*(x_1), \ldots, a, b, \ldots, Sf^*(x_k)$ or $x_r = ((y, ab), s_et_e)$ where (a, s_e) and (b, t_e) lie in some $A_{\lambda e} \times I$, with $Sf^*(V) = Sf^*(x_1), \ldots, (a, s_e), (b, t_e), \ldots, Sf^*(x_k)$.

In the first case let y_1 be a lift of $g_v(a)$ with $\lambda y_1 = \lambda y$, and put $y_2 = y_1^{-1}y$. Let $V' = x_1, \ldots, (y_1, a), (y_2, b), \ldots, x_k$.

In the second case, let y_1 be a lift of $g_{\lambda e}(a)$ with $\lambda y_1 = \lambda y$, and put $y_2 = y_1^{-1}y$. Then

$$V' = x_1, \ldots, ((y_1, a), s_e), ((y_2, b), t_e), \ldots, x_k .$$

Contraction of type (iii):

$$x_r, x_{r+1} = (y_1, a), (y_2, b) \text{ with } ab \text{ defined in } A_v ,$$

$$Sf^*(V) = Sf^*(x_1), \ldots, ab, \ldots, Sf^*(x_k) \quad ; \quad \text{or}$$

$$x_r, x_{r+1} = ((y_1, a), s_e), ((y_2, b), t_e) \text{ with } (ab, s_e t_e) \in A_{\lambda e} \times I_e ,$$
$$Sf^*(V) = Sf^*(x_1), \ldots, (ab, s_e t_e), \ldots, Sf^*(x_k) .$$

In the first case we can simply let $V' = x_1, \ldots, (y_1 y_2, ab), \ldots, x_k$.

In the second case we must prove that $y_1 y_2$ is defined. As elements of Q, the right hand end of x_r is the left hand end of x_{r+1}. For $\rho s_e = \lambda t_e = 0_e$, we have $\rho x_r = (\rho y_1, \rho a) = \lambda x_{r+1} = (\lambda y_2, \lambda b)$; hence $\rho y_1 = \lambda y_2$. For $\rho s_e = \lambda t_e = 1_e$, we have $\rho x_r = B_e(\rho y_1, \rho a) = (\rho W(\rho y_1, \rho a, e), \rho a)$ and $\lambda x_{r+1} = \lambda B_e(y_2, b) = (\rho W(\lambda y_2, \lambda b, e), \lambda b)$. But since f is conservative and $\rho a = \lambda b$, it follows that $\rho y_1 = \lambda y_2$. Thus we may put
$V' = x_1, \ldots, ((y_1 y_2, ab), s_e t_e), \ldots, x_k$.

Expansion of type (iv):

$$x_r = ((y, j), *_e) \text{ where } j \in \mathrm{Id}(A_{\lambda e}) , \quad e = e'e'' ,$$
$$Sf^*(V) = Sf^*(x_1), \ldots, (j, *_{e'}), (A_{e'}(j), *_{e''}), \ldots, Sf^*(x_k) .$$

Then

$$U \sim x_1, \ldots, ((y, j), 0_e), ((y, j), *_e), \ldots, x_k$$
$$\sim x_1, \ldots, (y, j), ((\rho y, j), *_{e'}), (B_{e'}(\rho y, j), *_{e''}), \ldots, x_k$$
$$\sim V' = x_1, \ldots, ((y, j), *_{e'}), (B_{e'}(\rho y, j), *_{e''}), \ldots, x_k .$$

Contraction of type (iv):

$$x_r, x_{r+1} = ((y_1, j), *_{e'}), ((y_2, A_{e'}(j)), *_{e''}) \text{ where } e = e'e'' ,$$
$$Sf^*(V) = Sf^*(x_1), \ldots, (j, *_e), \ldots, Sf^*(x_k) .$$

Then

$$U \sim x_1, \ldots, ((y_1, j), 0_{e'}), ((\rho y_1, j), *_{e'}), ((\lambda y_2, A_{e'}(j)), *_{e''}),$$
$$((y_2, A_{e'}(j), 1_{e''}), \ldots, x_k$$
$$\sim x_1, \ldots, (y_1, j), ((\rho y_1, j), *_{e'}), (B_{e'}(\rho y_1, j), *_{e''}), B_{e''}(y_2, A_{e'}(j)), \ldots, x_k .$$

Now let $\underline{i} \in \mathrm{Id}(Y)$ be such that $\rho W(\underline{i}, j, e') = \lambda y_2$ (note that $\rho y_2 = \lambda y_2$ since f is conservative). Let $z = W(\underline{i}, j, e').y_2.W(\underline{i}, j, e')^{-1}$. Then $f(z) = g_{\lambda e}(j)$ and $B_{e'}(z, j) = (y_2, A_{e'}(j))$. Thus $B_e(z, j) = B_{e''}(y_2, A_{e'}(j))$.
So

$$U \sim x_1, \ldots, \big(\{y_1, j\}, 0_e\big), \big(\{\rho y_1, j\}, *_e\big), \big((z, j), 1_e\big), \ldots, x_k .$$

Finally it follows from the argument used in the case of a contraction of type (iii) that $y_1 z$ is defined in Y. Thus $U \sim V' = x_1, \ldots, \big(\{y_1 z, j\}, *_e\big), \ldots, x_k$, proving Lemma 2.

We can now prove that γ is one-to-one. Since γ is proper, it suffices to prove that if $\gamma(x) = \gamma(\lambda x)$, then $x = \lambda x$.

Let U be a sequence in $S(Q)$ with $\nu^B(U) = x$ (refer to the diagram preceding Lemma 2). Let V be the one-term sequence (i, j), where $\lambda x = m_v^B(i, j)$. Then $mf^*\nu^B(U) = \beta\gamma(x) = \beta\gamma(\lambda x) = mf^*\nu^B(V)$, so $\nu^A Sf^*(U) = \nu^A Sf^*(V)$, that is, $Sf^*(U) \sim Sf^*(V)$. By Lemma 2, there is a sequence V' in $S(Q)$ such that $U \sim V'$ and $Sf^*(V') = Sf^*(V)$. But Sf^* leaves invariant the number of terms in a sequence, so V' has just one term, that is, $V' = (y, j)$ for some $y \in Y$. Then

$$\begin{aligned} y &= Sg^*(V') = g''\nu^V(V') = \alpha\gamma\nu^B(U) = \alpha\gamma(x) = \alpha\gamma(\lambda x) \\ &= g''m_v^B(i, j) = g_v^*(i, j) = i . \end{aligned}$$

Thus $V' = V$, and $x = \nu^B(U) = \nu^B(V) = \lambda x$. //

This result established part 1) of the program outlined at the beginning of this section, and provides a subgroupnet theorem sufficient for the results of Section 10. However to recover the usual theorem for groups, we must replace the vertex groupnets B_v by disjoint unions of groups. It seems this can only be done when C is the free category on a directed graph.

Let (D, A) be a groupnet diagram, with D a directed graph. For each vertex v of D, let A'_v be a strong deformation retract of A_v, that is, there are morphisms $s_v : A'_v \to A_v$ and $r_v : A_v \to A'_v$ with $r_v s_v = 1 : A'_v \to A'_v$ and a homotopy $G_v : A_v \times I \to A_v$ from the identity map on A_v to $s_v r_v$ which is constant on $s_v(A'_v)$, that is, $G_v\big(s_v(a), *\big) = s_v(a)$. For each edge e of D, define $A'_e : A'_{\lambda e} \to A'_{\rho e}$ by $A'_e = r_{\rho e} A_e s_{\lambda e}$; this uniquely determines a functor A' from $C(D)$, the free category on D, to the category of groupnets. Thus we have a groupnet diagram (D, A'). Define homotopies $s_e : A'_{\lambda e} \times I \to A_{\rho e}$ and $r_e : A_{\lambda e} \times I \to A'_{\rho e}$ for each edge e of D by $s_e = G_e\big(A_e s_{\lambda e}, 1\big) : A_e s_{\lambda e} \simeq s_{\rho e} A'_{\rho e}$ and

$$\left.\begin{aligned} r_e(a, 0) &= A'_e r_{\lambda e}(a) \\ r_e(i, *) &= r_{\rho e} A_e G_{\lambda e}\left(i, *^{-1}\right).r_{\rho e} G_{\rho e}\left(A_e(i), *^{-1}\right) \end{aligned}\right\} r_e : A'_e r_{\lambda e} \simeq r_{\rho e} A_e .$$

These maps define morphisms $s : (C, A') \to (C, A)$ and $r : (C, A) \to (C, A')$.

THEOREM 8.4. *$m(C, A')$ is a strong deformation retract of $m(C, A)$ under the inclusion ms and retraction mr; the deformation $1 \simeq ms \circ mr$ is given by $mG : m(C, A) \times I \to m(C, A)$ with $mG\left(m_v^A(i), *\right) = m_v^A G_v(i, *)$.*

9. The subgroup theorem for mapping cylinders

Let G be a group which has the homotopy type of a mapping cylinder $m(D, A)$, where D is connected directed graph and each vertex groupnet A_v is a group, and let H be a subgroup of G. We show that H has the homotopy type of a mapping cylinder $m(D, K)$, where K_v is a disjoint union of subgroups of A_v; if G is a graph product (each A_e is one-to-one), then so is H.

Let 1_v denote the identity in A_v, and choose a distinguished vertex 0 in D. G may be identified with the group of loops in $m(D, A)$ at 1_0. Let T be a maximal tree in D; T determines a maximal acyclic subgroupnet T^* of $m(D, A)$ with $m_e\left(1_{\lambda e}, *\right) \in T^*$ if and only if $e \in T$. The retraction $r : m(D, A) \to G$ is determined by $r(T^*) = 1_0$. Let H^* be the subgroupnet of $m(D, A)$ generated by $H \cup T^*$; then r retracts H^* onto H, and H^* is a connected wide subgroupnet of $m(D, A)$ of the homotopy type of H.

Let w_v be the unique element of T^* joining 1_0 to 1_v. Let $\overline{A}_v$ be the subgroup of G defined by $\overline{A}_v = w_v.m_v\left(A_v\right).w_v^{-1}$; if G is a graph product of (D, A), then $\overline{A}_v$ is isomorphic to A_v.

Let D_v be the set of double cosets $D_v = G/H : \overline{A}_v$, and choose double coset representatives $u_d \in d \subset G$ for each $d \in D_v$, where $u_d = 1_0$ if $d = H.1_0.\overline{A}_v$. Let K_d be the subgroup $m_v^{-1}\left(w_v^{-1}u_d^{-1}Hu_d w_v\right)$ of A_v; if G is a graph product, then K_d is isomorphic to $\overline{K}_d = \left(u_d^{-1}Hu_d\right) \cap \overline{A}_v$ via rm_v. Let K_v be the disjoint union of the groups K_d, $d \in D_v$.

Let $t_e = w_{\lambda e}.m_e\left(1_{\lambda e}, *\right).w_{\rho e}^{-1} \in G$; note that $t_e = 1_0$ if $e \in T$. Define a map $D_e : D_{\lambda e} \to D_{\rho e}$ by $D_e(d) = H.u_d t_e.\overline{A}_{\rho e}$, for each edge e . Since $H.u_{d'}.\overline{A}_{\rho e} = H.u_d t_e.\overline{A}_{\rho e}$ for $d' = D_e(d)$, we may choose $\alpha_{d,e} \in \overline{A}_{\rho e}$ such that $Hu_{d'} = Hu_d t_e \alpha_{d,e}$.

Choose a coset representative $c(\sigma) \in \sigma \subset A_v$ for each coset $\sigma \in A_v/K_d$, such that $c(K_d) = 1_0$. The map rm_v induces a one-to-one correspondence between A_v/K_d and $\overline{A}_v/\overline{K}_d$; let $\sigma_{d,e}$ be the coset in $A_{\rho e}/K_{d'}$ corresponding to $\overline{K}_{d'}\alpha_{d,e}^{-1}$, where $d' = D_e(d)$.

Define $K_e : K_{\lambda e} \to K_{\rho e}$ by $K_e(a) = c(\sigma_{d,e}).A_e(a).c(\sigma_{d,e})^{-1} \in K_{d'}$ for $a \in K_d$, $d \in D_{\lambda e}$. This defines a diagram (D, K) .

THEOREM 9.1 (the Subgroup Theorem for mapping cylinders). H *has the homotopy type of* $m(D, K)$.

PROOF. Let $\pi_H : G_H \to G$ be the covering morphism corresponding to H . We apply Theorems 8.3 and 8.4 to obtain the commutative diagram

$$\begin{array}{ccccc}
(D, K) & \xrightarrow{s} & (D, B) & \xrightarrow{\pi^*} & (D, A) \\
\downarrow{\scriptstyle m^K} & & \downarrow{\scriptstyle m^B} & & \downarrow{\scriptstyle m} \\
m(D, K) & \xrightarrow{ms} & m(D, A)_{H^*} & \xrightarrow{\pi} & m(D, A) \\
& & \downarrow{\scriptstyle r^*} & & \downarrow{\scriptstyle r} \\
H \simeq & & G_H & \xrightarrow{\pi_H} & G
\end{array} .$$

The vertex groupnet B_v is the covering groupnet of A_v corresponding to the collection of subgroupnets $\{K_d \mid d \in D_v = G/H : rm_v(A_v) = G/H : \overline{A}_v\}$, where K_d is the subgroup $m_v^{-1}r^{-1}\left(u_d^{-1}Hu_d\right) = m_v^{-1}\left(w_v^{-1}u_d^{-1}Hu_d w_v\right)$ (see Theorem 5.2).

Thus B_v may be identified with the disjoint union $\cup B_d$, where $B_d = \left(A_v/K_d\right) \times A_v$, with $\pi^*(\sigma, a) = a$ and

$$\begin{aligned} r^*m_v^B(\sigma, a) &= \left(Hu_d.rm_v(c(\sigma)), rm_v(a)\right) \\ &= \left(Hu_d w_v m_v(c(\sigma))w_v^{-1}, w_v m_v(a)w_v^{-1}\right) . \end{aligned}$$

Furthermore, it is straightforward to prove that $B_e(\sigma, a) = \left(\sigma_{d,e}A_e c(\sigma), A_e(a)\right)$.

According to Theorem 8.4, the morphism s is defined by $s_v(a) = \left(K_d, a\right)$ for $a \in K_d \subset A_v$, $s_e\left(1_{d,\lambda e}, *\right) = \left(K_{d'}, c(\sigma_{d,e})\right)^{-1}$ where $1_{d,\lambda e}$ denotes the identity of

$K_d \subset K_{\lambda e}$.

The map $K_e = s_{\rho e}^{-1}(s_e(-, 1))$ is then given by $K_e(a) = c(\sigma_{d,e}).A_e(a).c(\sigma_{d,e})^{-1}$.

By Theorems 8.4 and 5.1, r^*ms is a homotopy equivalence between $m(D, K)$ and G_H . //

The diagram (D, K) may of course be replaced by a diagram (D', K') in which each vertex group $K'_{d,v} \approx K_d$ is connected, where the vertices of D' are $\cup D_v$, and edges $\{(d, e) \mid d \in D_{\lambda e}\}$ with (d, e) joining d to $d' = D_e(d)$. If A_e is one-to-one for all e , then $K'_{d,e}$ will be one-to-one.

Finally, let us consider the case of an HNN construction or free product with amalgamation G . This is of the homotopy type of a graph product of a diagram

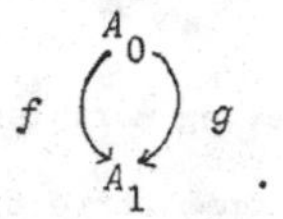

(In the case of a free product, A_1 is actually a disjoint union of two groups, but the argument is essentially the same). If H is a subgroup of G , then H is of the homotopy type of a graph product of

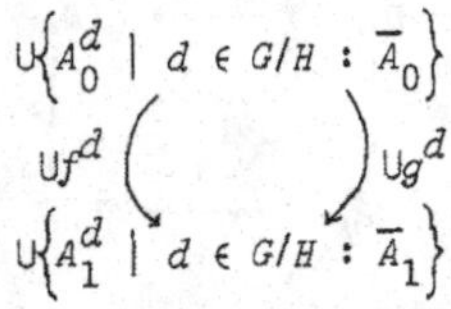

where $A_0^d \approx \{u_d^{-1}Hu_d\} \cap \bar{A}_0$, and so forth. The graph can be separated out so that the vertex groupnets are connected, and then a maximal tree chosen in this graph. Form the graph product P over this tree. The original graph product can then be written as a graph product of the form

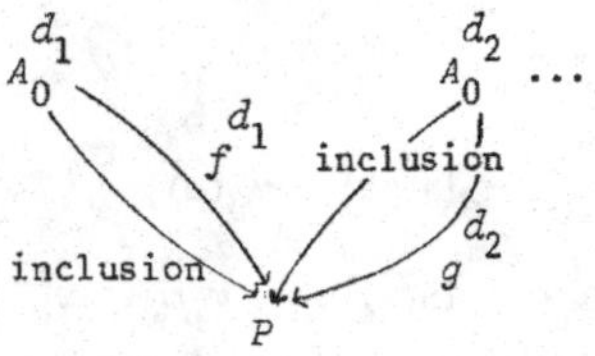

that is, a number of simple HNN extensions of a tree product (*cf.* Karrass and Solitar [4]).

THEOREM 9.2 (The Subgroup Theorem for Colimits). *Let G be the colimit of a group diagram (C, A) , and let H be a subgroup of G . Then H has the homotopy* *type of the colimit of the pullback diagram* (C, B) *in*

$$\begin{array}{ccc} (C, B) & \xrightarrow{\pi^*} & (C, A) \\ \downarrow m^B & & \downarrow m^A \\ m(C, B) & \xrightarrow{\pi} & m(C, A) \\ \downarrow \eta^* & & \downarrow \eta \\ H \simeq \quad G_H & \xrightarrow{\pi_H} & G \end{array} .$$

PROOF. By Theorem 6.3, G is the quotient of $m(C, A)$ by the normal subgroupnet generated by its core K^A (so that η is a monotone surjection). Let K^B be the core of $m(C, B)$, generating the normal subgroupnet $\overline{K}^B$.

Now

$$\begin{aligned} \pi m_e^B\{((Hg, 1), 1_{\lambda e}), *\} &= m_e^A(\pi_{\lambda e}^*((Hg, 1), 1_{\lambda e}), *).m_{\rho e}^A \pi^*{}_e\{((Hg, 1), 1_{\lambda e}), *\} \\ &= m_e^A(1_{\lambda e}, *) \end{aligned}$$

since π_e^* is the constant homotopy. Hence $\pi(\overline{K}^B) \subset \overline{K}^A = \ker \eta$, and hence $\overline{K}^B \subset \ker \eta^*$.

Moreover π is a covering map, and the lift of an element of K^A at any identity of $m(C, B)$ is in K^B; and if k is a loop in K^A, with lift k^* in $m(C, B)$, then $\eta^*(k^*)$ lifts $\eta(k) \in \mathrm{Id}(G)$, so that $\eta^*(k^*)$ is an identity, and is also a loop. Thus any conjugate aka^{-1} of a loop in K^A lifts to a conjugate $a^*k^*(a^*)^{-1}$; hence $\pi^{-1}(\overline{K}^A) = \overline{K}^B$. If $x \in \ker \eta^*$, then $\pi(x) \in \ker\eta = \overline{K}^A$ so $x \in \overline{K}^B$.

Thus G_H is the colimit of (C, B). //

It must be noted that H is not necessarily of the homotopy type of a diagram in which the vertex groupnets B_v are replaced by disjoint unions of groups (*cf.* Theorem 9.1), essentially because the inclusion morphism $s : (C, K) \to (C, B)$ does not have s_e the constant homotopy. The example which follows illustrates this point. This makes it difficult to compare the strength of Theorem 9.1 with the results of Hanna Neumann [6] on generalized free products, since it is not expressed in the language of group theory, but presumably they are essentially the same. (Note that in the case of a generalized free product G, each vertex group A_v is embedded in G. It follows easily that each B_v is embedded in G_H, and the loops in B_v are therefore embedded in H.)

We close this section with a simple example of the application of these two theorems. Let $G = Z_2 * Z_2$ with generators x and y , and let $H = G'$, the commutator subgroup. G has the homotopy type of the mapping cylinder of (D, A) ,

$$\begin{array}{ccc} & A_0 & \\ {}^{f}\swarrow & & \searrow^{g} \\ A_1 & & A_2 \end{array} \qquad \text{where } A_0 = \{1_0\} , \quad A_1 = \{1_1, x\} , \quad A_2 = \{1_2, y\} .$$

(G is also the colimit of (D, A) .)

Then $D_0 = G/H : A_0 = \{\alpha, \beta, \gamma, \delta\}$, with double coset representatives $1, x, y, xy$; $D_1 = G/H : A_1 = \{\varphi, \psi\}$ with representatives $1, y$; and $D_2 = G/H : A_2 = \{\mu, \nu\}$ with representatives $1, x$. $T = \{f, g\}$, and $t_e = 1$ for both edges $e = f, g$.

The maps D_f and D_g are given by

$$\begin{aligned} D_f : \alpha &\mapsto H.1.A_1 = \varphi , & D_g : \alpha &\mapsto H.1.A_2 = \mu , \\ \beta &\mapsto H.x.A_1 = \varphi , & \beta &\mapsto H.x.A_2 = \nu , \\ \gamma &\mapsto H.y.A_1 = \psi , & \gamma &\mapsto H.y.A_2 = \mu , \\ \delta &\mapsto H.xy.A_1 = \psi , & \delta &\mapsto H.xy.A_2 = \nu . \end{aligned}$$

K_d is the subgroup $H \cap A_0 = \{1_0\}$, for $d = \alpha, \beta, \gamma, \delta$. Similarly $K_\varphi = H \cap A_1 = \{1_1\} = y^{-1}Hy \cap A_1 = K_\psi$, and $K_\mu = K_\nu = \{1_2\}$. Thus B_1 consists of two copies of I , the universal covering of Z_2 , as does B_2 .

Finally,

$$\begin{aligned} \alpha_{\alpha,f} &= 1 , & \sigma_{\alpha,f} &= K_\varphi , & \alpha_{\alpha,g} &= 1 , & \sigma_{\alpha,g} &= K_\mu , \\ \alpha_{\beta,f} &= x , & \sigma_{\beta,f} &= K_\varphi x , & \alpha_{\beta,g} &= 1 , & \sigma_{\beta,g} &= K_\nu , \\ \alpha_{\gamma,f} &= 1 , & \delta_{\gamma,f} &= K_\psi , & \alpha_{\gamma,g} &= y , & \sigma_{\gamma,g} &= K_\mu y , \\ \alpha_{\delta,f} &= x , & \sigma_{\delta,f} &= K_\psi x , & \alpha_{\delta,g} &= y , & \sigma_{\delta,g} &= K_\nu y , \end{aligned}$$

and hence

$$B_f(K_\alpha, 1_0) = (K_\varphi, 1_1) , \quad B_g(K_\alpha, 1_0) = (K_\mu, 1_2)$$

and so forth, giving the diagram

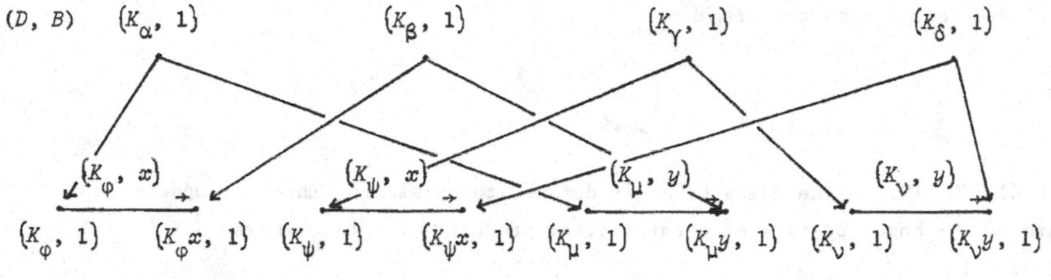

with associated diagram (D, K) **of groups**

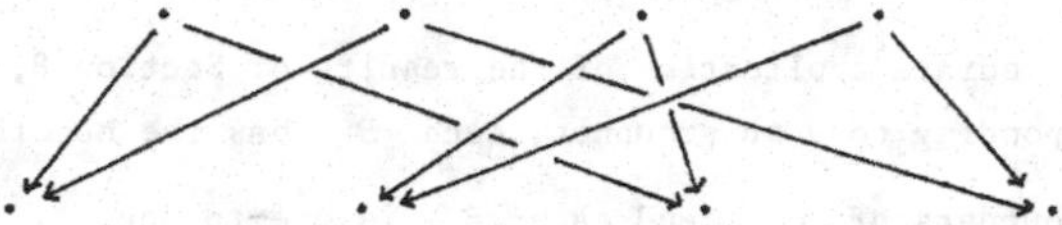

Thus $m(D, B)$ **is clearly of infinite cyclic homotopy type, generated by the element**

$$m^B_f((K_\alpha, 1), *).m^B_1(K_\varphi, x).m^B_f\left[(K_\beta, 1), *^{-1}\right].m^B_g((K_\beta, 1), *).m^B_2(K_\nu, y).$$
$$.m^B_g\left[(K_\delta, 1), *^{-1}\right].m^B_f((K_\delta, 1), *).m^B_1(K_\psi, x)^{-1}.m^B_f\left[(K_\gamma, 1), *^{-1}\right].$$
$$.m^B_g((K_\gamma, 1), *).m^B_2(K_\mu, y)^{-1}.m^B_g\left[(K_\alpha, 1), *^{-1}\right].$$

Thus H **is infinite cyclic, generated by the image of this element in** G **, that is,** $xyxy$.

The colimit of (D, B) **, obtained by collapsing the core of** $m(D, B)$ **, is clearly also infinite cyclic in homotopy type, but the colimit of** (D, K) **is trivial.**

10. A class of groupnets

We construct a class of groupnets which have some interest in the topology 3-manifolds. Hopefully the methods discussed in this work will be of use in their study.

DEFINITION 10.1. **A class of groupnets** G **will be said to be** *admissible* **if**

(i) **if** $G \simeq G' \in G$ **, then** $G \in G$ **,**

(ii) **if** $G \subset G' \in G$ **, then** $G \in G$ **,**

(iii) **if** G **is a disjoint union of members of** G **then** $G \in G$ **.**

For example, the class T **of acyclic groupnets is admissible; also the class of groupnets of the homotopy type of a disjoint union of subgroups of a fixed group** G **, or of a group in some fixed variety of groups, is admissible.**

Let C **be a class of small categories; in particular we shall be interested in**

HNN, the free category on the graph

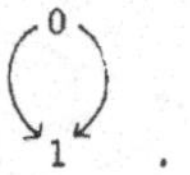

DEFINITION 10.2. The class $C(G)$ is defined to consist of those groupnets which are of the homotopy type of a categorical product $m(C, A)$, where $C \in C$ and each $A_v \in G$.

THEOREM 10.1. $C(G)$ *is an admissible class if* G *is an admissible class.*

(The proof is an immediate application of the results of Section 8, since in the pullback diagram corresponding to a subgroupnet, each B_v has the homotopy type of a disjoint union of subgroupnets of A_v , and each B_e is one-to-one.)

In particular we define admissible classes $C_n(G)$ by $C_0(G) = G$, $C_n(G) = C(C_{n-1}(G))$. For the class $G = T$, we let $C^n = C_n(T)$.

For example, HNN^1 consists of free groupnets, HNN^2 contains free products of free groups amalgamated over subgroups, and also HNN-constructions with base a free group. Groups with one defining relation without torsion appear in some HNN^n , as do the fundamental groups of sufficiently large 3-manifolds (see Waldhausen [9]). Actually such fundamental groups belong to an even more restrictive class, formed by taking categorical products over HNN for which A_0 is the fundamental groupnet of an orientable surface. This class is also closed under the operation of taking subgroupnets. As an analogue of Waldhausen's conjecture we

CONJECTURE. *The Whitehead group of any group in* HNN^n *is* 0 .

DEFINITION 10.3. Let H and K be admissible classes. Define HK to be the class of proper extensions of H by K , with $H \in H$, and $K \in K$. (A groupnet G is said to be a proper extension of H by K if there is a proper surjection of G onto K with kernel H ; H is then necessarily a disjoint union of groups.)

THEOREM 10.2. HK *is an admissible class.*

(Conditions *(ii)* and *(iii)* are obviously satisfied. The proof that HK is closed under homotopy equivalence is left to the reader.)

Note that $TG = G = GT$. The first equality is obvious, since if G is a proper extension of an acyclic H by K , then $H = \mathrm{Id}(G)$ and $G \approx K$. For the second, suppose G is an extension of H by an acyclic K . Then H contains all the loops of G , and each component of G is homotopically equivalent to some component of H .

THEOREM 10.3. $H.C_n(K) \subset C_n(HK)$.

PROOF. The statement is trivial for $n = 0$. Suppose it is true for $n \leq k-1$. Let $G \in HC_k(K)$. Then we have

$$\begin{array}{ccc} & & (C, A) \\ & & \downarrow \\ & & m(C, A) \\ & & \downarrow {\scriptstyle g} \\ G & \xrightarrow{\eta} & K \end{array}$$

where g is a homotopy equivalence, η is a proper surjection, (and therefore a conservative fibration on each component of G), each $A_v \in C_{k-1}(K)$, each A_e is one-to-one, and $H = \ker \eta \in H$.

For the pullback diagram

$$\begin{array}{ccc} (C, B) & \xrightarrow{N} & (C, A) \\ \downarrow & & \downarrow \\ m(C, B) & \xrightarrow{\eta^*} & m(C, A) \\ \downarrow {\scriptstyle g^*} & & \downarrow {\scriptstyle g} \\ G & \xrightarrow{\eta} & K \end{array} .$$

Then g^* is a homotopy equivalence, η^* is a proper surjection, and so is each $N_v : B_v \to A_v$. Let H_v be the kernel of N_v . We have a commutative diagram

$$\begin{array}{ccccc} H_v & \longrightarrow & B_v & \xrightarrow{N_v} & A_v \\ \downarrow & & \downarrow {\scriptstyle f_v^*} & & \downarrow {\scriptstyle f_v} \\ H & \longrightarrow & G & \xrightarrow{\eta} & K \end{array}$$

in which the right hand square is a pullback square.

Thus $\ker N_v = \{(y, j) \in G \times A_v \mid \eta(y) = f_v(j),\ j \in \mathrm{Id}(A_v)\}$. So the component of H_v consisting of loops at (i, j) is isomorphic to the subgroup of H consisting of loops at i . Hence $H_v \in H$.

Thus $G \in C(HC_{k-1}(K)) \subset C(C_{k-1}(HK)) = C_k(HK)$. //

In particular $HC^n \subset C_n(H)$ and $C^m.C^n \subset C^{m+n}$.

References

[1] R.H. Crowell and N. Smythe, *The theory of groupnets* (in preparation, partial preprint available).

[2] Philip J. Higgins, *Notes on categories and groupoids* (Van Nostrand Reinhold Mathematical Studies, 32. Van Nostrand, London, 1971). Zbl.226.20054.

[3] A. Karrass and D. Solitar, "The subgroups of a free product of two groups with an amalgamated subgroup", *Trans. Amer. Math. Soc.* 150 (1970), 227-255. MR41#5499.

[4] A. Karrass and D. Solitar, "Subgroups of *HNN* groups and groups with one defining relation", *Canad. J. Math.* 23 (1971), 627-643. Zbl.232.20051.

[5] Hanna Neumann, "Generalized free products with amalgamated subgroups. Part I. Definitions and general properties", *Amer. J. Math.* 70 (1948), 590-625. MR10,233.

[6] Hanna Neumann, "Generalized free products with amalgamated subgroups. Part II. The subgroups of generalized free products", *Amer. J. Math.* 71 (1949), 491-540. MR11,8.

[7] Edward T. Ordman, "On subgroups of amalgamated free products", *Proc. Cambridge Philos. Soc.* 69 (1971), 13-23. Zbl.203,323.

[8] John R. Stallings, "A topological proof of Grushko's theorem on free products", *Math. Z.* 90 (1965), 1-8. MR32#5723.

[9] F. Waldhausen, "Whitehead groups of generalized free products", preprint.

Dartmouth College,
New Hampshire 03755, USA.

Australian National University,
Canberra.

PROC. SECOND INTERNAT. CONF. THEORY OF GROUPS, CANBERRA 1973, pp. 281-284.

20H25 (20C10)

LINEAR REPRESENTATIONS OVER PRINCIPAL IDEAL DOMAINS

John D. Dixon

Let D be a principal ideal domain and F be its field of fractions. Then it is well known that a linear representation of a finite group over F is equivalent to a representation over D. This is proved by Burnside in an appendix to [1] for the case $D = \mathbb{Z}$ (the integers); and a complete proof appears, for example, in Theorem (75.3) of [3]. The object of this note is to prove a generalization for infinite groups.

THEOREM 1. *Let D be a principal ideal domain and F its field of fractions. Let $\rho : G \to \mathrm{GL}(n, F)$ be an irreducible representation of an arbitrary group G over F. Suppose that there exists a finite normal separable extension E of F such that ρ splits into absolutely irreducible representations over E. If for each $x \in G$, $\rho(x)$ is conjugate in $\mathrm{GL}(n, F)$ to an element of $\mathrm{GL}(n, D)$, then ρ is equivalent to a representation $\sigma : G \to \mathrm{GL}(n, D)$ over D.*

REMARK 1. $\rho(x)$ is conjugate to an element of $\mathrm{GL}(n, D)$ if and only if the eigenvalues of $\rho(x)$ are integral over D. The condition that each $\rho(x)$ have this property is clearly a necessary condition for the existence of a representation over D. In the case G is finite, this condition is automatically satisfied.

2. In the finite case it is not required that ρ should be irreducible, but some such condition is necessary in the general case. For example, if $G = (\mathbb{Q}, +)$ (the additive group of rationals), $D = \mathbb{Z}$, and $F = \mathbb{Q}$, then $\rho : G \to \mathrm{GL}(2, \mathbb{Q})$ defined by

$$\rho(x) = \begin{pmatrix} 1 & x \\ 0 & 1 \end{pmatrix}$$

is not equivalent to a representation over $\mathbb{Z}$.

3. It is not known whether the existence of the separable normal extension E is a necessary hypothesis. But certainly this hypothesis holds in the cases:

(i) ρ is absolutely irreducible; or

(ii) F is perfect (see [3], §68).

In the case where ρ is absolutely irreducible, an examination of the proof below shows that we do not need to assume each $\rho(x)$ is conjugate to an element of $GL(n, D)$, merely that the trace $\operatorname{tr} \rho(x) \in D$.

PROOF OF THE THEOREM. Let $\theta = \theta_1$ be an irreducible constituent of $\rho^E = \rho \otimes_F E$, and let $\theta_1, \ldots, \theta_s$ represent the inequivalent conjugates of θ under the Galois group $\mathrm{Gal}(E/F)$. Since E is separable over F, $\rho^E = m_1\theta_1 + \ldots + m_s\theta_s$ where each θ_i occurs as a constituent m_i times (see [3], Theorem (70.15), or [4], Theorem 1.2 for more details). We can extend the definitions of ρ and θ_i so that they are linear representations of the group algebra $F[G]$.

Define the F-linear functional $\tau : F[G] \to E$ by $\tau(a) = \sum_{i=1}^{s} \operatorname{tr} \theta_i(a)$ where tr denotes the trace. Note that the values of τ actually lie in F since the θ_i are the conjugates of θ over $\mathrm{Gal}(E/F)$. Moreover, if $x \in G$, then the eigenvalues of $\rho(x)$ are all integral over D by hypothesis; therefore $\tau(x)$ is integral over D. Since a principal ideal domain is integrally closed in its field of fractions, this shows that $\tau(x) \in D$ for each $x \in G$. We also observe that if $a \in F[G]$ and $\tau(ax) = 0$ for all $x \in G$, then $\rho(a) = 0$. Indeed, since the θ_i are mutually inequivalent absolutely irreducible representations of G, a theorem of Frobenius and Schur implies that their coordinate functions are linearly independent (see [3] Theorem (27.8)). Thus $0 = \tau(ax) = \sum \operatorname{tr}\{\theta_i(a)\theta_i(x)\}$ for all $x \in G$ implies $\theta_i(a) = 0$ for all i and so $\rho(a) = 0$.

Now choose $x_1, \ldots, x_m$ in G so that $\rho(x_1), \ldots, \rho(x_m)$ is an F-basis for the F-algebra spanned by $\rho(G)$. For each $x \in G$ we have $\lambda_j \in F$ such that

$$\rho(x) = \sum_{j=1}^{m} \lambda_j \rho(x_j) . \tag{1}$$

Then, for each k, $\rho(xx_k) = \sum \lambda_j \rho(x_j x_k)$ and so

$$\tau(xx_k) = \sum_{j=1}^{m} \lambda_j \tau(x_j x_k) \quad \text{for } k = 1, \ldots, m . \tag{2}$$

We claim that the $m \times m$ matrix $[\tau(x_j x_k)]$ is nonsingular. In fact, otherwise there would exist $a = \sum \alpha_j x_j \in F[G]$, such that $\rho(a) \neq 0$ and $0 = \sum \alpha_j \tau(x_j x_k) = \tau(ax_k)$ for $k = 1, \ldots, m$. But this would imply $\tau(ax) = 0$ for all $x \in G$ because $F\rho(G)$

is spanned by $\rho(x_1), \ldots, \rho(x_m)$; and we saw above that $\tau(ax) = 0$ for all $x \in G$ implies $\rho(a) = 0$. Thus $[\tau(x_i x_j)]$ is nonsingular and we can solve the system (2) of equations for the λ_i using Cramer's rule. Since we have already shown that $\tau(y) \in D$ for all $y \in G$, we conclude that there exist μ_j, υ in D such that

$$\lambda_j = \mu_j/\upsilon \quad \text{for } j = 1, \ldots, m . \tag{3}$$

The remainder of the proof is now very similar to the proof of the classical case of the theorem. Since D is a principal ideal domain we can find $\delta \neq 0$ in D such that each of the matrices $(\delta/\upsilon)\rho(x_j)$ $(j = 1, \ldots, m)$ has all its entries lying in D .

Let D^n denote the free D-module of all n-vectors over D and define

$$M = \{v \in D^n \mid v\rho(x) \in D^n \text{ for all } x \in G\} .$$

Clearly M is a D-module, and it follows from (1) and (3) and the choice of δ that $\delta D^n \subseteq M \subseteq D^n$. Since D is a principal ideal domain, a D-submodule of a free D-module of rank r is free of rank $\leq r$ (see [3], Theorem (16.1)).

Because δD^n and D^n are both free D-modules of rank n , this means M is free of rank n . Let $v_1, \ldots, v_n$ be a free D-basis of M and let $c \in \mathrm{GL}(N, D)$ be the matrix with rows $v_1, \ldots, v_n$. Since for all $x \in G$ and all $v \in M$ we have $v\rho(x) \in M$ by the definition of M , therefore $c\rho(x)c^{-1} \in \mathrm{GL}(n, D)$. Hence $\sigma : G \to \mathrm{GL}(n, D)$ defined by $\sigma(x) = c\rho(x)c^{-1}$ is the required representation of G . □

It is also possible to prove an analogous theorem where the integral domain D is merely assumed to be a (commutative) Bezout domain; that is, all its finitely generated ideals are principal. The class of Bezout domains includes all principal ideal domains as well as the following interesting domains: the ring of all algebraic integers, any valuation domain, and the ring of all analytic functions from C into itself (see [5], p. 72).

THEOREM 2. *The conclusion of Theorem* 1 *remains true under the same general hypotheses if it is merely assumed* D *is a Bezout domain, provided we add the condition that* G *is finitely generated.*

The proof of Theorem 2 requires only minor modifications of the earlier proof. Specifically, since G is finitely generated the entries of the matrices $\rho(x)$ $(x \in G)$ all lie in the field F_1 of fractions of some finitely generated (and hence Noetherian) subring D_1 of D . We can then construct the D_1-module

$$M_1 = \left\{ v \in D_1^n \mid v\rho(x) \in D_1^n \text{ for all } x \in G \right\}$$

and show as above that $\delta D_1^n \subseteq M_1 \subseteq D_1^n$ for some $\delta \neq 0$ in D_1. Because D_1 is Noetherian, M_1 is a finitely generated D_1-module, and so $M = DM_1$ is a finitely generated D-module. However it is known that for a Bezout domain D, any finitely generated submodule of a free D-module of rank r is free of rank $\leq r$ (see [2], p. 46). Since $\delta D^n \subseteq M \subseteq D^n$, M is a free D-module of rank n, and the proof is now completed precisely as in the case above.

REMARK. It is not known whether the added hypothesis that G be finitely generated is strictly necessary in Theorem 2.

References

[1] W. Burnside, *Theory of groups of finite order*, 2nd ed. (Cambridge University Press, Cambridge, 1911). FdM42,151.

[2] P.M. Cohn, *Free rings and their relations* (London Math. Soc. Monographs, 2. Academic Press, London, New York, 1971). Zbl.232.16003.

[3] Charles W. Curtis, Irving Reiner, *Representation theory of finite groups and associative algebras* (Pure and Applied Mathematics, 11. Interscience [John Wiley & Sons], New York, London, 1962). MR26#2519.

[4] Burton Fein, "The Schur index for projective representations of finite groups", *Pacific J. Math.* 28 (1969), 87-100. MR39#5717.

[5] Irving Kaplansky, *Commutative rings* (Allyn and Bacon, Boston, 1970). Zbl.203,346.

Carleton University,
Ottawa, Ontario K1S 5B6, Canada.

PROC. SECOND INTERNAT. CONF. THEORY OF GROUPS, CANBERRA 1973, pp. 285-287.

20G99

A PROPERTY COMMON TO THE WEYL AND ORTHOGONAL GROUPS

Gordon Bradley Elkington

Let W be the Weyl group associated with a root system Σ in an n-dimensional Euclidean space V. In his paper 'Conjugacy classes in the Weyl group' [1], Carter shows that every element $w \in W$ may be written in the form

(a) $$w = w_{r_1} \ldots w_{r_j} w_{r_{j+1}} \ldots w_{r_k}$$

where $r_1, \ldots, r_k \in \Sigma$ are linearly independent, and where each of the sets $\{r_1, \ldots, r_j\}$, $\{r_{j+1}, \ldots, r_k\}$ consists of mutually orthogonal roots. The set $\{r_1, \ldots, r_k\}$ gives rise to a graph, or diagram, similar in form to the Dynkin diagrams; the graphs so arising are useful in describing the conjugacy classes in W, and have certain other significance [2], [3].

The decomposition (a) is not canonical, and is proved simply by counting the elements having such a decomposition. In particular, setting $w' = w_{r_1} \ldots w_{r_j}$, $w'' = w_{r_{j+1}} \ldots w_{r_k}$, the formula

(b) $$w = w'w''$$

displays w as a product of two involutions (the identity being regarded as involutory).

The following analogous result in the real orthogonal group $O(V)$ gives a motivation for the formula (b):

Every element $X \in O(V)$ may be written in the form

(c) $$X = X'X''$$

where $X', X'' \in O(V)$ are involutions.

We first note that V has an orthogonal decomposition

(d) $$V = V_1 \perp V_2 \perp \ldots \perp V_s$$

where each V_i is an irreducible X-invariant subspace of V of dimension 1 or 2 . For the elementary divisors of X have the form ϕ^r , where ϕ is a real irreducible, that is to say, a linear or quadratic polynomial. In either case, V has an irreducible X-invariant subspace V_1 of dimension 1 or 2 , affording a decomposition

(e) $$V = V_1 \perp V_1^{\perp}$$

of V into X-invariant subspaces. The result follows by applying induction on the dimension to the pair $\left(V_1^{\perp}, X|V_1^{\perp}\right)$.

Write $X_i = X|V_i$ $(i = 1, \ldots, s)$. Now each V_i is a Euclidean space of dimension 1 or 2 , and $X_i \in O(V_i)$.

If $\dim V_i = 1$, then $X_i = \pm 1$. Writing $X_i' = X_i$, $X_i'' = 1$, we see that $X_i'^2 = X_i''^2 = 1$, and $X_i = X_i'X_i''$.

If $\dim V_i = 2$, then with respect to an orthonormal basis of V_i , the matrix of X_i has the form

$$\begin{bmatrix} \cos\theta, & \sin\theta \\ -\sin\theta, & \cos\theta \end{bmatrix}.$$

Writing X_i', X_i'' respectively for the orthogonal transformations represented by

$$\begin{bmatrix} 0, & 1 \\ 1, & 0 \end{bmatrix}, \begin{bmatrix} -\sin\theta, & \cos\theta \\ \cos\theta, & \sin\theta \end{bmatrix},$$

we see that $X_i'^2 = X_i''^2 = 1$, and $X_i = X_i'X_i''$.

Lastly, write $X' = X_1' \perp \ldots \perp X_s'$, $X'' = X_1'' \perp \ldots \perp X_s''$. Then $X', X'' \in O(V)$, $X'^2 = X''^2 = 1$ and $X = X'X''$.

References

[1] R.W. Carter, "Conjugacy classes in the Weyl group", *Compositio Math.* 25(1972), 1-59.

[2] R.W. Carter and G.B. Elkington, "A note on the parametrization of conjugacy classes", *J. Algebra* 20 (1972), 350-354. Zbl.239.20053.

[3] Gordon Bradley Elkington, "Centralizers of unipotent elements in semi-simple algebraic groups", *J. Algebra* 23 (1972), 137-163. Zbl.247.20053.

[4] G.E. Wall, "On the conjugacy classes in the unitary, symplectic and orthogonal groups", *J. Austral. Math. Soc.* 3 (1963), 1-62. MR27#212.

University of Sydney,
Sydney, NSW 2006.

PROC. SECOND INTERNAT. CONF. THEORY OF GROUPS, CANBERRA 1973, pp. 288-297. 20B15

PRIMITIVE EXTENSIONS OF MAXIMAL DIAMETER OF MULTIPLY TRANSITIVE PERMUTATION GROUPS

Hikoe Enomoto

Primitive extensions of multiply transitive permutation groups have been determined in case of rank three or four ([1], [2], [5], [6], [8]). In this paper we shall consider primitive extensions of maximal diameter of arbitrary rank.

THEOREM 1. *There exists no primitive permutation group* (G, Ω) *of rank* r (≥ 3) , *having a self-paired orbital* Δ *of length* k (≥ 5) *which satisfies the following conditions:*

(i) *the action of* G_a *on* $\Delta(a)$ *is quadruply transitive,*

(ii) G_a *is transitive on* $\Gamma_i(a)$ *for* $i \leq 4$, *where* $\Gamma_0(a) = \{a\}$, $\Gamma_1(a) = \Delta(a)$ *and*

$$\Gamma_{i+1}(a) = \Gamma_i \circ \Delta(a) - \Gamma_i(a) - \Gamma_{i-1}(a) ,$$

(iii) $|\Delta(a) \cap \Delta(b)| = 1$ *for* $b \in \Gamma_2(a)$, *and*

(iv) G_{ab} $(b \in \Gamma_2(a))$ *fixes two points in* $\Delta(a)$.

Note that $\Gamma_i(a)$ is the set of points of Ω of distance i from a in the orbital graph with respect to Δ . The condition *(i)* implies that G_a is transitive on $\Gamma_2(a)$ and $|\Delta(a) \cap \Delta(b)|$ is 1 or 2 for $b \in \Gamma_2(a)$ ([2, Proposition (2.3)]). The conditions *(i)* and *(iii)* imply that G_{ab} $(b \in \Gamma_2(a))$ fixes one or two points in $\Delta(a)$ ([2, Proof of Proposition (2.3)]). It seems difficult to treat systematically the case when G_{ab} $(b \in \Gamma_2(a))$ fixes only one point in $\Delta(a)$.

Note also that (G, Ω) is of maximal diameter with respect to Δ (in the sense of Higman [4]) if and only if G_a is transitive on every $\Gamma_i(a)$. Hence if (G, Ω) satisfies the condition *(i)* above and the rank of (G, Ω) is 3 or 4 , then (G, Ω) is of maximal diameter.

THEOREM 2. *Let* (G, Ω) *be a primitive permutation group of rank* r (≥ 3) *of maximal diameter with respect to a self-paired orbital* Δ *of length* k (≥ 5) *such that* G_a *acts faithfully on* $\Delta(a)$ *and is isomorphic to the symmetric group* Σ_k *or the alternating group* A_k *of degree* k . *Then one of the following holds:*

(1) k *is odd,* $r = \frac{k+1}{2}$ *and* G *has a regular normal subgroup of order* 2^{k-1} ;

(2) $k = 7$, $r = 3$ *and* G *is isomorphic to either* $U_3(5)$ *or* $\widehat{U_3(5)}$.

Cameron [3] proved that the case (1) holds in Theorem 2 if $|\Delta(a) \cap \Delta(b)| = 2$ for $b \in \Gamma_2(a)$. Therefore we have to consider only the case $k = 7$ and the case when G_a is isomorphic to A_5 , because otherwise the conditions of Theorem 1 are satisfied. The proof of Theorem 2 consists of tedious calculations, and we shall not give it here.

Originally Bannai and the author proved Theorem 2 (for sufficiently large k) making use of the knowledge about the subgroups of Σ_k or A_k . Later, the author gave a simpler proof, which leads to Theorem 1.

PROOF OF THEOREM 1. Let (G, Ω) be a primitive permutation group with a self-paired orbital Δ which satisfies the conditions of Theorem 1. The intersection numbers (relative to Δ) are defined by

$$\mu_{ij} = |\Gamma_i(a) \cap \Delta(b)| \ , \quad b \in \Gamma_j(a) \ .$$

These numbers depend only on i and j for $0 \leq i$, $j \leq 4$ by the condition *(ii)*. Fix a point ∞ in Ω , and let

$$\Delta(\infty) = \{x_1, \ldots, x_k\} \ .$$

Ordered s-ple $\left(x_{\alpha_1}, \ldots, x_{\alpha_s}\right)$ is denoted by $[\alpha_1, \ldots, \alpha_s]$, and unordered s-ple $\left\{x_{\alpha_1}, \ldots, x_{\alpha_s}\right\}$ is denoted by $\langle \alpha_1, \ldots, \alpha_s \rangle$. $\{a_1, \ldots, a_m\}$ $(a_i \in \Omega)$ is an m-gon if $a_i \in \Delta(a_{i+1})$, $1 \leq i \leq m-1$, and $a_m \in \Delta(a_1)$.

LEMMA 1. *The number of points in* $\Gamma_2(\infty)$ *is* $k(k-1)$ *and* $\Gamma_2(\infty)$ *may be identified with the set of ordered pairs* $\{[\alpha, \beta] \mid 1 \leq \alpha, \beta \leq k, \alpha \neq \beta\}$ *so that* x_α *and* $[\alpha, \beta]$ *are adjacent.*

PROOF. First, we obtain

$$|\Gamma_2(\infty)| = |\Gamma_1(\infty)| \cdot \mu_{21}/\mu_{12} = k(k-1) \ ,$$

because $\mu_{21} = k - 1$ by the condition *(i)* ([2, Proposition (2.2)]) and $\mu_{12} = 1$ by the condition *(iii)*. Let b be a point in $\Gamma_2(\infty)$. Then $G_{\infty,b}$ fixes two points, say x_α and x_β, in $\Delta(\infty)$ by the condition *(iv)*. One of them, say x_α, is in $\Delta(\infty) \cap \Delta(b)$. Then the lemma follows immediately by identifying b with the ordered pair $[\alpha, \beta]$.

LEMMA 2. *$\mu_{22} = 1$, and $[\alpha, \beta]$ is adjacent to $[\beta, \alpha]$.*

PROOF. Let $b = [\alpha, \beta]$ be a point in $\Gamma_2(\infty)$. Then

$$\mu_{22} = |\Delta(\infty) \cap \Gamma_2(b)|$$

and $\Delta(\infty) \cap \Gamma_2(b)$ is a $G_{\infty,b}$-invariant subset of $\Delta(\infty)$. On the other hand, $G_{\infty,b}$-orbits in $\Delta(\infty)$ are $\{x_\alpha\}$, $\{x_\beta\}$ and $\Delta(\infty) - \{x_\alpha, x_\beta\}$. Hence $\mu_{22} = 0, 1, k - 2$ or $k - 1$, because x_α is a point in $\Gamma_1(b)$. Without loss of generality, we may assume that $b = [1, 2]$. Suppose $\mu_{22} = k - 2$ or $k - 1$, or $\{x_3, \ldots, x_k\}$ is contained in $\Gamma_2(b)$. Then $\Delta(x_3) \cap \Delta(b) \cap \Gamma_2(\infty)$ consists of a single point, say $[3, \beta]$. This point is fixed by $G_{\infty,b,x_3} = G_{\infty,x_1,x_2,x_3}$. Hence $\beta = 1$ or 2. Suppose $\beta = 1$. This means that $[\alpha, \beta]$ and $[\gamma, \alpha]$ are adjacent if $\beta \neq \gamma$. Then we obtain a quadrangle $\{[1, 2], [3, 1], [1, 4], [5, 1]\}$, which contradicts the condition *(iii)*. Suppose $\beta = 2$. This implies that $[\alpha, \beta]$ and $[\gamma, \beta]$ are adjacent if $\alpha \neq \gamma$. Then we obtain a triangle $\{[1, 2], [3, 2], [4, 2]\}$, which is a contradiction. Next, suppose $\mu_{22} = 0$, or $\{x_2, \ldots, x_k\}$ is contained in $\Gamma_3(b)$. Choose an element $g \in G$ which interchanges b with ∞, and put $c = x_2^g \in \Gamma_3(\infty)$. Then

$$|G_{\infty,b,c}| = |G_{\infty,b,x_2}| = |G_{\infty,b}| .$$

Since $G_{\infty,b,c}$ is a subgroup of $G_{\infty,b}$, we obtain $G_{\infty,b,c} = G_{\infty,b}$. On the other hand

$$|G : G_{\infty,c}| = |\Gamma_3(\infty)| = |\Gamma_2(\infty)| \cdot \mu_{32}/\mu_{23}$$
$$= |G_\infty : G_{\infty,b}| \cdot (k-1)/\mu_{23} ,$$

and therefore

$$|G_{\infty,c} : G_{\infty,b,c}| = \mu_{23}/(k-1) .$$

This implies that $\mu_{23} = k - 1$ and that

$$G_{\infty,c} = G_{\infty,b,c} = G_{\infty,b} = G_{\infty,x_1,x_2} .$$

Hence $\Gamma_3(\infty)$ may be identified with the set of ordered pairs $\{[\alpha, \beta]_3 \mid 1 \le \alpha, \beta \le k, \alpha \ne \beta\}$ so that $[\alpha, \beta]$ and $[\alpha, \beta]_3$ are adjacent. (To distinguish the points of $\Gamma_3(\infty)$ from those of $\Gamma_2(\infty)$, we use the convention that the points of $\Gamma_i(\infty)$ are subscripted with i if necessary.) Put $c' = [1, 2]_3$. Then

$$\Delta(\infty) \cap \Gamma_2(c') = \Delta(\infty) - \{x_2\},$$

because $\mu_{23} = k - 1$. Hence

$$\Delta(x_3) \cap \Delta(c') = \{[3, \beta]\}$$

for some β. $[3, \beta]$ is fixed by $G_{\infty,c',x_3} = G_{\infty,x_1,x_2,x_3}$, hence $\beta = 1$ or 2. Suppose $\beta = 1$. Then we obtain a quadrangle $\{[3, 1], [1, 2]_3, [4, 1], [1, 5]_3\}$, which is a contradiction. Suppose $\beta = 2$. Then we obtain a quadrangle $\{[3, 2], [1, 2]_3, [4, 2], [5, 2]_3\}$, a contradiction.

We have proved that $\mu_{22} = 1$. Therefore

$$\Delta(\infty) \cap \Gamma_2(b) = \{x_2\}.$$

Suppose $\Delta(x_2) \cap \Delta(b) = \{[2, \alpha]\}$. Then $[2, \alpha]$ is fixed by G_{∞,b,x_2}, which implies that $\alpha = 1$. This completes the proof of Lemma 2.

COROLLARY 3. *For any pair of points of distance two, there exists one and only one pentagon that contains these two points.*

LEMMA 4. *Let* $b = [1, 2] \in \Gamma_2(\infty)$, *and* $c \in \Delta(b) \cap \Gamma_3(\infty)$. *Then* $d(c, x_2) = 3$. *Incidentally, this implies that* $\mu_{33} \ge 1$.

PROOF. Suppose $d(c, x_2) = 2$. Then there exists a point $b' \in \Delta(c) \cap \Delta(x_2)$. By the property of triangle-free, $b' \ne b$ nor $[2, 1]$. Then we obtain two pentagons which contain $\{b, x_2\}$. This contradicts Corollary 3.

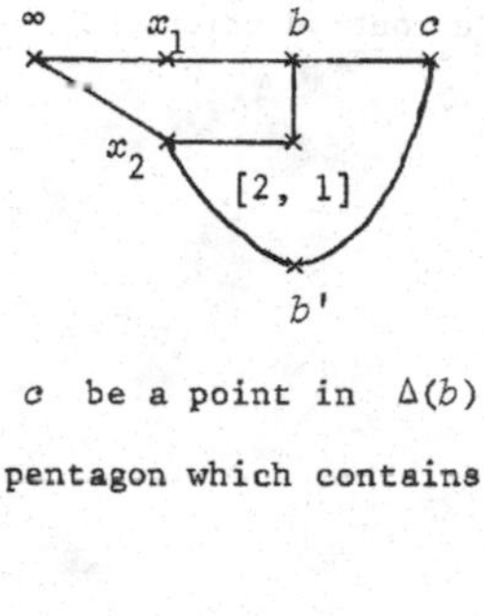

Let $b = [1, 2] \in \Gamma_2(\infty)$, c be a point in $\Delta(b) \cap \Gamma_3(\infty)$, and $\{x_1, b, c, c', [1, \alpha]\}$ be the pentagon which contains x_1 and c.

LEMMA 5. *c' is a point in* $\Gamma_3(\infty)$.

PROOF. Suppose c' is not in $\Gamma_3(\infty)$. Then c' is a point in $\Gamma_2(\infty)$. Hence $c' = [\alpha, 1]$, and we obtain two pentagons which contain x_1 and $[\alpha, 1]$.

LEMMA 6. $G_{\infty,b,c} = G_{\infty,x_1,x_2,x_\alpha}$.

PROOF. Let g be an element of G which interchanges ∞ with b . Then $x_1^g = x_1$, $x_2^g = [2, 1]$ and $[2, 1]^g = x_2$ by Corollary 3 . Hence c^g is in $\Delta(\infty) - \{x_1, x_2\}$. Therefore

$$|G_{\infty,b} : G_{\infty,b,c}| = |G_{\infty,b} : G_{\infty,b,c^g}| = k - 2 .$$

On the other hand, $G_{\infty,b,c}$ is contained in $G_{\infty,x_1,x_2,x_\alpha}$, because $G_{\infty,b,c}$ fixes $[1, \alpha]$.

LEMMA 7. $\mu_{23} = 1, 2, k - 2$ *or* $k - 1$.

PROOF. $\mu_{23} = |\Delta(\infty) \cap \Gamma_2(c)|$ and $\Delta(\infty) \cap \Gamma_2(c)$ is a $G_{\infty,b,c}$-invariant subset of $\Delta(\infty)$. On the other hand, $G_{\infty,b,c}$-orbits in $\Delta(\infty)$ are $\{x_1\}$, $\{x_2\}$ and $\{x_\alpha\}$ and $\Delta(\infty) - \{x_1, x_2, x_\alpha\}$.

LEMMA 8. $x_\alpha \notin \Gamma_2(c)$. *Hence* $\mu_{23} \neq 2, k - 1$.

PROOF. Suppose $x_\alpha \in \Gamma_2(c)$ and let $[\alpha, \beta]$ be in $\Delta(x_\alpha) \cap \Delta(c)$. Then x_β is fixed by $G_{\infty,x_1,x_2,x_\alpha}$, hence $\beta = 1$ or 2 .

Suppose $\beta = 1$. Then there exists a quadrangle $\{[\alpha, 1], c, c', [1, \alpha]\}$, which is not the case. Suppose $\beta = 2$. By the quadruple transitivity of G_∞ , there exists an element g in G_∞ which fixes x_1 and interchanges x_2 with x_α . Then g interchanges $[1, \alpha]$ with $[1, 2] = b$, $[\alpha, 2]$ with $[2, \alpha]$ and c with c' . Hence c' is adjacent to $[2, \alpha]$, and then we obtain a quadrangle $\{[\alpha, 2], c, c', [2, \alpha]\}$, which is a contradiction.

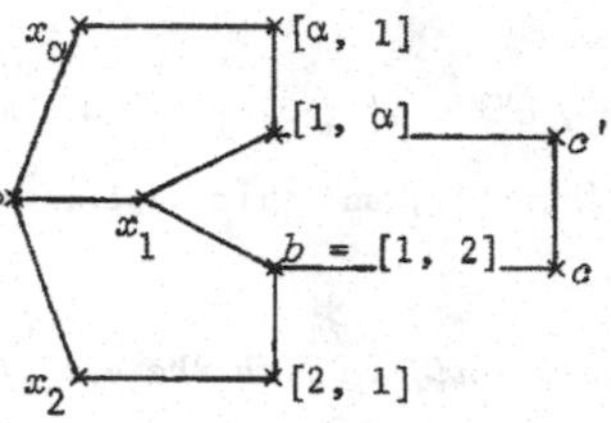

Without loss of generality, we may assume that $\alpha = 3$.

LEMMA 9. $\mu_{23} = 1$.

PROOF. Suppose $\mu_{23} = k - 2$. Then

$$|\Gamma_3(\infty)| = |\Gamma_2(\infty)|\cdot\mu_{32}/\mu_{23} = |\Gamma_2(\infty)| = k(k-1) ,$$

and therefore

$$|G_{\infty,c}| = |G_{\infty,b}| .$$

On the other hand, $\mu_{33} = 1$ or 2 , because

$$\mu_{33} = |\Delta(\infty) \cap \Gamma_3(c)| ,$$

and $\Delta(\infty) \cap \Gamma_3(c)$ is a $G_{\infty,b,c}$-invariant subset of $\Delta(\infty)$. Suppose $\mu_{33} = 1$. Then

$$\Delta(\infty) \cap \Gamma_3(c) = \{x_2\} ,$$

$$\Delta(\infty) \cap \Gamma_4(c) = \{x_3\} ,$$

and $G_{\infty,c}$ fixes x_2 and x_3 , which implies that $G_{\infty,c}$ is contained in G_{∞,x_2,x_3} . Hence $G_{\infty,c} = G_{\infty,x_2,x_3}$, because $|G_{\infty,c} : G_{\infty,b,c}| = k - 2 = |G_{\infty,x_2,x_3} : G_{\infty,b,c}|$. $\Gamma_3(\infty)$ may be identified with the set of ordered pairs $\{[\beta, \gamma]_3 \mid 1 \le \beta, \gamma \le k, \beta \ne \gamma\}$ so that $[\alpha, \beta]$ and $[\beta, \gamma]_3$ are adjacent if $\alpha \ne \gamma$. Then we obtain a quadrangle $\{[1, 2], [2, 3]_3, [4, 2], [2, 5]_3\}$, and this is a contradiction. Suppose $\mu_{33} = 2$, and let $\Delta(\infty) \cap \Gamma_3(\infty) = \{c', c''\}$. Then

$$|\Delta(\infty) \cap \Gamma_3(c)| = \{x_2, x_3\} ,$$

and

$$|\Delta(c) \cap \Gamma_2(x_2)| = |\Delta(c) \cap \Gamma_2(x_3)| = \mu_{23} = k - 2 .$$

This implies that

$$\Delta(c) \cap \Gamma_2(x_3) = \Delta(c) - \{b, c'\} ,$$

and hence

$$\Delta(c) = \{b, c', c''\} \cup \{[\alpha, 3] \mid 4 \le \alpha \le k\} .$$

Then $\Delta(c) \cap \Gamma_2(x_2)$ is at most $\{b, c''\}$, and this implies $k - 2 \le 2$, which is a contradiction.

COROLLARY 10. *$\Gamma_3(\infty)$ may be identified with the set of ordered triples $\{[\alpha, \beta, \gamma] \mid \alpha \ne \beta \ne \gamma \ne \alpha\}$ so that*

$$\Delta([\alpha, \beta, \gamma]) \cap \Gamma_2(\infty) = \{[\alpha, \beta]\} ,$$

and $[\alpha, \beta, \gamma]$ is adjacent to $[\alpha, \gamma, \beta]$.

LEMMA 11. $\mu_{33} = 1, 2, k-2$ *or* $k-1$.

PROOF. Put $c = [1, 2, 3] \in \Gamma_3(\infty)$. Then

$$\mu_{33} = |\Delta(\infty) \cap \Gamma_3(c)|$$

and $\Delta(\infty) \cap \Gamma_3(c)$ is a $G_{\infty,c} = G_{\infty,x_1,x_2,x_3}$-invariant subset of $\Delta(\infty)$. On the other hand $G_{\infty,c}$-orbits in $\Delta(\infty)$ are $\{x_1\}$, $\{x_2\}$, $\{x_3\}$ and $\{x_4, \ldots, x_k\}$.

LEMMA 12. $\mu_{33} \ne k-2, k-1$.

PROOF. Suppose $\mu_{33} = k-2$ or $k-1$, or $\{x_4, \ldots, x_k\}$ is contained in $\Gamma_3(c)$. Let

$$\Delta(c) \cap \Gamma_2(x_4) = \{[4, \beta, \gamma]\} .$$

Then x_β and x_γ are fixed by $G_{\infty,x_1,x_2,x_3,x_4}$. Hence $\beta, \gamma < 4$ or G_∞ is isomorphic to one of Σ_5, A_6 or M_{11} by a theorem of Nagao [7]. Suppose c is adjacent to $[4, 1, 2]$. Then we obtain a quadrangle $\{[1, 2, 3], [4, 1, 2], [3, 4, 1], [2, 3, 4]\}$. Suppose c is adjacent to $[4, 1, 3]$. Then we obtain a triangle $\{[1, 2, 3], [4, 1, 3], [2, 4, 3]\}$. Similarly, we obtain a triangle or a quadrangle if $\beta, \gamma < 4$. Suppose G_∞ is isomorphic to Σ_5, and one of β or γ is 5. Suppose c is adjacent to $[4, 5, 1]$. Then $[4, 5, 1]^g = c$ and $c^g = [3, 5, 4]$, where $g = (1, 3, 4)(2, 5) \in \Sigma_5$. Hence $\Delta(c)$ contains $[1, 3, 2]$, $[4, 5, 1]$, $[5, 4, 1]$, $[3, 5, 4]$ and $[3, 4, 5]$, which is a contradiction, because $\mu_{33} \le 4$. We can derive a similar contradiction for other cases. Suppose G_∞ is isomorphic to A_6 and at least one of β or γ is greater than 4. Suppose, for example, c is adjacent to $[4, 5, 3]$. Then we obtain a hexagon

$$\{[1, 2, 3], [1, 3, 2], [6, 5, 2], [6, 2, 5], [4, 3, 5], [4, 5, 3]\} .$$

We can derive a similar contradiction for other values of β and γ. By a little more complicated computation, we can derive a similar contradiction for the case when G_∞ is isomorphic to M_{11}.

LEMMA 13. $\mu_{33} \neq 2$.

PROOF. Suppose $\mu_{33} = 2$. Then $\Delta(\infty) \cap \Gamma_3(c) = \{x_2, x_3\}$. Let

$$\Delta(c) \cap \Gamma_2(x_3) = \{[3, \beta, \gamma]\} .$$

Then β and γ are fixed by $G_{\infty, x_1, x_2, x_3}$, which means that $\{\beta, \gamma\} = \{1, 2\}$. Suppose c is adjacent to $[3, 1, 2]$. Then we obtain a triangle $\{[1, 2, 3], [3, 1, 2], [2, 3, 1]\}$, which is a contradiction. Suppose c is adjacent to $[3, 2, 1]$. Then we obtain a hexagon

$$\{[1, 2, 3], [3, 2, 1], [3, 1, 2], [2, 1, 3], [2, 3, 1], [1, 3, 2]\} .$$

LEMMA 14. $\mu_{33} \neq 1$.

PROOF. Suppose $\mu_{33} = 1$. Then

$$\Delta(\infty) \cap \Gamma_2(c) = \{x_1\} ,$$

$$\Delta(\infty) \cap \Gamma_3(c) = \{x_2\} ,$$

and

$$\Delta(\infty) \cap \Gamma_4(c) = \{x_3, \ldots, x_k\} .$$

We can find an element g of G, which interchanges ∞ with c. Then g interchanges x_1 with b, x_2 with $[1, 3, 2]$, and $[2, 1]$ with $[1, 3]$.

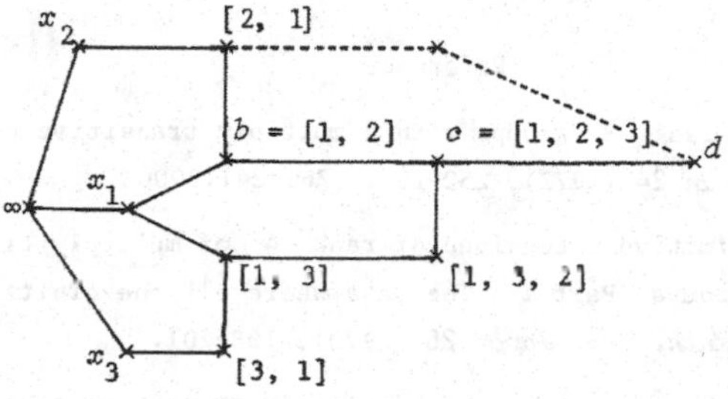

Let d be the image of x_3 by g. Then

$$d = x_3^g \in \Delta(c) \cap \Gamma_4(\infty) ,$$

and

$$|G_{\infty,c,d}| = |G_{\infty,c,x_3}| = |G_{\infty,c}| .$$

Hence

$$G_{\infty,c,d} = G_{\infty,c} = G_{\infty,x_1,x_2,x_3} .$$

On the other hand,

$$|G_\infty : G_{\infty,d}| = |\Gamma_4(\infty)| = (k-1)(k-2)^2/\mu_{34} .$$

Hence

$$|G_{\infty,d} : G_{\infty,c,d}| = \mu_{34}/(k-2) ,$$

which implies that $\mu_{34} = k - 2$, and

$$G_{\infty,d} = G_{\infty,c,d} = G_{\infty,c} = G_{\infty,x_1,x_2,x_3} .$$

Therefore $\Gamma_4(\infty)$ may be identified with the set of ordered triples $\{[\alpha, \beta, \gamma]_4 \mid \alpha \neq \beta \neq \gamma \neq \alpha\}$ so that $[\alpha, \beta, \gamma] \in \Gamma_3(\infty)$ is adjacent to $[\alpha, \beta, \gamma]_4 \in \Gamma_4(\infty)$. Without loss of generality we may assume that $d = [1, 2, 3]_4$. $\Delta(d) \cap \Gamma_3(\infty)$ contains $[3, 1]^g$, which is a point in $\Delta([2, 1]) \cap \Gamma_3(\infty)$. So, $[3, 1]^g$ is of the form $[2, 1, \gamma]$. But γ must be 3 , because x_γ is fixed by $G_{\infty,c,d} = G_{\infty,x_1,x_2,x_3}$. Then we obtain a quadrangle

$$\{[1, 2, 3], [1, 2, 3]_4, [2, 1, 3], [2, 1, 3]_4\} ,$$

which is a contradiction.

References

[1] Eiichi Bannai, "On rank 3 groups with a multiply transitive constituent", *J. Math. Soc. Japan* 24 (1972), 252-254. Zbl.241.20003.

[2] Eiichi Bannai, "Primitive extensions of rank 4 of multiply transitive permutation groups (Part I. The case where all the orbits are self-paired)", *J. Math. Soc. Japan* 25 (1973), 188-201.

[3] P.J. Cameron, "Permutation groups with multiply transitive suborbits", *Proc. London Math. Soc.* (3) 25 (1972), 427-440. Zbl.247.20004.

[4] D.G. Higman, "Intersection matrices for finite permutation groups", *J. Algebra* 6 (1967), 22-42. MR35#244.

[5] Shiro Iwasaki, "A note on primitive extensions of rank 3 of alternating groups", *J. Fac. Sci. Hokkaido Univ. Ser. I* 21 (1970), 125-128. MR42#3160.

[6] Shiro Iwasaki, "A note on primitive extensions of rank 4 of alternating groups", *Proc. Japan Acad.* 48 (1972), 5-8. MR46#1888.

[7] Hirosi Nagao, "On multiply transitive groups IV", *Osaka J. Math.* 2 (1965), 327-341. MR34#231d.

[8] Tosiro Tsuzuku, "On primitive extensions of rank 3 of symmetric groups", *Nagoya Math. J.* 27 (1966), 171-177. MR33#1358.

Tokyo University,
Tokyo, 113, Japan.

PROC. SECOND INTERNAT. CONF. THEORY OF GROUPS, CANBERRA 1973, pp. 298-314. 20C30

SOME CHARACTERS OF ORTHOGONAL GROUPS OVER THE FIELD OF TWO ELEMENTS

J.S. Frame

1. Introduction

Let G_n, or G denote the $(2n+1)$-dimensional orthogonal group $O_{2n+1}(2)$ over the field F_2 of two elements, which is isomorphic with the symplectic group $Sp_{2n}(2)$, let A_n or A and B_n or B denote the maximal full orthogonal subgroups $O_{2n}(2,-)$ and $O_{2n}(2,+)$ of G_n, and let A'_n or A' and B'_n or B' denote the subgroups of A_n or B_n of index 2 which are simple commutator subgroups (with the exception of B'_2 of order 36). Let 1_A^G and $\bar{1}_A^G$ denote the characters of G induced by the trivial 1-character 1_A of A and by the alternating character $\bar{1}_A$ of A whose value is $+1$ in the subgroup A' and -1 in its second coset $A'\tau$. Let $A_n \cap B_n = D_n$ or D, and $A'_n \cap B'_n = D'_n$ or D'. Then D'_n is isomorphic with G_{n-1} and has index $4^n(4^n-1)/2$ in G_n. We denote certain factors of this index by $a_n = 2^n+1$, $b_n = 2^n-1$.

We define the level 1 characters of G, A, and B to be the irreducible components of 1_A^G, $\bar{1}_A^G$, 1_B^G, $\bar{1}_B^G$ and of their restrictions to A and B, except for the trivial and alternating characters which are assigned level 0. Level k characters are irreducible characters of G, A, or B contained in products of k but not of $k-1$ level 1 characters. In this paper we prove some results conjectured in a recent paper of Frame and Rudvalis [4] concerning these characters.

In §2 we determine the degrees of the five level 1 characters of G_n and of the irreducible components of their restrictions to A_n, B_n and G_{n-1}. In §3 we describe a pairing of characters of A_n with those of B_n and of characters of G_n with their mates which interchanges the roles of A_n and B_n in the Frobenius induce-restrict tables of G_n and preserves the addition and multiplication in the character ring of G_n. Symbols are introduced to describe the degrees of characters of a given level k for all dimensions n. In §4 we characterize the 2^n classes of elements of odd order in G_n and determine the character values on these classes of all the five level 1 characters. Values of many higher level characters are immediately derivable from these (see Appendix 2). Finally in §5 we describe the symbols and degrees of all the 2^{n+1} characters of odd degree (for $n \geq 2$), half of which are self associated. The other half belong to 2^{n-2} tetrads related to the four paired characters at level 1.

The author has completed the table of 81 characters of $0_9(2)$ of order $2^{16}3^5 5^2 7\cdot 17$ and used this to discover many patterns which appear to be valid for all the groups $0_{2n+1}(2)$. Proofs of additional relationships should be forthcoming in the future.

2. Subgroup splitting of level 1 characters

The group $G = G_n = 0_{2n+1}(2)$ is representable by $(2n+1)$-dimensional matrices M over F_2, with bottom row $(0^{2n}, 1)$ that map row vectors Z so as to preserve the quadratic form

$$(2.1) \qquad \Omega_n(Z) = \sum_{i=1}^{n} z_{2i-1}z_{2i} + z_{2n+1}^2 = \Omega_n^+(Z) + z_{2n+1}^2 .$$

Matrices of the orthogonal subgroups $A = A_n = 0_{2n}(2, -)$ or 0_{2n}^-, and $B = B_n = 0_{2n}(2, +)$ or 0_{2n}^+ have the additional invariant $ZC^- = z_{2n-1} + z_{2n} + z_{2n+1}$ and $ZC^+ = z_{2n+1}$ respectively where $C^- = (0^{2n-2}, 1^3)^T$, and $C^+ = (0^{2n}, 1)^T$. Thus $M^-C^- = C^-$, $M^+C^+ = C^+$ for M^- in A and M^+ in B.

Matrices M of G permute the 4^n column vectors C with last entry 1 in two transitive sets. One set contains $[(3+1)^n-(3-1)^n]/2 = 2^{n-1}(2^n-1) = 1 + x_n$ vectors with $\Omega_n^+(C) = 1$, which have an odd number of products $z_{2k-1}z_{2k}$ equal to 1, and the other contains $[(3+1)^n+(3-1)^n]/2 = 2^{n-1}(2^n+1) = 1 + y_n$ vectors with $\Omega_n^+(C) = 0$, which have an even number of such products 1. These permutation representations of G are both doubly transitive, and are isomorphic to the representations induced in G by the 1-representations of A and B so we designate their characters

(2.2) $$1_A^G = 1 + X_n \ , \quad 1_B^G = 1 + Y_n \ ,$$

where X_n and Y_n are irreducible level 1 characters of degrees

(2.3) $$x_n = (2^n+1)(2^{n-1}-1) = a_n b_{n-1} \ , \quad y_n = (2^n-1)(2^{n-1}+1) = b_n a_{n-1} \ .$$

Column vectors with last entry 0 are also permuted by matrices M of G in two transitive sets, but one of these sets consists of the null vector alone. The remaining $2^{2n} - 1$ vectors $\neq 0$ correspond to points in the projective geometry $PG(2n-1, 2)$. Since the mapping is linear, the fixed points under any element g of G form a subspace of $2^m - 1$ points. Thus g fixes 2^m column vectors with last entry 0 , including the null vector.

THEOREM 2.1. *The sum of the permutation characters* $1_A^G = 1 + X_n$ *and* $1_B^G = 1 + Y_n$ *induced in* G_n *by its maximal orthogonal subgroups* A_n *and* B_n *has values on all classes which are powers of* 2 . *Every class of* G_n *is represented in* A_n *or in* B_n *or in both.*

PROOF. Since half the column vectors fixed by matrices M of G have last coordinate 1 and half have last coordinate 0 the two character values $(1+X_n) + (1+Y_n)$ and 2^m are equal. Since $1 + X_n$ and $1 + Y_n$ cannot both be 0 , every class of G_n is represented in A_n, B_n or both.

The group G has two (A, A) double cosets, two (B, B) double cosets and just one (A, B) double coset $AB = G$, since the characters 1_A^G and 1_B^G each have two irreducible constituents with one in common. Hence the characters $\bar{1}_A^G$ and $\bar{1}_B^G$ induced in G by the alternating characters of A and B also have two irreducible constituents each, with one in common. We write

(2.4) $$\bar{1}_{A_n}^{G_n} = \bar{S}_n + \bar{U}_n \ , \quad \bar{1}_{B_n}^{G_n} = \bar{S}_n + \bar{V}_n$$

and denote the degrees of $\bar{S}_n$, $\bar{U}_n$, $\bar{V}_n$ by s_n, u_n, v_n . A bar over the symbol for a character indicates that it has a negative value on the class of transpositions τ including the matrix M of $D = A \cap B$ that interchanges z_{2n-1} and z_{2n} .

THEOREM 2.2. *The restrictions to* G_{n-1} *of the five level* 1 *characters of* G_n *split into irreducible components as follows:*

(2.5) $$(1+X_n)_{n-1} = 2(1+X_{n-1}) + (1+Y_{n-1}) + (\bar{S}_{n-1}+\bar{U}_{n-1}) \ ,$$

$$\left(1+Y_n\right)_{n-1} = \left(1+X_{n-1}\right) + 2\left(1+Y_{n-1}\right) + \left(\bar{S}_{n-1}+\bar{V}_{n-1}\right) ,$$

(2.6) $$\left(\bar{S}_n\right)_{n-1} = \left(1+X_{n-1}\right) + \left(1+Y_{n-1}\right) + \bar{S}_{n-1} ,$$

(2.7) $$\left(\bar{U}_n\right)_{n-1} = \left(1+X_{n-1}\right) + \bar{U}_{n-1} ,$$

$$\left(\bar{V}_n\right)_{n-1} = \left(1+Y_{n-1}\right) + \bar{V}_{n-1} ,$$

where the subscripts $n-1$ *on the left indicate restrictions to* G_{n-1} .

PROOF. When restricted to the subgroup D'_n isomorphic with G_{n-1} , the doubly transitive permutation representation of G_n on the vectors C with $\Omega^+_n(C) = 1$ and $z_{2n+1} = 1$, with character $1 + X_n$, splits into three transitive constituents. The vector $C^- = \left(0^{2n-2}, 1^3\right)^T$ is fixed. The remaining $2^{2n-2}-1$ vectors with $z_{2n-1} = z_{2n}$ are permuted transitively by D'_n like the points of $PG(2n-3, 2)$ and yield the character $1 + X_{n-1} + Y_{n-1}$. The third transitive set contains the $\left(2^{n-1}-1\right)2^{n-1}$ vectors C with $z_{2n-1} + z_{2n} = 1$ which are permuted transitively like the cosets of A'_{n-1} in G_{n-1} and yield the character $1 + X_{n-1} + \bar{S}_{n-1} + \bar{U}_{n-1}$. We add these three permutation characters to obtain $\left(1+X_n\right)_{n-1}$ in (2.5). We obtain $\left(1+Y_n\right)_{n-1}$ by similar reasoning involving $\Omega^-_n(C)$.

Next we note the relations

(2.8) $$\left(1+X_n\right)_A = \left(\bar{S}_n+\bar{U}_n\right)_A\bar{1}_A , \qquad \left(1+Y_n\right)_B = \left(\bar{S}_n+\bar{V}_n\right)_B\bar{1}_B ,$$

(2.9) $$\left(1+X_n\right)_{n-1} = \left(\bar{S}_n+\bar{U}_n\right)_{n-1} , \quad \left(1+Y_n\right)_{n-1} = \left(\bar{S}_n+\bar{V}_n\right)_{n-1} .$$

We subtract equations (2.9) and use (2.5) to obtain

(2.10) $$\left(\bar{V}_n-\bar{U}_n\right)_{n-1} = \left(Y_n-X_n\right)_{n-1} = \left(Y_{n-1}+\bar{V}_{n-1}\right) - \left(X_{n-1}+\bar{U}_{n-1}\right) .$$

From (2.10) it follows that $\left(\bar{U}_n\right)_{n-1}$ contains $X_{n-1} + \bar{U}_{n-1}$. Both these and $\left(\bar{S}_n\right)_{n-1}$ contain the 1-character, since $\bar{S}_n$, $\bar{U}_n$, $\bar{V}_n$ are all contained in $\bar{1}^G_D$ and $1^G_{D'}$. After extracting $1 + X_{n-1} + \bar{U}_{n-1}$ from $\left(\bar{S}_n+\bar{U}_n\right)_{n-1} = \left(1+X_n\right)_{n-1}$ in (2.5), we have at most $1 + X_{n-1} + 1 + Y_{n-1} + \bar{S}_{n-1}$ left for $\left(\bar{S}_n\right)_{n-1}$. A check of values in character tables shows this to be exact for $n = 2, 3, 4$. Furthermore a transfer of X_{n-1} or Y_{n-1} or both from $\left(\bar{S}_n\right)_{n-1}$ to $\left(\bar{U}_n\right)_{n-1}$ yields degrees s_n, s_{n-1}, u_n, u_{n-1} which do not all divide the group order for $n \geq 3$. Hence formulas (2.6) and (2.7) are valid.

Since 1^B_D is a rank 3 permutation character of degree $2^{n-1}\left(2^n-1\right)$ with orbit lengths 1, $4^{n-1}-1$, and $4^{n-1}-2^{n-1}$, it contains two irreducible characters besides 1_B. The character $1_{D'}^B = \left(1_{D'}^{B'}\right)^B$ contains these three and their products with $\bar{1}_B$, and is the restriction to B of $1_{A'}^G = 1 + X_n + \bar{S}_n + \bar{U}_n$. Since $\bar{S}_n$ is a component of 1_B^G, its restriction to B contains $\bar{1}_B$. Hence $(\bar{S}_n)_B - \bar{1}_B$ and $(\bar{U}_n)_B$ are irreducible and $(X_n)_B$ is the sum of their associates which are positive on the transposition class. A similar argument shows that $(\bar{S}_n)_A - \bar{1}_A$ and $(\bar{V}_n)_A$ are irreducible, and $(Y_n)_A$ is the sum of their associates, which we denote by S' and V. Further analysis which we omit here shows that $\bar{U}_A - \bar{1}_A$ is irreducible in A and $\bar{V}_B - \bar{1}_B$ is irreducible in B. We write the decomposition in A and B of level 1 characters of G as follows:

(2.11)
$$X_A = S' + 1 + U', \quad Y_A = S' + V, \quad \bar{S}_A = \bar{1} + \bar{S}', \quad \bar{U}_A = \bar{1} + \bar{U}', \quad \bar{V}_A = \bar{V},$$
$$X_B = S' + U, \quad Y_B = S' + 1 + V', \quad \bar{S}_B = \bar{1} + \bar{S}', \quad \bar{U}_B = \bar{U}, \quad \bar{V}_B = \bar{1} + \bar{V}'.$$

THEOREM 2.3. *The dimensions of the irreducible level* 1 *characters* $\bar{S}_n, \bar{U}_n, \bar{V}_n$ *of* G_n *and of their components in* A_n *and* B_n *are*

(2.12) *for* G_n : $s_n = a_n b_n/3$,
$$u_n = b_n b_{n-1}/3, \quad v_n = a_n a_{n-1}/3;$$
for A_n : $s_n - 1 = 4a_{n-1}b_{n-1}/3$,
$$u_n - 1 = 2a_n b_{n-2}/3, \quad v_n = a_n a_{n-1}/3;$$
for B_n : $s_n - 1 = 4a_{n-1}b_{n-1}/3$,
$$u_n = b_n b_{n-1}/3, \quad v_n - 1 = 2b_n a_{n-2}/3;$$
where the values of a_n *and* b_n *are defined by*

(2.13)

n :	5	4	3	2	1	0	-1	-2
$a_n = 2^n + 1$:	33	17	9	5	3	2	3/2	5/4
$b_n = 2^n - 1$:	31	15	7	3	1	0	-1/2	-3/4 .

PROOF. Since $s_1 = 1$ and $s_n - s_{n-1} = 1 + x_{n-1} + 1 + y_{n-1} = 4^{n-1}$ by (2.6) and (2.3) it follows that

(2.14) $$s_n = 4^{n-1} + 4^{n-2} + \ldots + 4 + 1 = (4^n-1)/(4-1) = a_n b_n/3 .$$

The equations $1 + x_n = s_n + u_n$ and $1 + y_n = s_n + v_n$ derived from (2.8) determine u_n and v_n.

In this notation the order ${}^0G_n$ of G_n can be written

(2.15) $${}^0G_n = 2^{n^2} \prod_{k=1}^{n} a_k b_k .$$

3. Character symbols and pairing

Noting that the degree of each level 1 character is a product of two factors of the form $a_{n-k} = 2^{n-k} + 1$ and $b_{n-k} = 2^{n-k} - 1$ and a numerical factor with 2-power numerator and odd denominator d_j independent of n, we adopt the symbols in (3.1) below for level 1 characters of G_n, A_n, and B_n, in which 1's in row 1 correspond to factors $a_n, a_{n-1}, \ldots$, and 1's in row 2 correspond to factors $b_n, b_{n-1}, \ldots$. Since b_n is always missing in A_n characters and a_n in B_n characters we can and do place a + or − sign in these positions to distinguish an A_n or B_n character which is positive on the transposition class from its associate which is negative there. For $n \le 4$ we describe these characters by their degrees with subscripts for 0_8^+ as in [3], for 0_7 and 0_6^- as in [2], and describe the characters of symmetric groups $S_8 = 0_6^+$, $S_6 = 0_5$, $S_5 = 0_4^-$, $S_3 = 0_3$ by partition symbols corresponding to Young diagrams.

(3.1) For G_n

	Symbol	Degree	0_9	0_7	0_5	0_3
X_n	$\begin{pmatrix}10\\01\end{pmatrix}$	$a_n b_{n-1}$	119_i	27_a	$5[51]$	0
$\bar{U}_n$	$\frac{1}{3}\begin{pmatrix}00\\11\end{pmatrix}$	$b_n b_{n-1}/3$	$\overline{35}_i$	$\bar{7}_a$	$\bar{1}[1^6]$	0
$\bar{\bar{S}}_n$	$\frac{1}{3}\begin{pmatrix}1\\1\end{pmatrix}$	$a_n b_n/3$	$\overline{85}_i$	$\overline{21}_b$	$\bar{5}[21^4]$	$\bar{1}[1^3]$
$\bar{V}_n$	$\frac{1}{3}\begin{pmatrix}11\\00\end{pmatrix}$	$a_n a_{n-1}/3$	$\overline{51}_i$	$\overline{15}_a$	$\bar{5}[2^3]$	$\bar{2}[21]$
Y_n	$\begin{pmatrix}01\\10\end{pmatrix}$	$b_n a_{n-1}$	135_i	35_b	$9[42]$	$2[21]$

(3.2) For $A_n = O_{2n}^-$

	Symbol	Degree	O_8^-	O_6^-	O_4^-
V_n	$\frac{1}{3}\begin{pmatrix}11\\+0\end{pmatrix}$	$a_n a_{n-1}/3$	51_u	15_q	$5[32]$
U'_n	$\frac{2}{3}\begin{pmatrix}100\\+01\end{pmatrix}$	$2a_n b_{n-2}/3$	34_u	6_p	0
S'_n	$\frac{4}{3}\begin{pmatrix}01\\+1\end{pmatrix}$	$4a_{n-1}b_{n-1}/3$	84_u	20_p	$4[41]$

(3.3) For $B_n = O_{2n}^+$

			O_8^+	O_6^+	O_4^+
S'_n	$\frac{4}{3}\begin{pmatrix}+1\\01\end{pmatrix}$	$4b_{n-1}a_{n-1}/3$	84_x	$20[62]$	4
V'_n	$\frac{2}{3}\begin{pmatrix}+01\\100\end{pmatrix}$	$2b_n a_{n-2}/3$	50_x	$14[4^2]$	4
U_n	$\frac{1}{3}\begin{pmatrix}+0\\11\end{pmatrix}$	$b_n b_{n-1}/3$	35_x	$7[71]$	1

Degrees of level k factors are factors of the order (2.15) of G which are observed to be products of $2k$ factors a_{n-i} or b_{n-i} without repetitions, times a numerical factor with 2-power numerator and odd denominator d_i . We denote these characters by 2-rowed arrays of 0's and 1's , where a 1 in position $(1, i)$ or $(2, i)$ denotes a factor a_{n+1-i} or b_{n+1-i} , prefixed with the numerical factor independent of n . We find for all n that the alternating Kronecker squares of $\bar{U}$ and $\bar{V}$ are irreducible level 2 characters, whereas the symmetrized Kronecker squares and alternating Kronecker cubes split as follows:

(3.4) $$U^{[1^2]} = \frac{1}{9}\begin{pmatrix}100\\111\end{pmatrix}, \qquad U^{[2]} = 1 + \begin{pmatrix}10\\01\end{pmatrix} + \frac{2}{9}\begin{pmatrix}1100\\1001\end{pmatrix},$$

$$V^{[1^2]} = \frac{1}{9}\begin{pmatrix}111\\100\end{pmatrix}, \qquad V^{[2]} = 1 + \begin{pmatrix}01\\10\end{pmatrix} + \frac{2}{9}\begin{pmatrix}1001\\1100\end{pmatrix},$$

(3.5) $$U^{[1^3]} = \frac{1}{81}\begin{pmatrix}111\\111\end{pmatrix} + \frac{2}{27}\begin{pmatrix}1100\\1111\end{pmatrix}, \quad V^{[1^3]} = \frac{1}{81}\begin{pmatrix}111\\111\end{pmatrix} + \frac{2}{27}\begin{pmatrix}1111\\1100\end{pmatrix},$$

(3.6) $$VU^{[1^2]} = \frac{1}{9}\begin{pmatrix}111\\111\end{pmatrix} + \frac{2}{27}\begin{pmatrix}1100\\1111\end{pmatrix}, \quad UV^{[1^2]} = \frac{1}{9}\begin{pmatrix}111\\111\end{pmatrix} + \frac{2}{27}\begin{pmatrix}1111\\1100\end{pmatrix}.$$

The decomposition of the product of any two level 1 characters is given in Appendix 2.

We note that all the degree formulas involve polynomials of even degree in the quantity $t = 2^n$, and that a change in the sign of t interchanges the two rows in a character symbol while interchanging $\bar{U}_n$ and $\bar{V}_n$, X_n and Y_n in G_n and exchanging characters of A_n and B_n for their mates in the other subgroup. This automorphism of the character ring of G_n interchanges pairs of characters whose arrays have unequal rows. It fixes the self paired characters whose rows are the same, and it preserves multiplication in the character ring of G while interchanging the Frobenius induce-restrict table of G_n and A_n with that of G_n and B_n, and of A_n and G_{n-1} with that of B_n and G_{n-1}. There is one defect, however, that spoils the complete symmetry. If for a given n a character symbol has the degree factor $a_0 = 2$, this character, called a widow, has a mate with degree factor $b_0 = 0$. It vanishes identically and is called a ghost character. Each widow has two distinct symbols, obtained one from the other by interchanging the two rows except in the $(n+1)$th column involving a_0 and b_0, and then modifying the constant factor to obtain the same degree. In higher dimensions these two symbols and their mates represent a tetrad of distinct characters.

4. Characters of odd order elements

Any element g of G_n can be factored uniquely as a product rs of two of its powers, where r is 2-regular of odd order, and s is 2-singular of 2-power order or 1. All eigenvalues of s are 1, and the eigenvalues of g are those of r. There are exactly 2^n monic reciprocal polynomials $\varphi_{2n}(z)$ of degree $2n$ over F_2, and to each of these corresponds a unique class C_j of G_n with odd order elements. The 2^{n-1} polynomials with middle coefficient 0 have even coefficient sum, and hence are divisible by the factor $(z-1)^2 \equiv z^2 + 1 \pmod 2$ whose suppression reduces the polynomial to one for G_{n-1}. If $z^2 + 1$ factors out exactly d times we say the class has depth d and level $k = n - d$. The residual polynomial $\varphi_{2k}(z)$ has middle coefficient 1, and is identified by its first k coefficients which serve as a binary code for the class. We call the class negative or positive and assign it a $-$ or $+$ sign according as the class is represented in $A_k = 0^-_{2k}$ or $B_k = 0^+_{2k}$. It cannot be represented in both. A more convenient, but not always unique, symbol for a class is a product of odd integers $2^j + 1$ or $2^j - 1$ each corresponding to an irreducible factor φ_{2j} of φ_{2k} and such that a_j or b_j is a multiple of the order of the element corresponding to φ_{2j}. Here 3 represents $a_1 = 2 + 1$ rather than $b_2 = 2^2 - 1$, since the irreducible polynomial for 3 is $\varphi_2 = z^2 + z + 1$ with $j = 1$. The class belongs to 0^-_{2k} or 0^+_{2k} according as an odd or even number of factors in its symbol are a_j's . Algebraically conjugate classes are not distinguished in this symbolism, although they have distinct binary code designations.

THEOREM 4.1. *For elements of odd order in class* C_j *of level* k *in* G_k ,

$$X_{k,j} = 1 + Y_{k,j} = 0 \quad \text{if } A_k \cap C_j \neq 0 , \tag{4.1}$$

$$1 + X_{k,j} = Y_{k,j} = 0 \quad \text{if } B_k \cap C_j \neq 0 .$$

PROOF. The permutation characters $1_B^G = 1 + Y_k$ and $1_A^G = 1 + X_k$ vanish on C_j respectively when C_j is not represented in B_k or A_k . Since their sum is a power of 2 by Theorem 2.1, they cannot both vanish, and the non-vanishing one is a power of 2 . If $1 + X_{k,j} > 1$, then the permutation on column vectors representing an element g_j of C_j would fix more than one column vector, and some conjugate of g_j would be in G_{k-1} , contrary to the assumption that C_j has level k .

THEOREM 4.2. *For odd order elements in classes* C_j *of level* k *in* G_n ,

$$\left(2\bar{S}_n+\bar{U}_n+\bar{V}_n\right)_j = \left(1+X_n+1+Y_n\right)_j = 4^{n-k} , \tag{4.2}$$

$$\left(\bar{V}_n-\bar{U}_n\right)_j = \left(Y_n-X_n\right)_j = \begin{cases} -2^{n-k} & \text{if } A_k \cap C_j \neq 0 , \\ +2^{n-k} & \text{if } B_k \cap C_j \neq 0 . \end{cases} \tag{4.3}$$

PROOF. For such elements equations (2.8) yield

$$\left(\bar{S}_n+\bar{U}_n\right)_j = \left(1+X_n\right)_j , \quad \left(\bar{S}_n+\bar{V}_n\right)_j = \left(1+Y_n\right)_j . \tag{4.4}$$

From (2.5) we obtain

$$1 + X_n + 1 + Y_n = 4\left(1+X_{n-1}+1+Y_{n-1}\right) = 4^{n-k}\left(1+X_k+1+Y_k\right) , \tag{4.5}$$

$$\bar{V}_n - \bar{U}_n = Y_n - X_n = 2\left(Y_{n-1}-X_{n-1}\right) = 2^{n-k}\left(Y_k-X_k\right) . \tag{4.6}$$

From Theorem 4.1 we see that $1 + X_k$ and $1 + Y_k$ have the values 1, 0 or 0, 1 according as C_j is represented in A_k or B_k . Substitution of these values in (4.5) and (4.6) completes the proof of (4.2) and (4.3).

THEOREM 4.3. *If* m_j *is the multiplicity of the factor* $z^2 + z + 1$ *in the characteristic polynomial of an element* g_j *of odd order of level* k *, and hence is the exponent of* 3 *in the symbol for its class* C_j *, then*

(4.7) $$\left(\bar{U}_n+\bar{V}_n-\bar{S}_n\right)_j = (-2)^{m_j} = 1 - 3c_j .$$

PROOF. Under the given hypotheses, g_j^2 is in the same class C_j as g_j and g_j^3 is in a class $C_{j'}$ of level $k - m_j$, whose signature is the same if and only if m_j is even. Dropping subscripts n in the proof, we have by (4.3),

(4.8) $$(\bar{V}-\bar{U})_{j'} = (-2)^{m_j}(\bar{V}-\bar{U})_j .$$

Equations (3.5) and (3.6) yield the identities

(4.9) $$\bar{V}^{[1^3]} - \bar{U}^{[1^3]} = \overline{UV}^{[1^2]} - \overline{VU}^{[1^2]} ,$$

(4.10) $$\left(\bar{V}^3-\bar{U}^3\right)_j - 3\left(\bar{V}^2-\bar{U}^2\right)_j + 2(-2)^{m_j}(\bar{V}-\bar{U})_j = 3\left(\overline{UV}(\bar{V}-1)-\overline{VU}(\bar{U}-1)\right)_j ,$$

$$(\bar{V}-\bar{U})_j^3 - 3\left(\bar{V}^2-\bar{U}^2\right)_j + 2(-2)^{m_j}(\bar{V}-\bar{U})_j = 0 ,$$

(4.11) $$(\bar{V}-\bar{U})_j^2 - 3(\bar{V}+\bar{U})_j + 2(-2)^{m_j} = 0 .$$

Substituting for $(\bar{V}-\bar{U})_j^2$ in (4.3) and (4.2) we obtain

(4.12) $$\left(2\bar{S}+\bar{U}+\bar{V}\right)_j - 3(\bar{V}+\bar{U})_j = -2(-2)^{m_j} .$$

Dividing (4.12) by -2, we obtain (4.7).

THEOREM 4.4. *The values for odd order elements of* G_n *in class* C_j *of level* k, *of all level* 1 *characters are*

(4.13) $\bar{S}_{n,j} = s_{n-k} + c_j$, *where* $c_j = \left[1-(-2)^{m_j}\right]/3$,

(4.14) $X_{n,j} = y_{n-k}$, $Y_{n,j} = x_{n-k}$, $\bar{U}_{n,j} = v_{n-k} - c_j$, $\bar{V}_{n,j} = u_{n-k} - c_j$
if $A_k \cap C_j \neq 0$,

(4.15) $X_{n,j} = x_{n-k}$, $Y_{n,j} = y_{n-k}$, $\bar{U}_{n,j} = u_{n-k} - c_j$, $\bar{V}_{n,j} = v_{n-k} - c_j$
if $B_k \cap C_j \neq 0$.

PROOF. Values of X_n and Y_n are obtained for all j by solving (4.2) and (4.3). They depend only on the level k, the sign of the class, and the dimension n. Values of $\bar{S}_{n,j}$, $\bar{U}_{n,j}$ and $\bar{V}_{n,j}$ are obtained by solving (4.2), (4.7), and (4.3).

Numerical values for these characters for $n = 2, 3, 4$ are given in Appendix 1.

5. Characters of odd degree

The group $G_n = 0_{2n+1}(2)$ appears to have exactly 2^{n+1} characters of odd degree for $n \geq 2$, of which half are self-paired characters $S_n^{(j)}$ of degree $s_n^{(j)}$ in G_n. Each character $S_n^{(j)}$ can be assigned to one of the 2^n classes C_j of elements of odd order in such a way that the degree $s_n^{(j)}$ is the odd factor of the class size of C_j in G_n. This correspondence is one-to-one, except for the assignment of a family of algebraically conjugate characters of the same odd degree to an equal number of algebraically conjugate classes of odd order elements on which these characters have irrational values. The numerical factor in the symbol for $S_r^{(j)}$ of level k appears to be the reciprocal of the odd factor f_j in the order of the centralizer of an element of C_j in G_k. This factor appears to be

(5.1) $$f_j = (2^i-\varepsilon)(2^{2i}-1)(2^{3i}-\varepsilon) \ldots (2^{ri}-\varepsilon^r), \quad \varepsilon = \pm 1$$

if C_j is a "primitive" class of level k whose polynomial $\varphi_{2k}(z)$ is a power of an irreducible polynomial, and whose symbol is $(2^i-\varepsilon)^r$. In general it is the product of such factors, one for each such power factor in the symbol for C_j. The array part of the symbol for a self paired character $S^{(j)}$ of level k is always a $2 \times k$ matrix of 1's.

The remaining 2^n odd degree characters of G_n for $n \geq 2$ belong to 2^{n-2} tetrads, each consisting of two pairs of paired characters, labeled $U^{(j)}$ and $V^{(j)}$, $X^{(j)}$ and $Y^{(j)}$. Each of the 2^{k-1} tetrads of level $k+1 \leq n-1$ is associated with a self paired character $S^{(j)}$ of odd degree and level k. The symbols for such a tetrad are obtained by placing the 2 columns for $\bar{U}$, $\bar{V}$, X or Y respectively (which correspond to $S^{(j)} = 1$) at the right of the k columns of 1's for $S^{(j)}$ (except when $S^{(j)} = 1$) and multiplying the numerical factor $1/f_j$ for $S^{(j)}$ by 1/3 for $U^{(j)}$ and $V^{(j)}$ or by 1 for $X^{(j)}$ and $Y^{(j)}$.

When $S^{(j)}$ has level $n-1$ however, the 2^{n-2} tetrads have degrees 0 for $U^{(j)}$ and $X^{(j)}$ and twice an odd number for their widowed pairs $V^{(j)}$ and $Y^{(j)}$ which are two representations of the same widowed character $W^{(j)}$. For example if $n = 4$ and $G = 0_9(2)$ we have

(5.2)
$$C^{(j)} = \quad 3\cdot5, \quad 7, \quad 3^3, \quad 9,$$

$$V^{(j)} = \frac{1}{45}\begin{pmatrix}11111\\11100\end{pmatrix}, \frac{1}{21}\begin{pmatrix}11111\\11100\end{pmatrix}, \frac{1}{243}\begin{pmatrix}11111\\11100\end{pmatrix}, \frac{1}{27}\begin{pmatrix}11111\\11100\end{pmatrix},$$

$$Y^{(j)} = \frac{1}{3\cdot5}\begin{pmatrix}11101\\11110\end{pmatrix}, \frac{1}{7}\begin{pmatrix}11101\\11110\end{pmatrix}, \frac{1}{3\cdot3\cdot9}\begin{pmatrix}11101\\11110\end{pmatrix}, \frac{1}{9}\begin{pmatrix}11111\\11100\end{pmatrix},$$

$$W_4^{(j)} = \quad 32130_j, \quad 68850_i, \quad 5950_i, \quad 53550_i .$$

Just half of the characters of odd degree are "positive characters" with positive values on the transposition class, namely the characters $S^{(j)}$, $X^{(j)}$ and $Y^{(j)}$ corresponding to the positive classes (represented in B_k at level k, but not in A_k) and the characters $U^{(j)}$ and $V^{(j)}$ corresponding to negative classes (in A_k but not in B_k). The other half are "negative characters", which may be distinguished by placing a bar over their symbols.

For A_n or B_n the 2^n characters of odd degree consist of 2^{n-1} associated pairs. Symbols for those of A_n are obtained from symbols of self paired characters $S^{(j)}$ of level $k \le n-1$ by replacing the 1 representing b_n by a + or - sign, and adjoining a 1 for b_{n-k} if $S^{(j)}$ is positive or a 1 for a_{n-k} if $S^{(j)}$ is negative. By interchanging the two rows we obtain corresponding symbols for B_n.

APPENDIX I

Values for odd order elements of level 1 characters of $0_{2n+1}(2)$

Class	$0_{11}(2)$					$0_9(2)$					$0_7(2)$					$0_5(2)$					$0_3(2)$				
	$\bar{U}_5$	$\bar{V}_5$	$\bar{S}_5$	X_5	Y_5	$\bar{U}_4$	$\bar{V}_4$	$\bar{S}_4$	X_4	Y_4	$\bar{U}_3$	$\bar{V}_3$	$\bar{S}_3$	X_3	Y_3	$\bar{U}_2$	$\bar{V}_2$	$\bar{S}_2$	X_2	Y_2	$\bar{U}_1$	$\bar{V}_1$	$\bar{S}_1$	X_1	Y_1
1 +	155	187	341	495	527	35	51	85	119	135	7	15	21	27	35	1	5	5	5	9	0	2	1	0	2
3 −	50	34	86	135	119	14	6	22	35	27	4	0	6	9	5	1	-1	2	2	0	0	-1	1	0	-1
3^2 +	8	16	20	27	35	2	6	4	5	9	1	3	0	0	2	1	2	-1	-1	0					
5 −	15	7	21	35	27	5	1	5	9	5	2	0	1	2	0	1	0	0	0	-1					
3·5 +	0	4	6	5	9	-1	1	2	0	2	-1	0	1	-1	0										
7 +	1	5	5	5	9	0	2	1	0	2	0	1	0	-1	0										
3^3 −	2	-2	8	9	5	-1	-3	4	2	0	-2	-3	3	0	-1										
9 −	5	1	5	9	5	2	0	1	2	0	1	0	0	0	-1										
3^4 +	5	7	-4	0	2	5	6	-5	-1	0															
3·9 +	-1	1	2	0	2	-1	0	1	-1	0															
5^2 +	0	2	1	0	2	0	1	0	-1	0															
15 +	0	2	1	0	2	0	1	0	-1	0															
$3^2 5$ −	3	1	0	2	0	2	1	-1	0	-1															
3·7 −	1	-1	2	2	0	0	-1	1	0	-1															
17 −	2	0	1	2	0	1	0	0	0	-1															
17 −	2	0	1	2	0	1	0	0	0	-1															
$3^3 5$ +	-3	-2	3	-1	0																				
$3^2 7$ +	1	2	-1	-1	0																				
3·17 +	-1	0	1	-1	0																				
3·17 +	-1	0	1	-1	0																				
5·9 +	0	1	0	-1	0																				
31 +	0	1	0	-1	0																				
31 +	0	1	0	-1	0																				
31 +	0	1	0	-1	0																				
3^5 −	-10	-11	11	0	-1																				
$3^2 9$ −	2	1	-1	0	-1																				
$3·5^2$ −	0	-1	1	0	-1																				
3·15 −	0	-1	1	0	-1																				
5·7 −	1	0	0	0	-1																				
11 −	1	0	0	0	-1																				
33 −	1	0	0	0	-1																				
33 −	1	0	0	0	-1																				

Symbols and degrees for level 1 characters of $O_{2n+1}(2)$ and their values in positive and negative classes of level k whose class symbols have the factor 3^m are:

Symbol	Degree	Value for: + classes,	- classes
$\bar{U}_n = \frac{1}{3}\begin{pmatrix}00\\11\end{pmatrix}$,	$u_n = \frac{1}{3}b_n b_{n-1}$,	$u_{n-k} - c_m$	$v_{n-k} - c_m$
$\bar{V}_n = \frac{1}{3}\begin{pmatrix}11\\00\end{pmatrix}$,	$v_n = \frac{1}{3}a_n a_{n-1}$,	$v_{n-k} - c_m$	$u_{n-k} - c_m$
$\bar{S}_n = \frac{1}{3}\begin{pmatrix}1\\1\end{pmatrix}$,	$s_n = \frac{1}{3}a_n b_n$	$s_{n-k} + c_m$	$s_{n-k} + c_m$
$X_n = \begin{pmatrix}10\\01\end{pmatrix}$,	$x_n = a_n b_{n-1}$,	x_{n-k}	y_{n-k}
$Y_n = \begin{pmatrix}01\\10\end{pmatrix}$,	$y_n = b_n a_{n-1}$	y_{n-k}	x_{n-k}

where $a_n = 2^n + 1$, $b_n = 2^n - 1$, $c_m = (1-(-2)^m)/3$.

APPENDIX II

Characters of $O_{2n+1}(2)$ of levels 1 and 2 contained in products of two of the level 1 characters $\bar{U}_n, \bar{V}_n, \bar{S}_n, X_n, Y_n$

$O_{2n+1}(2)$		$\bar{U}_n = \frac{1}{3}\begin{pmatrix}00\\11\end{pmatrix}$,					$\bar{V}_n = \frac{1}{3}\begin{pmatrix}11\\00\end{pmatrix}$,					$\bar{S}_n = \frac{1}{3}\begin{pmatrix}1\\1\end{pmatrix}$,					$X_n = \begin{pmatrix}10\\01\end{pmatrix}$,					$Y_n = \begin{pmatrix}01\\10\end{pmatrix}$,				
	$O_9(2)$	$\overline{35}_i$					$\overline{51}_i$					$\overline{85}_i$					119_i					135_i				
		$\bar{U}$	$\bar{V}$	$\bar{S}$	X	Y	$\bar{U}$	$\bar{V}$	$\bar{S}$	X	Y	$\bar{U}$	$\bar{V}$	$\bar{S}$	X	Y	$\bar{U}$	$\bar{V}$	$\bar{S}$	X	Y	$\bar{U}$	$\bar{V}$	$\bar{S}$	X	Y
$\begin{pmatrix}0\\0\end{pmatrix}$	1_i	u						v						s						x						y
$\frac{1}{3}\begin{pmatrix}00\\11\end{pmatrix}$	$\overline{35}_i$				x										x		u		s							
$\frac{1}{3}\begin{pmatrix}11\\00\end{pmatrix}$	$\overline{51}_i$										y					y							v	s		
$\frac{1}{3}\begin{pmatrix}1\\1\end{pmatrix}$	$\overline{85}_i$				x						y				x	y	u		s				v	s		
$\begin{pmatrix}10\\01\end{pmatrix}$	119_i	u		s								u		s						xx	y				x	y
$\begin{pmatrix}01\\10\end{pmatrix}$	135_i							v	s				v	s						x	y				x	yy
$\frac{1}{9}\begin{pmatrix}11\\11\end{pmatrix}$	1785_i		v				u														y				x	
$\frac{1}{5}\begin{pmatrix}11\\11\end{pmatrix}$	$\overline{3213}_j$					y				x					x	y		v	s			u		s		
$\frac{2}{9}\begin{pmatrix}11\\11\end{pmatrix}$	3570_i													$\bar{s}$						$\bar{x}$	y				x	$\bar{y}$
$\frac{8}{9}\begin{pmatrix}101\\101\end{pmatrix}$	3400_i													s						x	y				x	y
$\frac{1}{9}\begin{pmatrix}100\\111\end{pmatrix}$	595_i	$\bar{u}$																		$\bar{x}$						
$\frac{1}{9}\begin{pmatrix}111\\100\end{pmatrix}$	1275_i							$\bar{v}$																		$\bar{y}$
$\frac{1}{3}\begin{pmatrix}110\\101\end{pmatrix}$	$\overline{2295}_i$				x										x	y	u		s					s		
$\frac{1}{3}\begin{pmatrix}101\\110\end{pmatrix}$	$\overline{2975}_j$										y				x	y			s				v	s		
$\frac{2}{9}\begin{pmatrix}1100\\1001\end{pmatrix}$	510_i	u																		x						
$\frac{2}{9}\begin{pmatrix}1001\\1100\end{pmatrix}$	1190_i							v																		y
$\frac{2}{15}\begin{pmatrix}1000\\1101\end{pmatrix}$	$\overline{238}_i$				x												u									
$\frac{2}{15}\begin{pmatrix}1101\\1000\end{pmatrix}$	$\overline{918}_i$										y												v			
$\frac{8}{15}\begin{pmatrix}010\\111\end{pmatrix}$	$\overline{1512}_i$				x	y									x		u		s			u				
$\frac{8}{15}\begin{pmatrix}111\\010\end{pmatrix}$	$\overline{2856}_j$									x	y					y		v					v	s		
$\frac{8}{9}\begin{pmatrix}110\\011\end{pmatrix}$	2856_i			s																$x\bar{x}$	y				x	
$\frac{8}{9}\begin{pmatrix}011\\110\end{pmatrix}$	4200_i								s												y				x	$y\bar{y}$

Each column in the table above indicates the decomposition of the product of two level 1 characters $\bar{U}_n, \bar{V}_n, \bar{S}_n, X_n, Y_n$ of $O_{2n+1}(2)$. For example, the x's in column 4 and the u's in column 16 are placed in rows opposite the irreducible components of $U_n \times X_n$ and $X_n \times U_n$. The three $\bar{x}$'s in column 19 indicate the components of the alternating Kronecker square of X_n, whereas the x's in column 19 indicate the components of the symmetrized square of X_n, in which X_n itself appears twice. Numerical degrees are given for the level 1 characters of $0_9(2)$. Paired characters are bracketed together. Note that an interchange of paired rows results from an interchange of U_n and V_n, X_n and Y_n in the column headings.

APPENDIX III

Characters of odd degree for $0_{2n+1}(2)$, $n \le 4$

Self paired characters

Class	Symbol	$0_9(2)$	$0_7(2)$	$0_5(2)$	$0_3(2)$
1 +	$\begin{Bmatrix}0\\0\end{Bmatrix}$	1_i	1_a	1[6]	1[3]
3 −	$\frac{1}{3}\begin{Bmatrix}1\\1\end{Bmatrix}$	$\overline{85}_i$	$\overline{21}_b$	$\bar{5}[2\cdot1^4]$	$\bar{1}[1^3]$
3^2 +	$\frac{1}{3^2}\begin{Bmatrix}11\\11\end{Bmatrix}$	1785_i	105_b	$5[3^2]$	
5 −	$\frac{1}{5}\begin{Bmatrix}11\\11\end{Bmatrix}$	$\overline{3213}_j$	$\overline{189}_b$	$\bar{9}[2^2 1^2]$	
$3\cdot5$ +	$\frac{1}{3\cdot5}\begin{Bmatrix}111\\111\end{Bmatrix}$	16065_h	189_a		
7 +	$\frac{1}{7}\begin{Bmatrix}111\\111\end{Bmatrix}$	34425_i	405_a		
3^3 −	$\frac{1}{3\cdot3\cdot9}\begin{Bmatrix}111\\111\end{Bmatrix}$	$\overline{2975}_i$	$\overline{35}_a$		
9 −	$\frac{1}{9}\begin{Bmatrix}111\\111\end{Bmatrix}$	$\overline{26775}_j$	$\overline{315}_a$		
3^4 +	$\frac{1}{3^2 9\cdot15}\begin{Bmatrix}1111\\1111\end{Bmatrix}$	595_j			
$3\cdot9$ +	$\frac{1}{3\cdot9}\begin{Bmatrix}1111\\1111\end{Bmatrix}$	26775_i			
5^2 +	$\frac{1}{5\cdot15}\begin{Bmatrix}1111\\1111\end{Bmatrix}$	9639_i			
15 +	$\frac{1}{15}\begin{Bmatrix}1111\\1111\end{Bmatrix}$	48195_i			
$3^2 5$ −	$\frac{1}{3^2 5}\begin{Bmatrix}1111\\1111\end{Bmatrix}$	$\overline{16065}_k$			
$3\cdot7$ −	$\frac{1}{3\cdot7}\begin{Bmatrix}1111\\1111\end{Bmatrix}$	$\overline{34425}_j$			
17 −	$\frac{1}{17}\begin{Bmatrix}1111\\1111\end{Bmatrix}$	$\overline{42525}_i$			
17 −	$\frac{1}{17}\begin{Bmatrix}1111\\1111\end{Bmatrix}$	$\overline{42525}_j$			

Paired characters

Class	Symbol	$0_9(2)$	$0_7(2)$	$0_5(2)$
1 +	$\frac{1}{3}\begin{Bmatrix}00\\11\end{Bmatrix}$	$\overline{35}_i$	$\bar{7}_a$	$\bar{1}[1^6]$
	$\frac{1}{3}\begin{Bmatrix}11\\00\end{Bmatrix}$	$\overline{51}_i$	$\overline{15}_a$	$\bar{5}[2^3]$
	$\begin{Bmatrix}10\\01\end{Bmatrix}$	119_i	27_a	$5[5\cdot1]$
	$\begin{Bmatrix}01\\10\end{Bmatrix}$	135_i	35_b	$9[4\cdot2]$
3 −	$\frac{1}{9}\begin{Bmatrix}100\\111\end{Bmatrix}$	595_i	21_a	
	$\frac{1}{9}\begin{Bmatrix}111\\100\end{Bmatrix}$	1275_i	105_c	
	$\frac{1}{3}\begin{Bmatrix}110\\101\end{Bmatrix}$	$\overline{2295}_i$	$\overline{105}_a$	
	$\frac{1}{3}\begin{Bmatrix}101\\110\end{Bmatrix}$	$\overline{2975}_j$	$\overline{189}_c$	
3^2 +	$\frac{1}{27}\begin{Bmatrix}1100\\1111\end{Bmatrix}$	$\overline{1785}_j$		
	$\frac{1}{27}\begin{Bmatrix}1111\\1100\end{Bmatrix}$	$\overline{8925}_j$		
	$\frac{1}{9}\begin{Bmatrix}1110\\1101\end{Bmatrix}$	8925_i		
	$\frac{1}{9}\begin{Bmatrix}1101\\1110\end{Bmatrix}$	16065_i		
5 −	$\frac{1}{15}\begin{Bmatrix}1100\\1111\end{Bmatrix}$	3213_i		
	$\frac{1}{15}\begin{Bmatrix}1111\\1100\end{Bmatrix}$	16065_j		
	$\frac{1}{5}\begin{Bmatrix}1110\\1101\end{Bmatrix}$	$\overline{16065}_q$		
	$\frac{1}{5}\begin{Bmatrix}1101\\1110\end{Bmatrix}$	$\overline{28917}_i$		

Classes are designated + or − according as they appear first in $0^+_{2k}(2)$ or $0^-_{2k}(2)$. A bar over the degree of a character indicates a negative value for the class of transpositions.

References

[1] Leonard Eugene Dickson, *Linear groups: With an exposition of the Galois field theory* (Teubner, Leipzig, 1901; reprinted, Dover, New York, 1958). MR21#3488.

[2] J.S. Frame, "The classes and representations of the groups of 27 lines and 28 bitangents", *Ann. Mat. Pura Appl.* (4) 32 (1951), 83-119. MR13,817.

[3] J.S. Frame, "The characters of the Weyl group E_8 ", *Computational problems in abstract algebra;* Proceedings, pp. 111-130 (Pergamon Press, Oxford, New York, Toronto, 1970). MR40#5374.

[4] J.S. Frame and A. Rudvalis, "Characters of symplectic groups over F_2 ", *Finite Groups. Proc. Gainesville Conf.*, 1972, pp. 41-54 (American Elsevier, New York, 1973).

Michigan State University,
East Lansing, Mich. 48823, USA.

PROC. SECOND INTERNAT. CONF. THEORY OF GROUPS, 20K99
CANBERRA 1973, pp. 315-317.

TYPE SETS AND ADMISSIBLE MULTIPLICATIONS

B.J. Gardner

The study of *nil groups*, abelian groups admitting only the trivial $(xy = 0)$ ring multiplication, was initiated by Szele [3]. More recently, Wickless [4] has considered groups which admit only nilpotent multiplications, and some related results for T-nilpotence have been obtained by the author [1]. Wickless' paper also contains some information about torsion-free groups on which only prime radical rings can be defined.

We shall be concerned here with groups whose admissible multiplications are subject to these various restrictions. Only torsion-free groups are of interest in this context: a torsion group has any one of the properties mentioned if and only if it is divisible, and there are no mixed nil groups [3], while a group G satisfies one of the other conditions precisely when its torsion subgroup G_t is divisible and G/G_t satisfies the condition (*cf.* [4], Theorem 1.1).

The completely decomposable nil groups are characterized by the following result.

THEOREM 1 (Ree and Wisner [2]). *Let* $G = \bigoplus_{\alpha\in A} X_\alpha$, *where* X_α *is a rational group of type* τ_α. *Then* G *is a nil group if and only if* $\tau_\alpha + \tau_\beta \nleq \tau_\gamma$ *for all* $\alpha, \beta, \gamma \in A$.

Thus a sufficient condition for G to be a nil group is that the τ_α be nil and pairwise incomparable, that is, $\{\tau_\alpha \mid \alpha \in A\}$ have no chains of length 2. It is therefore of some interest to see how other chain conditions on the type set of a torsion-free group affect the admissible multiplications.

THEOREM 2 (Wickless [4]; Gardner [1]). *Let* G *be a torsion-free group in which every non-zero element has nil type.*

(i) If the type set $\mathrm{T}(G)$ *of* G *has ascending chain condition, then every ring defined on* G *is prime radical.*

(ii) If $\mathrm{T}(G)$ *has ascending chain condition and descending chain condition,*

then every ring defined on G *is* T*-nilpotent.*

(iii) If every chain in $\mathsf{T}(G)$ *is of length* $\leq n$ *for some positive integer* n *, then every ring defined on* G *is nilpotent.*

(i) is proved in [4]; for another proof see [1]. *(ii)* is proved in [1], *(iii)* in [4].

There is no ambiguity in *(ii)*: every ring on a group G is left T-nilpotent if and only if every ring on G is right T-nilpotent. (Consider opposite rings.)

Examples 3.6 and 3.7 of [1] show that the ascending chain condition alone does not force T-nilpotence and that the ascending chain condition and the descending chain condition do not force nilpotence. It is perhaps worth noting that in both of these counterexamples the group concerned is completely decomposable.

The converse to each part of Theorem 2 is false, as the following result shows.

THEOREM 3. *There exists a torsion-free group* G *with the following properties:*

(i) G *is completely decomposable;*

(ii) every non-zero element of G *has nil type;*

(iii) $\mathsf{T}(G)$ *has neither the ascending chain condition nor the descending chain condition;*

(iv) G *is a nil group.*

PROOF. Let $p_1, p_2, \ldots$ be the natural enumeration of the primes. The types τ_n of the height sequences

$$H^{(n)} = (h_1, h_2, \ldots) \text{ , where } h_i = \begin{cases} i \text{ , if } i = p_1^k, \ldots, p_n^k \text{ ,} \\ 0 \text{ , otherwise,} \end{cases}$$

form an infinite ascending chain while the types σ_n of the height sequences

$$(\underbrace{0, \ldots, 0}_{n-1}, 1, 2, 3, \ldots)$$

form an infinite descending chain. Let X_n, Y_n be rational groups with types τ_n, σ_n respectively, and let $G = \bigoplus_{n=1}^{\infty} X_n \oplus \bigoplus_{n=1}^{\infty} Y_n$. Using Theorem 1 one can show that G is a nil group.

References

[1] B.J. Gardner, "Some aspects of *T*-nilpotence", *Pacific J. Math.* (to appear).

[2] Rimhak Ree and Robert J. Wisner, "A note on torsion-free nil groups", *Proc. Amer. Math. Soc.* 7 (1956), 6-8. MR17,1051.

[3] Tibor Szele, "Zur Theorie der Zeroringe", *Math. Ann.* 121 (1949), 242-246. MR11,496.

[4] William J. Wickless, "Abelian groups which admit only nilpotent multiplications", *Pacific J. Math.* 40 (1972), 251-259. MR46#3641.

University of Tasmania,
Hobart, Tasmania 7001.

PROC. SECOND INTERNAT. CONF. THEORY OF GROUPS, CANBERRA 1973, pp. 318-329.

20E10, 20E35

POWER SERIES AND MATRIX REPRESENTATIONS OF CERTAIN RELATIVELY FREE GROUPS

C.K. Gupta and N.D. Gupta

1. Introduction

Let ZF be the free integral group ring of the free group F generated by $y_1, y_2, \ldots$. Let

(1.1) $$\underline{\underline{f}} = \ker(\varepsilon : ZF \to Z)$$

be the fundamental (augmentation, basic) ideal of ZF , where $\varepsilon\left(\sum n_g g\right) = \sum n_g$. If R is a normal subgroup of F , let

(1.2) $$\underline{\underline{r}} = \ker\left(\theta : ZF \to Z(F/R)\right) ,$$

where θ is the natural map of F onto F/R linearly extended to ZF . For each $n \geq 1$, $\underline{\underline{f}}^n\underline{\underline{r}}$ is a free ideal of ZF and can be identified as:

(1.3) $$\underline{\underline{f}}^n\underline{\underline{r}} = \text{ideal}_{ZF}\{(y_{i_1}-1) \ldots (y_{i_n}-1)(w_k-1);\ i_j = 1, 2, \ldots \text{ and } w_1, w_2, \ldots \text{ are free generators of } R \text{ as a subgroup of } F\} .$$

Since $\underline{\underline{f}}^n\underline{\underline{r}}$ is a two-sided ideal,

(1.4) $$F(n, R) = \left(1+\underline{\underline{f}}^n\underline{\underline{r}}\right) \cap F$$

is a normal subgroup of F (see for instance Gruenberg [3], Chapters 3, 4 and Fox [2]). Further it should be noted that if R is a fully invariant subgroup of F , then $F(n, R)$ is a fully invariant subgroup of F .

For each $i = 1, 2, \ldots$ define a (right) Fox derivative D_i of ZF by:

Research supported by the National Research Council of Canada.

(1.5) $$D_i(u+v) = D_i(u) + D_i(v) \quad \text{for all } u, v \in ZF ;$$

$$D_i(uv) = D_i(u)v + \varepsilon(u)D_i(v) \quad \text{for all } u, v \in ZF ;$$

$$D_i(y_i) = 1 \text{ and } D_i(y_j) = 0 \quad \text{for } j \neq i .$$

The higher order Fox derivatives may be defined inductively as,

(1.6) $$D_{i_1,\ldots,i_k}(u) = D_i\left(D_{i_1,\ldots,i_{k-1}}(u)\right) .$$

From (4.5) of Fox [2], it follows that

(1.7) $$F(n, R) = \left\{w \in F;\ D_{i_1,\ldots,i_k}(w) \in \underline{f}^{n-k}\underline{r} \text{ for all } D_{i_1,\ldots,i_k}\ (0 \le k \le n)\right\} .$$

The identification of $F(n, R)$ is a difficult and outstanding problem of Fox [2] (see page 557). We note that $F(n, \{1\}) = (1+\{0\}) \cap F = \{1\}$ and $F(n, F) = \left(1+\underline{f}^{n+1}\right) \cap F = \gamma_{n+1}(F)$, the $(n+1)$-st term of the lower central series of F (see [2], (4.6)). Further $F(1, R) = [R, R]$ is a well-known result of Magnus [7] (see also [2], (4.9)). Recently, Enright [1] has shown that $F(2, R) = [R_2, R_2][R, R, R]$ and $F(3, R) = [R_3, R_3]\left[[R, R]_3, R_2\right][R, R, R, R]$, where $R_i = \gamma_i(F) \cap R$. Beyond this the only other known results are:

$$F(4, F_2) = [F_4, F_4][F_2, F_2, F_2] \quad \text{(Hurley [5]);}$$

$$F(5, F_2) = [F_5, F_5][F_3, F_2, F_3][F_2, F_2, F_2, F_2]$$

and

$$F(6, F_2) = [F_6, F_6][F_4, F_2, F_4][F_3, F_3, F_3][F_2, F_2, F_2, F_2]$$

(C.K. Gupta (unpublished); see also [4] for $F(3, F_2)$).

In this paper we shall identify $F(n, F_2)$ for all $n \ge 1$. Our main result is stated in Section 4.

2. Matrix representation of $F/F(n, R)$

For each $n \ge 1$, let $\Lambda_n = \left\{\lambda^{(k)}_{i,i-1};\ n+1 \ge i \ge 2,\ k = 1, 2, \ldots\right\}$ be a set of independent and commuting indeterminates and consider the polynomial ring $Q_n = Z(F/R)[\Lambda_n]$ in $\lambda^{(k)}_{i,i-1}$'s over the group ring $Z(F/R)$. Let $M(n, R)$ be the multiplicative group of $(n+1) \times (n+1)$ matrices over Q_n generated by all

$$(2.1)\qquad \begin{bmatrix} \underset{\sim}{y}_k & 0 & 0 & \cdots & 0 & 0 \\ \lambda_{21}^{(k)} & 1 & 0 & \cdots & 0 & 0 \\ 0 & \lambda_{32}^{(k)} & 1 & \cdots & 0 & 0 \\ \vdots & \vdots & \vdots & & \vdots & \vdots \\ 0 & 0 & 0 & \cdots & \lambda_{n+1,n}^{(k)} & 1 \end{bmatrix} = \langle y_k \rangle_n$$

for $k = 1, 2, \ldots$, where $\underset{\sim}{w}$ denote the coset wR for $w \in F$. For $w = y_{i_1}^{\varepsilon_1} \ldots y_{i_l}^{\varepsilon_l}$ define $\langle w \rangle_n = \langle y_{i_1} \rangle_n^{\varepsilon_1} \ldots \langle y_{i_l} \rangle_n^{\varepsilon_l}$. Enright [1] has observed that $\langle w \rangle_n = I$, the identity matrix if and only if $\theta\left(D_{i_1,\ldots,i_k}(w)\right) = 0$ for all $k = 1, \ldots, n$, where θ is defined by (1.2). Thus by (1.7) we have,

$$(2.2)\qquad F(n, R) = \ker\left(F \to M(n, R)\right) .$$

3. Power series representation of $F/F(n, F_2)$

Let $\underline{\underline{P}}$ denote the free associative power series ring over Z in the independent and non-commuting indeterminates $x_1, x_2, \ldots$. Define a mapping φ of F into $U(\underline{\underline{P}})$, the group of units of $\underline{\underline{P}}$, by

$$(3.1)\qquad \varphi(y_i) = 1 + x_i \quad \text{for } i = 1, 2, \ldots .$$

Then φ defines an isomorphism of F into $U(\underline{\underline{P}})$ (see Theorem 5.6 of Magnus, Karrass and Solitar [8]).

For each $n \geq 1$, define ideals $\underline{\underline{X}}^n$, $\underline{\underline{A}}$ and $\underline{\underline{X}}^n\underline{\underline{A}}$ of $\underline{\underline{P}}$ as follows:

$$(3.2)\qquad \underline{\underline{X}}^n = \text{ideal}_{\underline{\underline{P}}}\{x_{i_1} \ldots x_{i_n} \mid i_j = 1, 2, \ldots\} ;$$

$$(3.3)\qquad \underline{\underline{A}} = (\underline{\underline{X}}, \underline{\underline{X}}) = \text{ideal}_{\underline{\underline{P}}}\{(\alpha, \beta) \mid \alpha, \beta \in \underline{\underline{X}} \ (= \underline{\underline{X}}^1)\} ;$$

and

$$(3.4)\qquad \underline{\underline{X}}^n\underline{\underline{A}} = \text{ideal}_{\underline{\underline{P}}}\{x_{i_1} \ldots x_{i_n}(x_{i_{n+1}}, x_{i_{n+2}}) ; i_j = 1, 2, \ldots\} .$$

Using the ring monomorphism $\delta : ZF \to \underline{\underline{P}}$ defined by

$$(3.5)\qquad \delta(u) = \varepsilon(u) + \sum \varepsilon\left(D_i(u)\right)x_i + \sum \varepsilon\left(D_{i,j}(u)\right)x_i x_j + \ldots$$

(see Section 4.4 of [3]; *cf.* Magnus [6]), Hurley [5] has shown that $\underline{\underline{f}}^n\underline{\underline{a}}$ and $\underline{\underline{X}}^n\underline{\underline{A}}$ determine the same subgroups of F (here $\underline{\underline{a}} = \ker\left(F \to Z(F/F_2)\right)$). In other words

(3.6) $$F(n, F_2) = (1+\underline{\underline{f}}^n\underline{\underline{a}}) \cap F = \{w \in F;\ \varphi(w)-1 \in \underline{\underline{X}}^n\underline{\underline{A}}\} .$$

We shall use power series techniques to identify $F(n, F_2)$.

4. The main theorem

For each $n \geq 1$, $m \geq 2$, define a finite set $S_{n,m}$ of m-tuples of integers as follows:

(4.1) $S_{n,m} = \{\mathbf{r} = (r_1, \ldots, r_m);\ r_i \geq 2$ are least integers satisfying $r_1 + \ldots + \hat{r}_i + \ldots + r_m \geq n$ for all $i = 1, \ldots, m$ where $\hat{r}_i$ indicates r_i missing$\}$;

(for example, $S_{6,3} = \{(4, 2, 4), (2, 4, 4), (4, 4, 2), (3, 3, 3)\}$) .

Define

(4.2) $$S_n = \bigcup_{m=2}^{l} S_{n,m} , \text{ where } l = \left[\frac{n-1}{2}\right] + 2 .$$

Our main result can now be stated as:

THEOREM. *The subgroup* $F(n, F_2)$ *is given by*

(4.3) $$F(n, F_2) = \prod_{\mathbf{r}} [F_{r_1}, \ldots, F_{r_m}] ,$$

where $F_k = \gamma_k(F)$ = *k-th term of the lower central series of* F *and the product is taken over all elements* $\mathbf{r} \in S_n$.

It should be noted that $F(n, F_2)$ as given by (4.3) is not in general a minimal product of its factors. However, by repeated application of Hall's Three Subgroup Lemma (see Lemma 5.1 of Magnus, Karrass, Solitar [8]), it can be simplified to a minimal product. As an illustration we list below minimal representation of $F(n, F_2)$ for $n = 7, 8$ (for $n = 1, \ldots, 6$, $F(n, F_2)$ is listed among the known results in Section 1).

$$F(7, F_2) = [F_7, F_7][F_5, F_2, F_5][F_4, F_3, F_4][F_3, F_2, F_2, F_3][F_3, F_2, F_3, F_2][F_2, F_2, F_2, F_2, F_2] ;$$

$$F(8, F_2) = [F_8, F_8][F_6, F_2, F_6][F_5, F_3, F_5][F_4, F_4, F_4][F_4, F_2, F_2, F_4][F_4, F_2, F_4, F_2][F_3, F_3, F_3, F_2][F_3, F_2, F_3, F_3][F_2, F_2, F_2, F_2, F_2] .$$

Finally we remark that with $R = F_2$, $Z(F/R)[\Lambda_n]$ in Section 2 is a commutative integral domain. Thus together with (2.2) our theorem gives a series of relatively free linear groups. This contributes towards the solution of a general problem of Wehrfritz ([9] Question 3, page 35): Which relatively free groups are linear?

5. Notation and preliminaries

Our main reference is Chapter 5 of Magnus, Karrass, Solitar [8]. The commutator notation is $[g_1, g_2] = g_1^{-1}g_2^{-1}g_1g_2$ for groups and $(\alpha, \beta) = \alpha\beta - \beta\alpha$ for rings. The commutators $[g_1, \ldots, g_n]$ and $(\alpha_1, \ldots, \alpha_n)$ are left-normed commutators of weight n ; $F_n = \gamma_n(F)$ is the n-th term of the lower central series of F and $\gamma_m\gamma_n(F) = \gamma_m(\gamma_n(F))$. If $\underline{S}, \underline{T}$ are ideals in the power series $\underline{P}$, then

(5.1) $$(\underline{S}, \underline{T}) = \text{ideal}_{\underline{P}}\{(\alpha, \beta);\ \alpha \in \underline{S},\ \beta \in \underline{T}\} ,$$

(5.2) $$\underline{S} + \underline{T} = \text{ideal}_{\underline{P}}\{\alpha+\beta;\ \alpha \in \underline{S},\ \beta \in \underline{T}\} .$$

For $\alpha, \beta \in \underline{X}$ (the basic ideal of $\underline{P}$),

(5.3) $$(1+\alpha)^{-1} = 1 - \alpha + \alpha^2 - \alpha^3 + \ldots + (-1)^n\alpha^n + \ldots ,$$

(5.4) $$[1+\alpha, 1+\beta] = 1 + \sum_{i,j=0}^{\infty} \alpha^i\beta^j(\alpha, \beta) ;$$

(see Magnus, Karrass, Solitar [8], page 314).

Let φ be as defined in (3.1). If $\varphi(S) \subseteq 1 + \underline{S}$ and $\varphi(T) \subseteq 1 + \underline{T}$ for some subgroups S and T of F , then (5.4) gives

(5.5) $$\varphi([S, T]) \subseteq 1 + (\underline{S}, \underline{T}) .$$

Using induction, repeated application of (5.5) gives

(5.6) $$\varphi(F_m) \subseteq 1 + \sum_{i=0}^{m-2} \underline{X}^i\underline{\underline{A}}\underline{X}^{m-2-i} \quad \text{for all} \quad m \geq 2 .$$

Using (5.6) with $m = 2$, (5.5) and the fact that $\underline{\underline{A}} \subseteq \underline{X}^2$ we get $\varphi(\gamma_2\gamma_2(F)) \subseteq 1 + (\underline{\underline{A}}, \underline{\underline{A}}) \subseteq 1 + \underline{X}^2\underline{\underline{A}}$; and inductively

(5.7) $$\varphi(\gamma_m\gamma_2(F)) \subseteq 1 + \underline{X}^{2(m-1)}\underline{\underline{A}} \quad \text{for all} \quad m \geq 2 .$$

Similarly,

(5.8) $$\varphi([F_{r_1}, \ldots, F_{r_m}]) \subseteq 1 + \underline{X}^n\underline{\underline{A}} \quad \text{for all} \quad \mathbf{r} = (r_1, \ldots, r_m) \in S_{n,m}$$

(see (4.1)).

We now record some technical but elementary facts about the free associative algebra $\underline{P}$.

(5.9) *If* $\underline{S}$ $(\subseteq \underline{X})$ *is an ideal of* $\underline{P}$, *then* $\sum x_i\alpha_i \in \underline{XS}$ *implies* $\alpha_i \in \underline{S}$ *for all* i .

PROOF of (5.9). $\sum x_i\alpha_i \in \underline{XS}$ implies $\sum x_i\alpha_i = \sum x_j\beta_j$ (where $\beta_j \in \underline{S}$) and in turn $x_i\alpha_i = x_i\beta_i$, $x_i(\alpha_i-\beta_i) = 0$, $\alpha_i - \beta_i = 0$, $\alpha_i = \beta_i \in \underline{S}$.

(5.10) *If* $\mu_1, \ldots, \mu_p$ *are distinct monomials of length* $n - 1$ *and if* $\lambda_1, \ldots, \lambda_p$ *are in* $\underline{S}$ *such that* $\sum \mu_i\lambda_i \in \underline{X}^n\underline{S}$ *then* $\lambda_i \in \underline{XS}$ *for all* $i = 1, \ldots, p$.

PROOF of (5.10) (Corollary to (5.9)). Let $\beta_1, \ldots, \beta_p$ be distinct left normed basic Lie elements of weights $r_1, \ldots, r_p$ respectively $(r_i \geq 2)$. Let $n \geq 3$ be a fixed positive integer. For each p-tuple $(s_1, \ldots, s_p)$ $(s_i \geq 0)$ of integers with $\sum_{i=1}^{p} s_ir_i = n - 1$, let $\lambda(s_1, \ldots, s_p)$ denote a linear combination of products of the form $\beta_{j_1} \ldots \beta_{j_m}$ with precisely s_i occurrences of β_i for each $i = 1, \ldots, p$ (note that $m = s_1 + \ldots + s_p$). A straight forward application of Theorem 5.8 of Magnus, Karrass, Solitar [8] gives,

(5.11) *If* $(s_{1_1}, \ldots, s_{1_p}), \ldots, (s_{k_1}, \ldots, s_{k_p})$ *are distinct* p-*tuples satisfying* $\sum_{i=1}^{p} s_{j_i}r_i = n - 1$, *then* $\lambda(s_{1_1}, \ldots, s_{1_p}), \ldots, \lambda(s_{k_1}, \ldots, s_{k_p})$ *are linearly independent.*

Using techniques in Section 5.6 of Magnus, Karrass, Solitar [8], the following can be proved.

(5.12) *Let* $\beta_1, \ldots, \beta_p$ *be left-normed basic Lie elements (not necessarily distinct) and let* $\lambda_1, \ldots, \lambda_q$ *be distinct products of the form* $\beta_{1\sigma} \ldots \beta_{p\sigma}$, *where* σ *is a permutation of* $\{1, \ldots, p\}$. *Then* $\lambda_1, \ldots, \lambda_q$ *are linearly independent.*

Finally,

(5.13) *Let* $d = [b_1, b_2, \ldots, b_m]$, *where* $b_1, \ldots, b_m$ *are left-normed basic commutators of weights* $r_1, \ldots, r_m$ *respectively* $(r_i \geq 2)$ *with* $r_1 + \ldots + \hat{r}_i + \ldots + r_m \geq n - 1$ $(n \geq 3)$ *for each* i . *Then modulo*

$\underline{\underline{X}}^n\underline{\underline{A}}$, $\varphi(d) - 1 = \mu_1(\varphi(b_1)-1) + \dots + \mu_m(\varphi(b_m)-1)$, *where* μ_i *is a linear combination of products of the form* $\beta_{1\sigma} \dots \hat{\beta}_{i\sigma} \dots \beta_{m\sigma}$ *for each* $i = 1, \dots, m$ *and* β_j *is the Lie basic reflection of the group basic commutator* b_j .

PROOF of (5.13).

$$\varphi(d) - 1 = [1+(\varphi(b_1)-1), \dots, 1+(\varphi(b_m)-1)] - 1 \equiv (\varphi(b_1)-1, \dots, \varphi(b_m)-1) \bmod \underline{\underline{X}}^n\underline{\underline{A}}$$

(see (5.4)).

A straight forward expansion gives the desired result (note that $\varphi(b_i) - 1 = \beta_i$ + terms of degree $> r_i$).

6. Proof of the main theorem

We recall that for each $n \geq 1$, $F(n, F_2) = \{w \in F;\ \varphi(w)-1 \in \underline{\underline{X}}^n\underline{\underline{A}}\}$. For each $n \geq 1$, $m \geq 2$, let

(6.1) $$H_{n,m} = \prod_{r} [F_{r_1}, \dots, F_{r_m}] \quad (r \in S_{n,m}) ;$$

and

(6.2) $$H_n = H_{n,2} \dots H_{n,l} \quad \text{where } l = \left[\frac{n-1}{2}\right] + 2 .$$

By (4.2) and (4.3), the proof of the theorem consists in showing that $H_n = F(n, F_2)$. By (5.8), $\varphi(H_{n,m}) - 1 \subseteq \underline{\underline{X}}^n\underline{\underline{A}}$ for all $m = 2, \dots, l$, showing that $H_n \leq F(n, F_2)$. For the reverse inclusion we note that for $n = 1$, $H_1 = [F_2, F_2]$, which is the same as $F(1, F_2)$ by the theorem of Magnus metnioned earlier in Section 1. Let $n > 1$ and assume that $F(n-1, F_2) \leq H_{n-1}$. Let w be an arbitrary element of $F(n, F_2)$. Then $\varphi(w) - 1 \in \underline{\underline{X}}^n\underline{\underline{A}} \subseteq \underline{\underline{X}}^{n-1}\underline{\underline{A}}$ implies $w \in F(n-1, F_2)$ so that by the induction hypothesis $w \in H_{n-1}$. By (6.2), w may be written as

(6.3) $$w = w_2 \dots w_{l'} , \text{ where } w_i \in H_{n-1,i} \text{ and } l' = \left[\frac{n-2}{2}\right] + 2 .$$

If each $w_i \in H_{n,i}$, then $w \in H_n$ by (6.2) and there is nothing to prove. Thus we may assume

(6.4) $w = w_m w_{m+1} \dots w_{l'}$, where $w_i \in H_{n-1,i}$, $l' = \left[\frac{n-2}{2}\right] + 2$ and m is the least integer such that $w_m \notin H_{n,m} \cdot H_{n-1,m+1}$.

If $[d_1, \dots, d_m]$ is a factor of w_m with $d_i \in F_{r_i}$ for $i = 1, \dots, m$ and if for some j , $d_j = [d_{j_1}, d_{j_2}]$ $(j_1, j_2 \geq 2)$, then by the arithmetic conditions on $S_{n-1,m}$, it is easily seen that $(r_1, \dots, r_{j-1}, r_{j_1}, r_{j_2}, r_{j+1}, \dots, r_m) \in S_{n-1,m+1}$ and consequently $[d_1, \dots, d_m] \in H_{n-1,m+1}$. Thus we may assume that in (6.4), the following holds.

(6.5) *If* $[d_1, \dots, d_m]$ *is a factor of* w_m *then each* $d_i \in F_2 \backslash [F_2, F_2]$.

Let $[d_1, \dots, d_m]$ be a factor of w_m , with $d_i \in F_{r_i}$. If for any j, k , $j \neq k$, $d_j \in F_{r_j+1}$ and $d_k \in F_{r_k+1}$, then since $r_1 + \dots + \hat{r}_i + \dots + r_m \geq n - 1$ for all i , it follows that $[d_1, \dots, d_m] \in H_{n,m}$. Thus we may write $[d_1, \dots, d_m]$ as a product of commutators (left-normed) of the form $[b_{i_1}, \dots, b_{i_m}] \in H_{n-1,m}$, where at most one b_{i_j} is non-basic and all others are basic commutators. (This can be done by writing each d_k as a product of basic commutators modulo F_{r_k+1} .) By repeated application of Jacobi identity we may write $[d_1, \dots, d_m]$ as a product of commutators of the form $[b_0, b_1, \dots, b_{m-1}]$, where $b_1, \dots, b_{m-1}$ are basic commutators of weights $r_1, \dots, r_{m-1}$ respectively with $r_1 + \dots + r_{m-1} = n - 1$. Thus in addition to (6.5), we can assume that w in (6.4) satisfies the following conditions.

(6.6) w_m *is a product of commutators of the form* $[b_0, b_1, \dots, b_{m-1}]$ *, where* $b_1, \dots, b_{m-1}$ *are basic commutators of weights* $r_1, \dots, r_{m-1}$ *respectively with* $r_1 + \dots + r_{m-1} = n - 1$ *. Further* $b_0 \in F_2 \backslash [F_2, F_2]$.

We recall the hypothesis that $\varphi(w) - 1 \in \underline{\underline{X}}^n\underline{\underline{A}}$ so that $\sum_{i=m}^{l'} \varphi(w_i) - 1 \equiv 0$.

Using (5.13), it follows that $\varphi(w_i) - 1 \equiv \sum_j \mu_{ij}\lambda_{ij}$, where μ_{ij} is a linear combination of products of at least $i - 1$ basic Lie elements and $\lambda_{ij} \in \underline{\underline{A}}$ for $i = m, \dots, l'$. Thus by (5.11) and (5.10) we get,

(6.7) *In the representation* (6.6) *of* w_m , $\varphi(w_m) - 1 \in \underline{\underline{X}}^n\underline{\underline{A}}$.

Let $[b_0, b_1, \dots, b_{m-1}]$ be a factor of w_m , with $b_i \in F_{r_i}$ and $r_1 + \dots + r_{m-1} = n - 1$. We first of all show that $\varphi[b_0, b_1, \dots, b_{m-1}] - 1 \notin \underline{\underline{X}}^n\underline{\underline{A}}$.

Since b_0 may or may not occur in $\{b_1, \ldots, b_{m-1}\}$, we rewrite $[b_0, b_1, \ldots, b_{m-1}]$ as

(6.8) $d = [b_0, b_{1_1}, \ldots, b_{1_{s(1)}}, b_0, b_{2_1}, \ldots, b_{2_{s(2)}}, \ldots, b_0, b_{k_1}, \ldots, b_{k_{s(k)}}]$

where $k \geq 1$, $s(1) \geq 1$, $s(j) \geq 0$ for $j = 2, \ldots, k$, $b_0 \neq b_{i_j}$ for any i_j and if weight of $b_{i_j} = r_{i_j}$ (and weight $b_0 = r_0$) then $(k-1)r_0 + \sum_{i=1}^{k} \sum_{j=1}^{s(i)} r_{i_j} = n-1$ and $(k-1) + \sum_{i=1}^{k} s(i) = m-1$. A straight forward expansion using (5.13) shows that in the expansion of $\varphi(d) - 1$ modulo $\underline{\underline{X}}^n\underline{\underline{A}}$, there is a unique term

(6.9) $$(\varepsilon_1\mu_k\xi_{k-1}+\varepsilon_2\mu_k\beta_0\mu_{k-1}\xi_{k-2} + \ldots + \varepsilon_{k-1}\mu_k\beta_0 \ldots \mu_2\xi_1 + \varepsilon_k\mu_k\beta_0\mu_{k-1}\beta_0 \ldots \mu_2\beta_0\mu_1)(\varphi(b_0)-1),$$

where $\mu_i = \beta_{i_{s(i)}} \ldots \beta_{i_1}$ for $i = 1, \ldots, k$,

$$\xi_i = (\beta_0, \beta_{1_1}, \ldots, \beta_{1_{s(1)}}, \beta_0, \beta_{2_1}, \ldots, \beta_{2_{s(2)}}, \ldots, \beta_0, \beta_{i_1}, \ldots, \beta_{i_{s(i)}})$$

for $i = 1, \ldots, k-1$, $\mu_1\xi_0 = \mu_1$ and $\varepsilon_i = (-1)^{q_i}$ with $q_i = \sum_{j=0}^{j-1} (s(k-j)+1) - 1$, for $i = 1, \ldots, k$. By Theorem 5.8 of Magnus, Karrass, Solitar and (5.10) it follows that if $\varphi(d) - 1 \in \underline{\underline{X}}^n\underline{\underline{A}}$, then in particular

$$(\mu_k\beta_0 \ldots \mu_2\beta_0\mu_1)(\varphi(b_0)-1) \in \underline{\underline{X}}^n\underline{\underline{A}},$$

but $\mu_k\beta_0 \ldots \mu_2\beta_0\mu_1 \notin \underline{\underline{X}}^n$ so that by (5.9), (5.10) we must have $\varphi(b_0) - 1 \in \underline{\underline{XA}}$, which, by the theorem of Magnus, is possible only if $b_0 \in \gamma_2\gamma_2(F)$, a contradiction. Thus we have shown that for each individual factor d of w_m, $\varphi(d) - 1 \notin \underline{\underline{X}}^n\underline{\underline{A}}$. We now proceed to show that under these circumstances $\varphi(w_m) - 1 \notin \underline{\underline{X}}^n\underline{\underline{A}}$ and this contradiction to (6.7) establishes the final result.

CASE I. $d = [b_0, b_1, \ldots, b_{m-1}]$ is a factor of w_m with $b_0 \notin \{b_1, \ldots, b_{m-1}\}$ and weight of $b_i = r_i$, $r_1 + \ldots + r_{m-1} = n-1$.

As remarked before, in $\varphi(d) - 1$ there is a unique term $(-1)^{m-1}\beta_{m-1} \ldots \ldots \beta_1(\varphi(b\ ) -1) \notin \underline{\underline{X}}^n\underline{\underline{A}}$. Since $\varphi(w_m) - 1 \in \underline{\underline{X}}^n\underline{\underline{A}}$ (by 6.7), by 5.12 and 5.11, there is a factor $d(1)^{a_1} \ldots d(s)^{a_s}$, $(a_i \in Z)$ of w_m which contributes

$$\beta_{m-1} \dots \beta_1\lambda \quad (\text{with } \lambda \in \underline{\underline{XA}})$$

in the expansion of $\varphi\left(d(1)^{a_1} \dots d(s)^{a_s}\right) - 1$. Clearly

$$d(1)^{a_1} \dots d(s)^{a_s} = [b_{0_1}, b_1, \dots, b_{m-1}]^{a_1} \dots [b_{0_s}, b_1, \dots, b_{m-1}]^{a_s} =$$

$$= \left[b_{0_1}^{a_1} \dots b_{0_s}^{a_s}, b_1, \dots, b_{m-1}\right]$$

and $\lambda = \varphi\left(b_{0_1}^{a_1} \dots b_{0_s}^{a_s}\right) - 1$. Since $\lambda \in \underline{\underline{XA}}$, it follows that $b_{0_1}^{a_1} \dots b_{0_s}^{a_s} \in \gamma_2\gamma_2(F)$ and consequently $d(1)^{a_1} \dots d(s)^{a_s} \in \gamma_{m+1}\gamma_2(F) \cap H_{n-1,m} \leq H_{n-1,m+1} \cdot H_{n,m}$, contrary to the choice of w_m.

CASE II. $d = [b_0, b_1, \dots, b_{m-1}]$ is a factor of w_m with $b_0 \in \{b_2, \dots, b_{m-1}\}$. Note in particular that in this case b_0 is also basic. Let s be the largest integer such that $s \in \{2, \dots, m-1\}$ and $b_s = b_0$. Write $d = [b_0, b_1, \dots, b_{s-1}, b_0, b_{s+1}, \dots, b_{m-1}]$. By repeated application of Jacobi identity (recall that we are working modulo $\gamma_{m+1}\gamma_2(F) \cap H_{n-1,m} \leq H_{n,m} \cdot H_{n-1,m+1}$), d can be written as a product of commutators of the form

$$d' = [[b_0, b_1, \dots, b_{s-1}], b_0, c_k, \dots, c_1],$$

where $c_1 \leq \dots \leq c_k$ are basic commutators made of $b_{s+1}, \dots, b_{m-1}$ (and consequently c_i does not involve b_0 for any $i = 1, \dots, k$). By further applications of Jacobi identity we may write $[b_0, b_1, \dots, b_{s-1}]$ as a product of basic commutators so that d' can be written as a product of commutators of the form

$$d'' = [b_0, c(b_0), c_k, \dots, c_1], \tag{6.10}$$

where $c(b_0)$ is a basic commutator involving at least one b_0, and in view of (5.12) we may, if necessary, further define $c_1 \leq \dots \leq c_k \leq c(b_0)$.

Thus we may assume that w_m is a product of distinct commutators of the form (6.10).

Let $d'' = [b_0, c(b_0), c_k, \dots, c_1]$ be a factor of w_m. Then in the expansion of $\varphi(d'') - 1$, the coefficient of $\varphi(b_0) - 1$ comes precisely from

$$(-1)^{k+1}\xi_1 \dots \xi_k\xi(\beta_0)\left(\varphi(b_0)-1\right) + (-1)^k\xi_1 \dots \xi_k\beta_0\left(\varphi(c(b_0))-1\right), \tag{6.11}$$

where $\xi_1 \le \dots \le \xi_k \le \xi(\beta_0)$ are basic Lie commutators corresponding to the basic group commutators $c_1 \le \dots \le c_k \le c(b_0)$. Since $\varphi(w_m) - 1 \in \underline{X}^n\underline{A}$ and $\varphi(b_0) - 1 \notin \underline{XA}$, by (5.10) and by Theorem 5.8 of Magnus, Karrass, Solitar [8], it follows that either

(i) there is another factor of w_m contributing

$$-(-1)^{k+1}\xi_1 \dots \xi_k\xi(\beta_0)(\varphi(b_0)-1)$$

which is clearly not possible since w_m is a product of distinct factors of the form (6.10); or

(ii) $\xi_1 \dots \xi_k\xi(\beta_0)$ cancels in (6.11) itself, in which case $\xi_1 \dots \xi_k\xi(\beta_0) = \xi_1 \dots \xi_k\beta_0\lambda$, for some λ , where λ is a summand of the coefficient of $\varphi(b_0) - 1$ in $\varphi(c(b_0)-1)$.

The last equation implies that $\xi(\beta_0) = \beta_0\lambda$, which is possible only if $\xi(\beta_0) = \beta_0$ (and $\lambda = 1$). Thus $c(b_0) = b_0$ and consequently $d'' = 1$, contrary to the choice of d'' . This completes the proof of the theorem.

7. Concluding remarks

By a slight modification of our argument, $F(n, F_c)$ can be identified for all $c \ge 2$. Some progress has been made towards the general problem of identifying $F(n, R)$, $R \Delta F$. The details will be discussed in a subsequent paper.

References

[1] Dennis E. Enright, "Triangular matrices over group rings", (Doctoral dissertation, New York University, 1968).

[2] Ralph H. Fox, "Free differential calculus. I. Derivation in the free group ring", *Ann. of Math.* 57 (1953), 547-560. MR14,843.

[3] Karl W. Gruenberg, *Cohomological topics in group theory* (Lecture Notes in Mathematics, 143. Springer-Verlag, Berlin, Heidelberg, New York, 1970). MR43#4923.

[4] Chander Kanta Gupta, "On free groups of the variety $\underline{AN}_2 \wedge \underline{N}_2\underline{A}$ ", *Canad. Math. Bull.* 13 (1970), 443-446. MR43#322.

[5] T.C. Hurley, "Representations of some relatively free groups in power series rings", *Proc. London Math. Soc.* (3) 24 (1972), 257-294. Zbl.232.20056.

[6] Wilhelm Magnus, "Beziehungen zwischen Gruppen und Idealen in einem speziellen Ring", *Math. Ann.* 111 (1935), 259-280. FdM61,102.

[7] Wilhelm Magnus, "On a theorem of Marshall Hall", *Ann. of Math.* 40 (1939), 764-768. MR1,44.

[8] Wilhelm Magnus, Abraham Karrass, Donald Solitar, *Combinatorial group theory* (Pure and Appl. Math. 13. Interscience [John Wiley & Sons], New York, London, Sydney, 1966). MR34#7617.

[9] B.A.F. Wehrfritz, *Infinite linear groups* (Ergebnisse der Mathematik und ihrer Grenzgebiete, Band 76. Springer-Verlag, Berlin, Heidelberg, New York, 1973).

University of Manitoba,
Winnipeg, Manitoba R3T 2N2, Canada.

PROC. SECOND INTERNAT. CONF. THEORY OF GROUPS, CANBERRA 1973, pp. 330-332. 20D15

THE NILPOTENCY CLASS OF FINITELY GENERATED GROUPS OF EXPONENT FOUR

N.D. Gupta and M.F. Newman

Wright (Theorem 3 of [6]) has shown that for every positive integer n an n-generator group of exponent 4 is nilpotent of class at most $3n - 1$. When n is 2 this result is best possible (see, for example, §3 of [4]). On the other hand, if for some positive integer m the nilpotency class of every m-generator group of exponent 4 were $3m - 3$, then groups of exponent 4 would all be soluble [3] and there would be a positive integer k such that for every positive integer n every n-generator group of exponent 4 would have nilpotency class at most $n + k$ (Theorem A of [2]). The gap between these two results is intriguingly small. In this paper we report on some further narrowing of this gap which makes the problem of trying to close the gap still more attractive. We prove the following result.

THEOREM. *For n a positive integer greater than 2 every n-generator group of exponent 4 is nilpotent of class at most $3n - 2$.*

Some closing from the other side of the gap is also possible. Words $u_1, u_2, \ldots\ v_2, v_3, \ldots$ are defined, recursively, as follows: (left-norming convention is used for commutators) $u_1 = [x_1, x_2]$, $u_n = \left[u_{n-1}, x_{n+1}, x_n^2\right]$, $v_n = \left[u_{n-1}, x_n^2\right]$. In groups of exponent 4 the word v_n is in effect a commutator of weight $3n - 2$. *If for some positive integer n the word v_n is a law in groups of exponent 4, then groups of exponent 4 are soluble.* We do not give a proof of this here (in part because the result does not acquire real significance till its hypothesis has been shown to hold and in part because our proof consists of calculations similar to those in [3]).

PROOF OF THE THEOREM. Every finitely generated group of exponent 4 is finite (see, for example, Theorem 5.25 of [5]). It therefore suffices to prove that every n-generator group of exponent 4 which is nilpotent of class at most $3n - 1$ is

The work for this paper was supported by a Canadian NRC grant which enabled the second author to spend July-August, 1972 at the University of Manitoba.

nilpotent of class at most $3n - 2$. (Of course Theorem 3 of Wright's paper [6] guarantees that every n-generator group of exponent 4 is nilpotent of class at most $3n - 1$ but that fact is not needed here. On the other hand we will make heavy use of the details of Wright's proof. Some of these details can be simplified. We hope to report on such matters in notes we are preparing on groups of exponent 4 .) For n is 3 the result is given in the paper [1] by Bayes, Kautsky and Wamsley in these proceedings. (Our original proof consisted of lengthy commutator calculations.) For n greater than 3 assume the result has been proved for $n - 1$; in particular that $[x_1, x_3, x_1, x_3, x_{i(5)}, \ldots, x_{i(3n-4)}]$ is a law in groups of exponent 4 where $i(5), \ldots, i(3n-4)$ belong to $\{3, \ldots, n\}$. Let G be a group of exponent 4 which is nilpotent of class $3n - 1$ and has an n-element generating set $A = \{a_1, \ldots, a_n\}$. We will show that every left-normed commutator of weight $3n - 1$ with entries from A has value e (the identity) in G . We show first that if c is a left-normed commutator of weight $3n - 2$ with entries from A only one of which is a_1 and the first six of which are, in order, $a_1, a_2, a_2, a_2, a_3, a_3$, then c has value e . In the law of the form given by the inductive hypothesis in which $i(j)$ is the subscript on the $(j+2)$-th entry of c , substitute $[a_1, a_2, a_2]a_2$ for x_1 and $a_3, \ldots, a_n$ for $x_3, \ldots, x_n$. Expanding the resulting commutator using the usual identities (Theorem 5.3 of [5]) and using Theorem 1 of [6] (which makes all commutators with at least four entries a_2 trivial) gives

$$[[a_1, a_2, a_2], a_3, a_2, a_3, \ldots][a_2, a_3, [a_1, a_2, a_2], a_3, \ldots] = e .$$

Hence using the Jacobi identity gives

$$[a_1, a_2, a_2, a_2, a_3, a_3, \ldots] = e$$

as claimed. We now turn to proving that every left-normed commutator of weight $3n - 1$ with entries from A has trivial value in G . By Theorem 1 of [6] there is nothing to prove if some element of A occurs at least four times as an entry. Therefore we may assume that one element of A , without loss of generality a_1 , occurs twice as an entry and the others occur three times each as entries. We first prove by induction on i running through $\{3, \ldots, n\}$, that if c is a left-normed commutator of the above kind whose first $3i - 2$ entries include both a_1s and involve only i elements of A has trivial value in G . For i is 3 the commutator of the first seven entries of c either has trivial value or value equal to that of the commutator $[a_1, a_{i_2}, a_{i_2}, a_{i_2}, a_{i_3}, a_{i_3}]$ (see the table of Bayes, Kautsky and Wamsley [1]). It follows from our first result that c has value e . For i greater than 3 assume the result for $i - 1$. There is an element, a_j say, of A

which occurs three times amongst the 3rd to $(3i-2)$-th entries of c. By Lemma 2 of [6] c can be written as a product of left-normed commutators in which a_j occurs as $(3i-2)$-th entry and as two of $(3i-5)$-th, $(3i-4)$-th, $(3i-3)$-th entries. If the $(3i-5)$-th entry of such a commutator is not a_j, the inductive hypothesis applies to give that its value is e. If the $(3i-5)$-th entry is a_j, applying (9) (and if necessary (8)) of [6] yields a commutator to which the inductive hypothesis can be applied. The result follows. This completes the whole proof (taking $i = n$) except for commutators whose last entry is a_1. In that case an argument similar to the above enables such a commutator to be written as a product of commutators whose last entry is not a_1. Since these all have trivial value, the proof is complete.

References

[1] A.J. Bayes, J. Kautsky and J.W. Wamsley, "Computation in nilpotent groups (application)", these Proc.

[2] C.C. Edmunds and N.D. Gupta, "On groups of exponent four IV", *Conf. on Group Theory, University of Wisconsin-Parkside*, 1972, pp. 57-70 (Lecture Notes in Mathematics, 319. Springer-Verlag, Berlin, Heidelberg, New York, 1973).

[3] N.D. Gupta and R.B. Quintana, Jr., "On groups of exponent four. III", *Proc. Amer. Math. Soc.* 33 (1972), 15-19. MR45#2000.

[4] Marshall Hall, Jr., "Notes on groups of exponent four", *Conf. on Group Theory, University of Wisconsin-Parkside*, 1972, pp. 91-118 (Lecture Notes in Mathematics, 319. Springer-Verlag, Berlin, Heidelberg, New York, 1973).

[5] Wilhelm Magnus, Abraham Karrass, Donald Solitar, *Combinatorial group theory* (Pure and Appl. Math. 13. Interscience [John Wiley & Sons], New York, London, Sydney, 1966). MR34#7617.

[6] C.R.B. Wright, "On the nilpotency class of a group of exponent four", *Pacific J. Math.* 11 (1961), 387-394. MR23#A927.

University of Manitoba,
Winnipeg, Manitoba R3T 2N2, Canada.

Australian National University,
Canberra, ACT.

PROC. SECOND INTERNAT. CONF. THEORY OF GROUPS, 20E10
CANBERRA 1973, pp. 333-336.

HYPOCRITICAL AND SINCERE GROUPS

L.F. Harris

A group G is defined to be *hypocritical* if, whenever G is in a locally finite variety $\underline{V}$ of groups and $\mathcal{G}$ is a section closed class of groups which generates $\underline{V}$, then $G \in \mathcal{G}$. The critical groups which for some prime p are an extension of an abelian p-group by a p'-group are considered from the standpoint of hypocriticality.

1. Introduction

A *class of groups* is defined to be a union of isomorphism classes of groups. In particular $[H]$ denotes the class of groups isomorphic to the group H. For a set or class $\mathcal{G}$ of groups, $s\mathcal{G}$ and $Q\mathcal{G}$ denote respectively the classes of groups isomorphic to subgroups and factor groups of groups in $\mathcal{G}$. A class $\mathcal{G}$ of groups is said to be *section closed* if $s\mathcal{G} \subseteq \mathcal{G}$ and $Q\mathcal{G} \subseteq \mathcal{G}$. It is easy to see that $Qs\mathcal{G}$ is section closed. If $\mathcal{G}$ consists of a single group G we write sG and QG for $s\mathcal{G}$, and $Q\mathcal{G}$ respectively, and elements of QsG are called *sections* of G; $(Qs-1)G$ denotes the class of proper sections of G. A group G is said to be *critical* if it is not in the variety $\text{var}(Qs-1)G$ generated by $(Qs-1)G$.

The *skeleton* $S(\underline{V})$ *of a variety* $\underline{V}$ is defined to be the intersection of the section closed classes of groups which generate $\underline{V}$. Skeletons have some interest in their own right, as Brisley and Kovács [1] and Bryant and Kovács [2] have shown, and may be used to discuss distributivity properties of lattices of varieties, as Cossey [4] and Harris [6] have done. As part of the skeleton of a locally finite variety $\underline{V}$ one naturally finds the spine $T(\underline{V})$, that is, the intersection of the skeletons of the locally finite varieties which contain $\underline{V}$. A group G is said to be *hypocritical* if, whenever it is in a locally finite variety $\underline{V}$ and $\mathcal{G}$ is a section closed class of groups which generates $\underline{V}$, then $G \in \mathcal{G}$. Thus G is hypocritical if and only if G in a locally finite variety $\underline{V}$ implies $G \in S(\underline{V})$. Equivalently, G is hypocritical if $G \in T(\text{var } G)$. Clearly a hypocritical group is critical. It is easy to see that a critical group with nonabelian monolith is hypocritical.

It follows immediately from the definition that a locally finite variety generated by any class of hypocritical groups is generated by its spine. One reason for our interest in varieties generated by their spines, and hence in hypocritical groups, is the following. If a variety $\underline{V}$ generated by its spine is contained in a locally finite join, $\bigvee_{\lambda} \underline{V}_{\lambda}$, of a possibly infinite number of varieties, then a consideration of the finite free groups of the $\underline{V}_{\lambda}$ shows $\underline{V} = \bigvee_{\lambda} (\underline{V} \wedge \underline{V}_{\lambda})$. In particular if $\underline{V}$ and all its subvarieties are generated by their spines then

$$\underline{U} \wedge \left(\bigvee_{\lambda} \underline{V}_{\lambda}\right) = \bigvee_{\lambda} (\underline{U} \wedge \underline{V}_{\lambda})$$

whenever $\underline{U} \subseteq \underline{V}$ and $\bigvee_{\lambda} \underline{V}_{\lambda}$ is locally finite.

A finite group which is not hypocritical is said to be *sincere*. A variety generated by a single sincere critical group is not generated by its spine. Furthermore if the skeleton $S(\underline{V})$ of a locally finite variety $\underline{V}$ contains a sincere group H which is not in $\text{Qs}\big(S(\underline{V})\backslash[H]\big)$ then a routine calculation shows $\underline{V}$ is not generated by its spine.

Most of the results which follow are proved in my PhD Thesis [5] which was supervised by Dr R.M. Bryant and Dr L.G. Kovács whom I thank for many helpful discussions.

2. The theorems

Let p be a prime, Z_p the ring of integers modulo p, and G^* an irreducible linear p'-group of degree k over the field of p elements. Let α be a positive integer, S a k-generator homocyclic group of exponent p^{α} and ΦS the Frattini subgroup of S. Then $S/\Phi S$ becomes an irreducible $Z_p G^*$-module in an obvious way. Because S is a relatively free p-group and G^* a p'-group, an action of G^* on S can be defined such that the action induced on $S/\Phi S$ is the original action, and the split extension $G(p^{\alpha}, G^*)$ of S by G^* is unique up to isomorphism. A finite group G is said to be *monolithic* if it has a unique minimal normal subgroup, which is then called the *monolith* σG of G. Obviously $G(p^{\alpha}, G^*)$ is monolithic and by [7, 1.65] it is critical. Bryant and Kovács have shown (in unpublished work) that if $\alpha = 1$ or G^* has degree one then $G(p^{\alpha}, G^*)$ is hypocritical. On the other hand, we have the following result.

THEOREM 1. *If G^* has degree greater than one then there is an α such that $G(p^{\alpha}, G^*)$ is sincere. If $G(p^{\alpha}, G^*)$ is sincere then so is $G(p^{\alpha+1}, G^*)$.*

The obvious problem is to find the smallest smallest α such that $G(p^{\alpha}, G^*)$ is sincere. This problem is not solved, but there are solutions in some special cases. For G^* cyclic of order n we write $G(p^{\alpha}, n)$ for $G(p^{\alpha}, G^*)$; it is well defined by [9]. Let $a(p, n)$ be the smallest positive integer such that $G(p^{\alpha}, n)$ is sincere for all $\alpha \geq a(p, n)$.

THEOREM 2. *If* p *does not divide* n *and* n *does not divide* $p - 1$ *then* $2 \leq a(p, n) \leq 3$. *Let* k *be the smallest positive integer such that* n *divides* $p^k - 1$. *If either*

(a) *there is a nonconstant sequence* $a(1), \ldots, a(r)$ *with* $r \leq p$, $0 \leq a(i) < k$ *for all* i *and* $p^{a(1)} + \ldots + p^{a(r)} \equiv 1 \pmod{n}$, *or*

(b) n *is prime and some prime divisor of* k *is less than* $p - 1$, *or*

(c) *there exist integers* $a(1), a(2), a(3), a(4)$ *such that* $0 \leq a(i) < k$ *for all* i , $a(1) < a(2) < a(3)$ *and* $p^{a(1)} + p^{a(2)} + p^{a(3)} + p^{a(4)} \equiv 1 \pmod{n}$,

then $a(p, n) = 2$. *However* $a(2, 3) = 3$.

For a positive integer m let $\underline{\underline{A}}_m$ denote the variety of abelian groups of exponent dividing m . Since a variety generated by a single sincere critical group is not generated by its spine, and since, by Cossey [3], $G(p^{\alpha}, n)$ generates the product variety $\underline{\underline{A}}_{p^{\alpha}}\underline{\underline{A}}_n$, we have a corollary.

COROLLARY. *Let* $\alpha \geq 3$ *and* p *be a prime not dividing* n . *The variety* $\underline{\underline{A}}_{p^{\alpha}}\underline{\underline{A}}_n$ *is generated by its spine if and only if* n *divides* $p - 1$.

To describe group theoretic conditions which determine the sincerity of a group we need some further notation and terminology. The concept of similar normal subgroups is defined in [8, 53.11]. For a group H , $F(H)$ denotes the Fitting subgroup of H , H' the derived group, and, if H is monolithic, σ^*H the centralizer of σH in H .

THEOREM 3. *The group* $G = G(p^{\alpha}, G^*)$ *is sincere if and only if there is a monolithic group* H *such that* σH *in* H *is similar to* σG *in* G , $\sigma^*H = F(H)$, $\sigma H \leq F(H)'$, $F(H)/\Phi H$ *is similar in* $H/\Phi H$ *to* σG *in* G *and* p^{α} *does not divide the exponent of* $F(H)/F(H)'$.

In view of the last theorem it would be nice to have a general method for constructing a group H , as described there, when it exists. There is an algorithm, due to L.G. Kovács, which will give H in a finite number, $m(G)$ say, of steps if

H exists, but unfortunately it is not known if there is a bound on $m(G)$ independent of $G = G(p^{\alpha}, G^*)$.

References

[1] Warren Brisley and L.G. Kovács, "On soluble groups of prime power exponent", *Bull. Austral. Math. Soc.* 4 (1971), 389-396. MR43#4913.

[2] R.M. Bryant and L.G. Kovács, "The skeleton of a variety of groups", *Bull. Austral. Math. Soc.* 6 (1972), 357-378. Zbl.237.20028.

[3] P.J. Cossey, "On varieties of A-groups", (PhD thesis, Australian National University, Canberra, 1966).

[4] John Cossey, "Critical groups and the lattice of varieties", *Proc. Amer. Math. Soc.* 20 (1969), 217-221. MR38#1147.

[5] L.F. Harris, "Varieties and section closed classes of groups", (PhD thesis, Australian National University, Canberra, 1973; see also, Abstract, *Bull. Austral. Math. Soc.* 9 (1973), 475-476).

[6] L.F. Harris, "A product variety of groups with distributive lattice", *Proc. Amer. Math. Soc.* (to appear).

[7] L.G. Kovács and M.F. Newman, "On critical groups", *J. Austral. Math. Soc.* 6 (1966), 237-250. MR34#233.

[8] Hanna Neumann, *Varieties of groups* (Ergebnisse der Mathematik und ihrer Grenzgebiete, Band 37. Springer-Verlag, Berlin, Heidelberg, New York, 1967). MR35#6734.

[9] M.F. Newman, "On a class of metabelian groups", *Proc. London Math. Soc.* (3) 10 (1960), 354-364. MR22#8074.

8085 King George Highway,
Surrey, B.C., Canada.

PROC. SECOND INTERNAT. CONF. THEORY OF GROUPS, CANBERRA 1973, pp. 337-346.

20E20

SYLOW SUBGROUPS OF LOCALLY FINITE GROUPS

B. Hartley

1. Introduction

The theorems of Sylow are among the most basic in the theory of finite groups, and Hall's theorems on the existence and conjugacy of Hall π-subgroups occupy a similarly central position in the theory of finite soluble groups. It is therefore natural to ask for what kinds of infinite groups results like them are true, and to what extent other parts of finite group theory can be extended to such groups. The answer to the first question has probably turned out to be more disappointing than was at one time expected, in that it is only under rather severe restrictions that sensible analogues of Sylow's and Hall's theorems can be obtained; but this paradoxically rekindles interest in the question in that one may now hope for a reasonably complete classification of the circumstances under which theorems like Sylow's and Hall's are true. As regards the second question, there are nevertheless some interesting classes of groups with very civilized Sylow structure, and quite a lot of progress has been made in extending such things as formation theory to these groups. I want mainly to discuss the first question from the classification point of view, but first some background may be of interest.

DEFINITION. Let π be a set of primes. A *π-group* is a periodic group in which the order of every element is a product of primes in π. A *Sylow π-subgroup* of a group is a π-subgroup which is maximal under set-theoretic inclusion among all the π-subgroups of that group.

If π consists of a single prime then, by Sylow's theorem, this definition coincides with the usual one for finite groups. A Sylow π-subgroup of a finite group H need not be a Hall π-subgroup in general, although it is if H is soluble, and more generally if the Sylow π-subgroups of H are conjugate. Our usage of the term "Sylow p-subgroup" here differs from that introduced by Wehrfritz [26] and used in Kegel and Wehrfritz [15]. They require a Sylow p-subgroup of G not only to be maximal but to contain an isomorphic copy of every p-subgroup of G. In this sense, Sylow p-subgroups need not exist in a general locally finite group ([15], p. 85) but they exist and are of great utility in groups with min-p. However we shall not discuss them further here.

The following happy fact is a direct consequence of Zorn's Lemma:

PROPOSITION 1.1. *Let* G *be any group and* π *any set of primes. Then every* π*-subgroup of* G *is contained in a Sylow* π*-subgroup of* G *. In particular,* G *possesses Sylow* π*-subgroups.*

On enquiring about the relationship between the Sylow π-subgroups of a group, however, one usually discovers chaos.

2. Sylow pathologies

We now restrict our attention to locally finite groups, and use the term "group" to mean "locally finite group". One of the milder forms of respectable behaviour which one might hope for is that the Sylow p-subgroups of a group should all be isomorphic. It is, however, part of the folklore that the restricted symmetric group on a countably infinite set has continuously many pairwise non-isomorphic Sylow p-subgroups for each prime p . Much less complicated groups than this can exhibit similar behaviour. Kovács, Neumann and de Vries [16] have given examples of a countable metabelian group of exponent 6 which has continuously many pairwise non-isomorphic Sylow 3-subgroups, an uncountable metabelian group of exponent 6 in which one Sylow 2-subgroup is countable and another uncountable, a similar uncountable group in which all the Sylow p-subgroups are countable, and so on. Although it was at one time suspected that conjugacy of the Sylow p-subgroups might follow from the condition min-p , Wehrfritz [25] has shown that this condition does not even imply their isomorphism, and has given other examples of strange behaviour in groups with min-p .

Some general methods of constructing groups with many isomorphism types of Sylow subgroups are given in Hartley [10] and Heineken [13] (see also [14]). An entertaining consequence of [10] is

THEOREM 2.1. *There exists a countable group which contains, for each prime* p *, a copy of every countably infinite* p*-group as a Sylow* p*-subgroup.*

We mention in passing that this group involves the well known "universal" group of P. Hall [5] and so it is not locally soluble. In fact, I know of no countable locally soluble group in which every countable locally finite p-group can be embedded.

PROBLEM 1. *Is there a locally soluble group like the group of Theorem 2.1? To what extent can* p *be replaced by* π *?*

3. Sylow integrated groups

Section 2 describes the worst aspects of the picture. On the positive side, it has been known for some time that Sylow's and Hall's theorems extend in a natural way to the class of locally normal groups, in which every finite set of elements lies in a finite normal subgroup, and even such things as formation theory extend satisfactorily to these groups. The main change necessary is that conjugacy under the group of inner automorphisms has to be replaced by conjugacy under the group of locally inner automorphisms. See for example [21] and the literature there mentioned. However we wish to concentrate our attention on genuine conjugacy.

DEFINITION. A group G is *S_π-integrated* if, for every $H \leq G$, the Sylow π-subgroups of H are conjugate in H. A group which is S_π-integrated for all π is called *completely Sylow integrated.*

By a theorem of P. Hall, every finite subgroup of a completely Sylow integrated group is soluble, so that *completely Sylow integrated groups are locally soluble.* The class of completely Sylow integrated groups has often been denoted by $\underline{\underline{U}}$ (see Theorem 4.5).

We summarize below a number of properties which imply Sylow integration:

THEOREM 3.1. *If G has a normal Sylow π-subgroup then G is S_π-integrated. If G has a finite Sylow p-subgroup then G is S_p-integrated. If G is locally soluble and has a finite Sylow π-subgroup then G is S_π-integrated.*

This is well known and rather trivial.

THEOREM 3.2 (Wehrfritz [23], [24]). *Periodic linear groups are S_p-integrated for all primes p. Periodic soluble linear groups are completely Sylow integrated.*

THEOREM 3.3 (McDougall [17]). *Metabelian groups satisfying the minimal condition on normal subgroups are completely Sylow integrated.*

THEOREM 3.4 (Šunkov [20]). *Groups which satisfy* min-p *for all p are S_p-integrated for all p.*

These results are supplemented by the following information about closure properties. Notice that homomorphic images of S_π-integrated groups are S_π-integrated, a fact which would have been built into the definition had it not been automatic.

THEOREM 3.5 [8]. 1. *The class of S_π-integrated groups is closed under forming subgroups, homomorphic images, and direct products with finitely many factors. It is countably recognizable.*

2. *If G has an S_p-integrated subgroup of finite index then G is S_p-integrated.*

3. *If* G *is* π*-separable and has an* S_π*-integrated subgroup of finite index, then* G *is* S_π*-integrated.*

By a π-*separable* group we mean one having a finite series with π- or π'-factors. More generally we shall need to refer to *upper* π-*separable* groups, the obvious generalization in which the series is well-ordered ascending.

4. Sylow sparse groups

We now come to our main topic, which is the analysis of Sylow integrated groups. In fact, it turns out to be more convenient to work with an apparently weaker property:

DEFINITION. A group G is S_π*-sparse* if $|\mathrm{Syl}_\pi H| < 2^{\aleph_0}$ for every countable $H \leq G$. A group which is S_π-sparse for all π is called *completely Sylow sparse.*

Obviously we have

PROPOSITION 4.1. *Every* S_π*-integrated group is* S_π*-sparse.*

I know of no counterexample whatever to the converse, and, at any rate, the following conjecture seems very plausible:

CONJECTURE 1. *If* G *is locally* S_π*-integrated and* S_π*-sparse then* G *is* S_π*-integrated.*

Some evidence for this will be found below. Notice that local S_π-integration is no restriction if $\pi = \{p\}$, and follows from local solubility for general π. The advantages of working with S_π-sparseness are that its closure properties are rather transparent and that a countable group G is actually S_π-sparse if $|\mathrm{Syl}_\pi G| < 2^{\aleph_0}$. Thus, for example, it is a consequence of the next result that *a countable group* G *is* S_p*-integrated provided only that the Sylow* p*-subgroups of* G *itself are conjugate.*

THEOREM 4.2. *Let* G *be a group containing an* S_p*-sparse subgroup of finite index. Then*

(i) G *is* S_p*-integrated* (Asar [1], Hartley [7]),

(ii) if p *is odd and* G *is locally* p*-soluble, then* $O_{p,p',p}(G)$ *has finite index in* G *and* $G/O_p(G)$ *satisfies* min-p (Hartley [9]).

The proof of *(i)* is rather easy, but we should like to enlarge upon the proof of *(ii)* since it ties up with a number of apparently rather remote questions. The main difficulty lies in showing that G is p-separable. Now it is not hard to see that an S_p-sparse group G contains a finite p-subgroup P which lies in only one Sylow p-subgroup of G. This makes it rather easy to deduce p-separability if G is also locally p-soluble, if the following conjecture of Thompson can be established.

CONJECTURE 2 (Thompson). *There exists an integer-valued function f such that, whenever a finite p-soluble group X contains a p-subgroup Q of order p^l which lies in only one Sylow p-subgroup of X, then the p-length of X is at most $f(l)$.*

The proof of Theorem 4.2 (ii) given in [9] in fact avoids Conjecture 2 by working directly with the infinite group, using methods based on the Fitting chains of Dade [2] which were introduced to establish another conjecture of Thompson about Fitting heights and Carter subgroups. However, Rae [18] has since shown that these techniques may be improved to yield a proof of Conjecture 2 for odd p, thereby furnishing a more satisfactory proof of Theorem 4.2 (ii).

PROBLEM 2. *Prove Theorem 4.2 (ii) and/or Conjecture 2 for $p = 2$.*

Presumably this can be done by sufficiently refining and improving the methods used for the odd p case, possibly bringing in the more powerful representation - theoretic techniques being developed by Berger.

We point out that Conjecture 2 is closely bound up with questions about automorphism groups of p-soluble groups, because of the rather easy

LEMMA [12]. *Let G be a finite p-soluble group and let P be a subgroup of a Sylow p-subgroup P_0 of G. Then the following statements are equivalent:*

(i) P_0 *is the only Sylow p-subgroup of G containing P;*

(ii) $C_{U/V}(P) = C_{U/V}(P_0)$ *for every p'-section U/V of G normalised by P_0.*

Once the locally p-soluble case has been completely settled, there remains the case of S_p-sparse groups in general, and the simple ones in particular. Here it is by no means clear how to make progress. The only simple S_p-sparse groups I know are linear, but because of Theorem 3.4, proving a theorem to that effect is more difficult than proving that a simple group with min-p for all primes p is necessarily finite. This has not yet been done. We state the following conjecture without very much hard evidence, or idea how to make a start on it:

CONJECTURE 3. *If G is S_p-integrated and $O_{p'p}(G) = 1$ then G is an* *extension of a linear group by a group with finite Sylow p-subgroups.*

The case of general π

It seems natural, in trying to analyse S_π-sparse groups for general π, to impose some order on the finite subgroups of the groups considered by means of the condition of local π-separability. The problem then falls into two parts:

(i) proving (global) π-separability or at any rate upper π-separability, and

(ii) analysing the upper π-separable case.

We state straight away

CONJECTURE 4. *If G is locally π-separable and S_π-sparse, then $G/O_{\pi,\pi',\pi,\pi'}(G)$ is finite, and in particular, G is π-separable.*

This conjecture embodies both parts of the problem as just described, but because of the next result, the only difficulty remaining is in the first part.

THEOREM 4.3 (see [8]). *If G is upper π-separable and S_π-sparse, then*

(i) G is S_π-integrated,

(ii) $G/O_{\pi,\pi',\pi,\pi'}(G)$ is finite.

There are two situations in which progress has been made in establishing π-separability. The first is established by methods like those used for Theorem 4.2.

THEOREM 4.4 [9]. *Let π be a set of odd primes and let G be a locally π-separable and S_π-sparse group. Suppose there exists an integer $n \geq 0$ such that every chief factor of every finite π-subgroup of G is elementary abelian of rank $\leq n$. Then G is π-separable.*

In particular, Theorem 4.4 holds if the π-subgroups of G are locally supersoluble.

In the second situation we have stronger hypotheses.

THEOREM 4.5 [8]. *Let G be locally soluble and completely Sylow sparse. Then*

(i) G is completely Sylow integrated,

(ii) $G/\rho(G)$ is a metabelian-by-finite group of finite rank, where $\rho(G)$ is the Hirsch-Plotkin radical of G.

The proof of Theorem 4.3, and any kind of analysis of the upper π-separable case, rests on careful investigation of the *Schur-Zassenhaus situation*:

(*) $G = HK$, $H \triangleleft G$, *where H is a π'-group and K is a π-group* and once one has understood what S_π-sparseness means in this situation, results about the general upper π-separable case can be read off by considering successive layers of the upper π-series. What one then obtains are various refinements of Theorem 4.3, with stronger conclusions which we shall not go into.

The first move in analysing the Schur-Zassenhaus situation is to notice the following fact:

LEMMA. *Suppose that (*) holds. Then G is S_π-sparse if and only if, for each subgroup $L \le K$, there exists a finite subgroup $F \le L$ such that $C_H(L) = C_H(F)$.*

To fix our ideas, let us consider an example. Let $p_1, p_2, \ldots; q_1, q_2, \ldots$ be two disjoint infinite sequences of distinct primes such that $p_i | q_i - 1$ $(i = 1, 2, \ldots)$. Then there exists, for each $i = 1, 2, \ldots$, a group $K_i = A_i B_i$ containing a normal subgroup A_i of order q_i and a cyclic subgroup B_i of order p_i^2 such that $\left|C_{B_i}(A_i)\right| = p_i$. Let $K = K_1 \times K_2 \times \ldots$, let $C_i = C_{B_i}(A_i)$, and let $M = A_1 \times C_1 \times A_2 \times C_2 \times \ldots$. Then M is a locally cyclic subgroup of K which contains all the elements of prime order of K. Since M is locally cyclic it has a faithful irreducible module U over Z_r, where r is a prime different from any p_i and q_i; we may assume such a prime exists. Let H be the induced module $H = U^K$. Then every non-trivial element of K acts fixed point freely on H, and the above lemma shows that the split extension $G = HK$ is S_π-sparse, with $\pi = \{p_1, q_1, p_2, q_2, \ldots\}$.

This simple example in fact typifies the general situation, and it is interesting to see from the following results that there is no such example in which K_i is non-abelian of order $p_i q_i$.

DEFINITION. A group X is *pinched* if

(i) X contains a normal locally cyclic subgroup A such that X/A is abelian and A contains every element of prime order of X, and

(ii) every 2-subgroup of X is abelian.

Condition (ii) is just a technicality which ensures that every p-subgroup of a pinched group is locally cyclic, even for $p = 2$. It is easy to see that a pinched group is an extension of one locally cyclic group by another.

THEOREM 4.6. (see [11]). *Suppose that (*) holds, that G is S_π-sparse and that $C_K(H) = 1$. Then K has a subgroup of finite index which is a subdirect* *product of a finite number of pinched groups.*

This describes the structure of the top group in (*), and is best possible in that any group of the type described can occur as K, with $C_K(H) = 1$ and G S_π-sparse. The hypotheses of local solubility made in [11] Theorem B are actually superfluous, because of a recent result of Šunkov [20] stating that a group, all of whose abelian subgroups have finite rank, has a locally soluble subgroup of finite index.

Regarding what can be at the bottom in (*), we have a recent theorem of Rae [19].

THEOREM 4.7. *Suppose that (*) holds, that G is S_π-sparse, and that H is locally soluble. Then K contains a subgroup K_0 of finite index such that $[H, K_0]$ has a finite series with locally nilpotent factors.*

PROBLEM 3. *Can the hypothesis of local solubility on H be removed? If (*) holds, G is S_π-sparse and H is simple, does it follow that $K/C_K(H)$ is finite?*

Finally we mention an improvement of Theorem 4.5.

THEOREM 4.8 [11]. *Suppose that G is locally soluble and completely Sylow sparse. Then $G/\rho(G)$ has a subgroup of finite index which is a subdirect product of finitely many pinched groups.*

This, suitably interpreted, is best possible.

5. Formation theory

The best behaved class we have discussed here is the class $\underline{\underline{U}}$ of completely Sylow integrated groups. Its properties are very similar to those of the class of finite soluble groups, though usually more difficult to establish. For example, *every $\underline{\underline{U}}$-group has a unique conjugacy class of Sylow bases* [15], the appropriate analogues of the Sylow systems of a finite soluble group. One can then introduce locally defined formations, $\underline{\underline{F}}$-normalizers, $\underline{\underline{F}}$-projectors and so on, and develop most of the standard theory as in finite soluble groups. See for example [3], [6], [4].

6. Two more questions

PROBLEM 4. *Describe the countable locally soluble groups which admit a group of automorphisms which*

(i) permutes the Sylow bases transitively,

(ii) permutes transitively the Sylow π-subgroups for each π,

(iii) normalizes every normal subgroup.

Among them are all countable locally soluble groups which are either completely Sylow integrated or are locally normal.

So far I have said nothing about the influence of elements of infinite order. The reason is that I know of no results of the type discussed here which work in the presence of elements of infinite order. It certainly happens that groups containing elements of infinite order have their Sylow p-subgroups conjugate, for example full linear groups [22] and certain wreath products [10], but then such groups often seem to have badly behaved subgroups.

PROBLEM 5. *Prove something about* S_p*-integrated groups containing elements of infinite order.*

Perhaps one should try to prove that the elements of infinite order are only involved in a very trivial way.

References

[1] A.O. Asar, "A conjugacy theorem for locally finite groups", *J. London Math. Soc.* (2) 6 (1973), 358-360.

[2] E.C. Dade, "Carter subgroups and Fitting heights of finite solvable groups", *Illinois J. Math.* 13 (1969), 449-514. MR41#339.

[3] A.D. Gardiner, B. Hartley and M.J. Tomkinson, "Saturated formations and Sylow structure in locally finite groups", *J. Algebra* 17 (1971), 177-211. MR42#7778.

[4] C.J. Graddon, "Formation theoretic properties of certain locally finite groups" (PhD thesis, University of Warwick, Coventry, 1971).

[5] P. Hall, "Some constructions for locally finite groups", *J. London Math. Soc.* 34 (1959), 305-319. MR29#149.

[6] B. Hartley, "$\underline{\underline{F}}$-abnormal subgroups of certain locally finite groups", *Proc. London Math. Soc.* (3) 23 (1971), 128-158. MR46#3622.

[7] B. Hartley, "Sylow subgroups of locally finite groups", *Proc. London Math. Soc.* (3) 23 (1971), 159-192. MR46#3623.

[8] B. Hartley, "Sylow theory in locally finite groups", *Compositio Math.* 25 (1972), 263-280. Zbl.248.20036.

[9] B. Hartley, "Sylow p-subgroups and local p-solubility", *J. Algebra* 23 (1972), 347-369.

[10] B. Hartley, "Complements, baseless subgroups and Sylow subgroups of infinite wreath products", *Compositio Math.* 26 (1973), 3-30.

[11] B. Hartley, "A class of modules over a locally finite group I", *J. Austral. Math. Soc.* 16 (1973), 431-442.

[12] B. Hartley and A. Rae, "Finite p-groups acting on p-soluble groups", *Bull. London Math. Soc.* 5 (1973), 197-198.

[13] H. Heineken, "Maximale p-Untergruppen lokal endlicher Gruppen", *Arch. Math.* (to appear).

[14] C.H. Houghton, "Ends of groups and baseless subgroups of wreath products", *Compositio Math.* (to appear).

[15] Otto H. Kegel and Bertram A.F. Wehrfritz, *Locally finite groups* (North-Holland Mathematical Library, 3. North-Holland, Amsterdam, 1973).

[16] L.G. Kovács, B.H. Neumann and H. de Vries, "Some Sylow subgroups", *Proc. Roy. Soc. London Ser. A* 260 (1961), 304-316. MR29#4803.

[17] David McDougall, "Soluble groups with the minimal condition for normal subgroups", *Math. Z.* 118 (1970), 157-167. Zbl.194,36.

[18] A. Rae, "Sylow p-subgroups of finite p-soluble groups", *J. London Math. Soc.* (2) 7 (1973), 117-123.

[19] A. Rae, "Local systems and Sylow subgroups in locally finite groups II", *Proc. Cambridge Philos. Soc.* (to appear).

[20] В.П. Шунков [V.P. Šunkov], "О локально конечных конечного ранга" [Locally finite groups of finite rank], *Algebra i Logika* 10 (1971), 199-225; *Algebra and Logic* 10 (1971), 127-142. MR45#2002.

[21] M.J. Tomkinson, "Formations of locally soluble FC-groups", *Proc. London Math. Soc.* (3) 19 (1969), 675-708. MR41#5501.

[22] Р.Т. Вольвачев [R.T. Vol'vačev], "p-подгруппы Силова полной линейной группы" [Sylow p-subgroups of the general linear group], *Izv. Akad. Nauk SSSR Ser. Mat.* 27 (1963), 1031-1054; *Amer. Math. Soc. Transl.* (2) 64 (1967), 216-243. MR28#4037.

[23] B.A.F. Wehrfritz, "Sylow theorems for periodic linear groups", *Proc. London Math. Soc.* (3) 18 (1968), 125-140. MR36#3893.

[24] B.A.F. Wehrfritz, "Soluble periodic linear groups", *Proc. London Math. Soc.* (3) 18 (1968), 141-157. MR36#3894.

[25] B.A.F. Wehrfritz, "Sylow subgroups of locally finite groups with min-p", *J. London Math. Soc.* (2) 1 (1969), 421-427. MR40#1480.

[26] B.A.F. Wehrfritz, "On locally finite groups with min-p", *J. London Math. Soc.* (2) 3 (1971), 121-128. MR43#4920.

University of Warwick, Coventry CV4 7AL, England.

PROC. SECOND INTERNAT. CONF. THEORY OF GROUPS, 20F05
CANBERRA 1973, pp. 347-356. (20-04)

A REIDEMEISTER-SCHREIER PROGRAM

George Havas

1. Introduction

The Reidemeister-Schreier method yields a presentation for a subgroup H of a group G when H is of finite index in G and G is finitely presented. This paper describes the implementation and application of a FORTRAN program which follows this method. The program has been used satisfactorily for subgroups of index up to several hundred.

Following the theory of Reidemeister and Schreier (see for example Magnus, Karrass, Solitar [4], 2.3, p. 86), we see that we require the coset table of H in G. The program described is implemented as a set of subroutines called by the Todd-Coxeter program described in Cannon, Dimino, Havas and Watson [1] and we shall consider it in that context.

2. The procedure in detail

The Reidemeister-Schreier program commences by finding the coset table of H in G. Directly following the theory of Reidemeister and Schreier, the program finds Schreier generators for H and next finds a set of Reidemeister relators in terms of the Schreier generators.

At this stage the presentation is usually not in a useful form. The number of Schreier generators is of the order of $n_g[G : H]$ where n_g is the number of generators of G; the number of Reidemeister relators is of the order $n_r[G : H]$ where n_r is the number of relators in the presentation of G. In view of this, the program goes on further to improve the presentation using obvious but *ad hoc* techniques. In particular, this is done by using a canonical form for the relators, by eliminating redundant generators and by attempting relator simplification.

A somewhat more detailed description is given in the following paragraphs.

Supported by the Commonwealth Scientific and Industrial Research Organization.

2.1 Set up the coset table

Given $G = \langle g_1, \ldots, g_{n_g} \mid R_1 = R_2 = \ldots R_{n_r} = 1\rangle$ and $H = \langle h_1, \ldots, h_{n_h}\rangle$ the coset table of H in G is computed. (The actual subgroup generators, the h_m, are required only by the Todd-Coxeter part of the program and are not required subsequently.) The subscript i will be used to run through the cosets of H in G.

2.2 Compute coset representatives

Minimal Schreier coset representatives are computed by constructing a minimal spanning tree for the coset table. For each coset i, C_i will represent the corresponding minimal Schreier coset representative. Further K will designate a coset representative function. This means that K maps words in the g_j onto a coset representative system for G mod H. In this case $K(w)$ will be the minimal Schreier representative of the coset of w.

2.3 Compute Schreier generators

The Schreier generators $S_{i,j}$ are computed using the formula

$$S_{i,j} = C_i g_j \left(K\left(C_i g_j\right)\right)^{-1} .$$

The Schreier generators are freely reduced by the program. Some may be trivial and any such are noted.

2.4 Compute Reidemeister relators

The Reidemeister relators $r_{i,k}$ are computed using a Reidemeister rewriting process t.

If $w = g_{j_1}^{\varepsilon_1} g_{j_2}^{\varepsilon_2} \ldots g_{j_n}^{\varepsilon_n}$ then

$$t(w) = S_{i_1,j_1}^{\varepsilon_1} S_{i_2,j_2}^{\varepsilon_2} \ldots S_{i_n,j_n}^{\varepsilon_n}$$

where i_k is the coset of the initial segment of w preceding $g_{j_k}^{\varepsilon_k}$ if $\varepsilon_k = 1$ and i_k is the coset of the initial segment of w up to and including $g_{j_k}^{\varepsilon_k}$ if $\varepsilon_k = -1$. Using this rewriting we have

$$r_{i,k} = t\left(C_i R_k C_i^{-1}\right) .$$

The program eliminates all occurrences of trivial Schreier generators (previously noted for this purpose). Then the relators are freely and cyclically reduced.

All relators are converted to a canonical form as they are computed, and the relators are maintained in this form throughout. The canonical form is based on an ordering $<$ of the generators $S_{i,j}$. (The program uses the lexicographic ordering for the $S_{i,j}$.)

If x_k, y_k are Schreier generators we define our ordering $<$ on the relators by

$$x_1^{\varepsilon_1} < y_1^{\delta_1} \quad \text{if } x_1 < y_1 \text{ or } x_1 = y_1 , \ \varepsilon_1 = 1 , \ \delta_1 = -1 ;$$

$$x_1^{\varepsilon_1} x_2^{\varepsilon_2} \dots x_m^{\varepsilon_m} < y_1^{\delta_1} y_2^{\delta_2} \dots y_n^{\delta_n} \quad \text{if } m < n ;$$

and inductively

$$x_1^{\varepsilon_1} x_2^{\varepsilon_2} \dots x_m^{\varepsilon_m} < y_1^{\delta_1} y_2^{\delta_2} \dots y_m^{\delta_m}$$

if

$$x_1^{\varepsilon_1} x_2^{\varepsilon_2} \dots x_{m-1}^{\varepsilon_{m-1}} < y_1^{\delta_1} y_2^{\delta_2} \dots y_{m-1}^{\delta_{m-1}}$$

or

$$x_1^{\varepsilon_1} x_2^{\varepsilon_2} \dots x_{m-1}^{\varepsilon_{m-1}} = y_1^{\delta_1} y_2^{\delta_2} \dots y_{m-1}^{\delta_{m-1}} \quad \text{and} \quad x_m^{\varepsilon_m} < y_m^{\delta_m} .$$

We choose as our canonical relator the least (with respect to this ordering) of the set of relators made up of all cyclic rotations of the given relator and its formal inverse.

As each relator is computed its canonical representative is inserted into a relator list, provided it is not a repeat of a relator already there. The canonical form notion substantially reduces the number of relators in the presentation.

2.5 Eliminate redundant generators

When we have completed the above steps we have a presentation for H, namely

$$H = \langle S_{i,j} \mid r_{i,k} = 1 \rangle ,$$

with duplicate relators removed.

This presentation is still highly redundant. In particular there are usually many redundant generators in the sense that there are many relators containing exactly one occurrence of a particular generator. Such generators can be removed by substitution.

The most obvious approach is to remove one redundant generator at a time from each relator. However this turns out to be very time-consuming in computer implementation. Three possible techniques are discussed below.

2.5.1 ELIMINATION TECHNIQUE 1

This is the technique actually implemented for regular use in the Reidemeister-Schreier program. The eliminations are batched together to save computing time.

Each relator in the relator list is examined for any generator occurring exactly once (provided the relator is independent of previously discovered but not yet eliminated redundant generators). The shorter relators are examined first so that the generators selected for elimination will tend to have short generator strings as their equivalents.

Whenever a redundant generator is found the associated relator is removed from the relator list, the generator is marked redundant, and the value of the generator and its inverse computed. On the completion of such a pass through the relator list all remaining relators are examined for occurrences of redundant generators. Each such occurrence is eliminated by substituting for the redundant symbol its computed value. The new relators are again freely and cyclically reduced and a new relator list is formed.

On completing the redundant generator elimination we repeat this step (for new redundancies may now have appeared in the relators). We continue repeating this step till no further redundancies are found, when we go to the last step, 2.6.

2.5.2 ELIMINATION TECHNIQUE 2

Eliminate one generator at a time. Heuristically this seems superior in that we may select for elimination the generator with shortest equivalent string at each stage. Unfortunately the Reidemeister-Schreier method frequently gives us presentations with hundreds of generators and relators, and hundreds of redundancies, and this approach takes too long.

2.5.3 ELIMINATION TECHNIQUE 3

A compromise between techniques 1 and 2 is to eliminate redundant generators with values of the same length at the one time, doing this elimination for increasing length. This has the advantage of ensuring that all length zero and one eliminations are done as soon as possible. Length zero and one eliminations are most desirable for they do not increase relator lengths, whereas higher length eliminations may well increase relator lengths, and usually do.

In §3 comparisons of the above three techniques in terms of execution times and "niceness" of ensuing presentation are given.

2.6 Simplify the presentation

When no obviously redundant generators remain we resort to other attempts to simplify the presentation. There are an unlimited number of possibilities for this. The most obvious is looking for relator substrings which have shorter equivalent strings. At this stage only the following technique is included in the program.

2.6.1 SIMPLIFICATION TECHNIQUE

All relators are checked to see if they are of the form $S^n_{i,j}$. If any such relators are found then all other relators are processed for strings in the $S_{i,j}$ of known order. All long strings (that is of length exceeding $n/2$) are replaced by their shortest equivalent counterparts. Further, for even n , strings $S^{-n/2}_{i,j}$ are replaced by $S^{n/2}_{i,j}$.

After this simplification the program returns to step 2.5, and if no further redundancies appear, the program terminates. Otherwise steps 2.5 and 2.6 are repeated till no further improvement is obtained.

3. Examples

In this section some applications in determining previously unknown Macdonald groups (see [3]) are presented. Also some other test examples are presented to give an indication of the performance of the different possible techniques and the program as a whole.

Macdonald group coset enumerations are notoriously difficult (see [1]), and this forced us to resort to the techniques described here to determine them by coset enumeration based methods.

Wamsley [5] describes a technique for constructing the largest finite nilpotent p-factors of groups. Such an approach is in fact much more suitable for determining Macdonald groups and enables their determination more easily.

3.1 Determination of a Macdonald group

First let us consider in detail a reasonably easy application of the Reidemeister-Schreier program. The Macdonald group $G(-2, -2)$ is defined

$$G(-2, -2) = \langle a, b \mid b^{-1}a^{-1}bab^{-1}aba^2 = a^{-1}b^{-1}aba^{-1}bab^2 = 1 \rangle .$$

In practice it is impossible to do the coset enumeration $G|\langle 1 \rangle$ using Todd-Coxeter programs. This is not surprising for $[G : \langle a \rangle] = 729$ (by Todd-Coxeter), and it is easy to prove that the order of a is either 27 or 81 , making the order of G either 19683 or 59049 .

The subgroup $H = \langle [a, b], [a^{-1}, b], [b, a], [b^{-1}, a] \rangle$ is of index 9 in G

(hence H is the commutator subgroup), and the enumeration of cosets is easy. Using the Reidemeister-Schreier program the following presentation for H was found

$$H = \langle x, y, z \mid x^2y^{-1}xy^{-2} = y^2z^2yz = xz^2yxz^{-1}xz^2y^{-1} = xyz^{-2}x^{-1}(yz^{-2})^2 = xz^2y^{-1}xz^3y^{-1}xz^4y^{-1} = 1 \rangle .$$

Further, the enumeration $H|\langle 1 \rangle$ was done, revealing that H has order 6561, whence G has order 59049 and a has order 81.

3.2 Test examples

A set of test examples was run using each of the three techniques described in 2.5, and also with the final program. The following groups and subgroups (taken from Coxeter and Moser [2]) were used.

(a) $G_1 = \langle a, b \mid a^3 = b^6 = (ab)^4 = (ab^2)^4 = (ab^3)^3 = a^{-1}b^{-2}a^{-2}b^{-2}a^{-1}b^{-2}ab^2a^2b^2ab^2 = 1 \rangle$.

(G_1 is a presentation for PSL(3, 3), of order 5616.)

$H_1 = \langle a, b^2 \rangle$ is the Hessian group of order 216, $[G_1 : H_1] = 26$.

(b) $G_2 = \langle a, b \mid a^4 = b^4 = (ab)^4 = (a^{-1}b)^4 = (a^2b)^4 = (ab^2)^4 = (a^2b^2)^4 = [a, b]^4 = (a^{-1}bab)^4 = 1 \rangle$.

(G_2 is a presentation for $B_{2,4}$, of order 4096.)

$H_2 = \langle a, b^2 \rangle$, $|H_2| = 64$, $[G_2 : H_2] = 64$.

(c) $G_3 = \langle a, b, c \mid a^{11} = b^5 = c^4 = (bc^2)^2 = (abc)^3 = (a^4c^2)^3 = b^2c^{-1}b^{-1}c = a^4b^{-1}a^{-1}b = 1 \rangle$.

(G_3 is a presentation for M_{11}, of order 7920.)

$H_3 = \langle a, b, c^2 \rangle$ is PSL(2, 11) of order 660, $[G_3 : H_3] = 12$.

(d) $G_4 = \langle a, b, c \mid a^{11} = b^5 = c^4 = (ac)^3 = b^2c^{-1}b^{-1}c = a^4b^{-1}a^{-1}b = 1 \rangle$.

(G_4 is a "better" presentation for M_{11}.)

$H_4 = \langle c, b, c^2 \rangle$ is again PSL(2, 11) of order 660, $[G_4 : H_4] = 12$.

(e) $G_5 = \langle a, b, c \mid a^3 = b^7 = c^{13} = (ab)^2 = (bc)^2 = (ca)^2 = (abc)^2 = 1 \rangle$.

(G_5 is $G^{3,7,13}$, a presentation for PSL(2, 13) , of order 1092 .)

$H_5 = \langle ab, c \rangle$ is dihedral of order 26 , $[G_5 : H_5] = 42$.

(f) $G_6 = \langle a, b, c \mid a^3 = b^7 = c^{14} = (ab)^2 = (bc)^2 = (ca)^2 = (abc)^2 = 1 \rangle$.

(G_6 is $G^{3,7,14}$, of order 2184 .)

$H_6 = \langle ab, c \rangle$ is dihedral of order 28 , $[G_6 : H_6] = 78$.

In fact it is easy to read presentations for H_5 and H_6 off from the presentations for G_5 and G_6 , but it is still interesting to observe the behaviour of the algorithm in these cases.

Each of the three elimination techniques was used on each of the test groups. These runs were actually made at an early stage of program development, before the introduction of the canonical form or the simplification technique. The following table indicates the nature of the presentations obtained at that time.

Column a indicates the number of generators in the presentation,
Column b indicates the number of relators, and
Column c indicates the length of the longest relator.

	Technique 1			Technique 2			Technique 3		
Subgroup	a	b	c	a	b	c	a	b	c
H_1	3	35	80	3	37	114	3	37	114
H_2	2	85	264	2	76	160	2	81	132
H_3	4	27	157	3	19	42	4	25	115
H_4	3	13	84	3	11	24	3	11	24
H_5	3	12	182	3	13	134	3	13	142
H_6	3	16	56	3	16	80	3	16	80

We might say one presentation is better than another if it has fewer generators, fewer relators and/or shorter relators. The table indicates that sometimes the heuristically better technique produces a "worse" presentation.

The fact that the better technique sometimes leads to a worse presentation is troubling. The reason for this is that redundant generators may be eliminated in different orders when different techniques are used, leading to different presentations.

Other factors may also affect the final presentation. Obviously the nature of the initial group presentation is very relevant to the nature of the subgroup presentation deduced. The presentation for H_4 is much "nicer" than the presentation for H_3. This is because the presentation for G_4 is better than the presentation for G_3.

It is interesting to note that if we take different generators for the subgroup we may get significantly better or worse presentations.

The reason for this is that the coset table may be generated in a different order, leading to a different ordering on the Schreier generators. When given a choice of a number of redundant subgroup generators to eliminate in one relator, the elimination procedure uses the rule of selecting the largest with respect to the canonical ordering. Thus different Schreier generators are eliminated under different conditions.

Sample timings for each of the three techniques are as follows. For H_1 technique 3 took 1.5 times as long as technique 1 while technique 2 took 2.4 times as long. For H_5 technique 3 took 1.6 times as long as technique 1 while technique 2 took 8.5 times as long. These timing considerations justify the selection of technique 1 for the final program implementation.

It is also interesting to consider the nature of the presentation yielded by the original Reidemeister-Schreier method and the nature of the presentation obtained by the final program implementation. Columns a, b and c have the same significance as before.

	Original Reidemeister-Schreier Presentation			Final output Presentation		
Subgroup	a	b	c	a	b	c
H_1	27	156	14	2	29	168
H_2	65	576	16	2	67	204
H_3	25	96	15	4	22	111
H_4	25	72	11	3	10	21
H_5	157	546	14	3	13	65
H_6	85	294	13	2	12	66

At first sight, it may seem that neither of these presentations is of much use. The former has too many generators and too many relators while the latter has relators which are too long, and perhaps too many relators. But looking at the final output presentation shows us otherwise. Of the examples given, the presentation for

H_2 looks the worst. However the first four relations in the presentation produced for H_2 are $b^2 = a^4 = (ab)^4 = (a^2b)^4 = 1$ and these alone suffice for a presentation of H_2. In a similar fashion three of the first four relations of the presentation produced for H_6 are $a^2 = (ab)^2 = b^{14} = 1$, indeed a presentation for H_6.

If we can successfully select a likely presentation for H from the relators, it is an easy matter, using coset enumeration, to show whether the remaining relators are consequences of the selected set. So even when we get an apparently ungainly presentation, we may well be able to extract a useful presentation from it.

The final programmed algorithm, including the coset enumeration, took 6.5 seconds CPU time on a CDC 6600 to find the presentation for H_1.

3.3 The Reidemeister-Schreier abelianized

In certain cases (the Macdonald groups are again a case in point) there are abelian subgroups of a group of which we do not know the structure. In such cases it is useful to perform an abelianized Reidemeister-Schreier. What we do is abelianize each relator at each stage of the computation, and this greatly simplifies the ensuing presentation.

Let us consider the use of the abelianized Reidemeister-Schreier in the context of two other previously unknown Macdonald groups. Consider $G = G(3, 5)$. Again the enumeration $G|\langle 1 \rangle$ is too difficult. However, the enumeration $G|H$ where $H = \langle b \rangle$ can be done easily enough, to get $[G : H] = 8$.

Unfortunately we do not know the order of b.

If we try to use the straight Reidemeister-Schreier program to find a presentation for H we get a nasty presentation (3 generators, 10 relators, longest relator length 66). However, H is abelian of course. Using the abelianized Reidemeister-Schreier we obtain the one relator presentation $H = \langle b \mid b^{16} \rangle$, whence $|G| = 128$.

The second example is $G(-3, -5)$. $[G : H = \langle b \rangle] = 32$, and a presentation for H given by the abelianized Reidemeister-Schreier is $H = \langle b \mid b^{12} \rangle$, whence $|G| = 384$. However, using the straight Reidemeister-Schreier program we get a presentation with 3 generators, 34 relators and longest relator length 684.

4. Data structures and computation techniques

For those interested in the actual computer implementation, brief details of some of the methods used are given in this section.

After computing the coset table we need to compute coset representatives, Schreier generators and Reidemeister relators for the subgroup. All of these are of unpredictable length so a list structure is most appropriate. In order to compute Schreier coset representatives we need two locations per coset of the table. In addition we need n_g locations per coset for pointers to each of the n_g Schreier generators associated with each coset.

Thus, in addition to the data structures used in [1], we must augment the coset table by n_g columns and we require a reasonably large amount of list space. The list processing is done in-line in the FORTRAN code.

In order to perform the conversion of relators to canonical form, the relators are made into circular lists after free and cyclic reduction. This facilitates the examination of all rotations of a relator without the use of additional storage.

Finally, the relator list, in which all the canonical form relators for the subgroup are stored, is kept sorted according to the previously defined order. This makes the search for duplication easier and enables the program to simply process the relators in relator list order when applying generator elimination and relator simplification techniques.

References

[1] John J. Cannon, Lucien A. Dimino, George Havas and Jane M. Watson, "Implementation and analysis of the Todd-Coxeter algorithm", *Math. Comp.* 27 (1973), 463-490.

[2] H.S.M. Coxeter and W.O.J. Moser, *Generators and relations for discrete groups* (Ergebnisse der mathematik und ihrer Grenzgebiete, Band 14. Springer-Verlag, Berlin, Göttingen, Heidelberg, New York; 1st edition, 1957; 2nd edition, 1965). MR19,527 (1st edition); MR30#4818 (2nd edition).

[3] I.D. Macdonald, "On a class of finitely presented groups", *Canad. J. Math.* 14 (1962), 602-613. MR25#3992.

[4] Wilhelm Magnus, Abraham Karrass, Donald Solitar, *Combinatorial group theory* (Pure and Appl. Math., 13. Interscience [John Wiley & Sons], New York, London, Sydney, 1966). MR34#7617.

[5] J.W. Wamsley, "Computation in nilpotent groups (Theory)", these Proc.

Canberra College of Advanced Education,
Canberra City, ACT 2601.

PROC. SECOND INTERNAT. CONF. THEORY OF GROUPS, 20E15
CANBERRA 1973, pp. 357-360.

NON-NILPOTENT GROUPS WITH NORMALIZER CONDITION

H. Heineken and I.J. Mohamed

In [2] we proved the following result

THEOREM 1. *For each prime* p *there is a countable metabelian* p*-group* G *such that*

I $(G')^p = 1$ *and* $G/G' \simeq C_{p^\infty}$,

II *every proper subgroup of* G *is nilpotent and subnormal,*

III $Z(G) = 1$.

Some of the consequences of this theorem, and of the proof of the theorem, are shown in the following diagram of classes of soluble groups

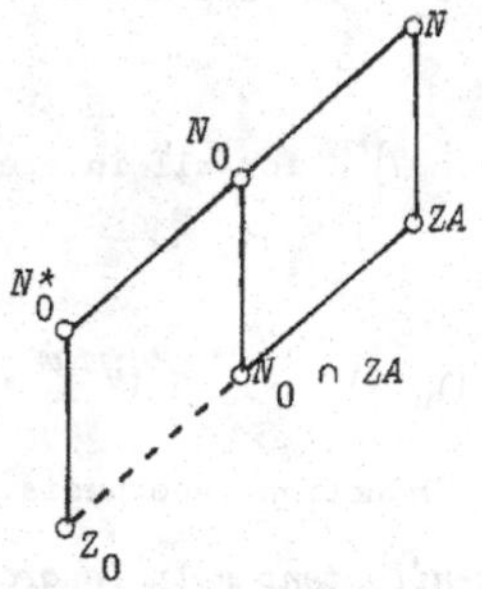

where

N is the class of groups with normalizer condition,

N_0 is the class of groups with all subgroups subnormal,

N_0^* is the class of groups with all proper subgroups nilpotent and subnormal,

ZA is the class of hypercentral groups,

Z_0 is the class of nilpotent groups.

It was also established in [2] that for soluble groups $N_0^* \cap ZA = Z_0$, but it is not known if $N_0 \cap ZA = Z_0$.

The groups G of Theorem 1 all contain elements of arbitrarily large p-power order. That this has to be the case for non-nilpotent soluble groups satisfying II is shown in Lemma 1 of [2]. It is now natural to ask if there are non-nilpotent (non-hypercentral) soluble groups of finite exponent satisfying all subgroups subnormal (the normalizer condition). It is clear from Lemma 1 of [2] that such a group must contain a proper subgroup of the same type. It is the object of this note to reduce this problem to a more easy situation. We shall appeal to the following result of Hall [1].

LEMMA 1. *If* $N \triangleleft G$ *and if* G/N' *and* N *are nilpotent, then so is* G .

In addition to this we shall also make use of the following simple analogue.

LEMMA 2. *If* $N \triangleleft G$, N *of exponent* m^k *and if* G/N^m *and* N *are nilpotent, then so is* G .

PROOF. The proof is by induction on k . The result is trivially true for $k = 1$. Without loss of generality we may assume that N is abelian, for G/N^m nilpotent implies $\dfrac{G/N'}{(N/N')^m}$ is nilpotent and so this special case gives G/N' nilpotent and hence by Lemma 1, G is nilpotent.

Assume $k > 1$ and let $M = N^m$. Then $\exp M = m^{k-1} < \exp N$. If $\gamma_n(G) \leq N^m$ a simple induction shows that

$$\gamma_{n+r}(G) \leq \left(\gamma_{r+1}(G) \cap N\right)^m \text{ for all integers } r \geq 0 .$$

Thus

$$\gamma_{2n-1}(G) \leq \left(\gamma_n(G) \cap N\right)^m \leq \left(N^m\right)^m = M^m$$

and hence the result follows by the induction hypothesis.

THEOREM 2. *If there is a non-nilpotent soluble group of finite exponent with all subgroups subnormal, then there is a countable such* p*-group* G *, for some prime* p *, such that*

(i) $\left(G'G^p\right)' = 1 = \left(G'G^p\right)^p$,

(ii) *if* $N \geq G'$ *and if* N *is nilpotent, then* N *and* G/N *are both infinite.*

Conversely, if G *is any group which satisfies (i) and (ii) then*

(iii) $G \nleq Z_0$,

(iv) $G \in N_0$ *if and only if for all subgroups* L *such that* $G'L$ *is non-nilpotent and* L *of infinite index in* $G'L$, $G'L/G' \cap L$ *is nilpotent.*

PROOF. It is clear that if any group of the required type exists, then there is a countable such group.

(i) If the class of non-nilpotent soluble groups of finite exponent with all subgroups subnormal is not empty then there is a group G such that G has minimal derived length and G has minimal exponent. Since an N_0-group is locally nilpotent and since a periodic N_0-group is the direct product of p-groups an N_0-group of finite exponent is the direct product of a finite number of p-groups. Thus we may assume that G is a group of exponent p^n , for some prime p . By Lemma 1 and Lemma 2 we may assume that G' is abelian of exponent p . Hence

$$\exp\left(G^{p^2}\right) \le \frac{\exp(G/G')}{p^2} \exp G' \le p^{n-1} < \exp G .$$

By the minimality of G we have G^{p^2} is nilpotent and hence G^{p^2} has exponent p by Lemma 2. In particular, G has exponent p^3 at most.

Assume the exponent of G is p^3 . Since G^{p^2} and G' are nilpotent so is $G^{p^2}G'$ and Lemma 1 allows us to assume $G^{p^2}G'$ is abelian. But we have shown that G^{p^2} and G' are both of exponent p and hence $G^{p^2}G'$ is of exponent p . Now $\left(G^p\right)^p \le G^{p^2}G'$ and therefore G^p has exponent p^2 at most. By the minimality of G we have that G^p is nilpotent. But if $G/\left(G^p\right)^p$ is nilpotent, then G is nilpotent by Lemma 2 and consequently the exponent of G must be p^2 .

We want to show further that we can assume that G^p is abelian of exponent p . If G^p is nilpotent, then G^p is abelian by Lemma 1 hence also of exponent p . On the other hand, if G^p is not nilpotent, let $H = G^p$. Then

$$H^p = \left(G^p\right)^p \le G^{p^2}\left(G^p\right)' = \left(G^p\right)' = H'$$

and

$$(H')^p = \left(\left(G^p\right)'\right)^p \le (G')^p = 1$$

by what we have shown earlier. In this case we can therefore take H as the required group. Finally $G'G^p$ is nilpotent and so Lemma 1 allows us to assume $G'G^p$

is abelian and hence also of exponent p .

(ii) For any $g \in G$, $N\langle g\rangle$ must be nilpotent since N and $\langle g\rangle$ are and since $N \triangleleft G$ and $\langle g\rangle \triangleleft\triangleleft G$. Thus any finite extension of N must be nilpotent and so G/N must be infinite.

If N is finite, then $G = C(N)K$, where K is finite. Therefore by what we have just shown NK is nilpotent and hence N is contained in some finite term of the upper central series of G . Hence N must be infinite.

For the converse assume G is a group which satisfies *(i)* and *(ii)*. Then *(iii)* is clear.

(iv) For any $g \in G$,

$$1 = [G', g^p] = [G', {}_pg]$$

and so $\langle g\rangle \triangleleft\triangleleft G'\langle g\rangle$ and $G'\langle g\rangle$ is nilpotent. Hence for any subgroup K of G such that NK/N is finite NK is nilpotent and hence $K \triangleleft\triangleleft G$.

Let L be a subgroup of G . If for any $N \geq G'$, N nilpotent, either $N/N\cap L$ or $L/N\cap L$ is finite then $L \triangleleft\triangleleft G$. For in the first case L has finite index in $NL \triangleleft G$, and the second case follows from what we have just shown. Thus $G \in N_0$ if and only if all subgroups L of G such that $N/N\cap L$ and $L/N\cap L$ are both infinite, are subnormal in G . But $L/N\cap L$ is infinite for all $N \geq G'$, N nilpotent, if and only if $G'L$ is non-nilpotent. Hence $G \in N_0$ if and only if all subgroups L of G such that $G'/G'\cap L$ is infinite and $G'L$ non-nilpotent are subnormal in G . Since $G'L/G'\cap L$ splits over $G'/G'\cap L$ the result follows.

ACKNOWLEDGEMENT. The second author gratefully acknowledges financial support from the Australian Academy of Science and the Australian National University as well as of the University of Botswana, Lesotho and Swaziland.

References

[1] P. Hall, "Some sufficient conditions for a group to be nilpotent", *Illinois J. Math.* 2 (1958), 787-801. MR21#4183.

[2] H. Heineken and I.J. Mohamed, "A group with trivial centre satisfying the normalizer condition", *J. Algebra* 10 (1968), 368-376. MR38#3347.

Mathematisches Institut der Universität,
87 Würzburg, Germany.

University of Lesotho,
PO Roma, Lesotho.

PROC. SECOND INTERNAT. CONF. THEORY OF GROUPS, 20D05
CANBERRA 1973, pp. 361-365.

THE INFLUENCE OF 2-SYLOW INTERSECTIONS ON THE STRUCTURE OF FINITE SIMPLE GROUPS

Marcel Herzog

During the last decade, finite simple groups were characterized mainly by the structure of a centralizer of one of their involutions and by the structure of their 2-Sylow subgroups. In this talk I would like to describe another line of investigation of finite simple groups, based on the structure of intersections of two or more of their 2-Sylow subgroups.

The first result in this direction was obtained by Suzuki in 1964. In a fundamental paper [12], which serves as a basic reference for all results to follow, Suzuki classified all finite groups which satisfy the following condition:

TI All S_2-intersections are trivial,

where by S_2-intersections we mean intersections of two distinct 2-Sylow subgroups. Denoting a group satisfying condition X as an X-group, Suzuki proved:

THEOREM 1 (Suzuki [12]). *Let* G *be a finite* TI*-group. Then one of the following holds:*

(A) G *is* 2*-closed;*

(B) S_2*-subgroup of* G *is of* 2*-rank* 1 *(that is, either cyclic or generalized quaternion);*

(C) there exists a normal series

$$G \supset G_1 \supset G_2 \supset 1$$

such that $[G : G_1]$ *and* $[G_2 : 1]$ *are odd, and* G_1/G_2 *is isomorphic to one of the following simple groups:* $\mathrm{PSL}(2, q)$, $\mathrm{PSU}(3, q)$ *or* $\mathrm{SZ}(q)$, $q = 2^n$.

Suzuki and Bender [3] generalized this result by showing that no new simple groups are involved if the TI-condition is replaced by a weaker one:

(SB) - there exists a nontrivial partition of $\mathrm{Syl}_2(G) = B_1 \cup B_2$ such that whenever $S_1 \in B_1$ and $S_2 \in B_2$, then $S_1 \cap S_2 = 1$.

The above mentioned collection of simple groups is usually referred to as the simple Bender-groups.

Generalizations of Theorem 1 of a different type appeared in 1971 and 1972. In 1971 Mazurov [9] classified simple groups satisfying condition

CI - all S_2-intersections are cyclic,

and in 1972 Aschbacher [2] classified all simple groups subject to the following condition:

OI - S_2-intersections are of 2-rank at most one.

Their lists were identical, and in addition to simple Bender-groups they also included $\mathrm{PSL}(2, q)$, $q \equiv 3$ or $5 \pmod 8$ and J_1, Janko's smallest simple group. As a matter of fact, all these simple groups could have been classified by the simple condition

WI - S_2-intersections are of order at most two.

Finally, Stroth has informed me that he had classified all simple groups satisfying condition

AI - all S_2-intersections are abelian.

He has shown [11] that the S_2-subgroups of an AI-group are either dihedral or of class two, however he has not specified the groups which could actually occur.

In 1971, while dealing with similar problems, I noticed the following consequence [4] of Shult's far reaching theorem [10]: if G is a non-TI simple group, then every involution in the center of an S_2-subgroup of G (so called central involution) lies in an S_2-intersection. This observation enabled me to prove [5] that the same simple groups as mentioned above occur if one requires only that G satisfies condition

C*I - all central S_2-intersections are cyclic,

where a central S_2-intersection is one containing a central involution. Finally, simple groups satisfying condition

O*I - all central S_2-intersections are of 2-rank one,

were classified by Shult and myself [7]. The list remains the same.

Suppose now that a single maximal (by inclusion) S_2-intersection $V = S_1 \cap S_2$

of the simple group G is of 2-rank 1. By adding the additional assumption that $V \triangleleft S_1$ I was able to classify the groups involved.

THEOREM 2 [6]. *Let G be a simple group and suppose that V is a maximal S_2-intersection of G satisfying the following conditions:*

(i) V *is cyclic, and*

(ii) $|G : N_G(V)|$ *is odd.*

Then G is isomorphic to one of the following groups:

(A) $PSL(2, q)$, $q > 3$ *and* $q \neq 2^n \pm 1$ *where* $n > 2$;

(B) $PSL(3, q)$, $q \equiv -1 \pmod 4$ *and* $q \neq 2^n - 1$ *where* $n \geq 2$;

(C) $PSU(3, q)$, $q \equiv 0$ *or* $1 \pmod 4$ *and* $q \neq 2^n + 1$ *where* $n \geq 2$;

(D) $SZ(q)$ *and*

(E) J_1.

Conversely, all groups mentioned satisfy the assumptions of the theorem.

THEOREM 3. *Replace (i) in Theorem 1 by*

(i') V *is generalized quaternion.*

Then $N_G(V)$ is solvable and an S_2-subgroup of G is either wreathed or quasidihedral.

The simple groups with either wreathed or quasidihedral S_2-subgroups are well known [1]. However I was not able to decide which groups actually occur in Theorem 3, if any.

I would like now to explain in a few lines the basic structure of the proof of Theorem 3. A complete proof will be published elsewhere.

Let $H = N_G(V)$; then H/V is a non-2-closed TI-group. If H is solvable, then by Theorem 1 an S_2-subgroup of H/V is of 2-rank 1. Consequently, an S_2-subgroup of H, hence of G, is of 2-rank 2. It follows by [1] that an S_2-subgroup of G is either wreathed or quasidihedral.

Let now H be nonsolvable and let S be the maximal solvable normal subgroup of H. Then V is the S_2-subgroup of S and H/S is a TI-group. It follows then by Theorem 1 that there exists a subgroup L of H of odd index, such that $S \subset L$ and L/S is isomorphic to a simple Bender-group. Let R denote an S_2-subgroup of L, and hence of G; then R/V is isomorphic to an S_2-subgroup of a simple Bender-group. Analysis of the structure of R yields that $R = VC_R(V)$ and $C_R(V)$ is quaternion. Hence R is of 2-rank 2 and a contradiction is reached by [1].

I would like also to report here on interesting results, which were obtained by my student, Mr Ariel Ish-Shalom. Let $G \in I(k)$ denote that the intersection of any distinct $k+1$ S_2-subgroups of G is trivial. Thus $I(1)$ is the collection of TI-groups. It is easy to see that $G \in I(2)$ implies $G \in I(1)$ and vice-versa. In general, it can be shown that $G \in I(2^n)$ iff $G \in I(2^{n-1})$. Ish-Shalom proved the following two theorems [8]:

THEOREM 4. *Let G be a simple group and suppose that $G \in I(4)$. Then G is isomorphic to one of the following groups: a simple Bender-group,* PSL(2, 11) *or* PSL(2, 13).

THEOREM 5. *Let G be a simple group and suppose that $G \in I(6)$. Then G is isomorphic to one of the following groups: a simple Bender-group, J_1,* PSL(3, 3), *M_{11} and* PSL(2, q) *for $q = 7, 9, 11, 13$ or 19.*

References

[1] J.L. Alperin, R. Brauer and D. Gorenstein, "Finite simple groups of 2-rank two", submitted.

[2] M. Aschbacher, "A class of generalized *TI*-groups", *Illinois J. Math.* 16 (1972), 529-532. MR45#6916.

[3] Helmut Bender, "Finite groups having a strongly embedded subgroup", submitted.

[4] Marcel Herzog, "On 2-Sylow intersections", *Israel J. Math.* 11 (1972), 326-327. MR46#251.

[5] Marcel Herzog, "Simple groups with cyclic central 2-Sylow intersections", *J. Algebra* 25 (1973), 307-312.

[6] Marcel Herzog, "On simple groups with a cyclic maximal 2-Sylow intersection", submitted.

[7] Marcel Herzog and E.E. Shult, "Groups with central 2-Sylow intersections of rank at most one", *Proc. Amer. Math. Soc.* 38 (1973), 465-470.

[8] A. Ish-Shalom, "On 2-Sylow intersections", submitted.

[9] В.Д. Мазуров [V.D. Mazurov], "Конечные простые группы с циклическими пересечениями силовских 2-подгрупп" [Finite simple groups with cyclic intersections of Sylow 2-subgroups], *Algebra i Logika* 10 (1971), 188-198. MR45#5218.

[10] E.E. Shult, "On the fusion of an involution in its centralizer", submitted.

[11] G. Stroth, "Gruppen mit abelschen 2-Sylow Durchschnitten", submitted.

[12] Michio Suzuki, "Finite groups of even order in which Sylow 2-groups are independent", *Ann. of Math.* 80 (1964), 58-77. MR29#145.

Tel-Aviv University, Israel.

PROC. SECOND INTERNAT. CONF. THEORY OF GROUPS,
CANBERRA 1973, p. 366.

RANK FOUR CONFIGURATIONS

D.G. Higman

EDITORIAL NOTE: Professor Higman has not submitted the text of his address for these Proceedings because he feels there is too much overlap with some recently published results.

PROC. SECOND INTERNAT. CONF. THEORY OF GROUPS, CANBERRA 1973, pp. 367-376.

20D05

SOME NONSIMPLICITY CRITERIA FOR FINITE GROUPS

Graham Higman

1. Introduction

If G is a finite group and P a Sylow p-subgroup of G then we can write

(1) $$N_G(P)/O_{p'}(N_G(P)) \simeq H = PT$$

where the p'-group T normalises P and acts faithfully on it. In order to be able to discuss such situations concisely we shall reserve the letter P for a p-group, T for a p'-group of automorphisms of P, and H for the semidirect product PT; and we shall say that H is *realised* in G if (1) holds. We are interested in the question whether H can be realised in a simple group, or perhaps in a perfect group, particularly in cases when P is abelian. (If the answer is yes, one should, of course, go on to ask for a list of the simple groups in which H is realised; but for odd p we do not have such a list even in the simplest cases.) This is a situation in which transfer says something. For instance, the Hall-Wielandt Theorem implies that if the class of P is less than p and $O^p(H) \neq H$ then H cannot be realised in a perfect group. Our theorems at least include cases that go beyond this, that is, where $O^p(H) = H$.

If $p = 2$ and P is abelian, we can answer the question by working back from the known list of simple groups with abelian Sylow 2-subgroups. We can also answer the question if P is cyclic. For T must then be a cyclic group of order n dividing $p - 1$. If $n = 1$, H cannot be realised in a perfect group, by transfer. If $n > 1$, H is realised in $\mathrm{PSL}(n, q)$ for prime powers q in suitable arithmetic progressions. Once beyond these two cases we quickly run into questions to which the answer appears not to be known. For instance, if H is the direct product of two dihedral groups of order 6, H can certainly be realised in a perfect group, but I do not know whether it can be realised in a simple group. Again, if P is elementary abelian of order 49, and T is the cyclic group $\langle t\rangle$ of order 3, where $t^{-1}xt = x^2$ for all x in P, I do not know whether H can be realised in a perfect group.

I want here to mention three theorems, each of some generality, and to prove one of them. The first was proved by S.D. Smith, with the collaboration in part of A.P. Tyrer (see [2, 3, 4]).

THEOREM 1 (Smith-Tyrer). *Suppose that* p *is odd, that* P *has class at most* 2, *and that* T *is cyclic of order* 2. *Then* H *cannot be realised in a perfect group unless* P *is cyclic.*

The second is an unpublished result of D.W. Garland.

THEOREM 2 (Garland). *Suppose that* $p \geq 5$, *that* P *is abelian, and that* T *is the cyclic group* $\langle t \rangle$ *of order* 3. *Then* H *cannot be realised in a perfect group unless for some integer* k, $k \not\equiv 1(p)$, $t^{-1}xt = x^k$ *for all* x *in* P.

This implies in particular that H cannot be realised if $p \not\equiv 1(3)$. The third theorem is the one I propose to prove here.

THEOREM 3. *Suppose that* p *is odd, that* P *is elementary abelian, and that* T *is a cyclic group of odd order acting semiregularly on the nontrivial elements of* P. *Suppose further that* H *has a* p*-fold covering group* $\hat{H} = \hat{P}T$, *where* $\hat{P}$ *is an extraspecial* p*-group. Then* H *cannot be realised in a perfect group.*

This applies, for instance, if P has order p^{n-1} where n is prime, being generated by elements $x_1, \ldots, x_n$ which are permuted cyclically by a generator of T and satisfy $x_1 \ldots x_n = 1$. The common features in the proofs of these theorems are the use of the principal-block methods of R. Brauer, and the application of a theorem of R.G. Swan [5], for which see Section 3 below. This is very much work in progress. It should be possible to prove a theorem generalising both Theorem 2 and Theorem 3, and possibly the abelian case of Theorem 1 as well. In the meantime I make this interim report.

2. Preliminaries. The principal block of G

For the rest of this paper, G will be a group realising H, where H satisfies the hypotheses of Theorem 3. Since factoring out a normal p'-subgroup does not affect the hypotheses we can and will assume that $O_{p'}(G) = 1$. The notations we now introduce will remain in force throughout the paper.

Since P has the extraspecial cover $\hat{P}$, its order is a square. We put

$$|P| = q^2, \quad |T| = n.$$

Since T acts semiregularly on the nontrivial elements of P, n divides $q^2 - 1$, so we write

$$q^2 - 1 = nr.$$

But furthermore, because a generator t of T induces a symplectic transformation on P, n divides either $q+1$ or $q-1$. For suppose that the order of the field of characteristic p generated by the n-th roots of unity is p^m. Because the transformation induced by t is semiregular, each of its characteristic roots is a primitive n-th root of unity, and the degree of its characteristic equation is therefore divisible by m; that is, $q^2 = p^{lm}$ for some integer l. But because the transformation is symplectic, any n-th root ζ has the same multiplicity as a characteristic root as its inverse ζ^{-1}. If ζ and ζ^{-1} are not conjugate, this implies that l is even, $l = 2l'$ say. Then $q = p^{l'm}$, and since n divides $p^m - 1$, n divides $q-1$. If ζ and ζ^{-1} are conjugate m is even, $m = 2m'$ say, and the automorphism of the field of order p^m interchanging ζ and ζ^{-1} is the map $x \to x^{p^{m'}}$. It follows that n divides $p^{m'}+1$; and since $q = p^{lm'}$, n divides $q+1$ if l is odd, and n divides $q-1$ if l is even. Thus in any case we can write

$$\varepsilon q - 1 = ns , \quad \varepsilon = \pm 1 .$$

Observe that, since n is odd, this implies $n \le \frac{1}{2}(q+1)$, and hence $n < r$.

We now consider the principal block of G, following Brauer [1]. Then a *column* over the principal block is a complex valued function on the (ordinary irreducible) characters in the block. The value of the column b at the character χ is written b_χ; and the columns form an inner product space if we set

$$(b, c) = \sum b_\chi \overline{c}_\chi$$

where the sum is over all characters χ in the block. The fact that T acts semiregularly on the nontrivial elements of P implies that for $x \ne 1$ in P, $C_G(x)$ is p-nilpotent. It follows that for any p'-element k in $C_G(x)$, and any character χ in the principal block, $\chi(xk) = \chi(x)$. Moreover, if we denote also by x the column defined by $x_\chi = \chi(x)$, we have $(x, x) = |P|$, and $(x, y) = 0$ if x, y are not conjugate in H, and hence not conjugate in G. To take care of the facts that x is a column of integers in a certain algebraic number field, and that these columns satisfy certain congruences, we introduce, for each non-principal character φ of P, a new column a_φ, by

$$a_{\varphi\chi} = \left(\chi|_P, 1-\varphi\right) = \frac{1}{|P|} \sum_x \chi(x) \overline{(1-\varphi(x))} .$$

Thus a_φ is a column of rational integers. If φ, ψ belong to the same orbit of the natural action of T on char(P), that is, if $\psi(x) = \varphi\left(t^i x t^{-i}\right)$ for some i, then $a_\varphi = a_\psi$, so that we only get r different columns a_φ . Using the orthogonality relations for the characters φ , and the inner products of the columns x , one checks that

$$(a_\varphi, a_\psi) = n + \delta(\varphi, \psi)$$

where $\delta(\varphi, \psi)$ is 1 or 0 according as ψ is or is not in the same orbit as φ under the action of T . Because $n < r$, essentially the only solutions of these equations are those in the following table.

	1	k	a_{φ_1} ... a_{φ_r}
χ_0	1	1	1 ... 1
χ_1	x_1	ξ_1	f_1 ... f_1
	⋮	⋮	⋮
χ_m	x_m	ξ_m	f_m ... f_m
	y	η	-1 ... 0
	⋮	⋮	⋮
	y	η	0 ... -1

Here "essentially" means that possibly some rows should have a change of sign throughout, and that we ignore rows consisting entirely of zeros. We have added to the table names for the first $m + 1$ rows, and columns for $\chi(1)$ and $\chi(k)$, where k is some p-regular element. These are obtained by writing down twice, with different symbols for the unknowns involved, the most general column orthogonal to all the a_φ and satisfying $\chi_0(1) = \chi_0(k) = 1$ (since χ_0 is the principal character). This procedure is correct, since the defining equations for the a_φ can be solved to give

$$\chi(x) = -\sum_\varphi a_{\varphi\chi}\varphi(x)$$

so that the columns a_φ span the same space as the columns x , *i.e.* the space of character values for p-singular elements. We have used the abbreviations

$$y = \sum f_\alpha x_\alpha , \quad \eta = \sum f_\alpha \xi_\alpha ;$$

and the equation

$$n = \sum f_\alpha^2$$

also holds. Here and subsequently, subscripts α, β range from 0 to m .

Recall next that the space spanned by the Brauer characters of modular representations in the principal block is spanned also by the restrictions to p-regular elements of the ordinary characters in the block. As a basis for this space we can clearly take the characters $\chi_0, \chi_1, \ldots, \chi_m$. Relative to this basis we have the decomposition matrix

$$\begin{bmatrix} 1 & 0 & \cdots & 0 \\ 0 & 1 & \cdots & 0 \\ \cdot & \cdot & \cdots & \cdot \\ \cdot & \cdot & \cdots & \cdot \\ \cdot & \cdot & \cdots & \cdot \\ 0 & 0 & \cdots & 1 \\ 1 & f_1 & \cdots & f_m \\ \cdot & \cdot & \cdots & \cdot \\ \cdot & \cdot & \cdots & \cdot \\ \cdot & \cdot & \cdots & \cdot \\ 1 & f_1 & \cdots & f_m \end{bmatrix}$$

and hence the Cartan matrix $[\delta_{\alpha\beta} + r f_\alpha f_\beta]$.

3. An application of a theorem of Swan

It would be perfectly reasonable at this point to try to finish the proof off by counting dihedral subgroups in the manner of Section 5. The reader may verify that what emerges is a single equation which, at least on the face of it, seems to be insufficient for the purpose. We need something extra, and this is provided by the following theorem of Swan [5].

The subgroup X containing the Sylow p-subgroup P of the finite group K is said to *control strong fusion* in P , relative to K , if an isomorphism between two subgroups of P is a restriction of an inner automorphism of K only if it is a restriction of an inner automorphism of X . A *covering group* of K is a group $\hat{K}$ with an epimorphism $\pi : \hat{K} \to K$ such that $\ker(\pi)$ is contained both in $Z(\hat{K})$ and in $\hat{K}'$; a *p-cover* is a covering group such that $\ker(\pi)$ is a p-group. The theorem then says that if X controls strong fusion in the Sylow p-subgroup P relative to K , any p-cover of X can be extended to a p-cover of K . (Swan actually stated the theorem in a less general form, though one quite wide enough for our application;

it is both easy to see and well-known that his argument gives the result I have stated.) All I shall say here about the proof is that anyone who wants to abstract from the homological algebra a self-contained proof may find the following elementary result useful.

LEMMA. *If* $P \leq X \leq K$ *, where* P *is a Sylow p-subgroup of* K *, then* X *controls strong fusion in* P *if and only if it admits a P-invariant transversal.*

It is well-known that if a Sylow p-subgroup P is abelian, then its normaliser controls strong fusion. So $N_G(P)$ controls strong fusion. Now by hypothesis (writing temporarily N for $N_G(P)$) we have the epimorphism $\rho : N \to H$, whose kernel is $O_{p'}(N)$, and the covering map $\pi : \hat{H} \to H$. If we define $\hat{N} = \{(g, h),\ g \in N,\ h \in \hat{H},\ g\rho = h\pi\}$ and define $\sigma : \hat{N} \to N$ by $(g, h)\sigma = g$, then $\hat{N}$ is easily seen to be a p-cover of N . By Swan's Theorem, G has a p-cover $\hat{G}$ such that $\hat{G}$ realises $\hat{H}$.

4. The principal block of $\hat{G}$

Since G is a homomorphic image of $\hat{G}$, a character of G can be lifted to a character of $\hat{G}$, and the lifted character will be in the principal block if the original was. Thus the table constructed in Section 2 can be thought of as part of the principal block table of $\hat{G}$. From this point of view, φ is to be thought of as a non-principal linear character of $\hat{P}$, and the columns a_φ are redefined by

$$a_{\varphi\chi} = \left(\chi|_{\hat{P}},\ 1-\varphi\right) = \frac{1}{|\hat{P}|} \sum_{x \in \hat{P}} \chi(x)\overline{(1-\varphi(x))} \ .$$

This gives the same value as before for characters with $Z(\hat{G})$ in the kernel. For other characters it gives zero. This can be seen directly, for such a character is afforded by a representation in which a generator of $Z(\hat{G})$ is represented by a non-trivial scalar matrix, so that its restriction to $\hat{P}$ can contain no linear character. Alternatively one can argue that, for an element x of $\hat{P}$ not in $Z(\hat{P})$, $C_{\hat{G}}(x)$ is p-nilpotent and has Sylow p-subgroup of order $|P|$, and hence that the inner products of columns x, y , and so also of columns a_φ, a_ψ , are the same in the principal block of $\hat{G}$ as in that of G .

To complete the p-singular part of the principal block table, we have to consider elements whose p-part is in $Z(\hat{G})$. The expression for the character values at such an element involves generalised decomposition numbers indexed by the irreducible modular representations of $\hat{G}$, or by some other basis for the space spanned by their Brauer characters. But such representations have $Z(\hat{G})$ in the kernel and are therefore representations of G . Thus as such a basis we may take the characters χ_α , $\alpha = 0, \ldots, m$ introduced in Section 2. The Cartan matrix of $\hat{G}$

with respect to this basis is p times the Cartan matrix of G, that is, is $[p(\delta_{\alpha\beta}+rf_\alpha f_\beta)]$. Thus if k is a p-regular element of $\hat{G}$, and z a generator of $Z(\hat{G})$, we can write

$$\chi(z^i k) = \sum_\alpha d^{(i)}_{\alpha\chi}\chi_\alpha(k) ,$$

where the columns $d^{(i)}_\alpha$ satisfy

$$\left(d^{(i)}_\alpha, d^{(j)}_\beta\right) = \delta_{ij}p(\delta_{\alpha\beta}+rf_\alpha f_\beta) .$$

Here and subsequently, indices i, j range from 1 to $p-1$. The columns $d^{(i)}_\alpha$ are columns of integers in the field $Q(\omega)$, where ω is a primitive p-th root of unity. Because z and k have coprime orders, $z^j k$ is a power of $z^i k$, and hence $\chi(z^j k)$ is obtained from $\chi(z^i k)$ by applying an automorphism of the field of the characters, which maps ω^i on ω^j, and fixes each $\chi_\alpha(k)$. It follows that we can write

$$d^{(i)}_\alpha = \sum_j b^{(j)}_\alpha \omega^{-ij}$$

where the $b^{(j)}_\alpha$ are columns of rational integers. These equations can be solved to give

$$b^{(i)}_\alpha = \frac{1}{p}\sum_j d^{(j)}_\alpha\left(\omega^{ij}-1\right)$$

whence

$$\left(b^{(i)}_\alpha, b^{(j)}_\beta\right) = \left(1+\delta_{ij}\right)\left(\delta_{\alpha\beta}+rf_\alpha f_\beta\right) .$$

Also, since the a_φ are linear combinations of columns x with x in $\hat{P}$ not in $Z(\hat{P})$, we have

$$\left(a_\varphi, b^{(i)}_\alpha\right) = 0 .$$

There is a generalised character θ of $\hat{P}$ such that $\theta(x) = q$ if x does not belong to $Z(\hat{P})$, $\theta\left(z^j\right) = q\left(1-\omega^{ij}\right)$ and $\theta(1) = 0$. If we write down the condition that $\left(\chi|_{\hat{P}}, \theta\right)$ is an integer, and use the obvious fact that

$$\chi_\alpha(1) \equiv f_\alpha \pmod{q^2}$$

we obtain the congruence

$$\sum_{\varphi} a_{\varphi} \equiv \sum_{\alpha} f_{\alpha} b_{\alpha}^{(i)} \pmod{q} .$$

(The sum on the left is over all characters φ, not over all distinct columns a_{φ}.)

Now we know the values of the columns $b_{\alpha}^{(i)}$ at the characters with $Z(\hat{G})$ in the kernel. By definition, $d_{\alpha\chi_{\beta}}^{(i)} = \delta_{\alpha\beta}$ for all i and hence $b_{\alpha\chi_{\beta}}^{(i)} = -\delta_{\alpha\beta}$; and the values at the other characters of G follow by the orthogonality with a_{φ}. Let $c_{\alpha}^{(i)}$ denote a column which is zero at the characters with $Z(\hat{G})$ in the kernel, and agrees with $b_{\alpha}^{(i)}$ elsewhere. Using the known value of $\left(b_{\alpha}^{(i)}, b_{\beta}^{(j)}\right)$ and the known values of the $b_{\alpha}^{(i)}$ on characters with $Z(\hat{G})$ in the kernel we find

$$\left(c_{\alpha}^{(i)}, c_{\beta}^{(j)}\right) = \delta_{ij}\left(\delta_{\alpha\beta}+rf_{\alpha}f_{\beta}\right) .$$

Also the congruence of the previous paragraph gives

$$\sum_{\alpha} f_{\alpha} c_{\alpha}^{(i)} \equiv 0 \pmod{q} .$$

Thus if we define columns $e_{\alpha}^{(i)}$ by

$$q e_{\alpha}^{(i)} = \sum_{\beta} \left(q\delta_{\alpha\beta}-\varepsilon s f_{\alpha}f_{\beta}\right) c_{\beta}^{(i)}$$

the columns we get are columns of integers. (For ε and s see Section 2.) We calculate

$$\left(e_{\alpha}^{(i)}, e_{\beta}^{(j)}\right) = \delta_{ij}\delta_{\alpha\beta} .$$

Thus there is essentially only one possibility for the columns $e_{\alpha}^{(i)}$, and since the defining equations of the $e_{\alpha}^{(i)}$ can be solved for the $c_{\alpha}^{(i)}$, to give

$$c_{\alpha}^{(i)} = \sum_{\beta} \left(\delta_{\alpha\beta}+s f_{\alpha}f_{\beta}\right) e_{\beta}^{(i)}$$

there is essentially only one possibility for the $c_{\alpha}^{(i)}$. That is, once the principal block of G is known, and hence the integers f_{α}, there is essentially only one possibility for the principal block of $\hat{G}$.

This can be written down in the following table.

	1	k	a_φ	$b_\beta^{(j)}$
χ_α	x_α	ξ_α	f_α	$-\delta_{\alpha\beta}$
ψ_θ	y	η	$-\delta_{\theta\varphi}$	$-f_\beta$
$\zeta_\alpha^{(i)}$	$x_\alpha+sf_\alpha y$	$\xi_\alpha+sf_\alpha\eta$	0	$\delta_{ij}(\delta_{\alpha\beta}+sf_\alpha f_\beta)$

Here the indices α, β range from 0 to m, the indices θ, φ over a set of representatives of the orbits of the natural action of T on the non-principal characters of P, and the indices i, j from 1 to $p-1$. Thus the block contains $m+1$ characters χ_α, of which χ_0 is the principal character, r characters ψ_θ, and $(m+1)(p-1)$ characters $\zeta_\alpha^{(i)}$. The table gives the value at each of these characters of each of the columns a_φ and $b_\beta^{(j)}$, in terms of the integers f_α, of which $f_0 = 1$. It gives also expressions for the degrees of the characters, in terms of unknown integers x_α, with $x_0 = 1$, using the abbreviation $y = \sum f_\alpha x_\alpha$, and for the value of each character at a general p-regular element k, in terms of unknowns ξ_α, with $\xi_0 = 1$, using the abbreviation $\eta = \sum f_\alpha \xi_\alpha$. These columns are simply the most general columns orthogonal to all a_φ and all $b_\beta^{(j)}$, and taking the value 1 at the principal character.

5. Counting dihedral groups

We shall complete the proof by deriving a contradiction from the assumption that G has even order: the Feit-Thompson Theorem then shows that G is not perfect.

If G has even order so has $\hat{G}$, and we can take k in the above table to be an involution. Now the Sylow p-subgroup P of G is abelian and $N_G(P)/C_G(P)$ is of odd order. Thus no element of P (except the identity) is conjugate to its inverse, in $N_G(P)$ or in G. Since the kernel of the covering map $\pi : \hat{G} \to G$ is in the centre, the same is true of $\hat{P}$. Thus no p-singular element of $\hat{G}$ is conjugate to its inverse. In particular

$$\#(k^\circ k^\circ = x) = 0$$

for all p-singular elements x of $\hat{G}$. This implies that the column $\chi^2(k)/\chi(1)$ over the principal block is orthogonal to each column $\chi(x)$, and hence to each column a_φ and $b_\beta^{(j)}$, since these span the same space. It follows that it can be obtained from the column $\chi(k)$ by replacing throughout ξ_α by $\xi_\alpha = \xi_\alpha^2/x_\alpha$, and hence η by $\eta^1 = \sum_\alpha f_\alpha \xi_\alpha^1$. This implies in particular that $\eta^2/y = \eta^1$ and $(1+s\eta)^2/(1+sy) = 1 + s\eta^1$, whence

$$(1+s\eta)^2/(1+sy) = 1 + s\eta^2/y$$

which implies $y = \eta$, and so $\eta^1 = y$. But also $(\xi_\alpha + sf_\alpha\eta)^2/(x_\alpha + sf_\alpha y) = \xi_\alpha^1 + sf_\alpha\eta^1$, whence $x_\alpha = \xi_\alpha = \xi_\alpha^1$, since $f_\alpha \neq 0$. But then we have $y = \eta$ and $x_\alpha = \xi_\alpha$ for all α, so that k is in the kernel of every representation in the principal block, and so is in $O_{p'}(\hat{G})$. But we are assuming $O_{p'}(G) = 1$, whence $O_{p'}(\hat{G}) = 1$, so this is the desired contradiction.

References

[1] Richard Brauer, "Some applications of the theory of blocks of characters of finite groups III", *J. Algebra* 3 (1966), 225-255. MR34#2716.

[2] Stephen D. Smith and A.P. Tyrer, "On finite groups with a certain Sylow normalizer. I", *J. Algebra* 26 (1973), 343-365.

[3] Stephen D. Smith and A.P. Tyrer, "On finite groups with a certain Sylow normalizer. II", *J. Algebra* 26 (1973), 366-367.

[4] Stephen D. Smith, "On finite groups with a certain Sylow normalizer. III", submitted.

[5] Richard G. Swan, "The p-period of a finite group", *Illinois J. Math.* 4 (1960), 341-346. MR23#A188.

The University of Oxford,
Oxford OX1 3LB, England.

PROC. SECOND INTERNAT. CONF. THEORY OF GROUPS, CANBERRA 1973, pp. 377-387. 22A99

FREE SUBGROUPS OF FREE TOPOLOGICAL GROUPS

David C. Hunt and Sidney A. Morris

SUMMARY

It is well known that every subgroup of a free group is a free group. However, it is not true in general that a subgroup of a free topological group is a free topological group.

It is shown that if X is the open unit interval $(0, 1)$ and Y is the closed unit interval $[0, 1]$ then the subgroup of the free topological group on Y generated by X is not a free topological group. On the other hand it is shown here that the commutator subgroup of the free topological group on Y is a free topological group.

1. Introduction

If G is a topological group then $|G|$ denotes the underlying abstract group and e always denotes the identity.

DEFINITION. If X is a topological space, with distinguished point e, the topological group $FG(X)$ is said to be the (Graev) *free topological group on* X *if*:

(a) $|FG(X)|$ is a free group with free basis $X\backslash\{e\}$ and identity e, and

(b) the topology of $FG(X)$ is the *finest* group topology on $|FG(X)|$ which induces the given topology on X.

NOTE. (i) Condition (b) can be replaced by (b'): for any continuous map γ of X into any topological group G such that $\gamma(e)$ is the identity element of G there exists a unique continuous homomorphism $\Gamma : FG(X) \to G$ such that $\Gamma|X = \gamma$.

(ii) $FG(X)$ is independent of the choice of the distinguished point e in X.

(iii) We define the free abelian topological group on X, $AG(X)$, by replacing "group" by "abelian group" in the above definition.

It is well known that every subgroup of a free group is a free group. However, it is not true in general that a subgroup of a free topological group is a free topological group. If X is compact and Y is a compact subspace of X, then the subgroup $gp(Y)$ of $FG(X)$ generated algebraically by Y is the free topological group on Y ; (Graev [2]). A recent as yet unpublished result of R. Brown and L. Hardy is that if X is compact then any *open* subgroup of $FG(X)$ is a free topological group. However, Graev showed that a *closed* subgroup of a free topological group need not be a free topological group. Another example is:

EXAMPLE. $gp(0, 1) \subset FG[0, 1]$. Let $b_i = \frac{1}{\pi i}$ and $a_i = \frac{1}{(i+\frac{1}{2})\pi}$, $i = 1, 2, \ldots$, and let $e = \frac{1}{\pi}$. Now $\gamma(x) = \sin\left(\frac{1}{x}\right)$ is a continuous function from $(0, 1)$ to R such that $\gamma(e) = 0$, $\gamma(b_i) = 0$ and $\gamma(a_i) = 1$ for all i . We claim that γ cannot be extended to a continuous homomorphism $\Gamma : gp(0, 1) \to R$ and hence (by condition (b') above) $gp(0, 1) \neq FG(0, 1)$. Suppose that Γ exists, then, since Γ is continuous and $a_i b_i^{-1}$ approaches e and $\Gamma(e) = 0$, $\Gamma\left(a_i b_i^{-1}\right)$ approaches 0 . However, $\Gamma(a_i)\Gamma\left(b_i^{-1}\right) = 1$, so Γ is not a homomorphism.

This does *not* show that $gp(0, 1)$ is not a free topological group, merely that it is not $FG(0, 1)$. In §3 we develop sufficient theory to prove this new result. In §2 we prove that the commutator subgroup of $FG[0, 1]$ is a free topological group.

We complete this section by giving some definitions, notation and preliminary results. Elementary properties of topological groups are assumed and can be found in Bourbaki [1].

DEFINITION. A Hausdorff topological space X is said to be a k_ω*-space* if $X = \bigcup_{n=1}^{\infty} X_n$ where

(i) X_n is compact for all n ,

(ii) $X_n \subseteq X_{n+1}$, for all n ,

(iii) a subset A of X is closed in X if and only if $A \cap X_n$ is compact for each n .

When we say $X = \cup X_n$ is a k_ω-decomposition we mean that the X_n have the properties (i), (ii) and (iii).

As examples of k_ω-spaces we have any compact space and any connected locally compact group. Note that $(0, 1)$ is homeomorphic to the topological group R , of real numbers, and hence $(0, 1) = \cup\left[\frac{1}{n}, 1 - \frac{1}{n}\right]$ is a k_ω-decomposition.

We will use the following properties of k_ω-spaces, (for further comments see

[5]):

(i) a closed subspace of a k_ω-space is a k_ω-space;

(ii) if Y is a compact subset of the k_ω-space $X = \cup X_n$, then there exists n such that $Y \subseteq X_n$;

(iii) if $X = \cup X_n$ is a k_ω-decomposition and $Y_1, Y_2, \ldots$ is a sequence of compact subsets of X such that each X_n is contained in some Y_m then $X = \cup Y_m$ is a k_ω-decomposition;

(iv) if G is a topological group and a k_ω-space then the X_n can be chosen such that $G = \cup X_n$ is a k_ω-decomposition and $X_n X_m \subseteq X_{n+m}$ for all n, m.

If G is a group and X is a subset of G, then $\text{gp}(X)$ denotes the subgroup of G generated algebraically by X. Further $\text{gp}_n(X)$ denotes the set of words in $\text{gp}(X)$ of length less than or equal to n with respect to X.

THEOREM A [5]. *If* $X = \cup X_n$ *is a* k_ω-*space then* $FG(X)$ *is a* k_ω-*space with decomposition* $FG(X) = \cup \text{gp}_n(X_n)$.

THEOREM B [5]. *Let* $X = \cup X_n$ *be a* k_ω-*space. Let* $Y \subset FG(X)$ *be a subset such that* $Y \backslash \{e\}$ *is a free algebraic basis for* $\text{gp}(Y)$. *Suppose* $Y_1, Y_2, \ldots$ *is a sequence of compact subsets of* Y *such that* $Y = \cup Y_n$ *is a* k_ω-*decomposition of* Y *inducing the same topology on* Y *that* Y *inherits as a subset of* $FG(X)$. *Put* $X^n = \text{gp}_n(X_n)$ *and* $Y^n = \text{gp}_n(Y_n)$. *If for each natural number* n *there is an* m *such that* $\text{gp}(Y) \cap X^n \subseteq Y^m$, *then* $\text{gp}(Y)$ *is the free topological group on* Y *and both* $\text{gp}(Y)$ *and* Y *are closed subsets of* $FG(X)$.

We note that the family $\{F_\alpha : \alpha \in I\}$ of sets is said to have the *finite intersection property* if for each $\alpha_1, \ldots, \alpha_n \in I$, $F_{\alpha_1} \cap \ldots \cap F_{\alpha_n} \neq \emptyset$. If X is a subset of a topological space, $\bar{X}$ denotes the closure of X.

DEFINITION. A *filterbase* on a set X is a set $\underline{F}$ of subsets of X such that

(i) every finite intersection of sets in $\underline{F}$ is a member of $\underline{F}$;

(ii) $\emptyset \notin \underline{F}$ and $\underline{F}$ is non-empty.

DEFINITION. Let X be a topological space and $\underline{F}$ a filterbase on X then $\underline{F}$ is said to be *convergent* if there exists x in X such that every neighbourhood of x contains an element of $\underline{F}$.

DEFINITION. A filterbase $\underline{F}$ on a topological group G is said to be a *Cauchy filterbase* (in the left uniformity) if for each neighbourhood U of e there exists F in $\underline{F}$ such that $F^{-1}F \subseteq U$.

DEFINITION. A topological group is said to be *complete* if every Cauchy filterbase is convergent.

PROPOSITION [1]. *If* G *is a topological group and* H *is a subgroup of* G *which is complete then* H *is closed in* G .

2. The commutator subgroup

THEOREM 1. *The commutator subgroup of* $FG([0, 1])$ *is a free topological group.*

PROOF. Let F' be the commutator subgroup. Then F' has a free basis equal to $Y\backslash\{e\}$, where

$$Y = \left\{x_1^{\varepsilon_1} \dots x_n^{\varepsilon_n} x_i x_n^{-\varepsilon_n} \dots x_i^{-\varepsilon_i - 1} \dots x_1^{-\varepsilon_1} \mid \varepsilon_i \text{ integers, } x_i \in [0, 1] ,\right.$$
$$\left. i = 1, \dots, n \text{ and } x_1 \le x_2 \le \dots \le x_n \right\} ; \text{ (see [4]).}$$

We will show that F' is the free topological group on Y .

Put $Y_m = Y \cap \mathrm{gp}_m[0, 1]$ and $Y^m = \mathrm{gp}_m(Y_m)$. By Theorem B of §1 it suffices to show

(a) for each n , there exists m such that $\mathrm{gp}(Y) \cap \mathrm{gp}_n[0, 1] \subseteq Y^m$, and

(b) Y_m is compact.

Let $F\{x_1, \dots, x_n\}$ be the algebraic free group on $x_1, \dots, x_n$. Define

$$Z = \left\{x_1^{\varepsilon_1} \dots x_n^{\varepsilon_n} x_i x_n^{-\varepsilon_n} \dots x_i^{-\varepsilon_i - 1} \dots x_1^{-\varepsilon_1} \mid \varepsilon_i \text{ integers, } x_i \in [0, 1] ,\right.$$
$$\left. i = 1, \dots, n \right\} .$$

Clearly $\mathrm{gp}(Z) \cap \mathrm{gp}_n\{x_1, \dots, x_n\}$ is finite and therefore is contained in $\mathrm{gp}_m(Z)$ for some m .

Now let $w \in \mathrm{gp}(Y) \cap \mathrm{gp}_n[0, 1]$.

Then there exists $x_1, x_2, \dots, x_n$ in $[0, 1]$, $x_i \ne e$, $i = 1, 2, \dots, n$, and $x_1 \le x_2 \le \dots \le x_n$ such that $w \in \mathrm{gp}_n\{x_1, \dots, x_n\}$. By the comment in the above paragraph there exists m such that $w \in \mathrm{gp}_m(Y)$. Since m is independent of

$x_1, \ldots, x_n$ we have (a) is true. Y_m is a *finite* union of sets

$$Y_{n,\varepsilon} = \left\{x_1^{\varepsilon_1} \ldots x_n^{\varepsilon_n} x_i x_n^{-\varepsilon_n} \ldots x_i^{-\varepsilon_i - 1} \ldots x_1^{-\varepsilon_1} \text{ where } x_i \in [0, 1] , \right.$$
$$\left. x_1 \le x_2 \le \ldots \le x_n \text{ for a fixed } n \text{ and fixed } \varepsilon = (\varepsilon_1, \ldots, \varepsilon_n)\right\} .$$

To prove (b) it suffices to show each $Y_{n,\varepsilon}$ is compact. Now $Y_{n,\varepsilon}$ is a continuous image of the set

$$K_n = \{(x_1, x_2, \ldots, x_n) \mid x_1 \le x_2 \le \ldots \le x_n\}$$
$$\subseteq [0, 1] \times [0, 1] \times \ldots \times [0, 1] \quad (n \text{ copies of } [0, 1])$$

under the obvious map f_ε. Clearly each K_n is compact. Hence $Y_{n,\varepsilon}$ is compact as required.

REMARK 1. Theorem 1 is also true if $[0, 1]$ is replaced by any totally ordered topological space which is k_ω in the order topology.

REMARK 2. On the other hand F. Clarke in a private communication stated that the commutator subgroup of $FG(S^n)$ is not a free topological group where S^n is the n sphere.

REMARK 3 . Theorem 1 is also true if the commutator subgroup is replaced by any verbal subgroup V_m containing the commutator subgroup. Of course V_m is the verbal subgroup defined by the words $[x, y]$ and x^m . The only difference in the proof is that Y is replaced by

$$Y(m) = \left\{x_1^{\varepsilon_1} \ldots x_n^{\varepsilon_n} x_i x_n^{-\varepsilon_n} \ldots x_i^{-\varepsilon_i - 1 + \delta} \ldots x_1^{-\varepsilon_1} \mid 0 \le \varepsilon_i \le m , \ x_i \in [0, 1] , \right.$$
$$i = 1, \ldots, n, \ x_1 \le x_2 \le \ldots \le x_n \text{ and } \delta = m \text{ if } \varepsilon_i = m - 1 \text{ and}$$
$$\left. \delta = 0 \text{ otherwise}\right\} .$$

This suggests the natural question:

QUESTION 1. *Which verbal subgroups and which characteristic subgroups of* $FG[0, 1]$ *are free topological groups?*

3. Completeness of k_ω-groups and its consequences

The following theorem generalizes Theorem 6 of Graev [2]. The proof given is essentially the same as Graev's.

THEOREM 2. *If* G *is a* k_ω *group then it is complete.*

PROOF. Let $G = \cup X_n$ be a k_ω-space decomposition of G such that $X_n X_m \subseteq X_{n+m}$

and $e \in X_1$.

Suppose G is not complete. Then there exists a Cauchy filterbase $\underline{\underline{F}}'$ in G which does not converge. Put $\underline{\underline{F}} = \{F_\alpha \mid \alpha \in I\}$ where each F_α is a finite intersection of closures of elements of $\underline{\underline{F}}'$. Clearly $\underline{\underline{F}}$ is a Cauchy filterbase which does not converge. Now $\bigcap_{\alpha \in I} F_\alpha = \emptyset$. For if $x \in \bigcap_\alpha F_\alpha$, let U be a neighbourhood of x then $x^{-1}U$ is a neighbourhood of e and $x^{-1}U \supseteq F_\alpha^{-1}F_\alpha \supseteq x^{-1}F_\alpha$, for some α , and hence $U \supseteq F_\alpha$ and the filterbase converges to x , a contradiction.

If $\{X_n \cap F_\alpha \mid \alpha \in I\}$ had the finite intersection property then, since X_n is compact, $\bigcap_\alpha (X_n \cap F_\alpha) \neq \emptyset$ which is false. Hence $\{X_n \cap F_\alpha \mid \alpha \in I\}$ does not have the finite intersection property. Thus for all $n \geq 2$ there exists $\alpha_n \in I$ such that $X_{n-1} \cap F_{\alpha_n} = \emptyset$.

We shall construct a sequence of sets U_i $(i = 1, 2, \ldots)$ in the group G with the following properties:

(a) $e \in U_1$;

(b) $U_i \subseteq X_i$ and U_i is open in X_i ;

(c) $U_j \subseteq U_i$ if $j \leq i$;

(d) $X_j\bar{U}_i \cap F_{\alpha_{2j}} = \emptyset$ if $j \leq i$.

Before continuing the proof we need a lemma.

LEMMA. *If M is a compact subset of G and N is a closed subset of G and $M \cap N = \emptyset$ then there exists an open neighbourhood U of e such that $M\bar{U} \cap N = \emptyset$.*

PROOF. For each $m \in M$ there exists an open neighbourhood $V(m)$ of e such that $mV^2(m) \cap N = \emptyset$. Since M is compact and $M \subseteq \bigcup_{m \in M} mV(m)$ then

$$M \subseteq \bigcup_{i=1}^{n} m_i V(m_i) , \quad m_i \in M .$$

Put $V = \bigcap_{i=1}^{n} V(m_i)$. Then $MV \cap N = \emptyset$. Using the regularity of a topological group we can see that there exists an open neighbourhood U of e such that $\bar{U} \subset V$ and hence $M\bar{U} \cap N = \emptyset$, as required.

Taking $M = X_1$ and $N = F_{\alpha_2}$ the lemma gives an open neighbourhood $U^{(1)}$ of e

such that $X_1\overline{U^{(1)}} \cap F_{\alpha_2} = \emptyset$. Put $U_1 = U^{(1)} \cap X_1$. Then (a), (b), (c), (d) are satisfied for U_1. Suppose $U_1, \ldots, U_n$ have been constructed with the required properties. We proceed to construct U_{n+1}.

By (d), $X_j\bar{U}_n \cap F_{\alpha_{2j}} = \emptyset$, $j = 1, \ldots, n$. Furthermore it is evident that $X_{n+1}\bar{U}_n \cap F_{\alpha_{2(n+1)}} = \emptyset$ as $X_{n+1}\bar{U}_n \subseteq X_{n+1}X_n \subseteq X_{2n+1}$.

Since $X_j\bar{U}_n$ is compact for $j = 1, \ldots, n+1$ there exists an open neighbourhood V of e such that $X_j\bar{U}_n\bar{V}_n \cap F_{\alpha_{2j}} = \emptyset$, $j = 1, \ldots, n+1$. Put $U_{n+1} = U_n.V \cap X_{n+1}$. Clearly U_{n+1} satisfies conditions (a), (b), (c) and since $U_{n+1} \subseteq \bar{U}_n.\bar{V}$ we have (d) for U_{n+1}.

Now set $U = \bigcup_{i=1}^{\infty} U_i$. Since G is a k_ω-space U is open in G. From condition (d), $X_jU \cap F_{\alpha_{2j}} = \emptyset$ for all j.

Since $\{F_\alpha\}$ is a Cauchy filterbase and U is an open neighbourhood of e, there exists F_α such that $F_\alpha^{-1}F_\alpha \subseteq U$. Let $x \in F_\alpha$ so $F_\alpha \subseteq xU$. Now $x \in X_n$ for some n and therefore $xU \cap F_{\alpha_{2n}} = \emptyset$ and $F_\alpha \cap F_{\alpha_{2n}} = \emptyset$, a contradiction. The proof is complete.

Since a free topological group on a k_ω-space is a k_ω-space (Theorem A) we have

COROLLARY 1. *If X is a k_ω-space then $FG(X)$ is a complete topological group.*

As an interesting application of Theorem 2 we prove the converse of Theorem B.

THEOREM 3. *Let $X = \cup X_n$ be a k_ω-space and $Y = \cup Y_m$ be a k_ω-space $\subseteq FG(X)$. Then* $\text{gp}(Y) = FG(Y)$ *if and only if for each n there exists an m such that* $\text{gp}(Y) \cap X^n \subseteq Y^m$ *and $Y \backslash \{e\}$ is a free algebraic basis for* $\text{gp}(Y)$.

PROOF. Suppose that there exists n such that for no m is $\text{gp}(Y) \cap X^n \subseteq Y^m$. Then there exists a sequence $\{a_m\}$ of elements of $\text{gp}(Y) \cap X^n$ such that $a_m \notin Y^m$, any m. Let $S = \{a_n\}$. Then $S \cap Y^m$ is compact (as it is finite) for each m. By Theorem A, $FG(Y) = \cup Y^m$ is a k_ω-decomposition. So S is closed in $FG(Y)$. Also $S - \{a_n\}$ for any n is closed in $FG(Y)$ and hence in S. Thus S has the discrete topology. Since $FG(Y)$ is a k_ω-space it is complete and therefore closed in $FG(X)$. So S is a closed infinite discrete subset of X^n, which is impossible as X^n is compact. The converse is Theorem B.

The following Corollary extends Theorem 7 of Graev [2].

COROLLARY 2. *Let* $X = \cup X_n$ *and* $Y = \cup Y_n$ *be* k_ω*-spaces and suppose* $FG(X) = FG(Y)$. *If each* X_n *is metric, then each* Y_n *is metric.*

We will need some information on the following question:

Is the closure of a free topological group necessarily a free topological group? (More precisely if $FG(X)$ *is a subgroup of a topological group* G, *is* $\overline{FG(X)}$ *a free topological group? Also is* $\mathrm{gp}(\bar{X})$ *a free topological group?)* As a technical proposition in this direction we have:

PROPOSITION 1. *If* $X \subset G$ *and* $\mathrm{gp}(X) = FG(X)$ *then* $\mathrm{gp}(\bar{X})$ *has the property that if* H *is any topological group with a completion and* ϕ *is any continuous map* $\bar{X} \to H$ *such that* $\phi(e) = e$ *then there exists a continuous homomorphism* $\Gamma : \mathrm{gp}(\bar{X}) \to H$ *such that* $\Gamma|_{\bar{X}} = \phi$.

PROOF. Firstly $\mathrm{gp}(X)$ is dense in $\mathrm{gp}(\bar{X})$. For if $a = x_1^{\varepsilon_1} \ldots x_n^{\varepsilon_n} \in \mathrm{gp}(\bar{X})$ and if U is a neighbourhood of a then there exist neighbourhoods $V_1, \ldots, V_n$ of $x_1, \ldots, x_n$ in G such that $V_1^{\varepsilon_1} V_2^{\varepsilon_2} \ldots V_n^{\varepsilon_n} \subseteq U$ and there exist elements $z_i \in V_i \cap X$ such that $z_1^{\varepsilon_1} z_2^{\varepsilon_2} \ldots z_n^{\varepsilon_n} \in U$.

Now $\phi|_X$ is continuous : $X \to H$ and $\phi(e) = e$. Hence there exists a continuous homomorphism $\Phi : FG(X) = \mathrm{gp}(X) \to H$ such that $\Phi|_X = \phi$. Let the topological group $\hat{H}$ be the completion of H. Since $FG(X)$ is dense in $\mathrm{gp}(\bar{X})$, there exists a continuous homomorphism $\Gamma : \mathrm{gp}(\bar{X}) \to \hat{H}$ such that $\Gamma|FG(X) = \Phi$, [1, page 246, Proposition 5]. Clearly $\Gamma|\bar{X} = \phi$. But $\Gamma(\bar{X}) \subseteq H$ and therefore $\Gamma\{\mathrm{gp}(X)\} \subseteq H$.

As every abelian topological group has a completion we have:

COROLLARY 3. *If* $X \subseteq G$ *and* $\mathrm{gp}(X) = AG(X)$ *then* $\mathrm{gp}(\bar{X}) = AG(\bar{X})$.

COROLLARY 4. *If* $X \subset FG(Y)$, *where* Y *is a* k_ω*-space, and* $\mathrm{gp}(X) = FG(X)$ *then* $\mathrm{gp}(\bar{X}) = FG(\bar{X})$.

PROOF. Firstly note that by Corollary 1, $FG(\bar{X})$ is a complete topological group. Consider the identity $\phi : \bar{X}\ (\subseteq \mathrm{gp}(\bar{X})) \to \bar{X}\ (\subseteq FG(\bar{X}))$. By Proposition 1, there exists a continuous homomorphism $\Gamma : \mathrm{gp}(\bar{X})\ (\subseteq G) \to FG(\bar{X})$ such that $\Gamma|\bar{X} = \phi$.

Since $|FG(\bar{X})|$ is the free group on $\bar{X}\backslash\{e\}$, $|\mathrm{gp}(\bar{X})|$ must be the free group on $\bar{X}\backslash\{e\}$ and Γ must be the identity map. Since Γ is continuous and $FG(\bar{X})$ has the finest group topology with respect to $\bar{X}$, Γ must be a homeomorphism; that is, $\mathrm{gp}(\bar{X}) = FG(\bar{X})$.

NOTE. Y is a k_ω-space would be an unnecessary restriction if it could be shown that every free topological group has a completion.

QUESTION 2. *Does every free topological group have a completion?*

As an extension of the Example in §1 we show:

THEOREM 4. *If* X *is metric and* $Y \subset X \subset FG(X)$ *then* $\mathrm{gp}(Y) = FG(Y)$ *implies* Y *is closed in* X.

PROOF. Suppose not and let a belong to $\bar{Y}\backslash Y$. Let $\{a_n\}$, $\{b_n\}$ be sequences in Y such that

(i) $a_n \to a$,

(ii) $b_n \to a$,

(iii) $a_n \neq b_m$ for all n and m,

(iv) $b_n \neq e$, $a_n \neq e$ for all n.

Put $S = \{a_n \mid n = 1, \ldots\} \cup \{b_n \mid n = 1, \ldots\} \cup \{e\}$. Define a continuous function $\phi : S \to R$ by

$$\phi(a_n) = 1 \quad \text{for all } n,$$

$$\phi(b_n) = 0 \quad \text{for all } n,$$

$$\phi(e) = 0 .$$

Since Y is normal and S is closed in Y by the Tietze Extension Theorem, ϕ can be extended to a countinuous map $\phi : X \to R$.

However ϕ cannot be extended to a continuous homomorphism $\Phi : \mathrm{gp}(Y) \to R$ as $a_n b_n^{-1} \to e$, implies $\Phi\left(a_n b_n^{-1}\right)$ would converge to 0 while $\Phi(a_n)\Phi\left(b_n^{-1}\right) = 1$.

Hence $Y = \bar{Y}$ and Y is closed in X.

THEOREM 5. *The subgroup* $\mathrm{gp}(0, 1)$ *of* $FG[0, 1]$ *is not a free topological group.*

PROOF. Suppose $\mathrm{gp}(0, 1) = FG(X)$ for some topological space X. Then by Corollary 4, $\mathrm{gp}(\bar{X}) = FG(\bar{X})$. By Theorem A, $FG[0, 1]$ is a k_ω-group and therefore the closed subset $\bar{X}$ is a k_ω-space. Therefore by Corollary 1, $FG(\bar{X})$ is complete and hence closed in $FG[0, 1]$. However $FG(\bar{X}) \supseteq \mathrm{gp}(0, 1)$ and $\mathrm{gp}(0, 1)$ is dense in $FG[0, 1]$. Therefore $FG(\bar{X}) = FG[0, 1]$. By Theorem 7 of Graev [2], $\bar{X}$ is compact and metrizable. Applying Theorem 4 with Y and X replaced by X and $\bar{X}$ respectively yields X is closed in $\bar{X}$; that is, $X = \bar{X}$. Hence $\mathrm{gp}(0, 1) = FG(X)$ and X is compact.

Therefore by Thoerem A, $\mathrm{gp}(0, 1)$ is a k_ω-group and hence by Theorem 2 is complete. Therefore $\mathrm{gp}(0, 1)$ is closed in $FG[0, 1]$, which is clearly false.

REMARK. The above theorem holds with $[0, 1]$ replaced by any compact metric space and $(0, 1)$ replaced by any dense subset.

QUESTION 3. *If* $Y \subset FG(X)$ *and* $\mathrm{gp}(Y) = FG(Y)$, *does this imply that*

(i) Y *is closed in* $FG(X)$ *?*

(ii) $FG(Y)$ *is closed in* $FG(X)$ *?*

(Note: *(ii)* implies *(i)*.)

Theorem 6 provides an answer in a special case.

THEOREM 6. *If* $X = \cup X_n$ *is metric and a* k_ω*-space and* $Y \subset X^n \subset FG(X)$ *then* $\mathrm{gp}(Y) = FG(Y)$ *implies* Y *is closed in* $FG(X)$. *Consequently* Y *is a* k_ω*-space and* $FG(Y)$ *is closed in* $FG(X)$.

PROOF. By Corollary 4, $\mathrm{gp}(\bar{Y}) = FG(\bar{Y})$. Noting that $\bar{Y} \subset X^n$ and X^n is metric we see that $\bar{Y}$ is metric. So Y is metric and $Y \subset \bar{Y} \subset FG(\bar{Y})$ and $\mathrm{gp}(Y) = FG(Y)$, hence by Theorem 4, Y is closed in $\bar{Y}$, that is, $Y = \bar{Y}$ as required.

ACKNOWLEDGEMENTS. The authors wish to thank both Dr M.F. Newman and Dr E.T. Ordman for helpful conversations. In particular they thank Dr Ordman for suggesting Theorem 4.

References

[1] Nicolas Bourbaki, *Elements of Mathematics: General topology*, Part I (Hermann, Paris, 1966). MR34#5044a.

[2] М.И. Граев [M.I. Graev], "Слободные топологические группы" [Free topological groups], *Izv. Akad. Nauk SSSR Ser. Mat.* 12 (1948), 279-324; *Amer. Math. Soc. Transl.* 35 (1951); reprinted *Amer. Math. Soc. Transl.* (1) 8 (1962), 305-364. MR10,11 (Russian); MR12,391 (Translation).

[3] John L. Kelley, *General topology* (Van Nostrand, New York, Toronto, London, 1969).

[4] A.G. Kuroš, *Theory of groups* (3rd augmented edition. Izdat "Nauka", Moscow, 1967). MR40#2740.

[5] J. Mack, Sidney A. Morris and E.T. Ordman, "Free topological groups and the projective dimension of a locally compact abelian group", *Proc. Amer. Math. Soc.* 40 (1973), 303-308.

University of New South Wales,
Kensington, NSW 2033.

PROC. SECOND INTERNAT. CONF. THEORY OF GROUPS 20E15
CANBERRA, 1973, pp. 389-394.

SOME QUESTIONS IN THE THEORY OF SOLUBLE GROUPS

M.I. Kargapolov

The aim of this lecture is to describe some recent results obtained by algebraists at Novosibirsk in the area of soluble groups.

1. Residual and algorithmic properties

Let G be a group and ρ a relation between elements and sets of elements defined on G and on all its homomorphic images. Let K be a class of groups. We say that G is residually in K relative to ρ if for every element and set of elements of G that are not related under ρ there exists a homomorphism of G onto a group in K such that the images of the element and the set of elements are also not related under ρ.

A special case is that of residual finiteness. Residual finiteness relative to ρ is denoted by $RF\rho$. If ρ runs over the predicates identity, conjugacy, inclusion in a subgroup, inclusion in a finitely generated subgroup, then the corresponding properties are denoted by RF, RFC, RFI, RFI_ω.

Mal'cev [1958] noted the connection between the residual finiteness of groups relative to identity, conjugacy, inclusion and the corresponding algorithmic problems. It turned out that the word (conjugacy, inclusion) problems are soluble in a finitely related group of a variety given by a single law if the group has the property RF $\left(RFC, RFI_\omega\right)$.

The following theorem of Remeslennikov [6] seems to be one of the more interesting results on residually finite groups.

THEOREM 1. *Every polycyclic group is residually finite relative to conjugacy*, that is, it is an RFC-group.*

By the same token, the conjugacy problem is soluble in the class of polycyclic groups. The fact that finitely generated nilpotent groups have property RFC was first proved by Blackburn, and Kargapolov did the same for supersoluble groups.

---

* Conjugacy-separable is more normal English usage.

In addition to the use of residual properties for a positive solution of algorithmic problems in group theory, there exist a number of other methods.

In [1] it was proved that the conjugacy problem has a positive solution in a free soluble group. The method of proof is based on a use of the Magnus embedding of a free soluble group of derived length n in the wreath product of a free abelian group by a free soluble group of derived length $n-1$.

It was proved in [8] that free soluble groups have property *RFC*. A theorem in [1] in a different direction has been generalized by Sarkisjan [12]. He proved the following:

THEOREM 2. *The conjugacy problem has a positive solution in a free polynilpotent group.*

The following question arises naturally in connection with Hall's Theorem on the residual finiteness of finitely generated metabelian groups: does every finitely generated metabelian group have the property *RFC* ? The answer is given by:

THEOREM 3 (see [2]). *There exists a finitely generated metabelian group which is not an RFC-group.*

The problems connected with residual properties and algorithmic questions for soluble groups are far from exhausted. Not even the following fundamental problem has been solved:

PROBLEM 1. *Does there exist an algorithm for solving the word problem in the class of soluble groups of derived length* $n \geq 3$ *?*

PROBLEM 2. *Does there exist an algorithm for solving the conjugacy problem in the class of metabelian groups?*

PROBLEM 3. *Does every free soluble group have property* RFI_ω *?*

We remark that free groups have RFI_ω (M. Hall) and that this property is preserved under free products (Romanovskiĭ).

Gruenberg and Remeslennikov have found necessary and sufficient conditions for the wreath product of *RF*-groups (*RFC*-groups) to be an *RF*-group (*RFC*-group). In this connection the following problem is of interest:

PROBLEM 4. *Find necessary and sufficient conditions for the wreath product of two* RFI_ω*-groups to be an* RFI_ω*-group.*

2. Linear representability and embedding theorems

The algebraists at Novosibirsk have obtained a fairly large number of results on the representability of groups as linear groups. These results have been published and can be found in a survey article of Merzljakov contained in one of the issues of

"Itogi Nauki" [4]. I want to state just one interesting result due to Merzljakov.

THEOREM 4 (see [3]). *The holomorph of every polycyclic group has a faithful representation by integral matrices.*

PROBLEM 5. *Is the following conjecture true? Every finitely generated metabelian group can be embedded in the direct product of a finite number of linear groups.*

The confirmation of this conjecture for torsion-free groups and groups with derived groups of prime exponents can be found in [5].

The theorems of G. Higman, B.H. Neumann and Hanna Neumann, and of P. Hall, on embedding countable groups of various types in 2-generator groups with pre-assigned properties have gained wide currency and have proved useful for the solution of a number of questions in group theory. The following results of Roman'kov are of the type just indicated.

THEOREM 5 (see [10]). *Every finitely generated nilpotent group can be embedded in a 2-generator nilpotent group.*

THEOREM 6 (see [11]). *Every polycyclic group can be embedded in a 2-generator polycyclic group.*

The method of proof of these theorems differs fundamentally from the previously known means for proving embedding theorems, which were based on the concepts of wreath product and free product with amalgamation. It is quite clear that the earlier methods are just not applicable to the situation considered by Roman'kov.

3. Groups that are finitely presented in soluble varieties

Very little is known about groups that are finitely presented in soluble varieties. Thanks to the work of Philip Hall, it is at least known that every finitely generated metabelian group is finitely presented in the variety of all metabelian groups, but that there is a finitely generated group that is not finitely presented in the variety of soluble groups of derived length $n > 2$.

The theory of groups with one defining relation is sufficiently developed to allow us the hope of getting some perspective for studying groups with one defining relation in varieties of soluble groups. However up to now only very preliminary and partial results have been obtained in this direction. Thus in [13] there is a positive solution on the word problem for groups with one defining relation in the variety of soluble groups of derived length n, under the additional assumption that the defining relation is in the last non-trivial term of the derived series of the free group. In [14] it is proved that the centre of a group with $n \geq 3$ generators and one defining relation in the class of metabelian groups is trivial.

Above all there are analogues, due to Romanovskiĭ, of Magnus' Freiheitssatz.

THEOREM 7 (see [9]). *Let* F *be a free nilpotent group with basis* $\{x, y_1, y_2, \ldots, y_m\}$. *For any* $r \in F_k \backslash F_{k+1}$, *let* $\dot{r}$ *be the element obtained from* r *by deleting* x .

Then $r \not\equiv \dot{r} \bmod F_{k+1}$ *if and only if* $y_1, y_2, \ldots, y_m$ *generate modulo* $\langle r^F \rangle$ *a free nilpotent group of the same nilpotency class as* F .

THEOREM 8 (see [9]). *Let* F *be a free soluble group with basis* $\{x, y_1, y_2, \ldots, y_m\}$, $m \geq 2$. *Suppose that* $r \in F^{(k-1)} \backslash F^{(k)}$ *and that* $r, \dot{r}$ *are not conjugate modulo* $F^{(k)}$.

Then $y_1, y_2, \ldots, y_m$ *generate modulo* $\langle r^F \rangle$ *a free soluble group of the same derived length as* F .

In the formulation of problems 6, 7, 8, we let H stand for a group with one defining relation in the variety of soluble groups of derived length n .

PROBLEM 6. *Does there exist an algorithm solving the word problem in* H ?

PROBLEM 7. *Find conditions under which* H *has trivial centre.*

PROBLEM 8. *Describe the abelian subgroups of* H .

4. Some more unsolved problems

The following is one of the more urgent and difficult unsolved problems.

PROBLEM 9. *Is the isomorphism problem in the class of nilpotent groups soluble?*

Ju.L. Eršov has shown that a finitely generated nilpotent group has soluble elementary theory if and only if its central factor-group is finite. In this connection we have the natural

PROBLEM 10. *Is the following conjecture true? The elementary theory of a finitely generated soluble group* G *is soluble if and only if* G *is almost abelian*.*

PROBLEM 11. *Describe the varieties* $\underline{\underline{M}}$ *such that* $\underline{\underline{M}} \subseteq \underline{\underline{A}}^n$, $\underline{\underline{A}}^k \nsubseteq \underline{\underline{M}}$, $k < n$.

PROBLEM 12. *Describe the finitely generated soluble groups in which the intersection of two finitely generated subgroups is always finitely generated.*

References

[1] М.И. Каргаполов, В.Н. Ремесленников [M.I. Kargapolov, V.N. Remeslennikov], "Проблема сопряженности для свободных разрешимых групп" [The conjugacy problem for free soluble groups], *Algebra i Logika* 5 no. 6 (1966), 15-25. MR34#5905.

* That is, has an abelian subgroup of finite index.

[2] М.И. Каргаполов, Е.И. Тимошенко [M.I. Kargapolov, E.I. Timošenko], "К вопросу о финитной аппроксимирувмости относительно сопряженности метабелевых групп" [On the problem of conjugacy separability for metabelian groups], *Fourth All-Union symposium on the theory of groups, Novosibirsk*, 1973, pp. 86-88 (Izdat. Sibirsk. Otdel. Akad. Nauk SSSR, Novosibirsk, 1973).

[3] Ю.И. Мерзляков [Ju.I. Merzljakov], "Целочисленное представление голоморфов полициклических групп" [Integral representation of the holomorphs of polycyclic groups], *Algebra i Logika* 9 (1970), 539-558; *Algebra and Logic* 9 (1970), 326-337. MR43#6298.

[4] Ю.И. Мерзляков [Ju.I. Merzljakov], "Линейные группы" [Linear groups], Алгебра. Топология. Геометрия [*Algebra. Topology. Geometry*], 1970, pp. 75-110 (Itogi Nauki. Ser. Mat. 23. Akad. Nauk SSSR Inst. Naucn. Tehn. Informacii, Moscow, 1971).

[5] В.Н. Ремесленнников [V.N. Remeslennikov], "Представление конечно порожденных метабелевых групп матрицами" [Representation of finitely generated metabelian groups by matrices], *Algebra i Logika* 8 (1969), 72-75; *Algebra and Logic* 8 (1969), 39-40. MR44#335.

[6] В.Н. Ремесленников [V.N. Remeslennikov], "Сопряженность в полициклических группах" [Conjugacy in polycyclic groups], *Algebra i Logika* 8 (1969), 712-725. MR43#6313.

[7] В.Н. Ремесленников [V.N. Remeslennikov], "О конечно-определенных группах" [Finitely presented groups], *Fourth All-Union symposium on the theory of groups, Novosibirsk*, 1973, 164-169 (Izdat. Sibirsk. Otdel. Akad. Nauk SSSR, Novosibirsk, 1973).

[8] В.Н. Ремесленников, В.Г. Соколов [V.N. Remeslennikov, V.G. Sokolov], "Некоторые свойства вложения Магнуса" [Some properties of the Magnus embedding], *Algebra i Logika* 9 (1970), 566-578; *Algebra and Logic* 9 (1970), 342-349. MR45#2001.

[9] Н.С. Романовский [N.S. Romanovskiĭ], "Теорема о свободе для групп с одним определяющим соотношением в многообразиях разрешимых и нильпотентных групп данных ступеней" [The Freiheitssatz for groups with one defining relation in varieties of soluble groups of given length and of nilpotent groups of given class], *Mat. Sb. (NS)* 89 (1972), 93-99.

[10] В.А. Романьков [V.A. Roman'kov], "Теоремы вложения для нильпотентных групп" [Embedding theorems for nilpotent groups], *Sibirsk. Mat. Ž.* 13 (1972), 859-867.

[11] В.А. Романьков [V.A. Roman'kov], "О вложении полициклических групп" [Embeddings of polycyclic groups], *Fourth All-Union symposium on the theory of groups, Novosibirsk*, 1973, pp. 178-179 (Izdat. Sibirsk. Otdel. Akad. Nauk SSSR, Novosibirsk, 1973).

[12] Р.А. Саркисян [R.A. Sarkisjan], "Сопряженность в свободных полинильпотентных группах" [Conjugacy in free polynilpotent groups], *Algebra i Logika* 11 (1972), 694-710.

[13] В.Г.Соколов [V.G. Sokolov], "Алгоритм тождества слов для одного класса разрешимых групп" [An algorithm for the equality of words in a class of soluble groups], *Sibirsk. Mat. Ž.* 12 (1971), 1405-1410.

[14] Е.И. Тимошенко [E.I. Timošenko], "Центр групп с одним определяющим соотношением в многообразии 2-ступенно разрешимых групп" [The centres of groups with one defining relation in the variety of all metabelian groups], *Fourth All-Union symposium on the theory of groups, Novosibirsk*, 1973, pp. 231-234 (Izdat. Sibirsk. Otdel. Akad. Nauk SSSR, Novosibirsk, 1973).

Steklov Math. Institute,
Vavilova 42,
Moscow V-333, USSR.

СССР
Москва В - 333
Математический институт имени В.А. Стеклова,
Академии Наук СССР,
Ул Вавилова, дом 42.

PROC. SECOND INTERNAT. CONF. THEORY OF GROUPS 20D05, 20E25
CANBERRA, 1973, pp. 395-400.

LOCALLY FINITE SIMPLE GROUPS

Otto H. Kegel

The group G is locally finite if each of its finitely generated subgroups is finite. Until rather recently the area of locally finite groups entirely belonged to the wilderness of counter-examples; and there absurdly wild behaviour is possible, indeed. What little progress has been made in cultivating some fringes of this wilderness is essentially due to the progress made in handling certain locally finite simple groups, and this in turn depends essentially on the advances achieved in the theory of finite simple groups.

Two natural examples of locally finite groups are the group S^{Ω} $\left(= S^{\Omega, <\aleph_0}\right)$ of all finitary permutations of the infinite set Ω and the group $GL(n, F)$ of all invertible n-by-n matrices over the infinite, absolutely algebraic field F of characteristic $c \neq 0$. In the first group every maximal elementary abelian p-subgroup is infinite, in the second the rank of such a subgroup is bounded by n, unless $p = c$. Consequently, for $p \neq c$, the p-subgroups of $GL(n, F)$ are abelian-by-finite (by Jordan's Theorem) and hence satisfy the minimum condition for subgroups. The c-subgroups of $GL(n, F)$ are nilpotent and of finite exponent. It may be observed that the simple group $PSL(n, F)$ has a natural embedding into the group $GL\left(n^2-1, F\right)$.

A locally finite p-group P satisfying the minimum condition for subgroups has a radicable abelian subgroup $R = R(P)$ of finite index, so the pair (rank R, $|P : R|$), the size of P, is an invariant of P which will play the rôle of order in finite groups. (Use lexicographical ordering.) Every proper subgroup of P has smaller size, so has every image of P under a homomorphism with infinite kernel. If the finite group G satisfies the minimum condition for p-subgroups for the fixed prime p, then in G there is a maximal p-subgroup P of maximal size, every p-subgroup of G is isomorphic to some subgroup of P and every finite p-subgroup of G is even conjugate in G to some subgroup of P, [11]. Thus the p-size is a useful invariant to try induction on in the proof of statements on locally finite groups satisfying the minimum condition on p-subgroups. One type of such a statement is:

1. *If the locally finite group* G *satisfies the minimum condition for p-subgroups for the (fixed) prime* p *together with some further property* P *, then* $G/O_{p',p}G$ *is finite.*

That some further property P is needed is shown by $\mathrm{PSL}(n, F)$ with $p \neq c$. One such property is that no infinite simple group containing elements of order p be involved in G, [5]. Other properties P with the same effect, but considerably more difficult to handle are "abelian subgroups of G have finite rank" [10], and "G satisfies the minimum condition for p-subgroups for every prime p and it also satisfies the minimum condition for centralisers", [6].

Considering a counter-example G to a statement of type 1, we may assume that it has minimal p-size and that it has a maximal normal subgroup M, so that G/M is infinite and simple containing an element of order p. Now it depends on the additional property P which we assume for G whether a reduction can be found to establish P also for the simple group G/M. In that case G/M will be a minimal counter-example to our conjecture, and the problem remains to wield from the information we have on this simple group a weapon lethal for G/M, a non-simplicity criterion.

QUESTION 1. *If the locally finite group* G *satisfies the minimum condition for p-subgroups for every prime* p*, does* G *then have a locally soluble subgroup of finite index?*

In this situation the reduction to a simple counter-example easily goes through as sketched, and Šunkov has shown that such a simple counter-example must have all its 2-subgroups finite.

Now we must take a look at the structure of locally finite groups in general. To simplify the presentation we shall assume that the locally finite simple group S is countably infinite. Then there exists an ascending sequence of finite subgroups F_i of S so that

$$(*)\quad 1 \subset F_1 \subset F_2 \subset \dots \subset F_i \subset F_{i+1} \subset \dots \subset S = \bigcup_{n=1}^{\infty} F_n \text{ and that } F_{i+1} \text{ has a}$$

maximal normal subgroup M_{i+1} with $M_{i+1} \cap F_i = \langle 1 \rangle$.

Then, in some sense, the simple group S is the limit of the sequence $\{S_i = F_i/M_i\}$ of finite simple groups. This sequence must contain substantial information on the group S.

Observing that the known infinite families of finite simple groups are distorted versions of the family {PSL(,)} and are finite in number, one finds, applying Dirichlet's pigeon hole principle, the following easy but startling result.

2. *If for the fixed prime* p *every p-subgroup of the countable locally finite*

simple group S *is either soluble or of finite exponent, then either* S *is linear, or almost all of the simple groups* S_i *obtained from* (*) *are "new".*

This result may be interpreted either as a linearity condition for locally finite groups (unfortunately, one does not yet understand linearity sufficiently well as an abstract property) or as evidence that there are still infinitely many "new" finite simple groups to be found.

If S is linear, that is, $S \subseteq GL(n, F)$, then - in fact - almost all of the subgroups F_i appearing in (*) are simple.

QUESTION 2. *Does every countable locally finite simple group* S *have such a sequence* (*) *consisting of finite simple groups?*

Assuming that the simple group S is linear and that the groups F_i appearing in (*) are all of known type then one needs an identification result of the following type.

3. *If the locally finite simple group* S *has an approximating sequence* (*) *with* $F_i \simeq C(n, q_i)$ *, where* $\underline{\underline{C}}(n, \ )$ *is a (twisted) Chevalley functor, say, then not only is there a locally finite field* F *so that* $S \subseteq GL(d(n), F)$ *but also* $S \simeq \underline{\underline{C}}(n, F)$.

This result has *not* been obtained for all (twisted) Chevalley functors yet, it holds for $PSL(n, \ )$, $PSU(3, \ )$, and $Sz(\ )$. Thus, in principle, one knows what one has to do if one can recognize the group S locally. If in the proof of a statement of type 1 one has identified the hypothetical simple minimal counter-example as one of the well known classical groups, then it is usually rather easy to decide whether it exists or not.

So the main problem has become to transform the additional information that we may be given about the simple group S into information of a local type which will allow us to apply theorems characterising certain finite simple groups to the groups F_i (or S_i) of (*) . Sometimes this transformation is immediate, for example, if the 2-subgroups of S have class less than γ - here we only have to wait for the corresponding theorem to be proved for finite groups.

There are other types of conditions which force a different sort of local, and hence of global information, namely embedding properties for certain subgroups. The prototype for such a result is Frobenius' Theorem. Bender's Theorem [2] appears as a variation of this.

4. *If the simple locally finite group* S *has a subgroup* $H \neq S$ *containing an involution and such that the intersection* $H \cap H^g$ *does not contain any involution for every* $g \in S \backslash H$ *, then* $S \simeq PSL(2, F)$ *,* $PSU(3, F)$ *, or* $Sz(F)$ *for a suitable locally finite field* F *of characteristic* 2 .

It should be pointed out that Bender's complete result in fact is more general, and hence much more useful - as it does *not* assume simplicity. Before Bender's Theorem was proved, Šunkov introduced in [8] a new embedding condition in the same spirit. I shall present it in the form that Wehrfritz has given it recently, [12].

5. *Let* G *be a locally finite group satisfying the minimum condition for 2-subgroups and* H *a subgroup of* G *containing an infinite 2-subgroup. Assume that for every involution* $i \in H$ *so that the centralizer* $\underline{\underline{C}}_H i$ *contains an infinite 2-subgroup one has* $\underline{\underline{C}}_H i = \underline{\underline{C}}_G i$ *and that* H *contains an involution* k *with* $\underline{\underline{C}}_G k \nsubseteq H$ *but so that* $\underline{\underline{C}}_G k$ *contains an infinite 2-subgroup. Then* $G/\underline{\underline{O}}G$ *has a unique minimal normal subgroup* $M/\underline{\underline{O}}G$ *so that* G/M *is an abelian 2'-group and* $M/\underline{\underline{O}}G \simeq \mathrm{PSL}(2, F)$ *for some quadratically closed, locally finite field* F *of odd characteristic.*

There is a difference in principle in the proofs of 4 and 5. In 4 one gets that suitable finite subgroups of S have a strongly embedded proper subgroup, then Bender's result applies in full force and gives us the structure of the 2-subgroups of S. In the proof of 5 one can push the maximal 2-subgroups about, manipulating involutions, until they have taken dihedral form (locally).

A new result of Asar is proved similarly, [1].

6. *Let* G *be a locally finite group satisfying the minimum condition for 2-subgroups and containing an infinite 2-subgroup. If* $\underline{\underline{O}}_2 G = \langle 1 \rangle$ *and any two distinct maximal 2-subgroups of* G *have bounded finite intersection then the radicable subgroup of finite index in a maximal 2-subgroup is locally cyclic.*

QUESTION 3. *Let* G *be a locally finite simple group satisfying the minimum condition for 2-subgroups and such that every radicable 2-subgroup of* G *is locally cyclic. Is it true that either every 2-subgroup of* G *is finite or* $G \simeq \mathrm{PSL}(2, F)$ *as in 5?*

In the proofs of 5 and 6 the inertia of the big radicable 2-subgroups is in fact a source of structural information. Things seem to be quite different if the 2-subgroups are all finite. In order to characterize infinite locally finite simple groups in this situation, one considers centralisers of involutions. These must be infinite in an infinite simple group, and one hopes that they contain subgroups with behaviour similar to that of the maximal radicable 2-subgroups in the proofs of 5 and 6. There exist such characterisations for structurally small classical groups, see [6], [9], or [7] Chapter 5.

For locally finite simple groups with finite 2-subgroups there is a result of Brauer's [3] which gives information on the structure of Sylow 2-subgroups.

7. *Let G be an infinite, locally finite, simple group with a finite Sylow 2-subgroup S. If for every involution $i \in S$ the factor group $\underline{C}_G i / \underline{OC}_G i$ of the centraliser $\underline{C}_G i$ is finite, then for every pair s, t of involutions of S and every element $x \in S$ there exist elements $a, b, c \in G$ such that*

$$s^a,\ t^b,\ x^c \in S,\quad s^a t^b = x^c,$$

and the order $|\underline{C}_S x^c|$ is maximal among the orders of centralisers in S of elements of S which are conjugate to x in G. The centre of S is elementary abelian.

This result will be useful in handling a proof of a statement of type 1 inductively, if P goes over to the simple factor group and the case of infinite 2-subgroups in this has been dealt with.

QUESTION 4. *Is every infinite locally finite simple group G satisfying the assumptions of 7 necessarily isomorphic to* PSL(2, F) *for some infinite, locally finite field F of odd characteristic?*

Well, we have mainly discussed situations which give rise to small, that is, linear simple groups. So to close that subject we ask the obvious question that goes with 2.

QUESTION 5. *Let G be an infinite, locally finite simple group such that for the fixed prime p the p-subgroups of G are either soluble or of finite exponent. Is G linear?*

Large groups exist in profusion. Here one can make many wild constructions. Let me mention one only which is due to Hall [4].

8. *Up to isomorphism there exists a single countable locally finite simple group which contains an isomorphic copy of every countable locally finite group and in which any two isomorphic finite subgroups are in fact conjugate.*

As in finite group theory the most useful characterisation results do not talk (in their hypotheses) about simple groups, and this is also the case for locally finite groups. Progress on simple locally finite groups will depend essentially on progress on finite simple groups and, possibly on new ideas of a geometrical nature as to how to use information on the embedding of subgroups. I believe that we shall see such progress within the next few years.

REFERENCES

[1] A.O. Asar, "On locally finite groups with min-2 ", *J. London Math. Soc.* (to appear).

[2] Helmut Bender, "Transitive Gruppen gerader Ordnung, in denen jede Involution genau einen Punkt festlässt", *J. Algebra* 17 (1971), 527-554. MR44#5370.

[3] Richard Brauer, "Some applications of the theory of blocks of characters of finite groups. II", *J. Algebra* 1 (1964), 307-334. MR30#4836.

[4] P. Hall, "Some construction for locally finite groups", *J. London Math. Soc.* 34 (1959), 305-319. MR29#149.

[5] М.И. Каргаполов [M.I. Kargapolov] "Локально конечные группы, обладающие нормальными системами с конечными факторами" [Locally finite groups possessing normal systems with finite factors], *Sibirsk. Mat. Ž.* 2 (1961), 853-873. MR26#193.

[6] Otto H. Kegel and Bertram A.F. Wehrfritz, "Strong finiteness conditions in locally finite groups", *Math. Z.* 117 (1970), 309-324. Zbl.195,38.

[7] Otto H. Kegel and Bertram A.F. Wehrfritz, *Locally finite groups* (North-Holland Mathematical Library, 3. North-Holland, Amsterdam, London, 1973).

[8] В.П. Шунков [V.P. Šunkov], "Об абстрактной характеризации простой проективной группы типа $PGL(2, K)$ над полем K характеристики $r \neq 0$, 2 " [Abstract characterization of a simple projective group of type $PGL(2, K)$ over a field K of characteristic $r \neq 0$ or 2], *Dokl. Akad. Nauk SSSR* 163 (1963), 837-840; *Soviet Math. Dokl.* 6 (1965), 1043-1047. MR32#7639.

[9] В.П. Шунков [V.P. Šunkov], "Об абстрактных характеризациях некоторых линейных групп" [Abstract characterizations for some linear groups], *Seminar algebraic systems*, pp. 5-54 (Krasnojarsk, 1970).

[10] В.П. Шунков [V.P. Šunkov], "О локально конечных группах конечного ранга" [On locally finite groups of finite rank], *Algebra i Logika* 10 (1971), 199-225. MR45#2002.

[11] Bertram A.F. Wehrfritz, "On locally finite groups with min-p ", *J. London Math. Soc.* (2) 3 (1971), 121-128. MR43#4920.

[12] Bertram A.F. Wehrfritz, "Divisible 2-subgroups of locally finite groups", *Quart. J. Math. Oxford Ser.* (to appear).

Queen Mary College,
Mile End Road,
London E.1, England.

PROC. SECOND INTERNAT. CONF. THEORY OF GROUPS 20D15
CANBERRA, 1973, pp. 401-408.

NORMAL SUBGROUPS OF GROUPS OF PRIME-POWER ORDER

Bruce W. King

In [5] Gaschütz noted that the dihedral and quaternion groups of order 8 cannot be Frattini subgroups of 2-groups. This, together with results in [6], raised the question of what limitations there may be on the structure of those normal subgroups which are contained in the Frattini subgroup of a p-group. Partial answers have been obtained in [1, 2, 3, 6, 7]. It has been shown that the subgroups considered cannot have cyclic centre, and that many such subgroups, if they have two generators, must be metacyclic. It is of interest to remove the restriction that the embedded normal subgroups have two generators. Under some weaker restrictions it may be shown that often such a subgroup N has the property $N' \leq N^p$, or $N' \leq N^4$ if $p = 2$. The parallel results obtained when the embedded normal subgroup has two generators strengthen the results mentioned above.

Notation and terminology

Groups G dealt with in this paper are always finite p-groups. The notation used is that of current texts, such as [9]. However, the following usage should be noted:

$G, \gamma_2(G), \gamma_3(G), \ldots$ refers to the descending central series of G ;

$\Phi(G), \Phi^2(G) = \Phi(\Phi(G)), \ldots$ refers to the Frattini series of G ;

$G^n = \langle g^n \mid g \in G \rangle$ is the subgroup generated by the n-th powers of elements of G ;

N_i denotes a normal subgroup of order p^i .

The term ordinary metacyclic group refers to a metacyclic p-group which has a presentation similar to those of the metacyclic p-groups of odd order. The remaining metacyclic p-groups, termed exceptional, are necessarily 2-groups. This class of groups includes the dihedral, semi-dihedral and quaternion groups. This terminology is elaborated in [10], in which it is shown that exceptional metacyclic groups also differ in group structure from ordinary metacyclic groups.

Embedding theorems

The key result is an elementary lemma dealing with the centralizing effect of elements of G raised to prime powers p^α on certain abelian normal subgroups N of G. The lemma may be proved using commutator calculations, but the proof takes its simplest form when expressed in terms of matrix transformations of a vector space.

LEMMA 1. *Let G be a p-group in which N_r is an elementary abelian normal subgroup of order p^r. If $r \leq p^\alpha$ then G^{p^α} centralizes N_r.*

PROOF. Regard G as a group of $r \times r$ matrices over $GF(p)$, acting on the additive vector space N_r of column vectors over $GF(p)$. Corresponding to some principal series of G passing through N_r, there is a series of G-invariant subspaces $0 < N_1 < \ldots < N_r$ of N_r. Hence a basis can be chosen in N_r such that every matrix $g \in G$ is in upper triangular form. As G is a p-group $g^{p^n} = I$ for some integer $n \geq 0$. Hence for any diagonal entry γ of g, $\gamma^{p^n} = 1$ and so $(\gamma-1)^{p^n} = 0$ {over $GF(p)$}. Hence $\gamma = 1$. Thus every matrix g is of the form $g = I + c$ where c is strictly upper triangular. But $g^{p^\alpha} = 1 + c^{p^\alpha}$ whence $c^{p^\alpha} = 0$ if $p^\alpha \geq r$ and so g^{p^α} acts as the identity transformation on N_r if $r \leq p^\alpha$.

The following corollary extends the lemma to deal with abelian normal subgroups of N of exponent greater than p.

COROLLARY 2. *Let G be a p-group in which N is an r-generator abelian normal subgroup of exponent $p^{\kappa+1}$. If $r \leq p^\alpha$ then $G^{p^{\alpha+\kappa}}$ centralizes N.*

PROOF. By application of Lemma 1 it may be assumed that $\kappa \geq 1$. It then follows by induction applied to N/N^{p^κ} that $\left[G^{p^{\alpha+\kappa-1}}, N\right] \leq N^{p^\kappa}$ and by induction applied to N^{p^κ} that $\left[G^{p^\alpha}, N^{p^\kappa}\right] = 1$. From the first of the relations above, if $n \in N$, $g \in G$, then $n^{g^{p^{\alpha+\kappa-1}}} = nk$ for some $k \in N^{p^\kappa}$. From the second relation above, $k^{g^{p^\alpha}} = k$. Hence $n^{g^{p^{\alpha+\kappa}}} = nk^p = n$. The result therefore follows.

The main embedding theorem may now be proved.

THEOREM 3. *Let* N *be an* r*-generator normal subgroup of a* p*-group* G *contained in the subgroup* $\mho^{p^\alpha}$, $\alpha \geq 1$, *generated by the* p^α*-th powers of elements of* G . *For odd primes* p , *if either* G *is regular or* $r \leq p^\alpha$ *then* $N' \leq N^p$. *For the even prime* $p = 2$, *if* $r \leq 2^{\alpha-1}$ *then* $N' \leq N^4$.

PROOF. (i) For $p \geq 3$, let N be a minimal counterexample to the assertion that $N' \leq N^p$. If $N^p \neq 1$ take $1 \lhd N_1 \lhd G$, $N_1 \leq N^p$ and consider G/N_1 . By minimality of N as a counterexample therefore $(N/N_1)' \leq (N/N_1)^p = N^p/N_1$ whence $N' \leq N^p$ (a contradiction). Therefore $N^p = 1$ and $N' = N_1$ is the unique G-invariant subgroup of N of order p . First assume that G is regular whence $\left[G^{p^\alpha}, N\right] = [G, N]^{p^\alpha}$ ([9], Kapitel 3, Satz 10.8). As $N \trianglelefteq G$ therefore $[G, N] \leq N$. Hence $[G, N]^{p^\alpha} = 1$ and so G^{p^α} centralizes N . As $N \leq Z\left(G^{p^\alpha}\right)$ therefore N is abelian (a contradiction). Next assume that N has at most p^α generators, whence $|N| \leq p^{\alpha+1}$. N possesses a maximal subgroup M such that $M \lhd G$ and $|M| \leq p^\alpha$. Hence by minimality of N , therefore $M' \leq M^p = 1$. Thus M is elementary abelian and so by Lemma 1, $M \leq Z(N)$, contradicting that N is not abelian. This completes the proof for the case of odd primes p .

(ii) For $p = 2$, an analogous argument to the first part of (i) shows it is is sufficient to assume that $N' = N_1$ is the unique G-invariant subgroup of order 2 . Again take M maximal in N , $M \lhd G$. If, first, $\Phi(M) \neq 1$ then $N_1 \leq \Phi(M)$ as $\Phi(M) \lhd G$. As N/N_1 is abelian with at most $2^{\alpha-1}$ generators the same is true of M/N_1 . Hence M has at most $2^{\alpha-1}$ generators. By minimality of N therefore $M' \leq M^4 = 1$ and so it follows that M is abelian of exponent 4 . By Corollary 2 therefore $\left[G^{2^\alpha}, M\right] = 1$ and so $M \leq Z(N)$. If, finally, $\Phi(M) = 1$ then M is elementary abelian with at most 2^α generators so that by Lemma 1, $M \leq Z(N)$. Thus, in either case, that N is not abelian is contradicted. This completes the proof.

By placing conditions on the nature of G or N , one can weaken the other hypotheses on N . For example, the following corollary shows that if G has class 2 then it suffices to take $N \leq \Phi(G)$.

COROLLARY 4. *Let* G *be a regular metabelian* p*-group,* $p \geq 3$, *in which* $\gamma_3(G)$

has exponent at most p . *If* N *is normally embedded in* G *and contained in the Frattini subgroup of* G , *then* $N' \leq N^p$.

PROOF. As in the proof of Theorem 3 suppose that $N^p = 1$ and $N' = N_1$ for some counterexample N . Thus $[N, \Phi(G)] = [N, G'G^p] = [N, G'][N, G^p]$. Now $[N, G^p] = [N, G]^p = 1$ by regularity, using the fact that $[N, G] < N$. But

$$[N, G'] \leq [G'G^p, G'] = [G', G'][G^p, G'] = [G, G']^p = \gamma_3(G)^p = 1 .$$

Hence $[N, \Phi(G)] = 1$ and so $N' = 1$ (a contradiction). This proves the corollary.

Note that from the corollary it can be seen that certain large subgroups $N \triangleleft G$, $N \leq \Phi(G)$, of a regular metabelian group G must be such that $N' \leq N^p$. This is the case, for example, if $\gamma_3(G) \leq N \leq \Phi(G)$.

COROLLARY 5. *Let* $p \geq 3$ *and* N, K *be normal subgroups of a* p*-group* G *such that* $N \leq K \leq G^p$. *If* K *has less than* p *generators then* K *is regular and* $N' \leq N^p$.

PROOF. By Theorem 3, $K' \leq K^p$ so that $\Phi(K) = K^p$ has index less than p^p in K . By [9], Kapitel 3, Sätze 10.12, 10.13, K is regular and N has less than p generators. Hence $N' \leq N^p$ by Theorem 3.

Theorem 7 deals with normal subgroups N which have two generators. Particular instances of this occur in [3, 6, 7] in which it is shown that N must be metacyclic, and in [2] is stated a general result which covers most cases. Theorem 7 of the present paper establishes a stronger result covering all cases. To complete the proof one needs the next theorem, which generalizes a theorem due to Hobby [1, 6]. The theorem gives a criterion for normal embedding in Frattini subgroups, which, like Theorem 3, but unlike the criterion in [5], depends only on the structure of the embedded subgroup.

THEOREM 6. *Let* S *and* N *be normal subgroups of a* p*-group* G *such that* $S \leq N \leq \Phi(G)$. *If* $C_S(N)$ *is cyclic then* S *is cyclic. In particular if* $Z(N)$ *is cyclic then* N *is cyclic, whence if* N *is not abelian then its centre cannot be cyclic.*

PROOF. Suppose that N is a counterexample of least order, containing some non-cyclic S with $C_S(N)$ cyclic. If $S < N$ then by minimality of N , $Z(S)$ is not cyclic. The elements of order p contained in $Z(S)$ generate an elementary abelian normal subgroup of G containing a subgroup N_2 of order p^2 which is normal in

G . By [3], $C_G(N_2) \geq \Phi(G) \geq N$ and so $N_2 \leq C_S(N)$, contradicting the assumption that $C_S(N)$ is cyclic. Hence $S = N$ and so $C_S(N) = Z(N)$ is cyclic. Also N is not abelian, else it is cyclic. If $S^* < N$ and $S^* \triangleleft G$ then $C_{S^*}(N) = S^* \cap Z(N)$ is cyclic and so S^* is cyclic. In particular every proper characteristic subgroup of N is cyclic and also N has a cyclic maximal subgroup, so is itself metacyclic, from which, by [9] Kapitel 3, Satz 13.10 it follows that N is either dihedral, semi-dihedral or a quaternion group. Obviously $N/\Phi^2(N)$ is dihedral or quaternion of order 8 , whence [5] applied to $G/\Phi^2(N)$ gives a contradiction.

Note that the proof also can be completed by elementary methods without recourse to [5, 9]. Theorem 7 can now be proved.

THEOREM 7. *Let N be a 2-generator normal subgroup of a p-group G , contained in the Frattini subgroup of G . $N' \leq N^p$ and when $p = 2$, $N' \leq N^4$.*

PROOF. (i) For the case $p \geq 3$, proceed as in (i) of Theorem 3. Thus N is a minimal counterexample, with $|N'| = p$, $N^p = 1$ and $|N| = p^3$. As N has cyclic centre this contradicts Theorem 6.

(ii) When $p = 2$, a similar procedure shows that $N^4 = 1$ and $N_1 = N'$ is the unique G-invariant subgroup of order 2 . By Theorem 6, $|N| \neq 2^3$. Now* N is a factor group of the group F which is free of rank 2 , exponent 4 and nilpotency class 2 . F contains a characteristic subgroup $S = \langle s, t, \ldots \rangle$ of order 4 consisting of elements of F^2 which are neither squares nor commutators, together with 1 . Now F' is cyclic and consists entirely of commutators, so that $S \cap F' = 1$. Thus if either $N \cong F$ or $N \cong F/\langle s \rangle$ then N would contain a normal subgroup $\bar{N}_1$ of G , such that $\bar{N}_1 \neq N_1$ (contradiction).

It follows that $N \cong F/\langle b^2 \rangle$ for some $b \in F\text{-}F^2$, whence up to isomorphism N has only one possible structure, with generators a, b such that $a^4 = b^2 = [a, b]^2 = 1$ (of class 2 and order 2^4). Note that $a^2 \in Z(N)$. As $|a| = |ab| = 4$ but $|b| = 2$ it follows that the automorphism group A of N contains no element of order 3 , whence $|A| = 2^n$, $n \leq 5$. Since $|\Phi(A)| \leq 2^3$ therefore by Theorem 6, $\Phi(A)$ is abelian, its elements being products of squares of elements of A . If $\alpha \in A$ then

* This argument, which is considerably shorter than the original one, follows a suggestion of M.F. Newman.

$$\alpha : \begin{cases} a \mapsto a^r b^s c^t , \\ b \mapsto a^{2u} b c^w , \\ c \mapsto c , \end{cases}$$

with $c = [a, b]$ and $s, t, u, w = 0$ or 1 , $r = \pm 1$. Hence

$$\alpha^2 : \begin{cases} a \mapsto (a^r b^s c^t)^r (a^{2u} b c^w)^s = a^{1+2us} c^k , \quad k = 0, 1 ; \\ b \mapsto (a^r b^s c^t)^{2u} a^{2u} b = b c^{us} \end{cases}$$

and so every element of $\Phi(A)$ maps $a \mapsto a^{1+2\Sigma us} c^{\Sigma k}$ and $b \mapsto b c^{\Sigma us}$ (summing over appropriate indices). Therefore the inner automorphism of N induced by the action of a , in which

$$\begin{cases} a \mapsto a , \\ b \mapsto b^a = bc \end{cases}$$

is not an element of $\Phi(A)$. By Gaschütz's criterion [5] it follows that $N \nleq \Phi(G)$. This contradiction now completes the proof.

Groups N of the type in Theorems 3, 7 which satisfy the conditions $N' \leq N^p$ $(p \geq 3)$ or $N' \leq N^4$ are in general not well-known (it is planned to discuss them more fully elsewhere). They are metacyclic provided that they have two generators. For odd primes p this was shown in [8]; that this is also true if $p = 2$ follows from results of Blackburn. For by a criterion in [4] it suffices to assume that N is non-abelian of class 2 and that $\Phi(N') = 1$. As N has 2 generators and class 2 therefore N' must be cyclic ([9], Kapitel 2 Hilfsatz 1.11); it follows that $|N'| = 2$. Let aN', bN' be canonical generators of the abelian group N/N' . Thus N has generators a, b such that $a^{2^m} = c^\alpha$, $b^{2^n} = c^\beta$, $[a, b] = c^\gamma$ for some non-negative integers $\alpha, \beta, \gamma, m, n$ such that $\alpha, \beta = 0, 1$. Now $N^2 = \langle g^2 \mid g \in N \rangle = \langle a^2, b^2, [a, b] \rangle$ by application of the usual power-commutator formula in a class 2 group. As $[a, b] \in N' \leq N^4 \leq \Phi(N^2)$ it follows that $N^2 = \langle a^2, b^2 \rangle$. N^2 is obviously abelian and so $c = a^{2\rho} b^{2\sigma}$ for some integers ρ, σ with $m > \rho \geq 0$, $n > \sigma \geq 0$. If $\alpha = \beta = 0$ in the defining relations for N above then $\rho \neq 0$, $\sigma \neq 0$ and so modulo N' there is a non-trivial relation $a^{2\rho} b^{2\sigma} \equiv 1$, contrary to the definition of a, b . Hence without loss of generality $\alpha \neq 0$ and so $[a, b] = a^{2^m \gamma}$, whence $a^b = a^k$ for some integer k and so N is metacyclic.

COROLLARY 8. *Under the hypotheses of Theorem 7, N is an ordinary metacyclic group.*

This follows since by [10] an exceptional metacyclic group N must have a non-abelian factor group N/K of order 8. Obviously $K \geq N^4$ and since $N^4 \geq N'$ therefore N/K is abelian (a contradiction). Corollary 8 is not the best result possible. After the exclusion from the ordinary metacyclic groups of those which have cyclic centre there still remain a few other metacyclic groups which cannot be embedded as the normal subgroup of another p-group contained in its Frattini subgroup. A full list of these groups is known but is not given here. Finally consider the problem of embedding N in G', not merely in $\Phi(G)$. It turns out that if

$$N = \left\langle a,\ b;\ a^{p^m} = 1,\ b^{p^n} = a^{p^{m-s}},\ a^b = a^{1+p^{m-c}} \right\rangle$$

then $s = 0$ and $m \geq m-c > n$. In [3] it is shown that if also $N = G'$ then

(i) $[G, N, N] = 1$, that is, $\gamma_3(G) \leq Z(G')$, and

(ii) $m \geq m-c > n > 2c$.

It seems unlikely that (i), (ii) need apply if only $N < G'$.

THEOREM 9. *Let N be a 2-generator p-group which is a normal subgroup of a p-group contained in its derived subgroup. N is an ordinary metacyclic group of class at most 2, and has a presentation*

$$N = \left\langle a,\ b;\ a^{p^m} = b^{p^n} = 1,\ a^b = a^{1+p^{m-c}} \right\rangle$$

in which $m \geq m-c > n > c$.

PROOF. Let $N \triangleleft G$ and $N \leq G'$. By Corollary 8, N is an ordinary metacyclic group, and following a remark in [7], N must have class at most 2. For as N' is cyclic therefore G induces an abelian group of automorphisms of N' so that $G/C_G(N')$ is abelian. Thus $C_G(N') \geq G' \geq N$, whence $N' \leq Z(N)$ as required. The remainder of the proof relies on [10] and proceeds by induction. Assume now that N is not abelian, and choose a, b so that in the notation of [10], $N = \langle a \parallel b;\ m, n, s, c, + \rangle$. As $1 = [a, b, b] = a^{p^{2(m-c)}}$ therefore $2(m-c) \geq m$ and so $m \geq 2c$. Since $a^{p^{m-1}} \in N'$ therefore $\left\langle a^{p^{m-1}} \right\rangle \triangleleft G$ and induction can be applied to $\bar{N} = N/\left\langle a^{p^{m-1}} \right\rangle$ in $G/\left\langle a^{p^{m-1}} \right\rangle$. By [10] (Lemma 4.5), $\bar{N} = \langle \bar{a} \parallel \bar{b} \rangle$ whence from the parameters of $\bar{N}$ it follows that $m-s \geq m-1 \geq m-c > n > c-1$ and $m-1 > 2(c-1)$. If $s = 1$ (so that N is not split) then by [10] it must be that $n > m-1$, contrary to the inequality obtained by induction. Thus $s = 0$ and so N

splits. By Theorem 6, $Z(N)$ cannot be cyclic so that by [10], $c < n$.

References

[1] Homer Bechtell, "Frattini subgroups and Φ-central groups", *Pacific J. Math.* 18 (1966), 15-23. MR33#5725.

[2] Я.Г. Беркович [Ja.G. Berkovič], "Нормальные делители конечной группы" [Normal subgroups in a finite group], *Dokl. Akad. Nauk SSSR* 182 (1968), 247-250; *Soviet Math. Dokl.* 9 (1968), 1117-1120. MR38#4562.

[3] N. Blackburn, "On prime-power groups in which the derived group has two generators", *Proc. Cambridge Philos. Soc.* 53 (1957), 19-27. MR18,464.

[4] N. Blackburn, "On prime-power groups with two generators", *Proc. Cambridge Philos. Soc.* 54(1958), 327-337. MR21#1348.

[5] Wolfgang Gaschütz, "Über die Φ-Untergruppe endlicher Gruppen", *Math. Z.* 58 (1953), 160-170. MR15,285.

[6] Charles Hobby, "The Frattini subgroup of a p-group", *Pacific J. Math.* 10 (1960), 209-212. MR22#4780.

[7] Charles R. Hobby, "Generalizations of a theorem of N. Blackburn on p-groups", *Illinois J. Math.* 5 (1961), 225-227. MR23#A209.

[8] Bertram Huppert, "Über das produkt von paarweise vertauschbaren zyklischen Gruppen", *Math. Z.* 58 (1953), 243-264. MR14,1059 and MR17,1436.

[9] B. Huppert, *Endliche Gruppen I* (Die Grundlehren der mathematischen Wissenschaften, Band 134. Springer-Verlag, Berlin, Heidelberg, New York, 1967). MR37#302.

[10] Bruce W. King, "Presentations of metacyclic groups", *Bull. Austral. Math. Soc.* 8 (1973), 103-131. Zbl.245.20016.

Riverina College of Advanced Education,
Wagga Wagga, NSW 2650.

PROC. SECOND INTERNAT. CONF. THEORY OF GROUPS 20F40
CANBERRA, 1973, pp. 409-416.

SOME RELATED QUESTIONS IN THE THEORY OF GROUPS AND LIE ALGEBRAS

A.I. Kostrikin

The aim of my talk is to give a survey of the present status of some old problems, which are to some extent common to the theory of groups and Lie algebras. Actually it is not so much that there are new results to report, nevertheless they provide grounds for the discussion. The connections between the theories of groups and Lie algebras are very numerous ([8], [14], [18]), but I shall mention only two in this lecture.

1. Several years ago I received a letter from James Wiegold. Let me quote it:

"Recently Dr L.G. Kovács and I had a conversation about the existence of a torsion-free locally nilpotent relatively free group which is not nilpotent. Because of Professor Shmelkin's work on varieties and your own work on the Burnside problem, this comes to the following question:

Let L *be a Lie algebra over a field of characteristic* 0 *satisfying an Engel condition* $[x, \underbrace{y, \ldots, y}_{n}] = 0$, *with* n *independent of* x, y . *Is it true that* L *is actually nilpotent? ... I would be very grateful for any comments you care to make. In view of Heineken's work my guess is 'yes'."*

There are many experts on Engel Lie algebras taking part in this conference, who would probably also say that the answer is "yes". But as far as I know, no solution has been found as yet.

For convenience we introduce the following notation. The symbol $E_{n,p}$ denotes the n-th Engel condition in Lie algebras L of characteristic p : $(\mathrm{ad}\, x)^n = 0 = px$ for all $x \in L$. We set $E_{n,\infty} = E_n$. It is possible that every Lie algebra L with $E_{n,p}$ is locally nilpotent, but this has been proved only for $n \leq p$ [9]. So let $L(d)$ be the "relatively free" Lie algebra on d generators satisfying $E_{n,p}$, $n < p$; let $c_{n,p}(d)$ be its nilpotency class $\left(c_{n,\infty}(d) = c_n(d)\right)$. The question is to study the behaviour of the function $c_{n,p}(d)$

(in particular, $c_n(d)$) as $d \to \infty$. If $c_{n,p} = \lim_{d\to\infty} c_{n,p}(d) = \infty$, then $L(\infty)$ is non-nilpotent. For sufficiently large p, we have $c_{n,p} = c_{n,\infty} \overset{\text{def}}{=} c_n$. Let d_n stand for the derived length of $L(\infty)$. As Higgins noted [6], $d_n < \infty \Rightarrow c_n < \infty$. In fact he showed that $d_4 \leq 9$, $c_4 \leq 87381$. Ten years later Heineken [5] improved this bound:

$$c_{3,p} \leq 5 \quad (p > 5)\ , \quad c_{4,p} \leq 29 \quad (p > 7)\ .$$

It can be verified that $c_{5,p} < \infty$ for $p > 7$. For arbitrary n I have been able only to prove the following assertion.

THEOREM 1. *Every Lie algebra* L *satisfying* $E_{n,p}$ *with* $p > n + [n/2]$ *contains a non-zero abelian ideal.*

COROLLARY. *There exists a sequence of ideals* $0 \subset L_1 \subset L_2 \subset \ldots \subset L_k \subset \ldots$ *such that* $M = \bigcup_{i=1}^{\infty} L_i$, $L_i < L$, $L_i^{n_i} = 0$ *for some nonnilpotent subalgebra* $M \subseteq L$ *(if* L *is not nilpotent).*

Some restriction on p is necessary, as a remarkable result of Razmyslov shows [15].

THEOREM 2. $c_{n,p} = \infty$ *for* $n \geq p-2 \geq 3$.

This result was found independently by Bachmuth and Mochizuki [1] in the case $p = 5$.

The case $n = p - 1$ is of particular interest since it has a direct application to the restricted Burnside problem.

COROLLARY. *The variety* $\underline{\underline{K}}_p$ *of all locally finite groups of prime exponent* $p > 3$ *is insoluble and non-Cross.*

Thus the situation in Lie algebras is somewhat more delicate than that in associative algebras, where the definitive result is due to G. Higman [7]: If A is a relatively free associative algebra satisfying $X^n = 0$ and having characteristic $p > n$, then A is nilpotent of class f_n where $(n/e)^2 < f_n \leq 2^n - 1$.

What must we do for Lie algebras? We consider the "relatively free" associative algebra $\underline{\underline{A}} = F_p\langle X, Y\rangle$ (where for F_p we take $\mathbb{Q}$ if the characteristic is 0) on two generators X, Y with the relations $U^n = 0$, $n << p$ (let us say $\frac{3}{2}n < p$) for all $U \in \underline{\underline{L}}$, where $\underline{\underline{L}}$ is the Lie algebra generated in $\underline{\underline{A}}$ by X, Y under bracket multiplication $[S, T] = ST - TS$. Of course, $\underline{A}$ is a nilpotent algebra, but that is not enough. It would be nice to prove the following conjecture:

CONJECTURE. $\underline{A}^{2n-2} = 0$.

If the conjecture is true, then $c_{n,p} < \infty$ (a rough bound would be $c_{n,p} < 3^{3^{n-2}}$), and we are done. It is enough to prove the identity $X^{n-1}Y^{n-1} = 0$, but in fact this identity is equivalent to the conjecture.

In connection with the restricted Burnside problem I must mention the result of G.E. Wall (see these Proceedings) which I learned at Canberra.

THEOREM 3. *The Lie algebra* $L(\overline{B}_{d,p})$ *associated with the finite Burnside group* $\overline{B}_{d,p}$ *is a proper factor-algebra of the relatively free Lie algebra* $\Lambda(d)$ *satisfying* $E_{p-1,p}$ *, at least for* $d \geq 3$ *and* $p = 5, 7, 11$.

More exactly, $\Lambda(d) = \sum_{i\geq 1} \Lambda_i(d)$ is a graded Lie algebra and $L(\overline{B}_{d,p}) \cong \Lambda(d)/W(d)$, where $W(d) = \sum_{i\geq 2p-1} W_i(d) \neq 0$ is a graded Wall ideal. If this result, which shows that the existence of $\overline{B}_{d,p}$ is not equivalent to the nilpotency of $\Lambda(d)$, had been known 15-20 years ago, then it is possible that a solution of the restricted Burnside problem would not have run on the purely ring-theoretical lines suggested by the work of Magnus. But now Theorem 3 provides a start to more detailed investigations on the structure of $\overline{B}_{d,p}$. In view of the Corollary to Theorem 2, not so many new relations arise from the fact that $W(d) \neq 0$. It is of interest to compare $c_{p-1,p}(d)$ with the nilpotency class $\overline{c}_p(d)$ of $\overline{B}_{d,p}$. This is apparently a difficult problem, since the existence of the Wall ideal $W(d)$ has been obtained up to now only with the aid of a computer. The question about the coincidence of $L(\overline{B}_{2,p})$ and $\Lambda(2)$ is still open, and we can only guess about the nilpotency class of $\overline{B}_{2,p}$ and its order. Probably $\overline{c}_p(2)$ is a quadratic function of p , but the only result in this direction known to me is that of Bomshik Chang [2]:

THEOREM 4. *For every positive integer* k *there exists a positive integer* m *such that* $c_{p-1,p}(2) > k\cdot p$ *if* $p > m$.

Before this it was known only that $\overline{c}_p(2) > 2p$ for all $p > 3$.

Finally, the most important unsolved problems (albeit special) are the following:

(a) Do the finite Burnside groups $\overline{B}_{2,8}$ *and* $\overline{B}_{2,9}$ *exist?*

Some approaches exist ([3], [13], [16]) but up to now they have not been good enough to make significant progress in this direction. The expected answer is positive.

(b) Set $[a, b] = aba^{-1}b^{-1}$, $[a, b; n+1] = [[a, b; n], b]$.

It is known that $[a, b; 6] = e$ in $\overline{B}_{2,5}$. Is the equation

$$(\dagger) \qquad [X, Y; 6] = \prod_{i=1}^{m} W_i(X, Y)^5$$

soluble in the free group $F(X, Y)$? If not, then the Burnside group $B_{2,5}$ is infinite. The solution of an auxiliary equation

$$[X, Y; 5] = \prod_{i=1}^{m} W_i(X, Y)^4$$

could provide some ideas for solving the equation (†).

(c) Is the set of finite simple groups of exponent 60 *finite?*

It is possible that a solution of this question, which is connected with the Hall-Higman reduction theorem [4], would be somewhat easier than the determination of all finite simple groups of orders $2^{\alpha}3^{\beta}5^{\gamma}$.

2. In its time the restricted Burnside problem gave the impetus to the investigation of finite-dimensional simple Lie algebras of characteristic $p > 0$ [10]. I would like now to discuss an aspect of this theory which has a direct application to the theory of groups.

If k is a field of characteristic $p \geq 3$ and

$$(*) \qquad L = H + \sum_{\omega\in\Omega} L_{\omega}$$

is the Cartan decomposition of a finite-dimensional simple Lie algebra of classical type with $L_{\omega} = \langle e_{\omega} \rangle$, then in most cases $(\text{ad } e_{\omega})^3 = 0$. The mapping $\exp(t \text{ ad } e_{\omega}) : L \to L$ $(t \in k)$ is an automorphism of L , and the group $\langle \exp(t \text{ ad } e_{\omega}) \mid t \in k, \omega \in \Omega \rangle$ is nothing else but a simple (or nearly simple) Chevalley group of normal type. The situation is not much changed if instead of ad we consider any irreducible p-representation of L .

There exist at least four additional infinite series of simple Lie p-algebras W_n, S_n, H_n, K_n over a field k (even in the algebraically closed case). The definitions can be found in the survey [11]. Let L be one of these algebras. Then there exists a Cartan decomposition of type (*), where generally speaking

$\dim L_\omega > 1$, and the space L_ω is spanned by elements e_ω such that $e_\omega^{[p]} = 0$ ($[p]$ is a unary operation in L). We consider an irreducible p-representation Γ of L and construct a finite group

$$G_q(L, \Gamma) = \langle \exp t\Gamma(e_\omega) \mid t \in \mathrm{F}_q, e_\omega \in L_\omega, \omega \in \Omega \rangle .$$

In the case of a non-classical Lie algebra, the influence of Γ on the structure of $G_q(L, \Gamma)$ is very strong. Let us consider as an example one of the simplest, namely a Witt algebra W_1 :

$$W_1 = \left\langle e_i, -1 \le i \le p-2, p > 3; [e_i, e_j] = \left\{\binom{i+j+1}{j} - \binom{i+j+1}{i}\right\} e_{i+j} \right\rangle .$$

The algebra W_1 has just one irreducible p-representation Γ_0 of dimension $p - 1$ and $p - 2$ irreducible p-representations Γ_j of dimension p . Suppose that $\Gamma_1 = \mathrm{ad}$.

THEOREM 5. *The following isomorphisms hold:*

(i) $G_q(W_1, \Gamma_0) \cong C_{(p-1)/2}(q)$, $q = p^m$,

(ii) $G_q(W_1, \Gamma_1) \cong \mathrm{SL}(p, q)$.

The corresponding Bruhat decompositions are obtained by simple but tedious calculation. We note that $G_q(W_1, \mathrm{ad})$ does not have any connection with the group of automorphisms of W_1 , since $\mathrm{Aut}(W_1)$ is soluble. I do not know what the groups $G_q(W_1, \Gamma_i)$ are. Are the groups $G_q(L, \Gamma)$ similar to the Chevalley groups of normal type? This is apparently a difficult question, because even the irreducible p-representations of classical Lie algebras are not completely known.

It would be useful to consider one more example connected with a number of interesting questions in the theories of Lie algebras and groups. There is a conjecture that every simple Lie p-algebra L over an algebraically closed field k of characteristic $p > 3$ can be written in the form $L = L_0 \otimes_{k_0} k$, where k_0 is the prime subfield of k . For $p \le 3$ this is false, and the example which we are interested in is connected with just this fact. One can construct (see [11] or [12]) a family of ten-dimensional simple Lie p-algebras $L(\varepsilon)$, $0 \ne \varepsilon \in k$, for which $L(\varepsilon) \cong L(\varepsilon') \Rightarrow \varepsilon' = \varepsilon^{\pm 1}$. In particular $L(-1) = B_2$, the classical orthogonal algebra. Here k is any field of characteristic 3 . The algebra $L(\varepsilon)$ admits a natural realization: the basis is to consist of differential operators

$$e_{-2\alpha-\beta} = -\partial_z ,$$

$$e_{-\alpha-\beta} = -\partial_x + y\partial_z ,$$

$$e_{-\alpha} = \partial y + x\partial_z ,$$

$$e_{-\beta} = y\partial_x ,$$

$$h_\beta = x\partial_x - y\partial_y ,$$

$$z = x\partial_x + y\partial_y - z\partial_z ,$$

$$e_\alpha = \{(1-\varepsilon)xy-\varepsilon z\}\partial_x + (\varepsilon-1)y^2\partial_y + \{(1+\varepsilon)xy^2+\varepsilon yz\}\partial_z ,$$

$$e_{\alpha+\beta} = (1-\varepsilon)x^2\partial_x + \{(\varepsilon-1)xy-\varepsilon z\}\partial_y + \{(1+\varepsilon)x^2y-\varepsilon xz\}\partial_z ,$$

$$e_{2\alpha+\beta} = \{-\varepsilon(1+\varepsilon)x^2y+\varepsilon^2xz\}\partial_x + \{\varepsilon(1+\varepsilon)xy^2+\varepsilon^2yz\}\partial_y + \{\varepsilon(1+\varepsilon)x^2y^2-\varepsilon^2z^2\}\partial_z .$$

These operators act on a 25-dimensional module $M = O'/k$, where $O' = O\backslash\{x^2y^2z^2\}$ $O = k[x, y, z]$ is *l'algèbra des puissances divisées* $x^iy^jz^k$, $0 \le i, j, k \le 2$ subject to the rules $t^it^j = \binom{i+j}{i}t^{i+j}$, $t = x, y, z$. Further $\Omega = \{u\alpha+v\beta\}$ is a root system of type B_2 ; $e_\omega^{[3]} = 0$ for all $0 \ne \omega \in \Omega$. Let Γ_0 be the irreducible representation corresponding to the module M . We set $k = \mathrm{F}_q$, $q = 3^m$. As we expected, $G_q(L(-1), \mathrm{ad}) \cong B_2(q)$. Moreover, $G_q(L(\varepsilon), \mathrm{ad}) \subseteq D_5(q)$, and I have almost finished checking that for $\varepsilon = 1$, the inclusion can be replaced by equality (actually by an isomorphism). The problem of the description of all irreducible p-representations of $L(\varepsilon)$ and of the corresponding groups is still open.

In the case of the representation Γ_0 , it is of interest to consider M an $\underline{\underline{L}}$-module, where $\underline{\underline{L}}$ is a Lie algebra generated by operators

$$\bar{c}_1 = e^2_{-2\alpha-\beta} , \quad \bar{c}_2 = e^2_{-\alpha-\beta} , \quad \bar{c}_3 = e^2_{-\alpha} ,$$

$$\bar{c}_{12} = e^2_{-2\alpha-\beta}e_{-\alpha-\beta} , \quad \bar{c}_{13} = e^2_{-2\alpha-\beta}e_{-\alpha} , \quad \bar{c}_{21} = e^2_{-\alpha-\beta}e_{-2\alpha-\beta} ,$$

$$\bar{c}_{23} = e^2_{-\alpha-\beta}e_{-\alpha} , \quad \bar{c}_{31} = e^2_{-\alpha}e_{-2\alpha-\beta} , \quad \bar{c}_{32} = e^2_{-\alpha}e_{-\alpha-\beta} ,$$

$$c_1 = e^2_{2\alpha+\beta} , \quad c_2 = e^2_{\alpha+\beta} , \quad c_3 = e^2_\alpha ,$$

$$c_{12} = e^2_{2\alpha+\beta}e_{\alpha+\beta} , \quad c_{13} = e^2_{2\alpha+\beta}e_\alpha , \quad c_{21} = e^2_{\alpha+\beta}e_{2\alpha+\beta} ,$$

$$c_{23} = e^2_{\alpha+\beta}e_\alpha , \quad c_{31} = e^2_\alpha e_{2\alpha+\beta} , \quad c_{32} = e^2_\alpha e_{\alpha+\beta} .$$

This $\underline{\underline{L}}$-module M is irreducible for $\varepsilon \ne 0, -1$, but it has an irreducible factor-module M_0 of dimension 24 for $\varepsilon = 1$. We set

$$\underline{\underline{G}} = \langle 1+tC \mid t \in F_9,\ C = \overline{C}_i,\ C_i,\ \overline{C}_{ij},\ C_{ij} \rangle .$$

It is easy to check that $C^2 = 0$, so that $\{\underline{\underline{G}}, M\}$ is a quadratic pair for $p = 3$ in the sense of Thompson [17]. Since the case of quadratic p-pairs for $p = 3$ has not been completed, it would seem of use to examine what the pairs $\{\underline{\underline{G}}, M\}$ give; in particular, what $\{\underline{\underline{G}}, M_0\}$ gives.

ACKNOWLEDGEMENT. I would like to express my thanks to the Academy of Sciences of USSR and the Australian National University for their support. I am grateful also to Dr L.G. Kovács for valuable comments.

References

[1] Seymour Bachmuth, Horace Y. Mochizuki and David Walkup, "A nonsolvable group of exponent 5 ", *Bull. Amer. Math. Soc.* 76 (1970), 638-640. MR41#1862.

[2] Bomshik Chang, "On Engel rings of exponent $p - 1$ over GF(p) ", *Proc. London Math. Soc.* (3) 11 (1961), 203-212. MR23#A3770.

[3] George Glauberman, Eugene F. Krause and Ruth Rebekka Struik, "Engel congruences in groups of prime-power exponent", *Canad. J. Math.* 18 (1966), 579-588. MR33#4138.

[4] P. Hall and Graham Higman, "On the p-length of p-soluble groups and reduction theorems for Burnside's problem", *Proc. London Math. Soc.* (3) 6 (1956), 1-42. MR17,344.

[5] Hermann Heineken, "Liesche Ringe mit Engelbedingung", *Math. Ann.* 149 (1963), 232-236. MR26#3754.

[6] P.J. Higgins, "Lie rings satisfying the Engel condition", *Proc. Cambridge Philos. Soc.* 50 (1954), 8-15. MR15,596.

[7] Graham Higman, "On a conjecture of Nagata", *Proc. Cambridge Philos. Soc.* 52 (1956), 1-4. MR17,453.

[8] Graham Higman, "Lie ring methods in the theory of finite nilpotent groups", *Proc. Internat. Congr. Mathematicians, Edinburgh*, 1958, pp. 307-312 (Cambridge University Press, Cambridge, 1960). MR22#6845.

[9] А.И. Кострикин [A.I. Kostrikin], "О проблеме Бернсайда" [The Burnside problem], *Izv. Akad. Nauk SSSR Ser. Mat.* 23 (1959), 3-34. MR23#A1947.

[10] А.И. Кострикин [A.I. Kostrikin], "Алгебры ли и конечные группы" [Lie algebras and finite groups], *Proc. Internat. Congr. Mathematicians, Stockholm*, 1962, pp. 264-269 (Inst. Mittag-Leffler, Djursholm, 1963; *Amer. Math. Soc. Transl.* (2) 31 (1963), 40-46). MR32#4167.

[11] A.I. Kostrikin, "Variations modulaires sur un theme de Cartan", *Actes Congr. Internat. Mathematiciens, Nice*, 1970, 1, pp. 285-292 (Gauthier-Villars, Paris, 1971).

[12] А.И. Кострикин [A.I. Kostrikin], "Параметрическое семейство простых алгебр" [A parametric family of simple Lie algebras], *Izv. Akad. Nauk SSSR Ser. Mat.* 34 (1970), 744-756. MR43#302.

[13] Eugene F. Krause, "On the collection process", *Proc. Amer. Math. Soc.* 15 (1964), 497-504. MR29#2299.

[14] R.C. Lyndon, "Burnside groups and Engel rings", *Finite groups*, pp. 4-14 (Proc. Symposia Pure Math. 1. Amer. Math. Soc., Providence, Rhode Island, 1959). MR22#9522.

[15] Ю.П. Размыслов [Ju.P. Razmyslov], "Об энгелевых алгебрах" [Lie algebras satisfying Engel conditions], *Algebra i Logika* 10 (1971), 33-44; *Algebra and Logic* 10 (1971), 21-29. MR45#3498.

[16] И.Н. Санов [I.N. Sanov], "О некоторой системе соотношений в периодических группах с периодом степенью простого числа" [On a certain system of relations in periodic groups with period a power of a prime number], *Izv. Akad. Nauk SSSR Ser. Mat.* 15 (1951), 477-502. MR14,722.

[17] J.G. Thompson, "Quadratic pairs", *Actes Congr. Internat. Mathématiciens, Nice*, 1970, 1, pp. 375-376 (Gauthier-Villars, Paris, 1971).

[18] Hans Zassenhaus, "On an application of the theory of Lie algebras to group theory", *Finite groups*, pp. 105-108 (Proc. Symposia Pure Math., 1. Amer. Math. Soc., Providence, Rhode Island, 1959). MR22#12168.

Steklov Math. Institute,
Vavilova 42,
Moscow V-333, USSR.

СССР
Москва В - 333
Математический институтимени В.А. Стеклова,
Академии Наук СССР,
Ул Вавилова, дом 42.

PROC. SECOND INTERNAT. CONF. THEORY OF GROUPS, 20E10
CANBERRA, 1973, pp. 417-431.

HANNA NEUMANN'S PROBLEMS ON VARIETIES OF GROUPS

L.G. Kovács and M.F. Newman

This is an informal report on the present status of the displayed problems in Hanna Neumann's book *Varieties of groups* [50]. The reader should have the book at hand, not only for notation and terminology but also because we do not re-state the problems nor repeat the comments available there. Our aim is to be up to date, not to present a complete historical survey; superseded references will be mostly ignored regardless of their significance at the time. Details of solutions will not be quoted from papers already published, unless needed to motivate further questions. The discussion and the problems highlighted in it reflect our personal interests rather than any considered value-judgement.

The preparation of this report was made much easier by access to the notes Hanna Neumann had kept on these problems. We are indebted to several colleagues who took part in a seminar on this topic and especially to Elizabeth Ormerod for keeping a record of these conversations. The report has also gained a lot from the response of conference participants; in particular, Professor Kostrikin supplied much useful information. Of course, all errors and omissions are our own responsibility: we shall be very grateful for information leading to corrections or additions.

PROBLEM 1 (page 6 in [50]).

As Hanna Neumann wrote, "this is of no great consequence". The answer is negative; see Kovács and Vaughan-Lee [47].

PROBLEMS 2, 3, 11 (pages 22, 92).

The celebrated "finite basis problem" asked whether every variety can be defined by a finite set of laws. Ol'šanskiĭ proved (in about 1968, unpublished) that this is equivalent to the problem: is the set of varieties countable? (See also Kovács [42].) A positive answer would have created a simple situation to report on; however, in general, the answer is negative. A comprehensive survey of the positive partial results is beyond the scope of this report, and the listing of open questions provoked by the complexity of the situation is also without any claim to completeness.

The negative answer was first obtained by Ol'šanskiĭ [55] in September 1969: he proved that there are continuously many locally finite varieties of soluble length at most 5 and exponent dividing $8pq$ whenever p, q are distinct, odd primes. This settled Problems 2 and 3. By December 1969, Vaughan-Lee [62] constructed (by entirely different means) continuously many varieties within $(\mathbf{B}_4 \wedge \mathbf{N}_2)^2$; and, early in 1970, Adyan ([1], see also [3]) gave an infinite independent set of very simple two-variable laws.

Given the negative solution, Zorn's Lemma yields the existence of at least one just non-finitely-based variety (a variety minimal with respect to not being definable by a finite set of laws). One may then ask:

QUESTION 1. *What is the cardinality of the set of just non-finitely-based varieties?*

Simplifying and extending Vaughan-Lee's construction, Newman [53] proved also that to each odd prime p there is at least one just non-finitely-based variety in $\left(\mathbf{A}_p^2 \wedge \mathbf{N}_p\right)\left(\mathbf{B}_p \wedge \mathbf{N}_2\right)$: so the answer to Question 1 is certainly 'infinite'.

Vaughan-Lee [64] was the first to show that the product of two finitely based varieties need not be finitely based.

Problem 11 was solved simultaneously and independently by Kleĭman [41] and Bryant [16]: $\mathbf{B}_4\mathbf{B}_2$ is not finitely based. This is still the easiest example to name. Further results of theirs (see also a forthcoming paper of Kleĭman), with little extra overlap between them, extend the scope of the work beyond what can be fully reported here. All we mention is that Bryant [16], starting from the method of Vaughan-Lee and Newman, produced varieties $\mathbf{U}$ and $\mathbf{V}$ such that there are continuously many varieties $\mathbf{W}$ with $\mathbf{U} \leq \mathbf{W} \leq \mathbf{V}$ and no such $\mathbf{W}$ can have a finite basis. This killed all hope that the set of finitely based varieties might in some sense be dense in the set of all varieties.

One of the questions provoked by the nature of these examples is the following.

QUESTION 2. *If all nilpotent groups in a (locally finite) variety are abelian, must the variety be finitely based?*

Another is prompted by the observation that elements of finite order appear to occur in the free groups of all these examples. Now it is easy to see that $\mathbf{AV}$ is always torsion-free (in the sense that its free groups are torsion-free) and that if $\mathbf{V}$ is not finitely based then neither is $\mathbf{AV}$: so torsion-free non-finitely-based varieties can also be made. However, here this was done at the cost of increasing soluble length and losing properties such as local nilpotence. Thus one may ask:

QUESTION 3. *Are all torsion-free metanilpotent varieties finitely based? Are all torsion-free subvarieties of* $\underline{\underline{A}}^3$ *(or even those of* $\underline{\underline{A}}^4$ *) finitely based? Are all torsion-free locally nilpotent varieties finitely based?*

The last of these questions is closely related, at least in one direction, to another problem mentioned by Professor Kostrikin in these Proceedings:

QUESTION 4. *Is every torsion-free locally nilpotent variety soluble (equivalently, nilpotent)?*

(The equivalence follows, for instance, from Lemma 4 of Groves [27].) While in general the meet of two torsion-free varieties need not be torsion-free, the intersection of a descending chain of torsion-free varieties is always torsion-free. Hence if the answer were negative, by Zorn's Lemma there would also exist minimal examples: torsion-free, insoluble, locally nilpotent varieties whose torsion-free proper subvarieties are all nilpotent.

We now turn to some of the positive results achieved since the publication of the book [50]. Many of these take the form that "all subvarieties of $\underline{\underline{V}}$ are finitely based": we shall paraphrase this as "$\underline{\underline{V}}$ is *hereditarily* finitely based". Perhaps the deepest result in this area, superseding many earlier ones and developing the technique initiated by Cohen [22] to its present limits, is due to Bryant and Newman [17]: $\underline{\underline{N}}_c\underline{\underline{A}} \wedge \underline{\underline{N}}_2\underline{\underline{N}}_c$ is hereditarily finitely based for every positive integer c. Others assert that the following varieties are hereditarily finitely based: $\underline{\underline{A}}_r\underline{\underline{A}}_s\underline{\underline{A}}_t$ provided $(s, rt) = 1$ and t is prime, Bryce and Cossey [19]; $\underline{\underline{A}}_m(\underline{\underline{B}}_n \wedge \underline{\underline{N}}_2)$ provided $(m, n) = 1$, Brady, Bryce, and Cossey [8]; and $\underline{\underline{B}}_6$, Atkinson [4].

QUESTION 5. *Which, if any, of the following varieties are hereditarily finitely based:* $\underline{\underline{A}}^3$, $\underline{\underline{A}}\underline{\underline{N}}_2$, $\underline{\underline{A}}_2(\underline{\underline{B}}_4 \wedge \underline{\underline{N}}_2)$, $\underline{\underline{N}}_c\underline{\underline{A}}$, $(\underline{\underline{B}}_4 \wedge \underline{\underline{N}}_3)\underline{\underline{A}}_2$, $(\underline{\underline{B}}_m \wedge \underline{\underline{N}}_2)(\underline{\underline{B}}_n \wedge \underline{\underline{N}}_2)$ *when* $(m, n) = 1$, $\underline{\underline{B}}_4$ *?*

Other positive results give ways of making new finitely based varieties from old. Abstracting the essence of an argument of Higman (34.23 in [50]), Brooks, Kovács, and Newman [10] introduced the concept of *strongly finitely based* variety and showed that if $\underline{\underline{U}}$ is strongly finitely based and $\underline{\underline{V}}$ is finitely based then $\underline{\underline{U}}\underline{\underline{V}}$ is finitely based. Indeed, if $\underline{\underline{V}}$ is also *strongly* finitely based, the same holds for $\underline{\underline{U}}\underline{\underline{V}}$. Bryant [14] proved that if $\underline{\underline{U}} \leq \underline{\underline{A}}\underline{\underline{N}}_c \wedge \underline{\underline{N}}_c\underline{\underline{A}}$ for some c and $\underline{\underline{V}}$ is (strongly) finitely based, then $\underline{\underline{U}} \vee \underline{\underline{V}}$ is (strongly) finitely based. On the other hand, a result of Vaughan-Lee [64] quoted above implies that not all finitely based varieties are strongly finitely based; in fact, Bryant pointed out in [16] that even $\underline{\underline{B}}_4 \wedge [\underline{\underline{A}}^2, \underline{\underline{E}}]$ fails to be strongly finitely based. Of the host of questions one might ask in this context, let us highlight just one.

QUESTION 6. *Are all Cross varieties strongly finitely based?*

One of the most intriguing general questions remains:

QUESTION 7. *Is the join of two finitely based varieties always finitely based?*

Deep positive partial results may be found in Bryant [14], [15]. We quote just one more: if $\underline{\underline{U}}$ is Cross while $\underline{\underline{V}}$ is locally finite and (hereditarily) finitely based, then $\underline{\underline{U}} \vee \underline{\underline{V}}$ is also (hereditarily) finitely based. This is derived in [15] by extending the method of proof of the Oates-Powell Theorem to its present limits. Finally, consider the following two propositions. If $\underline{\underline{U}} \leq \underline{\underline{W}}$ and $\underline{\underline{V}}$ is finitely based, then $\underline{\underline{UV}}$ is also finitely based. If $\underline{\underline{U}} \leq \underline{\underline{W}}$ and $\underline{\underline{V}}$ is finitely based, then $\underline{\underline{U}} \vee \underline{\underline{V}}$ is also finitely based. Both propositions are known to be valid provided $\underline{\underline{W}} \leq \underline{\underline{AN}}_c \wedge \underline{\underline{N}}_c\underline{\underline{A}}$ for some c, and neither is *known* to be valid otherwise. While this may be pure coincidence, the expectation is that the answer to Question 7 will be negative. This is further encouraged by the fact that Jónsson [37], [38] has shown the join of two finitely based varieties of lattices need not be finitely based.

PROBLEM 4 (page 23; insert "finitely generated" before "group").

No progress. One way towards a positive answer has been closed by the result that $\underline{\underline{K}}_p$ is not nilpotent (and hence not even soluble) if $p > 3$. This is due to Razmyslov [58]; for the case $p = 5$ it was found first by Bachmuth, Mochizuki, and Walkup [5]. In addition, Razmyslov identified [59] a just-non-Cross subvariety, satisfying the $(p-2)$th Engel condition, in each $\underline{\underline{K}}_p$ with $p > 3$.

PROBLEM 5 (page 42).

No progress: this is perhaps the most tantalizing problem of all. Many people feel there is a connection with the question (usually attributed to Tarski) concerning the existence of infinite groups in which all proper nontrivial subgroups are of prime order, but nobody seems to be able to prove even a one-way implication.

To facilitate discussion, call a variety *pseudo-abelian* if it is nonabelian but all its finite groups (equivalently, all its soluble groups) are abelian. In particular, a pseudo-abelian variety would not be generated by its finite groups. So far, the only way known for showing the existence of varieties not generated by their finite groups is still to point to $\underline{\underline{B}}_p$ with a large prime p and quote both Kostrikin's positive solution of the restricted Burnside problem and the negative solution by Novikov and Adyan of the unrestricted Burnside problem. This is one indication of the difficulties which would have to be overcome here.

At one time it was thought proved that all groups in a pseudo-abelian variety would have to be T-groups (groups in which normality is transitive; that is, all subnormal subgroups are normal). This claim survives as a conjecture supported by an unpublished partial result obtained independently by Kovács and Peter M. Neumann: if an element of squarefree order fails to normalize a subnormal subgroup of a group

G , then G has a metabelian, nonabelian factor and so cannot belong to any pseudo-abelian variety.

Any pseudo-abelian variety would have to contain a minimal pseudo-abelian variety (whose proper subvarieties are all abelian). This enables one to show that if a product contains a pseudo-abelian variety then at least one of the factors must also contain some (possibly different) pseudo-abelian variety. It follows that the class of those varieties which have no pseudo-abelian subvarieties, is closed with respect to all usual operations (meet, join, product, commutator) except (possibly) infinite joins, and contains, of course, all locally finite and all locally soluble varieties. One might therefore try to work within this class and prove theorems which would become generally valid if the existence of pseudo-abelian varieties were disproved. Unfortunately, this approach has failed in every case tried so far. As an example, let us note that if the existence of pseudo-abelian varieties were disproved one would wish to move on to questions like this:

QUESTION 8. *If all finite groups of a variety* $\underline{\underline{U}}$ *lie in a given (nonabelian) Cross variety* $\underline{\underline{V}}$ *, does* $\underline{\underline{U}} \leq \underline{\underline{V}}$ *follow?*

No counter-examples are known. However, assuming that $\underline{\underline{U}}$ has no pseudo-abelian subvarieties (or even that no pseudo-abelian varieties exist) seems to be no help at all, even if $\underline{\underline{V}}$ is taken as $\underline{\underline{B}}_p \wedge \underline{\underline{N}}_2$ for some large prime p . Apparently a positive answer to Question 8 cannot be derived from a dogma to the effect that there are no ghosts: what is needed is a surefire method for exorcising the pseudo-abelian ghost, and then an appropriate modification of the method might also yield the answer to Question 8.

The case of Question 8 with the special choice of $\underline{\underline{V}}$ mentioned above is vaguely related to another question: would the join of two pseudo-abelian varieties have to be pseudo-abelian? This could be answered positively if one knew that neither $\underline{\underline{B}}_4 \wedge \underline{\underline{N}}_2$ nor any $\underline{\underline{B}}_p \wedge \underline{\underline{N}}_2$ (with p an odd prime) can be contained in a join of two pseudo-abelian varieties (and used that $\underline{\underline{A}}_p\underline{\underline{A}}_q$, with p and q distinct primes, certainly cannot: see Kovács [45]).

Finally, we recall that Problem 5 arose in a discussion on what 'small' subvarieties must a variety possess. Theorem 21.4 of the book [50] could have been stated as follows: if $\underline{\underline{V}}$ is neither abelian nor pseudo-abelian, then it contains either an $\underline{\underline{A}}_p\underline{\underline{A}}_q$ (with p and q distinct primes) or $\underline{\underline{B}}_4 \wedge \underline{\underline{N}}_2$ or a $\underline{\underline{B}}_p \wedge \underline{\underline{N}}_2$ (with p some odd prime). Further, very much deeper, results of this kind are to be found in the contexts of just-non-Cross varieties (principal references: Brady [7], Ol'šanskiĭ [56]) and dichotomies (Groves [28], [29], [30], [31], Kargapolov and Čurkin [39]).

PROBLEM 6 (page 42).

No real progress. Peter M. Neumann (unpublished) improved the comment of the book [50]: a nonabelian variety other than $\underline{\underline{O}}$ in which verbal products with one normal amalgamation exist, would have to be pseudo-abelian. Meskin [49] noted that if $\underline{\underline{V}}$ is a variety of exponent 0 in which verbal products with one amalgamation exist, then the set of non-laws of $\underline{\underline{V}}$ is closed under those endomorphisms of the word group which map each variable to a nontrivial power of itself.

PROBLEM 7 (page 60).

The problem seems to have been asked, at least partly, in the hope that a positive answer would help make further examples of indecomposable varieties. This direction has been explored with some success through positive partial solutions of the second half of the problem by Brumberg [11] and Cossey [24]. However, the solution to both halves of the problem is, in general, negative: see Cossey [25].

The first half of the problem is easily seen to be equivalent to the following. Can a product of two nontrivial varieties ever be written as a join of two incomparable varieties, other than by writing the first factor as such a join and using the distributivity of right multiplication over joins? From this formulation the negative answer is almost evident; the example given by Cossey is a very simple one indeed.

The second part may be re-formulated similarly, with 'commutator' in place of 'join' (and one may as well omit 'incomparable'), but here the negative solution is far from obvious. Perhaps the most interesting positive result is due to Dunwoody [26] and Brumberg [11]; we describe it as a basis for an analogy to be drawn below. If $\underline{\underline{U}}, \underline{\underline{V}}, \underline{\underline{X}}, \underline{\underline{Y}}$ are varieties such that $\underline{\underline{X}} \neq \underline{\underline{E}}$ and $\underline{\underline{XY}} = [\underline{\underline{U}}, \underline{\underline{V}}] \neq \underline{\underline{O}}$, then there exist varieties $\underline{\underline{U}}', \underline{\underline{V}}'$ such that $\underline{\underline{U}}'\underline{\underline{Y}} \leq \underline{\underline{U}}$, $\underline{\underline{V}}'\underline{\underline{Y}} \leq \underline{\underline{V}}$, and $\underline{\underline{X}} = [\underline{\underline{U}}', \underline{\underline{V}}']$ (so $\underline{\underline{XY}} = [\underline{\underline{U}}'\underline{\underline{Y}}, \underline{\underline{V}}'\underline{\underline{Y}}] = [\underline{\underline{U}}, \underline{\underline{V}}]$). Thus if a product $\underline{\underline{XY}}$ (with $\underline{\underline{X}} \neq \underline{\underline{E}}$, $\underline{\underline{XY}} \neq \underline{\underline{O}}$) admits a commutator-decomposition, so does the first factor $\underline{\underline{X}}$; and in fact a commutator-decomposition of $\underline{\underline{X}}$ may be chosen so that the commutator-decomposition of the product obtained from it (by the appropriate distributive law) is 'smaller than or equal to' the original. In particular, if a commutator-decomposition $[\underline{\underline{U}}, \underline{\underline{V}}]$ of the product $\underline{\underline{XY}}$ is minimal (in the sense that $\underline{\underline{U}}_1 < \underline{\underline{U}}$, $\underline{\underline{V}}_1 < \underline{\underline{V}}$ imply $[\underline{\underline{U}}_1, \underline{\underline{V}}] < [\underline{\underline{U}}, \underline{\underline{V}}] > [\underline{\underline{U}}, \underline{\underline{V}}_1]$) then it comes from a minimal commutator-decomposition of the first factor $\underline{\underline{X}}$, and *vice versa*. Thus at least the minimal commutator-decompositions of $\underline{\underline{X}}$ and $\underline{\underline{XY}}$ are in one-to-one correspondence. It is easy to see that each commutator-decomposition $[\underline{\underline{U}}, \underline{\underline{V}}]$ of a variety is comparable to a minimal one: if U, V are the corresponding verbal subgroups of an absolutely free group F of infinite rank, let $C/[U, V]$ be the centralizer of $V/[U, V]$ in $F/[U, V]$, and $D/[U, V]$ the centralizer of $C/[U, V]$; then $C \geq U$, $D \geq V$, $[C, D] = [U, V]$, while C and D are verbal in F; so $[\text{var}F/C, \text{var}F/D]$ is a minimal commutator-decomposition of $[\underline{U}, \underline{V}]$, comparable to the original. Consequently, the negative solution demonstrates and exploits the existence of distinct, comparable, commutator-decompositions of certain varieties, while the positive partial solutions are obtained in situations where such ambiguities can be ruled out.

As this account shows, the context of the second half of Problem 7 is, by now, fairly well understood. By contrast, the situation surrounding the first half of the problem has remained largely unexplored. The first question suggested by analogy is whether a product can ever be a (proper, finite) join without the first factor being (trivial or such) a join. In other words:

QUESTION 9. *If* $\underline{X}$ *is a nontrivial join-irreducible and* $\underline{Y}$ *an arbitrary variety, is the product* $\underline{XY}$ *necessarily also join-irreducible?*

Kovács [45] shows that the answer is positive if either $\underline{X}$ is abelian or the infinite-rank free groups of $\underline{X}$ have no nontrivial abelian verbal subgroups. However, it is not known whether $(\underline{B}_4 \wedge \underline{N}_2)\underline{Y}$ or $(\underline{B}_p \wedge \underline{N}_2)\underline{Y}$ (for odd primes p) is join-irreducible for every $\underline{Y}$. (For $\underline{Y} = \underline{A}_m$ with m a divisor of $p - 1$, this problem has been settled positively by Woeppel [66]; see also [67].) If the answer were positive in general, one would proceed to ask whether each proper join-decomposition of a product is comparable to one obtained from one for the first factor:

QUESTION 10. *Does* $\underline{Y} \neq \underline{XY} = \underline{U} \vee \underline{V} \neq \underline{U} \nleq \underline{V}$ *imply that* $\underline{X} = \underline{U}' \vee \underline{V}'$ *for some* $\underline{U}'$, $\underline{V}'$ *with* $\underline{U}'\underline{Y} \leq \underline{U}$ *and* $\underline{V}'\underline{Y} \leq \underline{V}$ *? Equivalently, does every minimal proper join-decomposition* $\underline{U} \vee \underline{V}$ *of a product* $\underline{XY}$ *(with* $\underline{X} \neq \underline{E}$ *) come from a (necessarily minimal) proper join-decomposition* $\underline{X} = \underline{U}' \vee \underline{V}'$ *(in the sense that* $\underline{U} = \underline{U}'\underline{Y}$ *,* $\underline{V} = \underline{V}'\underline{Y}$ *)?*

Here $\underline{U} \vee \underline{V}$ is a minimal join-decomposition if $\underline{U}_1 < \underline{U}$, $\underline{V}_1 < \underline{V}$ imply $\underline{U}_1 \vee \underline{V} < \underline{U} \vee \underline{V} > \underline{U} \vee \underline{V}_1$. The equivalence claimed depends on the fact that every join-decomposition is comparable to a minimal one. (Prove this as the corresponding fact for commutator-decompositions, but instead of using centralizers choose $C/(U \cap V)$ maximal among the verbal subgroups of $F/(U \cap V)$ which avoid $V/(U \cap V)$, and so on: Zorn's Lemma makes this possible.) The 'converse' to Question 10 may be a separate question:

QUESTION 11. *If* $\underline{U} \vee \underline{V}$ *is a minimal join decomposition and* $\underline{Y} \neq \underline{U}$ *, is* $\underline{UY} \vee \underline{VY}$ *necessarily also minimal?*

Of course, a positive answer to Question 10 would imply one for Question 11, but it is not known whether the implication goes the other way as well.

PROBLEMS 8, 9 (page 69).

The only directly relevant work we know of is Baumslag's paper [6], which solved

Problem 8 partly and Problem 9 fully, and was already reported in the footnote on this page of the book [50].

One may well ask the question more generally (that is, not only for product varieties):

QUESTION 12. *Suppose* $F_k(\underline{\underline{V}})$ *generates* $\underline{\underline{V}}$, *and* $n > k$. *Is* $F_n(\underline{\underline{V}})$ *residually k-generator? Is* $F_n(\underline{\underline{V}})$ *residually* $F_k(\underline{\underline{V}})$? *Is* $F_n(\underline{\underline{V}})$ *residually* $F_{n-1}(\underline{\underline{V}})$?

Of course, these three questions are related, but it does not seem to be known whether any two of them are actually equivalent. Some answers are available for certain varieties defined via commutators: see the forthcoming paper [33] of Gupta and Levin. In particular, they show that $F_n([\underline{\underline{A}}^2, \underline{\underline{E}}])$ is residually $F_{n-1}([\underline{\underline{A}}^2, \underline{\underline{E}}])$ if and only if $n \neq 2$ and $n \neq 4$ (see also the comments after Question 13 below).

This leads on to residual properties of relatively free groups in general. The most interesting question seems to be:

QUESTION 13. *Are all soluble relatively free groups residually finite?*

The difficulties of progress beyond the results reported on in the book (26.31 in [50]) are best illustrated by the fact that even the residual finiteness of the $F_n([\underline{\underline{A}}^2, \underline{\underline{E}}])$ had not been conclusively established until recently, C.K. Gupta [32]. She showed also that for $n \leq 3$ these groups are torsion-free but for $n > 3$ they are not: indeed, $[\underline{\underline{A}}^2, \underline{\underline{E}}]$ is generated by its free group of rank 4 , and is the proper join of a nilpotent variety of 2-groups with the torsion-free variety generated by $F_2([\underline{\underline{A}}^2, \underline{\underline{E}}])$.

PROBLEM 10 (page 72).

No progress to report.

PROBLEM 11 (page 92).

The negative solution has been discussed with Problems 2 and 3.

PROBLEM 12 (page 101).

Unsolved. The very deep work of Ward mentioned in the remark preceding the problem was published in [65].

PROBLEM 13 (page 101).

The solution is in the negative. The classification of all subvarieties of $\underline{\underline{B}}_p \wedge [\underline{\underline{A}}^2, \underline{\underline{E}}] \wedge \underline{\underline{N}}_{p-1}$ by Stewart [61] yields this (see Stewart [60] for an explicit derivation), for instance, with $k = 3$, $c = 5$, $\underline{\underline{U}}_6 = \underline{\underline{B}}_7 \wedge [\underline{\underline{A}}^2, \underline{\underline{E}}] \wedge \underline{\underline{N}}_6$: the smallest free group to generate $\underline{\underline{U}}_6$ is $F_3(\underline{\underline{U}}_6)$, the centre of $F_6(\underline{\underline{U}}_6)$ contains the second derived group which is not in the last term of the lower central series, but $\underline{\underline{U}}_5$ $(= \underline{\underline{U}}_6 \wedge \underline{\underline{N}}_5)$ is also generated by its free group of rank 3 - indeed, even by its free group of rank 2 . On the other hand, the condition cannot be omitted altogether; Cossey had shown [23] that it cannot even be replaced by insisting that the varieties in question be torsion-free, or that they have prime-power exponent.

Hanna Neumann's lead-up to Problem 13 started with the comment: "As one might expect, the minimal rank of a generating group of a nilpotent variety is in general a non-decreasing function of the class". Her 35.21 is a specific instance of this general and intuitive statement. The nature of the negative solution of Problem 13 prompts one to look for other formulations. For example:

QUESTION 14. *If* $[\underline{\underline{U}}, \underline{\underline{E}}]$ *is generated by its free group of rank* k *, must the same be true of* $\underline{\underline{U}}$ *(at least if* $\underline{\underline{U}}$ *is nilpotent)? Does* $[\underline{\underline{U}}, \underline{\underline{E}}] = [\mathrm{var}F_k(\underline{\underline{U}}), \underline{\underline{E}}]$ *imply* $\underline{\underline{U}} = \mathrm{var}F_k(\underline{\underline{U}})$ *(at least if* $\underline{\underline{U}}$ *is nilpotent)?*

The first hypothesis implies the second. Note that Cossey has shown [25] that $[\underline{\underline{U}}, \underline{\underline{E}}] = [\underline{\underline{U}}_0, \underline{\underline{E}}]$ and $\underline{\underline{U}} \geq \underline{\underline{U}}_0$ need not imply $\underline{\underline{U}} = \underline{\underline{U}}_0$: however, in his example $\underline{\underline{U}}$ was not nilpotent, and $\underline{\underline{U}}_0$ was not generated by a free group of $\underline{\underline{U}}$. Note that a positive answer for the nilpotent case of Question 14 would, like 35.21, yield Corollary 35.22 of [50]. A positive answer to the following question would also be good enough for 35.22:

QUESTION 15. *If* $\underline{\underline{U}} \leq \underline{\underline{N}}_c$ *and* $[\underline{\underline{U}}, \underline{\underline{E}}]$ *is generated by its free group of rank* k *, must the same be true of* $\underline{\underline{N}}_c \wedge [\underline{\underline{U}}, \underline{\underline{E}}]$ *?*

PROBLEM 14 (page 102).

The suggestion that $d(c)$ might be $[c/2] + 1$ (and part of 35.35 of [50]) has to be replaced by $d(c) = c - 1$, established (for $c > 2$) by Kovács, Newman, Pentony [46] and Levin [48]. The latter paper contains also some further information, as does Vaughan-Lee [63]. Much work has been done on the varieties generated by groups of the form $F_k(\underline{\underline{N}}_c)$ by Chau [20], [21] and especially Pentony [57].

PROBLEM 15 (page 113).

No new information is available.

PROBLEM 16 (page 114).

Unpublished work of Peter M. Neumann together with the results of Groves [31] provide a positive solution for the case of metanilpotent varieties. In fact, they show that if in a metanilpotent variety all finitely generated groups are Hopf, then these groups are also residually finite and satisfy the maximum condition for normal subgroups.

PROBLEM 17 (page 125).

The negative solution is due to Adyan [2]; see also his paper [3] in these Proceedings.

PROBLEM 18 (page 128).

The positive solution was given by Bronšteĭn [9].

PROBLEMS 19, 20 (page 133).

No new information has come to our attention.

PROBLEMS 21, 22 (pages 141, 142).

For the exponent zero case, the solution to both problems is negative: it was obtained independently by Peter M. Neumann [51], [52] and Ol'šanskiĭ [54]. The counterexamples are made by joining $\underline{\underline{A}}$ to a variety of finite exponent which is known to be 'bad' and showing that the join remains 'bad'. One should replace 'exponent zero' by 'torsion-free' to revive these parts of the problem; the prime-power-exponent parts are open. Neumann and Ol'šanskiĭ give a lot of detailed information. Houghton's results on direct decompositions, partly reported in the book [50], have been published in [35] and [35a]. Bryant [13] and Bryce [18] have done much to explore splitting groups.

PROBLEM 23 (page 166).

Unsolved. Heineken and Peter M. Neumann claimed [34] and Jones eventually proved [36] that no variety other than $\underline{\underline{O}}$ contains infinitely many isomorphism types of the nonabelian finite simple groups which are now known. Related questions are whether any variety other than $\underline{\underline{O}}$ can contain an infinite simple locally finite group (Question IV.7 in the book [40] by Kegel and Wehrfritz), and whether any locally finite variety can contain infinite simple groups (Question IV.6 in [40]); see also Kovács [44].

PROBLEM 24 (page 171).

The paper of Burns quoted by Hanna Neumann in the lead-up to this problem had already answered positively the first half, and suggested the alternative which she presumably intended to put here. That, and the second half, have also been answered (at least for some small values of the parameters) in Kovács [43]: the first positively, the second negatively.

PROBLEM 25 (page 174).

The positive solution, and a lot more, was given by Bryant in [12].

References

[1] С.И. Адян [S.I. Adyan], "Бесконечные неприводимые системы групповых тождеств" [Infinite irreducible systems of group identities], *Izv. Akad. Nauk SSSR Ser. Mat.* 34 (1970), 715-734; *Math. USSR-Izv.* 4 (1970), 721-739 (1971). MR44#4078.

[2] С.И. Адян [S.I. Adyan], "О подгруппах свободных периодических групп нечетного показателя" [Subgroups of free groups of odd exponent], *Trudy Mat. Inst. Steklov.* 112 (1971), 64-72.

[3] S.I. Adyan, "Periodic groups of odd exponent", these Proc.

[4] M.D. Atkinson, "Alternating trilinear forms and groups of exponent 6 ", *J. Austral. Math. Soc.* 16 (1973), 111-128.

[5] Seymour Bachmuth, Horace Y. Mochizuki and David Walkup, "A nonsolvable group of exponent 5 ", *Bull. Amer. Math. Soc.* 76 (1970), 638-640. MR41#1862.

[6] Gilbert Baumslag, "Some theorems on the free groups of certain product varieties", *J. Combinatorial Theory* 2 (1967), 77-79. MR34#5902.

[7] J.M. Brady, "On the classification of just-non-Cross varieties of groups", *Bull. Austral. Math. Soc.* 3 (1970), 293-311. MR44#289.

[8] J.M. Brady, R.A. Bryce and John Cossey, "On certain abelian-by-nilpotent varieties", *Bull. Austral. Math. Soc.* 1 (1969), 403-416. MR41#1843.

[9] М.А. Бронштейн [M.A. Bronšteĭn], "О вербальных подгруппах свободных групп" [Verbal subgroups of free groups], *Dokl. Akad. Nauk SSSR* 177 (1967), 255-257; *Soviet Math. Dokl.* 8 (1967), 1386-1388. MR36#3861.

[10] M.S. Brooks, L.G. Kovács and M.F. Newman, "A finite basis theorem for product varieties of groups", *Bull. Austral. Math. Soc.* 2 (1970), 39-44. MR42#1882.

[11] Н.Р. Брумберг [N.R. Brumberg], "О взаимном коммутанте двух многообразий групп" [On the commutator of two varieties of groups], *Mat. Sb. (NS)* 79 (121) (1969), 37-58. MR39#5672.

[12] Roger M. Bryant, "On s-critical groups", *Quart. J. Math. Oxford Ser.* (2) 22 (1971), 91-101. MR44#2805.

[13] Roger M. Bryant, "Finite splitting groups in varieties of groups", *Quart. J. Math. Oxford Ser.* (2) 22 (1971), 169-172. MR44#5366.

[14] Roger M. Bryant, "On join varieties of groups", *Math. Z.* 119 (1971), 143-148. MR43#6293.

[15] Roger M. Bryant, "On locally finite varieties of groups", *Proc. London Math. Soc.* (3) 24 (1972), 395-408. MR45#8708.

[16] Roger M. Bryant, "Some infinitely based varieties of groups", *J. Austral. Math. Soc.* 16 (1973), 29-32.

[17] R.M. Bryant and M.F. Newman, "Some finitely based varieties of groups", *Proc. London Math. Soc.* (3) 28 (1974), 237-252.

[18] R.A. Bryce, "Projective groups in varieties", *Bull. Austral. Math. Soc.* 6 (1972), 169-174. MR45#3527.

[19] R.A. Bryce and John Cossey, "Some product varieties of groups", *Bull. Austral. Math. Soc.* 3 (1970), 231-264. MR42#4618.

[20] T.C. Chau, "The laws of some nilpotent groups of small rank" (PhD thesis, Australian National University, Canberra, 1968; Abstract: *Bull. Austral. Math. Soc.* 2 (1970), 277-279).

[21] T.C. Chau, "The laws of some nilpotent groups of small rank", *J. Austral. Math. Soc.* (to appear).

[22] D.E. Cohen, "On the laws of a metabelian variety", *J. Algebra* 5 (1967), 267-273. MR34#5929.

[23] John Cossey, "On a problem of Hanna Neumann", *Math. Z.* 106 (1968), 187-190. MR37#6357.

[24] John Cossey, "Some classes of indecomposable varieties of groups", *J. Austral. Math. Soc.* 9 (1969), 387-398. MR40#4341.

[25] John Cossey, "On decomposable varieties of groups", *J. Austral. Math. Soc.* 11 (1970), 340-342. MR42#3155.

[26] M.J. Dunwoody, "On product varieties", *Math. Z.* 104 (1968), 91-97. MR37#291.

[27] J.R.J. Groves, "On varieties of soluble groups", *Bull. Austral. Math. Soc.* 5 (1971), 95-109. MR45#3528.

[28] J.R.J. Groves, "Varieties of soluble groups and a dichotomy of P. Hall", *Bull. Austral. Math. Soc.* 5 (1971), 391-410. Zbl.217,76.

[29] J.R.J. Groves, "On varieties of soluble groups II", *Bull. Austral. Math. Soc.* 7 (1972), 437-441. Zbl.241.20023.

[30] J.R.J. Groves, "An extension of a dichotomy of P. Hall to some varieties of groups", *Arch. der Math.* 23 (1972), 573-580.

[31] J.R.J. Groves, "On some finiteness conditions for varieties of metanilpotent groups", *Arch. der Math.* 24 (1973), 252-268.

[32] Chander Kanta Gupta, "The free centre-by-metabelian groups", *J. Austral. Math. Soc.* 16 (1973), 294-299.

[33] Narain Gupta and Frank Levin, "Generating groups of certain soluble varieties", *J. Austral. Math. Soc.* (to appear).

[34] Hermann Heineken and Peter M. Neumann, "Identical relations and decision procedures for groups", *J. Austral. Math. Soc.* 7 (1967), 39-47. MR34#5931.

[35] C.H. Houghton, "Direct decomposability of reduced free groups", *J. London Math. Soc.* 43 (1968), 534-538. MR37#2840.

[35a] C.H. Houghton, "Directly decomposable finite relatively free groups", *J. London Math. Soc.* (2) 4 (1971), 381-384. MR46#7382.

[36] Gareth A. Jones, "Varieties and simple groups", *J. Austral. Math. Soc.* (to appear).

[37] Bjarni Jonsson, "The sum of two finitely based lattice varieties need not be finitely based", *Notices Amer. Math. Soc.* 21 (1974), A-2.

[38] Bjarni Jónsson, "Sums of finitely based lattice varieties", preprint.

[39] М.И. Каргаполов, В.А. Чуркин [M.I. Kargapolov, V.A. Curkin], "О многообразиях разрешимых групп" [On varieties of soluble groups], *Algebra i Logika* 10 (1971), 651-657.

[40] Otto H. Kegel and Bertram A.F. Wehrfritz, *Locally finite groups* (North-Holland Mathematical Library, 3. North-Holland, Amsterdam, London; American Elsevier, New York, 1973).

[41] Ю.Г. Клейман [Yu.G. Kleĭman], "О базисе произведения многообразий групп", [On bases for product varieties of groups], *Izv. Akad. Nauk SSSR Ser. Mat.* 37 (1973), 95-97.

[42] L.G. Kovács, "On the number of varieties of groups", *J. Austral. Math. Soc.* 8 (1968), 444-446. MR37#5277.

[43] L.G. Kovács, "A remark on critical groups", *J. Austral. Math. Soc.* 9 (1969), 465-466. MR39#5674.

[44] L.G. Kovács, "Varieties and finite groups", *J. Austral. Math. Soc.* 10 (1969), 5-19. MR40#1459.

[45] L.G. Kovács, "Inaccessible varieties of groups", *J. Austral. Math. Soc.* (to appear).

[46] L.G. Kovács, M.F. Newman and P.F. Pentony, "Generating groups of nilpotent varieties", *Bull. Amer. Math. Soc.* 74 (1968), 968-971. MR37#5276.

[47] L.G. Kovács and M.R. Vaughan-Lee, "A problem of Hanna Neumann on closed sets of group words", *Bull. Austral. Math. Soc.* 5 (1971), 341-342. MR45#3533.

[48] Frank Levin, "Generating groups for nilpotent varieties", *J. Austral. Math. Soc.* 11 (1970), 28-32. MR41#1844.

[49] Stephen Meskin, "Some varieties without the amalgam embedding properties", *Bull. Austral. Math. Soc.* 1 (1969), 417-418. MR41#1845.

[50] Hanna Neumann, *Varieties of groups* (Ergebnisse der Mathematik und ihrer Grenzgebiete, Band 37. Springer-Verlag, Berlin, Heidelberg, New York, 1967). MR35#6734.

[51] Peter M. Neumann, "Splitting groups and projectives in varieties of groups", *Quart. J. Math. Oxford Ser.* (2) 18 (1967), 325-332. MR36#3859.

[52] Peter M. Neumann, "A note on the direct decomposability of relatively free groups", *Quart. J. Math. Oxford Ser.* (2) 19 (1968), 67-79. MR36#6485.

[53] M.F. Newman, "Just non-finitely-based varieties of groups", *Bull. Austral. Math. Soc.* 4 (1971), 343-348. MR43#4891.

[54] А.Ю. Ольшанский [A.Ju. Ol'šanskiĭ], "Об одной задаче Ханны Нейман" [On a problem of Hanna Neumann], *Mat. Sb. (NS)* 76 (118) (1968), 449-453. MR37#2841.

[55] А.Ю. Ольшанский [A.Ju. Ol'šanskiĭ], "О проблеме конечного базиса тождеств в группах" [On the problem of a finite basis for the identities of groups], *Izv. Akad. Nauk SSSR* 34 (1970), 376-384; *Math. USSR-Izv.* 4 (1970), 381-389 (1971). MR44#4079.

[56] А.Ю. Ольшанский [A.Ju. Ol'šanskiĭ], "Разрешимые почти-кроссовы многообразия групп" [Soluble just non-Cross varieties of groups], *Mat. Sb. (NS)* 85 (127) (1971), 115-131; *Math. USSR-Sb.* 14 (1971), 115-129 (1972).

[57] Paul Pentony, "Laws in torsion-free nilpotent varieties with particular reference to the laws of free nilpotent groups" (PhD thesis, Australian National University, Canberra, 1970; Abstract: *Bull. Austral. Math. Soc.* 5 (1971), 283-284).

[58] Ю.П. Размыслов [Ju.P. Razmyslov], "Об Энгелевых алгебрах Ли" [On Engel Lie algebras], *Algebra i Logika* 10 (1971), 33-44; *Algebra and Logic* 10 (1971), 21-29. MR45#3498.

[59] Ю.П. Размыслов [Ju.P. Razmyslov], "Об одном примере неразрешимых почти кроссовых многообразий групп" [An example of an insoluble just-non-Cross variety of groups], *Algebra i Logika* 11 (1972), 186-205.

[60] A.G.R. Stewart, "On centre-extended-by-metabelian groups" (PhD thesis, Australian National University, Canberra, 1968).

[61] A.G.R. Stewart, "On centre-extended-by-metabelian groups", *Math. Ann.* 185 (1970), 285-302. MR41#6945.

[62] M.R. Vaughan-Lee, "Uncountably many varieties of groups", *Bull. London Math. Soc.* 2 (1970), 280-286. MR43#2054.

[63] M.R. Vaughan-Lee, "Generating groups of nilpotent varieties", *Bull. Austral. Math. Soc.* 3 (1970), 145-154. MR43#2091.

[64] M.R. Vaughan-Lee, "On product varieties of groups", *Bull. Austral. Math. Soc.* 5 (1971), 239-240. MR45#3529.

[65] M.A. Ward, "Basic commutators", *Philos. Trans. Roy. Soc. London Ser. A* 264 (1969), 343-412. MR40#4379.

[66] James Joseph Woeppel, "Finite groups generating a product variety" (PhD thesis, University of Illinois at Urbana-Champaign, 1970).

[67] James J. Woeppel, "Join-irreducible Cross product varieties of groups", *Trans. Amer. Math. Soc.* (to appear).

Australian National University,
Canberra.

PROC. SECOND INTERNAT. CONF. THEORY OF GROUPS, 20B99, 50D35
CANBERRA, 1973, pp. 432-436.

FINITE PROJECTIVE PLANES AND SHARPLY 2-TRANSITIVE SUBSETS OF FINITE GROUPS

Peter Lorimer

A sharply 2-transitive subset of a permutation group G acting on a set Σ is a subset R of G with the properties

(1) if $\alpha, \beta, \gamma, \delta \in \Sigma$, $\alpha \neq \beta$, $\gamma \neq \delta$, R contains a unique member r with $r(\alpha) = \gamma$, $r(\beta) = \delta$,

(2) $1 \in R$,

(3) the relation $\sim$ defined on R by $r \sim s$ if $r = s$ or $r^{-1}s$ fixes no symbol of Σ is an equivalence relation and each equivalence class is sharply transitive on Σ , that is, if $\alpha, \gamma \in \Sigma$ each class contains exactly one member r with $r(\alpha) = \gamma$.

Algebraically a sharply 2-transitive set can be interpreted as a ternary ring in the sense of Hall [10] and the existence of such a set is equivalent to the existence of a projective plane having order the cardinality of Σ . The set R determines a plane uniquely up to isomorphism and a plane together with two distinguished points determines a set R uniquely up to a type of isomorphism.

The conditions (1), (2) and (3) on R admit some simplification. If S is a subset of G satisfying (1) and $s \in S$ then the subset $s^{-1}S$ of G satisfies (1) and (2). If Σ is a finite set (3) follows from (1).

In the sequel we will only be concerned with the case that Σ (and hence R and G) are finite.

The classification problem for projective planes leads to a classification problem for permutation groups having a sharply 2-transitive subset. In terms of our present knowledge we can provide partial answers to this question in three directions. We will give here some answers to the following three questions.

(I) *Which permutation groups are known to have sharply 2-transitive subsets?*

(II) *Which permutation groups are known not to have sharply 2-transitive subsets?*

(III) *If a permutation group has a sharply 2-transitive subset what can be said about its structure?*

(I) The only finite symmetric groups known to have sharply 2-transitive subsets are those having degree a power of a prime number. (That is, the order of every known projective plane is a prime power.) If V is an elementary abelian group and A its group of automorphisms the semidirect product $V \times A$ acting as a permutation group on V always has a sharply 2-transitive subset (corresponding to a Desarguesian plane) and even more than one (corresponding to a non-Desarguesian translation plane) if V is not cyclic. If the two distinguished points are chosen correctly, the sharply 2-transitive set arising from any translation plane is a subset of $V \times A$. (See Dembowski [6].)

Information about permutation groups containing other known sharply 2-transitive sets does not seem to be readily available.

(II) Trivially, any permutation group which is not 2-transitive cannot have a sharply 2-transitive subset.

The only symmetric groups known not to have a sharply 2-transitive subset are those of degree n for those $n \equiv 1$ or $2 \bmod 4$ which cannot be written as the sum of two squares. This is the content of the Bruck-Ryser Theorem [5].

The present author has shown that the following 2-transitive groups do not have a sharply 2-transitive subset in their natural 2-transitive representation:

the semilinear projective groups $P\Gamma L(n, q)$, $n \geq 2$, $q \geq 5$ [19]. The most difficult case $n = 2$ has also been proved by Patrick D'Arcy using the methods of [22];

the automorphism groups of the projective unitary groups $PU(3, q^2)$ [20];

groups of Ree type [21] and their automorphism groups (unpublished);

the automorphism groups of the Suzuki groups [22]; the linear group $L_4(2)$ (unpublished).

In addition PSL(2, 11) acting on 11 symbols has no such subset (unpublished).

The position of other 2-transitive groups does not seem to be known.

(III) The following theorem is obtained by combining the results in (II) with characterization theorems for 2-transitive groups. In each case a reference to the characterization theorem follows the statement of the result. We use the common subscript notation for stabilizers.

THEOREM. *Let G be a 2-transitive group acting on a set Σ. If G contains a sharply 2-transitive subset but does not contain a regular normal subgroup then*

A. (1) G_α *does not have a normal subgroup regular on* $\Sigma - \{\alpha\}$ (Hering-Kantor-Seitz [14]).

(2) *If* G_α *has an abelian normal subgroup it is semiregular on* $\Sigma - \{\alpha\}$ *except possibly when* G *is isomorphic to a subgroup of* $P\Gamma L(n, q)$ *containing* $PSL(n, q)$ *and* $n > 2$, $q = 2, 3, 4$ (O'Nan [24]).

(3) $G_{\alpha\beta}$ *is not abelian* (Aschbacher [2]).

(4) $G_{\alpha\beta}$ *is not dihedral of order* $2p$, p *prime* (Kimura [15]).

(5) *If* t *is an involution of* G, *the centralizer of* t *contains a non-identity element stabilizing a different number of members of* Σ *from* t (Harada [11]).

(6) G *contains an involution fixing at least four points except possibly when* $|\Sigma| = 11$ *and* $G \simeq M_{11}$ (Bender [3, 4], Hering [13], King [16, 17], Harada [12]).

(7) *If* $|\Sigma| \not\equiv 0 \pmod 8$ *then* G *contains an involution fixing* 5 *points except possibly when* $|\Sigma| = 10$ *and* $G \simeq S_6$ *or when* $|\Sigma| = 12$ *and* $G \simeq M_{11}$ *or* M_{12} (Noda [23]).

(8) *If* $|\Sigma| = 1 + np$ *where* p *is a prime,* $n < p$, *and if* p^2 *divides* $|G|$ *then* G *contains the alternating group on* Σ. *If* $|\Sigma| = 29, 53, 149, 173, 269$ *or* 317 *then* G *contains the alternating group on* Σ (Appel-Parker [1]).

B. *If* $|\Sigma|$ *is even*

(1) $|G_{\alpha\beta} \cap G^x_{\alpha\beta}|$ *is even for some* $G^x_{\alpha\beta}$ *different from* $G_{\alpha\beta}$ (Harada [12]).

(2) *Either* $G_{\alpha\beta}$ *contains more than one nonidentity element fixing* 3 *symbols or* G *contains an involution fixing exactly* 0 *or* 2 *symbols* (Ree [25]).

C. *If* $|\Sigma| = 1 + p + p^2$ *where* p *is an odd prime and* p^3 *divides the order of* G *then* G *contains the alternating group on* Σ (Tsuzuku [26]).

In addition some purely group-theoretic characterization can be used to give information on these groups, for example the characterization of finite groups having dihedral Sylow subgroups by Gorenstein and Walter [7, 8, 9] can be used to deduce that a group satisfying the hypotheses of the Theorem cannot contain a dihedral Sylow 2-subgroup. This conclusion however depends on knowledge of the 2-transitive representations of the groups $P\Gamma L(2, q)$, q odd and A_7. In general knowledge of the different 2-transitive representation of the groups involved does not seem to be available so it seems best not to attempt to state anything like the previous theorem because of all the possible exceptions that would be involved in its statement. Also the results of Harada [12] when $|\Sigma|$ is odd could be used to give more information particularly about the Sylow 2-subgroups of G.

CORRECTION. The assertion that $V \times A$ has more than one sharply 2-transitive subset if V is not cyclic needs modification. There is only one projective plane of each of the orders 4 and 8 so the condition that V should have more than 8 members must be added.

References

[1] K.I. Appel and E.T. Parker, "On unsolvable groups of degree $p = 4q + 1$, p and q primes", *Canad. J. Math.* 19 (1967), 583-589. MR36#256.

[2] Michael Aschbacher, "Doubly transitive groups in which the stabilizer of two points is abelian", *J. Algebra* 18 (1971), 114-136. MR43#2059.

[3] Helmut Bender, "Endliche zweifach transitive Permutationsgruppen, deren Involutionen keine Fixtpunkte haben", *Math. Z.* 104 (1968), 175-204. MR37#2846.

[4] Helmut Bender, "Transitive Gruppen gerader Ordnung, in denen jede Involution genau einen Punkt festlässt", *J. Algebra* 17 (1971), 527-554. MR44#5370.

[5] R.H. Bruck and H.J. Ryser, "The nonexistence of certain finite projective planes", *Canad. J. Math.* 1 (1949), 88-93. MR10,319.

[6] P. Dembowski, *Finite Geometries* (Ergebnisse der Mathematik und ihrer Grenzgebiete, Band 44. Springer-Verlag, Berlin, Heidelberg, New York, 1968). MR38#1597.

[7] Daniel Gorenstein and John H. Walter, "The characterization of finite groups with dihedral Sylow 2-subgroups. I", *J. Algebra* 2 (1965), 85-151. MR31#1297a.

[8] Daniel Gorenstein and John H. Walter, "The characterization of finite groups with dihedral Sylow 2-subgroups - II", *J. Algebra* 2 (1965), 218-270. MR31#1297b.

[9] Daniel Gorenstein and John H. Walter, "The characterization of finite groups with dihedral Sylow 2-subgroups. III", *J. Algebra* 2 (1965), 354-393. MR32#7634.

[10] Marshall Hall, "Projective planes", *Trans. Amer. Math. Soc.* 54 (1943), 229-277. MR5,72.

[11] Koichiro Harada, "A characterization of the Zassenhaus groups", *Nagoya Math. J.* 33 (1968), 117-127. MR39#2854.

[12] Koichiro Harada, "On some doubly transitive groups", *J. Algebra* 17 (1971), 437-450. MR43#7499.

[13] Christoph Hering, "Zweifach transitive Permutationsgruppen, in denen 2 die maximale Anzahl von Fixpunkten von Involutionen ist", *Math. Z.* 104 (1968), 150-174. MR37#295.

[14] Christoph Hering and William M. Kantor and Gary M. Seitz, "Finite groups with a split *BN*-pair of rank 1 . I", *J. Algebra* 20 (1972), 435-475. MR46#243.

[15] Hiroshi Kimura, "On doubly transitive permutation groups of degree n and order $2p(n-1)n$ ", *Osaka J. Math.* 7 (1970), 275-290. MR44#2810.

[16] Jennifer D. King, "A characterization of some doubly transitive groups", *Math. Z.* 107 (1968), 43-48. Corrections: *Math. Z.* 112 (1969), 393-394.MR38#3329

[17] Jennifer D. King, "Doubly transitive groups in which involutions fix one or three points", *Math. Z.* 111 (1969), 311-321. MR40#5717.

[18] Peter Lorimer, "A note on doubly transitive groups", *J. Austral. Math. Soc.* 6 (1966), 449-451. MR34#7626.

[19] Peter Lorimer, "A property of the groups $P\Gamma L(m, q)$, $q \geq 5$ ", *Proc. Amer. Math. Soc.* 37 (1973), 393-396.

[20] Peter Lorimer, "A property of the groups Aut $PU(3, q^2)$ ", *Pacific J. Math.* 46 (1973), 225-230.

[21] Peter Lorimer, "A note on groups of Ree type", *Canad. Math. Bull.* (to appear).

[22] Peter Lorimer, "A property of the Suzuki group", *Nanta Math.* (to appear).

[23] Ryuzaburo Noda, "Doubly transitive groups in which the maximal number of fixed points of involutions is four", *Osaka J. Math.* 8 (1971), 77-90. MR45#3546.

[24] Michael O'Nan, "A characterization of $L_n(q)$ as a permutation group", *Math. Z.* 127 (1972), 301-314.

[25] Rimhak Ree, "Sur une famille de groupes de permutations doublement transitifs", *Canad. J. Math.* 16 (1964), 797-820. MR31#241.

[26] Tosiro Tsuzuku, "On doubly transitive permutation groups of degree $1 + p + p^2$ where p is a prime number", *J. Algebra* 8 (1968), 143-147. MR36#1527.

University of Auckland,
Auckland, New Zealand.

PROC. SECOND INTERNAT. CONF. THEORY OF GROUPS, CANBERRA, 1973, pp. 437-442.

20H10

NON EUCLIDEAN CRYSTALLOGRAPHIC GROUPS

Roger C. Lyndon

1.

The term *non Euclidean crystallographic group* (or NEC *group*) was introduced by Wilkie [1] for a discrete group of isometries of the hyperbolic plane with compact quotient space. If an NEC group contains no orientation reversing transformation it is a Fuchsian group; otherwise it contains a Fuchsian group as a subgroup of index two. Wilkie obtained presentations of all NEC groups, and a partial classification of these groups under isomorphism. Macbeath [8] completed the classification, using the theory of quasi-conformal mappings; Keller [5] established the same result by purely geometric methods.

Zieschang [11; see also 12] also obtained presentations for NEC groups. Hoare, Karrass, and Solitar [3, 4] used purely combinatorial and non geometric methods to characterize the subgroups of Fuchsian groups, and to obtain a Riemann-Hurwitz formula in the case of a subgroup of finite index in a Fuchsian group. Hoare extended these results to subgroups of finite index in an NEC group.

Here we obtain presentations for a class of groups that contains all subgroups of NEC groups (among which we include henceforth the Euclidean and spherical cases), including those of infinite index, which have non compact quotient spaces. Our general method is very close to that of Zieschang, while our argument in detail borrows heavily from Hoare, Karrass, and Solitar.

2.

We begin with some informal remarks. With every presentation $G = (X; R)$ of a group G by generators X and relators R a *Cayley complex* $C = C(X; R)$ is associated in a familiar way (Cayley, [1]; see also Lyndon and Schupp [6]). We shall call a group G a C *group* if it admits a presentation such that the Cayley graph C can be embedded in a 2-manifold, M, which we may take then to be either the plane or the sphere. Let P be a complex on M dual to C, that is, with its 1-skeleton dual to that of C. The natural action of G on C then induces an action of G on P that is regular on the faces of P. Conversely, if P is a complex in M, invariant under G and such that G acts regularly on the faces of P, and *provided that G preserves orientation*, by an argument known already to Poincaré [9; see also 7], a complex C dual to P is a Cayley graph for G and provides a presentation for G. By this means one can recognize that G is isomorphic to a Fuchsian group, and one concludes that the class of groups acting combinatorially in this manner on M coincides, abstractly, with that of groups acting conformally on M with respect to the usual geometry.

We relax the condition that G preserve orientation, and define a group G to be a C' *group* if G acts on a complex P in M, and regularly on faces. If G contains no *reflection*, that is no non trivial element fixing an edge pointwise, then G is a C group. Otherwise G contains a C group of index two. If G contains a reflection, then any complex C dual to P fails in a minor point from being a Cayley complex for G. The remedy for this is contained in a suggestion made already by Cayley; if a generator $x \in X$ represents an involution in G, Cayley suggested that one might replace the two edges at each vertex of C with labels (colours) x and x^{-1} by a single edge with ambiguous label $x^{\pm 1}$. Henceforth we modify the concept of a Cayley complex accordingly. This done, the C' groups are exactly those admitting a presentation with Cayley complex that can be embedded in the plane or the sphere.

We shall not give definitions amplifying the above concepts, but rather formulate explicitly the hypothesis that G is a C' group in the sense that it has a presentation with Cayley complex embedded in M, the plane or the sphere. Let F be the free product of infinite cyclic groups with generator x for all x in a set X and of groups of order two with generator y for all y in a set Y. For such an F we have $G = F/N$ where N is the normal closure in F of a set R of cyclically reduced elements of F. Now the vertices of C are in one-to-one correspondence with the elements of G, and will be identified with these elements. For each vertex g and each letter $z \in X \cup X^{-1} \cup Y$ there is an edge e from g to $g\bar{z}$, where $\bar{z}$ is the image of z in G, and to this edge we assign the *label* $\phi(e) = z$. The function ϕ on edges has a unique extension ϕ to a morphism from the groupoid of paths in C into the group F. For each $g \in G$ and $r \in R$ there is a unique (reduced) path p beginning and ending at g, and with label $\phi(p) = r$; then C contains a face $D(g, r)$ with boundary p. These are all the faces of C, and they are distinct except that if r has *root* s, that is, $r = s^m$ for m maximal, we take all the faces $D(gs^i, r)$, $0 \le i < m$, to be identical.

Let S be the set of all roots of elements in R. For each vertex g let $S(g)$ be the set of all roots of labels on the boundaries of faces, starting and ending at g, and in the positive sense. Then $S(g)$ is either the set of all cyclically reduced conjugates of elements of S or of S^{-1}; moreover,

$S(hg) = S(g)^{\pm 1}$ according as h preserves or reverses orientation. We may suppose that $S(1)$ consists of all cyclically reduced conjugates of elements of S. From local properties of the embedding of C in M we conclude that $S(1)$ contains at most two elements beginning with x or x^{-1} for each $x \in X$ and at most one element beginning with y for each $y \in Y$, and that no element of $S(1)$ belongs to Y.

We have stated all the consequences of the geometric hypotheses that are required for the theorem stated below. Let $S = T \cup U$ where T consists of all elements of S that contain a generator from Y and U of those that do not. A little elementary juggling shows that we may suppose that T consists of elements of the form $t = y_t a_t y'_t a_t^{-1}$ where y_t and y'_t are in Y and a_t contains no generators from Y. Let A be the set of a_t for all $t \in T$. We conclude that each $x \in X$ occurs at most twice in elements of the set $V = A \cup U$.

We have shown that G has a presentation

$$G = \left\{X \cup Y;\ Y^2 \cup R_T \cup R_U\right\}$$

where Y^2 is the set of all y^2 for $y \in Y$ and T and U are as described above, and R_T, R_U the set of all elements of R with roots in T, U.

3.

To obtain a more detailed description of presentations for C' groups it suffices to study such groups G that are not proper free products of C' groups. In this case S is *connected* in the sense of Hoare, Karrass, and Solitar, that is, S cannot be partitioned into two sets with no generator in common. Since free groups (of sufficiently small cardinality) are C' groups, we may assume also that S contains all the generators of G; under this assumption the presentation is fully described by listing the defining relators. Since the subgroups of Fuchsian groups and their analogs are well known, we may put these groups aside and suppose that Y and T are not empty.

THEOREM. *Let G be a C' group that is not a proper free product of C' groups and is not a G group (that is, not a subgroup of a Fuchsian group or of a Euclidean or spherical analog). Then G has a presentation with relators of one of the following forms.*

(I) y_{ij}^2 *for* $1 \le i \le p$, *some* $p \ge 1$ *and* $1 \le j \le n_i$, *some* $n_i \ge 1$;

$$(y_{i1}y_{i2})^{m_{i1}}, \ldots, \left(y_{in_i-1}y_{in_i}\right)^{m_{in_i-1}}, \left(y_{in_i}a_iy_{i1}a_i^{-1}\right)^{m_{in_i}},$$

where all $m_{ij} \geq 2$;

$b_k^{m_k}$ *for all* k , $1 \leq k \leq q$ *where* $q \geq 0$ *and each* $m_k \geq 2$;

$a_1 \ldots a_p b_1 \ldots b_q Q$ *where either* $Q = [c_1, c_2] \ldots [c_{2g-1}, c_{2g}]$ *or*

$Q = c_1^2 \ldots c_g^2$, $g \geq 0$.

(II) $y_1^2, \ldots, y_n^2$ *for some* $n \geq 1$;

$(y_1y_2)^{m_1}, \ldots, (y_{n-1}y_n)^{m_{n-1}}, \left(y_nay_1a^{-1}\right)^{m_n}$ *where each* $m_j \geq 2$;

(III) y_i^2 *and* $(y_iy_{i+1})^{m_i}$ *where* i *ranges over one of the sets*

$N_n = \{1, \ldots, n\}$ *for some* $n \geq 1$, $Z^+ = \{1, 2, \ldots\}$ *or*

$Z = \{\ldots, -1, 0, 1, \ldots\}$.

We make several comments on this theorem. The first is that, as is well known, the groups of type (I) are precisely the NEC groups. Second, an argument of Hoare, Karrass, and Solitar, involving the *coinitial graph*, shows that if a C' group G has a free factor of type (I) or (III), then G coincides with this single factor. Third, we note that a group of type (III) where $a = 1$ has a presentation of type (I), by taking the relators as in (III) with an additional relator aQ where $Q = 1$. We note also that type (II) is disjoint from (I) and (III) in that a occurs only once in S , and so the quotient space of P under G is not compact. Type (III) likewise differs from (I), except in trivial cases, in that such groups are generated by involutions. Finally, type (III) differs from type (II) in that a group G of type (III) abelianized yields a group of exponent two, while for G of type (II) the image of a has infinite order. As we have indicated, a complete classification of NEC groups, of type (I), is known. We do not know a similar classification of those of types (II) and (III).

We begin the proof of the theorem. Recall that we have assumed that T is not empty. Under the transitive closure of the relation among elements of T that they contain an element of Y in common, T splits into connected components T_i . Each component T_i consists of elements $t_{ij} = y_{ij}a_{ij}y_{ij+1}a_{ij}^{-1}$ where the set of indices j , for each i , runs through a set

(i) Z_{n_i} , cyclically ordered of cardinal n_i ; or

(ii) a set $N_{n_i} = \{1, \ldots, n_i\}$; or

(iii) a set infinite in one direction, which we may take to be Z^+ ; or

(iv) a set infinite in both directions which we may take to be Z .

In all cases except (i), by suitable Nielsen transformations, replacing the y_{ij} by conjugates, we may suppose that all $a_{ij} = 1$, and in case (i) we may suppose that all $a_{ij} = 1$ except $a_i = a_{in_i}$. In cases (ii), (iii), (iv) we conclude by connectedness that $p = 1$ and $S = T_1$, whence G has a presentation of type (III). Thus we may assume that all T_i are cycles, and, in view of a remark following the statement of the theorem, that all $a_i \neq 1$.

An argument of Hoare, Karrass, and Solitar shows that if the set V is infinite, or if some $x \in X$ occurs in it only once, then, after a Nielsen transformation of the set X of generators, we may suppose that $V \subseteq X$. Putting aside a free factor of G that is a free group, we may assume that $V = X$. Now a_1 , an element of X , occurs at most once in an element of S other that t_{1n_1} . If it has no such occurrence we have a presentation of type (II). Otherwise we may suppose either that $a_1 = a_2^{-1}$ and that G has a presentation with relators $S_1 \cup S_2 \cup \{a_1 a_2\}$ where T_i is the set of roots of S_i , or that $a_1 = b_1^{-1}$ and G has a presentation with relators $S_1 \cup \{b_1^{m_1}\} \cup \{a_1 b_1\}$, in either case of type (I).

In the remaining case that V is finite and contains each $x \in X$ exactly twice, another argument of Hoare, Karrass, and Solitar shows that a Nielsen transformation of the set X reduces us to the case that all but a single element v of V belong to X . If $v \in U$ and in fact $v \in R$ we are done, for the presentation is reducible by standard arguments to type (I); if $v \in U$ and $v^m \in R$ for $m \geq 2$, we accomplish the same by a Tietze transformation introducing a new generator b_0 and replacing the relator v^m by the two relators b_0^m and $b_0 v$. Similarly, if $v \in A$, say $v = a_1$, by a Tietze transformation we introduce a_1 as a new generator, and we replace the relator t_{1n_1} by the result of substituting a_1 for v , adjoining the further relator $a_1^{-1} v$. Again this leads to a presentation of type (I). This completes the proof.

References

[1] [A.] Cayley, "Desiderata and suggestions. No. 2. The theory of groups: graphical representation", *Amer. J. Math.* 1 (1878), 174-176. FdM10,105.

[2] A.H.M. Hoare, "Notes on NEC groups with reflections", preprint.

[3] A. Howard M. Hoare, Abraham Karrass and Donald Solitar, "Subgroups of finite index of Fuchsian groups", *Math. Z.* 120 (1971), 289-298. MR44#2837.

[4] A. Howard M. Hoare, Abraham Karrass and Donald Solitar, "Subgroups of infinite index in Fuchsian groups", *Math. Z.* 125 (1972), 59-69. MR45#2029.

[5] R. Keller, Thesis, Ruhr-Universität, Bochum, 1973.

[6] R.C. Lyndon and P.E. Schupp, *Combinatorial group theory* (Ergebnisse der Mathematik, to appear).

[7] A.M. Macbeath, *Fuchsian groups* (Notes, Summer School, Queen's College, Dundee, 1961).

[8] A.M. Macbeath, "The classification of non-euclidean plane crystallographic groups", *Canad. J. Math.* 19 (1967), 1192-1205. MR36#3890.

[9] H. Poincaré, "Théorie des groups fuchsiens", *Acta Math.* 1 (1882), 1-62.FdM14,338

[10] H.C. Wilkie, "On non-Euclidean crystallographic groups", *Math. Z.* 91 (1966), 87-102. MR32#2483.

[11] Х. Цишанг [H. Zieschang], "Дискретные группы движений плоскости и плоские групповые образы" [Discrete groups of motions of the **plane** and planar group diagrams], *Uspehi Mat. Nauk.* 21 (1966), 195-212. MR33#4150.

[12] H. Zieschang, E. Vogt and H.-D. Coldeway, *Flächen und ebene diskontinuierliche Gruppen* (Lecture Notes in Mathematics, 122. Springer-Verlag, Berlin, Heidelberg, New York, 1970). Zbl.204,240.

The University of Michigan,
Ann Arbor, USA.

PROC. SECOND INTERNAT. CONF. THEORY OF GROUPS, 20E15, 20F30
CANBERRA, 1973, pp. 443-445.

ON SUBNORMALITY IN SOLUBLE MINIMAX GROUPS

D.J. McCaughan

Finiteness conditions associated with subnormal subgroups are in general fairly difficult to handle. In this note we refer in particular to two restrictions of this type. The first is the so-called subnormal intersection property, which demands that the intersection of any family of subnormal subgroups should again be a subnormal subgroup. The second condition entails the existence of an upper bound for the subnormal indices of all subnormal subgroups. Since their introduction by Robinson almost ten years ago, these conditions have prompted a number of investigations aimed at elucidating the structure of groups (usually soluble) with such limitations on the behaviour of their subnormal subgroups. In his recent treatise [7, 8] - to which we refer the reader for background and terminology - Robinson remarks on the apparent difficulty of such investigations. Cases in point are Robinson's proof, in [5], that finitely generated soluble groups with the subnormal intersection property are finite-by-nilpotent, and McDougall's relatively incomplete description of soluble minimax groups with the subnormal intersection property ([4]).

In [5] Robinson shows that the second property is in general strictly stronger than the first, citing as an example the standard restricted wreath product of a p-cycle by a quasicyclic p-group, for any prime p. Incidental to his main result, quoted above, is the fact that for finitely generated soluble groups the two properties coincide. The results recorded in this note are concerned with proving a similar coincidence for the class of soluble minimax groups.

The fundamental theorem requires the following

DEFINITION. If N is a normal subgroup of a group G, we denote by $\sigma(G : N)$ the intersection of all subnormal subgroups of G which supplement N in G, that is,

$$\sigma(G : N) = \cap\{S : S \text{ is subnormal in } G, SN = G\} .$$

It is easily seen that $\sigma(G : N)$ is a normal subgroup of G, and that if M is any subgroup of G containing N, then $\sigma(G : N)$ contains $\sigma(M : N)$ (for if $SN = G$ then $(S \cap M)N = M$).

THEOREM. *Let* G *be a group with a nilpotent normal subgroup* N *such that* G/N *is finite and nilpotent. If* $\sigma(G : N) = 1$ *and* G *has the subnormal intersection property then* G *is nilpotent.*

To prove the theorem we first reduce, by the preceding remark, to the case where G/N is a finite p-group for some prime p. In this situation we know by Theorem A of [1] that G/P is nilpotent, where P is the maximal p-radicable subgroup of N. The proof is completed by an elementary but rather lengthy argument involving repeated use of the hypothesis $\sigma(G : N) = 1$. It may perhaps be of interest, in view of the results of [3], to point out that if the condition of finiteness on G/N is relaxed to periodicity we obtain the equivalent result that G is a Baer group.

Turning now to the main object of the investigation, we can use this theorem, together with a routine argument involving a "minimal" counterexample, to deduce the following result: *a soluble minimax group which is nilpotent-by-finite has the subnormal intersection property if and only if it has a bound on its subnormal indices.*

Now Theorem A of McDougall [4] states that any soluble minimax group with the subnormal intersection property has a radicable abelian normal subgroup satisfying the minimal condition, such that the relevant quotient group is nilpotent-by-finite. By the above, the quotient group, which of course inherits the subnormal intersection property, has bounded subnormal indices. Since, by Lemma 2.1 of [6], the presence of the radicable subgroup can only increase each subnormal index by a fixed integer, we have as an easy consequence our main result: *a soluble minimax group has the subnormal intersection property if and only if it has a bound on its subnormal indices.*

The class of soluble groups of finite rank, although still fairly restricted, is substantially wider than the class of soluble minimax groups. One can construct, by way of contrast, a metabelian group of rank two which has the subnormal intersection property but has no bound for its subnormal indices. It remains undecided, to my knowledge, whether for groups of finite total rank - a class intermediate between those just mentioned - the two properties coincide.

The details of this work will be found in [2].

References

[1] D.J. McCaughan, "Subnormal structure in some classes of infinite groups", *Bull. Austral. Math. Soc.* 8 (1973), 137-150. Zbl.248.20035.

[2] D.J. McCaughan, "Subnormality in soluble minimax groups", *J. Austral. Math. Soc.* (to appear).

[3] D.J. McCaughan and D. McDougall, "The subnormal structure of metanilpotent groups", *Bull. Austral. Math. Soc.* 6 (1972), 287-306. MR45#2023.

[4] David McDougall, "Soluble minimax groups with the subnormal intersection property", *Math. Z.* 114 (1970), 241-244. MR41#5490.

[5] Derek S. Robinson, "On finitely generated soluble groups", *Proc. London Math. Soc.* (3) 15 (1965), 508-516. MR31#253.

[6] Derek S. Robinson, "On the theory of subnormal subgroups", *Math. Z.* 89 (1965), 30-51. MR32#2481.

[7] Derek J.S. Robinson, *Finiteness conditions and generalized soluble groups*, Part 1 (Ergebnisse der Mathematik und ihrer Grenzgebiete, Band 62. Springer-Verlag, Berlin, Heidelberg, New York, 1972). Zbl.243.20032.

[8] Derek J.S. Robinson, *Finiteness conditions and generalized soluble groups*, Part 2 (Ergebnisse der Mathematik und ihrer Grenzgebiete, Band 63. Springer-Verlag, Berlin, Heidelberg, New York, 1972). Zbl.243.20033.

Otago University,
Dunedin, New Zealand.

PROC. SECOND INTERNAT. CONF. THEORY OF GROUPS, CANBERRA, 1973, pp. 446-447. 08A15

VARIETIES GENERATED BY FINITE ALGEBRAS

Sheila Oates Macdonald

This paper is a progress report (or perhaps more accurately, lack of progress report) on the problem:

"When does a finite algebra have a finite basis for its laws?"

As mentioned in [3] the following conjectures have been made:

CONJECTURE 1. *If* $\underline{V}$ *is any variety of algebras whose congruence lattices are modular, then any finite algebra in* $\underline{V}$ *has a finite basis for its laws.*

CONJECTURE 2. *If* $\underline{V}$ *is any variety of algebras whose congruence lattices satisfy some nontrivial lattice identity, then any finite algebra in* $\underline{V}$ *has a finite basis for its laws.*

In connection with the latter conjecture, Freese and Nation [2] have recently shown that if a variety of semigroups is such that some nontrivial lattice identity is satisfied by the congruence lattices of algebras in the variety, then the variety is a variety of groups.

The test case for Conjecture 1 would appear to be that of finite loops. Evans [1] has recently proved that nilpotent commutative Moufang loops (and this includes finite commutative Moufang loops) have a finite basis for their laws, and I have shown that a finite loop with no proper nontrivial subloops has a finite basis, [4]. However, Evans has conjectured that there is even a finite nilpotent loop whose laws are not finitely based.

The stumbling block in extending either method to a general loop would appear to be the fact that a normal subloop of finite index in a finitely generated loop is not in general finitely generated; in fact even in the free loop on one generator, the normal closure of the square of that generator is not finitely generated qua subloop. (However, it might be possible to extend the methods of [4] to slightly more general types of algebras with no proper nontrivial subalgebras.)

Attacking from another direction, perhaps the method used in Baker's Theorem (Conjecture 1 with modular replaced by distributive) could be adapted to the modular case, although the only proof of this which I have seen (Makkai, [5]) relies heavily on properties of algebras with distributive congruence lattices, which have no obvious analogue in the modular case. For instance, the initial reduction of the problem to: Let $\underline{\underline{V}}$ be a variety of algebras with distributive congruence lattices, then for any natural number M, the variety generated by all algebras in $\underline{\underline{V}}$ of cardinality $< M$ is finitely based, uses the fact that there are only finitely many critical algebras in such a variety which is generated by a finite algebra, and this in turn is a consequence of the fact that such a variety contains only finitely many subdirectly irreducible algebras, a property which is also used later in the proof.

To sum up: please would someone put us out of our misery by finding a finite loop without a finite basis for its laws; may I offer as a possible candidate the loop whose multiplication table is:

1	2	3	4	5	6	7	8	9	10
2	1	4	3	8	9	10	5	6	7
3	4	1	2	9	10	8	7	5	6
4	3	2	1	10	8	9	6	7	5
5	8	9	10	1	7	6	2	3	4
6	9	10	8	7	1	5	4	2	3
7	10	8	9	6	5	1	3	4	2
8	5	7	6	2	4	3	1	10	9
9	6	5	7	3	2	4	10	1	8
10	7	6	5	4	3	2	9	8	1

which is commutative but not Moufang.

References

[1] Trevor Evans, "Identities and relations in commutative Moufang loops", preprint.

[2] Ralph Freese and J.B. Nation, "Congruence lattices of semilattices", preprint.

[3] Sheila Oates Macdonald, "Various varieties", *J. Austral. Math. Soc.* 16 (1973), 363-367.

[4] Sheila Oates Macdonald, "Laws in finite strictly simple loops", *Bull. Austral. Math. Soc.* 9 (1973), 349-353.

[5] M. Makkai, "A proof of Baker's finite-base theorem on equational classes generated by finite elements of congruence distributive varieties", preprint.

PROC. SECOND INTERNAT. CONF. THEORY OF GROUPS, 20D05, 20-04
CANBERRA, 1973, pp. 448-452.

COMPUTING WITH FINITE SIMPLE GROUPS

John McKay

Leech [9] and Birkhoff and Hall [1] are standard references to computational group theory. The lesser-known Petrick [13] contains many articles on symbolic manipulation and group-theoretic work including a description by Sims of techniques he has developed to compute with very large degree permutation groups. These ideas have been used by him [14] most impressively to prove the existence and uniqueness of Lyons' simple group of order 51 765 179 004 000 000 = $2^8 3^7 5^6 7.11.31.37.67$ by constructing a permutation representation of it on the cosets of a subgroup $G_2(5)$ of index 8 835 156 . It should be added that Lyons' group has no proper subgroup larger than $G_2(5)$.

While it appears unlikely that the existence of Lyons' group could have been established without a computer, there are other smaller groups whose existence currently depends on constructing a permutation representation from a presentation by means of coset enumeration. Two such groups are Janko's group of order 50 232 960 = $2^7 3^5 5.17.19$ which has a subgroup SL*(2, 16) of index 6156 of rank 8 , and Held's group [6] of order 4 030 387 200 = $2^{10} 3^3 5^2 7^3 17$ which has a subgroup Sp*(4, 4) of index 2058 . The proof of the existence of J_3 appears in [8] but that for H has not yet appeared in print so I shall give here a presentation for it obtained by G. Higman in February 1969 which has been verified by computer.

Let

$$e = \begin{bmatrix} 0 & 0 & 0 & 1 \\ 0 & 0 & 1 & 0 \\ 0 & 1 & 0 & 0 \\ 1 & 0 & 0 & 0 \end{bmatrix}$$

be the matrix of a 4-dimensional symplectic form over F_4 . We identify generators with matrices over F_4 as follows:

$$a = \begin{bmatrix} 0 & 1 & 0 & 0 \\ 1 & 0 & 0 & 0 \\ 0 & 0 & 0 & 1 \\ 0 & 0 & 1 & 0 \end{bmatrix}, \quad b = \begin{bmatrix} \omega & 0 & 0 & 0 \\ 0 & \omega & 0 & 0 \\ 0 & 0 & \omega^2 & 0 \\ 0 & 0 & 0 & \omega^2 \end{bmatrix}, \quad c = \begin{bmatrix} 1 & 0 & 0 & 0 \\ 1 & 1 & 0 & 0 \\ 0 & 0 & 1 & 0 \\ 0 & 0 & 1 & 1 \end{bmatrix}$$

and the automorphism d acts as

$$d^{-1}ad = a^d = \begin{bmatrix} 1 & 0 & 0 & 0 \\ 0 & 0 & 1 & 0 \\ 0 & 1 & 0 & 0 \\ 0 & 0 & 0 & 1 \end{bmatrix}, \quad b^d = \begin{bmatrix} \omega^2 & 0 & 0 & 0 \\ 0 & 1 & 0 & 0 \\ 0 & 0 & 1 & 0 \\ 0 & 0 & 0 & \omega \end{bmatrix}, \quad c^d = \begin{bmatrix} 1 & 0 & 0 & 0 \\ 0 & 1 & 0 & 0 \\ 0 & 1 & 1 & 0 \\ 0 & 0 & 0 & 1 \end{bmatrix} .$$

Let

$$\begin{aligned} H^* = \langle a, b, c, d, e \mid a^2 = b^3 = c^2 = d^4 = e^2 = 1 ,& \\ (ad^2)^2 = (cd^2)^2 = (bd^2)^2 = (ad)^8 = (bdb^2d)^2 = abab^2 = adbdadb^2db^2 = 1,& \\ cbcb^2 = (d^3bdc)^3 = ((ad)^3cd)^2 = (ac)^3 = (adbdcdbd)^5 = (cadcda)^2 = 1,& \\ (cdad)^4 = (cd)^4adcadcdada = (de)^2 = (ce)^2 = (ad^2e)^3 = (adadae)^3 = 1,& \\ b^2dbdebdbde = 1\rangle ,& \end{aligned}$$

then the group $\langle a, b, c, d\rangle$ has index 2058 in H^* and Held's group is the group $\langle a, b, c, d^2, e, a^d\rangle$. The existence of the subgroup $Sp^*(4, 4)$ is assured by use of Brauer's restriction trick [2]. No more details are given as it is hoped that a full description will appear in due course.

As long as space restrictions can be met, that is, the product of the number of distinct generators and their inverses with the permutation degree is not too large (currently $\sim 10^5$), the above use of the computer can be regarded as standard and no special programs are likely to be needed. Sometimes existence proofs can be given by hand, even for large groups if they are of small rank, as they can for the Fischer groups F_{22}, F_{23}, and F_{24} which occur as automorphism groups of graphs.

Recently work (Conway [3], Magliveras [12], Finkelstein and Rudvalis [4, 5]) has been devoted to determining the subgroup structure of the sporadic simple groups. I give here a computational method for obtaining subgroups which are generated by two elements. It is expected that this method will be useful in determining non local subgroups. Apart from Higman [7] and the notable exception of Brauer [2], there appear to be few methods for proving the existence of non-local structure from character tables so first we examine necessary conditions before trying to find the subgroup.

We assume that we can compute inside G - this will usually mean that we have a permutation representation (or a matrix representation) of G. We also use character tables for G and for the subgroups we seek.

For H to be a subgroup of G there will be a permutation character of G of degree $[G : H]$ with well-known properties [11]. This character gives the classes of G containing those of H. The class structure constants are a further source of information. We observe that the number of solutions $(x, y) \in G \times G$ to $xy = z$ for x, y belonging to fixed classes of G and z a fixed element cannot be less than the number of solutions in $H \times H$ of the same equation. If there is no contradiction we try to find a pair of elements of G which satisfy a known presentation of H. For the known simple groups of order up to 10^6 we are in the process of obtaining presentations in terms of an involution x and an element y of least order such that $G = \langle x, y \rangle$. The presentations are complete in the sense that if $\langle x^*, y^* \rangle = G$ where x^* is an involution and y^* has the same order as y then there is a presentation in the list which will be satisfied on substituting x^* for x and y^{*k} (for some k) for y.

By using characters we find representatives of the G-classes to which x and y belong. If the class to which x belongs is small enough we may generate it and check if $H = \langle x^w, y \rangle$ by (for permutation representations) first checking orbit lengths then checking the relations. If H is simple then $H \subset G$ if the presentation is satisfied otherwise a factor group of H may have been obtained and this will need further investigation: if no presentation is satisfied then, of course, $H \not\subset G$. If we are unable to generate the G-class of x we may generate conjugates of x 'at random' in the hope of finding H. Upper bounds on the probability of success are given in [11] and should be evaluated before going ahead with the computing.

On finding $H \subset G$ it may be possible to find all K such that $H \subset K \subset G$ by obtaining a permutation representation of G on the cosets of H then constructing blocks of imprimitivity.

The methods described here were first used in showing the existence of the simple subgroups $L(2, 17)$ and $L(2, 19)$ in J_3 [8]; they have been refined by my student, K.C. Young, who has shown that $H \subset G$ for

$$G = {}^2F_4'(2) \text{ and } Sz \text{ and } H = L(3, 3), L(2, 25),$$

and

$$G = Ru, \quad H = L(2, 29), U(3, 5),$$

and

$$G = G_2(4), \quad H = L(2, 13).$$

The computations described here are 'numerical' in the sense that explicit individual groups are involved but there seems no reason why 'symbolic' techniques should not be used to investigate the structure of infinite families of groups such as the finite matrix groups parametrized by their matrix degree and the order of the finite field over which they are defined. Little has been done in this area.

References

[1] Garrett Birkhoff and Marshall Hall, Jr., *Computers in algebra and number theory*, Proc. Sympos. Appl. Math. Amer. Math. Soc. and the Soc. Indust. Appl. Math., New York City, 1970, 4 (Amer. Math. Soc., Providence, Rhode Island, 1971). Zbl.236.00006.

[2] Richard Brauer, "Representations of finite groups", *Lectures on Modern Mathematics* [Ed. T.L. Saaty], 1, pp. 133-175 (John Wiley & Sons, New York, London, 1963). MR31#2314.

[3] J.H. Conway, "Three lectures on exceptional groups", *Finite simple groups* [ed. M.B. Powell and G. Higman], pp. 215-247 (Academic Press, London and New York, 1971). Zbl.221.20014.

[4] Larry Finkelstein, "The maximal subgroups of Conway's group C_3 and McLaughlin's group", *J. Algebra* 25 (1973), 58-89.

[5] L. Finkelstein and A. Rudvalis, "Maximal subgroups of the Hall-Janko-Wales group", *J. Algebra* 24 (1973), 486-493.

[6] Dieter Held, "The simple groups related to M_{24} ", *J. Algebra* 13 (1969), 253-296. MR40#2745.

[7] G. Higman, "Construction of simple groups from character tables", *Finite simple groups* [ed. M.B. Powell and G. Higman], 205-214 (Academic Press, London and New York, 1971). Zbl.221.20014.

[8] Graham Higman and John McKay, "On Janko's simple group of order 50, 232, 960 ", *Bull. London Math. Soc.* 1 (1969), 89-94; Correction: *Bull. London Math. Soc.* 1 (1969), 219. MR40#224.

[9] John Leech (editor), *Computational problems in abstract algebra* (Oxford, 1967), (Pergamon Press, Oxford, 1970). MR40#5374.

[10] Richard Lyons, "Evidence for a new finite simple group", *J. Algebra* 20 (1972), 540-569. MR45#8722.

[11] John McKay, "Subgroups and permutation characters", *Computers in algebra and number theory* [ed. Garrett Birkhoff and Marshall Hall, Jr.], Proc. Sympos. Appl. Math. Amer. Math. Soc. and the Soc. Indust. Appl. Math., New York City, 1970, 4, pp. 177-181 (Amer. Math. Soc., Providence, Rhode Island, 1971).

[12] Spyros S. Magliveras, "The subgroup structure of the Higman-Sims simple group", *Bull. Amer. Math. Soc.* 77 (1971), 535-539. MR44#310.

[13] Stanley Roy Petrick (editor), *Second Symposium on symbolic and algebraic manipulation*, Special Interest Group on Symbolic and Algebraic Manipulation Proceedings (Assoc. for Computing Machinery, 1971).

[14] C.C. Sims, to appear.

School of Computer Science,
McGill University,
Montreal, Quebec, Canada.

PROC. SECOND INTERNAT. CONF. THEORY OF GROUPS, 22C05, 22D05
CANBERRA, 1973, pp. 453-462.

COMPACT TORSION GROUPS

John R. McMullen

Dedicated to the memory of Professor Hanna Neumann

1. Introduction

This paper is concerned exclusively with compact Hausdorff topological groups, all elements of which have finite order (compact torsion groups). It is well known that a compact torsion group is necessarily totally disconnected [5, (28.20)] and hence profinite.

The following are five unsolved problems concerning group-theoretic finiteness properties of compact torsion groups:

1. *Must a compact torsion group have finite exponent?*

This has a known (affirmative) answer only for solvable groups, and extension of this result is the main motiviation for the present paper (see Theorem 1 below).

2. *Must a compact torsion group be locally finite?*

This question is equivalent to the conjunction of the following two questions.

3. *Must a topologically finitely generated compact torsion group have finite exponent?*

4. *The restricted Burnside problem: must a topologically finitely generated profinite group of finite exponent be finite?*

The well-known difficulty of this problem gives little cause for hope for immediate solutions to problems 2 and 4.

5. *Must an infinite compact group have an infinite abelian subgroup?* (Of course, for nontorsion infinite compact groups the solution is trivial.)

Indeed owing to the work of Strunkov [9], Kargapolov [6], and Hall and Kulatilaka [2] an affirmative answer to question 2 would imply a like answer to this one. We have mildly improved Strunkov's result [9, Theorem 2] in the case of infinite compact groups, and this appears as Theorem 2 below.

Part of this work was supported by a National Science Foundation Grant.

TERMINOLOGY. In what follows, free use will be made of Sylow theory, as extended to apply to profinite groups (see [7] or [8]). In addition, if C is a class of groups, then by a *pro-C-group* we shall mean a group isomorphic to a projective limit of *finite* groups in C : for example, prosolvable group, pro-p-group. Other terminology and notation is standard.

The author would like to thank Roger Howe and Ralph McKenzie for their influence on this research.

2. Singular points

Let G be a topological torsion group. We say that the element $x \in G$ is *singular* if every neighbourhood of x contains elements of arbitrarily high order, otherwise x is *nonsingular*. The set of singular points of G we denote by $S(G)$. Clearly $S(G)$ is empty if and only if G has finite exponent.

PROPOSITION 1. *Let G be a compact torsion group. Then $S(G)$ is closed and nowhere dense in G. If $x \in S(G)$ and $n \in \mathbb{N}$, then $x^n \in S(G)$. In particular, $1 \in S(G)$ if $S(G) \neq \emptyset$.*

PROOF. All is clear save the fact that $S(G)$ is nowhere dense. Let N be a closed normal subgroup of finite index in G. Each coset xN $(x \in G)$ is compact and open. Let $G_k = \{x \in G : x^k = 1\}$, for $k = 1, 2, \ldots$. The sets $G_k \cap xN$ are closed subsets of xN and cover xN. By Baire's Theorem, there is a $k > 0$ for which $G_k \cap xN$ has nonvoid interior. Thus there is a coset $x'N'$, where N' is a closed normal subgroup of G of finite index, contained in $G_k \cap xN$ (such cosets form a base for the topology of G). Therefore $x'N' \subseteq xN$ and $x'N'$ avoids $S(G)$. Thus no open subset of G is contained in $S(G)$.

The following corollary is known. It follows from [4, (25.9)], for example.

COROLLARY. *A compact solvable torsion group has finite exponent.*

PROOF. We need only consider the abelian case. The above proof shows the existence of a coset xN with the properties $(xN)^n = 1$ $(G : N) < \infty$, whence the exponent of G divides $(G : N).n < \infty$.

Theorem 1 below generalizes this result.

The following property of $S(G)$ is evident:

LEMMA. *Let $\phi : G_1 \to G_2$ be a continuous surjection, where G_i are compact torsion groups $(i = 1, 2)$. Then $\phi^{-1}S(G_1) \subseteq S(G_2)$.*

We shall call a compact torsion group *vortical* if $S(G) = \{1\}$. The lemma shows that a vortical group can have no proper quotients of infinite exponent. Although the existence of a compact torsion group with a singular point is in doubt, vortical groups are certainly the simplest conceivable type. However, from the existence of a compact torsion group of infinite exponent we can, under certain assumptions, deduce the existence of a vortical one. We proceed as follows.

Let $M(G)$ be the set of all maximal subgroups of $S(G)$. Because of Proposition 1, $M(G)$ does not contain $\{1\}$ unless G is vortical. If U and V are open sets in G, let $N(U, V) = \{M \in M(G) : M \subseteq U$ and $M \cap V \neq \emptyset\}$. Then $M(G)$, with the topology generated by the sets $N(U, V)$ is a Hausdorff space, and G acts (properly) on $M(G)$ by conjugation.

PROPOSITION 2. *Suppose that the restricted Burnside conjecture holds. Let G be a compact torsion group with infinite exponent that is finitely generated topologically. Then only two cases are possible:*

(a) each orbit under G in $M(G)$ is a perfect set (hence of cardinality at least $2^{\aleph_0}$), or

(b) some orbit under G in $M(G)$ contains an isolated point M.

In case (b), Let N be the normalizer of M in G. Then N is closed, has finite index in G, and N/M is vortical.

PROOF. Since M is isolated, and since the space G/N is homeomorphic to the orbit of M in $M(G)$, G/N is Hausdorff and finite, so that N is closed of finite index.

Note that maximality of M in $S(G)$ implies that M is closed in G, hence also in N. If $S(N/M)$ contains any nontrivial element in N/M, then it contains some nontrivial subgroup P/M, with $P \neq M$. Then $P \subseteq S(N)$ by the lemma, and since $S(N) \subseteq S(G)$ this contradicts maximality of M in $S(G)$. Thus N/M is compact, Hausdorff, and either vortical or of finite exponent. In the latter case, since N is necessarily topologically finitely generated, being of finite index in G, our assumption entails that N/M is finite. Therefore M is open in G, contradicting the fact that $S(G)$ is nowhere dense. Thus N/M is vortical, as claimed.

3. Bounded *A*-groups

Let A be any group. An *order bound* on A is a function $\beta : A\backslash\{1\} \to \mathbb{N}^*$ such that $a^{\beta(a)} = 1$ for each $a \in A\backslash\{1\}$. For fixed A, the class $V(A, \infty)$ of all groups G on which A acts as a group of automorphisms is a variety of algebras, in the sense of universal algebra. Elements of $V(A, \infty)$ will be called *A-groups*. If β is a fixed order bound on A, we define a *β-bound A-group* to be an element G of $V(A, \infty)$ that satisfies the family of identities

$$P_a : x^{a^{\beta(a)-1}} x^{a^{\beta(a)-2}} \cdots x^{a^2} x^a x = 1 \quad (x \in G) ,$$

where a runs through $A \backslash \{1\}$. The class $V(A, \beta)$ of all such G forms a subvariety $V(A, \beta)$ of $V(A, \infty)$. By a *β-bound A-module* we shall mean a β-bound A-group whose group structure is abelian.

PROPOSITION 3. *Let* $A = \mathbb{Z}/(p) \times \mathbb{Z}/(p)$, p *prime, and let* β *be any order bound on* A . *Let* M *be a* β-*bound* A-*module written additively. Let*

$$k = \text{g.c.d.}\{\text{l.c.m.}\{\beta(a_1), \ldots, (a_{p+1})\} : a_1, \ldots, a_{p+1} \text{ generate distinct nontrivial cyclic subgroups of } A\}$$

Then $kM = 0$. *In particular, if* $\beta(a) = b$, *for all* $a \in A \backslash \{1\}$, *then* $bM = 0$.

PROOF. Let $a_1, \ldots, a_{p+1}$ generate distinct (hence all) cyclic nontrivial subgroups of A . We must prove that $k'M = 0$, where $k' = \text{l.c.m.}\{\beta(a_i)\}$.

Now $A = \bigcup_{i=1}^{p+1} \langle a_i \rangle$ and $\langle a_i \rangle \cap \langle a_j \rangle = 1 \quad (i \neq j)$. Hence in the group algebra $\mathbb{Z}A$, the identity

$$p.1 = \sum_{i=1}^{p+1} \sum_{a \in \langle a_i \rangle} a - \sum_{a \in A} a$$

holds. Let $m \in M$, and let $\mathbb{Z}A \xrightarrow{\phi} M$ be the unique A-module homomorphism satisfying $\phi(1) = m$. We have, for fixed i ,

$$\beta(a_i)p^{-1}\phi\left(\sum_{a \in \langle a_i \rangle} a\right) = m + a_i m + \ldots + a_i^{\beta(a_i)-1} m = 0$$

and if $\{h_1, \ldots, h_p\}$ is a transversal for $\langle a_i \rangle$ in A , then we have

$$\beta(a_i)p^{-1}\phi\left(\sum_{a \in A} a\right) = \sum_{j=1}^{p} \beta(a_i)p^{-1}\left(h_j m + a_i h_j m + \ldots + a_i^{p-1} h_j m\right)$$

$$= \sum_{j=1}^{p} \left(h_j m + a_i h_j m + \ldots + a_i^{\beta(a_i)-1} h_j m\right) = 0 .$$

Hence we obtain

$$k'm = \phi\left((k'p^{-1}).p.1\right) = \phi\left(k'p^{-1}\left(\sum_{i=1}^{p+1} \sum_{a \in \langle a_i \rangle} a - \sum_{a \in A} a\right)\right) = 0$$

which proves the claim.

THEOREM 1. *Let* G *be a vortical compact torsion group containing a subgroup* A *of the form* $\mathbb{Z}/(p) \times \mathbb{Z}/(p)$, p *prime. Then to each* $n > 0$ *there corresponds an* $m(n) > 0$ *such that every normal subquotient* M/N $(M, N \leq G)$ *of* G *which is solvable of length* n *has exponent dividing* $m(n)$.

REMARK. If G has an infinite abelian subgroup, then it certainly has a subgroup of the form required. See Problem 5 *supra* and Theorem 2 *infra*.

PROOF. Since G is vortical, each $a \in A\backslash\{1\}$ is nonsingular in G, and so there exists a neighbourhood U_a of $\{1\}$ such that the orders of the elements of U_a are bounded. Since G is profinite and A is finite there exists a normal subgroup N_0 of G, of finite index, contained in $\cap U_a$. Hence for some $c \in \mathbb{N}^+$, $(aN_0)^c = 1$ for $a \in A$. Thus for $n_0 \in N_0$ we have

$$1 = (an_0)^c = a^c n_0^{a^{c-1}} n_0^{a^{c-2}} \cdots n_0^a n_0 = n_0^{a^{c-1}} n_0^{a^{c-2}} \cdots n_0^a n_0$$

and so $N_0 \in V(A, \mathfrak{c})$ where $\mathfrak{c}$ is the order bound on A defined by $\mathfrak{c}(a) = c$ $(a \in A\backslash\{1\})$, and the action of A on N_0 is by conjugation.

Now let $M, N \leq G$ with M/N abelian. Then $(M \cap N_0)/(N \cap N_0)$ is a $\mathfrak{c}$-bound A-module, and so has exponent dividing c, by Proposition 3. Since $M/(M \cap N_0) \cong (MN_0)/N_0$ is finite of order dividing $(G : N_0)$, $M/(N \cap N_0)$ and hence M/N has finite exponent dividing $c.(G : N_0)$. It is now clear that for the required $m(n)$ in the statement of Theorem 1 we may take $c^n(G : N_0)$.

Proposition 3 also has the following consequence.

PROPOSITION 4. *A prosolvable infinite torsion group contains at least one infinite Sylow* p*-subgroup.*

PROOF. Let G be prosolvable, torsion, and infinite, and let p_i $(i = 1, 2, 3, \ldots)$ be the prime divisors of $|G|$. It can readily be shown that G has closed nontrivial subgroups P_i $(i = 1, 2, \ldots)$ such that P_i is a Sylow-p_i-subgroup, $P_iP_j = P_jP_i$ for all i, j, and the ascending union of the partial products $P_1P_2 \ldots P_k$ $(k = 1, 2, \ldots)$ is dense in G: in fact this is an easy extension of P. Hall's corresponding theorem for the finite solvable case.

Now suppose further that none of the P_i is infinite. We first observe that this implies that G has no infinite abelian subgroup. For if G_0 were such a subgroup (which we may assume to be closed), since its order divides that of G, it also has no infinite Sylow subgroup, and so is a direct product of infinitely many finite p-groups, for *distinct* primes. Such a group has infinite exponent, contradicting the Corollary of Proposition 1.

We now proceed to construct an infinite abelian subgroup of G , thus arriving at a contradiction. First we construct an infinite closed subgroup M_0 of G that is a projective limit of a sequence of quotients M_0/M_k and is a *split* extension of M_k for $k = 1, 2, \ldots$.

Since the P_i commute, we might as well assume that G has odd order, since we could otherwise arrange that P_1 be a Sylow 2-subgroup, and restrict attention to the closed subgroup generated by P_i $(i \geq 2)$. Because G is infinite and all P_i are assumed finite, there are infinitely many P_i . As P_1 is finite, there is an open normal subgroup N_1 of G of finite index avoiding P_1 . Set $i_1 = 1$, and let i_2 be the smallest index for which $P_{i_2} \subseteq N_1$ - such an i_2 exists, since G/N_1 is finite. Choose an open normal N_2 of finite index in G avoiding $\{P_{i_1} P_{i_2}\}$. This process leads to the selection of a subsequence $P_{i_1}, P_{i_2}, \ldots$ of pairwise commuting Sylow subgroups of G and a descending chain $N_1 \supseteq N_2 \supseteq \ldots$ of open normal subgroups of finite index in G such that $\{P_{i_1} P_{i_2} \ldots P_{i_k}\} \cap N_k = \{1\}$ and $\{P_{i_{k+1}} P_{i_{k+2}} \ldots\} \subseteq N_k$, for all k . For $k \geq 0$, let M_k be the closure in G of the subgroup $\{P_{i_{k+1}} P_{i_{k+2}} \ldots\}$. Then $M_k = M_0 \cap N_k$ is normal in M_0 , and M_0 is a split extension of M_k , for each k .

To show that M_0 is the projective limit of the inverse sequence of groups M_0/M_k , we invoke Bourbaki [1, Chapter III §7 Proposition 2] and the fact that M_0 is compact and Hausdorff, and thus restrict ourselves to showing that an arbitrary neighbourhood U of 1 in G contains one of the subgroups M_k . Indeed, we can assume that U is an open normal subgroup of G of finite index, and then since the order of $(M_0/M_0 \cap U)$ is finite, we have $\{P_{i_{k+1}} P_{i_{k+2}} \ldots\} \subseteq M_0 \cap U$ for some k , and hence $N_k \subseteq M_0 \cap U$, as required.

We now construct an infinite closed subgroup Q_0 of M_0 which is the projective limit of a sequence $\{Q_0/Q_k\}$ of its quotients, and in which all factors Q_i/Q_{i+1} are elementary abelian of pairwise distinct prime exponents. If P is a finite p-group, we write $m(P)$ for the last nontrivial term of its lower p-nilpotent series: $m(P)$ is elementary abelian and fully invariant in P. Let $R_0 = m\left(P_{i_1}\right)$. Let σ_k denote the natural isomorphism of M_0/M_k onto $P_{i_1}P_{i_2} \dots P_{i_k}$. Then $P_{i_2} = \sigma_2\left(M_1/M_2\right)$, and so is normalized by P_{i_1} in $P_{i_1}P_{i_2}$. Hence $m\left(P_{i_2}\right)$ is normalized by P_{i_1}, and thus by $R_0 \subseteq P_{i_1}$. Set $R_1 = R_0 . m\left(P_{i_2}\right) \subseteq P_{i_1}P_{i_2}$, and let $\phi_1 : R_1 \to R_0$ be the quotient map. Continuing we see that R_k normalizes $m\left(P_{i_{k+2}}\right)$ and define $R_{k+1} = R_k . m\left(P_{i_{k+2}}\right)$, with $\phi_{k+1} : R_{k+1} \to R_k$ the quotient map. Observe that ϕ_{k+1} is the restriction of the quotient map $\left(P_{i_1} \dots P_{i_{k+2}}\right) \to \left(P_{i_1} \dots P_{i_{k+1}}\right)$. Let Q_0 be the subgroup of M_0 corresponding to the inverse limit of the sequence $\left(R_k, \phi_k\right)$ (see Bourbaki [1, Chapter III §7 Proposition 3]) and let $Q_k = Q_0 \cap M_k$. Then $Q_k/Q_{k+1} = \left(Q_0 \cap M_k\right)/\left(Q_0 \cap M_{k+1}\right)$ is clearly isomorphic with $m\left(P_{i_{k+1}}\right)$, which is elementary abelian as claimed.

Let us agree to abuse the language and call a compact torsion group Q_0 *strictly proelementary* if it is a projective limit of quotients Q_0/Q_k, with all factors Q_i/Q_{i+1} finite and elementary abelian of pairwise prime exponent. We need two lemmas.

LEMMA 1. *Infinite closed subgroups and infinite quotient groups of infinite strictly proelementary groups by closed subgroups are strictly proelementary.*

The proof of this is easy, given Bourbaki [1, Chapter III §7 Proposition 3].

LEMMA 2. *The following statements are equivalent:*

(I) *every infinite strictly proelementary group contains an infinite abelian subgroup;*

(II) *every infinite strictly proelementary group* Q_0 *has at least one element* $x \neq 1$ *whose centralizer in* Q_0 *is infinite.*

The proof of Lemma 2 is identical, *mutatis mutandis*, with the proof of the corresponding statement for locally finite groups in [2] (this is because all subgroups in that proof are, in our case, closed, with the possible exception of the infinite abelian subgroup $H = \bigcup_{n=1}^{\infty} A_n$ constructed therein).

We are thus reduced to proving (II) for an arbitrary infinite strictly pro-elementary group Q_0. Such a group has a Hall decomposition $Q_0 = \left|m\left(P_{i_1}\right)m\left(P_{i_2}\right) \ldots\right|$ as before, with $\left\{m\left(P_{i_1}\right) \ldots m\left(P_{i_k}\right)\right\}$ a complement for Q_k, as is readily seen.

We consider two cases. Suppose first that all the factors $Q_k/Q_{k+1} \cong m\left(P_{i_{k+1}}\right)$ are cyclic (of prime order). Since the automorphism group of Q_k/Q_{k+1} is abelian, G/C_k is abelian, where C_k is the centralizer in Q_0 of Q_k/Q_{k+1}. Therefore $C_k \supseteq Q_0'$, and Q_0' centralizes all factors Q_k/Q_{k+1}. As Q_0/Q_0' is a compact abelian torsion group, it has finite exponent, and so at most finitely many of the subgroups $m\left(P_{i_k}\right)$ are not contained in Q_0'. Thus for sufficiently large j, $m\left(P_{i_j}\right)$ centralizes all $m\left(P_{i_k}\right)$ with $k > j$.

In the complementary case, suppose that one of the subgroups $m\left(P_{i_k}\right)$ is non-cyclic. Then since it is a p-group with $p \neq 2$, it contains a subgroup A of the form $\mathbb{Z}/(p) \times \mathbb{Z}/(p)$. Now A acts on each $m\left(P_{i_k}\right)$ $(j > k)$ by conjugation. If $a \in A$ and $x \in m\left(P_{i_j}\right)$, then a obviously centralizes the element $xx^a \ldots x^{a^{p-1}}$. Thus if A acts fixed-point-free on $m\left(P_{i_j}\right)$, then $xx^a \ldots x^{a^{p-1}} = 1$ for $x \in m\left(P_{i_j}\right)$ and $a \in A$, so that $m\left(P_{i_j}\right)$ is a p-bound A-module, and must have exponent p by Proposition 3 *supra*. Therefore A and P_{i_j} have the same exponent. This is absurd, and so for each $j > k$ there is an element $a_j \in A$ centralizing some element of $m\left(P_{i_j}\right)$. As A is finite, we conclude that some element of A must centralize elements of infinitely many of the $m\left(P_{i_j}\right)$ $(j > k)$. This proves that (II) holds, and Q_0 thus contains the promised infinite abelian subgroup.

We may now prove our modification of Strunkov's Theorem [9] for compact groups:

THEOREM 2. *A compact infinite group either contains an infinite abelian subgroup, or an infinite topologically-2-generator torsion pro-p-group, with* p *an odd prime.*

PROOF. Suppose G is compact and infinite, with no abelian infinite subgroup. Then G is a torsion group. In addition, Held's result [3] shows that G can contain no infinite 2-group. It follows that the order of G has the form $2^k n$ where k is finite and n is odd. Hence G has an infinite normal profinite (that is, closed) subgroup H of finite index, whose order is odd. Application of the Feit-Thompson Theorem to the finite quotients of H shows that H is a prosolvable infinite torsion group. We now invoke Proposition 4, to deduce the existence of an infinite Sylow p-subgroup P of H, with p odd. Strunkov's Theorem with our hypothesis implies that P contains an infinite 2-generator subgroup Q. The closure of Q fulfills the claims.

References

[1] Nicolas Bourbaki, *Elements of Mathematics: General Topology*, Part I (Hermann, Paris; Addison-Wesley, Reading, Massachusetts; London; Don Mills, Ontario; 1966). MR34#5044a

[2] P. Hall and C.R. Kulatilaka, "A property of locally finite groups", *J. London Math. Soc.* 39 (1964), 235-239. MR28#5111.

[3] Dieter Held, "On abelian subgroups of an infinite 2-group", *Acta Sci. Math. (Szeged)* 27 (1966), 97-98. MR33#4146.

[4] Edwin Hewitt and Kenneth A. Ross, *Abstract harmonic analysis. Volume I: Structure of topological groups. Integration theory, group representations* (Die Grundlehren der Mathematischen Wissenschaften, Band 115. Academic Press, New York; Springer-Verlag, Berlin, Göttingen, Heidelberg, 1963). MR28#158.

[5] Edwin Hewitt and Kenneth A. Ross, *Abstract harmonic analysis. Volume II: Structure and analysis for compact groups analysis on locally compact abelian groups* (Die Grundlehren der Mathematischen Wissenschaften, Band 152. Springer-Verlag, New York, Berlin, 1970). MR41#7378.

[6] М.И. Каргаполов [M.I. Kargapolov], "О проблеме О.Ю. Шмидта" [On a problem of O. Ju. Šmidt], *Sibirsk. Mat. Ž.* 4 (1963), 232-235. MR26#6241.

[7] Jean-Pierre Serre, *Cohomologie galoisienne* (Cours au Collège de France, 1962-1963. Seconde édition. Lecture Notes in Mathematics, 5. Springer-Verlag, Berlin, Heidelberg, New York, 1964). MR31#4785.

[8] Stephen S. Shatz, *Profinite groups, arithmetic, and geometry* (Annals of Mathematics Studies, 67. Princeton University Press; University of Tokyo Press, 1972). Zbl.M.236.12002.

[9] С.П. Стрyнков [S.P. Strunkov], "Подгруппы периодических групп" [Subgroups of periodic groups], *Dokl. Akad. Nauk SSSR* 170 (1966), 279-281. MR34#2705.

University of Sydney,
Sydney, NSW 2006.

PROC. SECOND INTERNAT. CONF. THEORY OF GROUPS, 55A25, 20E40
CANBERRA, 1973, pp. 463-487.

BRAID GROUPS: A SURVEY

Wilhelm Magnus

1. Introduction

The terms "*braid*" and "*braid groups*" were coined by Artin, 1925. In his paper, an n-braid appears as a specific topological object. We consider two parallel planes in euclidean 3-space which we call respectively the upper and the lower frame. We choose n distinct points U_ν $(\nu = 1, \ldots, n)$ in the upper frame and denote their orthogonal projections onto the lower frame by L_ν. Next, we join each U_ν with an L_μ by a polygon which intersects any plane between (and parallel to) the upper and lower frame exactly once. These polygons are called *strings*. We assume that they do not intersect anywhere, and that $\nu \to \mu(\nu)$ is a permutation of the symbols $1, \ldots, n$. By removing the strings from the slice of 3-space between upper and lower frame, we obtain an open subset of 3-space the isotopy class of which we call an n-braid. We define a composition between n-braids by hanging on one n braid to the other one. (This can be done by identifying the upper frame of the second braid with the lower frame of the first one, removing these two frames and compressing the slice of 3-space between the first upper and the second lower frame by an affine transformation to the same thickness as before.) Under this composition, the n-braids form a group B_n which has $n - 1$ generators σ_ν. These are represented respectively by braids which have a projection onto a plane perpendicular to the frames in which the ν-th and $(\nu+1)$-st string seem to cross once whereas all other strings go straight through as line segments orthogonal to both frames. The braid represented by n strings which go straight through is the representative of the unit element of B_n. (Figure 1 shows a representative of the particular 3-braid $\sigma_1\sigma_2^{-1}\sigma_1\sigma_2^{-1}$.) Of course, this rather vague definition of B_n can be made rigorous. See Artin 1947a, Burde 1963.

If we close a given n-braid in the manner indicated by the dotted lines in Figure 1 and then remove the frames we obtain a tame knot or linkage in 3-space. (For precise definitions see Crowell and Fox, 1963.) One of the applications of the

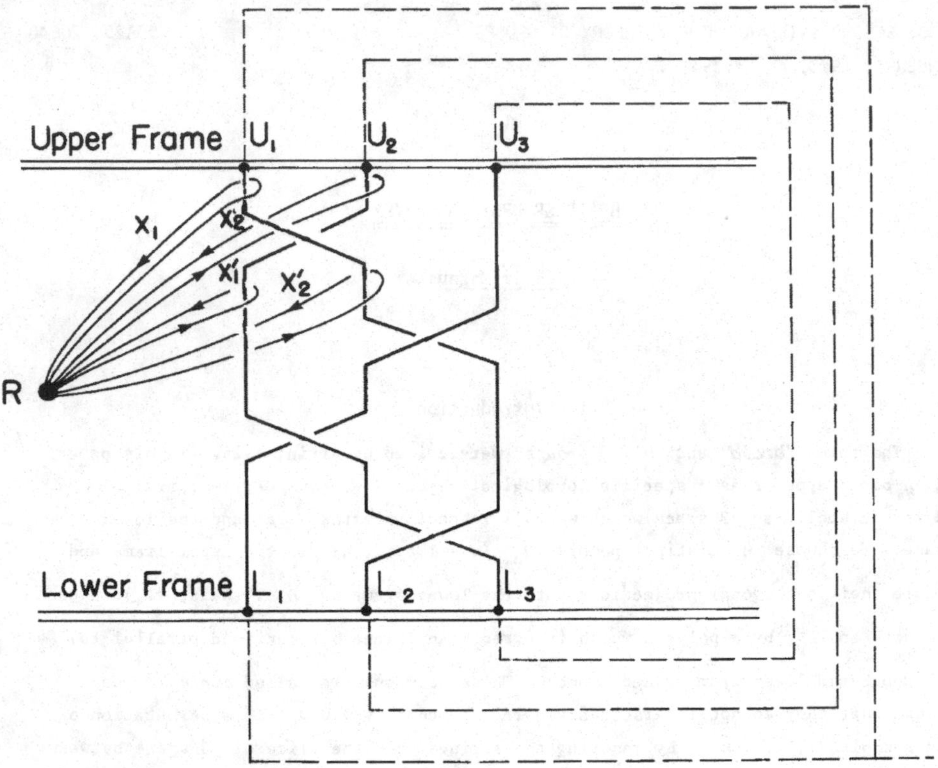

FIGURE 1. The 3-braid $\sigma_1\sigma_2^{-1}\sigma_1\sigma_2^{-1}$

theory of B_n given in Artin, 1925, refers to the theory of knots. A brief outline of some of the numerous results based on Artin's idea will be given in Section 4. A detailed exposition will appear in a monograph by Joan Birman, 1974a.

There exist other important aspects of B_n and they were discovered long before the appearance of Artin's paper. Fricke-Klein, 1897, indicated that B_n modulo its center is the mapping class group of the euclidean plane with n points removed. (For any orientable topological space we define the mapping class group as the group of all isotopy classes of orientation preserving selfhomeomorphisms of the space.) The connection between the definitions of B_n by Fricke and by Artin can be made obvious by a simple geometric argument (Magnus, 1934). We shall deal with the aspect of B_n and its recent generalizations in Section 5. Again, we shall be brief since more detailed information is available in an expository paper by Birman, 1974b.

Fricke's construction of the mapping class group of the euclidean plane with n punctures is implicit already in a paper by Hurwitz, 1891. But Hurwitz then proceeds to relate this group to the fundamental group of a higher dimensional space. Let $\mathbb{C}^n$ be the cartesian product of n replicas of the complex plane $\mathbb{C}$. A point in $\mathbb{C}^n$ is then defined uniquely by an ordered n-tuple of complex numbers z_ν , $(\nu = 1, \ldots, n)$. Consider the function

$$\Delta = \prod_{\nu<\mu} (z_\nu - z_\mu)$$

and remove the points on which $\Delta = 0$ from $\mathbb{C}^n$. The fundamental group of the remainder of the space is a normal subgroup $\overline{B}_n$ of B_n consisting of those elements for which the permutation $\nu \to \nu(\mu)$ is the identical one $\left(B_n/\overline{B}_n = \Sigma_n\right.$, the symmetric group on n symbols$\left.\right)$, and B_n itself is the fundamental group of the space whose coordinates are the elementary symmetric functions of the z_ν after removal of the points on which $\Delta^2 = 0$.

The interpretation of B_n itself as a fundamental group was rediscovered by Fox and Neuwirth 1962 and generalized: The n-th braid group $B_n(S)$ of any space S is the fundamental group of the space of unordered n-tuples of distinct points of S . B_n itself is then $B_n\left(E^2\right)$, the n-th braid group of the euclidean plane E^2 . This leads again to insights into the structure of mapping class groups, in particular of the groups $M(l, g)$ of mapping classes of two dimensional orientable manifolds of genus g with l boundary points (Birman, 1969a, b). However, the aim of Hurwitz was the investigation of classes of Riemann surfaces with a fixed type of ramification over the closed complex plane $\hat{\mathbb{C}}$. We shall give the necessary definitions and an outline of known results in Section 6. Right now we merely mention that Hurwitz' approach was again rediscovered and generalized, in this case by Arnol'd, 1968. We shall not go into the general theory resulting from the work of Arnol'd, but Section 6 will contain some references.

One of the "asides" in the paper by Artin, 1925, is a remark about the presentation of Σ_n in terms of generators and relations. Artin's presentation of Σ_n is exactly the presentation which shows that Σ_n is one of the groups "generated by reflections". Brieskorn and Saito, 1973, generalized Artin's result, starting with any group R generated by reflections, constructing from it a new group which has the same relationship to R as B_n has to Σ_n and proving that it has properties similar to those of B_n . For definitions and results see Section 7.

Sections 2 and 3 contain a report on the structure of Artin's groups B_n .

This survey stresses the group theoretical aspects of the theory of braid groups. The topological aspects are mentioned mainly for the purpose of motivation and are dealt with very sketchily.

2. The structure of B_n

Artin, 1925, 1947a, showed geometrically that B_n can be presented on $n-1$ generators σ_ν $(\nu = 1, \ldots, n-1)$ satisfying the defining relations

(1) $$\sigma_\nu\sigma_{\nu+1}\sigma_\nu = \sigma_{\nu+1}\sigma_\nu\sigma_{\nu+1} , \quad (\nu = 1, \ldots, n-2) ,$$

(2) $$\sigma_\nu\sigma_\mu = \sigma_\mu\sigma_\nu , \quad (|\nu-\mu| \geq 2) .$$

Introducing

$$\sigma = \sigma_1\sigma_2, \ldots, \sigma_{n-1} ,$$

B_n can be generated by σ_1 and σ because

(3) $$\sigma^\nu\sigma_1\sigma^{-\nu} = \sigma_{\nu+1} , \quad (\nu = 1, \ldots, n-2) .$$

The kernel $\overline{B}_n$ of the mapping $B_n \to \Sigma_n$ is the normal closure of σ_1^2 , and this leads to a presentation of Σ_n as a group generated by reflections (also called "Coxeter group"). See Coxeter and Moser, 1965, or Benson and Grove, 1971.

Artin also showed that B_n has a representation as a group of automorphisms of a free group F_n on free generators x_ν , $\nu = 1, \ldots, n$. Figure 1 indicates the geometric argument leading to this result. The space between the frames of the braid has, after removal of the braid, a fundamental group (with a reference point R) which is free of rank n with generators x_ν . These may be represented by simple loops around the strings. We may choose these loops on different levels of the braid and replace the x_ν by new generators x'_ν (as indicated in Figure 1, where the x_ν, x'_ν are drawn for $\nu = 1, 2$ in two special cases). Since the transition from one set of free generators of F_n to another one is effected by an automorphism, the following result emerges:

Let α_ν $(\nu = 1, \ldots, n-1)$ denote the automorphism

(4) $$\alpha_\nu : x_\nu \to x_{\nu+1} , \quad x_{\nu+1} \to x_{\nu+1}^{-1}x_\nu x_{\nu+1} , \quad x_\mu \to x_\mu \quad (\mu \neq \nu, \nu+1)$$

of F_n . Then $\sigma_\nu \to \alpha_\nu$ is a homomorphic mapping of B_n into the automorphism group of F_n .

It has been shown independently by several authors, (Magnus, 1934, Markoff, 1945, Bohnenblust, 1947, Chow, 1948) that the α_ν generate a group with defining relations

$$\alpha_\nu\alpha_{\nu+1}\alpha_\nu = \alpha_{\nu+1}\alpha_\nu\alpha_{\nu+1} , \quad \alpha_\nu\alpha_\mu = \alpha_\mu\alpha_\nu \quad (|\nu-\mu| \geq 2)$$

and therefore provide a faithful presentation of B_n . The automorphisms α in the group generated by the α_ν are exactly those for which

(5) $$\alpha : x_\nu \to T_\nu x_{\mu(\nu)} T_\nu^{-1} , \quad \prod_{\nu=1}^{n} x_\nu \equiv \prod_{\nu=1}^{n} T_\nu x_{\mu(\nu)} T_\nu^{-1} ,$$

where "$\equiv$" stands for "freely equal", and the T_ν are words in the generators of F_n .

We shall, from now on, not distinguish between an element of B_n and its representation as an automorphism of F_n .

The algebraic proof that $\sigma_\nu \to \alpha_\nu$ is an isomorphism reveals the structure of B_n and allows the solution of the *word problem* in two ways: For a group of automorphisms of F_n it is always solved, but in addition it can also be solved if we look upon B_n merely as a group with generators σ_ν and defining relations (1), (2). This follows from the existence of a normal form for the elements of B_n which is based on the existence of a normal series with quotient groups of a simple nature. We have:

Let F_n^* be the quotient group of F_n arising from the adjunction of the relation

(6) $$x_1 x_2 \ldots x_n = 1 .$$

Let $B_n^{(1)}$ be the subgroup of index n in B_n which consists of the automorphisms (5) in which $\mu(1) = 1$. Since the kernel of the mapping $F_n \to F_n^*$ is invariant under the action of the automorphisms α in (5), both B_n and $B_n^{(1)}$ act as automorphism groups on F_n^* . These automorphism groups will be denoted respectively by B_n^* and $B_n^{(1)*}$. Let $C(G)$ denote the center of G for any group G , and let

(7) $$\Theta = (\sigma_1\sigma_2 \dots \sigma_{n-1})^n = \sigma^n ,$$

(8) $$\begin{cases} \Theta_1 = (\sigma_2\sigma_3 \dots \sigma_{n-1})^{1-n} , \\ \Theta_\nu = (\sigma_1 \dots \sigma_{\nu-1})^\nu(\sigma_{\nu+1} \dots \sigma_{n-1})^{\nu-n} , \quad \nu = 2, \dots, n-2 , \\ \Theta_{n-1} = (\sigma_1\sigma_2 \dots \sigma_{n-2})^{n-1} . \end{cases}$$

Then $C(B_n)$ is generated by Θ which is of infinite order, and

(9) $$B_n^* = B_n/C(B_n) .$$

The elements $\Theta_1, \dots, \Theta_{n-1}$ in (8) freely generate a free group of rank $n - 1$. Its direct (cartesian) product with $C(B_n)$ is a normal subgroup K_n of both B_n and $B_n^{(1)}$. The Θ_ν in (8) represent generating inner automorphisms of F_n^*. We have

(10) $$B_n^{(1)}/K_n \simeq B_{n-1}/C(B_{n-1})$$

and the right-hand side in (10) can be represented as the group with generators $\sigma_2, \dots, \sigma_{n-1}$ and defining relations

(11) $$\begin{aligned} \sigma_\nu\sigma_{\nu+1}\sigma_\nu &= \sigma_{\nu+1}\sigma_\nu\sigma_{\nu+1} , \quad (\nu = 2, \dots, n-1) , \\ \sigma_\nu\sigma_\mu &= \sigma_\mu\sigma_\nu , \quad (|\nu-\mu| > 1) , \\ (\sigma_2\sigma_3 \dots \sigma_{n-1})^{n-1} &= 1 . \end{aligned}$$

The group K_n has a geometric interpretation: B_n/K_n is the mapping class group of the sphere with n points removed (Magnus, 1934).

It follows now that every element of $B_n^{(1)}$ can be represented in the form

$$W_1(\sigma_2, \dots, \sigma_{n-1})W_2(\Theta, \Theta_1, \dots, \Theta_{n-1})$$

where W_1, W_2 are words in the generators appearing in the parentheses, and since $B_n^{(1)}$ is of finite index in B_n this leads (by induction with respect to n) to a solution of the word problem in B_n.

Another solution of the word problem for B_n can be derived from Artin's theory of $\overline{B}_n$ (which is of index $n!$ in B_n). Artin, 1947a, showed the following: Let A_n be the subgroup of $\overline{B}_n$ (the unpermuted braid group with $\mu(\nu) = \nu$ for all ν) consisting of all braids in which only the n-th string does not go straight through in the projection of the braid on the plane. We may also say that A_n consists of those braids which become the trivial braid of $n - 1$ strings if we cut and take out the n-th string. The elements of A_n are called "n-pure" braids. Algebraically, the elements of A_n are automorphisms of F_n which degenerate into the identical automorphisms of the free group on free generators $x_1, \ldots, x_{n-1}$ if we add the relation $x_n = 1$. By "pinching" the frame of the braids so that the initial and terminal points of the n-th string coincide, one sees that A_n is simply the fundamental group of a frame with $n - 1$ strings and therefore indeed free of rank $n - 1$. Also, it is clear that A_n is a normal subgroup in $\overline{B}_n$ since the conjugate of an n-pure braid will be n-pure again. Artin showed that A_n is freely generated by elements a_ν, $\nu = 1, \ldots, n-1$ which are represented by the automorphisms of F_n:

(12) $$a_\nu : x_\mu \to x_\mu \,, \quad (\mu < \nu) \,, \quad x_\nu \to x_\nu C \,, \quad x_\lambda \to C^{-1} x_\lambda C \,, \quad (\nu < \lambda < n) \,,$$
$$x_n \to C^{-1} x_n \,, \qquad C = x_\nu^{-1} x_n^{-1} x_\nu x_n \,.$$

Every element in $\overline{B}_n$ can then be written in the form

(13) $$W_1(g_1, \ldots, g_m) W_2(a_1, \ldots, a_{n-1})$$

where $g_1, \ldots, g_m$ are words in $\sigma_1, \ldots, \sigma_{n-2}$ which generate $\overline{B}_{n-1}$. In fact, Artin gives a complete set of generators and defining relations for $\overline{B}_n$. In a purely algebraic way, Burau, 1936, had obtained an equivalent result. Shepperd, 1962, also found a presentation for $\overline{B}_n$ and for a geometrically defined subgroup of $\overline{B}_n$.

The *conjugacy problem* for B_n had withstood all attempts at a solution for quite a while, although partial results were found by Fröhlich, 1936. It has been solved by Garside, 1969, who based his solution on the following result:

A word $P(\sigma_1, \ldots, \sigma_{n-1})$ will be called positive if all the exponents of the σ_ν appearing in P are non-negative. The positive words form a semigroup under multiplication with defining relations (1), (2). This means that positive words which represent the same element in B_n can be changed into each other merely by replacing subwords which have the form of one side of the relations (1), (2) by the other side of the same relation. In particular, positive words of different length represent different elements of B_n. A positive word $\overline{P}$ is called *prime* to a particular positive word Δ if Δ does not appear anywhere as a subword in $\overline{P}$. Now define the *fundamental word* Δ as

(14a) $$\Delta = \Pi_{n-1}\Pi_{n-2} \cdots \Pi_2\Pi_1$$

where, for $\nu = 1, \ldots, n-1$,

(14b) $$\Pi_\nu = \sigma_1\sigma_2 \cdots \sigma_\nu .$$

Let W be any word in the σ_ν. Then there exists a unique integer $m = 0, \pm 1, \pm 2, \ldots$ and a uniquely defined positive word $\overline{P}$ prime to Δ such that

$$W = \Delta^m \overline{P} .$$

Conjugation by $\Delta^{\pm 1}$ has the following effect on the σ_ν:

(15) $$\Delta\sigma_\nu\Delta^{-1} = \sigma_{n-\nu} , \quad \Delta^{-1}\sigma_\nu\Delta = \sigma_{n-\nu} .$$

It follows that $\Delta^2 = \Theta = (\sigma_1 \ldots \sigma_{n-1})^n$ generates the center of B_n for $n > 1$, a result first obtained by Chow, 1948. The full solution of the conjugacy problem still requires some rather technical arguments for which we have to refer to Garside, 1969.

Fadell and Neuwirth, 1962, using topological methods, proved that the braid groups $B_n(S)$ are torsion free if S is the euclidean plane E^2 or a compact two dimensional manifold other than the two sphere S^2 or the projective plane P^2. (For these, see Section 5.) No purely group theoretical proof is known even for the fact that $B_n = B_n(E^2)$ is torsion free, although this is obvious for $\overline{B}_n$ because of (13).

Artin, 1947b, determined all homomorphic images of B_n which are transitive permutation groups on n symbols and showed that $\overline{B}_n$ is always a characteristic subgroup of B_n.

Artin's paper used the fact that there exists a prime number between $n/2$ and $n - 2$ for $n \geq 7$. D.I.A. Cohen, 1967, has shown how to prove Artin's results without using the existence of such a prime.

The commutator subgroup B_n' of B_n has been studied in some detail by Gorin and Lin, 1969.

Birman and Hilden, 1972b showed with topological methods that the action (4) of the α_ν on the group with generators x_ν and defining relations $x_\nu^k = 1$ $(k \in Z$, fixed, $\nu = 1, \ldots, n$) produces a group isomorphic with B_n for all $k > 1$.

3. Matrix representations of B_n

If a group B acts as a group of automorphisms on a group F, we may try to obtain a matrix representation of B as follows. Let C be a B-characteristic normal subgroup of F, and let C' be its commutator subgroup. Assume that the automrophisms in B induce the identity in B/C and let L be the group ring of B/C with the integers Z as ring of coefficients. Then C/C' is an L-module which is also a representation module for B. If C/C' has a finite basis as an L-module, we obtain thus a matrix representation of B where the entries of the matrices are elements of L.

In the case $B = B_n$, $F = F_n$, we can choose for C the normal closure of the set of elements $x_\nu x_{\nu+1}^{-1}$, $\nu = 1, \ldots, n-1$. The quotient group F_n/C is then infinite cyclic and its group ring is the ring of L-polynomials $L(v)$, that is of polynomials in $v^{\pm 1}$ with integral coefficients where v is an indeterminate corresponding to the generator of F_n/C. The group C/C' is then an $L(v)$-module with $n - 1$ basis elements t_ν ($\nu = 1, \ldots, n-1$) where the mapping from $C \to C/C'$ is given by

$$x_1^k x_\nu x_{\nu+1}^{-1} x_1^{-k} \to v^k t_\nu \tag{16}$$

and action of B_n on C/C' is described by the linear mappings

(17a) $\sigma_1 : t_1 \to -v^{-1} t_1 , \ t_2 \to v^{-1} t_1 + t_2 , \ t_\lambda \to t_\lambda \ (\lambda > 2)$

(17b) $\sigma_{n-1} : t_{n-2} \to t_{n-2} + t_{n-1} , \ t_{n-1} \to -v^{-1} t_{n-1} , \ t_\lambda \to t_\lambda \ (\lambda < n-2)$

and, for $1 < \rho < n-1$:

(17c) $\sigma_\rho : t_{\rho-1} \to t_{\rho-1} + t_\rho , \ t_\rho \to -v^{-1} t_\rho , \ t_{\rho+1} \to v^{-1} t_\rho + t_{\rho+1} ,$

$t_\lambda \to t_\lambda \ (\lambda \neq \rho-1, \rho, \rho+1)$.

The matrix representation (17) of degree $n - 1$ for B_n arises from a representation discovered (and stated without derivation) by Burau, 1936, by observing that his representation is reducible and splitting off a first-degree factor. We shall refer to (17) as the *Burau representation*. The derivation given above is taken from Magnus and Peluso, 1967.

A representation for $\overline{B}_n$ (the unpermuted braid group) of degree $n - 1$ with entries from the ring $L_n = L(v_1, \ldots, v_n)$ of polynomials in n indeterminates v_ν and their reciprocals v_ν^{-1} (with integral coefficients) can be obtained by replacing C by the commutator subgroup F_n' of F_n. Since $\overline{B}_n$ is of index $n!$ in B_n, standard methods of representation theory will then produce also a finite dimensional representation of B_n over L_n. The method to be used here has been described in detail by Bachmuth, 1965 and by Gassner, 1961.

It would be of interest for a question in the theory of knots (Section 4) to know whether the Burau representation of B_n is faithful or not. That this is true for $n = 3$ follows from results due to Lipschutz, 1961, who also showed that the matrices ocrresponding to the elements Θ_ν in (8) generate in pairs a free group of rank 2. The faithfulness of the Burau representation would follow if one could show that they generate a free group of rank $n - 1$. This is not known even for $n = 4$.

If we give numerical values to v in the Burau representations (17) we obtain other representations. For $v = 1$, we get a faithful representation of Σ_n. If we put $v = \varepsilon$, where ε is a primitive s-th root of unity and s is a divisor of n, we obtain a representation which becomes isomorphic with a subgroup of the symplectic group $\mathrm{Sp}(g, Z)$ of degree

(18) $$2g = (s-1)(n-2)$$

after splitting off a factor of degree 1. This emerges from the theory of monodromy groups of classes of Riemann surfaces developed by Hurwitz, 1891 and Arnol'd, 1968. (See Magnus and Peluso, 1969 and Section 6.) The representation with integral entries is obtained (for $s > 2$) by replacing ε by a matrix with integral entries of order s and of degree $s - 1$.

According to Hurwitz, 1891, there exists a representation of B_n as a transitive permutation group on $N(n, k)$ symbols where $k = 3, 4, 5, \ldots$, and where $N(n, k)$ denotes the number of compact Riemann surfaces which have k sheets and exactly one branch point of degree 2 over each one of n fixed points $P_1, \ldots, P_n$ of the Riemann sphere. In particular

$$N(n, 3) = \frac{1}{3!}\left(3^{n-1}-3\right), \quad N(n, 4) = \frac{1}{4!}\left(2^{n-2}-4\right)\left(3^{n-1}-3\right).$$

Hurwitz, 1902, also computed $N(n, k)$ for all k, using the theory of group characters of the symmetric groups. Cohen, 1973, showed that this permutation group is isomorphic to $\mathrm{PSp}(n/2-1, 3)$ for even $n \geq 4$ and $k = 3$, where PSp denotes the projective symplectic group of degree $n - 2$ over the Galois field of three elements and that, for even $n \geq 6$ and $k = 4$ it has $\mathrm{PSp}(n/2-1, 3)$ as a permutation homomorphic image.

4. Braids and links

It has been mentioned in the introduction and indicated in Figure 1 that we can close a braid in a well-defined manner and thus obtain a set of linked curves in 3-space. We shall call such a set a *linkage* or *link* even if it consists of only one curve. Should we wish to emphasize that the linkage consists of only one curve, we shall call it a *knot*. The *trivial knot* is the one which, in 3-space, is isotopic with a circle.

The fundamental group of the space arising from 3-space by removing the points of a link Λ is called the group of Λ. Let α be a braid which is represented by the automorphism (5) of F_n. Then the link $\Lambda(\alpha)$ arising from closing this braid has a group which can be presented in terms of n generators x_ν, $(\nu = 1, \ldots, n)$ and the defining relations

$$x_\nu = T_\nu x_{\mu(\nu)} T_\nu^{-1} \quad (\nu = 1, \ldots, n) \tag{19}$$

(see Artin, 1925). The number of cycles in the permutation $\nu \to \mu(\nu)$ is the number of distinct strings of the link. In particular, (19) is the group of a knot if and only if this permutation is an n-cycle.

Alexander, 1923, had shown that every tame link has a projection on a plane which also is the projection of a closed braid. Therefore, (19) provides a method of enumeration for the groups of all links, since the elements of B_n for all finite n can be enumerated (even recursively). Of course, different elements of the same or of different braid groups may produce, via (19), not only isomorphic link groups but even isotopic links. In particular, elements of B_n which are conjugate in B_n have this property as is obvious geometrically.

The characterization of links in terms of closed braids is a difficult and unsolved problem. The following partial results are known:

Assume that two projections of a link are given, both of which are also projections of closed braids. Markov, 1935, described a finite number of elementary changes of these projections each of which leads to a new projection of the same link and preserves the closed-braid character of the projections. A finite number of applications of these elementary changes will then carry one of the given projections into the other. Markov did not give a detailed proof of this statement. See Birman, 1974a, for a complete discussion.

The fact that conjugate elements of B_n lead, when closed, to isotopic links illustrates the importance of Garside's solution of the conjugacy problem in B_n.

However, there exist non-conjugate braids in the same B_n which, when closed, produce isotopic knots. A first example was given by Birman, 1969c. Murasugi and Thomas, 1972, showed that for $r = 1, 2, 3, \ldots$ the two elements γ_r and γ'_r of B_3 given by

$$\gamma_r = \sigma_1^{-1}\sigma_2^{2r}\sigma_1^{-2r}\sigma_2 \,, \quad \gamma'_r = \sigma_1\sigma_2^{-2r}\sigma_1^{2r}\sigma_2^{-1} \tag{20}$$

are not conjugate in B_3 but produce isotopic presentations of the same knot K when closed. In this example, K cannot be presented as a closed braid on fewer than 3 strings. For related questions see Murasugi, 1973.

For any $\alpha \in B_n$, we shall denote the group (19) by $G(\alpha)$ and its commutator subgroup by $G'(\alpha)$. It is obvious that the group $G(\alpha)/G'(\alpha)$ is free abelian and that its rank is the number of strings in $\Lambda(\alpha)$. From the proof of Dehn's Lemma by Papakyriakopoulos, 1957, it follows that $G(\alpha)$ is infinite cyclic if and only if $\Lambda(\alpha)$ is a knot isotopic with the trivial knot (a circle). In this case, $G'(\alpha) = 1$. For a non-trivial knot, in general $G'(\alpha) \neq G''(\alpha)$, which means that $G(\alpha)$ has solvable quotient groups which are not cyclic. A necessary and sufficient condition for $G'(\alpha) \neq G''(\alpha)$ can be obtained from the Burau representation (Burau, 1936, Magnus and Peluso, 1967). Let $M(\alpha)$ be the matrix corresponding to α in the representation defined by (17). Then $G'(\alpha) \neq G''(\alpha)$ if and only if the determinant of $M(\alpha) - I$ is not of the form

$$v^l\left(1-v^{-n}\right)\left(1-v^{-1}\right)^{-1} \tag{21}$$

where l is an integer and I denotes the unit matrix. For $n = 2, 3$, $G'(\alpha) = G''(\alpha)$ if and only if $G'(\alpha) = 1$ and the knot defined by α is trivial. This is not true any more for $n = 4$. Kinoshita and Terasaka, 1957, constructed a non-trivial knot which, according to Birman, has a projection as the closed four-braid belonging to

$$\alpha^* = \sigma_1^{-2}\sigma_2\sigma_3^{-1}\sigma_1^{-1}\sigma_2^{-1}\sigma_3\sigma_1^3\sigma_2^{-1}\sigma_3\sigma_2 \,, \tag{22}$$

and $G'(\alpha^*) = G''(\alpha^*)$. Infinitely many knots with the same property are known, but they are difficult to find. If the Burau representation of B_n $(n \geq 4)$ should turn out not to be faithful then for every element $\beta^* \in B_n$ for which $M(\beta^*) = I$, $\beta^* \neq 1$ we would have $G'\left(\alpha^*\beta^{*m}\right) = G''\left(\alpha^*\beta^{*m}\right)$ where $m = 0, \pm 1, \pm 2, \ldots$ and where α^* is now any element in B_n defining a knot with $G'(\alpha^*) = G''(\alpha^*)$. If, on the other hand, the Burau representation should be faithful one would have the standard information for B_n which goes with a finite matrix representation over a euclidean ring.

Levinson, 1973, showed how to find all n-braids which are decomposable in the following sense: Removal of any one string results in the trivial $(n-1)$-braid, (corresponding to the unit element of B_{n-1}). The decomposable n-braids form a normal subgroup D_n of $\overline{B}_n$ which is the intersection of Artin's normal subgroups A_ν for $\nu = 1, \ldots, n$ (where A_ν is defined analogously to A_n, see Section 2). For $n > 3$, the group theoretical description of D_n becomes increasingly difficult. However, Levinson describes a mapping μ which maps every element $\overline{\beta} \in \overline{B}_n$ onto an element of D_n and maps $\overline{\beta}$ onto itself if $\overline{\beta} \in D_n$. This leads to a constructive enumeration for the elements of D_n. Similar results are available for decomposable links and for generalizations (e.g. k-decomposable braids, that is, n-braids for which the removal of any k strings produces the trivial $(n-k)$-braid, $k > 1$). For these and numerous examples also see Levinson, 1973.

Braids can be closed in a different way and still produce all tame links. Following Reidemeister, 1960, a *plat* is defined as a closed $2m$-braid in which respectively the points U_{2j} and U_{2j+1} as well as the points L_{2j} and L_{2j-1} of the upper and of the lower frame have been identified for $j = 1, \ldots, m$. The resulting link then has a group defined by $2m$ generators x_{2j}, x_{2j-1} and defining relations

$$x_{2j-1}x_{2j} = 1 , \quad T_{2j-1}x_{\mu(2j-1)}T_{2j-1}^{-1}T_{2j}x_{\mu(2j)}^{-1}T_{2j}^{-1} = 1 \tag{23}$$

where the T_{2j}, T_{2j-1} replace the T_ν ($\nu = 1, \ldots, n = 2m$) in (19). That plats produce all tame links has been proved by Reidemeister, 1960. For a short proof and for motivation see Birman, 1973.

5. Braid groups and mapping class groups

Let $T(g, n)$ be an orientable two dimensional manifold of genus g with n points ("punctures") removed. The fundamental group $\pi_1 T(g, n)$ can then be presented on generators x_ν ($\nu = 1, \ldots, n$), a_i, b_i ($i = 1, \ldots, g$) and the defining relation

$$x_1 x_2 \ldots x_n \prod_{i=1}^{g} \left(a_i b_i a_i^{-1} b_i^{-1}\right) = 1 . \tag{24}$$

The mapping class group $M(T(g, n))$ can then be defined in a purely group theoretical manner as follows: Consider all automorphisms of $\pi_1 T$ which can be lifted to automorphisms of the free group on the x_ν, a_i, b_i which map each x_ν onto a conjugate of an $x_{\mu(\nu)}$ where $\nu \to \mu(\nu)$ is a permutation and which map the relator in (24) onto a conjugate of itself and not its inverse. This group A^* of automorphisms contains the inner automorphism I^*. Then $A^*/I^* = M(T)$. We note here that, in fact, each automorphism of $\pi_1 T$ can be lifted to an automorphism of the free group, but that this has not been proved group theoretically for $g > 1$. It is not even known how to find algebraically the generators of A^* if $g > 1$, and only topological methods are available here (see Birman, 1974b, for references). However, for $g = 0$ and $g = 1$ everything can be done algebraically, and for $g = 0$ there exists a close relationship between the mapping class groups $M(T(0, n))$, $M(T(0, n+1))$ and the braid groups $B_n(T(0, 0))$ and $B_n(T(0, 1))$, that is, the braid groups respectively of the 2-sphere and of the euclidean plane or a disk.

Using the notations of Section 2, we have (Magnus, 1935): The only inner automorphisms of F_n contained in B_n are the powers of Θ in (7). (According to Chow, 1948, Θ also generates the center of B_n.) We can define the mapping class group $M(E^2)$ of the euclidean plane as the subgroup of the mapping class group of the 2-sphere with $n + 1$ punctures where the neighborhood of one point (the infinite point of the plane) is kept fixed. This leads immediately to the result that $B_n(E^2)$ is the quotient group of B_n arising from an adjunction of the relation $\Theta = 1$ to (1) and (2). The relation between braids and selfmappings of the punctured plane becomes evident from the following remark (Magnus, 1935): An orientation-preserving selfmapping of the euclidean plane can be deformed continuously into the identical mapping. Considering now a selfmapping of the punctured plane, we can complete it to a selfmapping of the unpunctured plane. Deforming this continuously into the identical mapping means that the coordinates of each point P are now functions of a parameter t, $0 \leq t \leq 1$, such that, for any distinct points P_1, P_2, $P_1(t) \neq P_2(t)$. If we now move the plane parallel to itself and upward in the direction of a t-axis in 3-space, the orbits of the punctures during the deformation of the unpunctured plane become strings of an n-braid, the lower and upper frame of this braid being respectively the planes for $t = 0$ and $t = 1$. A modification of this argument has been formulated in a more complex situation by Birman, 1969b, to elucidate the relation between braid groups and mapping class groups.

The mapping class group $M(T(0, n))$ of the sphere with n punctures has been determined algebraically by Magnus, 1935. It is a quotient group of B_n which arises by adjunction of the relations

(25) $$\Theta = \Theta_1 = \ldots = \Theta_{n-1} = 1$$

(see (7), (8) for definitions). Θ defines the identical automorphism and

$\Theta_1, \ldots, \Theta_{n-1}$ define inner automorphisms of $\pi_1 T(0, n)$. If we define Ω by

(26) $$\Omega = \sigma_1\sigma_2 \ldots \sigma_{n-1}\sigma_{n-1} \ldots \sigma_2\sigma_1 = \Theta\Theta_1$$

we can define $M(T(0, n))$ also by (1), (2) and

(27) $$\Theta = \Omega = 1 .$$

This result is due to Fadell and Van Buskirk, 1962, who showed (with topological methods) that the n-th braid group $B_n(S^2) = B_n(T(0, n))$ of the 2-sphere S^2 arises from B_n by adjoining the single relation $\Omega = 1$, and that $\Omega = 1$ implies $\Theta^2 = \Theta_\nu^2 = 1$ for $\nu = 1, \ldots, n-1$. This shows that $B_n(S^2)$ is a non-splitting central extension of $M(T(0, n))$, a fact which is the topological analogue of the phenomenon that the three dimensional orthogonal group has a two valued representation. (See also Newman, 1942 and Fadell, 1962.)

Gillette and Van Buskirk, 1968 investigated $B_n(S^2)$ and $M(T(0, n))$ in detail, showing that Θ is the center of $B_n(S^2)$ and the only element of order 2 for $n > 1$ and that the center of $M(T(0, n))$ is trivial. They also showed that for $n > 2$ the orders m of elements of finite order in $B_n(S^2)$ are exactly the integers dividing $2n$, $2n - 2$ or $2n - 4$ and that for $n > 3$ and $m \geq 2$ an element of finite order exists in $M(T(0, n))$ if and only if m divides n, $n - 1$ or $n - 2$.

According to Fadell and Van Buskirk, 1962, the only compact 2-manifolds whose braid groups are not torsion free are S^2 and the projective plane P^2 . Van Buskirk, 1966, investigated $B_n(P^2)$ in detail, showing in particular that it has always non-trivial elements of finite order. As a presentation, he found:

Generators: σ_ν $(\nu = 1, \ldots, n-1)$, ρ_μ $(\mu = 1, \ldots, n)$.

(28) Defining relations:
$$\begin{cases} \sigma_\nu\sigma_\lambda = \sigma_\lambda\sigma_\nu , & (|\lambda-\nu| \geq 2) , \\ \sigma_\nu\sigma_{\nu+1}\sigma_\nu = \sigma_{\nu+1}\sigma_\nu\sigma_{\nu+1} , & (\nu = 1, \ldots, n-2) , \\ \sigma_\nu\rho_\mu = \rho_\mu\sigma_\nu , & (\mu \neq \nu, \nu+1) , \\ \rho_\nu = \sigma_\nu\rho_{\nu+1}\sigma_\nu , \quad \rho_{\nu+1}^{-1}\rho_\nu^{-1}\rho_{\nu+1}\rho_\nu = \sigma_\nu^2 , & \\ \rho_1^2 = \sigma_1\sigma_2 \ldots \sigma_{n-1}\sigma_{n-1} \ldots \sigma_2\sigma_1 & (= \Omega) . \end{cases}$$

From here on, the theory of braid groups and their relation to mapping class groups has been developed mainly by Birman. She showed first that for spaces S of a dimension > 2 the unpermuted braid group $\overline{B}_n(S)$ is the n-th cartesian power of the fundamental group $\pi_1 S$ (Birman, 1969a). In the same paper, she gave generators of the braid groups of orientable surfaces of arbitrary genus g and defining relations for $g = 1, 2$ with a procedure which indicates how to obtain defining relations for $g > 1$. Next, (Birman, 1969b) she described the relationship between the braid groups of orientable surfaces of arbitrary genus and the mapping class groups $M(T(g, n))$, deriving also a presentation of $M(T(1, n))$ for all n. (Algebraically, this had been done for $n = 1, 2$ by Magnus, 1935.) We shall not go into the details, partly because they are complex and partly because the methods used are strongly topological. A systematic presentation of the theory will be available in Birman, 1974a.

Bergau and Mennicke, 1960, had observed that $M(T(2, 0))$ is a quotient group of B_6. Their result appears as incidental. However, Birman and Hilden, 1972a, gave a presentation of $M(T(0, g))$ as a quotient group of B_6 which explains the relationship in a simple and natural manner. The surface $T(g, 0)$ is presented as a ramified two sheeted covering of the 2-sphere (Riemann surface) with branch points over $2g + 2$ points P_γ, $(\gamma = 1, \ldots, 2g+2)$. A particular selfmapping of $T(g, 0)$ is the involution J which exchanges the points of T covering the same point of the sphere. The normalizer of J in $M(T(g, 0))$ will be denoted by M_J. It consists of selfmappings of T which induce (by projection) a selfmapping of the sphere with punctures P_γ and, even more, induce all of these selfmappings. It turns out that the kernel of the homomorphism $M_J \to M(T(0, 2g+2))$ is the group generated by J. The final result is the following theorem:

$M(T(g, 0))$ contains a subgroup M_J which is an extension of the mapping class group $M(T(0, 2g+2))$ by a center of order 2. M_J is generated by elements σ_ν, $\nu = 1, \ldots, 2g+1$ with defining relations

$$(29)\quad \begin{cases} \sigma_\nu\sigma_\mu = \sigma_\mu\sigma_\nu, \quad |\nu-\mu| \geq 2 \\ \sigma_\nu\sigma_{\nu+1}\sigma_\nu = \sigma_{\nu+1}\sigma_\nu\sigma_{\nu+1} \qquad (\nu = 1, \ldots, 2g) \\ (\sigma_1\sigma_2 \cdots \sigma_{2g+1})^{2g+2} = 1, \\ \Omega^2 = 1, \quad \Omega\sigma_1 = \sigma_1\Omega, \quad \Omega = \sigma_1\sigma_2 \cdots \sigma_{2g+1}\sigma_{2g+1} \cdots \sigma_2\sigma_1. \end{cases}$$

For $g = 2$, the group M_J is the full mapping class group $M(T(g, 0))$.

The involution J is represented by the element Ω.

6. Braids and Riemann surfaces

A Riemann surface is a connected ramified covering of the sphere (represented for most purposes as the closed complex plane $\hat{\mathbb{C}}$) with a topology agreeing with that of $\hat{\mathbb{C}}$ under projection on S^2. We shall consider only finite, compact Riemann surfaces. Let P_ν ($\nu = 1, \ldots, n$) be the distinct points in $\hat{\mathbb{C}}$ over which the Riemann surface R is ramified. Let r denote the number of sheets and let Π_ν denote the permutation of the sheets induced by a simple closed loop with positive orientation around P_ν. (By this we mean the following: Lift the initial point of such a loop to the ρ-th sheet. Its endpoint, although coinciding with the initial point in $\hat{\mathbb{C}}$, is then lifted to the sheet with label $\tau(\rho)$. Now Π_ν is the permutation $\rho \to \tau(\rho)$.) We call the collection

(30) $$\{\Pi_1, \ldots, \Pi_n, P_1, \ldots, P_n\}$$

the *signature* of R, but we do not consider two signatures with permutations Π_ν and Π'_ν as distinct if there exists a permutation Π such that

$$\Pi'_\nu = \Pi^{-1}\Pi_\nu\Pi, \quad (\nu = 1, \ldots, n).$$

(Transition from Π_ν to Π_ν can be achieved by a relabeling of the sheets.)

The necessary and sufficient condition for the existence of a Riemann surface with signature (30) to exist are

(i) the Π_ν generate a permutation group which is transitive on the r symbols;

(ii) $\Pi_1\Pi_2 \ldots \Pi_n = 1$ (the identical permutation).

Condition (i) states that R is connected and (ii) states that, going around all branchpoints simultaneously gets us back into the sheet from which we started.

R is called a *regular* or *Galois-covering* if the following is true: If a closed loop in $\hat{\mathbb{C}}$ can be lifted to at least one closed loop in R, then all of its lifts into R are closed. A necessary and sufficient condition for this to happen is that

(iii) each Π_ν decomposes into cycles all of which have the same length.

(However, the cycle-length may depend on ν.) An equivalent condition is:

(iv) the Π_ν generate a group of order r.

Let c denote the total number of cycles occuring in all Π_ν. Then the genus g of R is given by the Hurwitz formula

(31) $$2g = r(n-2) - c + 2.$$

Hurwitz, 1891, showed: The mapping class group $M(T(0, n))$ of $\hat{\mathbb{C}}$ punctured in the points P_ν acts on the r-sheeted Riemann surfaces with signature (30) as follows: The generator σ_i $(i = 1, \ldots, n-1)$ has the effect:

$$(32) \qquad \sigma_i : \begin{cases} P_i \to P_{i+1} , \quad P_{i+1} \to P_i , \quad P_\mu \to P_\mu , & (\mu \neq i, i+1) \\ \Pi_i \to \Pi_{i+1} , \quad \Pi_{i+1} \to \Pi_{i+1}^{-1}\Pi_i\Pi_{i+1} , \quad \Pi_\lambda \to \Pi_\lambda , & (\lambda \neq i, i+1) . \end{cases}$$

He also showed: $M\big(T(0, n)\big)$ acts transitively on the Riemann surfaces with r sheets and exactly one branch point of degree 2 over every P_ν. (This means that each Π_ν consists of one two cycle and $r - 2$ one cycles.) We already mentioned in Section 3 that Cohen, 1973, has determined the quotient groups of $M\big(T(0, n)\big)$ resulting from this action for $r = 3, 4$. It should be noted that Hurwitz' proof for the transitivity is topological although the statement is purely group theoretical.

Arnol'd, 1968a, observed that the hyperelliptic algebraic curves of degree n,

$$(33) \qquad y^2 = x^n + s_1x^{n-1} + \ldots + s_{n-1}x + s_n \equiv Q_n(x) ,$$

which are non-degenerate (that is, for which $Q_n(x)$ has simple roots) form a fiber space with the hyperelliptic curve as fiber and a base space whose fundamental group is the n-th braid group $B_n(S^2)$ of the two-sphere. He used a topological theorem which states that the fundamental group of the base space acts on the homology groups of the fiber as a group of automorphisms to define these automorphism groups as the "*monodromy groups*" of the fiber. In the particular case of the space of hyperelliptic curves of degree n, he thus derives the following result:

The first monodromy group of the hyperelliptic curves of n-th degree is a subgroup of the symplectic group

$$\mathrm{Sp}(g, Z) , \quad g = [\tfrac{1}{2}(n-1)]$$

of $2g$ by $2g$ symplectic matrices with integral entries. Only for $n = 3, 4, 6$ is the first monodromy group the whole group $\mathrm{Sp}(g, Z)$.

Other topological investigations of a related nature may be found in Arnol'd, 1968b, 1970a, b, Brieskorn, 1971b and Gorin and Lin, 1969. The last paper contains a detailed investigation of the commutator subgroup of B_n.

The results found by Birman and Hilden, 1972a (see end of Section 5) show that Arnol'd, 1968a, obtains a two-valued representation of $B_n(S^2)$ as a subgroup of $\mathrm{Sp}(g, Z)$. The methods used by Arnol'd are topological. It is possible to derive his results and to generalize them by using essentially group theoretical methods.

Let R_n be a Riemann surface with r sheets and with branch points over n

points P_ν of $\hat{C}$. Let F_{n-1} be the fundamental group of the sphere $\hat{C}$ with the points P_ν removed. We present F_{n-1} on n generators x_ν (loops around P_ν) with the defining relation

$$x_1 x_2 \ldots x_n = 1 . \tag{34}$$

The automorphisms α_ν in (4), now again denoted by σ_ν, generate a group which we shall now denote by B_n^* (and which is actually the mapping class group of the euclidean plane with n points removed). According to Magnus, 1935, the defining relations of B_n^* are

$$\begin{cases} \sigma_\nu \sigma_{\nu+1} \sigma_\nu = \sigma_{\nu+1} \sigma_\nu \sigma_{\nu+1} & (\nu = 1, \ldots, n-2) \\ \sigma_\nu \sigma_\mu = \sigma_\mu \sigma_\nu & (|\nu-\mu| \geq 2,\ \nu, \mu = 1, \ldots, n-1) \\ (\sigma_1 \sigma_2 \ldots \sigma_{n-1})^n = 1 . & \end{cases} \tag{35}$$

It can be shown that a subgroup $H(R_n)$ of finite index in B_n^* acts as a group of automorphisms on the fundamental group $\pi_1(R_n)$ of the Riemann surface (Magnus, 1972). We consider the closed curves on R_n. They define (by projection) a subgroup C of F_{n-1}. Next we consider the curves on R_n which are closed and contractible to a point on R_n. They define a subgroup K of F_{n-1} which of course, is a subgroup of C. We can define both C and K algebraically merely by using the Π_ν in the signature of R_n. Then K is normal in C and $C/K = \pi_1(R_n)$. The group $H(R_n)$ is defined as the subgroup of B_n^* which maps C onto itself. It then maps also K onto itself and therefore acts as a group H' of automorphisms on $C/K = \pi_1(R_n)$. If we take the quotient group of $H'(R_n)$ with respect to the subgroup of inner automorphism which it may induce in $\pi_1(R_n)$ we obtain a subgroup $H^*(R_n)$ of the mapping class group of R_n which we shall call the *Hurwitz monodromy group* of R_n (which is now considered as being merely a closed two dimensional manifold of genus g). Magnus, 1972, proved: Given any β in B_n^*, there exists a Riemann surface R_n such that β is not in $H(R_n)$ if $n \geq 3$. Maclachlan, 1973, showed: Given any element β^* in B_n^* which is not an inner automorphism of F_{n-1}, there exists a regular Riemann surface R_n^* such that β^* is not in $H(R_n^*)$. For regular Riemann surfaces the inner automorphisms always belong to $H(R_n^*)$ and K is not only normal in C but even in F_{n-1}. However, inner automorphisms of F_{n-1} need not induce inner automorphisms of C or even of C/K. This explains why we have to use B_n^*

instead of $B_n(S^2)$ to derive subgroups of the mapping class group of R_n. Maclachlan's result also shows that the mapping class group of the sphere with punctures is residually finite.

In some cases $H(R_n)$ coincides with B_n^*. This is true in particular if R_n is the regular Riemann surface in whose signature all of the permutations Π_ν are r-cycles. If r is a divisor of n, this case has been investigated in detail by Magnus and Peluso, 1969. (The case $r = 2$ is the one considered by Arnol'd, 1968a.) The action of B_n^* on the abelianized group C/K can then be derived from the Burau representation (Section 3) by giving v a numerical value and splitting off a first degree factor. The resulting representation of B_n^* (and, therefore, of B_n) is isomorphic with a subgroup of $Sp(g, Z)$. That it cannot be all of $Sp(g, Z)$ for $g > 2$ follows then in a simple algebraic manner from our knowledge of isomorphisms (or, rather, non-isomorphisms) of the groups of finite order appearing as symplectic linear groups and the symmetric groups.

7. Artin groups

A *Coxeter matrix* M is defined as an n by n matrix with entries $m_{i,j}$, $i, j = 1, \ldots, n$ which are positive integers or ∞ and such that

$$m_{ij} = m_{ji}, \quad m_{ii} = 1, \quad (i, j = 1, \ldots, n).$$

An *Artin group* G is a group with the following presentation:

Generators a_i, $i = 1, \ldots, n$;

(36) Defining relations: $a_i a_j a_i \ldots = a_j a_i a_j \ldots$

where in (36) both sides contain m_{ij} factors which are alternatingly a_i and a_j. If $m_{i,j} = \infty$, the relation (36) is omitted.

If we add the relations

(37) $$a_i^2 = 1, \quad (i = 1, \ldots, n)$$

to (36) we obtain a *Coxeter group* $\overline{G}$. (For these groups, see Coxeter and Moser, 1965, or Benson and Grove, 1971.) The term "Artin groups" was coined by Brieskorn and Saito, 1972, because they are a natural generalization of Artin's braid groups B_n. For these, the corresponding Coxeter group is Σ_n.

Brieskorn, 1971a, had shown: If the Coxeter group $\overline{G}$ corresponding to G is an irreducible finite group, then G is the fundamental group of the space of regular orbits of the complex reflection group belonging to $\overline{G}$. (See Coxeter and

Moser, 1965, for definition of terms.) Brieskorn, 1971b, had also derived topological properties for these spaces in some cases, and Deligne, 1972, had generalized his results for all finite $\overline{G}$.

A few of the Artin groups other than B_n were first studied by Garside, 1969, who generalized his results for B_n . Garside's methods are group theoretical. Tits, 1968, gave an elegant solution of the word problem for Coxeter groups which, however, is based on geometric arguments. Brieskorn and Saito, 1972, solved the word problem and the conjugacy problem for the Artin groups G with a finite Coxeter group $\overline{G}$ in the same form in which this had been done by Garside. In particular, they construct a "fundamental word" for G which serves the same purposes as in Garside's paper, and they show that it will exist if and only if $\overline{G}$ is finite. They also show that, in the case of an irreducible $\overline{G}$, the center of G is infinite cyclic and generated by a power of $a_1 a_2 \dots a_n$.

ACKNOWLEDGEMENT. Work on this paper was supported in part by the United States National Science Foundation.

References

J.W. Alexander, II, (1923), "A lemma on systems of knotted curves", *Proc. Nat. Acad. USA* 9, 93-95. FdM49,408.

В.И. Арнольд [V.I. Arnol'd], (1968), "Замечание о ветвлении гиперэллиптических интегралов как функций параметров" [A remark on the branching of hyperelliptic integrals as functions of the parameters], *Funkcional. Anal. i Priložen.* 2, no. 3, 1-3; *Functional Anal. Appl.* 2, 187-189. MR39#5583.

В.И. Арнольд [V.I. Arnol'd], (1968), "О косах алгебраических функций и когомологиях ласточкиных хвостов" [Fibrations of algebraic functions and cohomologies of dovetails], *Uspehi Mat. Nauk* 23, no. 4, 247-248. MR38#156.

В.И. Арнольд [V.I. Arnol'd], (1969), "Кольцо когомологий группы крашеных кос" [The cohomology ring of the group of dyed braids], *Mat. Zametki* 5, 227-231.MR39#3529.

В.И. Арнольд [V.I. Arnol'd], (1970), "Топологические инварианты алгебраических функций. II" [Topological invariants of algebraic functions, II], *Funktional. Anal. i Priložen.* 4, no. 2, 1-9; *Functional Anal. Appl.* 4, 91-98. MR43#1991.

Emil Artin, (1925), "Theorie der Zöpfe", *Abh. Math. Sem. Univ. Hamburg* 4 (1926), 47-72. FdM51,450.

E. Artin, (1947a), "Theory of braids", *Ann. of Math.* (2) 48, 101-126. FdM8,367.

E. Artin, (1947b), "Braids and permutations", *Ann. of Math.* (2) 48, 643-649. MR9,6.

S. Bachmuth, (1965), "Automorphisms of free metabelian groups", *Trans. Amer. Math. Soc.* 118, 93-104. MR31#4831.

C.T. Benson and L.C. Grove, (1971), *Finite reflection groups* (Bogden and Quigley, Tarrytown on Hudson, New York).

P. Bergau und J. Mennicke, (1960), "Über topologische Äbbildungen der Brezelfläche vom Geschlecht 2", *Math. Z.* 74, 414-435. MR27#1960.

Joan S. Birman, (1969a), "On braid groups", *Comm. Pure Appl. Math.* 22, 41-72. MR38#2764.

Joan S. Birman, (1969b), "Mapping class groups and their relationship to braid groups", *Comm. Pure Appl. Math.* 22, 213-238. MR39#4840.

Joan S. Birman, (1969c), "Non-conjugate braids can define isotopic knots", *Comm. Pure Appl. Math.* 22, 239-242. MR39#6298.

Joan S. Birman, (1973), "Plat representations for link groups", *Comm. Pure Appl. Math.* 26,

Joan S. Birman, (to appear), "Braids, links and mapping class groups", *A research monograph* (Annals of Mathematics Studies).

Joan S. Birman, (to appear), "Mapping class groups of surfaces: a survey", *Proc. Third Conf. on Riemann Surfaces*, 1974 (Annals of Mathematics Studies).

Joan S. Birman and Hugh M. Hilden, (1971), "On the mapping class groups of closed surfaces as covering spaces", *Advances in the theory of Riemann surfaces*, pp. 81-115 (Proc. 1969 Stony Brook Conf. Annals of Mathematics Studies, 66. Princeton University Press and University of Tokyo Press, Princeton, New Jersey). MR45#1169.

Joan S. Birman and Hugh M. Hilden, (1972), "Isotopies of homeomorphisms of Riemann surfaces and a theorem about Artin's braid group", *Bull. Amer. Math. Soc.* 78, 1002-1004.

F. Bohnenblust, (1947), "The algebraical braid group", *Ann. of Math.* (2) 48, 127-136. MR8,367.

E. Brieskorn, (1971), "Die Fundamentalgruppe des Raumes der regulären Orbits einer endlichen komplexen Spiegelungsgruppe", *Invent. Math.* 12, 57-61. MR45#2692.

Egbert Brieskorn, (1973), "Sur les groupes de tresses d'après V.I. Arnol'd", *Seminaire Bourbaki*, Vol. 1971/72, Exposé 401 (Lecture Notes in Mathematics, 317. Springer-Verlag, Berlin, Heidelberg, New York).

Egbert Brieskorn und Kyoji Saito, (1972), "Artin-Gruppen und Coxeter-Gruppen", *Invent. Math.* 17, 245-271. Zbl.243.20037.

Werner Burau, (1936), "Über Verkettungsgruppen", *Abh. Math. Sem. Hansichen Univ.* 11, 171-178. FdM61,1021.

Gerhard Burde, (1963), "Zur Theorie der Zöpfe", *Math. Ann.* 151, 101-107. MR27#1942.

Wei-Liang Chow, (1948), "On the algebraical braid group", *Ann. of Math.* (2) 49, 654-658. MR10,98.

Daniel I.A. Cohen, (1967), "On representations of the braid group", *J. Algebra* 7, 145-151. MR38#220.

David B. Cohen, (1973), "The Hurwitz monodromy group", (PhD thesis, New York University, New York).

H.S.M. Coxeter, W.O.J. Moser, (1957), *Generators and relations for discrete groups* (Ergebnisse der Mathematik und ihrer Grenzgebiete, Band 14. Springer-Verlag, Berlin, Göttingen, Heidelberg). MR19,527.

Richard H. Crowell and Ralph H. Fox, (1963), *Introduction to knot theory* (Ginn & Co., Boston). MR26#4348.

Pierre Deligne, (1972), "Les immeubles des groupes de tresses généralisés", *Invent. Math.* 17, 273-302. Zbl.238.20034.

E. Fadell, (1962), "Homotopy groups of configuration spaces and the string problem of Dirac", *Duke Math. J.* 29, 231-242. MR25#4538.

Edward Fadell and Lee Neuwirth, (1962), "Configuration spaces", *Math. Scand.* 10, 111-118. MR25#4537.

Edward Fadell and James Van Buskirk, (1962), "The braid groups of E^2 and S^2 ", *Duke Math. J.* 29, 243-257. MR25#4539.

R. Fox and L. Neuwirth, (1962), "The braid groups", *Math. Scand.* 10, 119-126. MR27#742.

R. Fricke und F. Klein, (1965), *Vorlesungen über die Theorie der automorphen Functionen* (Band I. Die Gruppentheoretischen Grundlagen. Teubner, Leipzig, 1897; FdM28,334. Reprinted Johnson Reprint Corp., New York; B.G. Teubner Verlagsgesellschaft, Stuttgart). MR32#1348.

W. Fröhlich, (1936), "Über ein spezielles Transformationsproblem bei einer besonderen Klasse von Zöpfen", *Mh. Math. Phys.* 44, 225-237. FdM62,658.

F.A. Garside, (1969), "The braid group and other groups", *Quart. J. Math. Oxford Ser.* (2) 20, 235-254. MR40#2051.

Betty Jane Gassner, (1961), "On braid groups", *Abh. Math. Sem. Univ. Hamburg* 25, 10-22. MR24#A174.

Richard Gillette and James Van Buskirk, (1968), "The word problem and consequences for the braid groups and mapping class groups of the 2-sphere", *Trans. Amer. Math. Soc.* 131, 277-296. MR38#221.

Е.А. Горин, В.Я. Лин [E.A. Gorin, V.Ja. Lin], (1969), "Алгебраические уравнения с непрерывными коэффициентами и некоторые вопросы алгебраической теории кос" [Algebraic equations with continuous coefficients, and certain questions of the algebraic theory of braids], *Mat. Sb. (NS)* 78 (120), 579-610; *Math. USSR-Sb.* 7, 569-596. MR40#4939.

A. Hurwitz, (1891), "Ueber Riemann'sche Flächen mit gegebenen Verzweigungspunkten", *Math. Ann.* 39, 1-61. FdM23,429.

A. Hurwitz, (1902), "Ueber die Anzahl der Riemann'schen Flächen mit gegebenen Verzweigungspunkten", *Math. Ann.* 55, 53-66. FdM32,404.

S. Kinoshita and H. Terasaka, (1957), "On unions of knots", *Osaka Math. J.* 9, 131-153. MR20#4846.

H. Levinson, (1973), "Decomposable braids and linkages", *Trans. Amer. Math. Soc.* 178, 111-126.

Seymour Lipschutz, (1961), "On a finite matrix representation of the braid group", *Arch. Math.* 12, 7-12. MR23#A2469.

Seymour Lipschutz, (1963), "Note on a paper by Shepperd on the braid group", *Proc. Amer. Math. Soc.* 14, 225-227. MR26#4350.

Colin Maclachlan, (1973), "On a Conjecture of Magnus on the Hurwitz Monodromy Group", *Math. Z.* 132, 45-50.

Wilhelm Magnus, (1934), "Über Automorphismen von Fundamentalgruppen berandeter Flächen", *Math. Ann.* 109, 617-646. FdM60,91.

W. Magnus, (1972), "Braids and Riemann surfaces", *Comm. Pure Appl. Math.* 25, 151-161. Zbl.226.55002.

Wilhelm Magnus and Ada Peluso, (1967), "On knot groups", *Comm. Pure Appl. Math.* 20, 749-770. MR36#5930.

Wilhelm Magnus and Ada Peluso, (1969), "On a theorem of V.I. Arnol'd", *Comm. Pure Appl. Math.* 22, 683-692. MR41#8658.

Г.С. Маканин [G.S. Makanin], (1968), "Проблема сопряженности в группе кос" [The conjugacy problem in the braid group], *Dokl. Akad. Nauk SSSR* 182, 495-496; *Soviet Math. Dokl.* 9, 1156-1157. MR38#2195.

A. Markoff, (1936), "Über die freie Äquivalenz der geschlossenen Zöpfe", *Mat. Sb. (NS)* 1 (43), 73-78. FdM62,658.

А.А. Марков [A. Markoff], (1945), Основы алгеба ической теории кос [*Foundations of the algebraic theory of tresses*] (Trudy Mat. Inst. Steklov 16). MR8,131.

K. Murasugi, (1973), "On closed 3-braids", (Preprint).

K. Murasugi and R.S.D. Thomas, (1972), "Isotopic closed nonconjugate braids", *Proc. Amer. Math. Soc.* 33 , 137-139. MR45#1149.

M.H.A. Newman, (1942), "On a string problem of Dirac", *J. London Math. Soc.* 17, 173-177. MR4,252.

C.D. Papakyriakopoulos, (1957), "On Dehn's lemma and the asphericity of knots", *Ann. of Math.* (2) 66, 1-26. MR19,761.

Kurt Reidemeister, (1960), "Knoten und Geflechte", *Nachr. Akad. Wiss. Gottingen Math.-Phys. Kl.* 2, 105-115. MR22#1913.

G.P. Scott, (1970), "Braid groups and the group of homeomorphisms of a surface", *Proc. Cambridge Philos. Soc.* 68, 605-617. Zbl.203,563.

J.A.H. Shepperd, (1962), "Braids which can be plaited with their ends tied together at each end", *Proc. Roy. Soc. London Ser. A* 265, 229-244. MR24#A2959.

Jacques Tits, (1969), "Le problème des mots dans les groupes de Coxeter", *Symposia Mathematica, INDAM*, Rome, 1967/68, 1, 175-185 (Academic Press, London, New York). MR40#7339.

James Van Buskirk, (1966), "Braid groups of compact 2-manifolds with elements of finite order", *Trans. Amer. Math. Soc.* 122, 81-97. MR32#6440.

О.Я. Виро [O.Ja. Viro], (1972), "Зацепления, двулистные разветвленные накрытия и косы" [Linkings, 2-sheeted branched coverings and braids], *Mat. Sb. (NS)* 87 (129), 216-228; *Math. USSR-Sb.* 16 (1972), 223-236. MR45#7701.

Courant Institute of Mathematical Sciences,
New York University.

Present address:
11 Lomond Place,
New Rochelle, New York 10804, USA.

PROC. SECOND INTERNAT. CONF. THEORY OF GROUPS, CANBERRA, 1973, pp. 488-493.

20E40, 55A05, 55A25 (20E30, 20E15)

TWO-BRIDGE KNOTS HAVE RESIDUALLY FINITE GROUPS

E.J. Mayland, Jr.

0. Introduction

If C is a property of groups, then we say that a group is *residually-C* if the intersection of all its normal subgroups with quotient group possessing the property C is the identity subgroup. The residual finiteness of the groups of fibred knots or Neuwirth knots, that is, those knot groups with finitely generated and therefore free commutator subgroup, has been known for some time [5, p. 63]. It was shown in [4] that certain knots with infinitely generated commutator subgroup share with Neuwirth knots the property that their commutator subgroup is residually a finite p-group, therefore implying the groups of these knots are residually finite. In this paper we show that the class of knot groups with commutator subgroups which are residually a finite p-group includes the groups of all two-bridge knots [7]. We take this as support for the conjecture that the class includes the groups of all alternating knots. The proof depends on the one-relator structure of a two-bridge knot group and proceeds as in [4] by adjoining a countable sequence of roots to a free group, employing theorems of Baumslag.

1. Preliminaries

Schubert [7] describes and classifies knots with two bridges in terms of a normal form (α, β), where α may be an odd positive integer and $0 < \beta < \alpha$ an odd integer relatively prime to α. In fact two knots $K_1 = (\alpha, \beta_1)$ and $K_2 = (\alpha, \beta_2)$ have the same type if and only if $\beta_1 \equiv \pm\beta_2^{\pm 1} \pmod{2\alpha}$. In the following we will write a for the number $a + 2j\alpha$ such that $-\alpha < a + 2j\alpha < \alpha$, when a is not a multiple of α. Then, as in [6], Proposition 1 follows directly from Schubert's description of the form (α, β).

PROPOSITION 1. *If K is the two-bridge knot (α, β), then the group of K has the presentation*

Partially supported by a Canadian National Research Council Grant.

(1.1) $$G = \left\langle x_1,\ x_2;\ w^{-1}x_1wx_2^{-1}\right\rangle ,$$

in which

(1.2) $$w = x_2^{\eta_1}x_1^{\eta_2}x_2^{\eta_3} \ldots x_2^{\eta_{\alpha-2}}x_1^{\eta_{\alpha-1}} ,$$

(1.3) $$\eta_i = \mathrm{sign}(i\beta) \ ,\quad i = 1, 2, \ldots, \alpha-1 ,$$

(1.4) $$\eta_i = \eta_{\alpha-i} \ ,\quad i = 1, 2, \ldots, \alpha-1 .$$

Following the Magnus analysis of the structure of groups with one defining relator (see [3], Section 4.4), we will present the commutator subgroup γ_2G by a Reidemeister-Schreier rewriting process τ, using the powers of x_1 as coset representatives for the free cyclic quotient G/γ_2G. Thus γ_2G will be generated by symbols $S_k = x_1^kx_2x_1^{-k}$, $k = 0, \pm1, \pm2, \ldots$, and related by relators $R_k = \tau\left(x_1^k\cdot w^{-1}x_1wx_2^{-1}\cdot x_1^{-k}\right)$, $k = 0, \pm1, \pm2, \ldots$. We define M and m to be respectively the maximum and minimum subscript occuring in R_0.

If we denote the commutator $A^{-1}B^{-1}AB$ of elements A and B by $[A, B]$ and set $[g_1, \ldots, g_j] = [\ldots [[g_1, g_2], g_3] \ldots]$, then for a given group G, the terms of the *lower central series* of G are defined by $\gamma_jG = \mathrm{gp}\langle [g_1, \ldots, g_j];\ g_j \text{ in } G\rangle$. Also the *lower central sequence* of G is the sequence of quotients $G/\gamma_2G, G/\gamma_3G, \ldots$. The group G is *absolutely parafree* or *parafree* (in the variety of all groups) if G has the same lower central sequence as a free group F_r and G is residually nilpotent (see [1, 2]). Here r is the *rank* of G. Then the proof will contain repeated applications of the following special case of a theorem of Baumslag [1, 2].

PROPOSITION 2. *Let H be a finitely generated parafree group of rank r, let n be an integer, and suppose h is primitive in H modulo γ_2H (so that h is not a proper power modulo γ_2H and h generates its own centralizer in H). Then*

$$G = \langle H * \langle x\rangle;\ h = x^n\rangle$$

is parafree of rank r.

Finally, for G as in (1.1), let Y_d denote the subgroup of γ_2G generated by $S_m, S_{m+1}, \ldots, S_{m+d}$. (In fact all subgroups of γ_2G generated by $d + 1$ consecutive S-symbols are isomorphic.) Then Y_{M-1} is free, Y_M is a one-relator group, and the main effort will be expended on the following Theorem 1.

THEOREM 1. *The subgroup* Y_M *is parafree of rank* $M - m - 1$.

From the proof will follow the main results, namely:

COROLLARY 1. *The commutator subgroup* $\gamma_2 G$ *of a two-bridge knot group* G *is the union of parafree groups of rank* $M - m - 1$ *, hence residually a finite p-group.*

COROLLARY 2. *The group* G *of a two-bridge knot is residually finite.*

2. Proofs

The theorem will follow from a sequence of lemmas. We begin by showing that the occurrences of S_M are all contained within cyclic repetitions of a subword of R_0 .

Define γ to be the smaller of the representatives of β^{-1} and $-\beta^{-1}$ modulo α , so that $0 < \gamma < \alpha/2$. We will define $\kappa > 0$ by whichever of the following formulas is appropriate: $\beta\gamma \pm 1 = \kappa\alpha$. Then the following are immediate from the definition of the η_i .

LEMMA 1. *The exponents* $\eta_1, \eta_2, \ldots, \eta_{\alpha-1}$ *of* W *satisfy* $\eta_i = \eta_{i+\gamma}$ *for* $i = 1, 2, \ldots, \alpha-1-\gamma$ *, with the sign depending on whether* κ *is even (+) or odd (-).*

COROLLARY. $\eta_i = \eta_{i+2\gamma}$ *, for* $i = 1, 2, \ldots, \alpha-1-2\gamma$.

In fact, if $\gamma\beta \equiv 1$ modulo α , then $\eta_\gamma = \mp 1$ as κ is odd or even. Therefore

LEMMA 2. *If the relator* $w^{-1}x_1wx_2$ *is considered cyclically, the exponent cycle within* w *extends past* w *exactly* γ *symbols to the left or right as* $\gamma = \pm\beta^{-1}$.

If we now pair the X-symbols of w^{-1} and w symmetrically with respect to the x_1 symbol separating w^{-1} and w , we find that the corresponding S-symbol in $\tau(w)$ has subscript one greater than the subscript of the corresponding S-symbol of $\tau(w^{-1})$. Therefore all occurrences of the symbol S_M having maximal subscript are contained within $\tau\left(w \cdot x_2^{-1}\right)$, and not within $\tau\left(w^{-1} \cdot x_1\right)$. Together with the previous results this implies:

LEMMA 3. *If the relator* $R_D = \tau\left(w^{-1} \cdot x_1 \cdot w \cdot x_2^{-1}\right)$ *is considered cyclically, all occurrences of* S_M *are contained within a subword of the form* A^r *, where the length of* A *is* γ *. If the exponent sum of* $\tau^{-1}(A)$ *is not zero and if* κ *is even, then* $r = 1$ *. Otherwise* r *is the greatest integer in* $(\alpha-1+\gamma)/2\gamma$ *when* $\gamma = -\beta^{-1}$ *(and* *in* $(\alpha-2+\gamma)/2\gamma$ *when* $\gamma = \beta^{-1}$).

Now the subgroup Y_M of $\gamma_2 G$ generated by $S_m, S_{m+1}, \ldots, S_M$ is a one-relator group with relator R_0. If we adjoin to this group a new symbol t, together with the new relator tA^{-1}, then we can rearrange the relator R_0 to have the form $t^r = V$, for some non-empty word V in the generators $S_m, S_{m+1}, \ldots, S_{M-1}$. We consider the group (2.1) as the first stage of the construction of Y_M from Y_{M-1}, and note that Proposition 2 implies (modulo the proof of Lemma 4) that (2.1) is parafree of rank $M - m - 1$,

$$\left\langle S_m, S_{m+1}, \ldots, S_{M-1}, t;\ t^r = V\right\rangle. \tag{2.1}$$

We turn to an analogous consideration of the possibility of cyclically repeated subwords of the word A. Relabel the previously defined γ, κ, r, t, V and A respectively γ_1, κ_1, and so forth. For each integer $j = 2, 3, \ldots, (\alpha-1)/2$, consider in turn the smaller of the minimal positive representations modulo α of the two numbers $j\beta^{-1}$ and $-j\beta^{-1}$. Let γ_2 be the first of these numbers which is less than γ_1. We write, as $\gamma_2 \equiv \pm j_2 \cdot \beta^{-1}$, that $\gamma_2\beta = \kappa_2\alpha \pm j_2$. As above, the exponents in $\tau^{-1}(A_1)$ in w cycle with $\eta_i = \pm\eta_{i+\gamma_2}$ for $i = 1, 2, \ldots, \gamma_1-\gamma_2$, and this pattern continues γ_2 symbols to the left or right (and also with $\tau^{-1}(A_1)$ considered cyclically) as $\gamma_2 \equiv \pm j_2\beta^{-1}$. Thus A_1 will contain a subword of the form $A_2^{r_2}$, for $r_2 = 1$ if the exponent sum on $\tau^{-1}(A_1)$ is not zero and κ_2 is even, or for r_2 equal to the greatest integer in $(\gamma_1+\gamma_2)/2\gamma_2$, otherwise. Introducing a new symbol t_2, and rearranging the relator $t_1A_1^{-1}$ to the form $t_2^{r_2} = V_2$ (with V_2 clearly primitive in (2.1) modulo the commutator subgroup), we conclude (modulo Lemma 4) that $S_m, S_{m+1}, \ldots, S_{M-1}$, t_1 and t_2 generate a parafree group.

We repeat this reduction until a value of j has corresponding γ equal to $+1$. Then the repeated subword will be a power of S_M, so replacing the last t-symbol by S_M, and reducing the resulting presentation to that of Y_M, we complete the proof of Theorem 1 (again modulo Lemma 4).

LEMMA 4. *$V = V_1$ (in (2.1)) is primitive in Y_{M-1} modulo $\gamma_2 Y_{M-1}$.*

We will only sketch the algebraic proof of Lemma 4, which has many cases. First, no s-symbol occurs in $\tau\left(w^{-1}x_1wx_2\right)$ to both of the exponents $+1$ and -1. Thus the length of R_0 modulo $\gamma_2 Y_M$ is still α. By analogy with Lemma 3 define all occurrences of S_m in R_0 to occur within repetitions of the subword B (which will not involve the symbol S_M). It follows that V is primitive in Y_{M-1} modulo $\gamma_2 Y_{M-1}$ if and only if B is. Reducing the repeated word, the primitivity of V is eventually equivalent to that of S_m, proving Lemma 4.

The groups Y_d, $d = M+1, M+2, \ldots$, can be constructed inductively from the groups Y_{d-1} by adjoining roots to the words $\tau\left(x_1^{\upsilon}\cdot X\cdot x_1^{-\upsilon}\right)$, $\upsilon = d - M$, where X rewrites to V. If, in $Y_{d-1}/\gamma_2 Y_{d-1}$, we set $\tau\left(x_1^{\upsilon}\cdot X\cdot x_1^{-\upsilon}\right)$ equal to one, and use each relator R_k to delete whatever S-symbol it contains with smallest subscript, we obtain a quotient group which is free abelian of rank $M - m - 2$. Thus

LEMMA 5. *$\tau\left(x_1^{\upsilon}\cdot X\cdot x_1^{-\upsilon}\right)$ is primitive in $Y_{d-1}/\gamma_2 Y_{d-1}$.*

COROLLARY. *The groups Y_d are parafree of rank $M - m - 1$.*

Now Corollaries 1 and 2 follow as in [4, Section 2] from a theorem of Baumslag.

References

[1] Gilbert Baumslag, "Groups with the same lower central sequence as a relatively free group. I. The groups", *Trans. Amer. Math. Soc.* 129 (1967), 308-321. MR36#248.

[2] Gilbert Baumslag, "Groups with the same lower central sequence as a relatively free group. II. Properties", *Trans. Amer. Math. Soc.* 142 (1969), 507-538. MR39#6959.

[3] Wilhelm Magnus, Abraham Karrass, Donald Solitar, *Combinatorial group theory* (Pure and Appl. Math., 13. Interscience [John Wiley & Sons], New York, London, Sydney, 1966). MR34#7617.

[4] E.J. Mayland, Jr, "On residually finite knot groups", *Trans. Amer. Math. Soc.* 168 (1972), 221-232. MR45#4396.

[5] L.P. Neuwirth, *Knot theory* (Annals of Mathematics Studies, 56. Princeton University Press, Princeton, New Jersey, 1965). MR31#734.

[6] Robert Riley, "Parabolic representations of knot groups, I", *Proc. London Math. Soc.* (3) 24 (1972), 217-242. MR45#9313.

[7] Horst Schubert, "Knoten mit zwei Brücken", *Math. Z.* 65 (1956), 133-170. MR18,498.

York University,
Downsview, Canada.

PROC. SECOND INTERNAT. CONF. THEORY OF GROUPS, CANBERRA, 1973, pp. 494-498. 20E05, 20F55

PERIODIC AUTOMORPHISMS OF THE TWO-GENERATOR FREE GROUP

Stephen Meskin

Let $F_2 = \langle a, b \rangle$ be the free group of rank 2 . We show that Aut F_2 has four conjugacy classes of elements of order 2 and one conjugacy class each of elements of order 3 and 4 . Implicit in the proof of these statements is an algorithm for deciding in which conjugacy class a periodic element belongs. From an analysis of $\mathrm{GL}(2, Z) \cong \mathrm{Aut}\, F_2/\mathrm{Inn}\, F_2$, the orders of periodic elements are restricted to 2, 3, 4 and 6 , but Aut F_2 has no element of order 6 .

My curiosity in periodic automorphisms of free groups grew out of research into one-relator groups with center.

1. Background

We begin with a collection of results contained explicitly and implicitly in [1].

THEOREM 1.

$$\mathrm{Aut}\, F_2 = \langle P, X, Y, A, B;\ X^4 = P^2 = (PX)^2 = 1,\ (PY)^2 = B,\ X^2 = Y^3B^{-1}A,$$
$$A^P = A^X = A^Y = B,\ B^P = A,\ B^X = A^{-1},\ B^Y = A^{-1}B \rangle ,$$

where $a^P = a^X = a^Y = b$, $b^P = a$, $b^X = a^{-1}$, $b^Y = a^{-1}b$ *and* A *and* B *are the inner automrophisms induced by* a *and* b *respectively.*

Inn $F_2 = \langle A, B \rangle$.

$$\mathrm{Aut}\, F_2/\mathrm{Inn}\, F_2 \cong \mathrm{GL}(2, Z) = \langle p, x, y;\ x^4 = p^2 = (px)^2 = (py)^2 = 1,\ x^2 = y^3 \rangle ,$$

where $p = \begin{pmatrix} 0 & 1 \\ 1 & 0 \end{pmatrix}$, $x = \begin{pmatrix} 0 & 1 \\ -1 & 0 \end{pmatrix}$ *and* $y = \begin{pmatrix} 0 & 1 \\ -1 & 1 \end{pmatrix}$.

PROOF. Straightforward computation will show that the relations hold. To see that the indicated elements generate observe that $\mathrm{SL}(2, Z) = \langle x, y;\ x^4 = 1,\ x^2 = y^3 \rangle$ (see [1], p. 47, where it is denoted M) and $p \notin \mathrm{SL}(2, Z)$. Thus Aut F_2 maps onto $\mathrm{GL}(2, Z)$ by $P \mapsto p$, $X \mapsto x$, $Y \mapsto y$, $A \mapsto 1$ and $B \mapsto 1$ with kernel

Inn F_2 . //

The elements of Aut F_2 can be written uniquely in the form

$P^{\alpha}U(X, Y)X^{2\beta}W(A, B)$, where $\alpha, \beta = 0$ or 1 , $W(A, B)$ is a reduced word and $U(X, Y)$ is a reduced word where X, Y and Y^{-1} are the only powers of X and Y appearing. The elements of GL(2, Z) can be written in a similar manner with lower case symbols.

THEOREM 2. *A periodic element of* GL(2, Z) *is conjugate to exactly one of* p , px , x^2 , y^2 , x *and* y . *Their orders are* 2, 2, 2, 3, 4 *and* 6 *respectively.*

This is probably well-known. We include a proof for completeness.

PROOF. It is clear from the structure of SL(2, Z) as a generalized free product that a periodic element in SL(2, Z) is conjugate to exactly one of $x^2, y^2, y^{-2}, x, x^{-1}, y$ and y^{-1} . By conjugating with p , we see that y^2, x and y are all conjugate to their inverses. A periodic element of GL(2, Z) which is not in SL(2, Z) is of the form $pU(x, y, y^{-1})x^{2\beta}$. Now $(pU(x, y, y^{-1})x^{2\beta})^2 = U(x, y^{-1}, y)U(x, y, y^{-1})x^{2n}$ where n is the number of x's appearing in U . By conjugation we may assume that

$$U(x, y^{-1}, y)U(x, y, y^{-1})x^{2n} = 1, x^2, y^2 \ (= y^{-1}x^2), x \text{ or } y .$$

If there is no cancellation, then the only solution is $U = 1$ (and hence $n = 0$). If there is cancellation and if U begins x, y or y^{-1} , then U must also end x, y or y^{-1} respectively. Thus by conjugating $pUx^{2\beta}$ we can shorten U , unless $U = 1, x, y$ or y^{-1} . However, $(pyx^{2\beta})^y = (py^{-1}x^{2\beta})^{y^{-1}} = px^{2(\beta+1)}$, $px^2 = p^x$ and $px^3 = (px)^x$.

It only remains to show that p and px are not conjugate in GL(2, Z) (they are conjugate of course in GL(2, R)). However, note that in conjugating $pUx^{2\beta}$, n , the number of x's in U , changes only by an even number. //

2. The main result

THEOREM 3. *A periodic element of* Aut F_2 *is conjugate to exactly one of* P, PX, PXA, X^2, Y^2B^{-1} *and* X . *Their orders are* 2, 2, 2, 2, 3 *and* 4 *respectively.*

The proof follows from the following sequence of lemmas after a few observations. A computation will show that the given elements have the indicated orders. Every element of Aut F_2 of order n is of the form $U(P, X, Y)W(A, B)$ where $U(p, x, y)$

has order n in GL(2, Z) . We may restrict $U(p, x, y)$ to the elements listed in Theorem 2. Indeed, if $U(p, x, y) = S(p, x, y)^{T(p,x,y)}$ then

$$U(P, X, Y) = S(P, X, Y)^{T(P,X,Y)}V(A, B)$$

and

$$U(P, X, Y)W(A, B) = \left\{S(P, X, Y)V(A, B)^{T^{-1}} W(A, B)^{T^{-1}}\right\}^{T} .$$

Each lemma deals with one element from Theorem 2 and proceeds by induction on a word $W = W(A, B)$.

LEMMA 1. *A periodic element of the form* PW *is conjugate to* P .

PROOF. Let $C = A, A^{-1}, B$ or B^{-1} , then $(PW)^C = PC^{-P}WC$. Suppose $(PW)^2 = 1$ Since $(PW)^2 = W^PW$ we have $W = W^{-P}$. If W ends C^{-1} then W^P ends C^{-P} and thus W begins C^P . Since $C^P \neq C^{-1}$ we can, unless $W = 1$, shorten W by conjugation. //

LEMMA 2. *A periodic element of the form* PXW *is conjugate to* PX *or* PXA .

PROOF. We proceed exactly as in Lemma 1 with PX in place of P . The only difference is that $A^{PX} = A^{-1}$ so that we cannot shorten W by conjugation if $W = 1, A$ or A^{-1} . Note however that $PXA^{-1} = (PXA)^{A^{-1}}$.

LEMMA 3. *A periodic element of the form* X^2W *is conjugate to* X^2 .

PROOF. We proceed again as in Lemmas 1 and 2 with X^2 in place of P and PX respectively. In this case however $C^{X^2} = C^{-1}$ so that we cannot shorten W by conjugation if $W = 1, A, A^{-1}, B$ or B^{-1} . Note however that $X^2A^{-1} = X^{2Y^{-1}}$, $X^2A = X^{2Y^{-1}A}$, $X^2B = X^{2Y}$ and $X^2B^{-1} = X^{2YB^{-1}}$.

LEMMA 4. *A periodic element of the form* XW *is conjugate to* X .

PROOF. Let $(XW)^4 = 1$ then $(XW)^2 = X^2W^XW$ has order 2 . By conjugation in gp(A, B) we can assume from Lemma 3 that $W^XW = 1, A$ or B . But $W = W^{-X}A$ or $W = W^{-X}B$ are impossible by length considerations, for example, $|W^{-X}A| = |W| \pm 1$. Thus $W = W^{-X}$ and now proceeding exactly as in Lemma 1 we see that unless $W = 1$ we can shorten W by conjugation.

LEMMA 5. *A periodic element of the form* Y^2W *is conjugate to* Y^2B^{-1} .

PROOF. This lemma illustrates how much easier things can be if one is lucky enough to select the right conjugate. Indeed we will prove in fact that *a periodic element of the form* $XY^{-1}XW$ *is conjugate to* $XY^{-1}X$ and then denoting $XY^{-1}X$ by S note that $Y^2B^{-1} = S^{X^3}$.

Now $A^S = B$ and $B^S = A^{-1}B^{-1}$. If $W = W(A, B)$ then $W^S = W(B, A^{-1}B^{-1})$ and there is no A-cancellation in $W(B, A^{-1}B^{-1})$.

As in the previous lemmas $(SW)^C = SC^{-S}WC$. In this case it is not necessarily true that if $(SW)^3 = 1$ and W ends in C^{-1} then W begins with C^S. All is not lost, however, since we can modify our argument.

If W ends in A or A^{-1}, then by conjugation we can delete the ending while multiplying the beginning by B or B^{-1} respectively. In this process the length W does not increase.

Suppose now that $(SW)^3 = 1$. Then since $(SW)^3 = W^{S^2}W^SW$ we have $W = W^{-S}W^{-S^2}$.

If W ends in B, that is, $W = UB$, then $W^S = U^SA^{-1}B^{-1}$ and $W^{S^2} = U^{S^2}A$ and the A's do not cancel. Thus $W = UB = BAU^{-S}A^{-1}U^{-S^2}$. If U^{-S^2} does not cancel completely then U^{-S^2} also ends in B so $U^{S^2} = B^{-1}V$ and $U = U^{S^3} = BAV^S$ and W begins $BA = B^{-S}$. Even if U^{-S^2} does cancel completely, since W ends in B, $W \neq 1$ so W must at least begin with B. In either case conjugating by B^{-1} will shorten W unless $W = B$.

If W ends in B^{-1}, then as in the above argument we can show that W begins with A^{-1} so that conjugating by B will shorten W.

So we can shorten W by conjugation unless $W = 1, A, A^{-1}$ or B. However, W cannot be A^{-1}, $(SB) = S^{X^2}$ and $(SA) = S^{X^2A}$.

LEMMA 6. *There is no periodic element of the form* YW.

PROOF. If $(YW)^6 = 1$, then $(YW)^3 = X^2W_1$ has order 2 and $(YW)^2 = Y^2W_2$ has order 3. By conjugating X^2W_1 with an element of the form Y^nW_3 we can change it to X^2. So, for there to be an element of order 6, X^2 must commute with an element of order 3.

Now $(Y^2W)^{X^2} = Y^2B^{-1}AB^{-1}W^{X^2}$, $|W^{X^2}| = |W|$ and $|B^{-1}ABW^{X^2}| = |W| \pm 3$ or $|W| \pm 1$ depending on the amount of cancellation. Thus $W \neq B^{-1}AB^{-1}W^{X^2}$ and $(Y^2W)^{X^2} \neq Y^2W$ for all W .

To conclude the proof of Theorem 3 we need only

LEMMA 7. *PX and PXA are not conjugate.*

PROOF. If $PX^{U(P,X,Y)} = PXW(A, B)$, then px and $U(p, x, y)$ must commute so $U(p, x, y) = px, x^2$ or px^3 . However, in all these cases $U(P, X, Y)$ commutes with PX . Finally if $PX^{V(A,B)} = PXW(A, B)$, then W has even length and so $W \neq A$.

Reference

[1] Wilhelm Magnus, Abraham Karrass, Donald Solitar, *Combinatorial group theory* (Pure and Appl. Math. 13. Interscience [John Wiley & Sons], New York, London, Sydney, 1966). MR34#7617.

University of Connecticut,
Storrs, Connecticut 06268, USA.

PROC. SECOND. INTERNAT. CONF. THEORY OF GROUPS, 20D10, 20D15, 20E35
CANBERRA, 1973, pp. 499-503.

ON GROUPS OF EXPONENT FOUR: A CRITERION FOR NONSOLVABILITY

Horace Y. Mochizuki

It has been conjectured that $B(\infty, 4)$, the free infinitely generated group of exponent 4 is solvable (in particular, by Higman and Hall). In the references are given the many contributions toward a solution of this problem and related problems. I wish now to report on the recent joint work of Bachmuth and myself [1] and, in particular, to give evidence for the nonsolvability of $B(\infty, 4)$ and to give a ring-theoretic criterion for nonsolvability.

Using the results of Tobin [11], one can easily prove that if G is a solvable group of exponent 4, then the commutator subgroup $\delta_1(G)$ is nilpotent (see Edmunds and Gupta [3]). Thus, to prove nonsolvability, one can use the fact that there is a group G of exponent 4 such that $\delta_1(G)$ is non-nilpotent. We now proceed to exhibit what we believe to be such a group.

We use the following notation:

R is the free associative ring (with identity 1) of characteristic 2 generated by noncommuting indeterminates $x_1, x_2, \ldots$;

T is the ideal generated by all $T_3\left(x_{i_1}, \ldots, x_{i_m}\right)$, $m \geq 3$, where $T_3\left(x_1, \ldots, x_m\right)$ is the homogeneous component of degree one in each of $x_1, \ldots, x_m$ of $\left[\left(1+x_1\right) \ldots \left(1+x_m\right)-1\right]^3$;

$\Delta\left(x_i, x_j\right) = x_i x_j + x_j x_i$;

J is the ideal generated by all monomials in the x_i with a repeated indeterminate factor;

K is the ideal generated by x_i^3, $i \geq 1$;

Research supported by an NSF Grant.

$r_1, r_2, \ldots$ represent $x_1 + (T+J)$, $x_2 + (T+J), \ldots$, respectively;

H^* denotes the (multiplicative) group of $R/(T+J)$ generated by the $1 + r_i$, $i \geq 1$;

G is the group generated by the matrices $X_0 = \begin{pmatrix} 1 & 0 \\ 1 & 1 \end{pmatrix}$, $X_i = \begin{pmatrix} 1+r_i & 0 \\ r_i & 1 \end{pmatrix}$, $i \geq 1$;

$\delta_k(G)$ denotes the kth term of the derived series of G, $k \geq 0$;

$\gamma_k(G)$ denotes the kth term of the lower central series of G, $k \geq 1$.

We note that H^* satisfies the condition that $(g-1)^3 = 0$ for all $g \in H^*$ and that G is a group of exponent 4 satisfying for $i \geq 0$,

(I) $X_i^2 = 1$,

(II) $\langle X_i^G \rangle$ is abelian,

(III) $[X_i, Y, Y, Y] = 1$,

that is, G (as is H^*) is a homomorphic image of the group H introduced by Gupta and Weston in [8]. I also note that G was used in [5].

CONJECTURE 1. *$\delta_1(G)$ is not nilpotent, and hence G is not solvable.*

To translate this conjecture to a ring theoretic one, we note that

$$[[X_0, X_1], [X_2, X_3], \ldots, [X_{2n}, X_{2n+1}]] = \begin{pmatrix} 1 & 0 \\ r_1(r_2r_3+r_3r_2)\ldots(r_{2n}r_{2n+1}+r_{2n+1}r_{2n}) & 1 \end{pmatrix}$$

Thus, we need to prove

CONJECTURE 2. *$x_1\Delta(x_2, x_3) \ldots \Delta(x_{2n}, x_{2n+1})$ is not contained in $(T+J)$ or equivalently T.*

Let $S(n)$ be the homogeneous component of R of degree 3 in each of $x_1, x_2, \ldots, x_n$ and of degree zero in the other indeterminates.

THEOREM 1. *(i) In $S(n)$, $n \geq 2$, every monomial*

$$M \equiv \alpha x_{n-1}x_{n-2} \ldots x_2x_1x_nx_1^2x_2^2 \ldots x_n^2 \bmod (T+J) \quad \textit{for some } \alpha \in Z_2 .$$

Thus, $S(n)$ has dimension 1 or 0 modulo $(T+J)$.

(ii) In $S(2j+3)$, $j \geq 0$,

$$x_{2j+2}\Delta(x_{2j}, x_{2j+3}) \dots \Delta(x_4, x_7)\Delta(x_2, x_5)\Delta(x_1, x_3)x_1^2x_2^2 \dots x_{2j+3}^2 \equiv$$

$$x_{2j+2}x_{2j+1} \dots x_2x_1x_{2j+3}x_1^2x_2^2 \dots x_{2j+3}^2 \bmod (T+K) .$$

It is therefore clear that the truth of

CONJECTURE 3. $x_{n-1}x_{n-2} \dots x_2x_1x_nx_1^2x_2^2 \dots x_n^2 \notin (T+K)$, $n \geq 2$,

is sufficient for the truth of Conjecture 2 and hence of Conjecture 1.

As evidence for the validity of Conjectures 1, 2, and 3, we have

THEOREM 2. *(a)* $x_1x_2x_1^2x_2^2$ *and* $x_2x_1x_3x_1^2x_2^2x_3^2$ *are not contained in* $(T+K)$. *Thus,* $S(2)$ *and* $S(3)$ *each have dimension* 1 .

(b) $x_1\Delta(x_2, x_3) \dots \Delta(x_{2n}, x_{2n+1}) \notin T$, $1 \leq n \leq 4$.

(c) $\gamma_5(\delta_1(G)) \neq 1$. *In fact,* $[\delta_3(G), \delta_1(G)] \neq 1$.

Theorem 2 *(a)* was verified by computer computations, and *(a)* was used to prove parts *(b)* and *(c)*. For example, *(a)* implies

$$x_1\Delta(x_1, x_2)\Delta(x_2, x_3)\Delta(x_1, x_3)\Delta(x_2, x_3) \notin T + K$$

and thus

$$x_1\Delta(x_2, x_3)\Delta(x_4, x_5) \dots \Delta(x_8, x_9) \notin T .$$

We add that *(c)* implies

(1) $\gamma_{10}(B(5, 4)) \neq 1$.

These computer computations were also used to verify, for example, that

(2) $\gamma_9(B(4, 4)) \neq 1$;

(3) $\gamma_7(B(3, 4))$ has order 2^6 ;

(4) in $R/(T+J)$, there exist group elements $g_1, g_2 \in H^*$ such that $(g_1-1)(g_2-1)(g_1-1)^2(g_2-1)^2 \neq 0$;

(5) in $R/(T+J)$, there exist group elements $g_1, g_2, g_3, g_4 \in H^*$ such that $(g_1-1)(g_2-1)(g_3-1)^2(g_4-1)^2(g_2-1)^2 \neq 0$.

As another reason for the plausibility of Conjecture 3, the following inductive relation appears to be true; namely

$$Px_nx_jx_n^2 \notin (T+K) \text{ iff } Px_j \notin (T+K)$$

where P is a polynomial such that $Px_j \in S(n-1)$.

References

[1] Seymour Bachmuth and Horace Y. Mochizuki, "A criterion for non-solvability of exponent 4 groups", *Comm. Pure Appl. Math.* (to appear).

[2] S. Bachmuth, H.Y. Mochizuki and K. Weston, "A group of exponent 4 with derived length at least 4 ", *Proc. Amer. Math. Soc.* 39 (1973), 228-234.

[3] C.C. Edmunds and N.D. Gupta, "On groups of exponent four IV", *Conf. on Group Theory, University of Wisconsin-Parkside*, 1972, pp. 57-70 (Lecture Notes in Mathematics, 319. Springer-Verlag, Berlin, Heidelberg, New York, 1973).

[4] C.K. Gupta and N.D. Gupta, "On groups of exponent four. II", *Proc. Amer. Math. Soc.* 31 (1972), 360-362. MR44#6819.

[5] Narain D. Gupta, Horace Y. Mochizuki, and Kenneth W. Weston, "On groups of exponent four with generators of exponent two", *Bull. Austral. Math. Soc.* 10 (1974), 135-142.

[6] N.D. Gupta and M.F. Newman, *Monograph on groups of exponent four*, in preparation.

[7] N.D. Gupta and R.B. Quintana, Jr., "On groups of exponent four. III", *Proc. Amer. Math. Soc.* 33 (1972), 15-19. MR45#2000.

[8] Narain D. Gupta and Kenneth W. Weston, "On groups of exponent four", *J. Algebra* 17 (1971), 59-66. MR42#3176.

[9] Marshall Hall, Jr., "Notes on groups of exponent four", *Conf. on Group Theory, University of Wisconsin-Parkside*, 1972, pp. 91-118 (Lecture Notes in Mathematics, 319. Springer-Verlag, Berlin, Heidelberg, New York, 1973).

[10] И.Н. Санов [I.N. Sanov], "Решение проблемы Бернсайда для показателя 4 " [Solution of Burnside's problem for exponent four], *Leningrad. Gos. Univ. Ucen. Zap. Ser. Mat. Nauk* 10 (1940), 166-170. MR2,212.

[11] Seán Tobin, "On a theorem of Baer and Higman", *Canad. J. Math.* 8 (1956), 263-270. MR17,1182.

[12] A.L. Tritter, "A module-theoretic computation related to the Burnside problem", *Computational problems in abstract algebra* (Proc. Conf. Oxford, 1967, pp. 189-198. Pergamon Press, Oxford, 1970). MR41#6981.

[13] C.R.B. Wright, "On groups of exponent four with generators of order two", *Pacific J. Math.* 10 (1960), 1097-1105. MR22#6844.

[14] C.R.B. Wright, "On the nilpotency class of a group of exponent four", *Pacific J. Math.* 11 (1961), 387-394. MR23#A927.

University of California, Santa Barbara,
Santa Barbara, California 93106, USA.

PROC. SECOND INTERNAT. CONF. THEORY OF GROUPS, 22A05
CANBERRA, 1973, pp. 504-515.

THE TOPOLOGY OF FREE PRODUCTS OF TOPOLOGICAL GROUPS

Sidney A. Morris, Edward T. Ordman and H.B. Thompson

1. Introduction

In [3], Graev introduced the free product of Hausdorff topological groups G and H (denoted in this paper by $G \amalg H$) and showed it is algebraically the free product $G * H$ and is Hausdorff. While it has been studied subsequently, for example [4, 6, 7, 8, 11, 12], many questions about its topology remain unsolved. In particular, partial negative results about local compactness were obtained in [7, 11, 12]. In this paper we obtain a complete solution by showing that $G \amalg H$ is locally compact if and only if G, H and $G \amalg H$ are discrete. A similar line of reasoning allows us to show that $G \amalg H$ has no small subgroups if and only if G and H have no small subgroups.

We are able to obtain much stronger results when G and H are k_ω-spaces, a class of spaces which includes, for example, all compact spaces and all connected locally compact groups. In this case we are able to show that the cartesian subgroup, $\text{gp}[G, H] = \text{gp}\{g^{-1}h^{-1}gh : g \in G, h \in H\}$, of $G \amalg H$ is a free topological group, show that certain subgroups of $G \amalg H$ are themselves free products, and show that the topology of $G \amalg H$ depends only on the topologies and not on the algebraic structure of G and H .

2. Definitions and preliminaries

If X is a completely regular Hausdorff space with distinguished point e , the (Graev) *free topological group on* X, $FG(X)$, is algebraically the free group on $X \backslash \{e\}$, with e as identity element and the finest topology making it into a topological group and inducing the given topology on X ; by [2], $FG(X)$ is Hausdorff.

This research was done while the second author was a visitor at the University of New South Wales, partially supported by a Fulbright-Hays grant.

If G and H are topological groups, their *free product* $G \coprod H$ is a topological group whose underlying abstract group is the algebraic free product $G * H$ and whose topology is the finest topology making it into a topological group and inducing the given topologies on G and H; by [3], if G and H are Hausdorff then $G \coprod H$ is Hausdorff.

For the remainder of the paper all topological groups and spaces will be presumed Hausdorff.

A topological group is said to be NSS (*or to have no small subgroups*) if there is a neighbourhood of the identity e which contains no subgroup other than $\{e\}$. This property is most important for locally compact groups in that Hilbert's fifth problem yields that a locally compact group is a Lie group if and only if it is NSS.

We require the following algebraic preliminaries: The identity map $G \to G$ and the trivial map $H \to \{e\} \subset G$ extend simultaneously to a homomorphism $\pi_1 : G * H \to G$; by [3], this is also a continuous map from $G \coprod H$ to G. Similarly $\pi_2 : G * H \to H$ is a homomorphism and a continuous map on $G \coprod H$. The map $\pi_1 \times \pi_2 : G * H \to G \times H$ has kernel $\mathrm{gp}[G, H]$, where $[G, H] = \{g^{-1}h^{-1}gh : g \in G, h \in H\}$. Indeed $\mathrm{gp}[G, H]$ is a free group with free basis $[G, H]\backslash\{e\}$. We find it convenient below to introduce a map $c : G \times H \to [G, H]$ given by $c(g, h) = [g, h] = g^{-1}h^{-1}gh$. If w is any element of $G * H$ it has a unique representation $w = ghk$, where $g \in G$, $h \in H$ and $k \in \mathrm{gp}[G, H]$. We define a map $\pi_c : G * H \to \mathrm{gp}[G, H]$ by $\pi_c(w) = k = \pi_2(w)^{-1}\pi_1(w)^{-1}w$: notice that it is *not* a homomorphism. Finally we note that there is a bijection (not a homomorphism) $p : G \times H \times \mathrm{gp}[G, H] \to G * H$ given by $p(g, h, k) = ghk$. The inverse map is $p^{-1}(w) = \{\pi_1(w), \pi_2(w), \pi_c(w)\}$.

In §4 we use some additional machinery, that of k_ω-spaces; we rely heavily on [4]. A topological space X is said to be a k_ω-space with decomposition $X = \cup X_n$, if $X_1, X_2, \ldots$ are compact subsets of X, $X_n \subset X_{n+1}$ for all n, $X = \bigcup_{n=1}^{\infty} X_n$ and the X_n determine the topology on X in the sense that a subset A of X is closed if and only if $A \cap X_n$ is compact for all n. The decomposition $X = \cup X_n$ is essential, in that X may be a union of some other ascending chain of compact subsets which fail to determine the topology. If $X = \cup X_n$ and $Y = \cup Y_n$ where X_n and Y_n are ascending chains of compact sets, the two ascending chains determine the same topology on X provided each X_n is contained in some Y_k and each Y_n is contained in some X_m .

If G is a topological group and a k_ω-space the decomposition $G = \cup G_n$ may be chosen so that the G_n satisfy two additional conditions: if $g \in G_n$ then $g^{-1} \in G_n$, and if $g \in G_n$, $h \in G_k$ then $gh \in G_{n+k}$.

If X is any subset of a group G , we let $\mathrm{gp}_n(X)$ denote the set of elements of G which are products of at most n elements of X . Hence $\mathrm{gp}_n(G_n) \subset G_{n^2}$.

The class of topological groups which are k_ω-spaces is large enough to include many of the standard examples; in particular, every connected locally compact group is a k_ω-space [12].

We rely heavily on the following result of [4]:

PROPOSITION. *Let* G *be a topological group and* X *a subset which generates* G *algebraically. Let* $X = \cup X_n$ *be a* k_ω*-space. Then* G *has the finest group topology consistent with the original topology on* X *if and only if* G *is a* k_ω*-space with decomposition* $G = \cup \mathrm{gp}_n(X_n)$.

It follows that if $X = \cup X_n$ is a k_ω-space then $FG(X)$ is a k_ω-space with decomposition $FG(X) = \cup \mathrm{gp}_n(X_n)$. If $G = \cup G_n$ and $H = \cup H_n$ are k_ω-spaces then $G \coprod H$ is a k_ω-space with decomposition $G \coprod H = \cup \mathrm{gp}_n(G_n \cup H_n)$.

Finally note that when we say that a continuous map $f : X \to Y$ of topological spaces is *quotient map* we mean that Y has the finest topology for which f is continuous; this is equivalent to requiring that $A \subset Y$ is closed whenever $f^{-1}(A)$ is closed in X .

3. Results for general topological groups

We begin with a few words about Graev's proofs of the existence of free topological groups and free products of topological groups.

Let X be a completely regular space and e a distinguished point of X . Let $G(X)$ be the free group on the set $X \backslash \{e\}$, with e as the identity element of the group. Let $X' = X \cup X^{-1}$. Being completely regular, the topology of X is defined by a family of pseudometrics. Let ρ be a continuous pseudometric on X . Graev extended ρ to a two-sided invariant pseudometric on $G(X)$ as follows: Extend ρ to X' by setting $\rho(x^{-1}, y^{-1}) = \rho(x, y)$ and

$$\rho(x^{-1}, y) = \rho(x, y^{-1}) = \rho(x, e) + \rho(y, e)$$

for x and y in X. For u and v in $G(X)$ we have an infinity of representations $u = x_1 \dots x_n$, $v = y_1 \dots y_n$, where x_i and y_i X. Extend ρ to $G(X)$ by setting $\rho(u, v) = \inf\left[\sum_{i=1}^{n} \rho(x_i, y_i)\right]$, where the infimum is taken over all representations $u = x_1 \dots x_n$ and $v = y_1 \dots y_n$. The family of all such two-sided invariant pseudometrics on $G(X)$ yield a topological group $F_s(X)$. (It is shown elsewhere that $F_s(X)$ is the free topological SIN group on X.) Now $F_s(X)$ is Hausdorff; $FG(X)$ is the group $G(X)$ with the finest Hausdorff topology inducing the original topology on X. This topology $FG(X)$ is in general [9] a finer topology than $F_s(X)$.

Next we let G and H be topological groups. Graev defined a topology τ (not the free product topology, in general) on $G * H$ using the map $p : G \times H \times \text{gp}[G, H] \to G * H$. The method requires us to topologize $\text{gp}[G, H]$ in some way and then topologize $G * H$ to make the map p a homeomorphism. Since p is not a homomorphism it must be checked that this topology τ on $G * H$ is a group topology. (This is in fact quite difficult but our brief comments suppress this difficulty.) Let ρ_G and ρ_H be continuous right invariant pseudometrics on G and H respectively. Define a pseudometric ρ_{gh} on $[G, H]$ by

$$\rho_{GH}\left(g_1^{-1}h_1^{-1}g_1h_1, g_2^{-1}h_2^{-1}g_2h_2\right) = \min\left[\min\left(\rho_G(g_1, e), \rho_H(h_1, e)\right) + \min\left(\rho_G(g_2, e), \rho_H(h_2, e)\right); \rho_G(g_1, g_2)+\rho_H(h_1, h_2)\right].$$

The family of all such ρ_{GH} gives rise to a completely regular topology on $[G, H]$. Next, noting that $\text{gp}[G, H]$ is a free group on $[G, H]\backslash\{e\}$, we topologize $\text{gp}[G, H]$ by putting $(\text{gp}[G, H], \tau_1) = F_s[G, H]$. Finally we define the topology τ on $G * H$ by making

$$p : G \times H \times (\text{gp}[G, H], \tau_1) \to (G * H, \tau) \quad \text{a homeomorphism.}$$

Thompson [13] showed that $F_s(X)$ is NSS if and only if X admits a continuous metric. (Thompson's result is stronger than that of Morris and Thompson [10] which showed that $FG(X)$ is NSS if and only if X admits a continuous metric.)

Now if G is NSS, then G admits a continuous metric [10]; so if G and H are NSS, then $G \times H$ admits a continuous metric. Thus $[G, H]$ with the pseudometric topology described above admits a continuous metric. Hence $F_s[G, H]$ is NSS if G and H are NSS. We are now able to prove the following theorem:

THEOREM 1. *$G \amalg H$ is* NSS *if and only if G and H are* NSS.

PROOF. If $G \amalg H$ is NSS then any subgroup must be NSS. In particular, G and H must be NSS.

If G and H are NSS, then the above discussion yields that $F_s[G, H]$ is NSS. We shall prove that $(G * H, \tau)$ is NSS, as then $G \amalg H$ which has the same algebraic structure but a finer topology will also be NSS. Suppose that $(G * H, \tau)$, which is homeomorphic to $G \times H \times F_s[G, H]$, fails to be NSS. Let N and M be neighbourhoods of e in G and H, respectively, which contain no non-trivial subgroups. Then $\pi_1^{-1}(N) \cap \pi_2^{-1}(M)$ is a neighbourhood of e in $(G * H, \tau)$. Let A be a subgroup contained in $\pi_1^{-1}(N) \cap \pi_2^{-1}(M)$. Since π_1 is a homomorphism and $\pi_1(A) \subset N$ we must have $\pi_1(A) = e$. Similarly $\pi_2(A) = e$. Thus $A \subset F_s[F, G] \subset (G * H, \tau)$. Since $F_s[G, H]$ is NSS, $A = \{e\}$, as desired.

REMARKS. (1) This theorem generalizes the main result of [8] which says that if G and H are connected locally compact groups then $G \amalg H$ is NSS when and only when G and H are Lie groups.

(2) Note that the proof of Theorem 1 actually yields: $(G * H, \tau)$ is NSS if and only if G and H are NSS.

The fact that $(G * H, \tau)$ is homeomorphic to $G \times H \times \text{gp}[G, H]$ leads us to ask if a similar result is true for $G \amalg H$. It is!

THEOREM 2. *If* $\text{gp}[G, H]$ *is topologized as a subset of $G \amalg H$, then $G \amalg H$ is homeomorphic to $G \times H \times$* $\text{gp}[G, H]$ *(the homeomorphism is given by the map p).*

PROOF. Since $G \amalg H$ is a topological group, the product map $(G \amalg H) \times (G \amalg H) \times (G \amalg H) \to G \amalg H$, given by $(g, h, k) \to ghk$ is continuous, and so is its restriction $p : G \times H \times \text{gp}[G, H] \to G \amalg H$. We must show that the inverse map is continuous. The maps $\pi_1 : G \amalg H \to G$ and $\pi_2 : G \amalg H \to H$ are continuous, so $\pi_c(w) = \pi_2(w)^{-1}\pi_1(w)^{-1}w$ is a product of continuous maps and thus continuous. Hence the map $w \to (\pi_1(w), \pi_2(w), \pi_c(w)) = (g, h, k)$ is continuous, completing the proof.

THEOREM 3. *Suppose $G \neq \{e\}$ and $H \neq \{e\}$ are topological groups. Then $G \amalg H$ is not a locally compact space or a complete metric space unless G and H are both discrete.* (Of course if G and H are discrete, $G \amalg H$ is also discrete, and consequently locally compact and complete metric.)

PROOF. Suppose $G \amalg H$ is a locally compact space of a complete metric space; then so is the closed subgroup $\text{gp}[G, H]$. But as $\text{gp}[G, H]$ is algebraically a free group it follows from Dudley [1] that $\text{gp}[G, H]$ is discrete. Now G is also discrete: for if $\{g_\delta\}$ is a non-constant net converging to $g \in G$ and $h \in H\backslash\{e\}$, then $\{[g_\delta, h]\}$ is a non-constant net converging to $[g, h]$ in $gp[G, H]$, which is impossible. Similarly H is discrete. Finally we see $G \coprod H$, which is homeomorphic to $G \times H \times gp[G, H]$, is also discrete.

REMARK. Theorems 2 and 3 hold (with the same proofs) for any group topology μ on $G * H$ for which the projections $\pi_1 : (G * H, \mu) \to G$ and $\pi_2 : (G * H, \mu) \to H$ are continuous and which induce the given topologies on G and H. Thus it would be of interest to answer:

QUESTION 1.[1] *Is there any group topology* μ *on* $G * H$ *such that either projection* $\pi_1 : (G * H, \mu) \to G$ *or* $\pi_2 : (G * H, \mu) \to H$ *is discontinuous?*

If continuity of π_1 and π_2 could be shown even under the hypothesis that G, H and $(G * H, \mu)$ are locally compact, we could conclude that no group topology on an algebraic free product is locally compact (except trivially).

What is the topology that $gp[G, H]$ receives as a subset of $G \coprod H$? It is natural to hope that it has a free topological group topology, on an appropriate topology for $[G, H]$.

QUESTION 2. *(a) Does the topology induced on* $gp[G, H]$ *as a subgroup of* $G \coprod H$ *make it the free topological group* $FG[G, H]$ *?*

(b) Is the topology induced on $[G, H]$ *as a subset of* $G \coprod H$*, the same as the quotient topology under the map* $G \times H \to [G, H]$ *given by* $(g, h) \to [g, h]$ *?*

We have already noted that Graev's Topology $F_g[G, H]$ is not, in general, $FG[G, H]$. Example 1 in §5 shows that 2 *(b)* is also false for Graev's topology; that is, Graev does not give $[G, H]$ the quotient topology. On the other hand we will answer both 2 *(a)* and 2 *(b)* affirmatively when G and H are k_ω-groups.

4. Results for groups which are k_ω-spaces

We begin by answering Question 2 *(b)* for this case.

THEOREM 4. *Let* G *and* H *be topological groups which are* k_ω*-spaces. Then* $\sigma : G \times H \to [G, H] \subset G \coprod H$ *is a quotient map.*

PROOF. Let the k_ω-space decompositions of G and H be $G = \cup G_n$ and $H = \cup H_n$. In view of the Proposition stated in §2, $G \coprod H$ is a k_ω-space with decomposition $G \coprod H = \cup gp_n(G_n \cup H_n)$. (Thus a set A is closed in $G \coprod H$ if and only if $A \cap gp_n(G_n \cup H_n)$ is compact for all n, where $gp_n(G_n \cup H_n)$ is the set of

[1] This question has since been answered in the affirmative.

elements of $G \amalg H$ which are products of at most n elements of $G_n \cup H_n$; it is compact in $G \amalg H$.)

Now let $A \subset [G, H]$ be such that $c^{-1}(A)$ is closed in $G \times H$. We must show A is closed in $[G, H]$. It will suffice to show A is closed in $G \amalg H$. We shall prove that $A \cap \mathrm{gp}_n(G_n \cup H_n) = c\left(c^{-1}(A) \cap \left(G_{n^2} \times H_{n^2}\right)\right) \cap \mathrm{gp}_n(G_n \cup H_n)$ as the right hand side is the intersection of a continuous image of a compact set with a compact set it is compact.

If $n < 4$, both sides are trivial, so assume $n \geq 4$. Now if $w \in \mathrm{gp}_n(G_n \cup H_n)$, $w = x_1 \ldots x_n$, with $x_i \in G_n$ or H_n ; in reduced form $w = g^{-1}h^{-1}gh$, so clearly g is a product of at most n terms from G_n ; hence $g \in G_{n^2}$. Similarly $h \in H_{n^2}$. Since $w = c(g, h)$ we have that $w \in c\left(c^{-1}(A) \cap \left(G_{n^2} \times H_{n^2}\right)\right)$. The other inclusions needed are easy. Hence $A \cap \mathrm{gp}_n(G_n \cup H_n)$ is compact for all n, and A is closed, as required.

Note that it follows from the Proof of Theorem 4 that $[G, H]$ is closed in $G \amalg H$. We now turn to Question 2 *(a)*.

THEOREM 5. *Let* G *and* H *be topological groups which are* k_ω*-spaces. Then the topology on* $\mathrm{gp}[G, H]$ *as a subgroup of* $G \amalg H$ *is the free topological group topology* $FG[G, H]$.

PROOF. Again let $G = \cup G_n$ and $H = \cup H_n$ be k_ω-space decompositions. Then $G \amalg H = \cup \mathrm{gp}_n(G_n \cup H_n)$ and $[G, H] = \cup([G, H] \cap \mathrm{gp}_n(G_n \cup H_n))$ are k_ω-space decompositions.

Now from the Proposition given in §2, $FG[G, H]$ is a k_ω-space with decomposition $FG[G, H] = \cup \mathrm{gp}_n([G, H] \cap \mathrm{gp}_n(G_n \cup H_n))$. On the other hand, $\mathrm{gp}[G, H]$ is a closed subgroup of $G \amalg H$ and hence a k_ω-space with decomposition $\mathrm{gp}[G, H] = \cup(\mathrm{gp}[G, H] \cap \mathrm{gp}_n(G_n \cup H_n))$.

Clearly each $\mathrm{gp}_n([G, H] \cap \mathrm{gp}_n(G_n \cup H_n))$ is contained in $\mathrm{gp}[G, H] \cap \mathrm{gp}_k(G_k \cup H_k)$, for $k = n^2$; we must show for each n there is an m such that $\mathrm{gp}[G, H] \cap \mathrm{gp}_n(G_n \cup H_n) \subset \mathrm{gp}_m([G, H] \cap \mathrm{gp}_m(G_m \cup H_m))$.

Let $w \in \mathrm{gp}[G, H] \cap \mathrm{gp}_n(G_n \cup H_n)$. Without loss of generality suppose $n \geq 4$ and write $w = g_1h_2g_3 \ldots g_{n-1}h_n$, each $g_i \in G_n$ and each $h_i \in H_n$. We shall discuss a way of writing w as a product of commutators.

$$\begin{aligned} w &= g_1h_2g_3h_4 \cdots g_{n-1}h_n \\ &= \left[g_1^{-1}, h_2^{-1}\right]h_2(g_1g_3)h_4 \cdots g_{n-1}h_n \\ &= \left[g_1^{-1}, h_2^{-1}\right]\left[(g_1g_3)^{-1}, h_2^{-1}\right]^{-1}(g_1g_3)(h_2h_4)g_5 \cdots g_{n-1}h_n \\ &= \left[g_1^{-1}, h_2^{-1}\right]\left[(g_1g_3)^{-1}, h_2^{-1}\right]^{-1}\left[(g_1g_3)^{-1}, (h_2h_4)^{-1}\right] \cdots (g_1 \cdots g_{n-1})(h_2 \cdots h_n) . \end{aligned}$$

The last line has $n - 3$ commutators. Since $\pi_1(w) = \pi_2(w) = e$ we see that $g_1 \cdots g_{n-1} = h_2 \cdots h_n = e$. So w is a product of $n - 3$ commutators $[g, h]^{\pm 1}$, where each g is a product of at most n factors from G_n and hence lies in G_{n^2}. Similarly for h. So for any $m \geq n^2$ we have

$$[g, h] \in [G, H] \cap \mathrm{gp}_m(G_m \cup H_m)$$

and

$$w \in \mathrm{gp}_m([G, H] \cap \mathrm{gp}_m(G_m \cup H_m)) ,$$

as desired. Thus the topologies of $FG[G, H]$ and $\mathrm{gp}[G, H]$ are the same, completing the proof.

REMARK. It follows that if G and H are topological groups and k_ω-spaces, $G \coprod H$ contains a free topological group $FG[G, H]$ on a k_ω-space $[G, H]$. In this case we can draw somewhat stronger conclusions than Theorem 3; for instance, $G \coprod H$ is (except trivially) not metrizable and not SIN. (A topological group is said to be a SIN *group* if every neighbourhood of e contains a neighbourhood of the identity invariant under inner automorphisms of the group.) This leads us to ask

QUESTION 3. *If G and H are topological groups, at least one of which is not a discrete space, can $G \coprod H$ be*

(a) metrizable, or

(b) a SIN group?

By methods exactly similar to those used in Theorem 5 we obtain

THEOREM 6. *Let G and H be topological groups which are k_ω-spaces; let A be a closed subgroup of G and B be a closed subgroup of H. Then the subgroup of $G \coprod H$ generated by $A \cup B$ is closed and is (topologically and algebraically) $A \coprod B$.*

For general G and H, A and B closed does imply that the group generated by $A \cup B$ in $G \amalg H$ is closed; this however requires a careful examination of the Graev topology $(G * H, \tau)$ introduced before Theorem 1. It does not provide an answer to:

QUESTION 4. *Let* G *and* H *be topological groups and* A *and* B *closed subgroups of* G *and* H *respectively. Let* $\mathrm{gp}(A \cup B)$ *denote the subgroup of* $G \amalg H$ *generated by* $A \cup B$. *Algebraically it is* $A * B$. *Is* $\mathrm{gp}(A \cup B)$ *the topological free product* $A \amalg B$*?*

It is natural to ask whether the topology of $G \amalg H$ depends only on the topologies of G and H or also on the group structures. One may be inclined to conjecture that if $f_1 : G_1 \to H_1$ and $f_2 : G_2 \to H_2$ are hoemomorphisms, perhaps a homeomorphism $f_1 * f_2 : G_1 * G_2 \to H_1 * H_2$ can be constructed by letting $f_1 * f_2(r_1 s_1 \ldots r_n s_n) = f_1(r_1) f_2(s_1) \ldots f_1(r_n) f_2(s_n)$, where $r_i \in G_1$ and $s_i \in G_2$. This fails in general! For instance, if $\{s_\delta\}$ is a net converging to e in G_2, $f_2(e) = e$ and r_1 and r_2 are elements of G_1 with $f_1(r_1) f_1(r_2) \neq f_1(r_1 r_2)$, then

$$\lim f_1 * f_2(r_1 s_\delta r_2) = \lim f_1(r_1) f_2(s_\delta) f_1(r_2) = f_1(r_1) f_1(r_2)$$

while

$$f_1 * f_2(\lim r_1 s_\delta r_2) = f_1 * f_2(r_1 r_2) = f_1(r_1 r_2) \neq f_1(r_1) f_1(r_2),$$

so $f_1 * f_2$ is discontinuous.

In the k_ω-space case, another approach succeeds:

THEOREM 7. *Let* G_i *and* H_i *be topological groups which are* k_ω*-spaces, for* $i = 1, 2$. *If* G_i *is homeomorphic to* H_i, $i = 1, 2$ *then* $G_1 \amalg G_2$ *is homeomorphic to* $H_1 \amalg H_2$.

PROOF. As $G_1 \amalg G_2$ is homeomorphic to $G_1 \times G_2 \times FG[G_1, G_2]$ and $H_1 \amalg H_2$ is homeomorphic to $H_1 \times H_2 \times FG[H_1, H_2]$ and as $FG(X)$ and $FG(Y)$ are homeomorphic if X and Y are homeomorphic (independent of the choice of basepoints) it will suffice to show that $[G_1, G_2]$ is homeomorphic to $[H_1, H_2]$. Let $f_i : G_i \to H_i$ be a homeomorphism for $i = 1, 2$; since topological groups are homogeneous, we may assume that the f_i have been chosen so that $f_i(e) = e$ for each i. Hence the diagram

$$\begin{array}{ccc} G_1 \times G_2 & \xrightarrow{f_1 \times f_2} & H_1 \times H_2 \\ \Big\downarrow c & & \Big\downarrow c \\ [G_1, G_2] & \xrightarrow[j]{} & [H_1, H_2] \end{array}$$

is commutative, where $j([g_1, g_2]) = [f_1(g_1), f_2(g_2)]$, and as each vertical map is a quotient map, j is a homeomorphism. This completes the proof.

In view of this it appears that general solutions to Question 2 *(a)* and 2 *(b)* would allow a general solution of:

QUESTION 5. *Let* G_i *and* H_i *be topological groups for* $i = 1, 2$. *If* G_i *is homeomorphic to* H_i *for* $i = 1, 2$ *is* $G_1 \amalg G_2$ *necessarily homeomorphic to* $H_1 \amalg H_2$ *?*

It was shown in Ordman [12] that if G and H are arcwise connected topological groups, then the fundamental group

$$\pi(G \amalg H) = \pi(G \times H) \times L = \pi(G) \times \pi(H) \times L$$

for some group L . It was conjectured that L is always trivial. We now see that $\pi(G \amalg H) = \pi(G) \times \pi(H) \times \pi(\mathrm{gp}[G, H])$, where $\mathrm{gp}[G, H]$ has the induced topology from $G \amalg H$. Further if G and H are k_ω-spaces, then

$$\pi(G \amalg H) = \pi(G) \times \pi(H) \times \pi(FG[G, H]) .$$

So the group L has now been identified. However we have been unable to prove that $\pi(FG[G, H])$ is trivial in any case other than the one covered in [12]; that is, when G and H are countable CW-complexes with exactly one-zero-cell. It seems reasonable to conjecture that if G and H are simply connected then $\pi(G \amalg H) = \pi(G) \times \pi(H)$. However for this we need to answer

QUESTION 6. *If* X *is simply connected is* $FG(X)$ *necessarily simply connected? Is it true under the additional assumption that* X *is a* k_ω*-space?*

5. Examples

We conclude by giving two elementary examples which bear on the preceding.

EXAMPLE 1. The map $c : G \times H \to [G, H] \subset (G * H, \tau)$ is not a quotient map, in general, where τ is Graev's topology. Let $G = H = R$, the additive group of reals with the usual topology. Consider the sequence $a_n = (n, 1/n)$ in $R \times R$. Now $c(a_n)$ converges to e in $(R * R, \tau)$, for

$\rho(c(a_n), e) = \min(|n|, |1/n|) = 1/n \to e$,

where ρ is the metric (described in §3) arising from the usual metric on each copy of R. However $c(a_n)$ fails to converge to e in $R \coprod R$. To see this note that R is a k_ω-space with decomposition $R = \cup[-n, n]$. Since $\{c(a_k) : k = 1, 2, \ldots\}$ has finite intersection with each $\mathrm{gp}_n([-n, n] \cup [-n, n])$ (here the first $[-n, n] \subset R = G$, the second $[-n, n] \subset R = H$), it is a closed set in $R \coprod R$ and hence does not converge to e.

Since $c(a_n) \in [R, R]$ for all n and $e \in [R, R]$, it follows that $[R, R]$ is topologized differently in $(R * R, \tau)$ than in $R \coprod R$. Hence answering Question 2 will require more than an appeal to Graev's topology.

Incidentally the above argument also shows that the topology constructed in Ordman [11 (I)] also yields a topology on $R * R$ other than the free product topology.

EXAMPLE 2. While the free product of compact groups is a k_ω-space, it is very large. Although every discrete subgroup of a compact group is finite, the free product $T \coprod T$ of two circle groups contains a discrete subgroup which is not even finitely generated. Consider the subgroup $\{e, a\}$ of order 2 of the first factor and the subgroup $\{e, b, b^2\}$ of order 3 of the second factor. The free product $\{e, a\} \coprod \{e, b, b^2\}$ is discrete and by Theorem 6 it is a subgroup of $T \coprod T$. Hence its subgroup $\mathrm{gp}[\{e, a\}, \{e, b, b^2\}]$, the free group on the two generators $x = [a, b]$ and $y = [a, b^2]$ is discrete. This group in turn contains the free group on the countable set $\{x, yxy^{-1}, y^2xy^{-2}, \ldots\}$.

On the other hand, compact subgroups of $T \coprod T$ are very small. Every compact subset of $T \cup T$ is contained in some group $\mathrm{gp}_n(T \cup T)$; that is, has bounded word length. However the only subgroups of $T * T$ with bounded word length are those which are conjugates of subgroups of one of the two factors. Hence every compact subgroup $T \coprod T$ is either finite, or a conjugate of one of the two factors and hence itself a circle group.

QUESTION 7. *What are the locally compact subgroups of* $T \coprod T$ *?*

References

[1] R.M. Dudley, "Continuity of homomorphisms", *Duke Math. J.* 28 (1961), 587-594. MR25#141.

[2] М.И. Граев [M.I. Graev], "Свободные топологические группы" [Free topological groups], *Izv. Akad. Nauk SSSR Ser. Mat.* 12 (1948), 279-324; *Amer. Math. Soc. Transl.* 35 (1951), pp. 61; reprinted *Amer. Math. Soc. Transl.* (1) 8 (1962), 305-364. MR10,11.

[3] М.И. Граев [M.I. Graev], "О свободных произведениях топологических групп" [On free products of topological groups], *Izv. Akad. Nauk SSSR Ser. Mat.* 14 (1950), 343-354. MR12,158.

[4] J. Mack, Sidney A. Morris and E.T. Ordman, "Free topological groups and the projective dimension of a locally compact abelian group", *Proc. Amer. Math. Soc.* 40 (1973), 303-308.

[5] А.А. Марков [A.A. Markov], "О свободных топологических группах" [On free topological groups], *Izv. Akad. Nauk SSSR Ser. Mat.* 9 (1945), 3-64; *Amer. Math. Soc. Transl.* 30 (1950), 11-88; reprinted *Amer. Math. Soc. Transl.* (1) 8 (1962), 195-272. MR3,36.

[6] Sidney A. Morris, "Free products of topological groups", *Bull. Austral. Math. Soc.* 4 (1971), 17-29. MR43#410.

[7] Sidney A. Morris, "Local compactness and free products of topological groups I and II", submitted.

[8] Sidney A. Morris, "Free products of Lie groups", *Colloq. Math.* (to appear).

[9] Sidney A. Morris and H.B. Thompson, "Invariant metrics on free topological groups", *Bull. Austral. Math. Soc.* 9(1973), 83-88.

[10] Sidney A. Morris and H.B. Thompson, "Free topological groups with no small subgroups", *Proc. Amer. Math. Soc.* (to appear).

[11] E.T. Ordman, "Free products of topological groups with equal uniformities I and II", *Colloq. Math.* (to appear).

[12] E.T. Ordman, "Free products of topological groups which are k_ω-spaces", *Trans. Amer. Math. Soc.* (to appear).

[13] H.B. Thompson, "Remarks on free topological groups with no small subgroups", *J. Austral. Math. Soc.* (to appear).

University of New South Wales,
Kensington, NSW 2033.

University of Kentucky,
Lexington, Kentucky 40506, USA.

Flinders University of South Australia,
Bedford Park, South Australia 5042.

PROC. SECOND INTERNAT. CONF. THEORY OF GROUPS, 20-03
CANBERRA, 1973, pp. 516-519.

SOME GROUPS I HAVE KNOWN

B.H. Neumann

As this is the last lecture of our Conference, you will, I hope, permit me to make it less weighty, more autobiographical and anecdotal. No new results will occur, and even the problems I want to mention are not new. I want to say some nostalgic words about group theory as it was in the distant past when I was young, before many of you were born.

So I should like you to cast your minds back to the late 1920's, when I was a young student at the University of Berlin. Issai Schur was there, in his early or mid-fifties, Robert Remak, Heinz Hopf. Groups were finite, or they were groups of linear transformations, or Lie groups, or the fundamental groups of topological manifolds - groups were only rarely considered as interesting in their own right. The great success of the theory of finite groups was a thing of the past and a dream of the future; all finite simple groups were known! True, there were some stirrings: Remak was working on subdirect products of finite groups and presented his results, well before some of them were published, in a course of lectures, in which I was at the end the whole audience; an unknown young man at Cambridge, England, by name of Philip Hall, did interesting things with finite soluble groups and then even more interesting things with finite groups of prime-power order. G.A. Miller went on publishing hundreds of papers on groups, in a quaint, antiquated language.

But there were also stirrings of a new group theory, a theory of groups as groups. Though it was inspired by topology, both in its problems and its methods, it tended to cut loose from this connection. The great names were Jakob Nielsen, Max Dehn, Emil Artin, Otto Schreier, Kurt Reidemeister - all of them motivated by topology -; also Friedrich Levi, who was perhaps the most genuinely group-theoretical one among them, and who collaborated with a young man, Reinhold Baer; and Otto Juljewitsch Schmidt, whose book had been much before its time, and whose interests had then veered towards polar exploration. They began to look at free groups and presentations of groups, formulating many of the problems that kept later generations

---

This is not so much a record of the lecture as given, as a sketch of what it was intended to be.

busy. One of the greatest, William Burnside, had died in 1927, leaving a legacy of enormously fruitful problems.

That, very sketchily, is the scene on which I first started looking at groups. Jakob Nielsen had determined a presentation of the automorphism group of a free group of finite rank - even the term "automorphism" had not been invented yet when he did this, by deep topological methods; and the term "endomorphism" came much later. I had read Nielsen's paper [and some others - I even tried to translate his "Om regning med ikke-kommutative faktorer ..." from Danish into German, without any knowledge of Danish but with a dictionary, until some negation defeated me], and found that his automorphism groups of free groups could be generated by fewer (and much less useful) generators, and defined by slightly fewer defining relations - a piffling little result, especially when compared to Wilhelm Magnus' great Freiheitssatz. I won't bore you with the story of how my piffling little result became the main part of my doctoral dissertation at Berlin - I have bored enough people with the story, sufficiently many times.

Those were interesting times, when the free product and the free product with an amalgamated subgroup, or generalized free product, as it was to be called later, were introduced. Their great usefulness remained long unrecognized, except by Wilhelm Magnus. Heinz Hopf proposed the conjecture - very plausible at the time - that has given rise to so much fun with Hopf groups and non-Hopf groups. There were, in fact, three inter-related conjectures, one topological and two group-theoretical. The most famous one was that a finitely generated group can never be isomorphic to a proper factor group of itself, or, in terminology that has since become current, that every finitely generated group is a Hopf group. We all believed this to be so, and Wilhelm Magnus proved it to be true for an important class of groups; until, after the Second World War, Reinhold Baer rudely shattered our beliefs by publishing a counter-example. This counter-example did not live long, but our beliefs lay shattered, and thus the way lay open to making an example of a finitely generated non-Hopf group. The first example had 2 generators and infinitely many defining relations; but Graham Higman almost at once made a much better one, with 3 generators but only 2 defining relations. Then, about 10 years later, Gilbert Baumslag and Donald Solitar made a 2-generator, 1-relator non-Hopf group, thereby demolishing a theorem that Hanna and I had proved, but fortunately not yet published. Let me remind you in passing that Hanna's last substantial research paper was again concerned with Hopf groups, proving, with Ian M.S. Dey, that the free product of finitely many finitely generated Hopf groups is a Hopf group.

To go back to the groups of 40 years ago: there was Alexander Kurosch who proved a very general structure theorem on the subgroups of free products of groups, later to be extended by Hanna in her Oxford DPhil thesis to subgroups of generalized free products. This linked up with the theory of amalgams of groups, introduced by

Reinhold Baer, and with their embedding theory, in which Hanna and I worked together for some years - I might mention in passing that there is still an outstanding problem, probably difficult:

PROBLEM. *If a finite amalgam of groups is embeddable in a group, is it embeddable in a finite group?*

In trying to devise counter-examples to somebody's spurious argument trying to prove a positive solution, I have been playing with certain amalgams that may perhaps eventually lead to the construction of a new sporadic finite simple group; but so far our big computers have got nowhere with this.

To this part of our endeavours there also belongs the joint work with Graham Higman, Hanna, and myself on embeddings of groups that has been mentioned here a number of times, especially in Paul E. Schupp's fascinating survey: firstly, given a group and a partial automorphism, that is an isomorphism of a subgroup to another subgroup, one can embed the group in a bigger one in which the partial automorphism extends to a total one; and if the original group is finite, one can make the bigger group also still finite [this came a bit later, by different methods]. This led naturally to the construction of infinite groups, already mentioned by Paul Schupp, with only 2 classes of conjugate elements. Again let me mention an old unsolved problem:

PROBLEM. *To make a group with just 3 classes of conjugate elements, such that each of the classes is a semigroup, that is closed under the group multiplication, and the two classes that do not consist of the unit element are inverse to each other.*

Such a group would give a new example of a simple fully ordered group - the first examples, made quite differently, were constructed by C.G. Chehata and later generalised by Vlastimil Dlab.

Another result of the construction that is now often referred to by the initials HNN is the fact that every countable group can be embedded in a 2-generator group; or, in more modern parlance: the free group of rank 2 is SQ-universal [every countable group is isomorphic to a subgroup of a factor group of it]. Since then very many more SQ-universal 2-generator groups have been found. When, 10 years after the original HNN paper, Hanna and I returned to the subject and proved the embedding theorems by the use of wreath products rather than generalized free products, one of the incidental results was that the free product of two finite cyclic groups is SQ-universal if one has order 2 (or greater) and the other order 6 (or greater) ; and we conjectured then that "6" could be replaced by "3". In other words: the modular group, one of my favourites as it is a favourite of Wilhelm Magnus, is SQ-universal. This was proved later by Frank Levin and also, by a quite different method, by Philip Hall. There is a sequence of groups depending on a positive integral parameter: if the parameter is small, the groups are finite, some of them even trivial; but for quite moderate sizes of the parameter Graham Higman has shown the groups to be not only infinite, but SQ-universal.

As the fact has been mentioned (in Paul E. Schupp's lecture) that there are continuously many isomorphism classes of 2-generator groups, a fact first proved by an explicit construction more than a quarter of a century ago, I might add that among the spin-off of Hanna's and my "new", that is 1959, method of proving the embedding theorems is the fact that the cardinality of the set of isomorphism classes of 2-generator groups even in the variety of soluble groups of soluble length 3 is that of the continuum.

Having mentioned groups with only two classes of conjugate elements, I might mention, at the other end of the scale, groups in which the classes of conjugate elements are finite; Reinhold Baer introduced the name "FC-groups" for them. They provided much fun, especially also when the conjugacy classes were not only finite, but boundedly finite, the so called BFC-groups - James Wiegold and Ian D. Macdonald got much joy from these groups. They first occurred in a paper concerned with a different finiteness condition, namely groups covered by a finite set of cosets of subgroups - one can show that if a group is the set union of, say, n cosets of subgroups, then one of those subgroups has index no greater than n. On seeing this, Reinhold Baer observed that a group is covered by finitely many abelian subgroups if, and only if, its centre has finite index in it. At the recent International Colloquium on Finite and Infinite Sets held at Keszthely, Hungary, E.G. Straus asked what happens if "finite" is here replaced by "countable". It is not difficult to show that to an arbitrary infinite cardinal, say $\mathfrak{m}$, one can construct a group which is the union of countable many subgroups, but which has no proper subgroup of index less than $\mathfrak{m}$. But can one say anything about the index of the centre of a group covered by countably many abelian subgroups? I do not know the answer, so this is another problem:

PROBLEM. *Is there a bound on the index of the centre of a group that can be covered by countably many abelian subgroups?*

I should not be the least bit surprised if such a bound existed - perhaps the cardinal of the continuum.

Now I have talked all this time and not got round to saying much, or indeed anything, about some of the subjects dearest to my heart: wreath products, permutational products, and especially laws in groups and varieties of groups. But it is perhaps just as well that I did not get round to these things: they each deserve a full expository lecture to themselves, and varieties of groups were so treated, at our first Conference 8 years ago, by their very champion, Hanna.

PROC. SECOND INTERNAT. CONF. THEORY OF GROUPS, CANBERRA, 1973, 520-535. 20B10

TRANSITIVE PERMUTATION GROUPS OF PRIME DEGREE

Peter M. Neumann

1. Introduction

This paper is intended as a survey of what is now known about transitive permutation groups of prime degree. The topic arises from early work on the theory of equations. Over 200 years ago Lagrange was led to an interest in irreducible polynomial equations of prime degree by showing[1] that if every such equation were soluble in terms of root-extraction then polynomial equations of arbitrary degree would be. Even after Abel and Galois had shown that such solutions are impossible in general, Galois still devoted a good proportion of his work to equations of prime degree. It was of course he who emphasised the groups involved. Several 19th century mathematicians, notably Mathieu and Jordan, continued the work of Galois and provided foundations for the rich material that has been published since 1900. At present the problems concerning groups of prime degree remain near the centre of permutation group theory, retaining their interest partly as tests of the power and scope of techniques of finite group theory, partly as being typical of a range of similar problems concerning groups of degrees kp (with $k < p$) and p^m where p is prime.

2. Generalities about groups of prime degree

Throughout this paper

p will be a prime number;

Ω a set with p elements;

G a transitive group of permutations of Ω ;

and other notation will be mainly as in Wielandt [1964]. If $\alpha \in \Omega$ and G_α is its stabiliser then the index $|G : G_\alpha|$ is p . If Y is a subgroup of index p in a group X then X acts transitively on the set of right cosets of Y by right multiplication: this permutation representation of X is faithful provided that Y contains no non-trivial normal subgroup of X . Thus in terms of pure group theory

[1] This is not true: see the note at the end of this paper.

our interest is in the existence in G of a subgroup of prime index.

Since $p \,\big|\, |G|$ we know that G has a subgroup P of order p. In fact, since $|G| \,\big|\, p!$ and $p^2 \nmid p!$, P is a Sylow p-subgroup of G. We identify Ω with $\{0, 1, \ldots, p-1\}$, or with Z_p (the integers modulo p), in such a way that $P = \mathrm{gp}(a)$ where a is the p-cycle $(0, 1, \ldots, p-1)$. Thus P is the group of translations in the affine group $\mathrm{AGL}(1,p)$, which consists of all permutations $\pi_{a,b}$ of Z_p where $\pi_{a,b} : z \mapsto az + b$ $\left(a \in Z_p - \{0\},\ b \in Z_p\right)$. The following theorem is in the works of Galois ([1830], [1846]; *cf.* Écrits et Mémoires, pp. 65-69).

THEOREM 1 (Galois). *G is soluble if and only if* $G \le \mathrm{AGL}(1,p)$.

Since the subgroups of $\mathrm{AGL}(1,p)$, which is a metacyclic group of order $(p-1)p$, are very easy to enumerate, the difficulties in classifying groups of prime degree are all concentrated around insoluble groups.

The normal structure of G is always very simple:

THEOREM 2. *If S is the normal subgroup generated by the conjugates of P in G then*

(i) S is simple (and transitive on Ω);

(ii) G/S is cyclic and its order divides $p - 1$.

With this notation, if $S \ne P$, that is, if G is not soluble, then $C_G(S) = 1$ so that $G \le \mathrm{Aut}(S)$. Therefore we may usually restrict attention to the search for non-abelian simple groups which are permutation groups of prime degree. In terms of pure group theory we are seeking simple groups with conjugacy classes of subgroups of prime index.

Theorems 1 and 2 are quite superficial and very easy to prove (see, for example, Huppert [1967], Satz 21.1, p. 607). A very much deeper result, which is the main tool in studying groups of prime degree, is

THEOREM 3 (Burnside). *If G is insoluble then G is 2-fold transitive on* Ω.

This celebrated theorem was discovered by William Burnside in the course of his work on simple groups of odd order [1901]. His proof, an elegant character-theoretic argument, is reproduced in the second edition of his book [1911] and in Huppert [1967], (p. 609). In 1906 Burnside published a second proof, longer and less elegant, but free of character theory. Issai Schur [1908] gives a third proof which is elementary in the sense that, unlike both of Burnside's proofs, it makes no use of properties of complex numbers and p-th roots of 1. A beautiful proof of yet another kind is given by Wielandt in his notes [1969]. And the theorem can nowadays also be derived as an immediate corollary of deep representation-theoretic theorems of Brauer, Feit and others (see, for example, Feit [1967]).

3. The known groups

Table 1 below, describes some known examples of groups of prime degree. The first column is an approximate verbal description of the groups in question. The third column gives the unique largest group G_{max} for which S is the simple group described in Theorem 2; thus in general $S \leq G \leq G_{max}$. In the fourth column I have given the index of P in its normaliser (in S), a parameter of great significance which is to be the subject of §4, and in the fifth the number of inequivalent permutation representations of S of degree p, that is, the number of conjugacy classes of subgroups of index p.

Description	S	G_{max}	p	$t = \|N_S(P) : P\|$	r = number of representations
Cyclic	P	$AGL(1,p)$	p	1	1
Alternating	A_p	S_p	p	$\frac{1}{2}(p-1)$	1
Projective	$SL(d,q)$	$\Sigma L(d,q)$	$(q^d-1)/(q-1)$	d	1 if $d = 2$; 2 if $d \geq 3$
Exceptional	$PSL(2,5)$	$PGL(2,5)$	5	2	1
	$PSL(2,7)$	$PSL(2,7)$	7	3	2
	$PSL(2,11)$	$PSL(2,11)$	11	5	2
	M_{11}	M_{11}	11	5	1
	M_{23}	M_{23}	23	11	1

TABLE 1: Known examples

The first two rows of the table should be self-explanatory. The projective group $SL(d,q)$ indicated in the third row is the group of all $d \times d$ matrices of determinant 1 over a finite field F having q elements. This group acts naturally on the vector space F^d (of $1 \times d$ row vectors) and it acts transitively permuting the 1-dimensional subspaces of F^d, that is, the points of the geometry $PG(d-1,q)$, of which there are $(q^d-1)/(q-1)$. In general the kernel of this action is the set of scalar matrices included in $SL(d,q)$, and the factor group is known as $PSL(d,q)$, but the order of this kernel is the highest common factor $(d, q-1)$: and if $(q^d-1)/(q-1)$ is to be prime then d must be prime and must not divide $q-1$. If $q = q_0^m$ where q_0 is prime then $Aut(F)$ is a cyclic group A of order m. This acts in a natural way on F^d and hence on $PG(d-1,q)$. The group

$A.SL(d,q)$, which is the semi-direct product of $SL(d,q)$ by A acting in the obvious way, is $\Sigma L(d,q)$ and is G_{max} in this case. I have described these groups as acting transitively on the points of $PG(d-1,q)$: they also act transitively on the hyperplanes (that is, the $(d-1)$-dimensional vector subspaces of F^d) of which the number is the same, $(q^d-1)/(q-1)$. If $d = 2$ then points and hyperplanes are the same thing, but if $d \geq 3$ then the representations of $SL(d,q)$ on the set of points and on the set of hyperplanes are inequivalent (though they are "similar": they are interchanged by the 'inverse transpose' automorphism of the group).

I have already pointed out that if $(q^d-1)/(q-1)$ is to be prime then d must be prime. Also, if $q = q_0^m$ where q_0 (the characteristic of F) is prime then m must be a power of d . For $d = 2$ these numbers are familiar as the Fermat primes of which 3, 5, 17, 257 and 65537 are the only examples known. For $q = 2$ we have the Mersenne primes 3, 7, 31, 127, ... A list of all examples of primes $p = (q^d-1)/(q-1)$ up to approximately 4×10^8 is given by Bateman and Stemmler [1962]. Those less than 1000 are

3 5 7 13 17 31 127 257 307 757 .

The only prime known to have two such representations is 31 : both $SL(5,2)$ and $SL(3,5)$ act as groups of this degree.

The exceptional groups $PSL(2,5)$, $PSL(2,7)$ and $PSL(2,11)$ were described by Galois in his letter to Chevalier in 1832 (see Galois [1832], Ecrits et Memoires, pp. 179, 181), and the groups M_{11}, M_{23} are the Mathieu groups first described in Mathieu [1861], [1873]. Of course $PSL(2,5)$ of degree 5 is the same as A_5 (although Galois does not seem to have recognised this) and the same as $SL(2,4)$. And the two representations of $PSL(2,7)$ are the same as the representations of $SL(3,2)$ on points and lines of the Fano plane $PG(2,2)$. However, amongst the groups $PSL(2,p)$ only $PSL(2,5)$, $PSL(2,7)$ and $PSL(2,11)$ have permutation representations of degree p (Galois, *loc. cit.*) and as they arise frequently as special cases in this guise I find it convenient to list them with the other exceptional groups. Apart from the isomorphisms just mentioned, for $p \geq 5$ there are no further coincidences between groups described in the table.

The examples listed are the only known groups of prime degree. Therefore the central problem, which is to describe all such groups explicitly, may be put in the form of

PROBLEM 1. *Are there any transitive groups of degree* p *other than those listed in Table 1?*

Almost all the problems which follow are special cases. Since the question comes down to asking for a list of all those simple groups which contain subgroups of prime index perhaps it will not be answered until the search for all finite simple groups is complete. As a partial (and temporary) answer dealing with the simple groups known at present (see, for example, Carter [1973]) we have

THEOREM 4 (Praeger)[2]. *Amongst the known simple groups the only representations as transitive groups of prime degree are those given in Table* 1.

For the groups $PSL(d,q)$ this was proved by Ito [1960]. The complete theorem is due[2] to C.E. Praeger [1971], though her proof now needs to be supplemented with the observation (which is easy to justify) that the Rudvalis group and the O'Nan group (if it exists) do not have subgroups of prime index. Curtis, Kantor and Seitz [to be published] have recently completed a search for all 2-transitive permutation representations of the known simple groups, and of course, in view of Burnside's Theorem this gives a vast generalisation of Praeger's Theorem.

4. The influence of $N(P)$

Since the normaliser $N(P)$ is transitive it is a subgroup of $AGL(1,p)$ and its order is tp where $t|(p-1)$. The influence of $N(P)$, or equivalently of t since this determines the structure of $N(P)$, is illustrated by the following result from Neumann [1972a].

THEOREM 5. *If G is insoluble and t is even then G is 3-fold transitive.*

In unpublished work I have extended the method to show rather more, but to state the theorem we need some further terminology. I say that G is "a little generously 3-transitive" on Ω if whenever $\alpha, \beta, \gamma, \delta$ are distinct points of Ω there is a permutation in G whose expression as a product of disjoint cycles involves $(\alpha\beta)(\gamma\delta)$. It is not hard to show (Neumann [1973+]) that 3-fold transitivity follows if $|\Omega| > 4$.

THEOREM 6. *Suppose that G is insoluble and let $t = |N(P) : P|$.*

(i) If t is even then G is a little generously 3-transitive;

(ii) if t is odd and α, β are distinct points of Ω then $G_{\alpha\beta}$ has as most $(p-1)/t$ orbits in $\Omega - \{\alpha, \beta\}$.

I hope to submit an article containing the proof in the near future.

For small values of t one would hope to be able to find all the groups which occur. In fact we have

THEOREM 7. *(i) If $t = 1$ then $G = P$.*

(ii) (Itô). *If $t = 2$ and G is insoluble then $G = SL(2,q)$ where $q = p - 1$.*

The latter part of this theorem was proved by Noboru Ito in [1960] (see also

[2] Tsuzuku [1963]: see the notes at the end of this paper.

Huppert [1967], p. 613). Part *(i)* is an immediate consequence of Burnside's Transfer Theorem (Huppert [1967], p. 608), but it seems to have been known to Mathieu long before the transfer was available, and it can be proved by elementary means: if $P = N(P)$ then the number of conjugates of P is $|G|/p$ and since these conjugates intersect trivially in pairs there are $(p-1).|G|/p$, that is $|G| - |G|/p$, elements of order p; the stabiliser G_α however already accounts for all the remaining $|G|/p$ elements; therefore G_α is normal in G, it stabilises every element of and must be 1; consequently $G = P$.

PROBLEM 2. *Prove (or perhaps disprove) that if* G *is insoluble and* $t = 3$ *then either* G *is* A_7 *or* G *is* $SL(3,q)$ *where* q *is a prime-power and* $p = 1 + q + q^2$.

Much work has been done on this by T.P. McDonough [1972], and others, but although significant progress has been made, a complete proof is not yet foreseeable.

The data tabulated in Table 1 suggested to Noboru Ito the following question:

PROBLEM 3 (Ito). *If* G *is simple and neither cyclic nor alternating, is it true that* t *must be prime?*

Asking rather more:

PROBLEM 4. *If* G *is simple and* $1 < t < \frac{1}{2}(p-1)$ *is it true that* $G = SL(t,q)$ *for some prime-power* q *such that* $p = (q^t-1)/(q-1)$ *?*

And, as a special case which looks relatively accessible:

PROBLEM 5. *Prove that there is no simple group* G *with* $t = 4$.

If G does not contain odd permutations then t must divide $\frac{1}{2}(p-1)$. For the alternating and exceptional groups t is $\frac{1}{2}(p-1)$, and, of course, one would very much like to see a proof of the converse. I have proved a small extension of Theorem 6 *(ii)* in this case:

THEOREM 8. *If* G *is insoluble and* $t = \frac{1}{2}(p-1)$ *then*

(a) G *is 3-transitive; or*

(b) G_α *is primitive and of rank* 3 *on* $\Omega - \{\alpha\}$ *; or*

(c) $G = PSL(2,7)$ *of degree* 7.

With the further hypothesis that $\frac{1}{2}(p-1)$ is prime Ito has shown that G is 4-transitive unless $p \le 11$ (see Theorem 9 below).

PROBLEM 6. *Show that if* G *is insoluble and* $t = p - 1$ *then* G *is* S_p.

This is a question of Wielandt (see Huppert [1967], p. 618, see also §6.1 below, especially Problem 10). A variant of Wielandt's problem which, though weakened to the point where it looks feasible, is still sufficiently strong to be interesting, is

PROBLEM 7. *Show that if* G *is insoluble and* $t = p - 1$ *then* G *is* 4-*fold transitive.*

The main technique used in proving the theorems mentioned above is representation theory. In his famous papers [1942], [1943] Richard Brauer gives a general and satisfactorily detailed description of the characters of any group, such as G, whose Sylow p-subgroup is of order p. It is in this representation theory that the influence of t is most clearly visible. The papers of Thompson [1967], Feit [1966] and Green [1974] give further information about both the ordinary and the modular representations of G. To prove theorems one compares this knowledge with general information about the characters of permutation groups (Frobenius [1904] or Neumann [1973+].

5. Special primes

There are certain special prime numbers for which a great deal is known. The two main classes of these are (here, and in what follows q will always be prime):

(i) primes p for which $p = 2q + 1$;

(ii) "small" primes.

It was Mathieu who first drew attention to groups of degree p where $p = 2q + 1$ by suggesting ([1861], p. 242, and [1873]) that it is on just such an arithmetic accident as this that the existence of M_{11} and M_{23} depends. He introduced [1873] an elegant computational technique which was developed a little by Jordan [1874], Miller [1908], de Séguier [1912] (Chapter 5), and Fryer [1905] to show that other than soluble, alternating and symmetric groups there are no transitive groups of degrees 47, 59 and 83. In [1958] Parker and Nikolai report the results of a machine computation along these lines.

THEOREM 9 (Parker and Nikolai). *If* $p = 2q + 1$ *and* $23 < p \leq 4079$, *and if* G *is insoluble then* G *is* A_p *or* S_p.

On this evidence they make the conjecture embodied in

PROBLEM 8 (Parker and Nikolai). *Prove that if* $p = 2q + 1$ *and* $p > 23$, *and if* G *is insoluble then* G *must be* A_p *or* S_p.

Further evidence has been provided by Noboru Ito in his long and interesting papers [1963], [1964], [1965]. He proves

THEOREM 10 (Ito). *If* $p = 2q + 1 > 11$, *and if* G *is insoluble, then* G *is at least* 4-*fold transitive.*

In [1963] he adds further conditions to the prime p in an attempt to prove the conjecture in these special cases. His work was completed in Neumann [1972b] and we have

THEOREM 11 (Ito and П.M.N.). *If* $p = 2q + 1 = 4r + 3 > 23$ *, where* r *(as well as* q *) is also prime, and if* G *is insoluble, then* G *is* A_p *or* S_p .

Apparently the primes of this theorem form essentially the only class of numbers n for which all the transitive groups of degree n are known and the class itself probably is infinite. (However it does not appear to be known yet even whether there are infinitely many pairs p, q of primes with $p = 2q + 1$.)

The method of Parker and Nikolai has been modified by Appel and Parker [1967], and by M.D. Atkinson [unpublished] to compute with insoluble groups of degrees $p = 4q + 1$ and $p = 6q + 1$ respectively. They show that only alternating and symmetric groups occur for

$$13 < p = 4q + 1 \leq 317$$

and

$$31 < p = 6q + 1 \leq 139 .$$

Obstacles to extending the method to compute with groups of degree $mq + 1$ are first, that one requires enough knowledge of groups with t dividing $m/2$ to be able to assume that in an unknown group q divides t (which, recall, is $|N(P) : P|$); and second, that the number of operations required in the computation is of the order of magnitude $O(m!q^m)$, and this is far too large for feasibility if $m > 6$.

The other primes for which much is known are "small" primes. In the course of the last two years M.D. Atkinson and I have been checking primes $p < 100$. Many of these are covered by the computations of Parker and Nikolai, Appel and Parker, and Atkinson described in the foregoing paragraph. Indeed this was the main motivation for Atkinson's work. The remaining primes have to be treated 'by hand' with *ad hoc* methods: the main tools are Burnside's Theorem, the lemma of Jordan or Witt (see Wielandt [1964], Theorems 3.5 and 9.4), and theorems of W.A. Manning and others on elements of prime order in primitive permutation groups. The work is long and tedious and, since it has not been checked it is not very useful. However, we have the tentative theorem that for $p < 100$ no surprises occur. Some of the primes between 100 and 200 have also been treated. The situation is described in Table 2 (below). In this table only simple groups other than P and A_p are listed. A dash in the second column signifies that we know that no other groups exist. An entry in the column headed 'source' indicates that for this prime our knowledge is complete: the key to these entries is as follows:

$2q + 1$ refers to the computations of Parker and Nikolai [1958];

$4q + 1$ refers to the computation of Appel and Parker [1967];

$6q+1$ refers to the unpublished computation of M.D. Atkinson;

refers mainly to the unpublished work of M.D. Atkinson and Π.M.N.

However, many of the smaller primes were treated long ago by other authors; also C.C. Sims [1970] has published a table of the primitive groups of degrees $n \le 20$, and in unpublished work he has sought all primitive groups of degrees up to 50.

p	S	Source	p	S	Source
5	-	✓	101		
7	SL(3,2)	✓	103	-	$6q+1$
11	PSL(2,11)	✓	107	-	$2q+1$
	M_{11}	✓	109		
13	SL(3,3)	✓	113		
17	SL(2,16)	✓	127	SL(7,2)	
19	-	✓	131		
23	M_{23}	✓	137		
29	-	$4q+1$	139	-	$6q+1$
31	SL(3,5)	✓	149	-	$4q+1$
	SL(5,2)	✓	151		
37	-	✓	157		
41	-	✓	163		
43	-	$6q+1$	167	-	$2q+1$
47	-	$2q+1$	173	-	$4q+1$
53	-	$4q+1$	179	-	$2q+1$
59	-	$2q+1$	181		
61	-	✓	191		
67	-	$6q+1$	193	-	
71	-	✓	197		
73	SL(3,8)	✓	199		
79	-	$6q+1$	.		
83	-	$2q+1$	.		
89	-	✓	.		
97	-	✓	.		

TABLE 2: The "small" primes

6. Miscellany

6.1 Groups containing odd permutations

The following interesting problem comes from Ito [1963], p. 169, and may be viewed as an extension of Problem 6:

PROBLEM 9 (Ito). *Is it true that if* G *is insoluble and contains an odd permutation then* G *must be the symmetric group* S_p ?

The condition that G contain an odd permutation is equivalent, *via* the Frattini argument, to the condition that $N(P)$ contain an odd permutation. In particular then $|N(P)|$ must be even and by Theorem 5, G must be 3-fold transitive. Thus Ito's problem is covered by the more general conjecture:

if a 3-fold transitive group of odd degree contains an odd permutation then it is the full symmetric group.

This is known to be true in certain special cases (see Saxl [1973]), but since it would imply that a 5-fold transitive group of odd degree, and hence *any* 6-fold transitive group, is alternating or symmetric, it is presumably not going to be easy to prove.

Apart from the case where $p = 2q + 1 = 4r + 3$, which follows from Theorem 11 above, but which was settled rather earlier (in Ito [1963]: and indeed this led Ito to pose the question), the evidence is

THEOREM 12. *If* G *is insoluble and contains an odd permutation, and if*

(a) p *is a Fermat prime* (Ito [1967]), *or*

(b) $p = q + 2$ *where* q *is prime* (Neumann [1972a]), *or*

(c) $p = 3q + 2$ *where* q *is prime* (Saxl [1973]),

then G *is the symmetric group* S_p .

Only in case *(a)* is the primeness of p essential. In cases *(b)* and *(c)* the theorem holds for a triply transitive group of odd degree p .

Although Problems 6 and 9 are probably beyond the scope of techniques available at present, I believe that the following interesting special case should be within reach:

PROBLEM 10. *Prove that if* $p = 2q + 1$ *where* q *is prime, and if* G *is insoluble and contains an odd permutation, then* G *is the symmetric group* S_p .

6.2 Groups having several faithful representations

The last column of Table 1 lists the number r of distinct permutation representations of G as a group of degree p , that is, the number of conjugacy classes of subgroups of index p . Only PSL(2,7) , PSL(2,11) and the projective groups have as many as two such representations.

PROBLEM 11 *(i)* (Wielandt). *Is it true that* $r \leq 2$ *always?*

(ii) Is it true that if $r = 2$ *then* G *is* PSL(2,7) , PSL(2,11) *or a projective group?*

(For part *(i)* see Huppert [1967], p. 618.) In stating this problem I have shifted my ground slightly. Originally G was supposed to be given as a subgroup of the symmetric group S_p ; now we are thinking of G as an abstract group, and we are seeking all its representations as a transitive group of degree p . Of course, if these are not faithful then G may have many. However, if the Sylow p-subgroup of G has order p then every such representation has the largest normal subgroup whose order is not divisible by p as its kernel. Therefore we lose very little generality by seeking only the faithful representations.

Up to now the main attack on Problem 11 has used combinatorial methods. Since the soluble groups of degree p have only one representation we may assume that G is insoluble. Then, by Burnside's Theorem, G is 2-fold transitive in each of its representations. From this one shows quite easily that if $r \geq 2$ then G is a group of automorphisms of a non-trivial symmetric block design, that is, a configuration consisting of p "points", p "blocks" and a relation of incidence between points and blocks such that, for some numbers k and λ , each point is incident with k blocks, each block is incident with k points, and each pair of points is incident with λ blocks (see Ryser [1963], or Hall [1967]). Since this design has a cyclic group P of automorphisms it is derived from a 'cyclic difference set' (see, for example, Baumert [1971]). Using sometimes purely combinatorial arguments, sometimes algebraic arguments based on the adjacency matrix of the design the following theorems have been proved.

THEOREM 13. *(i) If* $|N(P) : P|$ *is even then* $r = 1$ *;*

(ii) if $p \leq 2\ 000\ 000$ *then* $r \leq 2$.

Part *(i)* is an unpublished theorem of Wielandt (see Huppert [1967], p. 618). Proofs have been published by Feit [1970] and by me [1972a]. Part *(ii)* comes from Cameron [1972]. In that paper Cameron shows that if $r \geq 3$ then taking the representations of G in pairs one obtains a family of "linked" symmetric block designs, and that this can be used to deduce a great many rather technical, but very suggestive, relations between the parameters of the designs. This approach gives elegant proofs (see Wielandt [1967], Cameron [1972]) of

THEOREM 14 (Wielandt). *If* G *is simple (and a transitive group of prime degree) then the outer automorphism class group of* G *is soluble.*

To conclude this subsection here is a problem about the combinatorial structures mentioned above;

PROBLEM 12. *Find the automorphism groups of the block designs derived from the known cyclic difference sets listed by* M. Hall, Jr *(see* Hall [1967] *or* Baumert [1971]*).*

The case of the Hadamard designs where $p \equiv 3 \pmod 4$, $k = \frac{1}{2}(p-1)$, $\lambda = \frac{1}{4}(p-3)$, and the difference set consists of the quadratic residues modulo p has been settled by Kantor [1969], p. 20 (or see Dembowski [1968], p. 98).

6.3 Concluding remarks

There are several facts about groups of prime degree which I have not mentioned above. These are (I hope) mainly statements which depend much more on the double transitivity of G than on the primeness of the degree. For example, M. O'Nan [unpublished] has proved that if G is 2-transitive of odd degree and the stabliser G_α is soluble then G is a known group. Or again, Kantor and McDonough [to be published] have shown that if $d \geq 3$, if $\mathrm{PSL}(d,q) \leq G \leq A_n$ where $n = (q^d-1)/(q-1)$ and if $G \nleq \mathrm{P\Sigma L}(d,q)$, then $G = A_n$. The interesting problem of finding all groups G such that $\mathrm{PSL}(2,q) \leq G \leq A_{q+1}$ is, as far as I know, still open. If we ask for subgroups rather then supergroups of the projective groups then little seems to be known beyond what is described in Perin [1970]. Therefore I offer

PROBLEM 13. *Prove that if* G *is insoluble and* $G \leq \mathrm{SL}(d,q)$ *where* $p = (q^d-1)/(q-1)$ *, then* $G = \mathrm{SL}(d,q)$.

There is some hope that Problem 13 is easier than the general problem of seeking 2-transitive subgroups of $\mathrm{PSL}(d,q)$ since, if the degree is prime, we know that t divides d (since $N_G(P) \leq N_{\mathrm{SL}(d,q)}(P)$), and so $t = d$ (since d is prime and $t \neq 1$); thus the representation theory of G may be related closely enough to the representation theory of $\mathrm{SL}(d,q)$ that one can exploit character theory in ways which are not in general possible.

NOTES ADDED DECEMBER 1973:

1. (Page 520). This is not true. But in [1771], §§56-58, 70-74, Lagrange concentrates on equations of prime degree. He treats equations of composite degree differently (see §§59-64, 75-85).
2. (Page 524). I owe to Professor M. Suzuki the information that this theorem was announced several years earlier by Toshiro Tsuzuku in "Transitive representations of finite simple groups", Proc. fourth algebra symposium (finite groups) at Hakone, Japan, 1963, 26-31, (in Japanese).

REFERENCES

K.I. Appel and E.T. Parker, [1967], "On unsolvable groups of degree $p = 4q + 1$, p and q primes", *Canad. J. Math.* 19, 583-589. MR36#256.

Paul T. Bateman and Rosemarie M. Stemmler, [1962], "Waring's problem for algebraic number fields and primes of the form $(p^r-1)/(p^d-1)$ ", *Illinois J. Math.* 6, 142-156. MR25#2059.

Leonard D. Baumert, [1971], *Cyclic difference sets* (Lecture Notes in Mathematics, 182. Springer-Verlag, Berlin, Heidelberg, New York). MR44#97.

Richard Brauer, [1942a], "On groups whose order contains a prime to the first power I", *Amer. J. Math.* 64, 401-420. MR4,1.

Richard Brauer, [1942b], "On groups whose order contains a prime to the first power II", *Amer. J. Math.* 64, 421-440. MR4,2.

Richard Brauer, [1943], "On permutation groups of prime degree and related classes of groups", *Ann. of Math.* (2) 44, 57-79. MR4,266. (MR6,334 - correction to MR4,266.)

W. Burnside, [1901], "On some properties of groups of odd order", *Proc. London Math. Soc.* 33, 162-185. FdM32,139.

W. Burnside, [1906], "On simply transitive groups of prime degree", *Quart. J. Math.* 37, 215-221. FdM37,172.

W. Burnside, [1911], *Theory of groups of finite order*, 2nd ed. (Cambridge University Press, Cambridge; reprinted, Dover, New York, 1955). FdM42,151; MR16,1086.

Peter J. Cameron, [1972], "On groups with several doubly-transitive permutation representations", *Math. Z.* 128, 1-14. Zbl.227.20001.

Roger W. Carter, [1972], *Simple groups of Lie type* (Pure and Appl. Math., 28. Interscience [John Wiley & Sons], London, New York). Zbl.248.20015.

P. Dembowski, [1968], *Finite geometries* (Ergebnisse der Mathematik und ihrer Grenzgebiete, Band 44. Springer-Verlag, Berlin, Heidelberg, New York). MR38#1597.

Walter Feit, [1966], "Groups with a cyclic Sylow subgroup", *Nagoya Math. J.* 27, 571-584. MR33#7404.

Walter Feit, [1967], "On finite linear groups", *J. Algebra* 5, 378-400. MR34#7632.

Walter Feit, [1970], "Automorphisms of symmetric balanced incomplete block designs", *Math. Z.* 118, 40-49. MR44#2809.

G. Frobenius, [1904], "Über die Charaktere der mehrfach transitiven Gruppen", *Sitzungsberichte der Königlich Preussischen Akademie der Wissenschaften zu Berlin*, 558-571; FdM35,154; *Gesammelte Abhandlungen III*, pp. 335-348 (Springer-Verlag, Berlin, Heidelberg, New York, 1968).

K.D. Fryer, [1955], "A class of permutation groups of prime degree", *Canad. J. Math.* 7, 24-34. MR16,793.

É. Galois, [1830], "Analyse d'un mémoire sur la résolution algébrique des équations", *Bulletin des Sciences mathématiques physiques et chimiques* (de M. Férussac), 13, 271-272; *J. Math. Pures Appl. (Liouville)* 11 (1846), 395-396; *Écrits et Mémoires Mathématiques d'Évariste Galois*, pp. 163-165 (Gauthier-Villars, Paris, 1962). MR27#21.

É. Galois, [1832], "Lettre à Auguste Chevalier (29 Mai 1832)", *Revue Encyclopédique* (Septembre 1832), 568-576; *J. Math. Pures Appl. (Liouville)* 11 (1846), 408-415; *Écrits et Mémoires Mathématiques d'Évariste Galois*, pp. 173-185 (Gauthier-Villars, Paris, 1962). MR27#21.

É. Galois, [1846], "Mémoire sur les conditions de résolubilité des équations par radicaux", *J. Math. Pures Appl. (Liouville)* 11, 417-433; *Écrits et Mémoires Mathématiques d'Évariste Galois*, pp. 43-71 (Gauthier-Villars, Paris, 1962). MR27#21.

J.A. Green, [1974], "Walking round the Brauer tree", *J. Austral. Math. Soc.*, (to appear).

Marshall Hall, Jr., [1967], *Combinatorial theory* (Blaisdell Publishing Co. [Ginn and Co.] Waltham, Massachusetts). MR37#80.

B. Huppert, [1967], *Endliche Gruppen I* (Die Grundlehren der mathematischen Wissenschaften, Band 134. Springer-Verlag, Berlin, Heidelberg, New York). MR37#302.

Noboru Itô, [1960], "Zur Theorie der Permutationsgruppen vom Grad p ", *Math. Z.* 74, 299-301. MR22#8064.

Noboru Ito, [1960], "Über die Gruppen $PSL_n(q)$, die eine Untergruppe von Primzahlindex enthalten", *Acta Sci. Math. (Szeged)* 21, 206-217. MR26#184.

Noboru Ito, [1963], "Transitive permutation groups of degree $p = 2q + 1$, p and q being prime numbers", *Bull. Amer. Math. Soc.* 69, 165-192. MR26#5050.

Noboru Ito, [1964], "Transitive permutation groups of degree $p = 2q + 1$, p and q being prime numbers. II", *Trans. Amer. Math. Soc.* 113, 454-487. MR30#3128.

Noboru Ito, [1965], "Transitive permutation groups of degree $p = 2q + 1$, p and q being prime numbers. III", *Trans. Amer. Math. Soc.* 116, 151-166. MR33#1355.

Noboru Ito, [1967], "On transitive permutation groups of Fermat prime degree", *Proc. Internat. Conf. Theory of groups*, (Canberra, 1965), pp. 191-202 (Gordon and Breach, New York). MR37#296.

C. Jordan, [1874], "Sur deux points de la théorie des substitutions", *C.R. Acad. Sci. Paris* 79, 1149-1151; FdM6,80; *Oeuvres de Camille Jordan I*, pp. 453-455 (Gauthier-Villars, Paris, 1961). MR24#A2526.

William M. Kantor, [1969], "2-transitive symmetric designs", *Trans. Amer. Math. Soc.* 146, 1-28. MR41#84.

J.-L. Lagrange, [1770/1771], "Réflexions sur la résolution algébrique des équations", *Nouveaux Mémoires de l'Academie royale des Sciences et Belles-Lettres de Berlin*; *Oeuvres de Lagrange*, 3, pp. 205-421 (Gauthier-Villars, Paris, 1869).

T.P. McDonough, [1972], "Some problems on permutation groups", (D. Phil. Thesis, Oxford).

Émile Mathieu, [1861], "Mémoire sur l'étude des fonctions de plusieurs quantités, sur la manière de les former et sur les substitutions qui les laissent invariables", *J. Math. Pures Appl. (Liouville)* (2) 6, 241-323.

Émile Mathieu, [1873], "Sur la fonction cinq fois transitive de 24 quantités", *J. Math. Pures Appl. (Liouville)* (2) 18, 25-46. FdM5,88-90.

G.A. Miller, [1908], "Transitive groups of degree $p = 2q + 1$, p and q being prime numbers", *Quart. J. Math.* 39, 210-216. FdM39,201.

Peter M. Neumann, [1972a], "Transitive permutation groups of prime degree", *J. London Math. Soc.* (2) 5, 202-208.

Peter M. Neumann, [1972b], "Transitive permutation groups of prime degree, II: a problem of Noboru Ito", *Bull. London Math. Soc.* 4, 337-339. Zbl.239.20005.

Peter M. Neumann, [1973+], "Generosity and characters of multiply transitive permutation groups", submitted for publication.

E.T. Parker and Paul J. Nikolai, [1958], "A search for analogues of the Mathieu groups", *Math. Tables Aids Comput.* 12, 38-43. MR21#450.

David Perin, [1972], "On collineation groups of finite projective spaces", *Math. Z.* 126, 135-142. Zbl.223.20051.

Cheryl E. Praeger, [1971], "Transitive permutation groups of prime degree", (MSc thesis, Mathematical Institute, Oxford).

Herbert John Ryser, [1963], *Combinatorial mathematics* (The Carus Mathematical Monographs, 14. Math. Assoc. Amer.; John Wiley & Sons, New York, 1963). MR27#51.

J. Saxl, [1973], "On triply transitive groups of odd degree", *J. London Math. Soc.* (2) 7, 159-167.

I. Schur, [1908], "Neuer Beweis eines Satzes von W. Burnside", *Jahresbericht Deutsche Math.-Verein.* 17, 171-176; FdM39,197; *Gesammelte Abhandlungen* 1, pp. 266-271 (Springer-Verlag, Berlin, Heidelberg, New York, 1973).

J.-A. de Séguier, [1912], *Théorie des groups finis. Éléments de la théorie des groupes de substitutions* (Gauthier-Villars, Paris). FdM43,197.

Charles C. Sims, [1970], "Computational methods in the study of permutation groups", *Computational problems in abstract algebra* (Oxford, 1967), pp. 169-183 (Pergamon, Oxford). MR41#1856.

John G. Thompson, [1967], "Vertices and sources", *J. Algebra* 6, 1-6. MR34#7677.

Helmut Wielandt, [1964], *Finite permutation groups* (Academic Press, New York). MR32#1252.

Helmut Wielandt, [1967], "On automorphisms of doubly transitive permutation groups", *Proc. Internat. Conf. Theory of Groups* (Canberra, 1965), pp. 389-393 (Gordon and Breach, New York).

Helmut Wielandt, [1969], *Permutation groups through invariant relations and invariant functions* (Lecture Notes, Ohio State University, Columbus, Ohio).

The Queen's College,
Oxford.

PROC. SECOND INTERNAT. CONF. THEORY OF GROUPS, CANBERRA, 1973, pp. 536-540.

20C05

IDEALS IN THE CENTRE OF A GROUP RING

M.F. O'Reilly

1.

Let G be a finite group, O a commutative ring with identity and $\Gamma(OG)$ the group ring of G over O. In this paper we obtain ideals of the centre $Z(OG)$ of $\Gamma(OG)$ which when $O = k$, a field of characteristic p, include the ideal of Reynolds ([2], Theorem 1).

Let A be a G-algebra over O (see [1]).

For $H \leq G$, $A_H = \{a \in A : a^h = a, \text{ all } h \in H\}$.

If $K \leq H \leq G$ then $A_H \leq A_K$.

For $b \in A_K$, $b_K^H = \sum_{h\in\Psi} b^h \in A_H$ where Ψ is a transversal of K in H.

$A_K^H = \left\{b_K^H, b \in A_K\right\}$.

From [1] we have

(i) for $K, L \leq H$, $a \in A_K$,

$$a_K^H = \sum_{x\in\Psi} \left(a^x\right)_{L\cap K^x}^L \in A_L$$

where Φ is a set of (L, K) double coset representatives in G,

(ii) for $H \leq K$, $a \in A_K$, $b \in A_H$,

$$ab_H^K = (ab)_H^K \text{ and } b_H^K a = (ba)_H^K ,$$

(iii) for $H \leq K \leq L$, $a \in A_H$,

$$\left(a_H^K\right)_K^L = a_H^L ,$$

(iv) for $H \leq K$, I an ideal of A_H then $I_H^K = \left\{a_H^K, a \in I\right\}$ is an ideal of A_K .

For $z \in \Gamma(OG)$, $H \leq G$, $St_H z = \{h \in H : z^h = z\}$. If $z \in A_K$, $K \leq H$ and $L = St_H z$, then $z_K^L = (L : K)z$ and so is zero if $0 = k$ and char $k = p$ divides $L : K$.

Constant use is made of the factorization $y = rs = sr$ of y into its p-regular part r and p-singular part s . The p'-section containing y which consists of all elements of G which have p-regular factors conjugate to r is denoted by S_r .

$N(H)$, $C(H)$ denote respectively the normalizer and centralizer of H in G ; for $g \in G$, $C(g)$ is the centralizer of g . The sum of the elements of a subset $S \subset \Gamma(OG)$ will be denoted by $\bar{S}$.

2.

$St_G \bar{H}y \leq N(H)$. Further

$$St_H \bar{H}y = H \cap H^y = St_H y\bar{H}$$

and so if $K = St_H y$ and $n_y = H \cap H^y : K$,

$$\bar{H}y_K^H = n_y (\bar{H}y)^H_{H\cap H^y} = n_y \overline{HyH} = y_K^H \bar{H} .$$

Thus the ideal $I(H) = A_H \bar{H} A_H$ has as basis the distinct double cosets sums $\overline{HyH}$, $y \in G$.

LEMMA 1. *There exist subgroups* $H_\alpha < H$ *and elements* $a_\alpha \in N(H_\alpha)$, $\alpha \in J$ *(index set) such that*

$$\overline{HyH} = \sum_J \left(\bar{H}_\alpha a_\alpha\right)^H_{H_\alpha} .$$

PROOF. The result is trivial when $H = \{1\}$ or $y \in N(H)$. We proceed by induction on $|H|$. $\overline{HyH} = (\bar{H}y)^H_R$ where $R = H \cap H^y$. As Hy is the union of (R, R) double cosets, $\bar{H}y = \sum \overline{RgR} = \sum \left(\bar{H}_\alpha a_\alpha\right)^R_{H_\alpha}$ where $a_\alpha \in N(H_\alpha)$ by hypothesis. the result follows from 1 (iii).

If $H < L$, $\{I(H)\}^L_H$ is an ideal of A_L by 1 (iv) with basis the non-zero elements $(\overline{HyH})_H^L$. From Lemma 1 we get

THEOREM 1. *Let* $R_L = \bigcup_{H\le L} \{I(H)\}_H^L$, $R'_L = \bigcup_{H<L} \{I(H)\}_H^L$. *Then* R_L, R'_L *are ideals of* A_L *with bases the sets of non-zero elements of the form* $(\bar{H}a)_H^L$, $a \in N(H)$ *and* $H \le L, H < L$ *respectively.*

3.

Restricting to the case where $O = k$, char $k = p$ and L is a p-subgroup we will show that R_L, R'_L have p'-section sums as bases.

LEMMA 2. *If* D *is a Sylow* p*-subgroup of* $C(r)$ *then* $n_r\bar{S}_r = (\bar{D}r)_D^G$ *where* $n_r = N(D) \cap C(r) : D$.

PROOF.

$$\begin{aligned} y_1 = r_1 s_1 \in S_r &\Longleftrightarrow r_1^h = r \text{ , some } h \in G \text{ ,} \\ &\Longleftrightarrow C(r_1) = C^h(r) \\ &\Longleftrightarrow s_1 = \tilde{s}^h \text{ , some } \tilde{s} \in C(r) \text{ ,} \\ &\Longleftrightarrow y_1 = (r\tilde{s})^h \text{ .} \end{aligned}$$

Conversely $\tilde{s} \in C(r)$ implies $(r\tilde{s})^h \in S_r$. Thus $S_r = \{(r\tilde{s})^h, h \in \Phi, s \in B\}$ where Φ is a right transversal of $C(r)$ in G and B is the set of all p-singular elements of $C(r)$. Thus $\bar{S}_r = (r\bar{B})_{C(r)}^G$.

$(\bar{D})_D^{C(r)} = n_r \sum \bar{D}^g$ where the sum is over the set T of $1 + np$ distinct conjugates of D in $C(r)$. For a fixed $x \in B$ partition T into classes of conjugates under powers of x . The order of the class containing D^g is a power of p and equals 1 only if $x \in N(D^g) \cap C(r)$ whence $x \in D^g$. Thus the number of conjugates of D not containing x is a multiple of p and so the number which contain x equals 1 mod p .

Thus $\sum \bar{D}^g = \bar{B}$ whence $n_r\bar{S}_r = (\bar{D}r)_D^G$.

For any p-group $P < G$, $\{I(P)\}_P^G$ has as basis the non zero elements $(\bar{P}a)_P^G$, $a \in N(P)$.

LEMMA 3. *Let* P *be a* p*-subgroup of* G , $y = rs \in N = N(P)$. *Then* *(a)* $Q < P$, $y \in N(Q)$ *implies* $(\bar{Q}y)_Q^P = \bar{P}y$ *or zero;*

(b) *if* $(\bar{P}y)_P^G \neq 0$ *then*

(i) $P \cap C(r) = D$, *a Sylow* p*-subgroup of* $C(r)$,

(ii) $(\bar{P}y)_P^G = (\bar{D}r)_D^G$.

PROOF. *(a)* Let $R = N(Q) \cap P > Q$ and let Ψ be a transversal of Q in R . Then $y \in N(R)$. For $u \in \Psi$, $u^{-1}Qyu = u^{-1}yuy^{-1}Qy = \tilde{u}Qy$ where $\tilde{u} = u^{-1}yuy^{-1} \in R$. For distinct $u, v \in \Psi$ if $(\bar{Q}y)_Q^R \neq 0$, $\tilde{u}Q$, $\tilde{v}Q$ are distinct cosets of Q in R and so $(\bar{Q}y)_Q^R = \sum_{u \in \Psi} \tilde{u}\bar{Q}y = \bar{R}y$. The result follows by induction on $P : Q$.

Each p-singular element $x \in N \cap C(y)$ is contained in some Sylow p-subgroup P_1 of N . Also $P \leq P_1$. If $x \notin P$, $(\bar{P}y)_P^{P_1} = 0$. So $P \cap C(y) = D_0$ is a normal Sylow p-subgroup of $N \cap C(y)$.

Now $s \in N \cap C(y)$ so $s \in P$ whence $Py = Pr$. So $(\bar{P}r)_P^G \neq 0$ and $P \cap C(r) = D$ is a normal Sylow p-subgroup of $N \cap C(r)$. $(\bar{D}r)_D^N$ is a scalar multiple of a p'-section sum in $\Gamma(kN)$ and is non-zero. So by *(a)*, $(\bar{D}r)_D^P = \bar{P}r$ giving $(\bar{P}y)_P^N = (\bar{D}r)_D^N$. If $D < D_1$, a Sylow p-subgroup of $C(r)$, then $(\bar{D}r)_D^{D_1} = 0$; so D is an Sylow p-subgroup of $C(r)$ and $(\bar{P}y)_P^G = (\bar{D}r)_D^G$.

From Lemmas 2 and 3 we see that the only non-zero elements $(\bar{P}y)_P^G$ of $\{I(P)\}_P^G$ are scalar multiples of p'-section sums. We thus obtain a corollary to Theorem 1.

COROLLARY 1. *If* P *is a* p*-subgroup of* G *the ideal* $(R_P)_P^G$ *has as basis the* p'*-section sums* $\bar{S}_r$ *for which a Sylow* p*-subgroup* D *of* $C(r)$ *lies in* P .

This is the intersection of Reynolds' ideal with the ideal of conjugacy class sums of defect group $\leq P$.

References

[1] James A. Green, "Some remarks on defect groups", *Math. Z.* 107 (1968), 133-150. MR38#2222.

[2] William F. Reynolds, "Sections and ideals of centres of group algebras", *J. Algebra* 20 (1972), 176-181. Zbl.232.20005.

University of Papua and New Guinea,
Boroko, Papua-New Guinea.

PROC. SECOND INTERNAT. CONF. THEORY OF GROUPS, 20E10
CANBERRA, 1973, pp. 541-549.

A NON-DISTRIBUTIVE METABELIAN VARIETY LATTICE

Elizabeth A. Ormerod

This paper presents an example to show that the lattice of subvarieties of $\underline{\underline{A}}_2\underline{\underline{A}}_4 \wedge \underline{\underline{N}}_6$ is not distributive.

1. Introduction

The notation and terminology used follows Hanna Neumann [7] with the addition of lat $\underline{\underline{V}}$ and lat G to denote respectively the lattice of subvarieties of a variety $\underline{\underline{V}}$ and the lattice of verbal subgroups of a group G.

In answering a question of Kovács at the First International Conference on the Theory of Groups, at Canberra, 1965, Higman [5] gave an example of three join-irreducible varieties of p-groups whose joins in pairs coincided. In particular, this showed that the lattice of subvarieties of $\underline{\underline{B}}_{p,c}$ (where $\underline{\underline{B}}_{p,c}$ denotes the variety of all groups of exponent p and class c) is non-distributive. At about the same time Bjarni Jónsson in some unpublished work showed that the lattice of subvarieties of $\underline{\underline{N}}_3$ is distributive. This raised the question of whether the subvariety lattice of a given variety is distributive or not.

Since then, Kovács and Newman [6] have shown that $\operatorname{lat}\left(\underline{\underline{A}}_{p^\alpha}\underline{\underline{A}}_p\right)$ is distributive for all primes p and all positive integers α. In contrast to this some unpublished work of the same authors demonstrated non-distributivity in $\operatorname{lat}\left(\underline{\underline{A}}_2\underline{\underline{A}}_8 \wedge \underline{\underline{N}}_6\right)$, thereby showing that $\operatorname{lat}\left(\underline{\underline{A}}_p\underline{\underline{A}}_{p^\alpha}\right)$ is generally non-distributive. Brooks [2] has given another example of non-distributivity in $\operatorname{lat}\left(\underline{\underline{A}}_p\underline{\underline{A}}_{p^\alpha}\right)$ with α as small as possible, namely $\alpha = 2$, and with $p = 3$. His result is that the lattice of subvarieties of $\underline{\underline{A}}_3\underline{\underline{A}}_9 \wedge \underline{\underline{N}}_{11}$ is non-distributive. He made the conjecture that $\operatorname{lat}\left(\underline{\underline{A}}_p\underline{\underline{A}}_{p^2} \wedge \underline{\underline{N}}_{p^2}\right)$ is not distributive for all primes $p \geq 3$. As supporting evidence he claimed that $\operatorname{lat}(\underline{\underline{A}}_3\underline{\underline{A}}_9 \wedge \underline{\underline{N}}_9)$ and $\operatorname{lat}(\underline{\underline{A}}_5\underline{\underline{A}}_{25} \wedge \underline{\underline{N}}_{25})$ are not distributive. For $p = 2$ his construction failed and he stated that the case of $\operatorname{lat}(\underline{\underline{A}}_2\underline{\underline{A}}_4)$ was still very much an open question. However, as has already been mentioned, $\operatorname{lat}(\underline{\underline{A}}_2\underline{\underline{A}}_8)$ is not distributive, and $\operatorname{lat}(\underline{\underline{A}}_4\underline{\underline{A}}_4)$ is not distributive because of a result of Bryce [4], who has shown that $\operatorname{lat}\left(\underline{\underline{A}}_{p^2}\underline{\underline{A}}_{p^2} \wedge \underline{\underline{N}}_{p+2}\right)$ is not distributive for any prime p.

In this paper the method used is a generalization of that of Brooks, but instead of considering the verbal subgroup lattice of the free group of rank two, we use the free group of rank three. The following result of Brooks [2] is used.

THEOREM 1.1. *Let* G *be a relatively free group. If* $\operatorname{lat} G$ *is not distributive, then neither is* $\operatorname{lat}(\operatorname{var} G)$. *In fact, if for some* $C, D_1, D_2 \in \operatorname{lat} G$,

$$C \cap D_1D_2 \neq (C \cap D_1)(C \cap D_2),$$

then

$$\underline{\underline{U}} \vee (\underline{\underline{M}}_1 \wedge \underline{\underline{M}}_2) \neq (\underline{\underline{U}} \vee \underline{\underline{M}}_1) \wedge (\underline{\underline{U}} \vee \underline{\underline{M}}_2),$$

where $\underline{\underline{M}}_i = \operatorname{var}(G/D_i)$, *and* $\underline{\underline{U}}$ *is any variety for which* $\underline{\underline{U}}(G) = C$.

We show that the verbal subgroup lattice of $G = F_3(\underline{\underline{A}}_2\underline{\underline{A}}_4 \wedge \underline{\underline{N}}_6)$ is not distributive. The example we provide occurs among the subgroups of the last term of the lower central series of G. In order to do this we first characterize the fully invariant subgroups of the last term of the lower central series of such a group.

2. Some fully invariant subgroups

The main result of this section is the following theorem.

THEOREM 2.1. *Let* $\underline{\underline{U}} = \underline{\underline{A}}_p\underline{\underline{A}}_{p^\alpha} \wedge \underline{\underline{N}}_c$ *where* $\alpha \geq 1$, $c > 2$. *Then the fully invariant subgroups in the last term of the lower central series of* $F = F_k(\underline{\underline{U}})$, $k > 1$, *are precisely those subgroups that are closed under the automorphisms and deletions of* F.

PROOF. Let M be the $\underline{\underline{A}}_p$ subgroup of F, and let η be any endomorphism of F. Denote by η/M the endomorphism of F/M induced by η. We note that F/M may be regarded as a k-dimensional vector space over the field of p elements, and that $\operatorname{Aut}(F/M) \cong \operatorname{GL}(k, p)$. Let S be a subgroup of $F_{(c)}$ that is closed under the automorphisms and deletions of F.

If $\ker \eta/M = \{1\}$, then $\eta/M \in \operatorname{Aut}(F/M)$ and η is in fact an automorphism of F since M is the Frattini subgroup $\Phi(F)$. Hence S admits η by hypothesis.

If $\ker \eta/M = F/M$, then $F\eta \subseteq M$, and $F_{(c)}\eta = \{1\}$ since $F_{(c)}$ has exponent p and $F_{(c+1)} = 1$. Hence S admits η.

We now consider the case $\{1\} < \ker \eta/M < F/M$. Let $\{f_1, \ldots, f_k\}$ be a free generating set for F and let the dimension of the space generated by $\{f_i\eta M;\ i = 1, \ldots, k\}$ be r, where $1 \le r < k$. Then r of the $f_i\eta M$ are linearly independent, say $f_{i_1}\eta M, \ldots, f_{i_r}\eta M$, where $\{i_1, \ldots, i_r\} \subset \{1, \ldots, k\}$. Then

$$f_j\eta = f_{i_1}^{\eta_{j1}} \ldots f_{i_r}^{\eta_{jr}} \bmod M, \text{ for } j \in \{1, \ldots, k\}\backslash\{i_1, \ldots, i_r\}.$$

Define an automorphism ϕ of F as follows:

$$f_j\phi = \begin{cases} f_j & \text{for } j \in \{i_1, \ldots, i_r\}, \\ f_j^{-1}f_{i_1}^{\eta_{j1}} \ldots f_{i_r}^{\eta_{jr}} & \text{for } j \in \{1, \ldots, k\}\backslash\{i_1, \ldots, i_r\}. \end{cases}$$

Then

$$f_j\phi\eta = \begin{cases} f_j\eta & \text{for } j \in \{i_1, \ldots, i_r\}, \\ 1 \bmod M & \text{for } j \in \{1, \ldots, k\}\backslash\{i_1, \ldots, i_r\}. \end{cases}$$

Define another automorphism $\hat{\eta}$ of F as follows:

$$f_j\hat{\eta} = \begin{cases} f_j\phi\eta & \text{for } j \in \{i_1, \ldots, i_r\}, \\ g_j & \text{for } j \in \{1, \ldots, k\}\backslash\{i_1, \ldots, i_r\}, \end{cases}$$

where the g_j are chosen in such a way that $\{f_j\hat{\eta}M;\ j = 1, \ldots, k\}$ forms a basis for F/M.

Now S admits $\hat{\eta}$ since it is an automorphism of F. Since S admits the deletions it is generated by homogeneous elements. If $x \in S$ is homogeneous then $x\phi\eta = 1$, or $x\phi\eta = x\hat{\eta}$, and so S admits $\phi\eta$. But $\phi \in \operatorname{Aut} F$, so $S\phi \subseteq S$, and in fact $S\phi = S$. Thus we have shown that $S \supseteq S\phi\eta = S\eta$, which is the required result.

To make this theorem useful we now look for a small set, P, of automorphisms of F with the property that closure under P implies closure under the automorphism group of F. To this end we define the automorphisms of F, α, ρ_j, $2 \le j \le k$, and $\varepsilon(a)$, where a generates the multiplicative group of the field of p elements, as follows:

$$f_i\alpha = \begin{cases} f_1f_2 , & i = 1 , \\ f_i , & 2 \le i \le k , \end{cases}$$

$$f_i\rho_j = \begin{cases} f_j , & i = 1 , \\ f_1 , & i = j , \\ f_i , & i \ne j,\ 2 \le i \le k , \end{cases}$$

and

$$f_i\varepsilon(a) = \begin{cases} f_i , & 1 \le i \le k-1 , \\ f_k^a , & i = k . \end{cases}$$

Let $P = \{\alpha;\ \rho_j,\ 2 \le j \le k;\ \varepsilon(a)\}$.

We have noted that $\mathrm{Aut}(F/M) = \mathrm{GL}(k, p)$. We make use of this to show that P is the required set. First we introduce some elements of $\mathrm{GL}(k, p)$.

For any $i \ne j$ and any $\lambda \in \mathrm{GF}(p)$, the field of p elements, we denote by $B_{ij}(\lambda)$ the matrix obtained from the unit matrix by replacing the element $a_{ij} = 0$ of the unit matrix by λ . For any $j \in \{2, \ldots, k\}$ we denote by P_j the matrix obtained from the unit matrix by replacing the elements $a_{1j} = a_{j1} = 0$ by 1 and $a_{11} = a_{jj} = 1$ by 0 . Let $D(\mu)$ be the diagonal matrix with diagonal entries $\{1, \ldots, 1, \mu\}$, where $\mu \in \mathrm{GF}(p)$.

We use the following theorem (4.1 of [1]) without proof.

THEOREM 2.2. *Every non-singular matrix* A *over* $\mathrm{GF}(p)$ *can be written in the form* $BD(\mu)$, *where* B *is a product of matrices* $B_{ij}(\lambda)$, *and* $\mu \in \mathrm{GF}(p)$, $\mu \ne 0$.

Thus we have a generating set for $\mathrm{GL}(k, p)$ but it is not yet in the form we require.

LEMMA 2.3. *The matrices* $B_{ij}(\lambda)$ *are contained in the group generated by* $\{B_{12}(1);\ P_j,\ 2 \le j \le k\}$.

PROOF. We first note that $B_{ij}(\lambda) = \left(B_{ij}(1)\right)^\lambda$, so it is sufficient to show that the matrices $B_{ij}(1)$ are in the group generated by $\{B_{12}(1);\ P_j,\ 2 \le j \le k\}$.

An easy calculation gives the following results.

$$B_{i2}(1) = P_iB_{12}(1)P_i , \quad \text{for } 3 \le i \le k ,$$

$$B_{i1}(1) = P_2B_{i2}(1)P_2 , \quad \text{for } 3 \le i \le k ,$$

$$B_{1j}(1) = P_j B_{j1}(1) P_j \ , \quad \text{for} \quad 3 \le j \le k \ ,$$

$$B_{ij}(1) = P_i B_{1j}(1) P_i \ , \quad \text{for} \quad 2 \le i \le k, \ 3 \le j \le k \ ,$$

and

$$B_{21}(1) = P_2 B_{12}(1) P_2 \ .$$

COROLLARY 2.4. *For* $p \neq 2$, $\mathrm{GL}(k, p)$ *is generated by* $\{B_{12}(1);\ P_j,\ 2 \le j \le k;\ D(\mu)$ *where* μ *generates the multiplicative group of* $\mathrm{GF}(p)\}$.

$\mathrm{GL}(k, 2)$ *is generated by* $\{B_{12}(1);\ P_j,\ 2 \le j \le k\}$.

We now have a generating set for $\mathrm{GL}(k, p)$, and we can show that P has the property mentioned earlier. First we prove a preliminary result.

LEMMA 2.5. *Let* η_1 *and* η_2 *be endomorphisms of* F *such that* $\eta_1/M = \eta_2/M$. *Then* $\eta_1\big|_{F_{(c)}} = \eta_2\big|_{F_{(c)}}$.

PROOF. The result follows because $F_{(c)}$ has exponent p and $F_{(c+1)} = \{1\}$.

THEOREM 2.6. *Let* S *be a subgroup of* $F_{(c)}$ *that admits the members of* P *and deletions of* F . *Then* S *is fully invariant in* F .

PROOF. It is sufficient to show that S admits the automorphisms of F .

F/M may be regarded as a k-dimensional vector space over the field of p elements. So with respect to a suitable basis of F/M , α/M is the linear transformation whose matrix is $B_{12}(1)$, ρ_j/M is the linear transformation whose matrix is P_j , where $2 \le j \le k$, and $\varepsilon(a)$ is the linear transformation whose matrix is $D(a)$. From Corollary 2.4 then $P/M = \{\alpha/M;\ \rho_j/M,\ 2 \le j \le k;\ \varepsilon(a)/M\}$ generates the automorphism group of F/M .

Let η be any automorphism of F . Then η/M is an automorphism of F/M and we can write η/M as a word in the generators, say

$$\eta/M = h\big(\alpha/M,\ \rho_j/M,\ \varepsilon(a)/M\big) \ .$$

Put

$$\nu = h\big(\alpha,\ \rho_j,\ \varepsilon(a)\big) \ .$$

Then $\eta/M = \nu/M$, and since S admits ν , it follows from Lemma 2.5 that S admits η .

3. Non-distributivity

We now consider $\underline{\underline{A}}_2\underline{\underline{A}}_4 \wedge \underline{\underline{N}}_6$. Let $\{g_1, g_2, g_3\}$ be a free generating set for $G = F_3(\underline{\underline{A}}_2\underline{\underline{A}}_4 \wedge \underline{\underline{N}}_6)$. $G_{(6)}$ is an elementary two-group, since it is a subgroup of G' A basis for $G_{(6)}$ is listed in Appendix A. The commutators listed are obviously in $G_{(6)}$. That they are a basis follows from 3.1 of [3].

From Theorem 2.6 to show that a subgroup of $G_{(6)}$ is fully invariant in G we have only to show that it is closed under the automorphisms α, ρ_2, ρ_3 , and the deletions of G . We now define certain subgroups of $G_{(6)}$ in terms of their generators.

Let

$w_1 = c_5c_{14}$, $w_2 = c_1c_7c_{11}$, $w_3 = c_3c_{10}c_{16}$, $w_4 = c_2c_8c_{12}c_{15}$, $w_5 = c_3c_5c_{12}c_{13}$, $w_6 = c_2c_4c_{11}c_{14}$, $w_7 = c_{25}$, $w_8 = c_{22}$, $w_9 = c_{19}$, $w_{10} = c_2c_5c_7$, $w_{11} = c_{12}c_{14}c_{16}$, $w_{12} = c_{24}c_{25}$, $w_{13} = c_{21}c_{22}$, $w_{14} = c_{19}c_{20}$, $w_{15} = c_{22}c_{23}$, $w_{16} = c_{18}c_{19}$, $w_{17} = c_{25}c_{26}$, $u_1 = c_5$, $u_2 = c_7$, $u_3 = c_3c_{10}$, $v_1 = c_5c_9$, $v_2 = c_6c_7$, $v_3 = c_3c_{10}c_{17}$.

Now put

$$V = \langle w_1, \ldots, w_{17} \rangle ,$$

$$D_1 = \langle u_1, u_2, u_3, V \rangle ,$$

$$D_2 = \langle v_1, v_2, v_3, V \rangle ,$$

and

$$C = \langle u_1v_1, u_2v_2, u_3v_3, V \rangle .$$

It can be shown, using Theorem 2.6, that V is a fully invariant subgroup of G , and it follows from Table B that D_1, D_2 , and C are fully invariant subgroups of G if V is fully invariant. It can also be seen that $C \cap D_1 = V$, $C \cap D_2 = V$ and $C \cap D_1D_2 = C$, which with Theorem 1.1 gives the required result.

We also give an example to show that it is necessary in Theorem 2.1 to include the condition that a subgroup be closed under the deletions.

Let $S = \langle w_1, w_2, w_3, w_4w_7, w_5w_8, w_6w_9 \rangle$. Then it can be seen from Table C that S is closed under the automorphisms α, ρ_2, ρ_3 , and it is therefore closed under the automorphism group of G , but inspection shows directly that S is not closed under the deletions of G , and hence S is not fully invariant.

The first example of non-distributivity I found in $\mathrm{lat}\left(\underline{A}_2\underline{A}_4 \wedge \underline{N}_6\right)$ was in the verbal subgroup lattice of the free group of rank four, and I thank Dr M.F. Newman for his help in reducing this to the free group of rank three. This result, however, is not necessarily the best possible, for the question of the distributivity of the lattices of subvarieties of $\underline{A}_2\underline{A}_4 \wedge \underline{N}_4$ and $\underline{A}_2\underline{A}_4 \wedge \underline{N}_5$ still remains open.

Finally, I would like to thank my supervisor Dr R.A. Bryce for his help in the preparation of this paper.

APPENDIX

Table A

This is a list of commutators that form a basis for $G_{(6)}$, where $G = F_3\left(\underline{A}_2\underline{A}_4 \wedge \underline{N}_6\right)$. The convention used is that g_i is represented by i .

$c_1 = [2, 1, 1, 2, 2, 3]$,	$c_9 = [3, 1, 1, 1, 1, 2]$,	$c_{18} = [2, 1, 1, 1, 1, 2]$
$c_2 = [2, 1, 1, 2, 3, 3]$,	$c_{10} = [3, 1, 1, 2, 3, 3]$,	$c_{19} = [2, 1, 1, 1, 2, 2]$
$c_3 = [2, 1, 1, 3, 3, 3]$,	$c_{11} = [3, 1, 1, 2, 2, 2]$,	$c_{20} = [2, 1, 1, 2, 2, 2]$
$c_4 = [2, 1, 1, 1, 2, 3]$,	$c_{12} = [3, 1, 1, 2, 2, 3]$,	$c_{21} = [3, 1, 1, 1, 1, 3]$
$c_5 = [2, 1, 1, 1, 3, 3]$,	$c_{13} = [3, 1, 1, 1, 2, 3]$,	$c_{22} = [3, 1, 1, 1, 3, 3]$
$c_6 = [2, 1, 2, 2, 2, 3]$,	$c_{14} = [3, 1, 1, 1, 2, 2]$,	$c_{23} = [3, 1, 1, 3, 3, 3]$
$c_7 = [2, 1, 2, 2, 3, 3]$,	$c_{15} = [3, 1, 2, 2, 2, 3]$,	$c_{24} = [3, 2, 2, 2, 2, 3]$
$c_8 = [2, 1, 2, 3, 3, 3]$,	$c_{16} = [3, 1, 2, 2, 3, 3]$,	$c_{25} = [3, 2, 2, 2, 3, 3]$
	$c_{17} = [3, 1, 2, 3, 3, 3]$,	$c_{26} = [3, 2, 2, 3, 3, 3]$

Table B

g	$g\alpha$	$g\rho_2$	$g\rho_3$
u_1	u_1u_2	u_2	u_3
u_2	u_2	u_1	w_2u_2
u_3	w_7u_3	w_3u_3	u_1
v_1	av_1v_2	v_2	v_3
v_2	v_2	v_1	w_2v_2
v_3	$w_{17}v_3$	w_3v_3	v_1

Note: $a = w_1w_2w_6w_{10}$.

The deletions act trivially on u_1, u_2, u_3, v_1, v_2 and v_3 .

Table C

g	$g\alpha$	$g\rho_2$	$g\rho_3$
w_1	w_1w_2	w_2	w_3
w_2	w_2	w_1	w_2
w_3	w_3	w_3	w_1
w_4w_7	w_4w_7	w_5w_8	w_6w_9
w_5w_8	$w_3w_4w_7w_5w_8$	w_4w_7	w_5w_8
w_6w_9	$w_2w_6w_9$	w_6w_9	w_4w_7

References

[1] E. Artin, *Geometric algebra* (Interscience, New York, London, 1957). MR18,553.

[2] M.S. Brooks, "On lattices of varieties of metabelian groups", *J. Austral. Math. Soc.* 12 (1971), 161-166. MR45#3526.

[3] M.S. Brooks, "On varieties of metabelian groups of prime-power exponent", *J. Austral. Math. Soc.* 14 (1972), 129-154.

[4] R.A. Bryce, "Metabelian groups and varieties", *Philos. Trans. Roy. Soc. London Ser. A* 266 (1970), 281-355. MR42#349.

[5] Graham Higman, "Representations of general linear groups and varieties of p-groups", *Proc. Internat. Conf. Theory of Groups* (Canberra, 1965), pp. 167-173 (Gordon and Breach, New York, London, Paris, 1967).

[6] L.G. Kovacs and M.F. Newman, "On non-Cross varieties of groups", *J. Austral. Math. Soc.* 12 (1971), 129-144. MR45#1966.

[7] Hanna Neumann, *Varieties of groups* (Ergebnisse der Mathematik und ihrer Grenzgebiete, Band 37. Springer-Verlag, Berlin, Heidelberg, New York, 1967). MR35#6734.

The Australian National University,
Canberra.

Present Address:
Ursuline College,
Armidale, NSW 2350.

PROC. SECOND INTERNAT. CONF. THEORY OF GROUPS, CANBERRA, 1973, pp. 550-561. 20F99

POLYNOMIAL MAPS

I.B.S. Passi

1. Introduction

Let M be a monoid (an associative multiplicative system with identity) and G an additive abelian group. A map $f : M \to G$ is called a polynomial map of degree $\leq n$ if the linear extension of f to $\mathbb{Z}(M)$, the integral monoid ring of M, to G vanishes on $\Delta_{\mathbb{Z}}^{n+1}(M)$, where $\Delta_{\mathbb{Z}}(M)$ is the augmentation ideal of $\mathbb{Z}(M)$. For example, if F is the free cyclic monoid $\{1, X, X^2, \ldots, X^n, \ldots\}$ and $f(t)$ is an integer-valued polynomial, then the map $\theta : F \to \mathbb{Z}$ given by $\theta(X^n) = f(n)$ is a polynomial map of degree $\leq$ degree of $f(t)$. The study of polynomial maps becomes more interesting when the monoid is, in fact, a group. In this case polynomial maps are an effective tool, particularly for the investigation of dimension subgroups. These maps also occur in wreath products. If A and B are groups, A abelian, B finite, then the elements of the nth term $\alpha_n(A \text{ wr } B)$ of the α-central series of $A \text{ wr } B$ can be seen to be exactly the polynomial maps $f : B \to A$ of degree $\leq n-1$.

This paper is a report on the present status of the theory and applications of polynomial maps.

2. Polynomial ideals in group rings

Let $f(x_1, x_2, \ldots, x_n)$ be an element of the integral group ring $\mathbb{Z}(F_n)$ of the free group F_n on $x_1, x_2, \ldots, x_n$. Let R be a commutative ring with unity, G a multiplicative group and $R(G)$ the group ring of G with coefficients in R. If $g_1, g_2, \ldots, g_n \in G$, then the expression $f(g_1, g_2, \ldots, g_n)$ can be regarded as an element of $R(G)$. We denote the 2-sided ideal of $R(G)$ generated by the elements $f(g_1, g_2, \ldots, g_n)$, $g_1, g_2, \ldots, g_n \in G$, by $\underline{\underline{A}}_{f,R}$.

DEFINITION 2.1 [7]. A 2-sided ideal $\underline{\underline{A}}$ of $R(G)$ is called a *polynomial ideal* it $\underline{\underline{A}} = \underline{\underline{A}}_{f,R}$ for some polynomial $f = f(x_1, x_2, \ldots, x_n) \in \mathbb{Z}(F_n)$ for some

$n \geq 1$.

EXAMPLES 2.2. (i) The augmentation ideal $\Delta_R(G)$ $\left(= \ker \varepsilon : R(G) \to R,\ \varepsilon\left(\sum_{\substack{r \in R \\ g \in G}} rg\right) = \sum r\right)$ is a polynomial ideal: take $f(x) = x - 1$. In fact, all powers $\Delta_R^i(G)$ of $\Delta_R(G)$ are polynomial ideals: $\Delta_R^i(G) = \underline{\underline{\Delta}}_{f_i,R}$, where $f_i = (x_1-1)(x_2-1) \ldots (x_i-1)$.

(ii) The Lie ideals $\Delta_R^{(n)}(G)$ defined inductively by $\Delta_R^1(G) = \Delta_R(G)$, $\Delta_R^{(n)}(G) = \left[\Delta_R(G),\ \Delta_R^{(n-1)}(G)\right]R(G)$, where $[M, N]$ denotes the R-submodule of $R(G)$ generated by $mn - nm$, $m \in M$, $n \in N$, are polynomial ideals [7].

DEFINITION 2.3 [7]. Let G be a group, R a commutative ring with unity. If $f(x_1, x_2, \ldots, x_n) \in Z(F_n)$, then a map $\theta : G \to M$, M an R-module, is called an *f_R-polynomial map* provided the linear extension of θ to $R(G)$ vanishes on $\underline{\underline{\Delta}}_{f,R}$, the polynomial ideal determined by f .

EXAMPLES 2.4. (i) Since $\Delta_R^i(G)$, $i \geq 1$, are polynomial ideals, maps $\theta : G \to M$ which vanish on $\Delta_R^i(G)$ are f_R-polynomial maps. Such maps were called *R-polynomial maps* of degree $\leq i-1$ in [8]. Z-polynomial maps have been studied in [2], [3], [6] and [8].

It may be observed that if M is an abelian group, N an R-module, $\theta : G \to M$ an f_Z-polynomial map, $\phi : M \to N$ a homomorphism, then the map $\phi \circ \theta : G \to N$ is an f_R-polynomial map [7].

(ii) Let $T : \underline{\underline{M}}_R \to \underline{\underline{M}}_R$ be a functor of degree $\leq n$ in the sense of Eilenberg-Maclane [4], where $\underline{\underline{M}}_R$ is the category of R-modules. Then, for every $M, N \in \underline{\underline{M}}_R$, $T_{MN} : \hom_R(M, N) \to \hom_R(T(M), T(N))$, $T_{MN}(\theta) = T(\theta)$, $\theta : M \to N$, is an R-polynomial map of degree $\leq n$.

(iii) Every homomorphism $\alpha : G \to M$ is a Z-polynomial map of degree ≤ 1 .

(iv) $\lambda_n : G \to Z(G)/\Delta_Z^{n+1}(G)$ given by

$$\lambda_n(x) = x + \Delta_Z^{n+1}(G) , \quad x \in G ,$$

is a Z-polynomial map of degree $\leq n$. λ_n is the universal Z-polynomial map of degree $\leq n$.

(iv) If $SP^n(G)$ denotes the nth symmetric product of an abelian group G, then

$$\theta_n : G \to SP^n(G)$$

given by

$$\theta_n(x) = \underbrace{x \hat{\otimes} x \hat{\otimes} \ldots \hat{\otimes} x}_{n \text{ times}}$$

is a $\mathbb{Z}$-polynomial map of degree $\leq n$ [18].

2.5 Dimension subgroups and Lie dimension subgroups

Let G be a group, R a commutative ring with unity. The nth *dimension subgroup* $D_{n,R}(G)$ and the nth *Lie dimension subgroup* $D_{(n),R}(G)$ are defined as follows:

$$D_{n,R}(G) = G \cap \left(1+\Delta_R^n(G)\right), \quad n \geq 1,$$

$$D_{(n),R}(G) = G \cap \left(1+\Delta_R^{(n)}(G)\right), \quad n \geq 1.$$

Let $i_R : \mathbb{Z}(G) \to R(G)$ be the ring homomorphism induced by $i_R : \mathbb{Z} \to R$, $i_R(n) = n1_R$, 1_R = identity of R. If $\underline{\underline{A}}_{f,R}$ is a polynomial ideal in $R(G)$, then using polynomial maps and techniques of Sandling [16], it has been shown [7] that $i_R^{-1}(\underline{\underline{A}}_{f,R})$ depends only on $\underline{\underline{A}}_{f,\mathbb{Z}}$, $i_{\mathbb{Z}_{p^n}}^{-1}\left(\underline{\underline{A}}_{f,\mathbb{Z}_{p^n}}\right)$ and the behaviour of the elements $p1_R$, p prime, where $\mathbb{Z}_{p^n}$ denotes the ring of integers and mod p^n.

THEOREM 2.6 [7]. *If characteristic of* $R = n > 0$,

$$i_R^{-1}(\underline{\underline{A}}_{f,R}) = i_{\mathbb{Z}_n}^{-1}\left(\underline{\underline{A}}_{f,\mathbb{Z}_n}\right).$$

If characteristic of $R = 0$,

$$i_R^{-1}(\underline{\underline{A}}_{f,R}) = \sum_{p\in\sigma(R)} \tau_p\left(\mathbb{Z}(G) \bmod \underline{\underline{A}}_{f,\mathbb{Z}}\right) \cap i_{\mathbb{Z}_{p^e}}^{-1}\left(\underline{\underline{A}}_{f,\mathbb{Z}_{p^e}}\right)$$

where $\sigma(R)$ *is the set of primes* p *for which there exists* n *such that* $p^nR = p^{n+1}R$, p^e *is the smallest power of* p *for which this holds.*

Here $\tau_p\left(\mathbb{Z}(G) \bmod \underline{\underline{A}}_{f,\mathbb{Z}}\right)$ *stands for the* p*-torsion subgroup of* $\mathbb{Z}(G) \bmod \underline{\underline{A}}_{f,\mathbb{Z}}$ *and if* $\sigma(R)$ *is empty, then the right hand side of the above equation is to be interpreted as* $\underline{\underline{A}}_{f,\mathbb{Z}}$.

Theorem 2.6 when applied to the ideals $\Delta_R^n(G)$ and $\Delta_R^{(n)}(G)$ yields the dimension subgroups and Lie dimension subgroups over arbitrary rings of coefficients.

THEOREM 2.7 [7]. *If characteristic of* $R = 0$, *then*

$$(*) \qquad D_{n,R}(G) = \prod_{p\in\sigma(R)} \left\{\tau_p\left(G \bmod D_{n,Z}(G)\right) \cap D_{n,Z_{p^e}}(G)\right\} \quad for \quad n \geq 1$$

and

$$(**) \quad D_{(n),R}(G) = \prod_{p\in\sigma(R)} G_2 \cap \left\{\tau_p\left(G \bmod D_{(n),Z}(G)\right) \cap D_{(n),Z_{p^e}}(G)\right\} \quad for \quad n \geq 2 .$$

If $\sigma(R)$ *is empty, then the right hand side of* (*) *is to be interpreted as* $D_{n,Z}(G)$ *and that of* (**) *as* $D_{(n),Z}(G)$.

If characteristic of $R = r \neq 0$, *then*

$$D_{n,R}(G) = D_{n,Z_r}(G) \quad and \quad D_{(n),R}(G) = D_{(n),Z_r}(G) \quad for\ all \quad n \geq 1 .$$

3. Z-polynomial maps

Let M be a monoid and G an additive abelian group. The set $P_n(M, G)$ of all Z-polynomial maps $\theta : G \to M$ of degree $\leq n$ is an abelian group: for $f_1, f_2 \in P_n(M, G)$, $(f_1+f_2)(x) = f_1(x) + f_2(x)$, $x \in M$. It is easy to check that $P_n(M, G) \cong \hom_Z\left[Z(M)/\Delta_Z^{n+1}(M), G\right]$. Let $P_n(M) = Z(M)/\Delta_Z^{n+1}(M)$. Then $P_n(M) \cong Z \oplus \overline{P}_n(M)$, where $\overline{P}_n(M) = \Delta_Z(M)/\Delta_Z^{n+1}(M)$. It is thus clear that the investigation of Z-polynomial maps depends on the structure of the groups $\overline{P}_n(M)$. These groups have been computed for several cases in [8].

Let A be an abelian group written additively and B an arbitrary group written multiplicatively. The following characterization is due to Buckley.

THEOREM 3.1 [3]. *A map* $f : B \to A$ *is a Z-polynomial map of degree* $\leq r$ *if and only if for each* $b_1, b_2, \ldots, b_k$ *in* B *there exist integer-valued polynomials* $p_1, p_2, \ldots, p_t$ *all of degree* $\leq r$ *and* $a_1, a_2, \ldots, a_t$ *in* A *such that*

$$f\left(b_1^{m_1} b_2^{m_2} \ldots b_k^{m_k}\right) = \sum_{i=1}^{t} p_i\left(m_1, m_2, \ldots, m_k\right) a_i$$

for all integers $m_1, m_2, \ldots, m_k$.

[An *integer-valued polynomial* $p\left(x_1, x_2, \ldots, x_k\right)$ is a polynomial with rational coefficients which takes integral values when integers are substituted for x_i's.]

3.2 Integer-valued polynomials

Let F_n = the set of integer-valued polynomials $f(x)$ of degree $\leq n$. With the obvious addition F_n is an abelian group. Let integers ρ, μ both ≥ 1, be given and let

$$F_{n,\rho,\mu} = \{f(x) \in F_n : \mu \mid f(t+\rho)-f(t) \quad \forall t \in Z\} .$$

Then $F_{n,\rho,\mu}$ is a subgroup of F_n and $\mu F_n \leq F_{n,\rho,\mu}$. Let $H_{n,\rho,\mu} = F_{n,\rho,\mu}/\mu F_n$.

THEOREM 3.3 [8]. $H_{n,\rho,\mu} \cong P_n(Z_\rho, Z_\mu)$.

3.4 Wreath products

Let $W = A \text{ wr } B$ be the standard restricted wreath product of two groups A and B. Let $F = A^{(B)}$ be the base group. Then W is the split extension of F by B. The α-*central series* of W is defined to be the series $\{\alpha_n(W)\}$ where $\alpha_n(W) = \zeta_n(W) \cap F$, $\{\zeta_n(W)\}$ being the upper central series of W. α-central series have been investigated by Meldrum [5]. In the case when A is abelian and B is finite, α-central series can be described in terms of Z-polynomial maps:

THEOREM 3.5 [3]. $\alpha_{n+1}(W) = P_n(B, A)$.

Thus the problem of calculating the α-central series is equivalent to calculating the 'polynomial groups' $\overline{P}_n(B)$. It may be remarked that even in the case when B is a cyclic group of prime power order, the structure of $\overline{P}_n(B)$ is fairly complicated. Let us denote by $Z_r^{(s)}$ the direct sum of s copies of the cyclic group Z_r of order r.

THEOREM 3.6 [8]. *If n and m are integers ≥ 1 and p is a prime, then*

$$P_n\left(Z_{p^m}\right) \cong Z_{p^{m+q_1}}^{(r_1)} \oplus Z_{p^{m+q_1-1}}^{(p-1-r_1)} \oplus Z_{p^{m+q_2-1}}^{(r_2)} \oplus Z_{p^{m+q_2-2}}^{\left(p^2-p-r_2\right)} \oplus \cdots$$

$$\oplus\, Z_{p^{q_m+1}}^{(r_m)} \oplus Z_{p^{q_m}}^{\left(p^m-p^{m-1}-r_m\right)} \quad \textit{for } n > p^{m-1} - 1 ,$$

and

$$\overline{P}_n\left(\mathbb{Z}_{p^m}\right) \cong \mathbb{Z}_{p^{m+q_1}}^{(r_1)} \oplus \mathbb{Z}_{p^{m+q_1-1}}^{(p-1-r_1)} \oplus \mathbb{Z}_{p^{m+q_2-1}}^{(r_2)} \oplus \mathbb{Z}_{p^{m+q_2-2}}^{\left(p^2-p-r_2\right)} \oplus \cdots$$

$$\oplus \mathbb{Z}_{p^{m-(s-1)+q_{s-1}+1}}^{r_{s-1}} \oplus \mathbb{Z}_{p^{m-(s-1)+q_{s-1}}}^{\left(p^{s-1}-p^{s-2}-r_{s-1}\right)} \oplus \mathbb{Z}_{p^{m-s+1}}^{\left(n-p^{s-1}+1\right)}$$

provided $p^{s-1} < n \le p^s-1$, $1 \le s \le m-1$.

Here q_i and r_i , $i = 1, 2, \ldots, m$ are integers satisfying

$$n - p^{i-1} + 1 = \left(p^i-p^{i-1}\right)q_i + r_i , \quad 0 \le r_i < p^i - p^{i-1} .$$

3.7 R-algebraic maps

Let M and N be two R-modules, R a commutative ring with unity. A map $f : M \to N$ is called an *R-algebraic map* of degree 0 if f is constant and of degree $\le n+1$ if for every $a \in M$ and $r \in R$, the maps $D_a f : M \to N$ and ${}_rD^{n+1}f : M \to N$ given by $D_a f(m) = f(a+m) - f(m)$, $m \in M$ and ${}_rD^{n+1}f(m) = f(rm) - r^{n+1}f(m)$ are each R-algebraic of degree $\le n$. R-algebraic maps have been studied by Müller [6]. It turns out that $\mathbb{Z}$-algebraic maps are the same as $\mathbb{Z}$-polynomial maps. R-algebraic maps of degree ≤ 1 , $f : M \to N$, between two R-modules which satisfy $f(0) = 0$ are precisely the R-homomorphisms. On the other hand, every homomorphism from M to N can be seen to be an R-polynomial map of degree ≤ 1 . Thus, in general, R-algebraic maps are not the same as R-polynomial maps.

4. Polynomial functors

Let C denote the category of abelian groups. Then $P_n : C \to C$, $G \longmapsto P_n(G) = \mathbb{Z}(G)/\Delta_{\mathbb{Z}}^{n+1}(G)$ is a functor. A homomorphism $\alpha : G \to H$ induces a ring homomorphism $\alpha^* : \mathbb{Z}(G) \to \mathbb{Z}(H)$. $P_n(\alpha) : P_n(G) \to P_n(H)$ is the homomorphism induced by α^* . Closely related to P_n is the functor $Q_n : C \to C$, $Q_n(G) = \Delta_{\mathbb{Z}}^n(G)/\Delta_{\mathbb{Z}}^{n+1}(G)$ and for $\alpha : G \to H$ a homomorphism, $Q_n(\alpha) = P_n(\alpha)|_{Q_n(G)}$. We call P_n and Q_n *polynomial functors*.

THEOREM 4.1 [10]. *P_n and Q_n are functors of degree exactly n for all $n \ge 1$.*

The rest of this section is based on some unpublished work of Rees. I am thankful to D. Rees for his permission to include these results in this report.

Let $\underline{\underline{M}}$ denote the category of monoids. Then for each $n \geq 1$ we have a functor $P_n : \underline{\underline{M}} \to C$.

4.2 Natural endomorphisms of polynomial functors

Let $M, N \in \underline{\underline{M}}$ and $\theta : M \to \mathbb{Z}(N)$ be a map (or by extension an additive homomorphism of $\mathbb{Z}(M) \to \mathbb{Z}(N)$). We shall say that θ is *linear* if the extended map $\theta : \mathbb{Z}(M) \to \mathbb{Z}(N)$ satisfies $\theta\left[\Delta^r_{\mathbb{Z}}(M)\right] \subseteq \Delta^r_{\mathbb{Z}}(N)$ for all $r \geq 1$. The linear maps clearly form an additive subgroup of $\hom_{\mathbb{Z}}(\mathbb{Z}(M), \mathbb{Z}(N))$. Now the latter is a ring if we define a product $\theta \circ \phi$ of two maps $\theta, \phi : M \to \mathbb{Z}(N)$ by $\theta \circ \phi(x) = \theta(x) \circ \phi(x)$, the product on the right being the ring product in $\mathbb{Z}(N)$. It can be shown that

if θ *and* ϕ *are linear, then* $\theta \circ \phi$ *is also linear*

Hence the linear maps form a subring of $\hom_{\mathbb{Z}}(\mathbb{Z}(M), \mathbb{Z}(N))$ with the product induced by the product in $\mathbb{Z}(N)$. Let us take $M = N$. Then the identity map and the constant maps are clearly linear. Hence any map

$$T_f : M \to \mathbb{Z}(M) \ , \quad T_f(x) = f(x) \ , \quad f(X) \in \mathbb{Z}[X] \ ,$$

determines a linear map. It is clear that a linear map $\theta : M \to \mathbb{Z}(N)$ induces an additive homomorphism $P_n(M) \to P_n(N)$ for each $n \geq 1$. In particular,

$$T_f : x \mapsto f(x)$$

induces a map

$$T_f^{(n)} : P_n(M) \to P_n(M)$$

which is a *natural endomorphism* of the functor P_n : for every homomorphism $\alpha : M \to N$ the diagram

$$\begin{array}{ccc} P_n(M) & \xrightarrow{T_f^{(n)}} & P_n(M) \\ {\scriptstyle P_n(\alpha)}\big\downarrow & & \big\downarrow{\scriptstyle P_n(\alpha)} \\ P_n(N) & \xrightarrow{T_f^{(n)}} & P_n(N) \end{array}$$

is commutative. By considering the free cyclic monoid, it can be shown that the maps $T_f^{(n)}$ are, in fact, the only natural endomorphisms of the functor P_n . Thus the ring of natural endomorphisms of P_n is a homomorphic image of the ring generated by the maps $T_f : \mathbb{Z}(M) \to \mathbb{Z}(N)$. The kernel consists of those maps T_f which map $\mathbb{Z}(M)$ into $\Delta^{n+1}_{\mathbb{Z}}(M)$ for all M . Again, by considering the free cyclic monoid, it is easy to see that the kernel consists of the maps T_f with $f(X)$ divisible by $(X-1)^{n+1}$.

These considerations show that *the ring of natural endomorphisms of* P_n *is isomorphic to the ring* $Z[X]^*/\{(X-1)^{n+1}\}$ *where* $Z[X]^*$ *is a ring whose elements are polynomials in* X *with integer coefficients but the composition is defined by*

$$\left(\sum a_r X^r\right)\left(\sum b_s X^s\right) = \sum a_r b_s X^{rs} ,$$

and $\{(X-1)^{n+1}\}$ *denotes the ideal consisting of multiples of* $(X-1)^{n+1}$ *(under ordinary multiplication).*

4.3 Homogeneous polynomial maps

A Z-polynomial map $f : M \to G$, $M \in \underline{M}$, $G \in C$ of degree $\leq n$ is said to be *homogeneous of order* r if $f(x^m) = m^r f(x)$, $x \in M$, $1 \leq m \in Z$.

The knowledge of the ring of natural endomorphisms of P_n yields a decomposition of a Z-polynomial map of degree $\leq n$ with values in an abelian group in which division by $n!$ is uniquely defined.

THEOREM 4.4 (Rees). *Let* G *be an additive abelian group in which division by* $n!$ *is uniquely defined. Let* $f(x) : M \to G$ *be a* Z*-polynomial map on a monoid* M *of degree* $\leq n$. *Then there is a natural decomposition*

$$f(x) = f_0(x) + f_1(x) + \ldots + f_n(x)$$

where $f_m(x) : M \to G$ *is a homogeneous polynomial of degree* $\leq n$ *and order* m .

REMARK 4.5. Theorem 4.5 applies, for example, to polynomial maps into abelian p-groups having degree $\leq p-1$.

5. An extension problem

A problem which arises in connection with the dimension subgroups ([9] and [12]) is the following:

Let G be a finite p-group of (nilpotency) class n , $G = G_1 > G_2 > \ldots > G_n > G_{n+1} = (1)$ its lower central series. *Can every homomorphism* $\alpha : G_n \to T$ *(*T *= additive group of rationals* mod 1 *) be extended to a polynomial map of degree* $\leq n$? In view of Rips' counterexample [15] the answer is: *not always.*

Most of the known results on dimension subgroups can be deduced from the following result.

THEOREM 5.1 [13]. *Let* $1 \to N \overset{i}{\to} \Pi \overset{\beta}{\to} G \to 1$ *be an exact sequence of groups with* N *abelian and* i *the inclusion map. Then a homomorphism* $\alpha : N \to M$, M *an abelian group, can be extended to a map* $\phi : \Pi \to M$ *whose linear extension to* $Z(\Pi)$ *vanishes on* $\Delta_Z^{n+1}(\Pi) + \Delta_Z(\Pi)\Delta_Z(N)$ *if and only if*

(i) there exists a transversal $\{\omega(g)\}_{g\in G}$, $\omega(1) = 1$, *for* G *in* Π *and a map* $\lambda : G \to M$, $\chi(1) = $, *such that*

$$\alpha\big(W\big(g_1, (g_2-1)(g_3-1) \ldots (g_{n+1}-1)\big)\big) = \chi\big((g_1-1)(g_2-1) \ldots (g_{n+1}-1)\big)$$

$g_i \in G$, $i = 1, 2, \ldots, n+1$, *where*

$$W(g_1, g_2) = W(g_1g_2)^{-1}W(g_1)W(g_2) : G \times G \to N$$

is the 2-*cocycle determined by* w , N *is regarded as a right* G-*module via conjugation in* Π *and both* W *and* χ *have been extended by linearity to* $\mathbb{Z}(G) \times \mathbb{Z}(G)$ *and* $\mathbb{Z}(G)$ *respectively;*

(ii) α *vanishes on* $\big[\underbrace{[\ldots[[N, \Pi], \Pi], \ldots]}_{n \text{ terms}}, \Pi\big]$.

6. Polynomial 2-cocycles

Let M be a monoid and G an additive abelian group. So far we have discussed $\mathbb{Z}$-polynomial maps $f : M \to G$ for one variable only. If we want to extend this notion to $\mathbb{Z}$-polynomial maps $f : \underbrace{M \times M \times \ldots \times M}_{n \text{ terms}} \to G$ of several variables, then we can talk of degrees in individual variables as well as total degree:

Extend f by linearity to $f : \mathbb{Z}(M) \times \mathbb{Z}(M) \times \ldots \times \mathbb{Z}(M) \to G$. f is said to be of *degree* $\leq n$ *in the ith variable* if $f\big|_{\mathbb{Z}(M)\times\ldots\times\Delta_{\mathbb{Z}}^{n+1}(M)\times\ldots\times\mathbb{Z}(M)} = 0$, where $\Delta_{\mathbb{Z}}^{n+1}(M)$ is in the ith coordinate. f is said to be of *total degree* $\leq n$ if $f\big|_{\Delta_{\mathbb{Z}}^{i_1}(M)\times\Delta_{\mathbb{Z}}^{i_2}(M)\times\ldots\times\Delta_{\mathbb{Z}}^{i_n}(M)} = 0$ whenever $i_1 + i_2 + \ldots + i_n \leq n + 1$.

An interesting case is that of maps $f : M \times M \to G$ of two variables which are normalized 2-cocycles, that is, satisfy the conditions

(i) $f(x, 1) = f(1, x) = 0$, $x \in M$,

(ii) $f(y, z) - f(xy, z) + f(x, yz) - f(x, y) = 0$, $x, y, z \in M$.

The condition (ii) can be written as

$$f\big((x-1)(y-1), (z-1)\big) = f\big((x-1), (y-1)(z-1)\big)$$

which shows that f is of degree $\leq n$ in the first variable if and only if it is of degree $\leq n$ in the second variable.

DEFINITION 6.1. A map $f : M \times M \to G$ is called a *polynomial* 2-*cocycle of degree* $\leq n$ if

(i) f is a normalized 2-cocycle, and

(ii) f is a $\mathbb{Z}$-polynomial map of degree $\leq n$ in the first variable.

Let $Z_n^2(M, G)$ denote the group of polynomial 2-cocycles of degree $\leq n$. These subgroups lead to a filtration

$$(0) = P_0H^2(M, G) \leq P_1H^2(M, G) \leq \ldots \leq P_nH^2(M, G) \leq \ldots$$

of $H^2(M, G)$, the second cohomology group of M with coefficients in G:

$$P_nH^2(M, G) = \left\{\xi \in H^2(M, G) : \xi \text{ has a representative 2-cocycle } f \in Z_n^2(M, G)\right\}.$$

This filtration has been studied for groups by Passi and Stammbach [14] when the coefficients are in T (= rationals mod 1) and by Sharma [17] when the coefficients are in $\mathbb{Z}_p$.

6.2 $P_nH^2(G, T)$

Let G be a group. $\left\{P_nH^2(G, T)\right\}$ is a filtration of Schur multiplicator $H^2(G, T)$ of G. $P_nH^2(G, T)$ can be characterized as

$$\ker\left[\operatorname{ext}_G^2(\mathbb{Z}, T) \to \operatorname{ext}_G^2\left[\mathbb{Z}(G)/\Delta_{\mathbb{Z}}^{n+1}(G), T\right]\right] \quad [14].$$

In the case of abelian groups, $P_1H^2(G, T) = H^2(G, T)$, [11]. For finite p-groups of class 2, $p \neq 2$, $P_2H^2(G, T) = H^2(G, T)$, [9]. Rips' example provides a 2-group of class 2 for which $P_2H^2(G, T) \neq H^2(G, T)$. These results suggest the the problem of investigating nilpotent groups G for which $P_nH^2(G, T) = H^2(G, T)$, n = class of G.

It is interesting to note [14] that the residually nilpotent groups G for which

(i) $P_nH^2(G, T) = 0$, $\forall n \geq 0$, and

(ii) G/G' is free abelian

are precisely the parafree groups in the sense of Baumslag [1].

6.3 $P_nH^2(G, \mathbb{Z}_p)$

The main result of Sharma [17] in this case is that for a finite p-group G of M-class c,

$$P_n H^2(G, Z_p) = H^2(G, Z_p) \quad \text{for} \quad n \geq pc-1 .$$

M-series of a group, relative to a prime p, is defined as follows:

$$M_1(G) = G , \quad M_i(G) = [G, M_{i-1}(G)]M^p_{(i/p)}(G) \quad \text{for} \quad i \geq 2 ,$$

where (i/p) is the least integer $\geq i/p$ and $[G, M_{i-1}(G)]$ denotes the subgroup generated by all commutators $[x, y] = x^{-1}y^{-1}xy$, $x \in G$, $y \in M_{i-1}(G)$. G is said to be M-class c if $M_{c+1}(G) = (1)$ and $M_c(G) \neq (1)$.

References

[1] Gilbert Baumslag, "Groups with the same lower central sequence as a relatively free group. I. The groups", *Trans. Amer. Math. Soc.* 129 (1967), 308-321. MR36#248.

[2] Joseph Thaddeus Buckley, "Polynomial mappings on groups", (PhD thesis, Indiana University, 1964).

[3] Joseph T. Buckley, "Polynomial functions and wreath products", *Illinois J. Math.* 14 (1970), 274-282. MR41#3609.

[4] Samuel Eilenberg and Saunders Mac Lane, "On the groups $H(\Pi, n)$. II. Methods of computation", *Ann. of Math.* (2) 60 (1954), 49-139. MR16,391.

[5] J.D.P. Meldrum, "Central series in wreath products", *Proc. Cambridge Philos. Soc.* 63 (1967), 551-567. MR35#4309.

[6] H. Müller, "Algebraische Abbildungen", (Dissertation, Universitat Bielefeld, 1971).

[7] M.M. Parmenter, I.B.S. Passi and S.K. Sehgal, "Polynomial ideals in group rings", *Canad. J. Math.* 25 (1973), 1174-1182.

[8] I.B.S. Passi, "Polynomial maps on groups", *J. Algebra* 9 (1968), 121-151. MR38#241.

[9] I.B.S. Passi, "Dimension subgroups", *J. Algebra* 9 (1968), 152-182. MR38#242.

[10] I.B.S. Passi, "Polynomial functors", *Proc. Cambridge Philos. Soc.* 66 (1969), 505-512. MR44#2831.

[11] I.B.S. Passi, "Induced central extensions", *J. Algebra* 16 (1970), 27-39. MR42#1902.

[12] I.B.S. Passi, "Dimension subgroup problem", (mimeographed notes, Queen Mary College, London, 1970).

[13] I.B.S. Passi, "Polynomial maps on groups. II", submitted.

[14] I.B.S. Passi and U. Stammbach, "A filtration of Schur multiplicator", submitted.

[15] E. Rips, "On the fourth integer dimension subgroup", *Israel J. Math.* 12 (1972), 342-346.

[16] Robert Sandling, "Dimension subgroups over arbitrary coefficient rings", *J. Algebra* 21 (1972), 250-265. Zbl.233.20002.

[17] S. Sharma, "A bound for the degree of $H^2(G, Z_p)$ ", *Canad. J. Math.* (to appear).

[18] L.R. Vermani, "On polynomial groups", *Proc. Cambridge Philos. Soc.* 68 (1970), 285-289. MR42#1895.

Kurukshetra University,
Kurukshetra, India.

PROC. SECOND INTERNAT. CONF. THEORY OF GROUPS, CANBERRA, 1973, pp. 562-564.

20F05, 20F10, 20H10 (20F10, 30A58, 50C15)

ÜBER ERZEUGENDE EBENER DISKONTINUIERLICHER GRUPPEN

N. Peczynski, G. Rosenberger und H. Zieschang

Besitzt eine endlich erzeugte, diskrete Gruppe G von Automorphismen der oberen Halbebene einen kompakten Fundamentalbereich, so läßt sich G darstellen durch

$$G = \left\{s_1, \ldots, s_r, a_1, b_1, \ldots, a_g, b_g \mid s_1^{n_1} = \ldots = s_r^{n_r} = s_1 \ldots s_r \prod_{i=1}^{g} [a_i, b_i] = 1\right\}$$

mit $\mu(G) = 2g - 2 + \sum_{i=1}^{r} \left(1 - \frac{1}{n_i}\right) > 0$, $n_i \geq 2$, $g \geq 0$, $r \geq 0$, wobei

$[a_i, b_i] = a_i b_i a_i^{-1} b_i^{-1}$.

Fragen wir nach einem Algorithmus, der ein beliebiges minimales Erzeugendensystem von G in endlich vielen Schritten in ein gegebenes minimales Erzeugendensystem überführt, so fragen wir natürlich als erstes nach dem Rang von G , das heisst, nach der kleinsten Zahl von Erzeugenden von G . Burns, Karrass, Pietrowski und Purzitsky zeigten uns durch ein Beispiel, daß ein früherer Satz von Zieschang über den Rang von G nicht in allen Fällen richtig ist [1]. Unser erstes Ziel war also, den Rang von G zu bestimmen. Für $r \geq 4$ falls $g = 0$ und $r \geq 2$ falls $g = 1$ läßt sich G schreiben als $G = H_1 \,{}_A{*}\, H_2$, das heisst, als freies Produkt von Gruppen H_1 und H_2 mit Amalgam $A = H_1 \cap H_2$. Sei im Augenblick $r \geq 4$ falls $g = 0$ und $r \geq 2$ falls $g = 1$. Ist nun m der Rang von G und $\{u_1, \ldots, u_m\}$ ein beliebiges minimales Erzeugendensystem von G , so läßt sich das System $\{u_1, \ldots, u_m\}$ durch freie Übergänge (Nielsen-Transformationen) in ein System $\{x_1, \ldots, x_m\}$ überführen, für das einer der folgenden Fälle erfüllt ist:

(1) jedes x_i liegt in H_1 oder H_2 ;

(2) wenigstens ein x_i liegt im Amalgam, und das Amalgam wird in einem der Faktoren echt von seinem Normalisator übertroffen;

(3) einige der x_i liegen in einer zu H_1 oder H_2 konjugierten

Untergruppe von G und ein Produkt in ihnen ist zu einem Element aus dem Amalgam konjugiert, [vergleiche 1].

Für die Lösung des Rangproblems sind also folgende Sätze von Bedeutung:

SATZ 1. *Sei* $F = \left\{s_1, \ldots, s_r \mid s_1^{n_1} = \ldots = s_r^{n_r} = 1\right\}$, $n_i \geq 2$, $r \geq 2$.

Seien $x_1, \ldots, x_n \in F$, $n \leq r$, *und* X *die von* $x_1, \ldots, x_n$ *erzeugte Untergruppe von* F.

Gilt $y^{-1}\left(\prod_{i=1}^{r} s_i\right)^{\alpha} y \in X$ *für ein* $\alpha \neq 0$ *und* $y \in F$, *so tritt einer der folgenden Fälle ein:*

(a) *es ist* $n = r$;

(b) *es gibt einen freien Übergang von* $(x_1, \ldots, x_n)$ *zu einem System, in dem ein Element zu einer Potenz von* $\prod_{i=1}^{r} s_i$ *oder von einem* s_i *konjugiert ist;*

(c) *es ist* $n = r - 1$, *alle* $n_i = 2$ *und* r *ungerade, und es gibt einen freien Übergang von* $(x_1, \ldots, x_n)$ *zu dem System* $(s_1 s_2, s_1 s_3, \ldots, s_1 s_r)$.

SATZ 2. *Sei* $F = \left\{s_1, s_2 \mid s_1^{n_1} = s_2^{n_2} = 1\right\}$ *mit* $2 \leq n_1 \leq n_2$. *Seien* $x_1, x_2 \in F$ *mit* $[x_1, x_2] = y(s_1 s_2)^{\alpha} y^{-1}$ *für ein* $\alpha \neq 0$ *und* $y \in F$. *Dann tritt einer der folgenden Fälle ein:*

(a) *es ist* $n_1 = n_2 = 2$, *und es gibt einen freien Übergang von* (x_1, x_2) *zu* $\left[(s_1 s_2)^{\gamma_1} s_1, s_2 (s_1 s_2)^{\gamma_2}\right]$;

(b) *es ist* $n_1 = 2$, $n_2 = 3$, *und es gibt einen freien Übergang von* (x_1, x_2) *zu* $\left[s_1 s_2 s_1 s_2^2, s_2^2 s_1 s_2 s_1\right]$;

(c) *es ist* $n_1 = 2$, $n_2 = 4$, *und es gibt einen freien Übergang von* (x_1, x_2) *zu* $\left[s_1 s_2^2, s_2^{-1} s_1 s_2^{-1}\right]$;

(d) *es ist* $n_1 = 3$, $n_2 = 3$, *und es gibt einen freien Übergang von* (x_1, x_2) *zu* $\left[s_1 s_2^2, s_2^2 s_1\right]$.

Die Lösung des Rangproblems wird durch den folgenden Satz gegeben:

SATZ 3.

$$G = \left\{ s_1, \ldots, s_r, a_1, b_1, \ldots, a_g, b_g \mid s_1^{n_1} = \ldots = s_r^{n_r} = \prod_{j=1}^{r} s_j \prod_{i=1}^{g} [a_i, b_i] = 1 \right\}$$

mit $r \geq 3$ *für* $g = 0$ *sowie* $n_i \geq 2$ *hat den Rang*

$2g$ *falls* $g \geq 1$, $r = 0$;

$2g + r - 1$ *falls* $g \geq 1$, $r > 0$ *oder* $g = 0$ *und mindestens zwei der* $n_i \geq 3$ *oder* $g = 0$ *und alle* n_i *gerade oder* $g = 0$ *und* r *ungerade;*

$r - 2$ *falls* $g = 0$, r *gerade, alle* $n_i = 2$ *bis auf eines und* $\sum_{j=1}^{r} n_j$ *ungerade.*

Die hier gemachten Aussagen erlauben einige Anwendungen, zum Beispiel auch auf einige spezielle Untergruppen-probleme.

Wir werden uns nun weiter mit dem Problem beschäftigen, für die Gruppen

$$G = \left\{ s_1, \ldots, s_r, a_1, b_1, \ldots, a_g, b_g \mid s_1^{n_1} = \ldots = s_r^{n_r} = s_1 \ldots s_r \prod_{i=1}^{g} [a_i, b_i] = 1 \right\}$$

einen Algorithmus anzugeben, der ein beliebiges minimales Erzeugendensystem in endlich vielen Schritten in ein gegebenes minimales Erzeugendensystem überführt.

Literatur

[1] Heiner Zieschang, "Über die Nielsensche Kürzungsmethode in freien Produkten mit Amalgam", *Invent. Math.* 10 (1970), 4-37. Zbl.185,52.

Mathematisches Seminar der Universität Hamburg,
2 Hamburg 13, West Germany.

PROC. SECOND INTERNAT. CONF. THEORY OF GROUPS, 20F99
CANBERRA, 1973, pp. 565-577.

CONTRIBUTION TO THE THEORY OF ABSTRACT GROUPS

Sophie Piccard

It is well known that in the theory of abstract groups an important role is played by the free groups of which the quotient groups make up an inexhaustible resource for the abstract groups. There exist, however, some more vast classes of abstract groups which contain the free groups as very particular case; they are the classes of P-groups. Any abstract multiplicative group can, as we know, be given by a set $A = \{a_i\}$, $i \in I$, of generators and an exhaustive set F of defining relations which connect these generators. Any relation between the elements of A results from the relations F and the trivial relations between the elements of A . There exist some properties named P-properties which can be common to all relations, including the trivial relations connecting the elements of certain sets of generators of multiplicative groups. At present, we know about thirty such properties, to each of them corresponds a vast class of P-groups which yields to an elegant general theory. Here, we present some important classes of P-groups, especially the quasi free groups, the quasi free groups modulo n , the quasi free groups moduli N , the free groups modulo n , the free groups moduli N and the P-symmetric groups. Next, we define the P-products of groups, products presenting some analogies with the free product, and we indicate some properties of these products. Then, we introduce the fundamental and the quasi fundamental groups as well as their bases and we define the rank, the essential invariant, of a fundamental group. To conclude, we review some curious properties of two fundamental groups of rank 2 , the first is given by a couple of generators bound by an exhaustive set of two defining relations and the second is given by a couple of generators bound by a single defining relation.

1. Some remarkable classes of P-groups

We start with a multiplicative group G generated by a set $A = \{a_i\}$, $i \in I$, of generators. Let

$$(1) \qquad a_{i_1}^{j_1} a_{i_2}^{j_2} \dots a_{i_k}^{j_k}$$

be a finite product of integer powers exponents $j_1, \ldots, j_k$, of elements $a_{i_1}, \ldots, a_{i_k}$ of A, not necessarily all distinct. Any element a of G can be written (and in general in many ways) in the form (1). A relation between elements of A is an equality of the form

$$a_{i_1}^{j_1} a_{i_2}^{j_2} \ldots a_{i_m}^{j_m} = a_{k_1}^{l_1} a_{k_2}^{l_2} \ldots a_{k_n}^{l_n}, \tag{2}$$

where the finite products of powers of elements of A which appear in the two sides of the equation represent a same element of G. The degree of the relation (2) with respect to any element a_i of A is equal to the difference between the degree of the left side and the degree of the right side of (2) with respect to this element a_i.

We consider the following three methods of reducing a product of the form (1).

(a) Trivial reduction, devised by taking into account the sole axioms of multiplicative groups. It leads to the trivially reduced form of this product which is either 1 * or a product of the form

$$a_{u_1}^{v_1} \ldots a_{u_{k'}}^{v_{k'}}, \tag{3}$$

where $\left\{a_{u_1}, \ldots, a_{u_{k'}}\right\} \subset \left\{a_{i_1}, \ldots, a_{i_k}\right\}$, $a_{u_i} \neq a_{u_{i+1}}$, $i = 1, 2, \ldots, k'-1$, and $v_1, \ldots, v_{k'}$ are integers different from zero.

The relation (2) is said to be trivially reduced if each of its two sides has been trivially reduced; it is said to be trivial if its two sides have the same trivially reduced form. Otherwise, it is not trivial.

(b) Reduction modulo n. Let n be a fixed integer ≥ 2. We modulo n reduce (1) by, firstly, trivially reducing it to the form (3), secondly, reducing all the exponents $v_1, \ldots, v_{k'}$ modulo n, and then proceeding once again to a trivial reduction and so on. After a finite number of operations, we arrive finally at the modulo n reduced form of (1) which is either 1 or a product of the form (3), where $1 \leq v_i \leq n-1$, $i = 1, 2, \ldots, k'$. The relation (2) is said to be trivial modulo n if its two sides have the same modulo n reduced form.

(c) Reduction moduli N. Let $N = \{n_i\}$, $i \in I$, be a set of rational integers (not necessarily all distinct), each of which is ≥ 2. We moduli N reduce (1) by, firstly, reducing it trivially, then modulo n_i reducing the exponent

---

* 1 = the identity of the group G.

of any element a_i of A ; after this, we reduce trivially the obtained product and start again the chain of operations: reduction moduli N of the exponents, trivial reduction, and so on, to arrive finally at the moduli N reduced form of (1) which is 1 or a product of the form (3) where $1 \leq v_r \leq n_{u_r}-1$, $r = 1, 2, \ldots, k'$. The relation (2) is said being moduli N reduced if each of its two sides has been moduli N reduced. It is said to be trivial moduli N if each of its two sides has the same moduli N reduced form.

The group G is free if it possesses at least one set of free generators bound by only trivial relations.

It is said to be free modulo n if it possesses at least one set of generators - said to be free modulo n - bound uniquely by some trivial relations modulo n .

Finally, G is said to be free moduli N if it possesses at least one set of generators - said to be free moduli N - which are bound uniquely by some trivial relations moduli N .

Any free group is also free modulo n , for any integer $n \geq 2$ and it is equally free moduli N for any set N of rational integers ≥ 2 associated with the elements of a set of free generators of the free group.

On the other hand, the class of free groups modulo n is contained in the class of the free groups moduli N .

We are now going to define three new types of P-groups.

The group G is said to be quasi free if it possesses at least one set A of generators - said to be quasi free - bound uniquely by some relations, each of which is of zero degree with respect to every element of A . G is said to be quasi free modulo n , where n denotes a fixed integer ≥ 2 , if it possesses at least one set of generators - said to be quasi free modulo n - bound by some relations, each of which is of degree congruent to zero modulo n with respect to any element of A . Finally G is said to be quasi free moduli $N = \{n_i\}$, $i \in I$, $n_i \in Z$, $n_i \geq 2$, if G possesses at least one set $A = \{a_i\}$, $i \in I$, of generators - said to be quasi free moduli N - such that any relation between the elements of A is of degree congruent to zero modulo n_i with respect to a_i , for any $i \in I$.

The free groups, free modulo n , free moduli N , quasi free, quasi free modulo n and quasi free moduli N , are all P-groups. In general, a P-group is a multiplicative abstract group given by a set $A = \{a_i\}$, $i \in I$, of generators bound by some relations which have a common property P - called P-property. A P-property is expressed by a certain character called P-character, proper to some, but in general not to all of the elements of a P-group. Any element of a P-group which possesses this P-character is said to be a P-element. For example, a P-element of a quasi free group is a finite product of integer powers of elements belonging to a set of quasi free generators of the group being considered, of zero degree with respect to each of these elements. For all of the P-properties known at this day, the set of P-elements of a P-group G form a subgroup of G. This subgroup may be invariant or not. It is invariant regarding the six classes of P-groups defined above. These six classes of P-groups are the subject of elegant general theories. Volume 2 of the series II of the published papers of "Seminaire de Geometrie de l'Universite de Neuchatel" is dedicated to quasi free groups. Volume I of the same collection treats the free and the quasi free modulo n groups. Many contributions to the P-groups have been published in the specialized reviews.

Let us point out some interesting properties of quasi free groups. Let G be a quasi free group and let $A = \{a_i\}$, $i \in I$, be a set of quasi free generators of G. Any element a of G has a fixed degree with respect to every element a_i of A, this degree being equal to the degree with respect to a_i of any finite product of integer powers of elements belonging to A, which represents a. The group G is fundamental. Any set of quasi free generators of G is irreducible in the strict sense and makes up a basis of G. An element of G is said to be quasi free if it belongs to at least one set of quasi free generators of G. Any quasi free element of a quasi free group is of infinite order. The cardinal of the set of quasi free elements of a non cyclic quasi free group G is equal to the cardinal of the set of non quasi free elements of G. The group G has only two quasi free elements if it is cyclic, but the set of quasi free elements is infinite in each non cyclic quasi free group. An element of G is said to be a zero element [zero modulo n, where n is an integer ≥ 2] if it is of degree zero [congruent to zero modulo n] with respect to every element of A. An element of G which is a zero element in a basis A of G is also a zero element of G in any other set of quasi free generators of G. The same hold true for the zero elements modulo n of G. The set of zero elements [zero elements modulo n] of a quasi free group G is a normal subgroup of G which coincides with the derived group G' of G. Any quasi free group G generated by a set of cardinal $\mathfrak{m}$ of quasi free generators possesses at least two distinct normal subgroups. Any subgroup of a quasi free group is not quasi free and a subgroup of a quasi free group generated by an infinite set of elements may be non fundamental. Any quasi free group G possesses an infinite set of quasi free subgroups and, if G is finitely generated, any quasi free subgroup of G is equally finitely generated. Any quasi free group G generated by a set of cardinal $\mathfrak{m}$ of quasi free elements has at least 2 distinct bases and a set of cardinal $> \mathfrak{m}$ of quasi free subgroups. For any cardinal number $\mathfrak{m}$, there exist quasi free groups whose bases are of cardinal $\mathfrak{m}$ and of which the set of outer automorphisms is of cardinal superior to $\mathfrak{m}$. The quasi free groups have many common properties with the free groups which constitute a particular case of quasi free groups.

The P-symmetry. We are now going to consider a new P-property named P-symmetry, of a totally different nature from those considered above. Let G be a multiplicative group generated by a set $A = \{a_i\}$, $i \in I$, of generators bound by a given family of defining relations and let S be the group of transformations, of the set A , formed by all the permutations of any finite number of any elements belonging to A . We say that the group G is P-symmetric, if for any relation $f\left(a_{i_1}, \ldots, a_{i_k}\right) = g\left(a_{i_1}, \ldots, a_{i_k}\right)$ relating between them the elements $a_{i_1}, \ldots, a_{i_k}$ of A and for any permutation of the group S , which maps the element a_{i_r} to the element a_{j_r} of A , $r = 1, 2, \ldots, k$, we have also the relation $f\left(a_{j_1}, \ldots, a_{j_k}\right) = g\left(a_{j_1}, \ldots, a_{j_k}\right)$, f and g being determined finite products of integer powers of the elements of A inside the parenthesis. If the group G is P-symmetric, A is said to be a set of P-symmetric generators of G . Any element a of a P-symmetric group G is represented by a product of the form (1) of integer powers of elements belonging to a set A of P-symmetric generators of G . The element a is said to be P-symmetric if any finite product of integer powers of elements of A , obtained by applying on (1) a permutation of the group S , represents equally a . The set of the P-symmetric elements of a P-symmetric group G is a subgroup of G , but this subgroup is in general not normal. The class of the P-symmetric groups is very rich, it contains numerous finite order groups, of which the symmetric group S_n and the alternative group A_n of any degree $n \geq 3$, the free groups, some abelian groups, and so on. A P-symmetric element of a P-symmetric group may be central and the set of the symmetric elements may in certain cases, coincide with the center of a P-symmetric group. It can also happen that 1 is the sole central element of G belonging to the subgroup of the P-symmetric elements of G . Such is, for instance, the case of the P-symmetric group of the permutations of the elements of the set $\{1, 2, 3, 4, 5, 6\}$ generated by the two elements $a = (1, 2)(4, 5, 6)$, $b = (1, 2, 3)(4, 5)$, group of order 36 of which the group of P-symmetric elements is formed by the six elements 1, (1, 2)(4, 5), (1, 3)(4, 6), (2, 3)(5, 6), (1, 2, 3)(4, 5, 6) and (1, 3, 2)(4, 6, 5) , and it is not a normal subgroup of the P-symmetric group being considered. Among the symmetric basis of the groups S_6 , we point out the couples of substitutions

(1, 2, 3, 4), (1, 3, 5, 6); (1, 2, 3)(4, 5), (1, 6)(2, 4, 3);
(1, 2, 3, 4, 5, 6), (1, 2, 6, 5, 4, 3); (1, 2, 3, 4, 5, 6), (1, 3, 2, 6, 5, 4);
(1, 2, 3, 4, 5, 6), (1, 4, 3, 2, 6, 5) .

The study of the class of symmetric P-groups is interesting apart. We indicate that the discovery of this class of P-groups has played a major role in the general theory of the P-groups and enabled us to resolve several difficult problems of this theory.

By an inventive procedure of relativisation which will be mentioned elsewhere, we obtain new P-properties from the P-properties already defined.

2. The P-products of groups

All the P-properties known at the present time are of such nature that if G is a multiplicative group and if A is a set of generators of G bound by some relations, of which some are P-relations, that is, they have a given P-property, then the product of two P-relations is always a P-relation and the inverse of every P-relation is also a P-relation, so that the set of all P-relations relating among them the elements of A forms a group with the usual composition law of the relations. We shall implicitely assume in what follows that all the P-properties being considered have this property. We distinguish the strong P-properties from the weak P-properties. A P-property is said to be *strong* if, for any set A of generators of a multiplicative group G and for any proper part A_1 of A, every P-relation between the elements of A_1 is also a P-relation between the elements of A. If it is not such a case, P is called a *weak* P-property. Each of the six first P-properties defined above is a strong P-property. On the contrary, the P-symmetry is an example of weak P-property. When we are talking about the P-products of groups, we shall only consider the strong P-properties.

A P-group G is a P-product of its subgroups G_i, $i \in I$, if there exist for any i of I, a set A_i of generators of G_i, bound by an exhaustive family F_i of defining relations, such that $A = \bigcup_{i \in I} A_i$ is a set of generators of G bound by a family F of defining relations such that F is the union of the set $\bigcup_{i \in I} F_i$ and a set F_0 of P-relations between elements of A, in each of which the elements of at least two sets A_i are presented in a non trivial way.

More generally, let G_i, $i \in I$, be a set of multiplicative groups; let, for any $i \in I$, A_i be a set of generators of G_i bound by the exhaustive family F_i of defining relations; let $A = \bigcup_{i \in I} A_i$ and let F_0 be a given set of defining P-relations between elements of A, in each of which the elements of at least two sets A_i are presented in an non trivial way, the set $F = \left(\bigcup_{i \in I} F_i\right) \cup F_0$ forming an exhaustive set of defining relations of A. The abstract group G given by the set A of generators and the family F of defining relations which bind the generators is called a P-product of the groups G_i. We obtain the group G by forming the free product $\prod^*_{i \in I} G_i$ of the groups G_i and by identifying in this product all equal elements by virtue of the relations between elements of A resulting from the defining relations F and the trivial relations between elements of A. A P-product of groups presents some analogies with the free product of multiplicative groups. It is liable to extension and to reduction. A P-product of groups G_i reduces to their free product when the set F_0 is empty.

Now, let G_i, $i \in I$, be a set of P-groups having a sole and same strong P-property and let, for any $i \in I$, A_i be a set of P-generators of the group G_i, bound by an exhaustive set F_i of defining P-relations. Any P-product of the groups G_i defined after the sets A_i is a P-group. This result makes the notion of P-product in the theory of P-groups very interesting.

The free product is a particular case of P-products.

3. The fundamental and quasi fundamental groups

Let $A = \{a_i, i \in I\}$, be a set of generators of a multiplicative group G. The set A is called irreducible in the large sense if, for any $i \in I$, the element a_i is not a finite product of integer powers of elements of the set $A - \{a_i\}$. It is said to be irreducible in the strict sense if, for any finite part $A^* = \{a_1, \ldots, a_k\}$ of A and for any finite subset $B = \{b_1, \ldots, b_l\}$ of G, such that $l \leq k$, $(A-A^*) \cup B$ is not a set of generators of G. Any set A of generators of a multiplicative group being irreducible in the strict sense is also irreducible in the large sense, however the converse is not true.

A multiplicative group G is said to be *fundamental* if it possesses at least one set of generators being irreducible in the strict sense. It is said to be *quasi fundamental* if it possesses at least one set of generators being irreducible in the large sense.

Any finite group, any finitely generated group, any free group and any quasi free group is fundamental. The group $S[A]$ of all the [even] permutations of any finite number of any rational integers is quasi fundamental, but it is not fundamental.

We call a *basis* of a fundamental [quasi fundamental] group any set of generators of this group which is irreducible in the strict sense [in the large sense].

Any set of free [quasi free] generators form a basis of a free group [of a quasi free group]. The set of transpositions $(0, n)$, $n \in Z-\{0\}$, makes up a basis of the quasi fundamental group S and the set of the cycles $(0, 1, n)$, $n \in Z-\{0, 1\}$, is a basis of the group A .

A group which is neither fundamental nor quasi fundamental is deprived of basis.

Any two bases of a fundamental group are two sets of the same cardinal. This cardinal is called the *rank* of the fundamental group.

The cardinal $\mathfrak{m}$ of any basis of a quasi fundamental group is $\geq \aleph_0$. This cardinal is an invariant of the quasi fundamental group. This invariant is also the cardinal of the set of all elements of the group considered. A fundamental group can have sets of generators irreducible in the large sense, which cardinal is superior to the rank of the group. A subgroup of a fundamental group may not be fundamental.

4. A study of two fundamental abstract groups of rank two

The group $*G$. This fundamental abstract group of rank 2 is generated by two elements $*t$ and $*t'$ bound by the exhaustive set of two defining relations

(i) $*t'^2 *t = *t *t'^2$ and

(ii) $*t *t'^{-1} = *t' *t^{-1}$.

This group possesses three maximal normal subgroups of index 2 , of these three, one is free abelian of rank 2 , the second is non abelian, non free, fundamental of rank 3 , the third is non free, non abelian, fundamental of rank 2 . For any odd prime integer $p \geq 3$, $*G$ possesses a maximal invariant subgroup of index p . We have determined all the basis of the group $*G$ as well as an irreducible exhaustive set of defining relations for each of these bases. Some of these bases are bound by an irreducible set of two defining relations while the others are bound by an irreducible set of three defining relations; this proves that the cardinal of an irreducible set of defining relations of a basis of a fundamental group is not an invariant of such a group. The center of $*G$ is the cyclic group generated by $*t'^2$, the commutator subgroup of $*G$ is the cyclic group generated by the element $*t^2 *t'^{-2}$. $*G$ possesses a set of the potency of the continuum of normal chains of subgroups in the form $G_0, G_1, G_2, \ldots$, where $G_0 = *G$ and G_i is an invariant and proper maximal subgroup of G_{i-1} for any $i = 1, 2, \ldots$. We have determined all the automorphisms of the group $*G$ which has a countable set of inner automorphisms as well as a countable set of outer automorphisms.

The group G . This fundamental abstract group of rank 2 is generated by two elements t and t' , bound by the sole defining relation

(iii) $tt' = t't^{-1}$.

This non cyclic, non abelian, countable group has an infinite set of basis which have all been discovered by now. We have also determined an irreducible exhaustive set of defining relations for each of these basis. A part of them are characterized by a single, the others by a irreducible set of two defining relations. A countable set of proper subgroups of G' are isomorphic to G ; some of these subgroups are normal, the others are not. The group G , like the group $*G$, has a countable set of inner automorphisms and the same quantity of outer automorphisms. Besides, G possesses a set of cardinal $2^{\aleph_0}$ of normal chains of subgroups in the form $G_0, G_1, G_2, \ldots$, where $G_0 = G$ and G_i is a proper normal subgroup of G , isomorphic to G , and is, on the other hand, a proper maximal normal subgroup of G_{i-1} , for any $i = 1, 2, \ldots$.

The group G_{s2} of all periodic transformations of period 2 of the set Z of rational integers is isomorphic to the group $*G$, while one of the maximal invariant subgroups of index 2 of G_{s2} is isomorphic to the group G .

Each of the two groups $*G$ and G is quasi free modulo 2 and metabelian. For each fixed integer $n \geq 2$, the group G_{sn} of all periodic transformations of period n of the set $\mathbb{Z}$ is fundamental, of rank 2 .

References

[1] А.Г. Курош [A.G. Kuroš], Теория групп [*Theory of groups*], Издание третье, дополненное. Издательство "Наука" Главная Редакция Физико-Математической Литературы, Москва, 1967; 3rd augmented edition. Izdat "Nauka", Moscow, 1967. MR40#2740.

[2] Sophie Piccard, *Sur les bases du groupe symétrique et les couples de substitutions qui engendrent un groupe régulier* (Mém. Univ. Neuchâtel, 19. Libraire Vuibert, Paris, 1946, 223 pp.). MR8,13.

[3] Sophie Piccard, *Sur les bases du groupe symétrique II* (Libraire Vuibert, Paris, 1948, 119 pp.). MR10,281.

[4] Sophie Piccard, *Sur les bases des groupes d'ordre fini. Avec une préface de Armaud Denjoy* (Mémoires de l'Université de Neuchâtel, 25. Sécretariat de l'Université, Neuchâtel, 1957. Gauthier-Villars, Paris, 1957, 242 pp.). MR20#902.

[5] Sophie Piccard, *Les groupes que peuvent engendrer trois éléments* a, b, c *générateurs d'un groupe multiplicatif, qui satisfont le système non exhaustif de relations fondamentales* $a^2 = 1$, $b^2 = 1$, $c^2 = 1$, $(ab)^3 = 1$, $(ac)^3 = 1$, $(bc)^3 = 1$ (Publ. Sém. Géom. Univ. Neuchâtel (1) 1, 1958, 42 pp.). MR21#683.

[6] Sophie Piccard, "Les groupes quasi libres", *C.R. Acad. Sci. Paris* 250 (1960), 3260-3262. MR22#5666.

[7] Sophie Piccard, "Les elements libres des groupes libres", *C.R. Acad. Sci. Paris* 251 (1960), 1328-1330. MR22#8050.

[8] Sophie Piccard, "Les groupes quasi libres", *C.R. Acad. Sci. Paris* 251 (1960), 2271-2273. MR22#8051.

[9] Sophie Piccard, "Les groupes fondamentaux et leur décomposition en produit quasi libre", *C.R. Acad. Sci. Paris* 251 (1960), 2450-2452. MR22#11025.

[10] Sophie Piccard, *Les groupes fondamentaux et leur décomposition en produit quasi libre* (Publ. Sém. Géom. Univ. Neuchâtel 1, no. 2, 1960, 20 pp.).MR24#A2624.

[11] Sophie Piccard, "Sur les elements libres des groups libres", *Enseignement Math.* (2) 6 (1960), 158-159.

[12] Sophie Piccard, "Théorie des groupes. Systemes irréductibles d'éléments d'un groupe. Les groupes fondamentaux, leurs bases et leurs éléments fondamentaux", *Verh. der Schweizerischen Naturforsch. Gesell.* (141. Versammlung in Biel, 1961): Wiss. Teil pp. 75-76. (Kommissionsverlag, Wabern-Bern, 1961.) Zbl.122,269.

[13] Sophie Piccard, "Les éléments quasi libres des groupes quasi libres", *C.R. Acad. Sci. Paris* 258 (1964), 1968-1970. MR29#1245.

[14] Sophie Piccard, "Les groupes pseudo-libres et les groupes fondamentaux", *C.R. Acad. Sci. Paris* 259 (1964), 24-26. MR29#5892.

[15] Sophie Piccard, "Dépendance et indépendance linéaire modulo n de vecteurs à composantes entièrs d'un espace vectoriel de dimension quelconque", *Enseignement Math.* (2) 10 (1964), 307-311.

[16] Sophie Piccard, "Les groupes libres modulo n", *C.R. Acad. Sci. Paris* 261 (1965), 2794-2797. MR33#175.

[17] Sophie Piccard, "Les groupes libres modulo n", *C.R. Acad. Sci. Paris* 261 (1965), 3016-3018. MR34#7619.

[18] Sophie Piccard, "Les groupes libres modulo n", *C.R. Acad. Sci. Paris* 261 (1965), 4303-4306. MR34#4342.

[19] Sophie Piccard, *Les groupes libres et les groupes quasi libres modulo n. Les P-produits et les P-groupes.* Publ. Sém. Géom. Univ. Neuchâtel (2) no. 1 (Librairie Gauthier-Villars, Paris, 1966, 215 pp.). Zbl.158,272.

[20] Sophie Piccard, "Les automorphismes et les endomorphismes modulo n d'un groupe libre modulo n", *C.R. Acad. Sci. Paris Ser. A-B* 263 (1966), A381-A384. MR34#7620.

[21] Sophie Piccard, "Quelques applications de la théorie des groupes quasi libres, des groupes quasi libres modulo *n* et des groupes libres modulo *n* à celle des groupes libres", *C.R. Acad. Sci. Paris Ser. A-B* 263 (1966), A403-A406. MR34#7621a.

[22] Sophie Piccard, "Quelques applications de la théorie des groupes quasi libres et libres modulo *n* à celle des groupes libres", *C.R. Acad. Sci. Paris Ser. A-B* 263 (1966), A440-A443. MR34#7621b.

[23] Sophie Piccard, "Dépendance et indépendance linéaire modulo *k* de vecteurs", *Bul. Inst. Politehn. Iasi (NS)* 12 (16) (1966), 39-45. MR35#4238.

[24] Sophie Piccard, "Quelques problèmes généraux de la théorie des groupes et les groupes libres modulo n ", *Proc. Internat. Conf. Theory of Groups* (Canberra, 1965), pp. 265-277. (Gordon and Breach, New York, London, Paris, 1967.) Zbl.157,349.

[25] Sophie Piccard, "Les groupes libres et quasi libres modulo *n* ", *Verh. der Schweizerischen Naturforsch. Gesell.* (146. Versammlung in Solothurn, 1966): Wiss. Teil pp. 114-117. (Kommissionsverlag, Berichthaus, Zürich, 1966.) Zbl.247.20033.

[26] Sophie Piccard, "Solutions de quelques problemes généraux de la théorie des groupes", *Verh. der Schweizerischen Naturforsch. Gesell.* (147. Jahresversammlung in Schaffhausen, 1967): Wiss. Teil pp. 98-100. (Kommissionsverlag Berichthaus, Zürich, 1967.) MR43#7283.

[27] Sophie Piccard, "Le groupe des relations reliant entre eux les éléments d'un ensemble de générateurs d'un groupe multiplicatif", *C.R. Acad. Sci. Paris Ser. A-B* 265 (1967), A504-A507. MR36#5205.

[28] Sophie Piccard, "Quelques propriétés générales des *P*-groupes au sens strict complets", *C.R. Acad. Sci. Paris Ser. A-B* 265 (1967), A553-A555. MR36#5206.

[29] Sophie Piccard, "Les *P*-propriétés et les *P*-groupes", *C.R. Acad. Sci. Paris Ser. A-B* 266 (1968), A705-A707. MR38#222a.

[30] Sophie Piccard, "Les *P*-propriétés et les *P*-groupes", *C.R. Acad. Sci. Paris Ser. A-B* 266 (1968), A760-A761. MR38#222b.

[31] Sophie Piccard, "Un théorème d'existence", *Publ. Sém. Géom. Univ. Neuchâtel* (1) 4 (1967), 1-6. MR38#1152.

[32] Sophie Piccard, "Sur les ensembles de générateurs de groupes multiplicatifs et les relations qui les lient", *Publ. Sém. Géom. Univ. Neuchâtel* (1) 4 (1967), 7-21. MR38#1153.

[33] S. Piccard, "Sur les groupes multiplicatifs définis par un ensemble irréductible de générateurs liés par un ensemble donne de relations fondamentales", *VII Österreichischer Mathematikerkongress*, 1968, 21-22 (Nachrichten der Österreichischen Mathematischen Gesellschaft, 1970).

[34] Sophie Piccard, "Relations entre générateurs d'un groupe multiplicatif", *Mém. Soc. Sci. Phys. Nat. Bordeaux*, Vol. *Special: Congress AFAS*, (1968), 25-27 (1969).

[35] Sophie Piccard, "Trois problèmes de la théorie générale des groupes", *Verh. der Schweizerischen Naturforsch. Gesell.* (148. Versammlung, Einsiedeln, 1968): Wiss. Teil pp. 82-85. (Kommissionsverlag Berichthaus, Zürich, 1968.) Zbl.247.20036.

[36] Sophie Piccard, "Groupes de relations", *Atti dell'VIII Congresso dell'Unione Matematica Italiana, Trieste*, 1969, 253-254.

[37] Sophie Piccard, "Quelques classes de *P*-groupes", *Atti dell'VIII Congresso dell'Unione Matematica Italiana, Trieste*, 1969, 254-255.

[38] Sophie Piccard, "Les groupes de transformations périodiques des entiers rationnels", *Publ. Sém. Géom. Univ. Neuchâtel* (1) 5 (1969), 10-55. MR41#6951.

[39] Sophie Piccard, "Quelques résultats de la théorie des groupes", *Verh. der Schweizerischen Naturforsch. Gesell.* (149. Versammlung, St. Gallen, 1969): Wiss. Teil pp. 87-89. (Kommissionsverlag Berichthaus, Zürich, 1969.) MR42#7765.

[40] Sophie Piccard, "Les groupes libres et les groupes quasi libres modulis N ", *J. Reine Angew. Math.* 242 (1970), 170-178. MR41#5474.

[41] Sophie Piccard, "Le groupe G_{s2} des transformations périodiques de période 2 des entiers rationnels", *C.R. Acad. Sci. Paris Ser. A-B* 270 (1970), A1222-A1225. MR44#6805.

[42] Sophie Piccard, "Les *P*-groupes", *Actes Congres Internat. des Math., Nice*, 1970. *Les 265 communications individuelles*, 8, pp. 24. (Gauthier-Villars, Paris, 1971.)

[43] Sophie Piccard, "Les groupes de transformations périodiques des entiers rationnels", *Universitatea "Alexandrou Ioan Cuza", Iaşi, Academia R.S.R., Sesuinea Jubilara A. Myller*, 20-25 Août, 1970. *Resumate Iaşi* (1970), p. 3.

[44] Sophie Piccard, "Les groupes de transformations périodiques des entiers rationnels", *C.R. Acad. Sci. Paris Ser. A-B* 271 (1970), A976-A979. MR44#6806.

[45] Sophie Piccard, "Structure du groupe multiplicatif engendré par deux éléments t, t' lies par la seule relation caractéristique $tt' = t't^{-1}$ ", *Publ. Sém. Géom. Univ. Neuchâtel* (1) 6 (1971), 1-40.

[46] Sophie Piccard, "Les groupes de transformations périodiques des entiers rationnels", *Sunti delle communicazioni, Congresso dell'Unione Matematica Italiana, Bari*, 23 Septembre - 3 Octobre, 1971, 8-9 (1972).

[47] Sophie Piccard, "Étude d'un groupe abstrait donné par un couple de générateurs liés par une relation caractéristique unique", *C.R. Acad. Sci. Paris Ser. A-B* 273 (1971), A565-A567. MR44#6801.

[48] Sophie Piccard, "Quelques questions choisies de la Théorie des groupes", *Verh. der Schweizerischen Naturforsch. Gesell.* (151. Versammlung in Freiburg, 1971): Wiss. Teil pp. 174-176. (Kommissionsverlag Berichthaus, Zürich, 1971.) Zbl.247.20034.

[49] Sophie Piccard, "Les groupes de transformations périodiques des entiers rationnels, II", *Publ. Sém. Géom. Univ. Neuchâtel* (1) 7 (1972), 1-40.

[50] Sophie Piccard, "Les bases et les éléments libres des groupes libres", *Sciences, Paris II*, 4 (1972), 203-206.

Université de Neuchâtel,
2002 Neuchâtel, Switzerland.

PROC. SECOND INTERNAT. CONF. THEORY OF GROUPS, CANBERRA, 1973, pp. 578-579. 20B99

SYLOW SUBGROUPS OF FINITE PERMUTATION GROUPS

Cheryl E. Praeger

If G is a transitive group of permutations on a set Ω of n points, and if P is a Sylow p-subgroup of G for some prime p dividing $|G|$, then our object is to obtain information about the structure of P as a permutation group on Ω. Questions like the following arise naturally.

How many fixed points does P have in Ω, that is, how large is $|\mathrm{fix}_\Omega P|$?

How many orbits does P have and what are their lengths?

If we restrict the lengths of the orbits of P, can we bound $|P|$ or $|\mathrm{fix}_\Omega P|$?

The only result I know about transitive groups in general is

THEOREM 1 [2]. *If G is a transitive permutation group on Ω, and if P is a Sylow p-subgroup of G for some prime p dividing $|G|$, then $|\mathrm{fix}_\Omega P| < \frac{1}{2}|\Omega|$.*

This is the best result possible for transitive groups, but if G is primitive and does not contain the alternating group, better results should be possible.

Jordan (c. 1873), Manning (1910 and later), and Weiss (1928) found strong bounds for the degree of a primitive non-alternating group G which contains an element of order p with a small number of cycles of length p (see [5], 13.10). Manning also showed that under these conditions, p^2 did not divide $|G|$, and so his results gave information about the Sylow p-subgroups of G. Instead of looking at p-elements of G, we shall consider the whole Sylow p-subgroup P in the case where P has no orbit of length greater than p.

THEOREM 2 [3]. *Suppose that G is a 2-transitive permutation group on a set Ω of degree n, such that G does not contain A_n, and let P be a Sylow p-subgroup of G for some prime p dividing $|G|$. Suppose further that P has no orbit of length greater than p. Then one of the following is true:*

(a) $|P| = p$,

(b) $|P| = 4$, *and* G *is* PSL(2, 5) *permuting the six points of the projective line,*

(c) $|P| = 9$, *and* G *is the Mathieu group* M_{11} *in its 3-transitive representation of degree* 12 .

This means that, in general, if p^2 divides $|G|$, then the Sylow p-subgroup P has an orbit of length at least p^2 . In the case where G is primitive but not 2-transitive and satisfies the conditions of the theorem, I believe, but cannot prove, that p^2 does not divide $|G|$. This result has been proved for special cases by Scott ([4], if $|\text{fix } P| \leq 1$), and O'Nan (1973, in unpublished work; if G is 2-closed, or if G contains a p-element with at most $p - 1$ cycles).

To prove the theorem we must consider three cases, namely, when P fixes no point, 1 point, or at least 2 points. The case in which P fixes only one point was dealt with by Appel and Parker [1], and the other two cases are considered separately in [3].

References

[1] K.I. Appel and E.T. Parker, "On unsolvable groups of degree $p = 4q + 1$, p and q primes", *Canad. J. Math.* 19 (1967), 583-589. MR36#256.

[2] Cheryl E. Praeger, "Sylow subgroups of transitive permutation groups", *Math. Z.* (to appear).

[3] Cheryl E. Praeger, "On the Sylow subgroups of a doubly transitive permutation group", (unpublished, 1972).

[4] Leonard L. Scott, "A double transitivity criterion", *Math. Z.* 115 (1970), 7-8. MR42#1886.

[5] Helmut Wielandt, *Finite permutation groups* (Academic Press, New York and London, 1964). MR32#1252.

Institute of Advanced Studies,
Australian National University, Canberra 2600.

PROC. SECOND INTERNAT. CONF. THEORY OF GROUPS, 20F05
CANBERRA, 1973, pp. 580-588.

ON THE NIELSEN EQUIVALENCE OF PAIRS OF GENERATORS IN CERTAIN HNN GROUPS

Stephen J. Pride

SUMMARY

This paper is concerned with obtaining information about the Nielsen equivalence classes and T-systems of certain two-generator HNN groups. The principal result (Theorem 1) states that the one-relator groups

$$\left\langle a,\ t;\ \left(a^{\alpha_1}t^{-1}a^{\varepsilon\beta_1}t \ldots a^{\alpha_r}t^{-1}a^{\varepsilon\beta_r}t\right)^n\right\rangle \qquad (r > 0)$$

where α_i, β_i $(i = 1, 2, \ldots, r)$ are positive integers, $|\varepsilon| = 1$, $n > 1$, have one Nielsen equivalence class of generating pairs. As a corollary of this result a counterexample to the converse of Corollary 4.13.1 of Magnus, Karrass and Solitar, "Combinatorial Group Theory" is obtained. The other main result of the paper (Theorem 2) gives a fairly detailed description of the Nielsen equivalence classes and T-systems of some HNN extensions of certain small cancellation groups.

The main idea behind the proofs of Theorems 1 and 2 can be explained in general terms as follows. Let G be an HNN group of the form

$$\left\langle a,\ b,\ t;\ R_1(a, b),\ R_2(a, b),\ \ldots,\ t^{-1}at = b^{\xi}\right\rangle$$

and suppose $\mathrm{sgp}(a)$ and $\mathrm{sgp}(b)$ are malnormal in the base group $H = \langle a, b; R_1, R_2, \ldots\rangle$. Then any generating pair of G is Nielsen equivalent to a pair of the form (th, a^{μ}) (see Theorem A), and h must be such that $(a, h^{-1}bh)$ is a generating pair for H (see Lemma 1). In certain situations (and in particular for the situations of Theorems 1 and 2) a and $h^{-1}bh$ generate H if and only if h is expressible in the form $b^{\beta}a^{\alpha}$, so that any generating pair of G is Nielsen equivalent to a pair of the form $(a^{\gamma}ta^{\alpha}, a^{\mu})$.

* * * * *

Let F_2 denote the free group freely generated by x_1, x_2, and let G be a two-generator group. Suppose that $\mathfrak{g} = (g_1, g_2)$ and $\mathfrak{g}' = (g_1', g_2')$ are generating pairs for G. Then $\mathfrak{g}$ and $\mathfrak{g}'$ are said to be *Nielsen equivalent* if there is an automorphism

$$x_i \mapsto Y_i(x_1, x_2) , \quad i = 1, 2 ,$$

of F_2 such that $g_i' = Y_i(g_1, g_2)$ for $i = 1, 2$. The generating pairs $\mathfrak{g}$ and $\mathfrak{g}'$ are said to lie in the same *T-system* [8] if there is an automorphism φ of G such that $\mathfrak{g}'$ is Nielsen equivalent to $\varphi\mathfrak{g} = (\varphi(g_1), \varphi(g_2))$. The concept of a T-system is of importance in classifying the different two-generator presentations of G. If $\langle x_1, x_2; R_1, R_2, \ldots\rangle$ is a presentation of G associated with the pair $\mathfrak{g}$, then $\mathfrak{g}'$ is in the same T-system as $\mathfrak{g}$ if and only if there is an automorphism ϕ of F_2 such that $\langle x_1, x_2; \phi(R_1), \phi(R_2), \ldots\rangle$ is a presentation of G associated with $\mathfrak{g}'$.

This paper is concerned with obtaining information about the Nielsen equivalence classes and T-systems of certain two-generator HNN groups, and in particular of certain two-generator one-relator groups with torsion. Use is made of the following reduction theorem, which shows that for a large family of two-generator HNN groups each Nielsen equivalence class has a representative with a particularly simple form. A subgroup B of a group C is *malnormal* in C if and only if for all elements c of C,

$$c^{-1}Bc \cap B \neq 1$$

implies $c \in B$.

THEOREM A. *Let H be a group having two non-trivial malnormal subgroups K_{-1} and K_1, and suppose K_{-1} and K_1 are isomorphic, where $\rho : K_{-1} \to K_1$ is a specified isomorphism. Any generating pair of the* HNN *group*

$$G = \left\langle H, t; \text{ rel } H, t^{-1}K_{-1}t \overset{\rho}{=} K_1\right\rangle$$

is Nielsen equivalent to a pair of the form (th, k) where $h \in H$ and $k \in K_{-1}$.

Proof to be published in a future paper.

The situation considered here is when H is generated by two elements a and b of equal order, and K_{-1} and K_1 are the subgroups generated by a and b respectively. In this case it is possible to relate a generating pair of $G = \left\langle H, t; \text{ rel } H, t^{-1}K_{-1}t \overset{\rho}{=} K_1\right\rangle$ of the form (th, k) to a generating pair of H.

LEMMA 1. *Let H be generated by two elements a and b of equal order, and let G be the* HNN *group*

$$\langle H, t; \text{rel } H, t^{-1}at = b^{\xi}\rangle$$

(where ξ is relatively prime to the order of b). If (th, a^{μ}) generates G then $(a, h^{-1}bh)$ generates H .

PROOF. If (th, a^{μ}) generates G then so does (th, a) . Since $(th)^{-1}ath$ is a power of $h^{-1}bh$ it follows that G is generated by the three elements $th, a, h^{-1}bh$. In particular, every element of the base group, H , can be written as a word in th, a and $h^{-1}bh$. To show that a and $h^{-1}bh$ generate H it suffices to prove that a word in th, a and $h^{-1}bh$ which is equal to an element of H is equal to a word in a and $h^{-1}bh$ alone.

The proof is by induction on the number of t-symbols appearing in the word. If no t-symbols appear the result is trivially true. Thus assume the result for all words with less than r t-symbols, where $r > 0$, and suppose Q is a word in th, $a, h^{-1}bh$ which involves r t-symbols and defines an element of H . It follows from Britton's Lemma (see [11]) that Q has a subword $t^{\varepsilon}qt^{-\varepsilon}$ $(|\varepsilon| = 1)$ where q is t-free and belongs to K_{ε} . This means that Q can be written as a product $Q = Q_1Q_2Q_3$ where Q_1, Q_2, Q_3 are words in $th, a, h^{-1}bh$ and *either* Q_2 has the form $thuh^{-1}t^{-1}$ where u is a word in a and $h^{-1}bh$, and $huh^{-1} \in K_1$ *or* Q_2 has the form $h^{-1}t^{-1}vth$ where v is a word in a and $h^{-1}bh$, and $v \in K_{-1}$. In the former case Q_2 can be replaced by a power of a , and in the latter case Q_2 can be replaced by a power of $h^{-1}bh$. In either case Q can be replaced by a new word Q' in $th, a, h^{-1}bh$, which is equal to Q in G and has less t-symbols. The inductive hypothesis can now be applied to give the desired result. //

Using Theorem A and Lemma 1, it is now possible to prove the main result of this paper.

THEOREM 1. *Let α_i, β_i , $i = 1, 2, \ldots, r$ $(r > 0)$ be positive integers. Any two generating pairs of the one-relator group*

$$G = \left\langle a, t; \left(a^{\alpha_1}t^{-1}a^{\varepsilon\beta_1}t \ldots a^{\alpha_r}t^{-1}a^{\varepsilon\beta_r}t\right)^n\right\rangle$$

where $n > 1$ and $|\varepsilon| = 1$, are Nielsen equivalent.

PROOF. The group G can be presented as an HNN group as follows:

$$G = \langle a, b, t; P^n(a, b), t^{-1}at = b^{\varepsilon} \rangle .$$

Here $P(a, b)$ is a positive word involving a and b non-trivially. Now the associated subgroups $K_{-1} = \mathrm{sgp}(a)$ and $K_1 = \mathrm{sgp}(b)$ are malnormal in the base group $H = \langle a, b; P^n(a, b) \rangle$ by Lemma 4 of [9], and so it follows from Theorem A that any generating pair of G is Nielsen equivalent to a pair of the form (th, a^{μ}). By Lemma 1, (th, a^{μ}) can generate G only if $(a, h^{-1}bh)$ generates the base group H. It will be shown that in order for $(a, h^{-1}bh)$ to generate H, h must be expressible as a product of the form $b^{\beta}a^{\alpha}$, where β and α are integers, possibly zero.

Let

$$W = \{w : w \text{ is a word in } a \text{ and } b \text{, and } b^{\xi}wa^{\theta} \text{ defines } h \text{ in } H \text{ for suitable integers } \xi \text{ and } \theta\} ,$$

and let w_0 be an element of W of minimal length. Suppose w_0 is not empty. It will be shown that $\mathrm{sgp}\{a, w_0^{-1}bw_0\}$ is free of rank 2. Consequently $\mathrm{sgp}(a, h^{-1}bh)$ is also free of rank 2 and is therefore not equal to H. For convenience, two consequences of the minimality of w_0 are noted here.

(1) w_0 starts with a or a^{-1} and ends with b or b^{-1}.

(2) w_0 does not contain more than half of any cyclic permutation of P^n or its inverse.

Consider a word

$$Q = a^{m_1}w_0^{-1}b^{n_1}w_0 \ldots a^{m_s}w_0^{-1}b^{n_s}w_0 \quad (s > 0)$$

where m_i, $n_i \neq 0$ $(i = 1, 2, \ldots, s)$. Then Q is freely reduced as a word in a and b by (1). If Q defines the identity of H then by the modification to Newman's Spelling Theorem recently announced in [3] (see Statement 1, page 1439), Q has a subword of the form $T^{n-1}T_1$ where T is a cyclic permutation of P or P^{-1}, $T \equiv T_1T_2$ and T_1 involves both a and b. Now since P is a positive word no subword of Q involving symbols from both an occurrence of w_0 and an occurrence of w_0^{-1} can be a subword of a cyclic permutation of P^n or its inverse. Thus using

(2), it follows that the only possible subwords of Q which could have the form $T^{n-1}T_1$ are of one of the types

$$b^k w_0 a^l \ , \quad k \ , \ l \neq 0 \ ,$$

$$a^{l'} w_0^{-1} b^{k'} \ , \quad k' \ , \ l' \neq 0 \ .$$

It suffices to consider a subword of the type $b^k w_0 a^l$ and show that it does not have the form $T^{n-1}T_1$. Now (2) implies that $b^k w_0 a^l$ can have the form $T^{n-1}T_1$ only if $n = 2$, and then w_0 must be a cyclic permutation of P or P^{-1}. But then there exist words U and V with $L(U) + L(V)$ less than $L(w_0)$ and $Ub^k w_0 a^l V$ a cyclic permutation of P^2 or P^{-2}. Consequently $w_0 = b^{-k}U^{-1}V^{-1}a^{-l}$ in H, and the minimality of w_0 is contradicted.

It has now been established that any generating pair of G is Nielsen equivalent to a pair of the form $(tb^{\beta}a^{\alpha}, a^{\mu})$, or equivalently (since $tb^{\beta} = a^{\varepsilon\beta}t$) a pair of the form $(a^{\gamma}ta^{\alpha}, a^{\mu})$. It will now be shown that if $(a^{\gamma}ta^{\alpha}, a^{\mu})$ generates G then $|\mu| = 1$. Once this has been verified it follows easily that any generating pair of G is Nielsen equivalent to (t, a). By Newman's Spelling Theorem, in order for $(a^{\gamma}ta^{\alpha}, a^{\mu})$ to generate G, α_i and β_i $(i = 1, 2, \ldots, r)$ must be divisible by μ, for otherwise $\mathrm{sgp}(a^{\gamma}ta^{\alpha}, a^{\mu})$ is a free subgroup of G. Now if $(a^{\gamma}ta^{\alpha}, a^{\mu})$ generates G then the factor group of G by the normal subgroup $\langle a^{\mu}\rangle^G$ generated by a^{μ} is cyclic. Since a presentation for $G/\langle a^{\mu}\rangle^G$ is $\langle a, t; a^{\mu}\rangle$ it follows that $|\mu| = 1$. //

COROLLARY 1 (Purzitsky and Rosenberger [10]). *The Fuchsian groups* $\langle a, t; [a, t]^n\rangle$, $n = 2, 3, \ldots$ *have the property that any two generating pairs are Nielsen equivalent.*

COROLLARY 2. *Let* G *be as in Theorem* 1, *and let* $\langle x_1, x_2; R\rangle$ *and* $\langle x_1, x_2; S\rangle$ *be two one-relator presentations for* G. *Then there is an automorphism* ϕ *of* F_2 *such that* $\phi(R) = S^{\pm 1}$.

PROOF. This follows from Theorem 1 and the conjugacy theorem for groups with one defining relation (Theorem 4.11 of [7]). //

Corollary 2 should be viewed in the light of Magnus' Conjecture, that if $\langle x_1, x_2, \ldots, x_\lambda; U\rangle$ and $\langle x_1, x_2, \ldots, x_\lambda; V\rangle$ are two one-relator presentations of a one-relator group then there is an automorphism of the free group on $x_1, x_2, \ldots, x_\lambda$ mapping U to V or its inverse. This conjecture has been disproved for torsion-free one-relator groups (see [2], [6]), but is still open for groups with torsion.

The next corollary provides a counterexample to the converse of Corollary 4.13.1 of [6]*.

COROLLARY 3. *There are two words* U *and* V *in* x_1 *and* x_2 *such that the groups* $L_n = \left\langle x_1, x_2; U^n\right\rangle$ *and* $M_n = \left\langle x_1, x_2; V^n\right\rangle$ $(n \geq 1)$ *are isomorphic if and only if* $n = 1$.

PROOF. Take

$$U \equiv x_1^{-1}x_2^2x_1x_2^{-3} ,$$

$$V \equiv x_2^{-1}[x_2, x_1]^2 .$$

It will first be shown that L_1 and M_1 are isomorphic:

$$\begin{aligned} L_1 &= \left\langle x_1, x_2; x_1^{-1}x_2^2x_1 = x_2^3\right\rangle \\ &\cong \left\langle x_1, x_2, y; y = x_2^2, x_1^{-1}x_2^2x_1 = x_2^2x_2\right\rangle \\ &\cong \left\langle x_1, y; y = [y, x_1]^2\right\rangle \\ &\cong M_1 . \end{aligned}$$

Now suppose $n > 1$. By Corollary 2, in order for L_n and M_n to be isomorphic there must exists an automorphism ϕ of F_2 mapping U^n to $V^{\pm n}$. This would imply (by Exercise 2, page 41 of [7]) that $\phi(U) = V^{\pm 1}$. However Brunner [2] has shown that neither V nor V^{-1} is an automorphic image of U . //

Theorem A and Lemma 1 can be used to obtain information about the Nielsen equivalence classes of other groups beside one-relator groups.

THEOREM 2. *Let* $H = \langle a, b; R_1, R_2, \ldots\rangle$ *satisfy the small cancellation condition* $C'\left(\frac{1}{8}\right)$ **, *and suppose that each relator is either a positive or negative*

---

* I thank B.B. Newman for suggesting the problem of finding such a counterexample.

** For the definition of $C'\left(\frac{1}{8}\right)$ see [5].

word. Assume further that a *and* b *have equal order in* H . *Any generating pair of the group*

$$G = \left\langle a,\ b,\ t;\ R_1,\ R_2,\ \ldots,\ t^{-1}at = b^{\xi}\right\rangle$$

(where ξ *is relatively prime to the order of* b *) is Nielsen equivalent to a pair of the form* $(a^{\gamma}ta^{\alpha},\ a^{\mu})$. *If* μ *is relatively prime to the order of* a *then* $(a^{\gamma}ta^{\alpha},\ a^{\mu})$ *is Nielsen equivalent to* $(t,\ a^{\mu})$. *Even if* μ *is not relatively prime to the order of* a *, the generating pair* $(a^{\gamma}ta^{\alpha},\ a^{\mu})$ *is in the same T-system as* $(t,\ a^{\mu})$.

SKETCH OF PROOF. It is convenient to treat two special cases separately. Firstly, if $H = \langle a, b; a, b\rangle$ then $G = \langle t;\ \rangle$ and it is easily shown that any generating pair of G is Nielsen equivalent to $(t, 1)$. Secondly, suppose H has a relator of length less than 8 involving both a and b . It is not difficult to show that then $H = \langle a, b; (ab)^k\rangle$ for some integer k with $1 \leq k \leq 3$. Thus $G = \langle a, b, t; (ab)^k, t^{-1}at = b^{\varepsilon}\rangle$ $(|\varepsilon| = 1)$, and the theorem holds. (For $k = 1$ see Section 2.5 of [1], and for $k = 2, 3$ use Theorem 1 of the present paper.) From now on it will be assumed that *H does not have a relator of length* 1 *and that every relator of H involving both a and b has length greater than* 7 . This assumption will not be repeated.

In order to be able to make use of Theorem A it must be shown that $K_{-1} = \mathrm{sgp}(a)$ and $K_1 = \mathrm{sgp}(b)$ are malnormal in H . Suppose for example that there is an element c of H and non-identity elements a^m and a^n of K_{-1} such that $c^{-1}a^mc = a^n$. By Theorems 1 and 1' of [4], $a^m = a^n$, and so it can be assumed that $m = n$. Furthermore, it is no loss of generality to assume that $|m|$ is less than or equal half the order of a . Write c as a word in a and b , say $c = a^{\xi}ua^{\theta}$, where u is either empty or starts and ends with b or b^{-1} . It can be assumed that u does not contain more than half of any cyclic permutation of a relator or its inverse. To prove that $c \in K_{-1}$ it suffices to show that if u is non-empty then $Q \equiv a^{-m}u^{-1}a^mu$ does not define the identity of H . Now by Corollary 4.1 (ii) of [5]*, Q can define the identity of H only if it has a subword identical with more than 3/4 of a cyclic permutation R_i^* of some relator R_i or its inverse. The only possible subwords of Q which can also be subwords of R_i^* have one of the forms:

---

* Corollary 4.1 (ii) is applicable - the triangle condition, T_3 , is satisfied because each relator of H is either a positive or negative word.

$a^j v_1^{-1}$, $j \neq 0$ and v_1^{-1} an initial segment of u^{-1} ;

$v_2^{-1} a^k$, $k \neq 0$ and v_2^{-1} a terminal segment of u^{-1} ;

$a^l v_3$, $l \neq 0$ and v_3 an initial segment of u .

It will be shown that $a^j v_1^{-1}$ is not identical with more than 3/4 of R_i^* . The fact that neither $v_2^{-1} a^k$ nor $a^l v_3$ is identical with more than 3/4 of R_i^* is proved similarly.

Suppose $a^j v_1^{-1}$ is a subword of R_i^* with $L\left(a^j v_1^{-1}\right) > \frac{3}{4} L\left(R_i^*\right)$. Since $L\left(v_1^{-1}\right) \leq \frac{1}{2} L\left(R_i^*\right)$ it follows that

(3) $$|j| > \frac{1}{4} L\left(R_i^*\right) .$$

Now $a^{|j|-1}$ is a piece and so

(4) $$|j| - 1 < \frac{1}{8} L\left(R_i^*\right) .$$

Hence from (3) and (4), $L\left(R_i^*\right) < 8$, which is a contradiction.

The proof of the first part of the present theorem now proceeds similarly to the proof of Theorem 1. The proof reduces to showing that if w is a word starting with b or b^{-1} and ending with a or a^{-1} and w does not contain more than half of a cyclic permutation of a relator of H or its inverse then $\operatorname{sgp}\left(a, w^{-1}bw\right)$ is the free product of $\operatorname{sgp}(a)$ and $\operatorname{sgp}\left(w^{-1}bw\right)$, provided that w is minimal (in the sense that w is not equal in H to a word of the form $b^{\beta} w' a^{\alpha}$ with $L(w') < L(w)$). The verification that $\operatorname{sgp}\left(a, w^{-1}bw\right) \cong \operatorname{sgp}(a) * \operatorname{sgp}\left(w^{-1}bw\right)$ uses Corollary 4.1 (ii) of [5] in a similar fashion to the way it was used above to establish that $Q \neq 1$.

The second part of the theorem is trivial. The third part is proved by showing that $\left(a^{\gamma} t a^{\alpha}, a^{\mu}\right)$ is a generating pair for G if and only if $\left(t, a^{\mu}\right)$ is, and in this case the mapping

$$t \mapsto a^{\gamma} t a^{\alpha} , \quad a^{\mu} \mapsto a^{\mu}$$

defines an automorphism of G . The verification of these facts uses Britton's Lemma [11]. //

References

[1] A.M. Brunner, "Fibonacci sequences and group theory", (PhD thesis, Australian National University, Canberra, 1973. See also abstract: *Bull. Austral. Math. Soc.* 9 (1973), 473-474).

[2] A.M. Brunner, "Generating sets of one-relator groups", these Proc.

[3] Г.А. Гуревич [G.A. Gurevič], "К проблеме сопряженности для групп с одним определяющим соотношением" [On the conjugacy problem for groups with one defining relation], *Dokl. Akad. Nauk SSSR* 207 (1972), 18-20; *Soviet. Math. Dokl.* 13 (1972), 1436-1439.

[4] Seymour Lipschutz, "On powers of elements in *S*-groups", *Proc. Amer. Math. Soc.* 13 (1962), 181-186. MR26#3760.

[5] Roger C. Lyndon, "On Dehn's algorithm", *Math. Ann.* 166 (1966), 208-228. MR54#5499.

[6] James McCool and Alfred Pietrowski, "On free products with amalgamation of two infinite cyclic groups", *J. Algebra* 18 (1971), 377-383. MR43#6296.

[7] Wilhelm Magnus, Abraham Karrass, Donald Solitar, *Combinatorial group theory* (Pure and appl. Math., 13. Interscience [John Wiley & Sons], New York, London, Sydney, 1966). MR34#7617.

[8] Bernhard H. Neumann und Hanna Neumann, "Zwei Klassen charakteristischer Untergruppen und ihre Faktorgruppen", *Math. Nachr.* 4 (1951), 106-125. MR12,671.

[9] B.B. Newman, "Some results on one-relator groups", *Bull. Amer. Math. Soc.* 74 (1968), 568-571. MR36#5204.

[10] Norman Puritsky and Gerhard Rosenberger, "Two generator Fuchsian groups of genus one", *Math. Z.* 128 (1972), 245-251. Zbl.232.20105.

[11] Joseph J. Rotman, *The theory of groups: An introduction* (Allyn and Bacon, Boston, Massachusetts, 1965). MR34#4338.

Australian National University,
Canberra, ACT 2600.

Present address:
The Open University,
Walton Hall, Milton Keynes, Bucks MK7 6AA, England.

PROC. SECOND INTERNAT. CONF. THEORY OF GROUPS, 20H15
CANBERRA, 1973, pp. 589-594.

THE COLLINEATION GROUPS OF FINITE GENERALIZED HALL PLANES

Alan Rahilly

1. Introduction

The aim of this paper is to present a number of results recently obtained by the author on the collineation groups of finite generalized Hall planes. The chief result (Theorem 1) appears in §3 with an outline of a proof. In §4 a classification of the finite generalized Hall planes of order greater than 16 which are not Hall planes, in terms of central collineations, appears as well as four theorems on central collineations. No proofs of these results are included. Some examples of generalized Hall planes are exhibited in §5 with information on their collineation groups.

2. Generalized Hall planes

DEFINITION. A projective plane π is a *generalized Hall plane* with respect to the line l_∞ and Baer subplane π_0 containing l_∞ if

(i) π is a translation plane with respect to l_∞, and

(ii) there exists a group $G(\pi_0)$ fixing π_0 pointwise and sharply transitive on $l_\infty \backslash \underline{M}$, where $\underline{M} = l_\infty \cap \pi_0$.

The subplane π_0 can be shown to be desarguesian (see [3] and [7]). If we coordinatize the *finite* plane π over $0, I, X, Y$ in π_0 such that $XY = l_\infty$ the coordinate system is a $V - W$ system F whose multiplication is given by

$$(Z\alpha+\beta)Z = Z\{f(\alpha)+h(\beta)\} + g(\alpha) + k(\beta)$$

for all $\alpha, \beta \in F_0$ and $Z \in F\backslash F_0$, where F_0 is the field coordinatizing π_0 and

(a) f, g, h and $k \in \text{end}(F_0(+))$,

(b) $h(1) = 1$, $k(1) = 0$,

(c) $h \in \operatorname{aut}(F_0(+))$, and

(d) $M_\lambda = g + (k-m_\lambda)h^{-1}(m_\lambda-f)$ is non-singular for all $\lambda \in F_0$, where

$m_\lambda(x) = \lambda x$ for all $\lambda, x \in F_0$.

Such a system F is called a *generalized Hall system* and the functions f, g, h and k are called *defining functions* for F . The Hall systems (see [1], p. 364) are given by $f = m_r$, $g = m_s$, $h = 1$ and $k = 0$, where $x^2 - rx - s$ is irreducible over F_0 .

3. Collineation groups

THEOREM 1. *If* π *is a generalized Hall plane and* $16 < |\pi| < \infty$ *then* $\underline{\underline{M}}$ *is fixed by all collineations of* π .

The proof that the author has for this result is quite long. The result was proved for the Hall planes by Hughes [2]. In the case of π not a Hall plane the basic steps are as follows.

STEP 1. The first step is to investigate collineations fixing a point $P \in \underline{\underline{M}}$, a point $0 \notin l_\infty$ and $\underline{\underline{M}}$. Such collineations fix π_0 and it is possible to find an explicit representation of them in terms of their action on π_0 .

STEP 2. Next it is possible to prove that if $\underline{\underline{M}}$ is shifted by a collineation and $|\pi| > 16$ then either

(a) there are two points A and B on l_∞ such that the group $G_{A,B}$ fixing A and B is transitive on $l_\infty \backslash \{A, B\}$ and permutes the images of $\underline{\underline{M}} \backslash \{A, B\}$ under $G_{A,B}$ as sets of imprimitivity, or

(b) there is a point $C \in l_\infty$ fixed by all collineations such that the full collineation group G is transitive on $l_\infty \backslash \{C\}$ and permutes the images of $\underline{\underline{M}} \backslash \{C\}$ under G as sets of imprimitivity.

STEP 3. The examination of the collineations which arise from cases (a) and (b) of Step 2.

Let $\underline{\underline{M}}(z)$ be the image of $\underline{\underline{M}}$ under G containing (z) where $z \in F \backslash F_0$, and 0 a point not on l_∞ .

(a) If case (a) of Step 2 applies then $\underline{\underline{M}}(z) = \{(z)^g \mid g \in H\} \cup \{A, B\}$, where H is a subgroup of order $q - 1$ of $G(\pi_0)$. (There are q such subgroups of

$G(\pi_0)$.) Further, there is a group of collineations fixing 0 and $\underline{M}(z)$ pointwise and transitive on $\underline{M}\backslash\{A, B\}$. These collineations are of the type investigated in Step 1. It is possible to show, using our knowledge of such collineations, that $|\pi| \le 16$.

(b) If case (b) of Step 2 applies then $\underline{M}(z) = \{(z)^g \mid g \in K\} \cup \{C\}$, where K is the unique subgroup of order q of $G(\pi_0)$ and there is a group of collineations of order q fixing 0 and $\underline{M}(z)$ pointwise and transitive on $\underline{M}\backslash\{C\}$. These collineations are of the type considered in Step 1 and it is possible to show $|\pi| = 4$.

4. Central collineations

An ordered pair of points (P, Q) on l_∞ is called a *central pair* if there exists a line l $(\ne l_\infty)$ such that $Q \in l$ and there is a non-trivial (P, l)-central collineation. The pair (P, Q) is a *special pair* if (P, Q) and (Q, P) are central pairs. Also, it is a *proper* central (or special) pair if $P \ne Q$ and *improper* otherwise.

With these preliminaries we may classify the generalized Hall planes of order greater than 16 which are not Hall planes. We consider the odd and even order cases separately.

$|\pi|$ IS ODD. Odd order generalized Hall planes have no improper central pairs. They fall into the disjoint classes:

O_1: no proper central pairs;

O_2: a unique proper special pair in $\underline{M}$ and no other central pairs;

O_3: more than one proper special pair in $\underline{M}$ and no other central pairs. (Here the special pairs are mutually disjoint.)

$|\pi|$ IS EVEN.

E_1: no central pairs proper or improper;

E_2: improper central pairs in $\underline{M}$ and no other central pairs;

E_3: proper central pairs in $\underline{M}$ and no other central pairs. (Here if (P, Q) and (R, S) are distinct central pairs then $P = S$ and $Q = R$ or $\{P, Q\} \cap \{R, S\} = \emptyset$.)

The Hall planes of odd (even) order are such that $\underline{M}$ is the disjoint union of proper (improper) special pairs (see Hughes [2]). The finite Hall planes also possess proper central pairs in $l_\infty \backslash \underline{\underline{M}}$ (see Ostrom [5], p. 359).

THEOREM 2. *Suppose* π *is a generalized Hall plane of odd order coordinatized over* $0, I, X, Y$ *by the generalized Hall system* F *with defining functions* f, g, h *and* k. *Then* (X, Y) *and* $\{(\alpha), (\beta)\}$, *where* $\alpha, \beta \in F_0^* = F_0 \backslash \{0\}$, *are central pairs if and only if*

(i) $f = k = 0$,

(ii) $hm_\lambda = m_\lambda h$, *where* $\lambda = (\alpha+\beta)(\alpha-\beta)^{-1}$, *and*

(iii) $g = m_{\alpha\beta\gamma} h^{-1} m_{\gamma-1}$, *where* $\gamma = 2(\alpha-\beta)^{-1}$.

THEOREM 3. *Suppose* π *is a generalized Hall plane of even order. Then,* π *possesses a* $(Y, 0Y)$*-elation taking* X *to* (δ), *where* $\delta \in F_0^*$, *and a* $\{(1), 0(1)\}$*-elation swapping* X *and* Y *if and only if*

(i) $h = 1$,

(ii) $f + k = m_\delta$,

(iii) $g + kf = 1$,

(iv) $m_\delta k = km_\delta$.

THEOREM 4. *Suppose* π *is a generalized Hall plane of even order possessing elations as in Theorem* 3. *Then,* π *possesses a* $\{(\lambda), 0(\lambda)\}$*-elation, where* $\lambda \in F_0$, *taking* Y *to* (λ') *if and only if* $\lambda \in \ker(F)$ *and* $\lambda'\delta = 1 + \lambda^2$.

THEOREM 5. *The finite generalized Hall plane* π *is a Hall plane if and only if every point of* $\underline{\underline{M}}$ *is the centre of a non-trivial central collineation with axis* $\neq l_\infty$.

Theorem 5 may be proved with the aid of Theorems 2, 3 and 4. Special cases of Theorem 5 are considered in [4] and [7].

5. Examples of generalized Hall planes

In this section some examples of generalized Hall planes will be given and some information on their collineation groups. Before proceeding to this we shall state a theorem.

THEOREM 6. *Suppose* π *is a generalized Hall plane of odd order coordinatized by the generalized Hall system with defining functions* f, g, h *and* k *over* $0, I, X, Y$. *If* $f = k = 0$ *and there is no* $\rho, \sigma \in F_0^*$ *nor* $\psi \in \mathrm{aut}(F_0)$ *such that* $g = m_\rho \psi h^{-1} \psi^{-1} m_\sigma$, *then the full collineation group* G *of* π *fixes* X *and* Y, *and so is generated by translations and* $0, X, Y$ *autotopism collineations.*

EXAMPLE 1. Let $|F_0|$ be odd and $f = k = 0$, $h = \theta$, $g = m\gamma\Phi$, where $\theta, \Phi \in \operatorname{aut}(F_0)$ and γ is a non-square of F_0.

If $\theta \neq \Phi^{-1}$ such systems satisfy the hypotheses of Theorem 6. The planes coordinatized by these systems are in 0_2.

If $\theta = \Phi^{-1}$ it is possible to use Theorem 2 to obtain (X, Y) and $\left((\mu), \left(\gamma\Phi(\mu)^{-1}\right)\right)$ as special pairs, where $\Phi^2(\mu) = \mu\theta(\gamma)\gamma^{-1}$. Thus, the planes coordinatized by these systems are in 0_3.

EXAMPLE 2. Let $F_0 = GF\left(3^2\right)$, $E = GF(3)$ and $t \in F_0$ be such that $t^2 = -1$. The endomorphisms $h = 1$, $g = m_t$, $k = 0$ and f the non-trivial involutory automorphism of F_0 are the defining functions for a generalized Hall system. The plane coordinatized by this system is in 0_1.

EXAMPLE 3. Let $F_0 = GF\left(2^{ud}\right)$, where d is even. Then $f = k = 0$, $h = \theta^{-1}$, and $g = m_\nu\theta$, where $\theta^{-1} : x \to x^{2u}$, are the defining functions of a generalized Hall system if $\theta(x)x - \nu$ is irreducible over F_0. The planes over these systems are in E_3.

EXAMPLE 4. Let $F_0 = GF\left(2^4\right)$, $E = GF\left(2^2\right)$ and $\{u, 1\}$ be a basis for $\left(F_0, E\right)$, where $u^2 = u + t$ and t is a primitive root of E. Then $f = \begin{pmatrix} 0 & 0 \\ 0 & t \end{pmatrix}$, $g = h = 1$ and $k = \begin{pmatrix} t & 0 \\ 0 & 0 \end{pmatrix}$ are the defining functions for a generalized Hall system which satisfies *(i)* to *(iv)* of Theorem 3 with $\delta = t$. Thus the plane over this system belongs to E_2. (Here matrices are to be understood as acting on the left.)

References

[1] Marshall Hall, Jr., *The theory of groups* (The Macmillan Co., New York, 1959). MR21#1996.

[2] D.R. Hughes, "Collineation groups of non-desarguesian planes, I. The Hall Veblen-Wedderburn systems", *Amer. J. Math.* 81 (1959), 921-938. MR22#920.

[3] P.B. Kirkpatrick, "Generalization of Hall planes of odd order", *Bull. Austral. Math. Soc.* 4 (1971), 205-209. MR43#1034.

[4] P.B. Kirkpatrick, "A characterization of the Hall planes of odd order", *Bull. Austral. Math. Soc.* 6 (1972), 407-415. Zbl.229.50026.

[5] T.G. Ostrom, "Collineation groups generated by homologies in translation planes", *Atti Convegno Geom. combinat. appl.* (Perqugia, 1970), pp. 351-366 (Instituto di Matematica, 1971). Zbl.225.50013.

[6] Alan Rahilly, "A class of finite projective planes", *Proc. First Austral. Conf. Combinatorial Mathematics* (Newcastle, 1972), pp. 31-37 (TUNRA, Newcastle, 1972).

[7] Alan Rahilly, "Generalized Hall planes of even order", *Pacific J. Math.* (to appear).

University of Sydney,
Sydney, NSW 2006.

PROC. SECOND INTERNAT. CONF. THEORY OF GROUPS, 20K30
CANBERRA, 1973, pp. 595-604.

ABELIAN GROUPS WITH SELF-INJECTIVE ENDOMORPHISM RINGS

K.M. Rangaswamy

It is shown that if the full ring $R = E(G)$ of all endomorphisms of an abelian group G is left self-injective, then $G = D \oplus H$, D torsionfree divisible and H is a pure fully invariant subgroup of $\prod_{p \in P} B_p$, where P is a set of primes and, for each $p \in P$, B_p is a direct sum of isomorphic cyclic p-groups. The converse also holds if $D = 0$. In the case when $D \neq 0$ and H has the required properties, it is shown that R/J is left self-injective (von Neumann) regular, idempotents lift modulo the Jacobson radical J and any left ideal of R isomorphic to a summand of the left R-module R is itself a summand of R. Finally, arbitrary (reduced) abelian groups H for which $E(H)$ is generalised left continuous (left continuous) are characterized. These results provide an 'almost' complete solution to a problem of Fuchs (see Problem 84d) of [1]).

All the rings that we consider here are associative rings with identity and all the modules are unitary left modules. A submodule S of a module M is said to be essential in M if $S \cap T \neq 0$ for every non-zero submodule T of M. We denote this by writing $S \subset' M$. A ring R is called generalised left continuous if any left ideal of R isomorphic to a summand of the left R-module R is itself a summand of R. If, in addition, every left ideal is essential in a direct summand of the left R-module R, then R is called left continuous. A ring R is called left self-injective if R is injective as a left R-module. Clearly a left self-injective ring is left continuous, but the converse is not true. An R-module M is called quasi-injective if, for any submodule S of M, every R-morphism $\alpha : S \to M$ extends to an endomorphism of M. For any abelian group G, $E(G)$ denotes the ring of all the endomorphisms of G, $T(G)$ the torsion part of G and G_p the p-component of G. Q denotes the field of rational numbers, Z the ring of integers, $Z/(p^n)$ the ring of integers modulo p^n and $Z(p^n)$ the cyclic group of order p^n. The symbol $\oplus A$ stands for the direct sum of arbitrary number of carbon copies of the group or ring A. The additive group of a ring R is denoted by R^+ . We follow mostly the notation and terminology of [1]. The only exception is: *all the endomorphisms are assumed to act on the right.*

We begin with the following simple result.

LEMMA 1. *Suppose A is an S-module and $R = \hom_S(A, A)$ is generalised left continuous. If B and C are isomorphic submodules of A and if one is a summand of A, then so is the other.*

Proof. We need only to observe that $B \cong C$ as S-modules implies that $\hom_S(A, B) \cong \hom_S(A, C)$ as left R-modules and that a submodule E of A is a summand of A if and only if $\hom_S(A, E)$ is a summand of the left R-module R.

PROPOSITION 2. *If $R = E(G)$ is generalised left continuous, then, for each prime p, the p-component G_p is a direct sum of isomorphic copies of a cyclic group and $G/T(G)$ is divisible.*

PROOF. By Lemma 1, all the cyclic summands of G_p (if they exist) are isomorphic. Hence G_p is either divisible or G_p is a direct sum of carbon copies of a cyclic group. In either case G_p is a direct summand: $G = G_p \oplus X$. Since multiplication by p is monic on X, $G_p \oplus pX \cong G$ and so, by Lemma 1, $G_p \oplus pX$ is a direct summand of G. Because $X[p] = 0$, we conclude that $X = pX$.

Now

$$\begin{aligned} G_p = pG_p &\Rightarrow G = pG \\ &\Rightarrow R[p] = 0 \\ &\Rightarrow (pR \cong R \text{ and so by Lemma 1) } pR = R \\ &\Rightarrow G_p = 0 . \end{aligned}$$

Thus, for every p, G_p is a direct sum of isomorphic copies of a cyclic group of order, say p^{n_p}. Then $G = G_p \oplus p^{n_p}G$ for every p and hence $G/T(G)$ is divisible.

PROPOSITION 3. *If, for an abelian group G, $E(G)$ is left self-injective, then $G = D \oplus H$, D torsionfree divisible, H reduced, $H/T(H)$ divisible, each H_p is a direct sum of isomorphic cyclic groups and H can be imbedded as a fully invariant pure subgroup of $\prod H_p$ containing $\oplus H_p$.*

PROOF. By Proposition 2, $G = D \oplus H$, D torsionfree divisible, H reduced and $H = H_p \oplus p^{n_p}H$ for each prime p. Now it is easy to see that $K = \bigcap_p p^{n_p}H$ is divisible and, since H is reduced, $K = 0$. Thus the projections $H \to H_p$ give an imbedding of H as a subgroup of $\prod_p H_p$ containing $\oplus_p H_p$. Since $H/T(H)$ is divisible, H is pure in $\prod_p H_p$ and every endomorphism of H extends uniquely to an endomorphism of $\prod_p H_p$. Thus we have an imbedding

$$\bigoplus_p E(H_p) \subseteq E(H) \subseteq E\left(\prod_p H_p\right) = \prod_p E(H_p) .$$

Now $\hom(G, H)$ is annihilated by the ideal $\hom(G, D)$ and so the left self-injectivity of R implies that of $E(H) \cong \hom(G, H)$. Hence $E(H)$ is a summand of $E\left(\prod_p H_p\right)$. But, as abelian groups, $\prod_p E(H_p) / \oplus_p E(H_p)$ is divisible and $\prod E(H_p)$ is reduced and so we conclude that $E(H) = E\left(\prod_p H_p\right)$; that is, H is fully invariant in $\prod_p H_p$.

COROLLARY 4. *If $E(G)$ is left self-injective, then G is a quasi pure injective abelian group.*

PROOF. Now $G' = D \oplus \prod_p H_p$ is algebraically compact and G is pure fully invariant in G' . The corollary follows immediately.

We shall first observe a few simple properties of the groups mentioned in Proposition 2.

5. *Properties of a reduced group H with $H/T(H)$ divisible, each p-component $H_p = \oplus Z\left(p^{n_p}\right)$, n_p being a positive integer depending on p .*

(i) For each p , every subgroup of H_p is an endomorphic image of H .

(ii) H is Hausdorff in its n-adic topology. In fact $\bigcap_p p^{n_p}H$, being divisible, is zero.

(iii) A subgroup S of H is pure if and only if $S/T(S)$ is divisible and $T(S)$ is a summand of $T(H)$.

(iv) If α is an endomorphism of H and $S = \operatorname{Im} \alpha \neq 0$, then $T(S) \neq 0$ and $S/T(S)$ is divisible.

For, $H = H_p \oplus p^{n_p}H$, for all primes p implies that

$S = H_p\alpha + p^{n_p}H\alpha = S_p + p^{n_p}S$ for all p so that $S/T(S)$ is divisible. Since H is reduced, $T(S) \neq 0$.

(v) The closure $\bar{S}$ of a pure subgroup S (in the n-adic topology of H) is again pure in H and is actually the closure $\overline{T(S)}$ of the torsion part $T(S)$. In particular, $T(\bar{S}) = T(S)$.

PROOF. By (iii), $\bar{S} = \overline{T(S)}$. Since H is pure in $Y = \prod H_p$, H is a subspace of Y in the n-adic topology and $\bar{S} = H \cap S^*$, where S^* is the closure of the pure subgroup S in Y. As Y is reduced algebraically compact, S^* is a summand of Y. Since $T(Y) = T(H)$, $H \cap S^* = \bar{S}$ is then pure in H.

(vi) H has no torsion free subgroups closed in the n-adic topology of H.

PROOF. It is enough to prove that $Y = \prod H_p$ has no non-zero torsionfree closed subgroups. Suppose S is a torsionfree closed subgroup of Y. Then $X = Y/S$ is reduced algebraically compact and $X/T(X)$ is divisible, so that $X \cong \prod X_p$. Now the natural map $\eta : Y \to Y/S = X$ is monic on $T(Y)$. Hence the map $\eta' : Y \to X = \prod X_p$ given by $\langle \ldots, y_p, \ldots \rangle \eta' = \langle \ldots, y_p \eta, \ldots \rangle$ is a monomorphism and, since X is reduced, $Y/T(Y)$ is divisible and $\eta - \eta'$ vanishes on $T(Y)$, we have $\eta = \eta'$ is a monomorphism: that is, $S = 0$.

(vii) If R is a ring with $R^+ = H$ and if I is a non-zero left ideal of R, then $I = I_p \oplus p^{n_p} I$ for all p and I^+ is not torsionfree.

(viii) If $\alpha : H \to H$, then $S = \operatorname{Im} \alpha$ is pure $\Leftrightarrow K = \ker \alpha$ is pure.

PROOF. Observe that $T(H)\alpha = T(\operatorname{Im} \alpha)$. Thus $T(S)$ is a summand of $T(H)$ if and only if $T(K)$ is a summand of $T(H)$. Then, by 5 (iii) and (iv), the purity of K implies that of S. Let α^* be the extension of α to $H^* = \prod H_p$. If S is pure, then $T(S)$ and hence $T(K)$ is a summand of $T(H)$. Then $K^* = \ker \alpha^* = \prod K_p$ is a summand of H^*, whence $K = H \cap K^*$ is pure in H.

THEOREM 6. *$R = E(G)$ is generalised left continuous $\Leftrightarrow G = D \oplus H$, D torsionfree divisible, H reduced, $H/T(H)$ divisible, each p-component $H_p = \oplus Z\left(p^{n_p}\right)$, n_p being a positive integer depending on p, and any subgroup isomorphic to a summand of G is itself a summand of G.*

For the sake of convenience, we shall single out a special case in the following lemma.

LEMMA 7. *Let H be a reduced abelian group satisfying the conditions of Theorem 6. Then $R' = E(H)$ is generalised left continuous.*

PROOF. Let L be a left ideal of R' and $\theta : R'e \to L$ an R'-isomorphism, where $e^2 = e \in R'$. Since $R'e$ is a summand, θ may be considered as an element of R' and its action given by right multiplication by θ. If $\alpha = e\theta$, then

$L = R'\alpha$. Now $H = \mathrm{Im}\, e \oplus \ker e$ and $K = \ker \alpha \supseteq \ker e$ so that $K = \ker e \oplus (K \cap \mathrm{Im}\, e)$. Since K is closed, $K \cap \mathrm{Im}\, e$ is closed and, by 5 (vi), $K \cap \mathrm{Im}\, e \neq 0$ implies that $(K \cap \mathrm{Im}\, e)_p \neq 0$ for some p . If $\delta : H \to (K \cap \mathrm{Im}\, e)_p$ is an epimorphism, then $0 \neq \delta \in R'e$ and $\delta\theta = \delta e\theta = \delta\alpha = 0$, contradicting the fact that θ is monic. Thus $\ker \alpha = \ker e$ is a summand and $\alpha|T$ is an isomorphism of the summand $T = \mathrm{Im}\, e$ to $S = \mathrm{Im}\, \alpha$ and so, by hypothesis, S is a summand. Then, by Lemma 121.1 of [1], $\alpha = \alpha\beta\alpha$ for some $\beta \in R'$ and $f = \beta\alpha$ is an idempotent. Clearly $L = R'f$ and so L is a summand of R' .

PROOF OF THEOREM 6. $\Rightarrow$. Follows from Lemma 1 and Proposition 2.

$\Leftarrow$. First observe that if $\eta : G \to D$ is a projection, then $R\eta$ is a two sided ideal being the divisible part of R^+ and, for any left ideal I of R with $I \subseteq R\eta$, I^+ is divisible. Suppose L is a left ideal of R and L is R-isomorphic to a summand U of R . Then $L_1 = L \cap R\eta$ and $U_1 = U \cap R\eta$ are the divisible parts of L and U and U_1 is a summand of R , say $U_1 = Re_1$, $e_1^2 = e_1$. Clearly we have an R-isomorphism $\theta : Re_1 \to L_1$, so that $L_1 = R\alpha$, where $\alpha = e_1\theta$. Let $S_1 = \mathrm{Im}\, \alpha$, $T_1 = \mathrm{Im}\, e_1$. Then $\theta' : T_1 \to S_1$, given by $t\theta' = te_1\theta$ for all $t \in T_1$, is an epimorphism. Now T_1 and S_1 are torsionfree divisible and so $K = \ker \theta'$ is a summand of T_1 . If $K \neq 0$, we have a non-zero $\delta : G \to K$ satisfying $\delta\theta = \delta e_1\theta = \delta\alpha = 0$ and this contradicts that θ is monic. Hence $\theta' : T_1 \to S_1$ is an isomorphism. Since S_1 is a summand, $(\theta')^{-1}$ extends to an endomorphism θ^* of G . Then $\eta^* = \theta^* e_1\theta \in L_1$, $\eta^{*2} = \eta^*$ and $\mathrm{Im}\, \eta^* = S_1$. Hence $L_1 = R\eta^*$ is a summand of R : $R = L_1 \oplus X = U_1 \oplus Y$. Then $L = L_1 \oplus L_2$, $U = U_1 \oplus U_2$, where $L_2 = L \cap X$, $U_2 = U \cap Y$ and $L_2 \cong U_2$. Now $U_2 = Re_2$, $e_2^2 = e_2$ and L_2, U_2 are reduced as abelian groups so that $S_2 = \sum_{\alpha \in L_2} \mathrm{Im}\, \alpha$, $T_2 = \mathrm{Im}\, e_2$ are reduced subgroups of G with T_2 a summand of G . Then the reduced component $H \supseteq S_2$ will contain a summand T isomorphic to T_2 . If $e : G \to T$ is a projection and $\chi : G \to H$ a projection with $\ker \chi = D$, then $L_2 \subseteq R\chi$, $Re \subseteq R\chi$ and $Re = \hom(G, T) \cong \hom(G, T_2) = Re_2 \cong L_2$. Since $(1-\chi)R\chi = 0$, $R\chi = \chi R\chi \cong \hom(H, H)$ and, by Lemma 7, $\chi L_2 = \chi L_2\chi = \chi R\chi f$, where $f^2 = f \in \chi R\chi$. Since $(1-\chi)L_2 = (1-\chi)L_2\chi = 0$, $L_2 = R\chi f = Rf$ is a summand. Then $L = L_1 \oplus L_2$ is a summand of R . This proves the theorem.

The next theorem characterizes reduced abelian groups H for which $E(H)$ is left continuous.

THEOREM 8. *Let H be a reduced abelian group. Then $R = E(H)$ is left continuous if and only if*

(a) $H/T(H)$ is divisible,

(b) for each prime p there exists a non-negative integer n_p such that

$$H_p = \oplus Z\left(p^{n_p}\right),$$

(c) the closure of any pure subgroup (in the n-adic topology of H) is a direct summand, and

(d) any endomorphic image which is pure in H is a direct summand of H.

PROOF. Let $R = E(H)$ be left continuous. Then, by Theorem 6, H fulfills the conditions *(a)* and *(b)*. Let S be the closure of a pure subgroup in H. By 5 (v), S is pure in H. Let $I = \{\alpha \in R \mid \operatorname{Im} \alpha \subseteq S\}$. Then I is a left ideal of R and, by hypothesis, $I \subset' Re$, $e^2 = e \in R$. If $U = \operatorname{Im} e$, then U is a summand of H and $S \subseteq U$. By the purity of S, S_p is a summand. If $U_p = S_p \oplus X_p$ with $X_p \neq 0$ and $\eta : G \to X_p$ a projection, then $0 \neq \eta \in Re$ so that $0 \neq \beta\eta \in I$ for some $\beta \in R$. But $\operatorname{Im} \beta\eta \subseteq S \cap X_p = 0$ so that $\beta\eta = 0$, a contradiction. Thus $T(U) = T(S)$. By 5 (v), $S = \overline{T(S)} = U$. Hence S is a summand. Next, let $\alpha : H \to H$ with $S = \operatorname{Im} \alpha$ pure in H. Then, by 5 (viii), $K = \ker \alpha$ is also pure. Since K is already closed, by what we have proved above, K is a summand of H : $H = K \oplus M$. Since $\alpha|M$ is an isomorphism of M to S, by Lemma 1, S is a summand of H.

Conversely, let H satisfy the conditions *(a)*-*(d)*. If S is isomorphic to a summand T of H, then it is easy to see that S is pure in H and hence, by condition *(d)*, S is a summand of H. Thus, by Theorem 6, R is generalised left continuous.

Let I be a left ideal of R and $S = \sum_{\alpha \in I} \operatorname{Im} \alpha$. By 5 (iv), $S/T(S)$ is divisible. Now $T(S)$ can be imbedded in a summand E of $T(H)$ such that $S[p] = E[p]$ for all primes p. Then the closure $C = \overline{E}$ of E is a summand of H and $S \subseteq C$, since $S/T(S)$ is divisible. Let $\eta : H \to C$ be a projection. Then $I \subseteq R\eta$. We claim $I \subset' R\eta$. Suppose $0 \neq f \in R\eta$. Since $H/T(H)$ is divisible, $f\,T(H) \neq 0$. Moreover $(\operatorname{Im} f)_p = \operatorname{Im}(f|H_p)$. Since $T(S) \subset' T(C) = E$, $S_p \cap (\operatorname{Im} f)_p \neq 0$ for some prime p. Let $0 \neq s \in S_p \cap (\operatorname{Im} f)_p$ be an element of order p, so that $s = zf$, for some $z \in H_p$. Let $\langle t \rangle$ be a cyclic summand of H_p containing s, and let $H = \langle t \rangle \oplus K$. Define $\alpha_2 : H \to H$ by $\alpha_2|K = 0$ and

$t\alpha_2 = z$. Then $0 \neq \alpha_2 f \in Rf$. Now $s = \sum_{i=1}^{n} x_i\beta_i$, where $\beta_i \in I$ and $x_i \in H_p$. Let $a \in H_p$ be an element of maximal order p^{n_p} , so that $H = \langle a\rangle \oplus M$. Let $\eta_i : H \to H$, $(i = 1, 2, \ldots, n)$ given by $\eta_i|M = 0$ and $a\eta_i = x_i$. Then $s = a\delta$, where $\delta = \sum_{i=1}^{n} \eta_i\beta_i \in I$. Define $\alpha_1 : H \to H$ by $\alpha_1|K = 0$ and $t\alpha_1 = a$. Then $\alpha_1\delta \in I$, and $\alpha_2 f = \alpha_1\delta$. Thus $I \subset' R\eta$. This completes the proof of Theorem 8.

We are now ready to characterize the reduced abelian groups possessing left self-injective endomorphisms.

THEOREM 9. *Let* G *be a reduced abelian group. Then* $R = E(G)$ *is left self-injective if and only if* G *is a fully invariant pure subgroup of* $\prod_{p\in P} B_p$ *, where* P *is a set of primes and for each* $p \in P$ *there exists a positive integer* n_p *such that* B_p *is a direct sum of isomorphic copies of* $Z\left(p^{n_p}\right)$.

PROOF. The necessity follows from Proposition 3. Conversely, let G satisfy the given conditions. Then $R = E(G) \cong E\left(\prod_p B_p\right) \cong \prod_{p\in P} E\left(B_p\right)$. Since a ring direct product of left self-injective rings is left self-injective, we need only show that $R_p = E\left(B_p\right)$ is left self-injective for each p . We may assume B_p is non-cyclic. By Theorem 8, R_p is left continuous. Let $B_p = \oplus Z\left(p^{n_p}\right)$. Write $B_p = A_1 \oplus \ldots \oplus A_n$ where $n > 1$ and $A_1 \cong A_2 \cong \ldots \cong A_n$. Let $\eta_i : G \to A_i$, $i = 1, \ldots, n$, be orthogonal projections: $\eta_i\eta_j = 0$ if $i \neq j$ and $\sum_{i=1}^{n} \eta_i = 1_{B_p}$. Since $R\left(1-\eta_i\right)$ contains an isomorphic copy of $R\eta_i$ for $i = 1, \ldots, n$, we conclude, by Theorem 7.1 of [4], that R_p (and hence R) is left self-injective.

COROLLARY 10. *Let* R *be left self-injective and* $\hom(Q, R) = 0$. *Then* $R \cong E(G)$ *for some abelian group* G *if and only if* $R = \prod_{i\in N} L_i$ *, where* N *is at most countable and, for each* $i \in N$, L_i *is the ring of all column-finite matrices over* $Z/\left(p_i^{n_i}\right)$ *with* p_i *primes satisfying* $p_i \neq p_j$ *if* $i \neq j$ *and* n_i *positive integers depending on* p_i .

REMARK 11. It is clear from Proposition 3, that if R^+ is torsionfree and R

is left self-injective, then $R \cong E(G)$ exactly when R is the ring of all linear transformations on a vector space over Q and, in this case, R therefore becomes von Neumann regular.

We now wish to investigate how far the converse of Proposition 3 is true.

Let G be an arbitrary abelian group with $G/T(G)$ divisible and each p-component G_p a direct sum of isomorphic copies of $Z\left(p^{n_p}\right)$, n_p a positive integer depending on p. Let A be a subring of $\prod_p Z/\left(p^{n_p}\right)$ containing $\bigoplus_p Z/\left(p^{n_p}\right)$ and the identity element $u = \langle \ldots, 1_p, \ldots \rangle$ such that $A/\bigoplus_p Z/\left(p^{n_p}\right)$ is the pure subring isomorphic to Q generated by the coset $u + \bigoplus Z/\left(p^{n_p}\right)$ (see page 306 of [1]). Proceeding the same way as in the proof of Theorem 125.3 of [1], we can make G an A-module and observe that if G is torsionfree, then for any $0 \neq x \in G$, $Ax = A/T(A)x = Qx$.

LEMMA 12. *If* $E(G)$ *is left self-injective, then* G *is quasi-injective as an* A*-module.*

PROOF. By Proposition 3, $G = D \oplus H$, D torsionfree divisible, H reduced, $H/T(H)$ divisible, $H_p = \bigoplus Z\left(p^{n_p}\right)$ and H is fully invariant and pure in $\prod H_p$. Now for any (left) ideal L of A, $\hom_A\left(L, H_p\right) = \hom_A\left(L_p, H_p\right) = \hom_{A_p}\left(L_p, H_p\right)$.

Since H_p is injective as a module over $Z/\left(p^{n_p}\right) = A_p$ and $A = A_p \oplus p^{n_p}A$, H_p is injective as an A-module for each prime p. Hence $\prod_p H_p$ is an injective A-module. We claim that D is injective as an A-module. Suppose $D \subset' E$, where E is injective as an A-module. Since D is torsionfree, E is also torsionfree as an abelian group. For any $0 \neq x \in E$, $Ax = A/T(A)x = Qx \cong Q$. Thus $Ax \cap D \neq 0$ implies that $(p/q)x = d \in D$. By the divisibility of D, $(p/q)x = (p/q)y$, $y \in D$. Since E is torsionfree $x = y \in D$. Thus $D = E$ is injective, so that $C = D \oplus \prod_p H_p$ is an injective A-module. Because D and H are fully invariant in C and $\prod H_p$ respectively, $G = D \oplus H$ is a fully invariant subgroup of C and hence is quasi-injective as an A-module.

PROPOSITION 13. *Suppose* $G = D \oplus H$, D *torsionfree divisible,* H *a fully invariant pure subgroup of* $\prod_{p \in P} H_p$ *containing* $\bigoplus_p H_p$ *where* P *is a set of primes and for each* $p \in P$ *there exists a positive integer* n_p *such that* H_p *is a direct sum of isomorphic copies of* $Z\left(p^{n_p}\right)$. *Then* $R = E(G)$ *is*

(i) generalised left continuous and

(ii) R/J *is left self-injective von Neumann regular and idempotents lift modulo* J *, where* J *is the Jacobson radical of* R .

PROOF. We shall first prove *(ii)*. Now

$$\hom_Z(G, H) = \hom_Z(H, H) = \hom_Z\left(\prod H_p, \prod H_p\right) = \hom_A\left(\prod H_p, \prod H_p\right) = \hom_A(G, H)$$

and since $G/T(G)$ is torsionfree divisible,

$$\hom_Z(G, D) = \hom_Z\left(G/T(G), D\right) = \hom_A\left(G/T(G), D\right) = \hom_A(G, D) .$$

Thus $\hom_Z(G, G) = \hom_A(G, G)$. By Lemma 10, G is quasi-injective as an A-module and so, by a theorem of Faith-Utumi-Osofsky (see [2]), R/J is left self-injective von Neumann regular and idempotents left mod J . Now *(i)* follows from the fact that G , being fully invariant and pure in the algebraically compact group $D \oplus \prod H_p$, is quasi-pure injective and hence satisfies the conditions of Theorem 6.

PROPOSITION 14. *If* R *is left self-injective and* $R \cong E(G)$ *for some abelian group* G *, then* $R/J \cong E(B)$ *for a suitable abelian group* B .

PROOF. By Proposition 3, $R \cong E\left(D \oplus \prod_{p \in P} H_p\right)$, where D is torsionfree divisible and $H_p = \oplus Z\left(p^{n_p}\right)$, n_p being a positive integer depending on p . As indicated in the proof of Proposition 13, $\hom_Z(X, X) = \hom_A(X, X)$ where $X = D \oplus \prod H_p$ and since X is injective as an A-module, by Utumi (see [2]), $J(R) = \{\alpha \in R \mid \ker \alpha \subset' X \text{ as } A\text{-modules}\}$. Let $\alpha \in J(R)$. Now $\ker \alpha \cap D \subset' D$ as A-modules and hence as Z-modules, which implies that $D\alpha$ is torsion. But $D\alpha \subseteq D$ so that $D\alpha = 0$. Also $H[p] \subseteq \ker \alpha$, for all primes p . Thus if $B = D \oplus \bigoplus_{p \in P} H[p]$, then $J(R) = \{\alpha \in R \mid \ker \alpha \supseteq B\}$. Now define $\theta : R \to E(B)$, by $\alpha \mapsto \alpha' = \alpha|B$. Clearly θ is a ring homomorphism whose kernel is precisely $J(R)$. Since $\hom_A(B, B) = \hom_Z(B, B)$ and every Z-endomorphism of B obviously extends to an endomorphism of X , we conclude that θ is an epimorphism. Thus $R/J \cong E(B)$.

Since R left self-injective implies that R/J is left self-injective von Neumann regular, we have the following corollary.

COROLLARY 15 ([3]). *Let* R *be left self-injective and von Neumann regular. Then for some abelian group* G ,

$$R \cong E(G) \iff R \cong E(\text{divisible + elementary}) \iff R = \prod_{i \in N} L_i ,$$

where N *is at most countable,* L_i *are full rings of linear transformations on vector spaces* V_i *over prime fields* F_i *and* $F_i \not\cong F_j$ *if* $i \neq j$.

References

[1] L. Fuchs, *Infinite abelian groups*, Vol. II (Academic Press, New York, 1973).

[2] B.L. Osofsky, "Endomorphism rings of quasi-injective modules", *Canad. J. Math.* 20 (1968), 895-903. MR39#184.

[3] K.M. Rangaswamy, "Representing Baer rings as endomorphism rings", *Math. Ann.* 190 (1970), 167-176. MR42#6105.

[4] Yuzo Utumi, "On continuous rings and self injective rings", *Trans. Amer. Math. Soc.* 118 (1965), 158-173. MR30#4793.

Institute of Advanced Studies,
Australian National University, Canberra, ACT 2600.

PROC. SECOND INTERNAT. CONF. THEORY OF GROUPS, 20D10, 20D30
CANBERRA, 1973, pp. 605-610.

LATTICE ISOMORPHISMS AND SATURATED FORMATIONS OF FINITE SOLUBLE GROUPS

Roland Schmidt

Let G and $\overline{G}$ be (finite) groups and let α be an isomorphism from the subgroup lattice $L(G)$ of G onto $L(\overline{G})$. Then in general G and $\overline{G}$ need not be isomorphic. So it is natural to ask which group theoretic properties groups with isomorphic subgroup lattices have in common. We call a class $\underline{\underline{K}}$ of groups invariant under lattice isomorphisms (short: lattice-invariant) if $G \in \underline{\underline{K}}$ implies $\overline{G} \in \underline{\underline{K}}$ for every group $\overline{G}$ with $L(\overline{G})$ isomorphic to $L(G)$. Then our question raised above is what classes of groups are lattice-invariant.

It is well known [4, Theorem 10, p. 46] that the class $\underline{\underline{S}}$ of all finite soluble groups has this property. In [3] we showed that also some subclasses of $\underline{\underline{S}}$ are lattice-invariant among them the classes $\underline{\underline{N}}^k$ $(k \geq 2)$ of groups of Fitting length at most k. It is the main purpose of this note to generalize this result in showing that a large number of saturated formations of finite soluble groups is lattice-invariant. For the definition and properties of saturated formations see [1] or [2, p. 696]. Our main result is the following

THEOREM. *For every prime* p *let* $\underline{\underline{F}}_p$ *be a formation of finite soluble groups satisfying*

(1) $\underline{\underline{F}}_p$ *contains some group* $F \neq 1$, *and*

(2) *if* $X \in \underline{\underline{F}}_p$ *and* $Z(X)$ *is cyclic, then* $Y \in \underline{\underline{F}}_p$ *for every group* Y *with* $L(Y)$ *isomorphic to* $L(X)$.

Then the saturated formation $\underline{\underline{F}}$ *defined locally by* $\{\underline{\underline{F}}_p\}$ *is invariant under lattice isomorphisms.*

If $\underline{\underline{K}}$ is a formation and $\underline{\underline{F}}_p = \underline{\underline{K}}$ for all primes p, then the saturated formation $\underline{\underline{F}}$ defined locally by $\{\underline{\underline{F}}_p\}$ is just $\underline{\underline{NK}}$, the class of groups whose $\underline{\underline{K}}$-residuum is nilpotent. Hence we get

COROLLARY 1. *If* $\underline{\underline{K}}$ *is a formation satisfying* (1) *and* (2) *with* $\underline{\underline{K}}$ *in place of* $\underline{\underline{F}}_p$ *, then* $\underline{\underline{F}} = \underline{\underline{NK}}$ *is lattice-invariant.*

We give some examples of formations $\underline{\underline{K}}$ satisfying (1) and (2). The formation $\underline{\underline{A}}$ of abelian groups satisfies (1). Since cyclic groups are exactly the groups with distributive subgroup lattices [4, Theorem 2, p. 4], $\underline{\underline{A}}$ also satisfies (2). Hence by Corollary 1, $\underline{\underline{NA}}$ is lattice-invariant (although $\underline{\underline{N}}$ and $\underline{\underline{A}}$ are not). Now, obviously, $\underline{\underline{NA}}$ satisfies (1) and (2). Hence $\underline{\underline{N}}^k\underline{\underline{A}}$ is lattice-invariant for every $k \geq 1$. Well known theorems of Suzuki [4, Theorem 4, p. 5 and Theorem 12, p. 12] show that if X is a nilpotent group with cyclic centre and $L(Y) \simeq L(X)$, then Y is nilpotent. Hence the formation $\underline{\underline{N}}$ of nilpotent groups satisfies (1) and (2), and as above we get that $\underline{\underline{N}}^k$ is lattice-invariant for $k \geq 2$. Thus the theorem contains and generalizes the result mentioned in the introduction. Other formations satisfying (1) and (2) are the class $\underline{\underline{M}}_L$ of supersoluble groups which induce automorphism groups of at most prime order in every chief factor [4, Theorem 10 and 11, pp. 10 and 11] and the classes $\underline{\underline{S}}_r$ of soluble groups of rank at most r $(r \geq 1)$ [3, Satz 3.1, p. 455]. We collect these examples in

COROLLARY 2. *The saturated formations* $\underline{\underline{N}}^k\underline{\underline{A}}, \underline{\underline{N}}^k\underline{\underline{M}}_L, \underline{\underline{N}}^k\underline{\underline{S}}_r$ $(k \geq 1,\ r \geq 1)$ *, and* $\underline{\underline{N}}^s$ $(s \geq 2)$ *are invariant under lattice-isomorphisms.*

For the proof of the theorem we need two lemmas. In [3] we have shown that a non-cyclic minimal normal subgroup N of a group G is mapped onto a minimal normal subgroup of G^α under every lattice-isomorphism α of G . In Lemma 1 we prove that, moreover, the centralizer of N is mapped onto the centralizer of N^α . This fact together with some information on cyclic minimal normal subgroups (Lemma 2) will imply the theorem.

LEMMA 1. *Let* G *and* $\overline{G}$ *be (arbitrary) finite groups,* α *an isomorphism from* $L(G)$ *onto* $L(\overline{G})$ *, and let* N *be a minimal normal subgroup of* G .

(a) If N *is not cyclic, then* N^α *is a minimal normal subgroup of* $\overline{G} = G^\alpha$ *and* $C_G(N)^\alpha = C_{G^\alpha}(N^\alpha)$.

(b) If N *is cyclic of order* p *, say, and if*

(b1) N^α *is normal in* G^α *, and*

(b2) α *and* α^{-1} *are* p*-regular (that is, map* p*-subgroups onto* p*-subgroups),* *then also* $C_G(N)^\alpha = C_{G^\alpha}(N^\alpha)$.

PROOF. The first statement in *(a)* is [3, Korollar 2.7, p. 454]. We prove that $C_G(N)^\alpha = C_{G^\alpha}(N^\alpha)$.

First suppose that N is non-abelian. Let $X \le C_G(N)$ such that $|X| = p^n$, p a prime. There exists a prime $q \ne p$ dividing $|N|$. If Q is a q-subgroup of N , $QX = Q \times X$ and $(|Q|, |X|) = 1$. By [4, Theorem 4, p. 5], $(QX)^\alpha = Q^\alpha \times X^\alpha$, that is, $X^\alpha \le C_{G^\alpha}(Q^\alpha)$. Since N is a minimal normal subgroup of G , N is generated by its q-subgroups Q and therefore N^α is generated by the Q^α . Hence $X^\alpha \le C_{G^\alpha}(N^\alpha)$. Since $C_G(N)$ is generated by its Sylow subgroups, $C_G(N)^\alpha \le C_{G^\alpha}(N^\alpha)$. Application of this result to α^{-1} and N^α yields the other inequality.

Now suppose that N is abelian of order p^n , say. If N is cyclic, then $|N^\alpha| = p$ by (b2). If N is not cyclic, then $|N^\alpha| = |N| = p^n$, since N and N^α have the same number of minimal subgroups. Furthermore, (b1) and (b2) are satisfied in this case, too. This is clear for (b1), and since G and G^α have non-cyclic p-chief factors, it follows for (b2) from [4, Propositions 2.8 and 2.9, pp. 43 and 44]. So in any case we have that $|N^\alpha| = |N|$ and (b1), (b2) hold. Let $L = O^p(C_G(N))$ be the group generated by all p'-subgroups of $C_G(N)$. If Q is any p'-subgroup of $C_G(N)$, then $NQ = N \times Q$, and again $(NQ)^\alpha = N^\alpha \times Q^\alpha$, $(|N^\alpha|, |Q^\alpha|) = 1$. Hence $L^\alpha \le O^p\left(C_{G^\alpha}(N^\alpha)\right)$. Considering α^{-1} we get $L^\alpha = O^p\left(C_{G^\alpha}(N^\alpha)\right)$, so $L^\alpha \triangleleft G^\alpha$ by (b1). Hence α induces an isomorphism $\bar{\alpha}$ from G/L onto G^α/L^α . By (b2), $\bar{\alpha}$ and $\bar{\alpha}^{-1}$ are p-regular. Hence

$$C_G(N)^\alpha/L^\alpha = \left(C_G(N)/L\right)^{\bar{\alpha}} \le O_p(G/L)^{\bar{\alpha}} = O_p\left(G^\alpha/L^\alpha\right) \le C_{G^\alpha}(N^\alpha)/L^\alpha ,$$

since $O_p\left(G^\alpha/C_{G^\alpha}(N^\alpha)\right) = 1$. So $C_G(N)^\alpha \le C_{G^\alpha}(N^\alpha)$ and considering α^{-1} we get the other inequality.

REMARK. Simple examples show that neither (b1) nor (b2) can be omitted in *(b)*.

LEMMA 2. *Let* G *and* $\bar{G}$ *be groups,* N *a normal subgroup of order* p *in* G ,

p a prime, and suppose that α is an isomorphism from $L(G)$ onto $L(\bar{G})$ such that N^{α} is not normal in $\bar{G}$. If α and α^{-1} are p-regular, then $N \leq Z(G)$.

PROOF. Suppose $C_G(N) \neq G$. Let $x \in G \backslash C_G(N)$, $X = \langle x \rangle$, and $H = NX$. Then $C_H(N) = NL$ where $L = C_H(N) \cap X = X \cap X^a$ for every $1 \neq a \in N$. Since any lattice isomorphism maps cyclic groups onto cyclic groups, $L^{\alpha} = X^{\alpha} \cap (X^a)^{\alpha}$ is centralized by X^{α} and $(X^a)^{\alpha}$ and is therefore contained in $Z(H^{\alpha})$. So if Q is any Sylow q-subgroup of H, Q^{α} normalizes N^{α}. For if $p = q$, $Q \leq NL$ and Q^{α} even centralizes N^{α}; if $q \neq p$, N^{α} is the only Sylow p-subgroup in $(NQ)^{\alpha}$, since α and α^{-1} are p-regular. Hence $N^{\alpha} \triangleleft H^{\alpha}$. Since $G = \langle x \mid x \in G \backslash C_G(N) \rangle$, $N^{\alpha} \triangleleft G^{\alpha}$, a contradiction. Hence $N \leq Z(G)$.

REMARK. It is easy to show that one can omit any one (but not both) of the conditions "α p-regular" and "α^{-1} p-regular" in Lemma 2.

PROOF OF THE THEOREM. Suppose that the Theorem is false and let $G \in \underline{\underline{F}}$ of minimal order such that $G^{\alpha} \notin \underline{\underline{F}}$ for some lattice-isomorphism α of G. By [4, Theorem 11, p. 46], G contains normal subgroups A and B such that A^{α} and B^{α} are normal in G^{α}, α induces an index-preserving isomorphism in $L(A)$, α^{-1} induces an index-preserving isomorphism in $L(B^{\alpha})$, and G/A and G^{α}/B^{α} are direct products of cyclic groups and p-groups of coprime orders. So if $D = A \cap B$, G^{α}/D^{α} induces in every chief factor an automorphism group of at most prime order. By (1), there is a group of prime order in $\underline{\underline{F}}_p$, and hence all groups of prime order are in $\underline{\underline{F}}_p$, by (2). Therefore $G^{\alpha}/D^{\alpha} \in \underline{\underline{F}}$. Since $G^{\alpha} \notin \underline{\underline{F}}$, $D \neq 1$. Let N be a minimal normal subgroup of G contained in D, $|N| = p^n$, say. By the construction of A and B given in [4, p. 46], it is clear that α and α^{-1} are p-regular. If $N^{\alpha} \triangleleft G^{\alpha}$, α induces an isomorphism $\bar{\alpha}$ from $L(G/N)$ onto $L(G^{\alpha}/N^{\alpha})$. By the choice of G, $G^{\alpha}/N^{\alpha} \in \underline{\underline{F}}$. Furthermore by Lemma 1, $C_G(N)^{\alpha} = C_{G^{\alpha}}(N^{\alpha})$. Hence α induces an isomorphism β from $L(G/C_G(N))$ onto $L\left[G^{\alpha}/C_{G^{\alpha}}(N^{\alpha})\right]$. Obviously $G/C_G(N) \in \underline{\underline{F}}_p$, and $Z(G/C_G(N))$ is cyclic by Schur's Lemma. By (2), $G^{\alpha}/C_{G^{\alpha}}(N^{\alpha}) \in \underline{\underline{F}}_p$ and hence $G^{\alpha} \in \underline{\underline{F}}$, a contradiction. So N^{α} is not normal in G^{α}, in particular $|N| = p$. By Lemma 2, $N \leq Z(G)$. Hence $H \neq 1$, if

$H = O_p(G) \cap C_G(O^p(G))$. Since α and α^{-1} are p-regular, p- and p'-subgroups of G centralize each other if and only if their images under α do, and hence $H^\alpha = O_p(G^\alpha) \cap C_{G^\alpha}(O^p(G^\alpha))$. Hence $H^\alpha \triangleleft G^\alpha$ and α induces an isomorphism from $L(G/H)$ onto $L(G^\alpha/H^\alpha)$. By the choice of G , $G^\alpha/H^\alpha \in \underline{\underline{F}}$. Since G^α induces trivial automorphisms in the chief factors contained in H^α , $G^\alpha \in \underline{\underline{F}}$, a final contradiction.

REMARK. (a) It is clear that with two formations $\underline{\underline{F}}_1$ and $\underline{\underline{F}}_2$ also $\underline{\underline{F}}_1 \cap \underline{\underline{F}}_2$ is lattice-invariant. Hence there exists a unique minimal lattice-invariant saturated formation $\underline{\underline{K}} \neq \{1\}$. One can easily prove that $\underline{\underline{K}}$ can be defined locally by $\{\underline{\underline{K}}_p\}$ where $\underline{\underline{K}}_p$ is the formation of abelian groups of square-free exponent dividing $p - 1$. Hence $\underline{\underline{K}}$ lies (strictly) between the (non-saturated) formation $\underline{\underline{M}}_L$ and the (saturated) formation of supersoluble groups.

(b) We conclude with an example showing that not every saturated formation $\underline{\underline{F}}$ containing $\underline{\underline{K}}$ is lattice-invariant. Let p, q, r be primes such that q^2r^2 divides $p - 1$ (for example $p = 37$), let $\underline{\underline{F}}_t = \underline{\underline{S}}$ for all primes $t \neq p$, and let $\underline{\underline{F}}_p$ be the formation of abelian groups of exponent dividing qv where v is the product of all (different) primes dividing $p - 1$. If $\underline{\underline{F}}$ is the formation defined locally by $\{\underline{\underline{F}}_p\}$, then $\underline{\underline{F}} \geq \underline{\underline{K}}$. If G and $\bar{G}$ are the semidirect product of the cyclic group of order p by its automorphism of order q^2, respectively r^2 , then $G \in F$, $\bar{G} \notin \underline{\underline{F}}$, but $L(G) \simeq L(\bar{G})$. Hence $\underline{\underline{F}}$ is not lattice-invariant.

References

[1] Wolfgang Gaschütz, "Zur Theorie der endlichen auflösbaren Gruppen", *Math. Z.* 80 (1963), 300-305. MR31#3505.

[2] B. Huppert, *Endliche Gruppen I* (Die Grundlehren der mathematischen Wissenschaften, Band 134. Springer-Verlag, Berlin, Heidelberg, New York, 1967). MR37#302.

[3] Roland Schmidt, "Verbandsisomorphismen endlicher auflösbarer Gruppen", *Arch. Math.* 23 (1972), 449-458.

[4] Michio Suzuki, *Structure of a group and the structure of its lattice of subgroups* (Ergebnisse der Mathematik und ihrer Grenzgebiete, Neue Folge, Heft 10. Springer-Verlag, Berlin, Göttingen, Heidelberg, 1956). MR18,715.

Mathematisches Seminar der Christian-Albrechts-Universität,
23 Kiel, (West) Germany.

PROC. SECOND INTERNAT. CONF. THEORY OF GROUPS, 20E30
CANBERRA, 1973, pp. 611-632.

SOME REFLECTIONS ON HNN EXTENSIONS

Paul E. Schupp

Two of the basic constructions of infinite group theory are the free product with amalgamated subgroup, introduced by Schreier [11] in 1927, and Higman-Neumann-Neumann extensions, introduced by G. Higman, B.H. Neumann, and H. Neumann [4] in 1949. The main point of this talk is to argue for the adoption of a certain point of view towards these constructions. We stress from the outset that these two constructions are very parallel, and are best viewed as each being half of a single basic concept. Indeed, we shall later give a single axiomatization (Stallings' concept of bipolar structure [13]) of both constructions.

We being by reviewing the definitions of the constructions. Let $G = \langle x_1, \ldots; r_1, \ldots \rangle$ and $H = \langle y_1, \ldots; s_1, \ldots \rangle$ be groups. Let $A \subseteq G$ and $B \subseteq H$ be subgroups, such that there exists an isomorphism $\varphi : A \to B$. Then the *free product of* G *and* H, *amalgamating the subgroups* A *and* B *by the isomorphism* φ is the group

$$\langle x_1, \ldots, y_1, \ldots; r_1, \ldots, s_1, \ldots, a = \varphi(a), a \in A \rangle .$$

We introduce the following convention on notation. If G is a group for which a presentation has been chosen, by the notation

$$\langle G, z, \ldots; u, \ldots \rangle$$

we mean the generators and defining relators of G together with whatever additional generators and relators are indicated. Any additional generators are understood to be disjoint from the generators of G. Thus we can write the free product with amalgamation as

$$\langle G * H;\ a = \varphi(a), a \in A \rangle ,$$

or even more simply as

$$\langle G * H;\ A = B, \varphi \rangle .$$

The basic idea of the free product with amalgamation is that the subgroup A is identified with its isomorphic image $\varphi(A) \subseteq H$. The free product with amalgamation

Research for this paper was supported by an NSF Grant.

depends on G, H, A, B and the isomorphism φ. The groups G and H are called the *factors* of the free product with amalgamation, while A and B are called the *amalgamated subgroups*.

From now on, we will shorten the term Higman-Neumann-Neumann extension to HNN extension. We now turn to the definition. Let G be a group, and A and B be subgroups of G with $\varphi : A \to B$ an isomorphism. *The* HNN *extension of* G *relative to* A, B *and* φ is the group

$$G^* = \langle G, t;\ t^{-1}at = \varphi(a),\ a \in A \rangle .$$

The group G is called the *base* of G^*, t is called the *stable letter*, and A and B are called the *conjugated subgroups*.

Note that both the free product with amalgamation and HNN constructions involve two subgroups and an isomorphism between them. In very loose language, the two constructions might be called the "disconnected case" and the "connected case" of one basic idea. In the free product with amalgamation, A and B are subgroups of separate groups G and H. In the HNN extension, A and B are already contained in a single group G.

As an aside, we discuss a topological situation which is often used as motivation for studying free products with amalgamation. All the spaces and subspaces which we mention are assumed to be arcwise connected. If X is a topological space, $\pi_1(X)$ will denote the fundamental group of X. Let X and Y be spaces, and let U and V be open arcwise connected subspaces of X and Y respectively such that there is a homeomorphism $h : U \to V$. Choose a basepoint $u \in U$ for the fundamental groups of U and X. Similarly, choose a basepoint $v \in V$. There is a homomorphism $\eta_u : \pi_1(U) \to \pi_1(X)$ defined by simply considering a loop in U as a loop in X. Suppose that η_u and the similarly defined homomorphism $\eta_v : \pi_1(V) \to \pi_1(Y)$ are both injections. The homeomorphism h induces an isomorphism $h^* : \pi_1(U) \to \pi_1(V)$. Suppose we identify U and V by the homeomorphism h to obtain a new space Z. Under the assumptions made, the Seifert-Van Kampen Theorem (*cf.* Massey [6]) says that

$$\pi_1(Z) = \langle \pi_1(X) * \pi_1(Y);\ \pi_1(U) = \pi_1(V),\ h^* \rangle$$

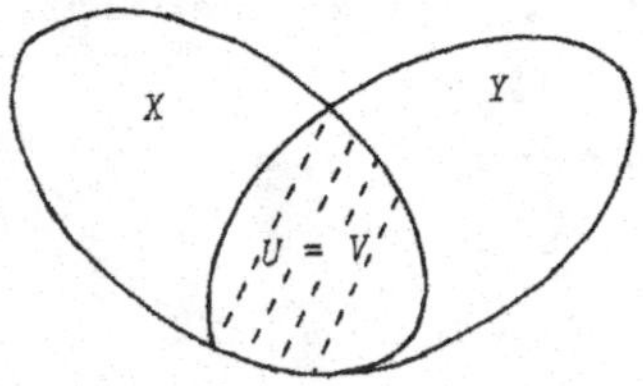

The HNN extension has a similar topological interpretation. Suppose that U and V are both subspaces of the arcwise connected space X. Assume the same hypothesis on U and V as above. Let I be the unit interval, and let $C = U \times I$. Identify $U \times \{0\}$ with U and identify $U \times \{1\}$ with V by the homeomorphism h. Let Z be the resulting space. (What we have done is to attach a handle to X.) The Seifert-Van Kampen Theorem can be used to show that

$$\pi_1(Z) = \left\langle \pi_1(X),\ t;\ t^{-1}\pi_1(U)t = \pi_1(V) \right\rangle$$

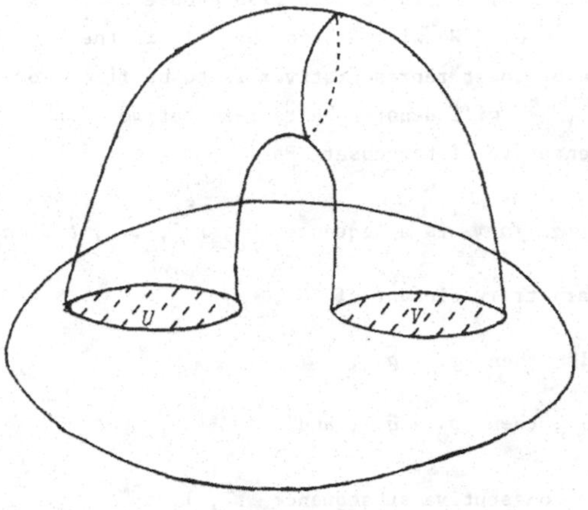

For any statement about general properties of free products with amalgamation or HNN extensions there is a corresponding statement about the other construction. Following the historical development of the subject, it is customary to first develop the basic properties of free products with amalgamation and then to study HNN extensions. We feel that there are good reasons for doing things the other way round, taking HNN extensions as the basic construction. This point of view often allows one to simplify proofs.

Let $G^* = \langle G,\ t;\ t^{-1}at = \varphi(a),\ a \in A \rangle$ be an HNN extension. We consider two definitions which will allow us to formulate a normal form theorem for HNN extensions. For the rest of this section, the letter g, with or without subscripts, will denote an element of G. If g is thought of as a word, it is a word on the generators of G, that is, g contains no occurrences of $t^{\pm 1}$. The letter ε, with or without subscripts, will denote 1 or -1.

DEFINITION. A sequence $g_0, t^{\varepsilon_1}, g_1, \ldots, t^{\varepsilon_n}, g_n$ $(n \geq 0)$ is said to be reduced if there is no consecutive subsequence t^{-1}, g_i, t with $g_i \in A$ or t, g_j, t^{-1} with $g_j \in B$.

In their original paper [4], Higman, Neumann, and Neumann proved that G is embedded in G^* by the map $g \to g$. The rest of the normal form theorem for HNN groups was proved by Britton [1], and is usually referred to as Britton's Lemma.

BRITTON'S LEMMA. *If the sequence* $g_0, t^{\varepsilon_1}, \ldots, t^{\varepsilon_n}, g_n$ *is reduced and* $n \geq 1$ *, then* $g_0 t^{\varepsilon_1} \ldots t^{\varepsilon_n} g_n \neq 1$ *in* G^* .

The products of the elements in two distinct reduced sequences may be equal in G^* . To actually get normal forms we need a further refinement. Choose a set of right coset representatives of A in G and also choose a set of right coset representatives of B in G . We will assume that 1 is the representative of both A and B . The choice of coset representatives is to be fixed for the rest of the discussion. If $g \in G$, $\bar{g}$ will denote the representative of the coset Ag , and $\hat{g}$ will denote the representative of the coset Bg .

DEFINITION. A *normal form* is a sequence $g_0, t^{\varepsilon_1}, \ldots, t^{\varepsilon_n}, g_n$ $(n \geq 0)$ where

(i) g_0 is an arbitrary element of G ,

(ii) if $\varepsilon_i = -1$, then $g_i = \bar{g}_i$,

(iii) if $\varepsilon_i = +1$, then $g_i = \hat{g}_i$, and

(iv) there is no consecutive subsequence $t^{\varepsilon}, 1, t^{-\varepsilon}$.

The following discussion will explain our definition of a normal form. The defining relations

(1) $$t^{-1}at = \varphi(a) \ , \quad a \in A$$

of the HNN extension, can be written as

(2) $$t^{-1}a = \varphi(a)t^{-1} \ .$$

By conjugating both sides of (1) by t , the relations can also be written as

(3) $$tbt^{-1} = \varphi^{-1}(b) \ , \quad b \in B \ ,$$

which are equivalent to

(4) $$tb = \varphi^{-1}(b)t \ .$$

We can view (2) and (4) as "quasi-commuting" relations. These relations allow us to move an element $a \in A$ to the left of a t^{-1} but changing a to $\varphi(a)$. Similarly, we can move $b \in B$ to the left of a t , changing b to $\varphi^{-1}(b)$. By working from right to left, we see that every element of G^* is equal to a product $g_0 t^{\varepsilon_1} \ldots t^{\varepsilon_n} g_n$ where $g_0, t^{\varepsilon_1}, \ldots, t^{\varepsilon_n}, g_n$ is a normal form.

The normal form theorem has two equivalent statements (I) and (II) below. Note that (I) is the combination of the theorem of Higman, Neumann, and Neumann, and Britton's Lemma.

THEOREM I (Normal Form Theorem for HNN extensions). *Let* $G^* = \langle G, t;\ t^{-1}at = \varphi(a),\ a \in A \rangle$ *be an* HNN *extension. Then*

(I) *the group* G *is embedded in* G^* *by the map* $g \to g$.

If $g_0 t^{\varepsilon_1} \ldots t^{\varepsilon_n} g_n = 1$ *in* G^* *where* $n \geq 1$, *then* $g_0, t^{\varepsilon_1}, \ldots, t^{\varepsilon_n}, g_n$ *is not reduced.*

(II) *Every element* w *of* G^* *has a unique representation as* $w = g_0 t^{\varepsilon_1} \ldots t^{\varepsilon_n} g_n$ *where* $g_0, t^{\varepsilon_1}, \ldots, t^{\varepsilon_n}, g_n$ *is a normal form.*

PROOF. It turns out that there are fewer details to verify in directly proving the normal form theorem for HNN extensions than in proving the normal form theorem for free products with amalgamation. In order to prove the theorem, we follow the Artin-van der Waerden idea of making G^* permute normal forms. Intuitively, the action of G^* will be "multiply on the left and then reduce to normal form". Let W be the set of all normal forms from G^*, and let $S(W)$ denote the group of all permutations of W. In order to define a homomorphism $\Psi : G^* \to S(W)$, it suffices to define Ψ on G and t, and then show that all defining relations go to 1.

If $g \in G$, define $\Psi(g)$ by

$$\Psi(g)\left(g_0, t^{\varepsilon_1}, \ldots, t^{\varepsilon_n}, g_n\right) = gg_0, t^{\varepsilon_1}, \ldots, t^{\varepsilon_n}, g_n .$$

Thus $\Psi(g)$ simply multiplies the first element of the sequence by g. Clearly, $\Psi(g'g) = \Psi(g')\Psi(g)$. In particular, $\Psi(g)\Psi(g^{-1}) = 1_W = \Psi(g^{-1})\Psi(g)$. Hence, $\Psi(g)$ is a permutation of W and Ψ is a homomorphism from G into $\Psi(g)$.

Define $\Psi(t)$ as follows. Let $g_0, t^{\varepsilon_1}, g_1, \ldots, t^{\varepsilon_n}, g_n$ be a normal form. If $\varepsilon_1 = -1$ and $g_0 \in B$,

$$\Psi(t)\left(g_0, t^{-1}, \ldots, t^{\varepsilon_n}, g_n\right) = \varphi^{-1}(g_0)g_1, t^{\varepsilon_2}, g_2, \ldots, t^{\varepsilon_n}, g_n .$$

Otherwise,

$$\Psi(t)\left(g_0, t^{\varepsilon_1}, \ldots, t^{\varepsilon_n}, g_n\right) = \varphi^{-1}(b), t, \hat{g}_0, t^{\varepsilon_1}, g_1, \ldots, t^{\varepsilon_n} g_n$$

where $g_0 = b\hat{g}_0$ with $b \in B$.

It is necessary to verify that $\Psi(t)$ is actually a permutation of W. We will show that $\Psi(t)$ has inverse $\Psi(t^{-1})$ defined as follows. Let $g_0, t^{\varepsilon_1}, \ldots, t^{\varepsilon_n}, g_n$ be a normal form. If $\varepsilon_1 = +1$ and $g_0 \in A$, then

$$\Psi(t^{-1})\left(g_0, t^1, \ldots, t^{\varepsilon_n}, g_n\right) = \varphi(g_0)g_1, t^{\varepsilon_2}, g_2, \ldots, t^{\varepsilon_n}, g_n .$$

Otherwise,

$$\Psi(t^{-1})\left(g_0, t^{\varepsilon_1}, \ldots, t^{\varepsilon_n}, g_n\right) = \varphi(a), t^{-1}, \bar{g}_0, t^{\varepsilon_1}, \ldots, t^{\varepsilon_n}, g_n$$

where $g_0 = a\bar{g}_0$ with $a \in A$.

In checking that $\Psi(t^{-1})\Psi(t) = 1_W$, there are two cases. Let $g_0, t^{\varepsilon_1}, \ldots, t^{\varepsilon_n}, g_n$ be a normal form. If the first clause in the definition of $\Psi(t)$ applies, then $\varepsilon_1 = -1$ and $g_0 \in B$. Note that it is then impossible to have $\varepsilon_2 = +1$ and $g_1 \in A$ since the sequence is a normal form. Now

$$\Psi(t)\left(g_0, t^{-1}, \ldots, t^{\varepsilon_n}, g_n\right) = \varphi^{-1}(g_0)g_1, t^{\varepsilon_2}, g_2, \ldots, t^{\varepsilon_n}, g_n .$$

By the preceding remark, the second clause of the definition of $\Psi(t^{-1})$ applies. Since g_1 is a coset representative for A and $\varphi^{-1}(g_0) \in A$, the coset representative of $A\varphi^{-1}(g_0)g_1$ is g_1. Therefore,

$$\Psi(t^{-1})\left(\varphi^{-1}(g_0)g_1, t^{\varepsilon_2}, g_2, \ldots, t^{\varepsilon_n}, g_n\right) = g_0, t^{-1}, g_1, t^{\varepsilon_2}, \ldots, t^{\varepsilon_n}, g_n .$$

If the second clause of the definition of $\Psi(t)$ applies, it is immediate that

$$\Psi(t^{-1})\Psi(t)\left(g_0, t^{\varepsilon_1}, \ldots, t^{\varepsilon_n}, g_n\right) = g_0, t^{\varepsilon_1}, \ldots, t^{\varepsilon_n}, g_n .$$

Thus $\Psi(t^{-1})\Psi(t) = 1_W$. A similar check shows that $\Psi(t)\Psi(t^{-1}) = 1_W$ and that if $b \in B$, $\Psi(b) = \Psi(t^{-1})\Psi(\varphi^{-1}(b))\Psi(t)$. Thus Ψ is indeed a homomorphism from G^* into $S(W)$.

To finish the proof it is necessary only to note that if $g_0, t^{\varepsilon_1}, \ldots, t^{\varepsilon_n}, g_n$ is a normal form, then

$$\Psi\left(g_0t^{\varepsilon_1} \ldots t^{\varepsilon_n}g_n\right)(1) = g_0, t^{\varepsilon_1}, \ldots, t^{\varepsilon_n}, g_n .$$

Thus the products of the elements in distinct normal forms represent distinct elements of G^* . //

For most purposes, there is no need to choose coset representatives. In practice, it is possible to almost exclusively use statement (I) of the normal form theorem. What is important is that G is embedded in G^* and that we have a criterion for telling when words of G^* do not represent the identity.

It is usual to be rather sloppy in formally distinguishing between a sequence $g_0, t^{\varepsilon_1}, \ldots, t^{\varepsilon_n}, g_n$ and the product $g_0 t^{\varepsilon_1} \ldots t^{\varepsilon_n} g_n$. It is clear from the context which is actually meant. The following lemma follows easily from the normal form theorem.

LEMMA. *Let* $u = g_0 t^{\varepsilon_1} \ldots t^{\varepsilon_n} g_n$ *and* $v = h_0 t^{\delta_1} \ldots t^{\delta_m} h_m$ *be reduced words, and suppose that* $u = v$ *in* G^* . *Then* $m = n$ *and* $\varepsilon_i = \delta_i$, $i = 1, \ldots, n$.

We assign a length to each element z of G^* as follows. Let w be any reduced word of G^* which represents z . If $w = g_0 t^{\varepsilon_1} \ldots t^{\varepsilon_n} g_n$, the *length* of z , written $|z|$, is the number of occurrences of $t^{\pm 1}$ in w . In view of the above lemma, $|z|$ is well-defined.

There is a natural notion of cyclically reduced in HNN extensions. An element $w = t^{\varepsilon_1} \ldots t^{\varepsilon_n} g_n$ is *cyclically reduced* if all cyclic permutations of the sequence $t^{\varepsilon_1}, \ldots, t^{\varepsilon_n}, g_n$ are reduced. Clearly, every element of G^* is conjugate to a cyclically reduced element.

The torsion theorem for HNN extensions is easily proved from the normal form theorem.

THEOREM II (The Torsion Theorem for HNN extensions). *Let* $G^* = \langle G, t; t^{-1}At = B, \varphi \rangle$ *be an HNN extension. Then every element of finite order in* G^* *is a conjugate of an element of finite order in the base* G . *Thus* G^* *has elements of finite order* n *only if* G *has elements of order* n .

Given a statement about HNN extensions, the corresponding statement about free products with amalgamation can usually be inferred almost immediately. We illustrate this procedure for the normal form theorem for free products with amalgamation, which we will state in the form not involving coset representatives.

Let G and H be groups with subgroups $A \subseteq G$ and $B \subseteq H$, and with $\varphi : A \to B$ an isomorphism. Form the group

$$P = \langle G * H;\ a = \varphi(a),\ a \in A \rangle .$$

We can view P as the quotient of the free product $G * H$ by the normal subgroup generated by $\{a\varphi(a)^{-1} : a \in A\}$. A sequence $c_1, \ldots, c_n$, $n \geq 0$, of elements of $G * H$ will be called *reduced* if

(1) each c_i is in one of the factors G or H,

(2) successive c_i, c_{i+1} come from different factors,

(3) if $n > 1$, no c_i is in A or B,

(4) if $n = 1$, $c_1 \notin A$.

It is clear that every non-trivial element of P is equal to the product of the elements in a reduced sequence. On the other hand we have

THEOREM III (Normal Form Theorem for free products with amalgamation). *If* $c_1, \ldots, c_n$ *is a reduced sequence,* $n \geq 1$, *then the product* $c_1 \ldots c_n \neq 1$ *in* P. *In particular,* G *and* H *are embedded in* P *by the maps* $g \to g$ *and* $h \to h$.

PROOF. The group $F^* = \langle G * H, t;\ t^{-1}at = \varphi(a),\ a \in A \rangle$ is an HNN extension of the free product $G * H$. Define $\psi : P \to F^*$ by

$$\begin{cases} \psi(g) = t^{-1}gt & \text{if } g \in G, \\ \psi(h) = h & \text{if } h \in H. \end{cases}$$

Now ψ is a homomorphism since the defining relations of P go to 1. But ψ sends a reduced sequence $c_1, \ldots, c_n$ of elements of P to a reduced sequence of elements in F^*. (For example, $\psi(h_1g_1h_2g_2) = h_1t^{-1}g_1th_2t^{-1}g_2t$ and each $g_i \notin A$, $h_i \notin B$.) The theorem thus follows from the normal form theorem for HNN extensions. //

The proof shows that ψ is, in fact, an embedding. Thus P is isomorphic to the subgroup of F^* generated by $t^{-1}Gt$ and H.

2. Some embedding theorems

We shall write down proofs of some famous theorems, stressing the HNN extension. In their original paper, Higman, Neumann, and Neumann proved the following celebrated result.

THEOREM IV. *Every countable group* C *can be embedded in a group* G *generated by two elements of infinite order. The group* G *has an element of finite order* n

iff C *does. If* C *is finitely presented then so is* G .

PROOF. Assume that $C = \langle c_1, c_2, \ldots; s_1, \ldots \rangle$ is given with a countable set of generators. Let $F = C * \langle a, b \rangle$. The set

$$\{a, b^{-1}ab, b^{-2}ab^2, \ldots, b^{-n}ab^n, \ldots\}$$

freely generates a free subgroup of $\langle a, b \rangle$ since it is Nielsen reduced. Similarly, the set

$$\left\{b, c_1a^{-1}ba, \ldots, c_na^{-n}ba^n, \ldots\right\}$$

freely generates a free subgroup of F . (To check this let π be the projection of F onto $\langle a, b \rangle$ defined by $a \to a$, $b \to b$, $c_i \to 1$ for all i . Since the images b, $a^{-i}ba^i$, $i \geq 1$, are free generators, so also is the indicated set.)

Hence, the group

$$G = \left\langle F, t; t^{-1}at = b, t^{-1}b^{-i}ab^it = c_ia^{-i}ba^i, i \geq 1\right\rangle$$

is an HNN extension of F . Thus C is embedded in G . That G is generated by t and a is immediate from the defining relations. That G has an element of order n iff C does follows from the torsion theorem for HNN extensions. Finally, suppose that $C = \langle c_1, \ldots, c_m; s_1, \ldots, s_k \rangle$ is finitely presented. The HNN relations, $t^{-1}at = b$, $t^{-1}b^{-i}ab^it = c_ia^{-i}ba^i$, can all be eliminated by Tietze transformations since each relation contains a single occurrence of some generator (namely, b or c_i). //

The next theorem was proved by B.H. Neumann [8] in 1937.

THEOREM V. *There are* $2^{\aleph_0}$ *non-isomorphic 2-generator groups.*

PROOF. Let S be any set of primes. Let $C_S = \sum_{p \in S} Z_p$ (direct sum), and embed each C_S in a 2-generator group G_S as in the previous theorem. Now C_S , and thus G_S , has an element of order p iff $p \in S$. Since there are $2^{\aleph_0}$ distinct sets of primes, the result follows //

One of the embedding theorems proved by Higman, Neumann, and Neumann in their original paper is the following.

THEOREM VI. *Any countable group* C *can be embedded in a countable group* G *in which all elements of the same order are conjugate.*

PROOF. The first thing is to embed C in a group C^* in which any two elements of C which have the same order are conjugate. To do this, let $\{\langle a_i, b_i\rangle : i \in I\}$ be the set of all ordered pairs of elements of C which have the same order. The group

$$C^* = \left\langle C, t_i, i \in I; t_i a_i t_i^{-1} = b_i, i \in I\right\rangle$$

has the desired property. To prove the theorem, let $G_0 = C$. Suppose inductively that G_i has been defined. Embed G_i in a group G_{i+1} in which any two elements of G_i which have the same order are conjugate. The group

$$G = \bigcup_{i=0}^{\infty} G_i$$

is the desired group. //

The next theorem is well-known and can be proved in several ways.

THEOREM VII. *Every countable group C can be embedded in a countable, simple, divisible group G.*

PROOF. First, embed C in a countable group K which has elements of all orders. (For example, the direct sum of C, $\mathbb{Z}$, and the groups $\mathbb{Z}_p$ for all primes p.) Embed the group $K * \langle x\rangle$ in a two generator group U in which both generators have infinite order. Finally, embed U in a countable group G in which all elements elements of the same order are conjugate. In summary, we have embeddings

$$C \to K \to K * \langle x\rangle \to U \to G .$$

We claim that the group G is both simple and divisible. Let $\{1\} \neq N \triangleleft G$. Since K contains elements of all orders and all elements of the same order are conjugate in G, N contains an element $1 \neq z \in K$. Now $x^{-1}z^{-1}x \in N$, and thus $x^{-1}z^{-1}xz$ is an element of N which has infinite order. Thus N contains the generators of U by the conjugacy property of G. Thus $K \subseteq U \subseteq N$. Since K contains elements of all orders and all elements of the same order are conjugate in G, $N = G$.

To check divisibility, let $g \in G$, and let n be any positive integer. Let g have order m (which may be ∞). Since G contains elements of all orders, G has an element z of order mn. Since z^n has order m, there is a $v \in G$ so that $g = v^{-1}z^n v$. Thus $g = (v^{-1}zv)^n$. //

In view of the last theorem and the fact that there are $2^{\aleph_0}$ 2-generator groups, there must be $2^{\aleph_0}$ non-isomorphic countable simple groups. (Any particular countable group has at most $\aleph_0$ 2-generator subgroups.) Most techniques for constructing infinite simple groups yield groups which are not finitely generated. Indeed, it was not until 1951 that G. Higman [3] proved the existence of finitely generated infinite simple groups. Nevertheless, a remarkable theorem of P. Hall [2] asserts that every countable group can be embedded in a 3-generator simple group. We shall give a different proof of Hall's Theorem (requiring, however, six generators) by using the construction of Rabin [10] which is well-known in showing that certain group theoretic decision problems are unsolvable.

THEOREM VIII. *Every countable group* C *can be embedded in a six-generator simple group.*

PROOF. First embed C in a countable simple group S. Embed the free product $S * \langle x \rangle$ in a two generator group U where the chosen generators u_1 and u_2 of U both have infinite order. Then the group

$$J = \left\langle U, y_1, y_2; y_1^{-1}u_1y_1 = u_1^2, y_2^{-1}u_2y_2 = u_2^2 \right\rangle$$

is an HNN extension of U with stable letters y_1 and y_2. The group

$$K = \left\langle J, z; z^{-1}y_1z = y_1^2, z^{-1}y_2z = y_2^2 \right\rangle$$

is an HNN extension of J.

We next consider the group

$$Q = \langle r, s, t; s^{-1}rs = r^2, t^{-1}st = s^2 \rangle .$$

Letting $P = \langle r, s; s^{-1}rs = r^2 \rangle$, P is an HNN extension of $\langle r \rangle$ and Q is an HNN extension of P with stable letter t. We claim that r and t freely generate a free subgroup of rank 2. For, let V be any non-trivial freely reduced word on r and t, and suppose that $V = 1$ in Q. If V does not contain t, then V is identically r^n for some $n \neq 0$. But r has infinite order in Q so this is impossible. Hence, V contains t. By Britton's Lemma applied to Q over P, V contains a subword $t^{\varepsilon}Rt^{-\varepsilon}$ where R does not contain t and is in the subgroup $\langle s \rangle$ or $\langle s^2 \rangle$ depending on the sign of ε. Since V is freely reduced and contains only t's and r's R must be r^n for some $n \neq 0$. Hence an equation $r^n = s^j$ with $n \neq 0$ holds in P. But this is impossible by Britton's Lemma applied to P. We thus have a contradiction and the claim is established.

Let w be an element of the group S with $w \neq 1$. The commutator $[w, x] = wxw^{-1}x^{-1}$ has infinite order in U. As in the proof of the previous claim, $[w, x]$ and z freely generate a free subgroup of K. Thus we can form the free product with amalgamation

$$D = \langle K * Q;\ r = z,\ t = [w, x] \rangle .$$

We claim that D has the property that if $N \lhd D$, then $N \cap S = \{1\}$ or $N = D$. For suppose that $N \cap S \neq 1$. Since S is simple, $w \in N$. Thus in the quotient D/N , $w = 1$. But we next see (and this is the main point of Rabin's construction) that $w = 1$ implies that $D/N = \{1\}$. For, if $w = 1$ then $[w, x] = 1$, and, using the defining relators, this successively implies that $t = 1$, $s = 1$, $r = 1$, $z = 1$, $y_1 = 1$, $y_2 = 1$, $u_1 = 1$, $u_1 = 1$. Thus $N = D$.

There exists a normal subgroup M of D which is maximal with respect to $M \cap S = \{1\}$. Thus S is embedded in D/M and, by maximality and the property above, D/M is simple. Since the generators r and t can be eliminated, we see that D has six generators. //

3. Bipolar structures

In his work on the theory of ends [13], Stallings shows that if G is a finitely generated group with infinitely many ends, then G has a decomposition as a non-trivial free product with amalgamation or as an HNN extension where the amalgamated or conjugated subgroups are finite. In his proof, Stallings introduces the concept of a bipolar structure. It turns out that bipolar structures provide a characterization for a group to be either a non-trivial free product with amalgamation or an HNN extension. We shall apply this characterization to prove Hanna Neumann's subgroup theorem [9] for finitely generated subgroups of free products with amalgamation and HNN extensions.

We turn to the definition of a bipolar structure on a group. (The reader familiar with Stallings' work will notice two slight differences in our definition. In working with ends, it is crucial that the subgroup F mentioned below be finite. Here, of course, we must allow F to be infinite. Also, we shall not need Stallings' set S .)

DEFINITION. A *bipolar structure* on a group G is a partition of G into five disjoint subsets F, EE, EE^*, E^*E, E^*E^* satisfying the following axioms. (The letters X, Y, Z will stand for the letters E or E^* with the convention that $(X^*)^* = X$, and so on.)

1. F is a subgroup of G .
2. If $f \in F$ and $g \in XY$, then $gf \in XY$.
3. If $g \in XY$, then $g^{-1} \in YX$, (inverse axiom).
4. If $g \in XY$ and $h \in Y^*Z$, then $gh \in XZ$, (product axiom).

5. **(Boundedness axiom).** If $g \in G$, there is an integer $N(g)$ such that, if there exist $g_1, \ldots, g_n \in G$ and $X_0, \ldots, X_n$ with $g_i \in X^*_{i-1}X_i$ and $g = g_1 \ldots g_n$, then $n \le N(g)$.

6. $EE^* \neq \emptyset$, (non-triviality axiom).

We next show that every non-trivial free product with amalgamation and every HNN extension possess bipolar structures. Let $G = \langle A * B;\ F = \varphi(F)\rangle$ be a free product with amalgamation where F and $\varphi(F)$ are proper subgroups of A and B respectively. (It may be that $F = \{1\}$, so that G is an ordinary free product.) We can define a bipolar structure on G as follows. The subgroup F of G is the subgroup F in the bipolar structure. Every element $g \in G-F$ has a representation as

$$g = c_1 \ldots c_n$$

where no $c_i \in F$, each c_i is in one of the factors A or B, and successive c_i, c_{i+1} come from different factors. The normal form theorem for free products with amalgamation ensures that the number n and what factors the c_i come from are the same for any representation satisfying the stated restrictions. We define the sets EE, EE^*, E^*E, and E^*E^* as follows:

$$g \in EE \quad \text{iff} \quad c_1 \in A \text{ and } c_n \in A\,,$$

$$g \in EE^* \quad \text{iff} \quad c_1 \in A \text{ and } c_n \in B\,,$$

$$g \in E^*E \quad \text{iff} \quad c_1 \in B \text{ and } c_n \in A\,,$$

$$g \in E^*E^* \quad \text{iff} \quad c_1 \in B \text{ and } c_n \in B\,.$$

Verification of the axioms is immediate. For example, if $g \in XY$ and $h \in Y^*Z$, then g and h have representations

$$g = c_1 \ldots c_n \quad \text{and} \quad h = d_1 \ldots d_m$$

where c_n and d_1 are in different factors. Thus the representation of gh is $c_1 \ldots c_n d_1 \ldots d_m$, and axiom 4 holds. The number $N(g)$ in axiom 5 is simply the free product with amalgamation length of g. Since F and $\varphi(F)$ are proper subgroups of A and B respectively, there are elements $c \in A-F$ and $d \in B-\varphi(F)$. Thus $cd \in EE^*$, verifying the non-triviality axiom.

We now consider the HNN case. Let

$$G = \langle H, t;\ tFt^{-1} = \varphi(F)\rangle$$

be an HNN extension. We can define a bipolar structure on G as follows. The subgroup F of G is the subgroup F of the bipolar structure. Each element $g \in G-F$ has a representation as a product of elements in a reduced sequence; say

$$g = g_0 t^{\varepsilon_1} g_1 \ldots t^{\varepsilon_n} g_n ,$$

where each $g_i \in H$.

We put

$$\begin{array}{lll} g \in EE & \text{iff} & g_0 \in H-F \text{ , or } g_0 \in F \text{ and } \varepsilon_1 = +1 \\ & \text{and} & g_n \in H-F \text{ , or } g_n \in F \text{ and } \varepsilon_n = -1 \text{ ,} \\ g \in EE^* & \text{iff} & g_0 \in H-F \text{ , or } g_0 \in F \text{ and } \varepsilon_1 = +1 \text{ ,} \\ & \text{and} & g_n \in F \text{ and } \varepsilon_n = +1 \text{ ,} \\ g \in E^*E & \text{iff} & g_0 \in F \text{ and } \varepsilon_1 = -1 \text{ ,} \\ & \text{and} & g_n \in H-F \text{ , or } g_n \in F \text{ and } \varepsilon_n = -1 \text{ ,} \\ g \in E^*E^* & \text{iff} & g_0 \in F \text{ and } \varepsilon_1 = -1 \text{ ,} \\ & \text{and} & g_n \in F \text{ and } \varepsilon_n = +1 \text{ .} \end{array}$$

In particular, all elements of $H - F$ are in EE .

It follows from the normal form theorem, that the above definition is independent of the choice of the reduced sequence for g .

We verify the product axiom (axiom 4). As with free products with amalgamation, the point is that if $g \in XY$ and $k \in Y^*Z$, then we can obtain a reduced sequence for gk by putting together the sequences for g and k . Let $g = g_0 t^{\varepsilon_1} \ldots t^{\varepsilon_n} g_n$, and let $k = k_0 t^{\delta_1} \ldots t^{\delta_m} k_m$. Suppose $g \in XE$ and $k \in E^*Z$. Then $g_n \in H-F$, or $g_n \in F$ and $\varepsilon_n = -1$, while $k_0 \in F$ and $\delta_1 = -1$. Thus the sequence

$$g_0 t^{\varepsilon_1} \ldots t^{\varepsilon_n} (g_n k) t^{\delta_1} \ldots t^{\delta_n} k_m$$

is reduced and $gk \in XZ$. A similar verification works for $g \in XE^*$ and $k \in EZ$.

To verify the boundedness axiom, consider a product $g = g_1 \cdots g_n$ where $g_i \in X^*_{i-1} X_i$. The condition $g_i \in X^*_{i-1} X_i$ implies that at most half the g_i can be in EE . Each $g_i \notin EE$ must have a t-symbol in its reduced form. Since we have seen that we can obtain a reduced sequence for g by putting together the reduced sequences for the g_i , merely consolidating elements of H at the beginning and end of the sequences for the g_i , we see that n is less than or equal to twice the number of t-symbols in the reduced form of g .

Finally, $t \in EE^*$, verifying non-triviality.

We note that a group may possess many different bipolar structures. Consider, for example, the fundamental group of the surface of genus 3 . The bipolar structures arising from viewing G as a free product with amalgamation, say

$$G = \langle a, b, c, d, e, f; [a, b][c, d] = [f, e] \rangle ,$$

and those arising from viewing G as an HNN extension, say

$$G = \langle a, b, c, d, e, f; f^{-1}ef = e[d, c][b, a] \rangle ,$$

are rather different.

The following sequence of lemmas follows Stallings [13]. Let G be a group with a bipolar structure. An element $g \in G$ is said to be *irreducible* if $g \in F$, or $g \in XY$ and g is not equal to a product $g = hk$ with $h \in XZ$ and $k \in Z^*Y$.

The boundedness axiom, axiom 5, immediately implies that G is generated by irreducible elements. It follows from the inverse axiom (axiom 3) and the fact that F is a subgroup, that g is irreducible if and only if g^{-1} is irreducible.

LEMMA 1. *If* $g \in XY$, $h \in YZ$, *and* h *is irreducible, then* $gh \in F$ *or* $gh \in XW$ *for some* W .

PROOF. If $gh \in F$, we are done. Suppose $gh \in X^*W$. Now $g^{-1} \in YX$ by the inverse axiom. But then $h = g^{-1}(gh)$, which contradicts h being irreducible. //

LEMMA 2. *If* $h \in ZX$, $g \in XY$, *and* h *is irreducible, then* $hg \in F$ *or* $hg \in WY$ *for some* Y .

PROOF. Similar to Lemma 1. //

LEMMA 3. *If* $g \in XY$ *and* $h \in YZ$ *are both irreducible, then* $gh \in F \cup XZ$ *and is irreducible.*

PROOF. By Lemmas 1 and 2, $gh \in F \cup XZ$. If gh is not irreducible, then $gh \in XZ$ and $gh = pq$ where $p \in XW$ and $q \in W^*Z$. Now $g = p\left(qh^{-1}\right)$. Since $h^{-1} \in ZY$ and is irreducible, we have $qh^{-1} \in F$ or $qh^{-1} \in W^*U$ by Lemma 1. We cannot have $qh^{-1} \in W^*U$, since this would contradict g being irreducible. Hence, $qh^{-1} \in F$. Thus $g = p\left(qh^{-1}\right) \in XW$ by axiom 2. Also, $h^{-1} = q^{-1}\left(qh^{-1}\right)$, so $h^{-1} \in ZW^*$ by the inverse axiom and axiom 2. Hence, $h \in W^*Z$. This contradicts $g \in XY$ and $h \in YZ$. //

LEMMA 4. *If* $g \in XY$, $f \in F$, *and* g *is irreducible, then* gf *and* fg *are irreducible elements of* XY .

PROOF. We have $gf \in XY$ by axiom 2. If gf is not irreducible, then $gf = pq$ where $p \in XW$ and $q \in W^*Y$ for some W . But then $g = p(qf^{-1})$. Since $p \in XW$ and $(qf^{-1}) \in W^*Y$ by axiom 2, this contradicts g being irreducible. Note that $g^{-1}f^{-1}$ is an irreducible element of YX by the inverse axiom and the first part of the proof. Thus fg is an irreducible element of XY . //

We can now prove Stallings' characterization theorem.

THEOREM IX. *If a group* G *has a bipolar structure, then* G *is either a non-trivial free product with amalgamation (possibly an ordinary free product) or an* HNN *extension.*

PROOF. Let G have a bipolar structure. Define

$$G_1 = F \cup \{x : x \text{ is an irreducible element of } EE\},$$

and

$$G_2 = F \cup \{x : x \text{ is an irreducible element of } E^*E^*\}.$$

We claim that G_1 and G_2 are subgroups of G . The inverse of an element of G_1 is in G_1 by axiom 1 and the inverse axiom. Consider a product hk of two elements of G_1 . If both h and k are in F , then $hk \in F$ since F is a subgroup. If both h and k are irreducible elements of EE , then $hk \in F$ or hk is an irreducible element of EE by Lemma 3. If exactly one of h, k is in F , then hk is an irreducible element of EE by Lemma 4. Similarly, G_2 is a subgroup of G .

Case 1. If there are no irreducible elements of EE^* , then $G = \langle G_1 * G_2;\ F = F \rangle$.

PROOF. Since the elements of E^*E are inverses of elements of EE^* , there are no irreducible elements in E^*E . Thus $G_1 \cup G_2$ contains all the irreducible elements of G . Since G is generated by irreducible elements, $G_1 \cup G_2$ generates G .

Take disjoint copies $\bar{G}_1$ and $\bar{G}_2$ of G_1 and G_2 . For any element $g_i \in G_i$, $i = 1, 2$, $\bar{g}_i$ will denote the corresponding element of $\bar{G}_i$. Let

$$\bar{G} = \langle \bar{G}_1 * \bar{G}_2;\ \bar{F} = \bar{F} \rangle .$$

Define $\psi : \bar{G} \to G$ by $\psi(\bar{g}_1) = g_1$ if $\bar{g}_1 \in \bar{G}_1$, and $\psi(\bar{g}_2) = g_2$, $\bar{g}_2 \in \bar{G}_2$. We claim that ψ is an isomorphism. First, ψ is onto G by the remark of the preceding paragraph. Now ψ is certainly one-to-one on $\bar{G}_1$ and $\bar{G}_2$. If $\bar{g}$ is an element not in a factor of $\bar{G}$, write $\bar{g} = \bar{c}_1 \ldots \bar{c}_n$ where each $\bar{c}_i$ is in a factor, no $\bar{c}_i \in \bar{F}$, and successive $\bar{c}_i, \bar{c}_{i+1}$ come from different factors. The images $c_i = \psi(\bar{c}_i)$ thus come alternatively from EE and E^*E^*. Thus, by the product axiom,

$$\psi(\bar{g}) = c_1 \ldots c_n$$

is in one of the sets XY and is thus not equal to 1. Hence, ψ is an isomorphism.

If F were not a proper subgroup of both G_1 and G_2, say $F = G_1$, then $G_1 \cup G_2 = G_2$. Since G_2 then generates G, we have $G = G_2 \subseteq F \cup EE$, which contradicts the non-triviality axiom $EE^* \neq \emptyset$.

Case 2. If there is an irreducible element $t \in EE^*$, then

$$G = \left\langle G_1, t;\ tFt^{-1} = \varphi(F) \right\rangle .$$

PROOF. We first show that $G_1 \cup \{t\}$ generates G. If g is an irreducible element of E^*E, then tg is an irreducible element of $F \cup EE$ by Lemma 3, and $g = t^{-1}(tg)$. The irreducible elements of EE^* are inverses of irreducible elements of E^*E. If g is an irreducible element of E^*E^*, then gt^{-1} is an irreducible element of E^*E. (Since G_1 is a subgroup, no product gt^{ε} or $t^{\varepsilon}g$ is in F.)

Certainly, conjugation by t induces an isomorphism φ between F and a subgroup $\varphi(F)$ of G. We check that $tFt^{-1} \subset G_1$. If $f \in F$, then tf is an irreducible element of EE^* by Lemma 4, and $(tf)t^{-1}$ is thus an irreducible element of $F \cup EE$ by Lemma 3.

Let $\bar{G}_1$ be an isomorphic copy of G_1, and let

$$G = \left\langle \bar{G}_1, \bar{t};\ \overline{tFt}^{-1} = \overline{\varphi(F)} \right\rangle .$$

Define $\psi : \bar{G} \to G$ by $\psi(\bar{g}_1) = g_1$ if $\bar{g}_1 \in \bar{G}_1$, and $\psi(\bar{t}) = t$. We know that ψ is onto G since $G_1 \cup \{t\}$ generates G. If $\bar{g} \in \bar{G}-\bar{G}_1$, write g as the product of elements in a reduced sequence, say

$$\bar{g} = \bar{g}_0 \bar{t}^{\varepsilon_1} \ldots \bar{t}^{\varepsilon_n} \bar{g}_n , \quad n \geq 1 ,$$

where each $\bar{g}_i \in \bar{G}_1$. We must show that

$$\psi(\bar{g}) = g_0 t^{\varepsilon_1} \ldots t^{\varepsilon_n} g_n \neq 1 .$$

In exhibiting a bipolar structure for an arbitrary HNN extension, we gave a scheme for putting elements into the sets XY. We claim that $\psi(\bar{g})$ is in XY according to the same scheme. (In the present context, G_1 plays the role of H.) The proof is by induction on n. We will first analyze the case where $n > 1$. The case $n = 1$ is easily seen to follow from the same analysis.

We consider the sign of ε_{n-1}. Suppose $\varepsilon_{n-1} = 1$. Then $g = wg_{n-1}t^{\varepsilon_n}g_n$ where $w = g_0 \ldots g_{n-2}t$. Now $w \in XE^*$ by the induction hypothesis. We cannot have both $g_{n-1} \in F$ and $\varepsilon_n = -1$ since the sequence for $\bar{g}$ is reduced. Suppose $\varepsilon_n = -1$. Then $g_{n-1} \in EE$. Regardless of whether $g_n \in EE$ or $g_n \in F$, we have $t^{-1}g_n \in E^*E$ by applying either Lemma 3 or axiom 2. Thus $wg_{n-1}\left(t^{-1}g_n\right) \in XE$ by the product axiom. Suppose $\varepsilon_n = 1$. Regardless of whether $g_{n-1} \in EE$ or $g_{n-1} \in F$, we have $g_{n-1}t \in EE^*$. If $g_n \in EE$, we have $w(g_{n-1}t)g_n \in XE$. If $g_{n-1} \in F$, we have $w(g_{n-1}tg_n) \in XE^*$.

Suppose $\varepsilon_{n-1} = -1$. Write $g = vt^{-1}g_{n-1}t^{\varepsilon_n}g_n$ where $v = g_0 \ldots g_{n-2}$. Suppose that v does not consist solely of $g_0 \in F$. Then we claim that $v \in XE$, for, otherwise, v would have to end in tg_{n-2} with $g_{n-2} \in F$. This contradicts the original sequence for $\bar{g}$ being reduced. Now $t^{-1}g_{n-1}$ is an irreducible element of E^*E regardless of whether $g_{n-1} \in EE$ or $g_{n-1} \in F$. (Apply either Lemma 3 or 4.) Suppose $\varepsilon_n = 1$. Then $t^{-1}g_{n-1}t \in F \cup E^*E^*$ by Lemma 3. Now $t^{-1}g_{n-1}t \in F$ is impossible since this implies $g_{n-1} \in \varphi(F)$ which contradicts the original sequence for $\bar{g}$ for being reduced. Thus $t^{-1}g_{n-1}t \in E^*E^*$. If $g_n \in EE$, $v\left(t^{-1}g_{n-1}t\right)g_n \in XE$, while, if $g_n \in F$, $v\left(t^{-1}g_{n-1}tg_n\right) \in XE^*$. Finally, suppose that $\varepsilon_n = -1$. Then $t^{-1}g_n \in E^*E$ and $v\left(t^{-1}g_{n-1}\right)\left(t^{-1}g_n\right) \in XE$. The result for the case $v \in F$ follows easily from the above analysis.

The case where $n = 1$ follows easily by analyzing the product $g_0 t^{\varepsilon_1} g_1$ along the lines above. Thus we see that $g = \psi(\bar{g})$ is in one of the sets XY and cannot be equal to 1. Hence, ψ is one-to-one and is thus an isomorphism. This concludes the proof of the theorem. //

We have seen that a group G has a bipolar structure if and only if G is a non-trivial free product with amalgamation or an HNN extension. We will now use the bipolar structure characterization to prove the Hanna Neumann subgroup theorem [9], stated below, for finitely generated subgroups of free products with amalgamation or HNN extensions.

THEOREM X (Hanna Neumann). *Let* $G = \langle A * B;\ F = \varphi(F) \rangle$ *be a non-trivial free product with amalgamation. (Respectively, let* $G = \langle A, t;\ t^{-1}Ft = \varphi(F) \rangle$ *be an* HNN *extension.) Let* H *be a finitely generated subgroup of* G *such that all conjugates of* H *intersect* F *trivially. Then*

$$H = K * \left(\underset{\alpha}{*} H_\alpha\right)$$

where K *is a free group and each* H_α *is a subgroup of a conjugate of a factor (respectively, the base) of* G.

The theorem remains true without the assumption that H is finitely generated. For a complete discussion of subgroup theorems for free products with amalgamation and HNN extensions see Serre [12] or Karrass and Solitar [5]. An immediate corollary of the theorem is the following.

COROLLARY. *Let* $G = \langle A * B;\ F = \varphi(F) \rangle$ *be a non-trivial free product with amalgamation. (Respectively, let* $G = \langle A, t;\ t^{-1}Ft = \varphi(F) \rangle$ *be an* HNN *extension.) If* H *is a finitely generated subgroup of* G *which has trivial intersection with all conjugates of the factors (respectively, the base) of* G, *then* H *is free.*

Before proving the theorem, we need a short technical lemma.

LEMMA. *Let* $G = \langle A * B;\ F = \varphi(F) \rangle$ *be a non-trivial free product with amalgamation. (Respectively, let* $G = \langle A, t;\ t^{-1}Ft = \varphi(F) \rangle$ *be an* HNN *extension.) Let* H *be a finitely generated subgroup of* G. *Then either some conjugate of* H *contains a cyclically reduced element of length at least two or* H *is contained in a conjugate of a factor (respectively, the base).*

PROOF. We prove the lemma for free products with amalgamation. The proof for HNN extensions is similar. The proof is by induction on s, the sum of the lengths of the elements in a finite set of generators for H. If $s = 1$, H is contained in a factor. Let $H = \langle h_1, \ldots, h_n \rangle$. If any of the h_i are cyclically reduced of length at least two we are done. Assuming that this is not the case, if we write any $h_i = c_1 \ldots c_k$ in reduced form (possibly $k = 1$), then c_1 and c_k are in the same factor. If two generators for H, say h_i and h_j, begin and end with elements from different factors, the product h_ih_j is cyclically reduced of length at least two. So suppose that reduced forms of all the h_i begin and end with elements from the same factor, say A. If all the h_i have length one, $H \subset A$. So assume that some $h_i = c_1 \ldots c_k$, $k > 1$, $c_1, c_k \in A$. We can then replace H by $c_1^{-1}Hc_1 = \langle h_1^*, \ldots, h_n^* \rangle$, where $h_j^* = c_1^{-1}h_j c_1^{-1}$. The sum of the lengths of the h_j^* is less than the sum of the lengths of the h_j, and the result follows. //

We now turn to the subgroup theorem. The reader familiar with Stallings' proof of Serre's conjecture in the finitely generated case, will note that the overall strategy of our proof, while not involving ends, is the same.

PROOF OF THEOREM. We use the notation in the statement of the theorem. If H is contained in a conjugate of a factor (respectively, the base) of G, the result holds trivially, so we assume that this is not the case. The proof is by induction on n, the minimum number of generators of H. If $n = 1$ and $H = \langle h_1 \rangle$ is not in a conjugate of a factor, then h_1 has infinite order, so H is free.

Assume the result true for $1 \leq n \leq k$, and prove for $n = k + 1$. Note that a subgroup H of G has a decomposition of the desired type stated in the theorem if and only if all conjugates of H have a decomposition of the desired type. By the lemma, some conjugate H^* of H contains a cyclically reduced element of length greater than one. We replace H by H^*.

Give G the bipolar structure associated with viewing G as the given free product with amalgamation or HNN extension. We put a bipolar structure on H^* by simply taking the set-theoretic intersection of H with the sets F, EE, and so on. Thus $F_{H^*} = H^* \cap F$, $EE_{H^*} = H^* \cap EE$, and so on. Note that axioms 1-5 for a bipolar structure are hereditary, and thus hold in the proposed bipolar structure for H. We must check the non-triviality axiom. But, H^* contains a cyclically reduced element h^* of length at least two. Hence, one of h^* or $(h^*)^{-1}$ is in EE^*. Thus $EE^*_{H^*} = H^* \cap EE^* \neq \emptyset$.

By the hypothesis on H, $H^* \cap F = \{1\}$. But $F_{H^*} = \{1\}$ means that H^* is a non-trivial free product, say $H^* = H_1 * H_2$. By Grushko's theorem, H_1 and H_2 have fewer generators than H^*. Thus H_1 and H_2 have decompositions of the desired type by the induction hypothesis. But it is immediate that the free product of two groups having a decomposition of the desired type again has a decomposition of desired type. This concludes the proof of the theorem. //

It seems surprising that this subgroup theorem, which has been proved only by using a rather large amount of machinery, should follow so easily from the bipolar structure characterization.

I would like to conclude by pointing out that bipolar structures may be useful in showing that various groups have nice decompositions as free products with amalgamations or HNN extensions. The following theorem is due to Nagao [7].

THEOREM XI. *Let* $K[x]$ *be the polynomial ring in one variable over a field* K . *Then*

$$GL_2(K[x]) = \langle GL_2(K) * T;\ U = U \rangle$$

where U *is the group of upper triangular matrices with entries from* K , *and* T *is the group of matrices of the form* $\begin{pmatrix} k_1 & f(x) \\ 0 & k_2 \end{pmatrix}$ *where* k_1 *and* k_2 *are non-zero elements of* K *and* $f(x) \in K[x]$.

For generalizations of Nagao's result see Serre [12]. The easiest way to prove Nagao's result is a direct proof. It turns out, however, that $GL_2(K[x])$ has a rather natural bipolar structure, which we now describe.

Let $F = U$, and put all matrices in $GL_2(K) - U$ into EE . Now consider any matrix $M = \begin{pmatrix} f_{11} & f_{12} \\ f_{21} & f_{22} \end{pmatrix}$ where at least one entry has degree greater than zero. If $f \in K[x]$, $\delta(f)$ will denote the degree of f . Now put

$$M \in EE \quad \text{iff} \quad \delta(f_{12}) \le \delta(f_{11}) \quad \text{and} \quad \delta(f_{12}) \le \delta(f_{22}) ,$$

$$M \in EE^* \quad \text{iff} \quad \delta(f_{12}) \le \delta(f_{11}) \quad \text{and} \quad \delta(f_{12}) > \delta(f_{22}) ,$$

$$M \in E^*E \quad \text{iff} \quad \delta(f_{12}) > \delta(f_{11}) \quad \text{and} \quad \delta(f_{12}) \le \delta(f_{22}) ,$$

$$M \in E^*E^* \quad \text{iff} \quad \delta(f_{12}) > \delta(f_{11}) \quad \text{and} \quad \delta(f_{12}) > \delta(f_{22}) .$$

In short, one compares the degree of f_{12} with the degrees of the entries on the main diagonal. To verify the axioms, it seems that one must multiply out all the various cases. The inverse axiom, however, is an immediate consequence of the formula for inverting 2×2 matrices: $M^{-1} = \frac{1}{\det M} \begin{pmatrix} f_{22} & -f_{12} \\ -f_{21} & f_{11} \end{pmatrix}$.

References

[1] John L. Britton, "The word problem", *Ann. of Math.* (2) 77 (1963), 16-32. MR29#5891.

[2] P. Hall, "On the embedding of a group in a join of given groups", *J. Austral. Math. Soc.* (to appear).

[3] Graham Higman, "A finitely generated infinite simple group", *J. London Math. Soc.* 26 (1951), 61-64. MR12,390.

[4] Graham Higman, B.H. Neumann and Hanna Neumann, "Embedding theorems for groups", *J. London Math. Soc.* 24 (1949), 247-254. MR11,322.

[5] A. Karrass and D. Solitar, "The subgroups of a free product of two groups with an amalgamated subgroup", *Trans. Amer. Math. Soc.* 150 (1970), 227-255. MR41#5499.

[6] W.S. Massey, *Algebraic topology: an introduction* (Harcourt, Brace and World, New York, 1967). MR35#2271.

[7] Hirosi Nagao, "On $GL_2(K[x])$ ", *J. Inst. Polytech. Osaka City Univ. Ser. A.* 10 (1959), 117-121. MR22#5684.

[8] B.H. Neumann, "Some remarks on infinite groups", *J. London Math. Soc.* 12 (1937), 120-127. FdM63,64.

[9] Hanna Neumann, "Generalized free products with amalgamated subgroups. Part II. The subgroups of generalized free products", *Amer. J. Math.* 71 (1949), 491-540. MR11,8.

[10] Michael O. Rabin, "Recursive unsolvability of group theoretic problems", *Ann. of Math.* (2) 67 (1958), 172-194. MR22#1611.

[11] Otto Schreier, "Die Untergruppen der freien Gruppen", *Abh. Math. Sem. Univ. Hamburg* 5 (1927), 161-183. FdM53,110.

[12] J.-P. Serre, *Groupes discrets* (Lecture Notes in Mathematics. Springer-Verlag, to appear).

[13] John Stallings, *Group theory and three-dimensional manifolds* (Yale Mathematical Monographs, 4. Yale University Press, New Haven and London, 1971).

University of Illinois,
Urbana, Illinois 61801, USA.

PROC. SECOND INTERNAT. CONF. THEORY OF GROUPS, 20E30
CANBERRA, 1973, pp. 633-640.

AMALGAMES ET POINTS FIXES

Jean-Pierre Serre

Disons qu'un groupe G est un *amalgame* s'il est isomorphe à une somme amalgamée $G_1 *_A G_2$, avec $G_1 \neq A \neq G_2$ (pour la définition des sommes amalgamées, voir par exemple [1], §7 ou [3], §4.2). Dans ce qui suit, je montre que certains groupes, notamment $SL_3(Z)$, *ne sont pas* des amalgames; comme on le verra ci-dessous, cela revient à prouver que, lorsque ces groupes opèrent sur des arbres, ils ont nécessairement des points fixes.

1. La propriété de point fixe pour les groupes opérant sur les arbres

Soit X un arbre, *i.e.* un graphe connexe non vide sans circuit, et soit G un groupe opérant sur X ; quitte à remplacer X par sa subdivision barycentrique, on peut supposer que G opère sans inversion ([4], p. I-37), autrement dit qu'il n'existe pas d'élément s de G et d'arête PQ de X tels que $s(PQ) = QP$. L'ensemble X^G des points fixes de G dans X est alors un sous-graphe de X ; si P et Q sont deux sommets de X^G, la géodésique joignant P à Q est fixe par G, donc contenue dans X^G ; il en résulte que, si X^G est non vide, c'est un sous-arbre de X. Nous nous intéressons aux groupes G ayant la propriété:

(FA) *Quel que soit l'arbre X sur lequel opère G, on a $X^G \neq \emptyset$.*

Cette propriété est "presque" équivalente à celle de ne pas être un amalgame. Plus précisément:

THÉORÈME 1. *On suppose G dénombrable. Pour que G ait la propriété* (FA), *il faut et il suffit que les conditions suivantes soient satisfaites:*

(i) G n'est pas un amalgame;

(ii) G n'a pas de quotient isomorphe à Z ;

(iii) G est de type fini.

DÉMONSTRATION. (FA) $\Rightarrow$ *(i)*: Si G est un amalgame $G_1 *_A G_2$, avec $G_1 \neq A$ et

$G_2 \neq A$, il existe un arbre X sur lequel G opère avec pour domaine fondamental un segment PQ , le stabilisateur de P (resp. de Q) étant G_1 (resp. G_2) , *cf.* [4], p. I-5 . Comme G est distinct de G_1 et G_2 , on a donc $X^G = \emptyset$, ce qui contredit (FA).

(FA) $\Rightarrow$ *(ii)*: Si G a un quotient isomorphe à $\mathbb{Z}$, on peut le faire opérer par translations sur une chaîne doublement infinie:

$$\ldots - \circ \longrightarrow \circ \overset{\rightarrow}{\longrightarrow} \circ \longrightarrow \circ \longrightarrow \circ - \ldots$$

et cela contredit (FA).

(FA) $\Rightarrow$ *(iii)*: Comme G est dénombrable, il est réunion d'une suite croissante $G_1 \subset G_2 \subset \ldots \subset G_n \subset \ldots$ de sous-groupes de type fini. Formons un graphe X dont l'ensemble des sommets est la somme disjointe des ensembles G/G_n , deux sommets étant joints par une arête si et seulement si ils appartiennent à deux ensembles consécutifs G/G_n et G/G_{n+1} et se correspondent par l'application canonique $G/G_n \rightarrow G/G_{n+1}$. On vérifie immédiatement que X est un arbre; de plus G opère de façon évidente sur X . Si G a la propriété (FA), il existe un sommet P de X invariant par G ; si $P \in G/G_n$, cela entraîne que $G_n = G$, donc G est de type fini.

Inversement, supposons que G ait les propriétés *(i)*, *(ii)* et *(iii)*, et qu'il opère (sans inversion) sur un arbre X . Si $T = X/G$ désigne le graphe quotient de X par G , on voit facilement que le groupe fondamental $\pi_1(T)$ est isomorphe à un quotient de G (*cf.* [4], I, §5). Comme $\pi_1(T)$ est un groupe libre, ce n'est possible, d'après *(ii)*, que si $\pi_1(T) = \{1\}$, et T est donc un arbre, et on peut le relever en un sous-arbre de X ([4], p. I-37). Le groupe G s'identifie alors au groupe $G_T = \varinjlim (G, T)$ limite de l' "arbre de groupes" défini par les fixateurs G_P et G_y des sommets P et des arêtes y de T , *cf.* [4], n°. 4.5. Le groupe G est donc réunion des groupes $G_{T'} = \varinjlim (G, T')$, où T' parcourt l'ensemble des sous-arbres *finis* de T . Comme G est supposé de type fini, il existe au moins un tel T' tel que $G = G_{T'}$; choisissons T' minimal pour cette propriété. Si T' est réduit à un seul sommet P , on a $G = G_P$ et G a un point fixe. Sinon, T' possède un sommet terminal P , et $T'' = T' - P$ est un arbre ([4], p. I-29); si y désigne l'unique arête qui joint P à T'' , on a

$$G = G_{T'} = G_{T''} *_A G_P \text{ , où } A = G_y \text{ .}$$

Vu l'hypothèse de minimalité faite sur T', on a $G_{T''} \neq G$ et $G_P \neq G$, ce qui montre que G est un amalgame et contredit l'hypothèse *(i)*.

REMARQUE. Lorsque G n'est pas dénombrable, le Théorème 1 reste valable à condition de remplacer *(iii)* par:

(iii') G *n'est pas réunion d'une suite strictement croissante de sous-groupes.*

Des exemples de groupes non dénombrables satisfaisant aux conditions *(i)*, *(ii)*, *(iii')* ont été construits par J. Tits et S. Koppelberg.

2. Conséquences de la propriété (FA)

PROPOSITION 1. *Soit* G *un groupe ayant la propriété* (FA). *Si* G *est contenu dans un amalgame* $G_1 *_A G_2$, G *est contenu dans un conjugué de* G_1 *ou de* G_2.

Cela traduit simplement le fait que G a un point fixe dans l'arbre associé à l'amalgame $G_1 *_A G_2$, *cf.* [4], I, §4.

PROPOSITION 2. *Soit* G *un groupe dénombrable ayant la propriété* (FA), *et soit* $\rho : G \to \mathrm{GL}_2(k)$ *une représentation linéaire de degré* 2 *de* G *sur un corps commutatif* k. *Alors, pour tout* $s \in G$, *les valeurs propres de* $\rho(s)$ *sont entières sur* $\mathbf{Z}$.

(Lorsque k est de caractéristique 0, ces valeurs propres sont donc des *entiers algébriques*; lorsque k est de caractéristique $\neq 0$, ce sont des *racines de l'unité*.)

Soit k_ρ le sous-corps de k engendré par les coefficients des matrices $\rho(s)$, pour $s \in G$. D'après le Théorème 1, G est de type fini, donc k_ρ est de type fini sur le corps premier. Soit v une valuation discrète de k_ρ, et soit $\mathcal{O}_v$ l'anneau de valuation correspondant. Notons X_v l'arbre associé à v ([4], II, §1), arbre sur lequel opère $\mathrm{GL}_2(k_\rho)$. Soit $\mathrm{GL}_2(k_\rho)^\circ$ le noyau de l'homomorphisme

$$v \circ \det : \mathrm{GL}_2(k_\rho) \to \mathbf{Z}, \ cf. \ [4], \text{p. II-8}.$$

La condition *(ii)* du Théorème 1 montre que $\rho(G)$ est contenu dans $\mathrm{GL}_2(k_\rho)^\circ$, donc opère sans inversion sur X_v. Puisque G a la propriété (FA), il existe un sommet de X_v invariant par G. Cela entraîne ([4], p. II-11) que $\rho(G)$ est contenu dans un conjugué de $\mathrm{GL}_2(\mathcal{O}_v)$. Ainsi, pour tout $s \in G$, les coefficients du polynôme caractéristique de s appartiennent à l'intersection des $\mathcal{O}_v$. Mais on sait que cette intersection est égale à l'ensemble des éléments de k_ρ qui sont entiers sur

$\mathbf{Z}$ (*cf.* par exemple [2], p. 140, Corollaire 7.1.8); les valeurs propres des $\rho(s)$ sont donc bien entières sur $\mathbf{Z}$.

3. Exemples

3.1. Un groupe *de torsion* de type fini a la propriété (FA). Vu le Théorème 1, il suffit de vérifier qu'un tel groupe ne peut pas être un amalgame $G_1 *_A G_2$. Or c'est clair, car si l'on prend $s_1 \in G_1 - A$ et $s_2 \in G_2 - A$, l'élément $s_1 s_2$ est d'ordre infini (cela résulte de la structure des "mots" d'un amalgame, *cf.* [1], p. I-81).

3.2. Si G a la propriété (FA), il en est de même de tout *quotient* de G .

C'est clair.

3.3. Soit H un *sous-groupe distingué* de G . Si H et G/H ont la propriété (FA), il en est de même de G .

En effet, si G opère sur un arbre X , le groupe G/H opère sur l'arbre X^H , donc a un point fixe.

3.4. Soit G' un *sous-groupe d'indice fini* de G . Si G' a la propriété (FA), il en est de même de G .

Soit X un arbre sur lequel opère G . Choisissons un sous-groupe distingué H d'indice fini de G contenu dans G' (on sait que c'est possible). Puisque G' a des points fixes dans X , il en est de même de H , et le groupe G/H opère sur l'arbre X^H ; comme G/H est fini, il a des points fixes.

3.5. Par contre, il ne faudrait pas croire que, si G a la propriété (FA), il en est de même de tout sous-groupe d'indice fini de G . Voici un contre-exemple: prenons pour G le groupe de Schwarz $G(a, b, c)$ défini par deux générateurs x, y liés par les relations $x^a = y^b = (xy)^c = 1$, où a, b, c sont des entiers ≥ 2 tels que $\frac{1}{a} + \frac{1}{b} + \frac{1}{c} \leq 1$. On peut montrer que $G(a, b, c)$ a la propriété (FA) , et qu'il contient un sous-groupe H d'indice fini isomorphe au groupe fondamental d'une surface orientable compacte de genre ≥ 1 ; comme H a un quotient isomorphe à $\mathbf{Z}$, il ne vérifie pas (FA).

3.6. On trouvera au n° 5 divers exemples de groupes ayant la propriété (FA).

4. Résultats auxiliaires

Dans ce n°, G désigne un groupe *de type fini* opérant sans inversion sur un arbre X . Si $X^G \neq \emptyset$, on dit que G "a un point fixe".

PROPOSITION 1. *Si tout élément de* G *a un point fixe, il en est de même de* G.

Supposons que ce ne soit pas le cas. D'après la Proposition 3.4 de [6], il existe alors un *bout* b de X qui est invariant par G (rappelons qu'un bout est une classe d'équivalence de chaînes simplement infinies, deux chaînes étant équivalentes si leur intersection est une chaîne simplement infinie). Soit $\{s_1, \ldots, s_n\}$ une famille génératrice finie de G, et soit P_i $(1 \leq i \leq n)$ un sommet de X fixe par s_i. La chaîne simplement infinie c_i joignant P_i à b est invariante par s_i ; l'intersection c des c_i est une chaîne infinie invariante par tous les s_i, donc par G, ce qui montre bien que G a un point fixe.

PROPOSITION 2. *On suppose* G *nilpotent. Deux cas seulement sont possibles (et s'excluent mutuellement):*

(a) G *a un point fixe;*

(b) il existe une chaîne doublement infinie C, *stable par* G, *sur laquelle* G *opère par translations au moyen d'un homomorphisme non trivial* $G \to \mathbf{Z}$.

Notons d'abord que, si l'on est dans le cas *(b)*, la chaîne C est unique (cela résulte de la Proposition 3.2 de [6], appliquée à un élément sans point fixe de G). Ceci étant, choisissons une suite de composition

$$1 = G_0 \subset G_1 \subset \ldots \subset G_n = G$$

telle que les quotients successifs G_i/G_{i-1} soient cycliques, et raisonnons par récurrence sur n, le cas $n = 0$ étant trivial. Supposons donc $n \geq 1$, et appliquons l'hypothèse de récurrence au groupe $H = G_{n-1}$. Si H a un point fixe, le groupe G/H opère sur l'arbre X^H ; comme G/H est cyclique, on peut lui appliquer la Proposition 3.2 de [6], d'où le résultat cherché. Si H n'a pas de point fixe, la chaîne doublement infinie C stable par H est stable par G (puisque H est distingué dans G), et l'on obtient ainsi un homomorphisme $G \to \mathrm{Aut}(C)$ dont l'image contient un groupe non trivial de translations. Cette image est donc, soit le groupe diédral infini, soit le groupe $\mathbf{Z}$. Mais le premier cas est impossible, le groupe diédral infini n'étant pas nilpotent. Il reste donc seulement le second cas, *i.e.* le cas *(b)*.

COROLLAIRE 1. *Si* G *est engendré par des éléments qui ont des points fixes, alors* G *a un point fixe.*

Supposons que l'on soit dans le cas *(b)*, et que G soit engendré par une famille $\{s_i\}$. Puisque $G \to \mathbf{Z}$ est non trivial, l'un au moins des s_i a une image $\neq 0$ dans $\mathbf{Z}$; d'après la Proposition 3.2 de [6], un tel élément s_i ne peut pas avoir de point fixe.

COROLLAIRE 2. *Soit G' le groupe des commutateurs de G, et soit s un élément de G tel que $s^n \in G'$ pour un entier $n \geq 1$. Alors s a un point fixe.*

C'est clair si G a un point fixe. Sinon, on est dans le cas *(b)*, et l'hypothèse faite sur s entraîne que son image par l'homomorphisme $G \to Z$ est nulle. L'élément s laisse donc fixe la chaîne C.

REMARQUE. On a un résultat analogue à la Proposition 2 chaque fois que G admet une suite de composition dont les quotients successifs sont cycliques, ou ont la propriété (FA).

5. Le cas de $SL_3(Z)$

THÉORÈME 2. *Le groupe* $SL_3(Z)$ *a la propriété* (FA).

On sait que $SL_3(Z)$ est engendré par les matrices élémentaires $1 + e_{ij}$, avec $i, j \in \{1, 2, 3\}$ et $i \neq j$. Il est commode d'indexer de façon circulaire ces six matrices:

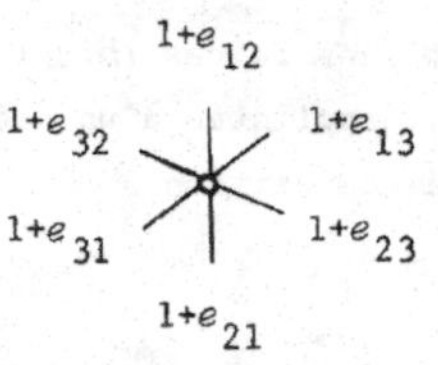

(cela correspond au fait que les "racines" de SL_3 forment un hexagone régulier).
On est ainsi conduit à poser:

$$z_0 = z_6 = 1 + e_{12}, \quad z_1 = 1 + e_{13}, \quad z_2 = 1 + e_{23}, \quad z_3 = 1 + e_{21},$$
$$z_4 = 1 + e_{31}, \quad z_5 = 1 + e_{32} \quad \text{et} \quad z_{i+6} = z_i \quad \text{pour tout } i .$$

On a les propriétés suivantes:

(i) z_i commute à z_{i+1} et z_{i-1},

(ii) le commutateur (z_{i-1}, z_{i+1}) est égal à z_i^{-1} ou z_i suivant que i est pair ou impair.

En particulier, $SL_3(Z)$ est engendré par $\{z_1, z_3, z_5\}$. De plus, pour tout $i \in Z/6Z$, les éléments z_{i-1} et z_{i+1} engendrent un groupe nilpotent B_i, et z_i appartient au groupe dérivé de B_i.

Supposons maintenant que $SL_3(Z)$ opère sans inversion sur un arbre X. Le Corollaire 2 de la Proposition 2, appliqué au groupe B_i, montre que z_i a un point fixe. Comme ceci est vrai pour tout $i \in Z/6Z$, on voit que B_i est engendré par des éléments qui ont des points fixes, donc a un point fixe d'après le Corollaire 1 de la Proposition 2. Si l'on note X_i l'ensemble des points fixes de z_i, on a donc $X_{i-1} \cap X_{i+1} \neq \emptyset$. En particulier, les trois arbres X_1, X_3, X_5 ont deux à deux des points communs. Or on vérifie facilement le lemme suivant:

LEMME. *Soit* $(X_i)_{i \in I}$ *une famille finie de sous-arbres d'un arbre* X. *Si* $X_i \cap X_j \neq \emptyset$ *pour tout couple* $(i, j) \in I \times I$, *l'intersection des* X_i *est non vide.*

On a donc $X_1 \cap X_3 \cap X_5 \neq \emptyset$. Mais $X^G = X_1 \cap X_3 \cap X_5$ puisque G est engendré par $\{z_1, z_3, z_5\}$. On a donc bien $X^G \neq \emptyset$.

REMARQUES. (1) Soit a un entier ≥ 1, et soit G_a le sous-groupe de $SL_3(Z)$ engendré par z_1^a, z_3^a et z_5^a. Un argument analogue à celui utilisé ci-dessus montre que G_a a la propriété (FA). Lorsque $a \geq 3$, le groupe G_a est sans torsion, et de dimension cohomologique 3.

(2) On démontre par un procédé analogue que les groupes $SL_n(Z)$, $n \geq 3$ et $Sp_{2n}(Z)$, $n \geq 2$, ont la propriété (FA). Il en est de même, plus généralement, des groupes $G(Z)$ où G est un groupe simple "de Chevalley" simplement connexe de rang ≥ 2 (utiliser le fait que $G(Z)$ est engendré par des sous-groupes cycliques correspondant aux différentes racines, *cf.* [5], p. 115). Ce résultat s'étend aux groupes $G(Z[1/N])$, $N \geq 1$.

6. Questions

Il serait intéressant de savoir quels sont les groupes arithmétiques (ou S-arithmétiques) qui ont la propriété (FA). Est-ce le cas, par exemple, pour les groupes suivants:

sous-groupes d'indice fini de $SL_n(Z[1/N])$, $n \geq 3$, $N \geq 1$?

sous-groupes d'indice fini de $SL_2(Z[\sqrt{d}])$, $d > 0$ non carré?

Références

[1] N. Bourbaki, *Éléments de Mathématique.* Livre II: *Algèbre,* Chapitre I: *Structures algébriques,* nouvelle edition (Hermann, Paris, 1970). MR43#2.

[2] A. Grothendieck, *Éléments de géométrie algébrique.* II: *Étude globale élémentaire de quelques classes de morphismes* (rédigés avec la collaboration de J. Dieudonné. Publications Mathématiques, 8. Institut des Hautes Études Scientifiques, Paris, 1961). Zbl.118,362-372.

[3] Wilhelm Magnus, Abraham Karrass, Donald Solitar, *Combinatorial group theory* (Pure and Appl. Math. 13. Interscience [John Wiley & Sons], New York, London, Sydney, 1966). MR34#7617.

[4] Jean-Pierre Serre, *Arbres, amalgames et* SL_2 , Collège de France 1968/69 (Notes rédigées avec la collaboration de Hyman Bass. Lecture Notes in Mathematics, Springer-Verlag, Berlin, Heidelberg, New York, à paraître).

[5] Robert Steinberg, *Lectures on Chevalley groups* (Notes prepared by John Faulkner and Robert Wilson. Yale University, 1968).

[6] Jacques Tits, "Sur le groupe des automorphismes d'un arbre", *Essays on topology and related topics: Mémoires dédiés à Georges de Rham,* pp. 188-211 (Springer-Verlag, Berlin, Heidelberg, New York, 1970). MR45#8582.

Collège de France,
75-Paris 16e, France.

PROC. SECOND INTERNAT. CONF. THEORY OF GROUPS, 20D99
CANBERRA, 1973, pp. 641-654.

GEOMETRIC CHARACTERIZATIONS IN FINITE GROUP THEORY

E.E. Shult

1. Introduction

Over the past several years one may be able to observe an increasing trend to use purely geometric arguments in the proofs of theorems in the theory of finite groups. The idea is a simple one. In the course of proving a theorem about finite groups, one displays some geometric configuration built out of a finite group G. He then proceeds to characterize the known configuration as being some very familiar geometric object. Because of this, the group G is a subgroup of the group of automorphisms of the geometric object, and this can frequently be used to characterize the group G. A good illustration of this principle would be the Suzuki-O'Nan characterization of the three dimensional projective unitary groups over a finite field by the centralizer of an involution.

There are two features which have made proof by geometric characterization especially rewarding. First, most of the characterization theorems are fairly simple-minded and are not of excessive length, certainly a non-trivial asset in present-day finite group theory. Second, the trend among group theorists to seek proofs utilizing the characterization of an incidence structure has already resulted in definite contributions by the group theorists to the subjects of combinatorial theory, graph theory and finite geometry.

No attempt will be made to survey these contributions. Instead I would like to illustrate the various ways in which characterizations of these structures have led to group theoretic results. (In some instances, of course, I will be raising more open questions than results.)

Three areas which seem most illuminating are:

(1) doubly transitive block designs,

(2) 2-graphs and the graph extension theorem, and

(3) more general graph-theoretic characterizations.

In order to show that these geometric characterizations lead to real results, I have listed below three theorems, corresponding one-by-one to the three areas I have mentioned. Each theorem is a purely group-theoretic theorem whose proof proceeds from the characterization of a geometric structure on combinatorial grounds alone.

THEOREM 1. *Let* G *be a doubly transitive finite permutation group with the property that the stabilizer of two letters fixes exactly three letters. If* G *contains an involution fixing fewer than seven letters, then either* G *contains a normal regular* 3*-group or* G *is a subgroup of* A_8 .

THEOREM 2. *Let* G *be a doubly transitive finite group. Suppose the one-point stabilizer,* G_α *, contains a normal subgroup* N_α *of index* 2 *in* G_α *and that no element in* N_α *is conjugate to an element in* $G_\alpha - N_\alpha$. *If* G *is triply transitive, then* G *contains a normal subgroup of index* 2 .

(Theorem 2 is the result of joint work with Hale, [8]. A version of this theorem with $[G_\alpha : N_\alpha]$ an odd prime also exists.)

THEOREM 3. *Let* K *be a union of classes of subgroups of order* p *in a finite group,* G . *Assume at least two members of* K *centralize one another and that whenever* X *and* Y *are members of* K *which centralize one another, then*

(i) *all subgroups of order* p *in* $\langle X, Y\rangle$ *lie in* K , *and*

(ii) $C(Z) \cap \langle X, Y\rangle$ *is non-trivial for all* Z *in* K .

Then G *contains a non-identity normal abelian* p*-subgroup.*

2. Doubly transitive block designs

The list of known doubly transitive groups is surprisingly short. Some of these admit block designs on which they may act, others do not. At times the design arises naturally from a group-theoretic setting, rather than being introduced *ad hoc*. For example, O'Nan considers doubly transitive groups in which the 2-point stabilizer $G_{\alpha\beta}$ contains a subgroup W weakly closed in $G_{\alpha\beta}$ with respect to G_α . The blocks are then the fixed points of the conjugates of W . Designs arise in other contexts in doubly transitive groups and Theorems 4 and 6 below are also illustrations of the use of characterization theorems and geometric arguments in designs in proving theorems from group theory.

THEOREM 4 (O'Nan [15]). *If* (G, Ω) *is a finite doubly transitive group whose one-point stabilizer* G_α *contains a normal abelian subgroup* N_α *which is not semi-regular on* $\Omega - (\alpha)$. *Then for some* n *and* q , $X \leq G \leq \mathrm{Aut}(X)$ *where* $X \simeq \mathrm{PSL}(n, q)$.

Theorem 4 follows from a general theorem of O'Nan characterizing the projective spaces among block designs with $\lambda = 1$:

THEOREM 5 (O'Nan [16]). *Suppose* (G, Ω) *is a finite doubly transitive group acting on a block design* Ω *with* $\lambda = 1$. *Let* K_α *be the normal subgroup of* G_α *of elements stabilizing every block through* α . *Then if* K_α *is not semiregular on* $\Omega - (\alpha)$, *we have* $X \le G \le \mathrm{Aut}(X)$, *where* $X \simeq \mathrm{PSL}(n, q)$ *and the design* Ω *is a projective space of dimension* $n - 1$ *with the projective lines as blocks.*

THEOREM 6 ([22]). *Suppose* (G, Ω) *is a finite doubly transitive group of even degree* n . *Suppose all involutions lying within a fixed 2-Sylow subgroup of* $G_{\alpha\beta}$ *have a common fixed point set. Then* $X \le G \le \mathrm{Aut}(X)$ *where* $X \simeq \mathrm{PSL}(2, q)$, $U(3, q)$ *or is a group of Ree type, or* (G, Ω) *is* A_6 *acting on six letters.*

The obvious design here has as its blocks the fixed point sets of $\Omega_1(S)$ where $S \in \mathrm{Syl}_2(G_{\alpha\beta})$, α, β ranging over Ω . (A special case of Theorem 6, namely the case when $|\Omega| \equiv 2 \bmod 4$, was proved earlier by Aschbacher [2].)

A theorem of Hering [11] classifies doubly transitive groups whose one point stabilizer G_α contains a normal subgroup N_α of even order which is semiregular on the remaining letters. This theorem can be combined with Theorem 5, to yield the following theorem.

THEOREM 7. *Let* G *be a finite doubly transitive group of automorphisms of a block design* Ω *having* $\lambda = 1$. *Suppose each block contains an odd number of points. If* G *contains an involution fixing exactly one letter, then either*

(i) G *contains a normal regular p-subgroup and the design is an affine space over a field of characteristic* p , *or*

(ii) $X \le G \le \mathrm{Aut}(X)$ *where* $X \simeq \mathrm{PSU}(3, q^2)$ *where* q *is a power of* 2 , *and the design is the unitary design with parameters* $(v, b, r, k, \lambda) = (1+q^3, q+q^4, q^2, q+1, 1)$.

It would be interesting to know what could be proved if the assumption of odd block size is dropped, or if one weakened "doubly transitive" to "primitive".

In the case $k = 3$, Ω is a Steiner system, and for this case the above theorem was proved by Hall [9]. Note that if $q = 2$, conclusion *(ii)* states that Ω is an affine plane of order 3 and is included under conclusion *(i)*. If S is a Steiner system, 3 points not lying in a block is called a triangle. Hall has also proved that if for each block in a Steiner system there exists an involution in the automorphism group of the Steiner system whose fixed points are that block, then every triangle of S generates either the projective plane $P(2)$ of order 2 , or the affine plane $A(3)$ of order 3 . We call any such Steiner system a Hall triple system. The following theorem on Hall triple systems is proved by purely geometric methods:

THEOREM 8 (Jonathan Hall. This theorem was also proved by R. Liebler and independantly proved in [23]). *Let* S *be a Hall triple system. Then either*

(i) every triangle of S *generates a* $P(2)$ *or*

(ii) every triangle of S *generates an* $A(3)$.

In the first case, S is a projective space of dimension d over GF(2) . In the second case Hall shows that the automorphism group of S contains a class of involutions K , the product of any two of which have order 3 . By Glauberman's Z^*-theorem $O_3(\langle K\rangle)$ is non-trivial [7].

We are now ready to extract from this a proof of Theorem 1. Immediately one obtains a triple system on Ω whose blocks are the fixed points of the 2-point stabilizers. By hypothesis there exists an involution t fixing fewer than seven letters. Since the fixed points of t must be a subsystem of the triple system, there are two possibilities:

(a) either t fixes exactly one point, or

(b) t fixes exactly a block.

In the first case since G is transitive, every letter is the unique fixed point of a conjugate of t and so every triangle generates an $A(3)$. By the above, $O_3\big(\mathrm{Aut}(\Omega)\big)$ is non-trivial. Since $G \leq \mathrm{Aut}(\Omega)$, $\mathrm{Aut}(\Omega)$ is doubly transitive on Ω and so has a regular elementary abelian 3-subgroup. It follows that Ω is an affine space over GF(3) , and since G is a doubly transitive group of collineations of Ω , G contains the group of translations on Ω (it is an easy lemma that the only exception to this rule occurs for $\Omega \simeq A_3(2)$ on eight points). Thus G contains a normal regular 3-group. Now consider the case that t fixes the three points of a block. Then by the theorem of Hall, Ω is a Hall triple system. By Theorem 8, $\Omega \simeq P_d(2)$ or every triangle generates an $A(3)$. We have just seen that the latter case leads to a normal regular 3-group for G . So suppose $\Omega \simeq P_d(2)$. Since involutions acting on $P_d(2)$ fix a subspace of dimension at least $(d-1)/2$, the assumption that t fixes 3 letters forces $d \leq 3$. Since G is doubly transitive either $d = 2$, $|\Omega| = 7$ and $G \simeq L(3, 2)$ or else $d = 3$, $|\Omega| = 15$ and $G \simeq A_7$ or $G \simeq L(4, 2)$. In all three cases, G is a subgroup of A_8 , proving Theorem 1.

3. 2-graphs and the graph extension theorem

Most of the results in this section were developed by two groups working independently of each other. On the one hand, was the graph extension theorem and subsequent work of Hale and myself on the theorem [8]. 2-graphs on the other hand were invented by Higman and the greatest part of our present knowledge of 2-graphs is developed in an important paper on regular 2-graphs by Taylor [25], and in a considerable amount of unpublished work of Taylor. The connection between the two is this: everything that results from the graph extension theorem is associated with a doubly transitive 2-graph, and conversely, every doubly transitive 2-graph comes out of the graph extension theorem. Observe, however, that 2-graphs are a more general incidence system since not all of them have doubly transitive automorphism groups. The 2-graph approach contains certain cohomological overtones, and in this connection mention should be made of the important work of Seidel and his concept of switching classes of graphs [18]. For purposes of achieving a fairly simple-minded presentation (and for the reason that I am most familiar with this unsophisticated approach) I will describe things from the point of view of graphs and the graph extension theorem. For the reader interested in more detailed information on 2-graphs I strongly recommend Taylor's excellent paper [25].

Let G be a graph, undirected and without loops. For any vertex x in G, let $A(x)$ denote the set of vertices adjacent to x and let $A'(x)$ denote the set of vectices distinct from x and not adjacent to x, so that we have a decomposition

$$\{x\} + A(x) + A'(x)$$

for every vertex x in G. The graph G is called *graph extendable* if the following two conditions are satisfied:

(a) $H = \mathrm{Aut}(G)$ is transitive on the vertices of G, and

(b) there exist automorphisms of the subgraphs $A(x)$ and $A'(x)$, say $h_1 \in \mathrm{Aut}(A(x))$ and $h_2 \in \mathrm{Aut}(A'(x))$, such that if (a, b) is in $A(x) \times A'(x)$, then (a, b) is an arc if and only if $(h_1(a), h_2(b))$ is *not* an arc.

Thus applying h_1 to $A(x)$ and h_2 to $A'(x)$ interchanges the set of arcs extending between $A(x)$ and $A'(x)$ with the set of non-arcs between $A(x)$ and $A'(x)$. Let us adjoin a new symbol, ∞, to the graph G. Let π denote the permutation on $\{\infty\} \cup G$ defined by

$$\pi = (\infty, x)(h_1 \text{ on } A(x))(h_2 \text{ on } A'(x)) .$$

We view $H = \mathrm{Aut}(G)$ as acting on $\{\infty\} \cup G$ with each element fixing ∞. The graph extension theorem then states that

$G = \langle \pi, H \rangle$ is doubly transitive on $\{\infty\} \cup G$ and H is the full subgroup of G stabilizing ∞. In particular, $|G| = (1+|G|)|H|$.

This theorem produces for us an interesting collection of doubly transitive groups. Indeed, with only two exceptions, it gives us all doubly primitive groups which are not triply transitive, although there is no apparent reason for this. The list of doubly transitive groups arising from the graph extension theorem is also interesting in that it seems to include nearly all non-triply transitive groups which still lie outside the present-day classification theorems (Bender [4], Bender [5], Hering [10], O'Nan [15], Shult [21], Hering [11], Hering, Kantor and Seitz [12], Kantor, O'Nan, and Seitz [13], Aschbacher [1], Aschbacher [3], Shult [22]). The groups are as follows:

Symmetric groups	n letters
$P\Sigma L(2, q)$, $q \equiv 1 \bmod 4$	$1 + q$ letters
$\mathrm{Aut}(U(3, q))$, q odd	$1 + q^3$ letters
$\mathrm{Aut}(X)$ where X is a group of Ree type	$1 + 3^{3m}$ letters, m odd
$Sp(2n, 2)V^+(2n, 2)$ (semidirect product)	2^{2n} letters
$Sp(2n, 2)$	$2^{n-1}(2^n-1)$ letters
$Sp(2n, 2)$	$2^{n-1}(2^n+1)$ letters
Higman-Sims group	176 letters
Conway groups (.3)	276 letters.

It is not really clear why this strange collection of groups should appear. The graph extension theorem itself cannot tell us, since it is merely a construction. It would therefore be desirable to find a group-theoretic property of a doubly transitive group which would tell us when the graph extension theorem operates to produce the group. There *is* such a group-theoretic setting. Consider

HYPOTHESIS (A). *Let G be a doubly transitive group on a finite set of letters. Suppose the one-point stabilizer G_α contains a normal subgroup of index 2, say N_α. It is also assumed that no element of N_α is conjugate to an element of $G_\alpha - N_\alpha$.*

The last statement of the hypothesis involves fusion in G_α. There is a very useful elementary lemma which tells us about fusion in the one-point stabilizer of a doubly transitive group [8]:

LEMMA (Fusion Lemma for doubly transitive groups). *Let (G, Ω) be a doubly transitive group of permutations. Suppose x and $g^{-1}xg$ are two elements of G_α. Then either x fixes exactly one letter in Ω and $g \in G$ or else $g = hn$ where $h \in G_\alpha$, x and x^h both fix two letters α and β and n is an element of $N(G_{\alpha\beta})$ transposing α and β.*

Then it is easy to see:

THEOREM 9. *If (G, Ω) is a doubly transitive group and if $N_\alpha \le G_\alpha$ with*

$[G_\alpha : N_\alpha] = 2$, *then* (G, Ω) *satisfies hypothesis* (A) *if and only if* $N_G(G_{\alpha\beta})$ *normalizes* $N_\alpha \cap G_{\alpha\beta}$.

Sufficient conditions that G *satisfy hypothesis* (A) *are*

(i) $G_{\alpha\beta}$ *is a complete group or*

(ii) $G_{\alpha\beta}$ *has no four-group as a factor.*

Thus one can see that if $G_{\alpha\beta}$ has a cyclic 2-Sylow subgroup as in the examples involving $\mathrm{PSL}(2, q)$ or $U(3, q)$, *(ii)* holds, so hypothesis (A) obtains. Similarly, *(ii)* holds for both representations of the groups $\mathrm{Sp}(2n, 2)$, or for the Conway group (.3) since in these examples $N_\alpha \cap G_{\alpha\beta}$ is a perfect group. Similarly, in the Higman-Sims group, $G_{\alpha\beta} \simeq \mathrm{Aut}(A_6)$ and is complete, so *(i)* holds.

Let (G, Ω) be a group satisfying hypothesis (A). Let $(-1)_\alpha$ denote the alternating character on G_α having kernel N_α . Then it is easy to show that the character of G induced from this alternating character is a sum of two irreducible complex characters which are afforded by real representations. Thus we write

$$(-1)_\alpha^G = \chi_1 + \chi_2 .$$

Let χ be one of the χ_i's and let V be the real module affording χ . Then V is a real Euclidean space. By Frobenius reciprocity, G_α stabilizes a one-dimensional subspace L affording the alternating character of G_α . If $L < V$, G is doubly transitive on the G-orbit, L^G , and so the system of lines L^G , is an equiangular set of lines in the sense of Lemmens and Seidel [14]. Thus if E is the collection of unit vectors in the union $\cup L^x$ $(L^x \in L^G)$, there exists a constant such that if two vectors lie on different lines of L^G , there inner product is $\pm c$. We then convert E into a graph $\mathcal{E}$ by declaring two vectors to be an edge if and only if their inner product is c . The definition of c can be chosen so that $\mathcal{E}$ is connected, and we assume this choice has been made. Then if $e \in \mathcal{E}$, $-e$ is the unique vertex in $\mathcal{E}$ at distance three from e . Two things can now be shown.

(1) *For* $e \in \mathcal{E}$, *the subgraph* $A(e)$ *is graph extendable* (as defined above).

(2) *The full automorphism group* $\mathrm{Aut}(\mathcal{E})$ *can be represented as a subgroup of* $\mathrm{GL}(V)$ *so that its restriction to* G *coincides with* V *as a* G*-module.*

(The first of these observations was essentially proved by Taylor in [25]; he gives it as a direct relation between equiangular lines and 2-graphs. The second was proved by Hale [8].)

We now have the machinery to prove Theorem 2. In Theorem (2), (G, Ω) satisfies Hypothesis (A), and we have the additional information that (G, Ω) is triply transitive. We can suppose χ chosen so $\dim \chi > 1$, and so a non-trivial graph E is defined, and G is triply transitive on the set of antipodal pairs $A = \{(e, -e) \mid e \in E\}$, of E. Then N_α is either doubly transitive or $1\frac{1}{2}$-transitive on $\Omega - (\alpha)$. In either event $G_\alpha = N_\alpha G_{\alpha\beta}$. Let the antipodal pair $(e, -e)$ correspond to the letter α and $(f, -f)$ correspond to the letter β, the notation f being chosen so that (e, f) and $(-e, -f)$ are edges of E. Then there is a decomposition

$$E = \{e\} + \{f\} + E_1 + E_2 + (-E_1) + (-E_2) + \{-f\} + \{-e\}$$

where $E_1 = A(e) \cap A(f)$, $E_2 = A(e) - (A(f)+\{f\})$, and (since E is connected) E_2 is non-empty. Then since $-E_i = \{-v \mid v \in E_i\}$ we see that $-E_1 = A(-e) \cap A(-f)$ and $-E_2 = A(-e) - (A(-f)+\{-f\})$. Suppose E_1 is non-empty. Then N_α is $1\frac{1}{2}$-transitive on $\Omega - (\alpha)$ and on $A(e)$, and so $N_\alpha \cap G_\alpha$ acts on $A(e)$ in two orbits E_1 and E_2. But we can choose $x \in G_{\alpha\beta} - N_\alpha$. Then x transposes the two orbits of $N_\alpha \cap G_\beta$ on Ω_α and so in its action on E, exchanges the two sets $(E_1 \cup -E_1)$ and $(E_2 \cup -E_2)$. But x induces the transpositions $(e, -e)$ and $(f, -f)$ on E. Thus x maps $E_1 = A(e) \cap A(f)$ into $A(-e) \cap A(-f) = -E_1$ and so stabilizes $E_1 \cup (-E_1)$. This is a contradiction. Hence E_1 is empty. Then E is determined up to isomorphism and $\mathrm{Aut}(E) \simeq \mathrm{Sym}(n) \times Z_2$.

Suppose by way of contradiction that G contains no normal subgroup of index 2. Then $G \le \mathrm{Alt}(n) \le \mathrm{Sym}(n) \times Z_2 \simeq \mathrm{Aut}(E)$. This means that G_α fixes pointwise the pair $(e, -e)$ corresponding in A to the letter α in Ω. But this is impossible since G_α acts on $\langle e \rangle = L$ with its alternating character. This contradiction proves Theorem 2.

A number of very interesting questions still remain. Starting with a group G satisfying Hypothesis (A) we obtain a connected graph E, whose automorphism group $\mathrm{Aut}(E)$ induces a doubly transitive group $E(G)$ on the set A of antipodal pairs of vertices of E. The kernel of this action is a central involution t of $\mathrm{Aut}(E)$ which transposes the members of each antipodal pair. It is easy to show that if $\mathrm{Aut}(E)$ splits over its central involution t, then $E(G)$ also satisfies Hypothesis (A). In that case, if $E(G)$ is not the symmetric group, $E(G)$ satisfies Hypothesis (A) in a way commensurate with G - that is, $E(G)_\alpha$ contains a strongly closed normal subgroup $NE(G)_\alpha$ of index 2 and $NE(G)_\alpha \cap G = N_\alpha$. In this case the alternating character for $E(G)_\alpha$ with kernel $NE(G)_\alpha$ induces to a sum of two irreducible real characters $\chi_1^* + \chi_2^*$ of $E(G)$ and the subscripts can be chosen so that

$$\chi_i^*|_G = \chi_i \ , \quad i = 1, 2 \ .$$

The doubly transitive groups $Sp(2n, 2)V^+(2n, 2)$ of the graph extension theorem do not satisfy Hypothesis (A) for $n \geq 3$. Nonetheless, there is a graph E with a system of antipodal pairs A such that $A(e)$ is graph extendable for all $e \in E$ and such that $\mathrm{Aut}(E)$ induces the group $Sp(2n, 2)V^+(2n, 2)$ on A. Thus $\mathrm{Aut}(E)$ is a non-split central extension of $Sp(2n, 2)V^+(2n, 2)$ by Z_2 and in fact is the centralizer of an involution u in the group $Sp(2n, 2)$. One can then realize the graph E from the equiangular set $E(2n+2)$ for $Sp(2n+2, 2)$ by picking up those lines corresponding to the subset of Ω centralized by u. Is $Sp(2n, 2)V^+(2n, 2)$ the only case in which $\mathrm{Aut}(E)/\langle t\rangle$ fails to satisfy Hypothesis (A)?

If $G \simeq U(3, 3)$ or $S\Gamma L(2, 8)$ on 28 letters, it is fairly easy to show that $E(G) \simeq Sp(6, 2)$ in its representation on 28 letters. In general if G is not normal in $E(G)$ and $A(e)$ is not totally disconnected for $e \in E$, then $E(G)$ is a larger doubly transitive group than G acting on the same set of letters as G. Moreover, these assumptions imply that $E(G)$ is not triply transitive, and except in the case of 28 letters mentioned above, must involve a new simple group as a composition factor. So far, the only known results comparing G and $E(G)$ for G ranging over the rest of the unitary groups, is a result of Taylors showing that if $G = U(3, 5)$, then $E(G) \leq \mathrm{Aut}(G)$, [26]. This is a very difficult problem and is intimately related to the problem of characterizing the groups of Ree type as Ree groups. In general we may ask whether there are further cases (other that $U(3, 3)$ and $S\Gamma L(2, 8)$) in which $E(G) \nleq \mathrm{Aut}(G)$ among groups satisfying Hypothesis (A)?

There is still one last feature of the equiangular line set associated with E. If the degree of the character χ is less than half of the degree of the permutation representation of G, then a lemma of Peter Neumann forces the Z-span of E in V to be a discrete lattice. For example, one of the characters χ_i appearing in the induced alternating character for $U(3, 3)$ has degree 7. Then the lattice $L = ZE$ spanned by E is isomorphic to the lattice spanned by the system of fundamental roots of type E_7. In that case $\mathrm{Aut}(L) \simeq Sp(6, 2)$, the Weyl group of E_7. Starting with a group G, satisfying Hypothesis (A), when does it fail to be true that

$$\mathrm{Aut}(ZE) \leq \mathrm{Aut}(G) \ ?$$

It is not even clear, for example, that E comprises all the unit vectors of

ZE ? If not, the possibility that Aut(ZE) is larger than Aut(E) certainly remains open.

4. General graph-theoretic characterizations

Let us return for a moment to the graph extension theorem. If the two mappings $h_1 \in \mathrm{Aut}(A(x))$ and $h_2 \in \mathrm{Aut}(A'(x))$ are composed with an automorphism g of G fixing vertex x, then the mappings $h_1 \circ g$ and $h_2 \circ g$ also serve as an h_1 and h_2 satisfying the hypothesis of the graph extension theorem. Classification of the groups arising from the graph extension theorem is an outstanding and probably difficult problem. There seemed to be some possibility of attacking this problem in the special case that, by composing with an automorphism g fixing x, one of the maps h_1 or h_2 could be taken to be the identity mapping. By replacing G with its dual, if necessary, one could assume without loss of generality that h_1 is the identity mapping on the subgraph $A(x)$. In that case the graph has the following property:

(P). *For each edge* (a, b) *and distinguished end point* a, *there exists a vertex* c *such that* (a, c) *is an arc, and that every vertex of* G *not arced to* a, *is arced to either* b *or* c *but not both.*

At this point one might choose to go shopping for theorems in the Journal of Combinatorial Theory, and if he were to do so, he would be doomed to a certain dissappointment. It becomes rather clear at once, that a relatively small number of regular graphs have been characterized by a relatively small but valiant band of combinatorial theorists, and in most cases one must explicitly know many of the graph-theoretic parameters, and so are frequently not in a form to be useful to group theorists. It is here that I believe that group theory has begun to make real contributions to combinatorial theory. (It is impossible to let the previous sentence stand without mentioning the names of Professors D. Higman, H. Enomoto and Bannai, just to name a few authors.) In terms of the problem presented above, one can prove the following

THEOREM 10. *Let* G *be an edge-regular graph with property* (P). *Suppose, for some edge* (a, b) *and vertex* a, *the hypothesized vertex* c *does not form an edge with* b. *Then* G *is either totally disconnected or is a pentagon.*

Otherwise, in graphs with property (P), one may assume that c always forms an edge with b. Then if G is regular it is easy to show that c is uniquely determined by the edge (a, b) and that every edge of G lies in a unique odd triangle - that is, a complete subgraph T of cardinality 3, with the property that any vertex not on an edge with an odd number of vertices of T. Graphs with this property are characterized in

THEOREM 11 ([20]). *Let* G *be a regular graph in which every edge lies in a unique odd triangle. Then either* G *is totally disconnected, or is complete or is isomorphic to the graphs* $O^-(2n, 2)$, $O^+(2n, 2)$ *and* $Sp(2n, 2)$ *which are the graphs whose vertices are the non-zero singular vectors in the two non-degenerate orthogonal geometries over* $GF(2)$ *or is all the non-zero vectors in a non-degenerate symplectic geometry over* $GF(2)$. *In all three cases an edge is a pair of vectors perpendicular to one another under the appropriate geometry.*

These geometric characterization theorems can then be returned to our (partially) group-theoretic problem to prove

THEOREM 12. *If in the graph extension theorem, one of the mappings* h_i $(i = 1, 2)$ *can be made trivial by composing with an automorphism of* G *fixing* x, *then the doubly transitive group* G *which results is either*

(i) *the symmetric group,*

(ii) *one of the two doubly transitive representations of* $Sp(2n, 2)$ *on* $2^{n-1}(2^n \pm 1)$ *letters,*

(iii) *the semidirect product* $Sp(2n, 2)V^+(2n, 2)$ *acting as a group of linear substitutions on* $V^+(2n, 2)$ *or*

(iv) $PSL(2, 5)$ *acting on six letters.*

Shortly after Theorem 11 was proved, Beukenhout and Seidel showed that the uniqueness of the triangle could be proved without the assumption of regularity. Seidel [19] then went on to prove Theorem 11 with both the words "regular" and "unique" removed - a considerably more general theorem. At about this time, Seidel's role as an intermediary stimulated an exchange of correspondence between Beukenhout and myself which resulted in a theorem far more general than any of us had thought possible. (Seidel's role as "Postillons des Sciences" should not be underestimated.) We state here the version of the theorem for finite graphs:

THEOREM 13 ([6]). *Let* G *be a finite graph. Assume that for each edge* (x, y) *there exists a complete subgraph* $C(x, y)$ *containing* x, y *and at least one further vertex, such that every vertex not in* $C(x, y)$ *is adjacent to exactly one or all of the vertices of* $C(x, y)$. *Then one of the following holds:*

(i) G *contains a vertex adjacent to all of the other vertices of* G;

(ii) G *is a totally disconnected graph;*

(iii) $G \simeq S(\pi)$, *where* π *is a polarity or a non-degenerate quadratic form on a projective space* P, *the vertices of* $S(\pi)$ *are the absolute or singular points of* P, *and edges are perpendicular pairs of points;*

(iv) *the* $C(x, y)$*'s are maximal cliques of* G, *are uniquely determined*

by x *and* y , *and* G *is a generalized* 4*-gon.*

We are now in a position to sketch the proof of Theorem 3. We convert K into a graph (also denoted K) by defining edges in K to be mutually commuting pairs of subgroups of K . If (X, Y) is an edge in K , $C(X, Y)$ will denote the $1 + p$ subgroups of order p in $\langle X, Y \rangle$. Then clearly K satisfies the hypothesis of Theorem 13. We then show that conclusions *(ii)-(iv)* fail. By legislation *(ii)* fails. If *(iii)* holds, $K \simeq S(\pi)$. One can then prove with only moderate difficulty that the subgroups of K induce the automorphisms on $S(\pi)$ given by the transvections of the relevant linear group. Since the product of two commuting transvections with distinct directions is not a transvection, hypothesis *(i)* of Theorem 3 yields a contradiction. If *(iv)* holds, it can be shown that $G = \mathrm{Aut}(G)$ is transitive on edges, from this in turn, that $\mathrm{Stab}_G\left(C(X, Y)\right)$ is doubly transitive on $C(X, Y)$, and (by the Hering-Kantor-Seitz Theorem) induces $\mathrm{PSL}(2, q)$ and no more than $\mathrm{P\Gamma L}(2, q)$ on it. Then the points not adjacent to a given point partition into totally disconnected subgraphs of size p , each vertex of one arced to exactly one vertex of each of the other partitional summands. From this it follows that $G \simeq S(\pi)$ for π a symplectic polarity on projective 3-space. This is a reduction to case *(iii)* which already fails. Thus case *(i)* holds. Then if $K_1 = \{X \in K \mid \text{each element of } K \text{ centralizes } X\}$ then K_i is a non-trivial normal set in G and so $N = \langle K_1 \rangle$ is a normal elementary p-subgroup of G .

Theorem 13 should be useful in other contexts of finite group theory since the graph-theoretic hypothesis are fairly crude and are simple enough to be duplicated within a group-theoretic structure, while at the same time include the graphs associated with all of the classical groups. Of course one drawback of the theorem is the set of open cases which occur at the level of generalized 4-gons. A survey of the known constructions of the generalized 4-gons will be found in a forthcoming paper of Thas [27]. From the class of generalized 4-gons, we may distinguish the 4-gons associated with the classical groups $\mathrm{Sp}(4, q)$, $O(5, q)$, $O^+(6, q)$, $U(4, q)$ and $U(5, q)$. Of these, those associated with $\mathrm{Sp}(4, q)$, $O(5, q)$ and $U(4, q)$ follow from the work of Benson (after Payne [17]) and Tallini [24]. No characterization of the 4-gons associated with $U(5, q)$ or $O^+(6, q)$ is known to us. In view of the fact that the 4-gons comprise an open case in Theorem 13, it might be useful to characterize the 4-gons associated with the classical groups from the remaining 4-gons. It may thus be meaningful to conclude this article with the following

CONJECTURE. *Let* G *be a generalized* 4*-gon and suppose* $\mathrm{Aut}(G)$ *is a rank-*3 *permutation group on its vertices. Then either* G *is a square grid, or else* G *is the generalized* 4*-gon associated with one of the classical groups,* $\mathrm{Sp}(4, q)$, $O(5, q)$, $O^+(6, q)$, $U(4, q)$ *or* $U(5, q)$.

References

[1] Michael Aschbacher, "Doubly transitive groups in which the stabilizer of two points is abelian", *J. Algebra* 18 (1971), 114-136. MR43#2059.

[2] M. Aschbacher, "On doubly transitive groups of degree $n \equiv 2 \bmod 4$ ", *Illinois J. Math.* 16 (1972), 276-279. MR45#8713.

[3] M. Aschbacher, "2-transitive groups whose 2-point stabilizer has 2-rank 2 ", submitted.

[4] Helmut Bender, "Endliche zweifach transitive Permutationsgruppen, deren Involutionen keine Fixpunkte haben", *Math. Z.* 104 (1968), 175-204. MR37#2846

[5] Helmut Bender, "Transitive Gruppen gerader Ordnung, in denen jede Involutionen genau einen Punkt festlasst", *J. Algebra* 17 (1971), 527-554. MR44#5370.

[6] F. Beukenhout and E. Shult, "On the foundations of polar geometry", submitted.

[7] George Glauberman, "Central elements in core-free groups", *J. Algebra* 4 (1966), 403-420. MR34#2681.

[8] M. Hale and E. Shult, "Equiangular lines, the graph extension theorem, and transfer in triply transitive groups", submitted.

[9] Marshall Hall, "Automorphisms of Steiner Triple systems", *IBM J. Res. Develop.* 4 (1960), 460-472. MR23#A1282.

[10] Christoph Hering, "Zweifach transitive Permutationsgruppen, in denen zwei die maximale Anzahl von Fixpunkten von Involutionen ist", *Math. Z.* 104 (1968), 150-174. MR37#295.

[11] Christoph Hering, "On subgroups with trivial normalizer intersection", *J. Algebra* 20 (1972), 622-629. Zbl.239.20026.

[12] Christoph Hering, and William M. Kantor, and Gary M. Seitz, "Finite groups with a split *BN*-pair of rank 1", *J. Algebra* 20 (1972), 435-475. Zbl.244.20003

[13] William M. Kantor, Michael E. O'Nan and Gary M. Seitz, "2-transitive groups in which the stabilizer of two points is cyclic", *J. Algebra* 21 (1972), 17-50.

[14] P.W.H. Lemmens and J.J. Seidel, "Equiangular lines", *J. Algebra* (to appear).

[15] M. O'Nan, "A characterization of $L_n(q)$ as a permutation group", *Math. Z.* 127 (1972), 301-314.

[16] M. O'Nan, "Normal structure of the one-point stabilizer of a doubly transitive permutation group, I", submitted.

[17] Stanley E. Payne, "Affine representations of generalized quadrangles", *J. Algebra* 16 (1970), 473-485. MR42#8381.

[18] J.J. Seidel, "Strongly regular graphs", *Recent Progress in Combinatorics*, 185-198 (Proc. Third Waterloo Conf. Combinatorics, 1968; Academic Press, New York, London, 1969). MR40#7148.

[19] J.J. Seidel, "On two-graphs and Shult's characterization of symplectic and orthogonal geometries over GF(2) ", Tech. Univ. Eindhoven T.H.-Report 73-WSK-02.

[20] Ernest E. Shult, "Characterizations of certain classes of graphs", *J. Combinatorial Theory Ser. B* 13 (1972), 142-167. Zbl.227.05110.

[21] Ernest Shult, "On a class of doubly transitive groups", *Illinois J. Math.* 16 (1972), 434-445. MR45#5211.

[22] E.E. Shult, "On doubly transitive groups of even degree", submitted.

[23] E.E. Shult, "Hall triple systems", University of Florida, mimeographed notes.

[24] G. Tallini, "Ruled graphic systems", *Atti. Conv. Geo. Comb. Perugia* (1971), 403-411.

[25] D. Taylor, "Regular 2-graphs", submitted.

[26] D. Taylor, Personal communication.

[27] J.A. Thas, "On 4-gonal configurations", *Geometrica Dedicata* (to appear).

University of Florida,
Gainesville, Florida 32601, USA.

PROC. SECOND INTERNAT. CONF. THEORY OF GROUPS, CANBERRA, 1973, pp. 655-666.

20E15 (20F15, 55A25)

TORSION-FREE METABELIAN GROUPS WITH INFINITE CYCLIC QUOTIENT GROUPS

H.F. Trotter

1. Introduction and statement of main results

For any group G, we write G' for the commutator subgroup $[G, G]$ and G'' for the second commutator subgroup $(G')' = [G', G']$. The quotient groups G/G' and G'/G'' are of course abelian. In this paper we are concerned with the problem of classifying the groups G/G'' when G is finitely generated and the conditions

(1) G/G' is infinite cyclic,

(2) G/G'' is torsion-free,

are both satisfied.

These conditions are satisfied when G is the fundamental group of the complement of a tame knot in the three-sphere, and our results are closely related to work of Crowell [2] and Rapaport [4]. Both these authors actually considered more general classes of groups than the titles of their articles suggest, and as will appear, groups satisfying our conditions also satisfy theirs.

Let t be a generator of the infinite cyclic group G/G', and identify the group ring $Z(G/G')$ with the ring $L = Z[t, t^{-1}]$ of Laurent polynomials. Then G'/G'' becomes an L-module in the usual way, with the group operation written additively and tm defined for m in $M = G'/G''$ by $tm = xmx^{-1}$, with x any element of G that maps into t in G/G'. Then G/G'' is given by an extension

$$0 \to M \to G/G'' \to G'/G \to 0 .$$

Because G/G' is free cyclic, the extension splits, and is completely determined by the structure of M as an L-module. The module structure of M actually depends on the choice of the generator t as well. To allow for this, we consider *oriented* groups, that is, groups with a distinguished generator for G/G'. (The terminology is consistent with that of algebraic topology, where an orientation distinguishes a

---

Research supported in part by a National Science Foundation grant.

generator for an homology group.) We then have the following simple statement (which is a special case of the situation discussed by Crowell in [1]).

If G_1 and G_2 are oriented groups satisfying (1) and (2), then there is an orientation-preserving isomorphism between G_1/G_1'' and G_2/G_2'' if and only if the associated L-modules M_1 and M_2 are isomorphic.

It will be convenient to introduce the ad hoc abbreviation "*-module" to stand for "L-module obtained as G'/G'' from some finitely generated group G satisfying conditions (1) and (2)". Our problem is thus equivalent to classifying *-modules.

Our first result describes a concrete representation for *-modules. Recall that a matrix Q over a ring R is a *relation matrix* for an R-module M (or *presents* M) if M is isomorphic to the quotient of the free module R^k of k-component "row vectors" (where k is the number of columns of Q) by the submodule generated by the rows of Q.

THEOREM 1. *An L-module M is a *-module if and only if it has a relation matrix of the form $tA + (I-A)$, with A a square matrix of integers. Except for the trivial case $M = 0$, it is always possible to choose A so that both A and $I - A$ have non-zero determinant.*

REMARK. If G is a knot group, then such an A is given by $V\left(V-V^T\right)^{-1}$ where V is a Seifert matrix for the knot. The idea for the present paper came from noticing that some of the methods applied to Seifert matrices in [6] could be used in a more general context.

The following definitions are needed to describe how matrices that describe isomorphic *-modules are related. Two square matrices A and B are *integrally similar* if $B = PAP^{-1}$ for some integral unimodular matrix P. They are *rationally similar* if the same equation holds for P some non-singular matrix of rationals. We call two non-singular integer matrices A and B *quasi-similar* if there exist integer matrices U and V such that $A = UV$ and $B = VU$. Since B is then equal to VAV^{-1}, quasi-similarity implies rational similarity. On the other hand, if $B = PAP^{-1}$ with P and P^{-1} integral, taking $U = P^{-1}$ and $V = PA$ shows that integral similarity implies quasi-similarity. We now define A and B to be *m-equivalent* if there is a finite sequence $A = A_0, A_1, \ldots, A_n = B$ such that for $i = 1, \ldots, n$ either A_{i-1} and A_i or $\left(I-A_{i-1}\right)$ and $\left(I-A_i\right)$ are quasi-similar. (Note: quasi-similarity is not an equivalence relation, because it is not transitive.)

THEOREM 2. *If A and B are integer matrices such that A, B, $(I-A)$, and $(I-B)$ are all non-singular, then the *-modules presented by $tA + (I-A)$ and*

$tB + (I-B)$ are isomorphic if and only if A and B are m-equivalent.

Theorems 1 and 2 translate the isomorphism problem for $*$-modules into concrete terms, but do not solve it in any satisfactory sense. In many cases, however, the problem can be solved, at least in principle.

Since similarity (integral or rational) of A and B is equivalent to similarity of $(I-A)$ and $(I-B)$ our earlier remarks about quasi-similarity yield the implications

$$\text{integral similarity} \Rightarrow m\text{-equivalence} \Rightarrow \text{rational similarity}.$$

This last implication allows us to define the *characteristic polynomial* $\varphi(\lambda)$ of a $*$-module M to be the characteristic polynomial $\det(A-\lambda I)$ of any matrix A such that $tA + (I-A)$ presents M and both A and $I - A$ are non-singular. It should be noted that φ determines and is determined by the *Alexander polynomial* $\Delta(t) = \det\big(tA+(I-A)\big)$, used in [2] and [4]. If k is the rank of A, then

$$\Delta(t) = (t-1)^k \det\big(A-(1-t)^{-1}\big) = (t-1)^k \varphi\big((1-t)^{-1}\big)$$

and

$$\varphi(\lambda) = (-\lambda)^k \det\big((1-\lambda^{-1})A+(I-A)\big) = (-\lambda)^k \Delta\big(1-\lambda^{-1}\big) .$$

There is a special case in which the classification of $*$-modules with a given characteristic polynomial is particularly simple. Suppose φ is irreducible, with root α, and let $R = Z\big[\alpha, \alpha^{-1}, (1-\alpha)^{-1}\big]$ be the subring of the algebraic number field $Q(\alpha)$ generated by α, α^{-1}, and $(1-\alpha)^{-1}$. There is a unique homomorphism $L \to R$ taking t into $-\alpha^{-1}(1-\alpha)$, and any R-module can be viewed as an L-module via this homomorphism. We claim that *$*$-modules with φ as characteristic polynomial are isomorphic to ideals of R*, and vice versa. In case $Z[\alpha]$ is integrally closed (that is, is the full ring of algebraic integers in $Q(\alpha)$) we can be more explicit. Let C be the ideal class group of $Z[\alpha]$, and C_0 the subgroup generated by representatives of ideals dividing $\alpha(1-\alpha)$. Then *the distinct isomorphism classes of $*$-modules with φ as characteristic polynomial correspond to the elements of the quotient group C/C_0*. Thus in this case the classification problem is "solved" in the sense of reducing it to a classical problem.

We can also in principle solve the classification problem whenever a given rational similarity class contains only a finite number of integral similarity classes. As will be shown in Section 4, the matrices quasi-similar to a given one fall into finitely many integral similarity classes, and it is easy to find a set of representatives (usually redundant) for these classes. Thus starting with a given matrix one can build up a family of integral similarity classes which is closed under quasi-similarity and constitutes an m-equivalence class. Actually carrying out the procedure of course requires a method for detecting integral similarity, which in general is easier said than done.

At the conference, Gilbert Baumslag pointed out to me that G/G'' may fail to be finitely presentable even if G is finitely presented, and raised the question of what happens for groups of the type considered in this paper. The answer turns out to be quite simple. If $tA + (I-A)$, with A and $I - A$ non-singular, presents the *-module associated with G/G'', then G/G'' is finitely presentable if and only if at least one of A and $I - A$ is unimodular. The proof is given in Section 5. This may be compared with Rapaport's result (Theorem 1 in [4]) which, restated in our terminology, asserts that G'/G'' is finitely generated if and only if *both* A and $I - A$ are unimodular. If G is a knot group, the symmetry of the Alexander polynomial implies that $\det(A) = \det(I-A)$. This yields the curious result that for a knot group G, G/G'' is finitely presented if and only if G'/G'' is finitely generated.

2. Proof of Theorem 1

The first step in proving Theorem 1 is to show that any *-module possesses a finite relation matrix. If G is any finitely generated group with G/G' the infinite cyclic group generated by t, we can choose a finite set of generators $x, a_1, \ldots, a_k$ such that x maps into t and the a_i all become trivial under the map into G/G'. Then the elements $x^i a_j x^{-i}$ for $i = 0, \pm 1, \pm 2, \ldots$ and $j = 1, \ldots, k$ generate G'. Since $x^i a_j x^{-i}$ becomes $t^i a_j$ in $M = G'/G''$ it follows that M is generated as an L-module by $a_1, \ldots, a_k$ and hence is finitely generated. Thus it is the quotient of some free module of finite rank by a submodule, which is also finitely generated because L is Noetherian. We get a finite relation matrix with one row for each member of any finite set of generators of the kernel.

The entries in this relation matrix may be arbitrary elements of L. Since powers of t are units in L, and multiplying rows of a relation matrix by units does not alter the module it presents [3], we may assume that only non-negative powers of t actually appear. We can get a new relation matrix in which only the zero and first powers of t appear as follows. Let t^m be the highest power that does appear in the given matrix. Suppose the matrix has k columns, corresponding to generators $e_1, \ldots, e_k$. The i-th row of the matrix gives a relation which may be written in the form

$$\sum_{j=1}^{k} \sum_{p=0}^{m} r_{ijp} t^p e_j = 0 ,$$

where the r_{ijp} are integers. Take new generators e_{pj} (to be identified with $t^p e_j$) for $j = 1, \ldots, k$ and $p = 0, \ldots, m$. We get a new relation matrix, with $k(m+1)$ columns, with rows corresponding to the old relations (with e_{pj} replacing $t^p e_j$) plus rows for the relations $e_{p,j} - te_{p-1,j} = 0$ for $p = 1, \ldots, m$ and $j = 1, \ldots, k$. The result is a finite relation matrix of the form $tU + V$, with U and V matrices of integers.

From a relation matrix of this form, with n columns, we obtain a presentation for G/G'' with generators $x, a_1, \ldots, a_n$, relations

$$x\left(\prod_{j=1}^{n} a_j^{u_{ij}}\right)x^{-1}\left(\prod_{j=1}^{n} a_j^{v_{ij}}\right) = 1$$

and further relations trivializing G''. Abelianization gives a presentation for G/G' with the same generators (now supposed to commute) and relations $\sum_{j=1}^{k} (u_{ij}+v_{ij})a_j$ (rewritten in additive notation). Thus G/G' is the direct sum of an infinite cyclic group and the abelian group having $U + V$ for a relation matrix over Z. Since G/G' is cyclic by assumption, it follows that $U + V$ *presents the trivial* Z*-module*. Hence $U + V$ can be converted to the identity matrix followed by rows of zeros, by elementary row and column operations. These same operations can be carried out on $tU + V$ without changing the module presented, so we may assume that our relation matrix has the form

$$\begin{bmatrix} I+(t-1)A \\ (t-1)B \end{bmatrix}$$

for some integer matrices A and B (with A square). From the first block of relations (those involving the matrix A) it follows that every generator is equal to $(t-1)$ times some combination of generators, and hence that *multiplication by* $(t-1)$ *maps* M *onto itself*.

If A is singular, it can be converted to have a row of zeros by elementary row operations on the relation matrix, and complementary column operations can be used to restore the identity matrix to its proper form without destroying the zero row in A. (Of course B will be modified in the process.) The result will be a row in the relation matrix which is zero except in one column, which contains a unit (1, in this case). Consequently, the row and column may be dropped [3]. Similarly, if $I - A$ is singular, the matrix can be converted to a form with a row which is zero except for a single entry of t, and again a row and column can be dropped.

Our next claim is that M *has finite rank as an abelian group*, that is, $Q \otimes A$ is finite-dimensional. We may suppose A and $I - A$ to be non-singular, so that

$T = -A^{-1}(I-A)$ is a non-singular rational matrix. Let K be the module presented by $tA + (I-A)$. The same matrix presents $Q \otimes K$ as a module over $Q \otimes L = Q[t, t^{-1}]$, and over this ring the matrix is equivalent to $tI + A^{-1}(I-A)$. As in Proposition 2.5 of [6], it can then be shown that K is a vector space over Q of dimension equal to the rank of T, with a basis such that multiplication by t is represented by multiplication by the matrix T. Since M is obtained from K by adding the relations corresponding to $(t-1)B$, its rank is no greater than that of K, and hence is finite.

Since multiplication by $(t-1)$ takes M onto itself, it does the same for $Q \otimes M$. Since the latter is a finite-dimensional vector space, a linear map is one-to one if it is onto. M is by assumption torsion-free, that is, the natural map $M \to Q \otimes M$ is an inclusion. It follows that for any m in M, $(t-1)m = 0$ only if $m = 0$. Thus *multiplication by* $(t-1)$ *is an automorphism on* M.

The relations represented by the rows of $(t-1)B$ therefore imply relations represented by the rows of B. Elementary column operations on B can be used to make rows with a single non-zero element. Such a row gives a relation asserting that some integer multiple of a generator is 0. Since M is torsion-free, the generator must be 0, and a row and column can be eliminated from the relation matrix. This process can be repeated until B is reduced to 0. It follows that we have a relation matrix for M in the form $tA + (I-A)$ with both A and $I - A$ non-singular (unless *all* rows and columns are eliminated, which can happen only when $M = 0$).

The converse part of Theorem 1, asserting that such a relation matrix always presents a *-module (for an arbitrary matrix A) is easily established by giving a presentation for a group which leads directly to the given matrix. (See the construction given above in the proof that $U + V$ presented the trivial module, and similar constructions in [4].)

3. Proof of Theorem 2

We showed that $1 - t$ acts as an automorphism on any *-module M, so multiplication by $(1-t)^{-1}$ is defined and we may extend the coefficient ring from $L = Z[t, t^{-1}]$ to $\Lambda = Z[t, t^{-1}, (1-t)^{-1}]$. It is convenient to write

$$z = (1-t)^{-1}$$

and note that

$$1 - z = -tz \quad \text{and} \quad t = -z^{-1}(1-z) .$$

These last equations show that Λ may alternatively be defined as

$Z[z, z^{-1}, (1-z)^{-1}]$. If $tA + (I-A)$ is a relation matrix for M over L then it is also a relation matrix for M over Λ .

Let $Q\Lambda = Q \otimes \Lambda = Q[t, t^{-1}, (1-t)^{-1}]$ be the ring like Λ , but with rational coefficients. Then $QM = Q \otimes M$ is a module over $Q\Lambda$, with $tA + (I-A)$ as relation matrix over $Q\Lambda$. Because M is torsion-free as an abelian group, the natural map $M \to QM$ is an injection and we may view M as a Λ-submodule of QM . We use this idea (as in Section 2 of [6]) to convert Theorem 2 into a proposition about lattices in a vector space.

A little more terminology will prove convenient. We shall call a matrix A *proper* if both A and $I - A$ are non-singular. If $tA + (I-A)$ is a relation matrix over Λ for M , we shall simply say that A *belongs to* M . If it is a relation matrix over Q for QM , we shall say that A *belongs rationally to* M . In this terminology, Theorem 2 asserts that two proper matrices belong to isomorphic modules if and only if they are m-equivalent.

LEMMA 3.1. *If A is a proper matrix of rank r , belonging to the module M , then $W = QM$ is a vector space of dimension r over Q . W possesses a basis X such that*

(a) the Λ-submodule generated by X is isomorphic to M ,

(b) if w in W is identified with the row vector of its coordinates with respect to X then

$$tw = -w(I-A)A^{-1}$$

and

$$zw = wA .$$

Conversely, if X is any basis for W such that the matrix A defined by condition (b) above has integer entries, then A is a proper matrix belonging to the Λ-submodule of W generated by X .

PROOF. Essentially identical with the proof of Proposition 2.5 in [6].

Now let $Y = \{y_1, \ldots, y_r\}$ be any other basis for W , and B the corresponding matrix which describes the action of z in terms of Y-coordinates. If P is the matrix whose i-th row consists of the X-coordinates of y_i then $B = PAP^{-1}$. It follows as a corollary of Lemma 3.1 that proper matrices belong rationally to isomorphic modules if and only if they are rationally similar.

With X, Y, A, B , and P all as above, let K and L be the Z-modules generated by X and Y respectively. (K and L are *lattices on* W .) The matrices A and B have integer entries if and only if $zK \subseteq K$ and $zL \subseteq L$. If

$K = L$ then P is unimodular and $B = PAP^{-1}$ is integrally similar to A. Conversely, given X (which determines A) and any B integrally similar to A, there exists a basis Y determining B and such that $L = K$. Thus integral similarity of matrices can be translated into a condition on lattices. The next lemma gives an analogous translation for quasi-similarity.

LEMMA 3.2. *If* $zK \subseteq L \subseteq K$ *then* A *and* B *are quasi-similar. Conversely, given* X *and the matrix* A *determined by it, and any* B *quasi-similar to* A, Y *can be chosen so that it determines* B *and* $zK \subseteq L \subseteq K$.

PROOF. Let P be the matrix giving Y in terms of X, as before. The condition $L \subseteq K$ holds if and only if P is integral, and $zK \subseteq L$ if and only if AP^{-1} is integral. If both conditions hold, the factorizations $A = (AP^{-1})P$ and $B = P(AP^{-1})$ show that A and B are quasi-similar. Conversely, given $A = UV$ and $B = VU$, take $P = V$ and use it to define Y in terms of X.

The same argument shows that quasi-similarity of $I - A$ and $I - B$ is similarly related to the condition $(1-z)K \subseteq L \subseteq K$.

LEMMA 3.3. *If* K *is a lattice on* W *such that* $zK \subseteq K$ *then*

$$\Lambda K = \bigcup_{m=0}^{\infty} z^{-m}(1-z)^{-m}K .$$

PROOF. Immediate from the fact that Λ is generated by z, z^{-1}, and $(1-z)^{-1}$.

COROLLARY. *If* $zK \subseteq L \subseteq K$ *or* $(1-z)K \subseteq L \subseteq K$, *then* $\Lambda K = \Lambda L$.

Combining the corollary with Lemmas 3.2 and 3.1 immediately gives half of Theorem 2, that is, m-equivalent matrices belong to isomorphic $*$-modules. For the converse, we observe that Lemmas 3.2 and 3.1 combine with the following lemma to complete the proof of Theorem 2.

LEMMA 3.4. *If* K *and* L *are lattices on* W *such that* $zK \subseteq K$, $zL \subseteq L$, *and* $\Lambda K = \Lambda L$ *then* K *and* L *can be converted to the same lattice in a finite number of steps, where each step consists in replacing* K *by a lattice* M *such that* $zK \subseteq M \subseteq K$ *or* $(1-z)K \subseteq M \subseteq K$ *or else replacing* L *by a lattice* M *such that* $zL \subseteq M \subseteq L$ *or* $(1-z)L \subseteq M \subseteq L$.

PROOF. We first note that by Lemma 3.3 (and the fact that K and L are finitely generated), the assumption $\Lambda K = \Lambda L$ implies the existence of a finite exponent s such that $K \subseteq z^{-s}(1-z)^{-s}L$ and $L \subseteq z^{-s}(1-z)^{-s}K$. We can replace K by $z^{s}(1-z)^{s}K$ in $2s$ steps, so there is no loss of generality in assuming $K \subseteq L$. Then for some non-negative m, n we have

$$z^{m}(1-z)^{n}L \subseteq K \subseteq L .$$

If both exponents are 0 then $K = L$ and we are done. Otherwise one of them, say m, is positive. Define $M = K + zL$. Obviously $zL \subseteq M \subseteq L$, so replacing L by M is a legitimate step. Furthermore,

$$z^{m-1}(1-z)^n M = z^{m-1}(1-z)^n K + z^m(1-z)^n L \subseteq K + K = K \subseteq M ,$$

and we have the same relation as before with the exponent reduced by one. Hence we can arrive at the desired equality in a finite number of steps.

4. Remarks on m-equivalence

We collect in this section a number of examples and remarks about m-equivalence and the classification of *-modules.

REMARK 4.1. Lemma 3.2 and the comments about integral similarity which precede it show that for each integral similarity class of matrices B that are quasi-similar to a given proper matrix A there is at least one lattice L satisfying $zK \subseteq L \subseteq K$. Such lattices are in one-to-one correspondence with subgroups of the finite abelian group K/zK, and finite in number. Hence, to within integral similarity there are only a finite number of matrices quasi-similar to a given one.

EXAMPLE 4.2. Let

$$A = \begin{bmatrix} 0 & 2 & 0 \\ 0 & 0 & 2 \\ 2 & 0 & 0 \end{bmatrix}, \quad U = \begin{bmatrix} 2 & 0 & 0 \\ 0 & 1 & 0 \\ 0 & 0 & 1 \end{bmatrix}, \quad V = \begin{bmatrix} 0 & 1 & 0 \\ 0 & 0 & 2 \\ 2 & 0 & 0 \end{bmatrix}, \quad B = \begin{bmatrix} 0 & 1 & 0 \\ 0 & 0 & 2 \\ 4 & 0 & 0 \end{bmatrix}, \quad C = \begin{bmatrix} 0 & 1 & 0 \\ 0 & 0 & 1 \\ 8 & 0 & 0 \end{bmatrix}.$$

Then $A = UV$ and $B = VU$, so A and B are quasi-similar. Via an analogous decomposition, B and C are quasi-similar. A and C are not quasi-similar, however, because C has rank 2 over the field $Z/2Z$ and therefore in any factorization $C = RS$, both R and S must have rank at least 2 over $Z/2Z$. Then SR has rank at least one over $Z/2Z$ and $SR = A$ is impossible. Thus *quasi-similarity is not transitive.*

REMARK 4.3. *If* $\det A$ *is* ± 1 *or a prime, and* B *is quasi-similar to* A, *then* B *is integrally similar to* A.

The reason is that in any decomposition $A = UV$, at least one of U and V must be unimodular, and then $B = VU = VAV^{-1} = U^{-1}AU$ is integrally similar to A.

EXAMPLE 4.4. The matrices A and B in Example 4.2 are quasi-similar but not integrally similar because they have different ranks over $Z/2Z$. Hence $I - A$ and $I - B$ are not integrally similar. Since $\det(I-A) = -7$, they cannot be quasi-similar, by Remark 4.3. The example shows that quasi-similarity of A and B does *not* imply that of $I - A$ and $I - B$.

We now turn to the connections with algebraic number theory indicated in the introduction. Let A be a proper matrix belonging to the module M, and suppose that its characteristic polynomial, $\varphi(\lambda)$, is irreducible. Then φ is also its minimal polynomial. Let $F = Q(\alpha)$ be the field obtained by adjoining a root α of φ to the rationals. There is a homomorphism $Q\Lambda \to F$, taking z to α , and by Lemma 3.1, the action of $Q\Lambda$ on QM can be factored through this homomorphism because $\varphi(A) = 0$. QM is then a one-dimensional vector space over F . The image of Λ in F is the integral domain $R = Z[\alpha, \alpha^{-1}, (1-\alpha)^{-1}]$, and M becomes a module of rank one over R . It is therefore isomorphic as an R-module to an ideal of R . If $Z[\alpha]$ is the full ring of integers in F , then R can be characterized as consisting of the elements of F with value less than or equal to 1 at all valuations corresponding to primes of $Z[\alpha]$ except those dividing $\alpha(1-\alpha)$. Then R is a Dedekind ring and the isomorphism classes of rank one modules correspond exactly with the ideal classes of R .

5. Finite presentability of G/G''

Suppose A is a given proper matrix of rank s (so A and $I - A$ are non-singular). A group G with associated $*$-module presented by $tA + (I-A)$ can be obtained as follows. Take $x, y_1, \ldots, y_s$ as generators and $R_1, \ldots, R_s$ as relations, where R_i is the relation

$$x\left(\prod_{j=1}^{s} y_j^{a_{ij}}\right)x^{-1} = y_i^{-1} \prod_{j=1}^{s} y_j^{a_{ij}} .$$

Define $y_{im} = x^m y_i x^{-m}$. Then the y_{im} generate G' and adding the commutator relations $[y_{im}, y_{jn}] = 1$ for $1 \le i, j \le s$ and all m, n gives a presentation for G/G'' .

Suppose first that A is unimodular, and let B be the integer matrix $-A^{-1}(I-A)$. Consider the result of adding only the $\frac{1}{2}s(s-1)$ commutator relations $[y_i, y_j]$, $1 \le i < j \le s$ to the presentation of G . From these and the R_i the relations

$$xy_i x^{-1} = \prod_{j=1}^{s} y_j^{b_{ij}}$$

follow, and more generally,

$$y_{i,m+1} = \prod_{j=1}^{s} y_{jm}^{b_{ij}} .$$

Thus by induction all y_{im} with $m \ge 0$ can be expressed in terms of the y_i and consequently commute.

Since $x^k[y_{im}, y_{jn}]x^{-k} = [y_{i,m+k}, y_{j,n+k}]$, having y_{im} and y_{jn} commute for $m, n \geq 0$ implies that they commute for all m and n .

Essentially the same argument works if $I - A$ is unimodular. We have proved the following.

THEOREM 3a. *If A and $I - A$ are non-singular of rank s , and at least one of them is unimodular, then there is a group G with associated *-module presented by $tA + (I-A)$ such that G/G'' has a presentation with $s + 1$ generators and $\frac{1}{2}s(s+1)$ relations.*

We claim that the following converse also holds. Its proof occupies the rest of the section.

THEOREM 3b. *Let A be such that neither A nor $(I-A)$ is singular or unimodular, and let G be a group with associated *-module presented by $tA + (I-A)$. Then G/G'' is not finitely presented.*

Since the conclusion of the theorem depends only on G/G'' , there is no loss of generality in assuming that G has the presentation given at the beginning of this section. Adjoining all the relations $[y_{im}, y_{jn}] = 1$ gives a presentation for G/G'' . Hence if G/G'' has any finite presentation, it has one obtained by adjoining finitely many of these commutator relations to the presentation of G .

Let G_K be the quotient group of G obtained by adjoining the relations $[y_{im}, y_{jn}] = 1$ for $1 \leq i, j \leq s$ and all m, n such that $|m-n| \leq K$. Any finite subset of the commutator relations will be included in the relations for G_K if K is large enough, so G/G'' is finitely presented only if G_K is metabelian for some K . We show that this is false (under the assumed hypotheses on A) by constructing a non-abelian homomorphic image of G_K' .

By the Reidemeister-Schreier Theorem, G_K' has a presentation with generators y_{im} , $1 \leq i \leq s$, $-\infty < m < \infty$, and relations

$$R_{im} : \prod_{j=1}^{s} y_{j,m+1}^{a_{ij}} = y_{im}^{-1} \prod_{j=1}^{s} y_{jm}^{a_{ij}} , \quad 1 \leq i \leq s , \quad -\infty < m < \infty$$

and the commutator relations

$$[y_{im}, y_{jn}] = 1 , \quad |m-n| \leq K , \quad 1 \leq i, j \leq s .$$

Let $U = (Q/Z)^s$ be the direct sum of s copies of Q/Z . Let $T = -A^{-1}(I-A)$, and define u_{im} to be the element of U obtained by reducing the ith row of T^m

modulo Z . Finally, let $i_1, i_2 : U \to U * U$ be the injections onto the first and second factor of the free product. We claim that the following formulas define a homomorphism from G'_K into $U * U$.

$$h(y_{im}) = \begin{cases} i_1(u_{im}) & , \ m < 0 , \\ 1 & , \ 0 \le m \le K , \\ i_2(u_{i,m-k}) & , \ m > K . \end{cases}$$

Compatibility with the relations R_{im} is trivial for $0 \le m < K$, and verified by direct computation for the separate cases $m < -1$, $m = -1$, $m = K$, and $m > K$. Compatibility with the commutator relations follows from the observation that if $|m-n| \le K$ then either $h(y_{im})$ and $h(y_{jn})$ lie in the same factor of the free product, or else at least one of them is trivial.

Since $T = I - A^{-1}$ and $T^{-1} = I - (I-A)^{-1}$, the assumption that neither A nor $I - A$ is unimodular implies that neither T nor T^{-1} is an integer matrix. Consequently, if m is 1 or -1 there is at least one value of i for which u_{im} is not zero. Hence the image of h has non-trivial intersection with each factor of $U * U$ and is therefore non-abelian.

References

[1] R.H. Crowell, "Corresponding group and module sequences", *Nagoya Math. J.* 19 (1961), 27-40. MR25#3977.

[2] R.H. Crowell, "The group G'/G" of a knot group G ", *Duke Math. J.* 30 (1963), 349-354. MR27#4226.

[3] Richard H. Crowell and Ralph H. Fox, *Introduction to knot theory* (Ginn and Co., Boston, New York, 1963). MR26#4348.

[4] Elvira Strasser Rapaport, "On the commutator subgroups of a knot group", *Ann. of Math.* 71 (1960), 157-162. MR22#6842.

[5] Olga Taussky, "Matrices of rational integers", *Bull. Amer. Math. Soc.* 66 (1960), 327-345. MR22#10994.

[6] H.F. Trotter, "On S-equivalence of Seifert matrices", *Invent. Math.* 20 (1973), 173-207.

Princeton University,
Princeton, New Jersey 68540, USA.

PROC. SECOND INTERNAT. CONF. THEORY OF GROUPS, CANBERRA, 1973, pp. 667-690. 20F40

ON THE LIE RING OF A GROUP OF PRIME EXPONENT

G.E. Wall

Certain properties of a p-group G are reflected in its Lie ring $L(G)$. For example, if the identity

(1.1) $$x^p = 1$$

holds in G, then the identities

(1.2) $$pu = 0 ,$$

(1.3) $$[\underbrace{[\dots [u, v], \dots], v}_{p-1 \text{ terms}}] = 0$$

hold in $L(G)$. Thus, we have an isomorphism of graded Lie rings

(1.4) $$L(B(n)) \cong \Lambda(n)/\Sigma(n) ,$$

where $B(n)$ is the n-generator free group of the variety of groups defined by (1.1) and $\Lambda(n)$ the n-generator free Lie ring of the variety of Lie rings defined by (1.2) and (1.3).

Let $\Sigma_m(n)$ denote the homogeneous component of $\Sigma(n)$ of degree m. Sanov [8] proved that

(1.5) $$\Sigma_m(n) = 0 \quad for \quad m \leq 2p-2$$

and asked whether $\Sigma(n) = 0$. Kostrikin [4] proved that

(1.6) $$\Sigma_{2p-1}(2) = \Sigma_{2p}(2) = 0 .$$

I will show, on the other hand, that

(1.7) $$\Sigma_{2p-1}(n) \neq 0 \quad when \quad n \geq 3 \quad and \quad p = 5, 7, 11 .$$

It seems likely that, in fact, (1.7) holds for all $p \geq 5$. Whether $\Sigma(2) = 0$ remains an open question.

Financial support by the Australian Research Grants Committee is gratefully acknowledged.

The component $\Sigma_m(n)$ has a naturally defined $SL(n, p)$-module structure. I prove that

(1.8) $\Sigma_{2p-1}(n)$ *is either zero or an irreducible* $SL(n, p)$*-module of specified structure.*

2. Preliminaries

It is largely a matter of taste whether one expresses the results below in terms of a Magnus algebra of formal power series or a suitable finite dimensional quotient algebra of it. We take the latter course.

Apart from supplying the necessary background material, the main purpose of the present section is to get a convenient representation of the group $B(n)$ - or rather, of $B(n, c)$, its largest nilpotent quotient group of class $\leq c$. This representation differs from those used by Magnus [6], Sanov [8] and Kostrikin [4] for similar purposes in that the coefficient domain is a field of characteristic p.

It is tacitly assumed that all associative algebras and rings introduced have unit elements. Homomorphisms, subalgebras, and so on, are interpreted accordingly.

2.1 THE ALGEBRA A

Let k be a commutative ring and n, c positive integers. Then we denote by $A = A(n, c; k)$ the associative k-algebra generated by (non-commuting) elements $x_1, \ldots, x_n$ subject to the following conditions:

(a) the monomials in $x_1, \ldots, x_n$ of total degree $\leq c$ form a k-basis of A;

(b) all monomials of degree $> c$ are zero.

Let $A^{(m)}$ denote the k-submodule of A spanned by the monomials of total degree m. If B is an additive subgroup of A, write $B^{(m)} = B \cap A^{(m)}$. We call B *graded* when $B = \sum B^{(m)}$.

Clearly, every element of A has a unique expansion

$$u = \sum u^{(m)}, \quad \left(u^{(m)} \in A^{(m)}\right).$$

The first of the components $u^{(0)}, u^{(1)}, \ldots$ to be nonzero is called the *leading term* of u. If $S \subseteq A$, then we denote by $\mathrm{gr}\, S$ the (graded) additive subgroup generated by the leading terms of the elements of S.

It is sometimes necessary to use the finer grading of A by partial degrees. Let $A^{(m_1, \ldots, m_n)}$ denote the k-submodule spanned by those monomials which have degrees $m_1, \ldots, m_n$ in $x_1, \ldots, x_n$ respectively and let

$$u = \sum u^{(m_1,\ldots,m_n)}, \quad \left(u^{(m_1,\ldots,m_n)} \in A^{(m_1,\ldots,m_n)}\right),$$

be the corresponding expansion of an element of A. The term $u^{(m_1,\ldots,m_n)}$ will be referred to as the $\left(x_1^{m_1} \ldots x_n^{m_n}\right)$-component of u.

Let us consider an endomorphism ϕ of A which stabilizes the ideal

$$\underline{\underline{a}} = \sum_{m>0} A^{(m)} .$$

(Notice that *every* endomorphism of A stabilizes $\underline{\underline{a}}$ when k contains no nilpotent elements.) Suppose that

$$x_i\phi = \sum a_{ij}x_j + \sum a_{ijk}x_jx_k + \ldots , \quad (i = 1, \ldots, n) .$$

Then ϕ is an automorphism if, and only if, $\left(a_{ij}\right) \in \mathrm{GL}(n, k)$. We say that ϕ is *graded* when it stabilizes every component $A^{(m)}$, that is, when only the linear terms in the equations above are present. The graded counterpart, $\mathrm{gr}\ \phi$, of ϕ is defined by

$$x_i(\mathrm{gr}\ \phi) = \sum a_{ij}x_j , \quad (i = 1, \ldots, n) .$$

We note the following simple functorial properties: if $S, S' \subseteq A$ and $S\phi \subseteq S'$, then $(\mathrm{gr}\ S)(\mathrm{gr}\ \phi) \subseteq \mathrm{gr}\ S'$; if ϕ' is another endomorphism of A which stabilizes $\underline{\underline{a}}$, then $\mathrm{gr}(\phi\phi') = (\mathrm{gr}\ \phi)(\mathrm{gr}\ \phi')$.

The bracket product $[u, v] = uv - vu$ is a Lie operation on A. We denote by $L = L(n, c; k)$ the Lie k-algebra generated by $x_1, \ldots, x_n$. The basic Lie products in $x_1, \ldots, x_n$ of total degree $\leq c$ form a k-basis of L (see Magnus, Karrass, Solitar [7], Section 5.6).

For $v \in A$, the mapping $\mathrm{ad}\ v : A \to A$ is defined, as usual, by $u\ \mathrm{ad}\ v = [u, v]$ $(u \in A)$. Let δ denote the homomorphism of A into the algebra of k-linear transformations on A defined by $x_i\delta = \mathrm{ad}\ x_i$ $(i = 1, \ldots, n)$. Then we write

$$[u|v] = u(v\delta) , \quad (u, v \in A) . \tag{2.1}$$

Since the Lie homomorphisms $\delta|_L$ and $\mathrm{ad}|_L$ agree on the generators x_i, we have

$$[u|v] = [u, v] , \quad (u \in A, v \in L) . \tag{2.2}$$

The elements $[u|v^m]$, $(u, v \in L)$, generate the *mth Engel ideal*, E_m, of L.

PROPOSITION 1. *Suppose* k *is a field and let* M *be a graded ideal of* L . *Then* $M^A \cap L = M$, *where* M^A *denotes the (associative) ideal of* A *generated by* M .

PROOF. The quotient Lie k-algebra $N = L/M$ is graded and nilpotent of class $\leq c$. Since k is a field, we may embed N in its universal associative algebra U . The essential point of the proof is that U is graded and that its grading is compatible with that of N - this follows, for example, from the explicit form of the Poincaré-Birkhoff-Witt Theorem (see Jacobson [3], Chapter 5).

Now in terms of U , the assertion of the Lemma is that

$$N \cap \left(U^{(c+1)} + U^{(c+2)} + \ldots\right) = 0 .$$

This is obvious because the homogeneous components of N of degree $> c$ are all zero.

2.2 GROUPS AND LIE RINGS (Lazard [5])

We shall use the following notation in a group G : (x, y) is the commutator $x^{-1}y^{-1}xy$ of elements x, y and (H, K) the commutator of subgroups H, K ; G^m denotes the subgroup generated by the mth powers of the elements of G ; G_r is the rth member of the descending central series of G ; ${}_pG_r$, for p a prime number, is the rth member of the descending central p-series of G , defined explicitly by

$$(2.3) \qquad {}_pG_r = G_r\left(G_{\{r/p\}}\right)^p\left(G_{\{r/p^2\}}\right)^{p^2} \ldots ,$$

where $\{r/p^s\}$ means the least integer $\geq r/p^s$.

A descending series of subgroups

$$N : G = N_1 \supseteq N_2 \supseteq \cdots$$

satisfying $(N_r, N_s) \subseteq N_{r+s}$ for all $r, s > 0$, determines a graded Lie ring $L(N)$ in the following way. Its underlying additive group is the direct sum of the abelian groups $L_r(N) = N_r/N_{r+1}$. The Lie product is determined by specifying its values for homogeneous elements: if $u = xN_{r+1} \in L_r(N)$ and $v = yN_{s+1} \in L_s(N)$, then $u \circ v = (x, y)N_{r+s+1} \in L_{r+s}(N)$. The *Lie ring*, $L(G)$, *of* G is, by definition, the Lie ring determined by the descending central series.

If the series N also satisfies

$$(N_r)^p \subseteq N_{rp} \quad \text{for all } r > 0 ,$$

then $L(N)$ carries the richer structure of a Lie p-ring and we will write

${}_pL(N)$, ${}_pL_r(N)$ instead of $L(N)$, $L_r(N)$. {By a Lie p-ring, I mean a "restricted Lie algebra of characteristic p" in the sense of Jacobson [3], Chapter 5, or equivalently a "Lie p-algebra" in the sense of Bourbaki [1], p. 106, *with the important proviso that the coefficient domain is the field,* F_p *, of* p *elements.*} Here, the pth power operation on ${}_pL(N)$ is defined by specifying its values on homogeneous elements: if $u = xN_{r+1} \in {}_pL_r(N)$, then $u^p = x^pN_{rp+1} \in {}_pL_{rp}(N)$. *The Lie p-ring,* ${}_pL(G)$ *, of* G is, by definition, the Lie p-ring determined by the descending central p-series.

JENNINGS-LAZARD THEOREM (Lazard [5], Theorem 6.10). *Let kG denote the group algebra of a group G over a field k of characteristic $p > 0$ and let $\underline{\underline{g}}$ be its augmentation ideal (spanned by the elements $g - 1$, $g \in G$). Then*

$$G \cap (1+\underline{\underline{g}}^r) = {}_pG_r, \quad (r = 1, 2, \ldots).$$

2.3 THE VARIETY ${}_p\underline{\underline{\mathrm{N}}}_c$

In the present section, k is a field of characteristic $p > 0$

Let $\underline{\underline{\mathrm{N}}}_c$ denote, as usual, the variety formed by the nilpotent groups of class $\le c$. Let ${}_p\underline{\underline{\mathrm{N}}}_c$ denote the corresponding variety formed by the groups G which satisfy

$${}_pG_{c+1} = 1$$

that is,

$$(G_r)^{p^s} = 1 \quad \text{whenever} \quad rp^s > c.$$

The formula $(1+u)^{-1} = 1 - u + u^2 - \ldots$ shows that the set $1 + \underline{\underline{a}} = \{1+u \mid u \in \underline{\underline{a}}\}$ is a group under multiplication. Let $F = F(n, c; k)$ denote the subgroup of $1 + \underline{\underline{a}}$ generated by the elements $1 + x_1, \ldots, 1 + x_n$.

PROPOSITION 2. *F is the n-generator free group of the variety* ${}_p\underline{\underline{\mathrm{N}}}_c$. *We have*

$$F \cap (1+\underline{\underline{a}}^r) = {}_pF_r, \quad (r = 1, 2, \ldots).$$

PROOF. Let G be a free group of rank n and let $\underline{\underline{g}}$ be the augmentation ideal of kG. The Proposition follows directly from the Jennings-Lazard Theorem once we have proved the existence of an isomorphism

$$(*) \qquad A \cong kG/\underline{\underline{g}}^{c+1}$$

under which $\underline{a}$ corresponds to $\underline{g}/\underline{g}^{c+1}$.

Let $\theta : G \to F$ be the homomorphism which maps a set of free generators $g_1, \ldots, g_n$ of G to the corresponding generators $1 + x_1, \ldots, 1 + x_n$ of F . Extend θ to an algebra homomorphism $\phi : kG \to A$. Then $\underline{g}^{c+1}\phi = 0$, so that we get an induced homomorphism $\alpha : kG/\underline{g}^{c+1} \to A$ such that $(g_i+\underline{g}^{c+1})\alpha = 1 + x_i$, $(i = 1, \ldots, n)$.

On the other hand, the definition of A shows that there is a homomorphism $\beta : A \to kG/\underline{g}^{c+1}$ such that $x_i\beta = g_i - 1 + \underline{g}^{c+1}$, $(i = 1, \ldots, n)$. Clearly, α and β are mutually inverse. This proves the existence of the isomorphism (*) and so gives the Proposition.

Let $P = P(n, c; k)$ denote the Lie p-ring generated by $x_1, \ldots, x_n$. The following analogues of (2.2) and Proposition 1 are proved by similar methods.

(2.4) $$[u|v] = [u, v] , \quad (u \in A, v \in P) .$$

PROPOSITION 3. *If* M *is a graded Lie* p*-ideal of* P *, then* $M^A \cap P = M$.

The next result is the group theoretical counterpart of Propositions 1 and 3.

PROPOSITION 4. *If* $N \triangleleft F$ *, then* $(N-1)^A \cap (F-1) = N - 1$.

PROOF. Write $G = F/N$ and let $\underline{g}$ denote the augmentation ideal of kG . By Proposition 3, ${}_pF_{c+1} = 1$ and therefore ${}_pG_{c+1} = 1$. Hence by the Jennings-Lazard Theorem, we may identify G with its canonical image in $kG/\underline{g}^{c+1}$. We now extend the canonical homomorphism $\theta : F \to G$ to an algebra homomorphism $\phi : A \to kG/\underline{g}^{c+1}$. Then

$$\begin{aligned}(N-1)^A \cap (F-1) &\subseteq (\ker \phi) \cap (F-1) \\ &= (\ker \theta) - 1 = N - 1 .\end{aligned}$$

Since the reverse inclusion is obvious, this proves the Proposition.

PROPOSITION 5. *If* $N \triangleleft F$ *, then* $\mathrm{gr}(N-1)$ *is a Lie* p*-ideal of* P *and*

$$_pL(F/N) \cong P/\mathrm{gr}(N-1) .$$

PROOF. Consider the series

$$\mathsf{F} : F = \mathsf{F}_1 \supseteq \mathsf{F}_2 \supseteq \ldots$$

$$\mathsf{F}/N : F/N = \mathsf{F}_1N/N \supseteq \mathsf{F}_2N/N \supseteq \ldots ,$$

where $\mathsf{F}_r = F \cap (1+\underline{a}^r)$. It is a straightforward matter to verify that $\mathrm{gr}(F-1)$ is a

Lie p-ring isomorphic to $_pL(F)$, that $gr(N-1)$ is a Lie p-ideal of $gr(F-1)$ and that $gr(F-1)/gr(N-1) \cong {}_pL(F/N)$. Moreover, by Proposition 2, $_pL(F) = {}_pL(F)$ and so $_pL(F/N) = {}_pL(F/N)$. Now, it is known that $_pL(F)$ is generated by its elements of degree 1 (Lazard [5], p. 135), so that $gr(F-1)$ is generated by $x_1, \ldots, x_n$, that is, $gr(F-1) = P$. Putting these results together, we get the Proposition.

PROPOSITION 6. *If N is a fully invariant subgroup of F , then $gr(N-1)$ is mapped into itself by every graded endomorphism of P .*

PROOF. Every endomorphism ω of F can be extended to an endomorphism ω^* of A . Since N is fully invariant in F , $gr(N-1)$ is mapped into itself by $gr(\omega^*)$. Thus, it will be sufficient to prove that every graded endomorphism θ of P has the form $gr(\omega^*)|_P$.

Now,

$$x_i\theta = \sum m_{ij}x_j \ , \quad (i = 1, \ldots, n) \ ,$$

where $m_{ij} \in \mathbb{Z}$. We may then take ω to be the endomorphism of F defined by

$$(1+x_i)\omega = (1+x_1)^{m_{i1}} \ldots (1+x_n)^{m_{in}} \ , \quad (i = 1, \ldots, n) \ .$$

This completes the proof.

An element of A is called *uniform* if all monomials which appear in it involve precisely the same variables x_i .

LEMMA. *Let N be a fully invariant subgroup of F . Let $1 + u$ be an element of N such that the leading term, $u^{(r)}$, of u is uniform. Then there is an element $1 + v$ in N such that v has leading term $u^{(r)}$ and v is uniform.*

PROOF. Since $A(n, c; k)/\underline{a}^c \cong A(n, c-1; k)$ we may assume, by induction on c , that $u^{(r)} + \ldots + u^{(c-1)}$ is uniform.

Let T be the set of variables x_i which appear in $u^{(r)}$. For $S \subseteq T$, define an endomorphism θ_S of F by

$$(1+x_i)\theta_S = \begin{cases} 1 + x_i \ , & (x_i \in S) \ , \\ 1 \ , & (x_i \notin S) \ . \end{cases}$$

If $w \in A$, let w_S denote the sum of those components $w^{(m_1, \ldots, m_n)}$ which involve only variables in S. Then, if S is a *proper* subset of T,

$$(1+u)\theta_S = 1 + u_S = 1 + \left(u^{(c)}\right)_S$$

is in the centre of F. It may be verified that the requirements of the Lemma are fulfilled by the element

$$1 + v = \prod_{S \subseteq T} \left[(1+u)\theta_S\right]^{\varepsilon_S},$$

where $\varepsilon_S = (-1)^{|T-S|}$.

PROPOSITION 7. *Let* N *be a fully invariant subgroup of* F. *Let* u *be an element of* $\mathrm{gr}(N-1)$ *which is multilinear in those variables which appear in it. Then* $u\theta \in \mathrm{gr}(N-1)$ *for every endomorphism* θ *of* P.

PROOF. Suppose, for simplicity of notation, that the variables appearing in u are $x_1, \ldots, x_r$ and write $u = u(x_1, \ldots, x_r)$. Since u is multilinear, it will be sufficient to prove that $u(y_1, \ldots, y_r) \in \mathrm{gr}(N-1)$ whenever $y_1, \ldots, y_r$ are homogeneous elements of P. We assume that $u(y_1, \ldots, y_r) \neq 0$, so that each $y_i \neq 0$.

Since $P = \mathrm{gr}(F-1)$, we may choose $1 + v \in N$ and $1 + z_1, \ldots, 1 + z_r \in F$ so that $u, y_1, \ldots, y_r$ are the leading terms of $v, z_1, \ldots, z_r$ respectively. Moreover, by the Lemma, we may assume that $v = v(x_1, \ldots, x_r)$ is uniform. Under these assumptions, $u(y_1, \ldots, y_r)$ *is the leading term of* $v(z_1, \ldots, z_r)$. For

$$v(z_1, \ldots, z_r) = u(z_1, \ldots, z_r) + \ldots,$$

where the omitted terms are monomials in the z_i which involve each z_i at least once and at least one z_i more than once; furthermore, it is obvious from the multilinearity of u that $u(y_1, \ldots, y_r)$ is the leading term of $u(z_1, \ldots, z_r)$.

Now, if ϕ is an endomorphism of F such that $(1+x_i)\phi = 1 + z_i$, $(i = 1, \ldots, n)$, then $(1+v)\phi = 1 + v(z_1, \ldots, z_r) \in N$ because $1 + v \in N$ and N is fully invariant in F. Therefore the leading term of $v(z_1, \ldots, z_r)$, namely, $u(y_1, \ldots, y_r)$, is in $\mathrm{gr}(N-1)$. This proves the Proposition.

Let $\underline{V}$ be a variety of groups. Let $V(G)$ denote the verbal subgroup of a group G determined by $\underline{V}$, so that $G/V(G)$ is the largest quotient group of G which lies in $\underline{V}$. Then, clearly, $F/V(F)$ is the n-generator free group of the variety $\underline{V} \cap \underline{N}_{p-c}$. We now make some simple observations about the dependence of $F/V(F)$ and its Lie p-ring $P/\mathrm{gr}\{V(F)-1\}$ on n (for fixed c, k and $\underline{\underline{V}}$).

Let us write $A = A(n)$, $F = F(n)$ and so on, and let $m \leq n$. Then there is the obvious embedding of $A(m)$ in $A(n)$ in which the generators $x_1, \ldots, x_m$ of $A(m)$ are regarded as the first m of the n generators of $A(n)$. It is readily verified that $F(m)$, $P(m)$, $V\{F(m)\}$ and $\mathrm{gr}\{V\{F(m)\}-1\}$ are the intersections of $F(n)$, $P(n)$, ... with $A(m)$. Now, $F(n)/V\{F(n)\}$ has the n generators

$$\theta_i = \{1+x_i\}V\{F(n)\} , \quad (i = 1, \ldots, n) ,$$

and $P(n)/\mathrm{gr}\{V\{F(n)\}-1\}$ the n generators

$$\xi_i = x_i + \mathrm{gr}\{V\{F(n)\}-1\} , \quad (i = 1, \ldots, n) .$$

The results just cited above imply that we may identify $F(m)/V\{F(m)\}$ with the subgroup of $F(n)/V\{F(n)\}$ generated by $\theta_1, \ldots, \theta_m$ and $P(n)/\mathrm{gr}\{V\{F(n)\}-1\}$ with the Lie p-subring of $P(n)/\mathrm{gr}\{V\{F(n)\}-1\}$ generated by $\xi_1, \ldots, \xi_m$. The practical effect of these remarks is that when dealing with m-variable laws (either in the group or Lie p-ring) we may use any larger number of variables that happens to be convenient.

2.4 APPLICATION TO GROUPS OF PRIME EXPONENT

In the present section, p *is a prime number and* $k = \mathrm{F}_p$ *, the field of* p *elements.*

Consider a group G of exponent p . Then the descending central series and p-series coincide, so that $L(G)$ is the underlying Lie ring of ${}_pL(G)$. Thus, ${}_pL(G)$ satisfies the $(p-1)$th Engel condition. Further, it is clear from the definition of pth powers in ${}_pL(G)$ that the pth powers of all its *homogeneous* elements are zero. It follows from the last two statements that the pth power of *every* element of ${}_pL(G)$ is zero, for, if u, v are elements of a Lie p-ring, then $(u+v)^p = u^p + v^p + f(u, v)$, where $f(u, v)$ is in the $(p-1)$th Engel ideal of the Lie ring generated by u, v .

In the discussion which follows we shall replace $B(n)$, $\Lambda(n)$, $\Sigma(n)$ (see §1) by their largest nilpotent quotients of class $\leq c$, say $B(n, c)$, $\Lambda(n, c)$, $\Sigma(n, c)$. Then (1.4) takes the modified form

$$(1.4)' \qquad L\{B(n, c)\} \cong \Lambda(n, c)/\Sigma(n, c) .$$

We wish to interpret this equation within the algebra $A = A\{n, c; \mathrm{F}_p\}$.

The group $B(n, c)$ is the n-generator free group of the variety $\underline{\underline{B}}_p \cap \underline{\underline{N}}_c$, where $\underline{\underline{B}}_p$ is the Burnside variety of exponent p . Since $\underline{\underline{B}}_p \cap \underline{\underline{N}}_c = \underline{\underline{B}}_p \cap {}_p\underline{\underline{N}}_c$, we have

(2.5) $$B(n, c) \cong F/F^p ,$$

(2.6) $$_pL(B(n, c)) \cong P/\mathrm{gr}(F^p-1) .$$

Then, since the pth power of every element of $_pL(B(n, c))$ is zero,

(2.7) $$P^{[p]} \subseteq \mathrm{gr}(F^p-1) ,$$

where $P^{[p]}$ is the Lie p-ideal of P generated by the pth powers of its elements.

On the other hand, since L is the n-generator free algebra of the variety of all nilpotent Lie algebras of class $\le c$, it follows that

(2.8) $$\Lambda(n, c) \cong L/E_{p-1} ,$$

where E_{p-1} is the $(p-1)$th Engel ideal of L. Now, it is known that

$$P^{[p]} + L = P , \quad P^{[p]} \cap L = E_{p-1} ,$$

so that we have the Lie algebra isomorphism

(2.9) $$L/E_{p-1} \cong P/P^{[p]} .$$

These considerations show that we may identify the terms $L(B(n, c))$, $\Lambda(n, c)$, $\Sigma(n, c)$ in (1.4)' with the underlying Lie algebras of $P/\mathrm{gr}(F^p-1)$, $P/P^{[p]}$, $\mathrm{gr}(F^p-1)/P^{[p]}$ respectively. In particular, $\Sigma(n, c)$ is nonzero if, and only if, $\mathrm{gr}(F^p-1) \neq P^{[p]}$.

We comment briefly on the $\mathrm{SL}(n, p)$-module structure of $\Sigma_m(n)$ referred to in §1. If $c \ge m$, we may identify $\Sigma_m(n)$ with the mth homogeneous component, say T_m, of $\mathrm{gr}(F^p-1)/P^{[p]}$. Now, Proposition 6 shows that T_m is stable under every graded endomorphism of P. The graded endomorphisms of P form an algebra isomorphic to the algebra, $M(n, p)$, of all $n \times n$ matrices over F_p. Thus, T_m is a module for the multiplicative semigroup $M(n, p)^\times$ of $M(n, p)$ and *a fortiori* a module for its special linear subgroup $\mathrm{SL}(n, p)$.

3. Identities

We have obtained suitable representations of the group $B(n, c)$ and its Lie ring and now set up the formal apparatus for carrying out calculations in them. Although our results are ultimately *applied* in characteristic p, they are mostly *proved* by passing to characteristic 0; it is only in this way that we can take direct advantage of the Baker-Campbell-Hausdorff formula.

The notation now to be introduced will remain in force for §3.1 - 3.3.

Let p be a prime. Let $\mathbb{Q}$ denote the rational field and $\mathbb{Q}^0$ its subring of p-integers (that is, rational numbers with denominator prime to p). Write

$$A = A(n, c; \mathbb{Q}) \quad , \quad L = L(n, c; \mathbb{Q}) \quad , \text{ and so on;}$$

$$A^0 = A(n, c; \mathbb{Q}^0) \quad , \quad L = L(n, c; \mathbb{Q}^0) \quad , \text{ and so on.}$$

We embed A^0 in A in the obvious way and call its elements p-integral. By the Poincaré-Birkhoff-Witt Theorem, $L^0 = L \cap A^0$.

3.1 POWER FORMULAE*

Let us write the multinomial expansion in the form

$$(3.1) \qquad (x_1 + \ldots + x_n)^m = \sum_{m_1+\ldots+m_n=m} \langle m_1x_1, \ldots, m_nx_n \rangle ,$$

where $\langle m_1x_1, \ldots, m_nx_n \rangle$ stands for the sum of all the different monomials having partial degrees $m_1, \ldots, m_n$ in $x_1, \ldots, x_n$ respectively.

The following simple identities are set down for reference:

$$(3.2) \qquad \langle (m-1)x_1, x_2 \rangle = \sum_1^m x_1^{m-i} x_2 x_1^{i-1}$$

$$= \sum_1^m \binom{m}{i} x_1^{m-i} \left[x_2 \mid x_1^{i-1}\right] ,$$

$$(3.3) \qquad \left[x_2, x_1^m\right] = \langle (m-1)x_1, [x_2, x_1] \rangle$$

$$= \sum_1^m \binom{m}{i} x_1^{m-i} \left[x_2 \mid x_1^i\right] ,$$

$$(3.4) \qquad \left[x_2 \mid x_1^m\right] = \sum_0^m \binom{m}{i} (-x_1)^{m-i} x_2 x_1^i .$$

We now consider pth powers.

LEMMA 1.

$$\langle x_1, \ldots, x_p \rangle = p!\,|x_1, \ldots, x_p| + [x_1 \mid \langle x_2, \ldots, x_p \rangle]$$

where $|x_1, \ldots, x_p| \in A^0$.

PROOF. Clearly,

$$\langle x_1, \ldots, x_s \rangle = \sum (x_1y_2 \ldots y_s + y_2x_1 \ldots y_s + \ldots + y_2 \ldots y_sx_1) ,$$

* *Cf.* the initial sections of Kostrikin [4].

where summation is over the permutations $y_2, \ldots, y_s$ of $x_2, \ldots, x_s$. A simple inductive proof shows that

$$[x_1 \mid \langle x_2, \ldots, x_s \rangle] =$$
$$\sum \left\{ x_1 y_2 \ldots y_s - \binom{s-1}{1} y_2 x_1 \ldots y_s + \ldots + (-1)^{s-1} \binom{s-1}{s-1} y_2 \ldots y_s x_1 \right\},$$

where summation is over the same range. Since

$$(-1)^i \binom{p-1}{i} \equiv 1 \pmod{p},$$

the Lemma follows.

COROLLARY. *If* $m_1, \ldots, m_n$ *are non-negative integers such that* $\sum m_i = p$ *and* $0 < m_1 < p$, *then*

$$\langle m_1 x_1, \ldots, m_n x_n \rangle = p! |m_1 x_1, \ldots, m_n x_n| + \frac{1}{m_1} [x_1 \mid \langle (m_1 - 1) x_1, \ldots, m_n x_n \rangle]$$

where

$$|m_1 x_1, \ldots, m_n x_n| = (m_1! \ldots m_n!)^{-1} |\underbrace{x_1, \ldots, x_1}_{m_1 \text{ terms}}, \ldots, \underbrace{x_n, \ldots, x_n}_{m_n \text{ terms}}| \in A^0 .$$

LEMMA 2.

$$\left(\sum_1^n x_i\right)^p - \sum_1^n x_i^p = p! R(x_1, \ldots, x_n) + Q(x_1, \ldots, x_n)$$

where

$$R(x_1, \ldots, x_n) = \sum |m_1 y_1, \ldots, m_r y_r| \in A^0 ,$$

$$Q(x_1, \ldots, x_n) = \sum \frac{1}{m_1} [y_1 \mid \langle (m_1 - 1) y_1, \ldots, m_r y_r \rangle] \in L^0 ,$$

summation in both cases being over the subsequences $y_1, \ldots, y_r$ *of* $x_1, \ldots, x_n$ *and over the integers* $m_1, \ldots, m_r$ *satisfying* $\sum m_i = p$ *and* $0 < m_i < p$ $(i = 1, \ldots, r)$.

PROOF. This follows from the Corollary and the multinomial expansion (3.1).

LEMMA 3. $R(x_1, \ldots, x_n)$ *is a symmetric function modulo* L^0.

PROOF. If $y_1, \ldots, y_n$ is a permutation of $x_1, \ldots, x_n$, then

$$R(y_1, \ldots, y_n) - R(x_1, \ldots, x_n) = \frac{1}{p!} (Q(x_1, \ldots, x_n) - Q(y_1, \ldots, y_n)) .$$

The left hand side is in A^0 and the right hand side in L, so that the common value is in $L \cap A^0 = L^0$. (This is a standard argument which we will use a number of times.)

LEMMA 4. *If* $\lambda_i, \mu_i \in Q^0$, *there exist* $\nu_i \in Q^0$ *such that*

$$R(\lambda_1 x_1, \ldots, \lambda_n x_n) + R(\mu_1 x_1, \ldots, \mu_n x_n) + R\left(\sum_1^n \lambda_i x_i, \sum_1^n \mu_i x_i\right)$$
$$\equiv R\left((\lambda_1+\mu_1)x_1, \ldots, (\lambda_n+\mu_n)x_n\right) + \sum_1^n \nu_i x_i^p \pmod{L^0+pA^0}.$$

PROOF. Similar to that of Lemma 3.

We define, for $u \in \underline{\underline{a}}^{\bullet}$,

$$e^u = \sum_0^\infty u^m/m!, \quad e_p(u) = \sum_0^{p-1} u^m/m!.$$

If $v \in \underline{\underline{a}}$, the equation $e^u = 1 + v$ has a unique solution $u \in \underline{\underline{a}}$. Similarly, if $w \in \underline{\underline{a}}^0$ $\left(= \underline{\underline{a}} \cap A^0\right)$, the equation $e_p(u) = 1 + w$ has a unique solution $u \in \underline{\underline{a}}^0$.

LEMMA 5. *If*

$$e^u = e^{x_1} \ldots e^{x_n}, \quad e_p(v) = e_p(x_1) \ldots e_p(x_n),$$

then $v^{(m)} = u^{(m)} \in L^0$, $(1 \le m \le p-1)$,

$$v^{(p)} - R(x_1, \ldots, x_n) = u^{(p)} + (p!)^{-1}Q(x_1, \ldots, x_n) \in L^0.$$

PROOF. We regard (1) as equations to determine the homogeneous components $u^{(1)}, u^{(2)}, \ldots$ and $v^{(1)}, v^{(2)}, \ldots$ by recurrence. This yields

(2) $$v^{(m)} = u^{(m)}, \quad (m < p),$$

(3) $$v^{(p)} = u^{(p)} + (p!)^{-1}\left\{\left(\sum x_i\right)^p - \sum x_i^p\right\},$$

and the latter becomes, after rearranging,

(4) $$v^{(p)} - R(x_1, \ldots, x_n) = u^{(p)} + (p!)^{-1}Q(x_1, \ldots, x_n).$$

In (2) and (4), the right hand sides are in L by the Baker-Campbell-Hausdorff formula and the left hand sides are in A^0; therefore both sides are in L^0.

3.2 MULTINOMIAL ELEMENTS

Given a *multilinear* element

$$f(x_1, \ldots, x_s) \in A(s, c; \mathrm{Q}) ,$$

we may, for each n , define a corresponding element

$$F_n(x_1, \ldots, x_n) \in A(n, c; \mathrm{Q})$$

by the equation

(3.5)

$$F_n(x_1, \ldots, x_n) = \sum_{\substack{m_1+\ldots+m_n=s \\ m_1\geq 0,\ldots,m_n\geq 0}} (m_1! \ldots m_n!)^{-1} f(\underbrace{x_1, \ldots, x_1}_{m_1 \text{ terms}}, \ldots, \underbrace{x_n, \ldots, x_n}_{m_n \text{ terms}}) .$$

We call F_n the *multinomial element* of $A(n, c; \mathrm{Q})$ with *generic term* f . For example $(x_1 + \ldots + x_n)^s$ and $Q(x_1, \ldots, x_n)$ are the multinomial elements with generic terms $\langle x_1, \ldots, x_s \rangle$ and $[x_1 \mid \langle x_2, \ldots, x_p \rangle]$ respectively.

We remark that $f(x_1, \ldots, x_s)$ is the $(x_1 \ldots x_s)$-component of $F_n(x_1, \ldots, x_n)$ when $s \leq n$ and that

$$F_m(x_1, \ldots, x_m) = F_n(x_1, \ldots, x_m, 0, \ldots, 0)$$

when $m \leq n$. For this reason, we shall omit the suffix n and write simply $F(x_1, \ldots, x_n)$.

LEMMA 6. *Let* $F(x_1, \ldots, x_n)$, $G(x_1, \ldots, x_n)$ *be multinomial elements with generic terms* $f(x_1, \ldots, x_s)$, $g(x_1, \ldots, x_t)$ *respectively. Then their product*

$$H(x_1, \ldots, x_n) = F(x_1, \ldots, x_n).G(x_1, \ldots, x_n)$$

is multinomial with generic term

$$h(x_1, \ldots, x_{s+t}) = \sum f(u_1, \ldots, u_s)g(v_1, \ldots, v_t) ,$$

where summation is over the partitions of the sequence $x_1, \ldots, x_{s+t}$ *into a subsequence* $u_1, \ldots, u_s$ *of* s *elements and the complementary subsequence* $v_1, \ldots, v_t$ *of* t *elements.*

We omit the proof, which is a straightforward verification.

3.3 FURTHER IDENTITIES

In the present section, we use the following abbreviations: $R = R(x_1, \ldots, x_n)$, $Q = Q(x_1, \ldots, x_n)$, $X = x_1 + \ldots + x_n$.

By Lemma 6, the element

$$-\frac{1}{p!}[Q|X^t] , \quad (t = 1, 2, \ldots) ,$$

is multinomial. In analogy with (3.1), we denote its generic term by

$$\langle\langle x_1, \ldots, x_{p+t}\rangle\rangle \tag{3.6}$$

and write

$$-\frac{1}{p!}[Q|X^t] = \sum_{\substack{m_1+\ldots+m_n=p+t \\ m_1\geq 0,\ldots,m_n\geq 0}} \langle\langle m_1x_1, \ldots, m_nx_n\rangle\rangle , \tag{3.7}$$

where

$$\langle\langle m_1x_1, \ldots, m_nx_n\rangle\rangle = (m_1! \ldots m_n!)^{-1}\langle\langle \underbrace{x_1, \ldots, x_1}_{m_1 \text{ terms}}, \ldots, \underbrace{x_n, \ldots, x_n}_{m_n \text{ terms}}\rangle\rangle . \tag{3.8}$$

LEMMA 7. *Let* $m_1, \ldots, m_n$ *be non-negative integers with sum* $p + t > p$.

(a) If $n \geq p+t$, $\langle\langle x_1, x_2, \ldots, x_{p+t}\rangle\rangle$ *is an element of* L^0 *which is symmetric modulo the* t*-th Engel ideal* E_t^0 *of* L^0 .

(b) If all $m_i < p$, *then* $\langle\langle m_1x_1, \ldots, m_nx_n\rangle\rangle$ *is the* $\begin{pmatrix} x_1^{m_1} & \ldots & x_n^{m_n}\end{pmatrix}$*-component of* $[R|X^t]$.

(c) If some $m_i \geq p$, *then*

$$\langle\langle \underbrace{x_1, \ldots, x_1}_{m_1 \text{ terms}}, \ldots, \underbrace{x_n, \ldots, x_n}_{m_n \text{ terms}}\rangle\rangle \in pA^0 + E_p^0 .$$

PROOF. By Lemma 2,

$$-\frac{1}{p!}[Q|X^t] = \left[R + \frac{1}{p!}\sum_1^n x_i^p \mid X^t\right] , \quad (t > 0) , \tag{3.9}$$

and, by definition, (3.8) is the $\begin{pmatrix} x_1^{m_1} & \ldots & x_n^{m_n}\end{pmatrix}$-component of the left hand side. Parts *(b)*, *(c)* of the Lemma follows by inspection (notice that $[x_i^p \mid X^t] \in pA^0 + E_p^0$ by (3.3)). Moreover if $n \geq p+t$, then (3.6), being the $(x_1 \ldots x_{p+t})$-component of $[R|X^t]$, is in $A^0 \cap L = L^0$; it is symmetric modulo E_t^0 because the same is true of $[R|X^t]$ by Lemma 3.

LEMMMA 8.

$$\langle\langle x_1, \ldots, x_{p+t}\rangle\rangle = -[x_1 \mid f_{t-1}(x_2, \ldots, x_{p+t})] , \quad (t > 0) ,$$

where $f_s(x_1, \ldots, x_{p+s})$, $(s \geq 0)$, *is the* $(x_1 \ldots x_{p+s})$*-component of*

$$F_s = \sum_{i=0}^{s} \binom{s+1}{i} (-X)^{s-i} R X^i , \quad (n \geq p+s) .$$

PROOF. Let us assume that, for some $t \geq 0$, the generic term of $(p!)^{-1}[Q|X^t]$ has the form $[x_1 \mid \phi_{t-1}(x_2, \ldots, x_{p+t})]$. Then, by Lemma 6, the generic term of $(p!)^{-1}[Q|X^{t+1}]$ is

$$\sum_{i=2}^{p+t+1} [x_1 \mid \phi_{t-1}(x_2, \ldots, \hat{x}_i, \ldots, x_{p+t+1})x_i] + [x_2 \mid \phi_{t-1}(x_3, \ldots, x_{p+t+1})x_1] =$$
$$= [x_1 \mid \phi_t(x_2, \ldots, x_{p+t+1})] ,$$

where

$$\phi_t(x_1, \ldots, x_{p+t}) = \sum_{i=1}^{p+t} \phi_{t-1}(x_1, \ldots, \hat{x}_i, \ldots, x_{p+t})x_i - [x_1 \mid \phi_{t-1}(x_2, \ldots, x_{p+t})] .$$

By Lemma 2, $\phi_{-1}(x_1, \ldots, x_{p-1}) = \frac{1}{p!}\langle x_1, \ldots, x_{p-1}\rangle$. Therefore

$$\phi_0(x_1, \ldots, x_p) = \frac{1}{p!}\left(\sum_{i=1}^{p} \langle x_1, \ldots, \hat{x}_i, \ldots, x_p\rangle x_i - [x_1 \mid \langle x_2, \ldots, x_p\rangle]\right)$$
$$= \frac{1}{p!}\left(\langle x_1, \ldots, x_p\rangle - [x_1 \mid \langle x_2, \ldots, x_p\rangle]\right)$$
$$= |x_1, \ldots, x_p| = f_0(x_1, \ldots, x_p) .$$

Let $t > 0$ and let us assume that $\phi_{t-1} = f_{t-1}$. Then $\phi_t(x_1, \ldots, x_{p+t})$ is the $(x_1 \ldots x_{p+t})$-component of

$$F_{t-1}X + [R|X^t] = \sum_{i=0}^{t-1} \binom{t}{i} (-X)^{t-1-i} R X^{i+1} + \sum_{j=0}^{t} \binom{t}{j} (-X)^{t-j} R X^j$$
$$= \sum_{j=0}^{t} \binom{t+1}{j} (-X)^{t-j} R X^j = F_t .$$

This proves the Lemma by induction.

COROLLARY. *Let* $\Omega(x_1, x_2, x_3)$ *be the* $\left(x_1 x_2^{p-1} x_3^{p-1}\right)$*-component of* $[R|X^{p-1}]$, $(n \geq 3)$, *and* $\Theta(x_1, x_2)$ *the* $\left(x_1^{p-1} x_2^{p-1}\right)$*-component of* $\langle R, (p-2)X\rangle$. *Then*

$$\Omega(x_1, x_2, x_3) \equiv [x_1 \mid \Theta(x_2, x_3)] \pmod{pA^0} .$$

PROOF. By the Lemma,

$$\Omega(x_1, x_2, x_3) = [x_1 \mid \Phi(x_2, x_3)] ,$$

where

$$\Phi(x_2, x_3) = -\frac{1}{(p-1)!^2} f_{p-2}(\underbrace{x_2, \ldots, x_2}_{p-1 \text{ terms}}, \underbrace{x_3, \ldots, x_3}_{p-1 \text{ terms}}) .$$

However, since

$$\sum_0^{p-2} \binom{p-1}{i} (-X)^{p-2-i} \left(R + \frac{1}{p!} \sum x_j^p\right) X^i = \sum_0^{p-2} \binom{p-1}{i} (-X)^{p-2-i} \frac{1}{p!} (X^p - Q) X^i$$

is multinomial, $\Phi(x_2, x_3)$ is the $\left(x_2^{p-1} x_3^{p-1}\right)$-component of

$$-\sum_0^{p-2} \binom{p-1}{i} (-X)^{p-2-i} R X^i .$$

Since $\langle R, (p-2)X \rangle = \sum_0^{p-2} X^{p-2-i} R X^i$ and $-\binom{p-1}{i} (-1)^{p-2-i} \equiv 1 \pmod p$, the Corollary follows.

3.4 SPECIALIZATION

In the present section, $k = \mathrm{F}_p$ *, where* p *is prime.*

Some of the p-integral formulae take a particularly simple form in the present case:

$$\langle x_1, \ldots, x_p \rangle = [x_1 \mid \langle x_2, \ldots, x_p \rangle] , \tag{3.10}$$

$$\left(\sum_1^n x_i\right)^p - \sum_1^n x_i^p = Q(x_1, \ldots, x_n) , \tag{3.11}$$

$$\langle (p-1)x_1, x_2 \rangle = \left[x_2 \mid x_1^{p-1}\right] , \tag{3.12}$$

$$\left[x_2, x_1^p\right] = \left[x_2 \mid x_1^p\right] . \tag{3.13}$$

These identities will be applied freely and usually without specific reference.

It is quite easily deduced from (3.10) that the Engel ideal E_{p-1} is spanned by the *Kostrikin elements*

$$\langle u_1, \ldots, u_p \rangle , \quad (u_i \in L) .$$

The group F was considered in §2.4. In the sequel, we shall deal not with F itself but rather with the group $\hat{F}$ generated by the elements $e_p(x_i)$, $(i = 1, \ldots, n)$. Clearly, F is mapped onto $\hat{F}$ by the automorphism ϕ of A

defined by $x_i\phi = e_p(x_i) - 1$, $(i = 1, \ldots, n)$. Since $\text{gr}\ \phi$ is the identity, the basic results (2.5) - (2.7) remain intact when F is replaced by $\hat{F}$.

LEMMA 9. *Suppose* $c = p$. *Then* $e_p(v) \in \hat{F}$ *if, and only if,*

$$v \equiv \sum_1^n \lambda_i x_i + R(\lambda_1 x_1, \ldots, \lambda_n x_n) \pmod{P \cap \underline{\underline{a}}^2} ,$$

where $\lambda_i \in \mathsf{F}_p$.

PROOF. Let S denote the set of elements v satisfying the condition in the Lemma and let $G = \{e_p(v) \mid v \in S\}$. Then

$$e_p(x_i) \in G , \quad (i = 1, \ldots, n) ,$$

and $|G| = |S| = |P| = |{}_pL(\hat{F})| = |\hat{F}|$. Thus, it is sufficient to prove that G is a group.

Let

$$u \equiv \sum \lambda_i x_i + R(\lambda_1 x_1, \ldots, \lambda_n x_n) ,$$

$$v \equiv \sum \mu_i x_i + R(\mu_1 x_1, \ldots, \mu_n x_n) ,$$

where $\equiv$ denotes congruence mod $P \cap \underline{\underline{a}}^2$. Let $e_p(u)e_p(v) = e_p(w)$. Then

$$w \equiv R(\lambda_1 x_1, \ldots, \lambda_n x_n) + R(\mu_1 x_1, \ldots, \mu_n x_n) + R\left(\sum \lambda_i x_i, \sum \mu_i x_i\right)$$
$$+ \sum (\lambda_i + \mu_i) x_i \quad \text{(by Lemma 5)}$$
$$\equiv \sum (\lambda_i + \mu_i) x_i + R((\lambda_1 + \mu_1) x_1, \ldots, (\lambda_n + \mu_n) x_n) \quad \text{(by Lemma 4)}.$$

This proves that G is a group, as required.

3.5. THE MODULE $\Delta^{(m)}$

We turn aside to consider the polynomial algebra

$$S = \mathsf{F}_p[X_1, \ldots, X_n]$$

in *commuting* indeterminates X_i . Then S has the natural grading by degree and its group of graded automorphisms is isomorphic to $\mathrm{GL}(n, p)$. We consider the mth homogeneous component $S^{(m)}$ as an $\mathrm{SL}(n, p)$-module.

Let T denote the graded ideal of S generated by the pth powers X_i^p . Then

$$\Delta = S/T$$

inherits the grading of S and we may identify $\Delta^{(m)}$ with $S^{(m)}/T^{(m)}$. Clearly $\Delta^{(m)} = 0$ when $m > n(p-1)$.

LEMMA 10. *If* $m \leq n(p-1)$, $\Delta^{(m)}$ *is an irreducible* $\mathrm{SL}(n, p)$*-module.*

PROOF. Write $\mathrm{SL} = \mathrm{SL}(n, p)$, $\mathrm{GL} = \mathrm{GL}(n, p)$. Suppose M is a finitely generated $\mathsf{F}_p G$-module, where $\mathrm{SL} \subseteq G \subseteq \mathrm{GL}$. Let N denote the subspace of M formed by the vectors fixed under the group, U , of all lower unitriangular matrices. Let B denote the group of all lower triangular matrices in G . Then the following results are easily proved:

(1) if $M \neq 0$, then $N \neq 0$;

(2) N is the sum of the 1-dimensional B-submodules of M .

Thus, in order to prove the irreducibility of $M = \Delta^{(m)}$, it suffices to show that

(3) $\dim N = 1$;

(4) N generates M as SL-module.

Let D denote the group of all diagonal matrices in GL . The 1-dimensional D-submodules of M are those generated by the elements

$$X_1^{\lambda_1} \ldots X_n^{\lambda_n} + T^{(m)}$$

with $\sum \lambda_i = m$ and all $\lambda_i < p$. The only such element fixed by U is

$$(X_1 \ldots X_t)^{p-1} X_{t+1}^{\mu} + T^{(m)} ,$$

where $m = t(p-1) + \mu$, $0 \leq \mu < p-1$. However, by looking at the action of upper triangular matrices, one easily sees that this vector generates M as SL-module.

4. The main results

Throughout §4, $k = \mathsf{F}_p$, *where* p *is prime.*

4.1. STATEMENT

We shall make the identifications

$$\Lambda(n, c) = P/P^{[p]} ,$$

$$\Sigma(n, c) = \mathrm{gr}(\hat{F}^{p}-1)/P^{[p]} ,$$

(see §2.4, §3.4). Then $\Lambda(n, c)$ is generated (as Lie ring) by the elements

$$\xi_i = x_i + P^{[p]} , \quad (i = 1, \ldots, n) .$$

Our main results are as follows. $\left(\langle\langle m_1x_1, \ldots, m_nx_n\rangle\rangle, \Theta(x_1, x_2)\right.$ were defined in §3.4; $\Delta^{(m)}$ was defined in §3.5.$\left.\right)$

THEOREM A. *$\Sigma(n, 2p-1)$ is spanned by the elements $\langle\langle m_1\xi_1, \ldots, m_n\xi_n\rangle\rangle$ with* $\sum_1^n m_i = 2p-1$ *and* $0 \le m_i < p$, $(i = 1, \ldots, n)$.

THEOREM B. *$\Sigma(n, 2p-1)$ is either zero or an irreducible $SL(n, p)$-module isomorphic to* $\Delta^{(2p-1)}$.

THEOREM C. *If $n \ge 3$, $\Sigma(n, 2p-1)$ is nonzero if, and only if, the following condition holds:*

$(C)_p$ *$\Theta(x_1, x_2) \notin J$, where J is the ideal of $A(2, 2p-2; \mathsf{F}_p)$ generated by the $(p-1)$th powers of the elements of $L(2, 2p-2; \mathsf{F}_p)$.*

These theorems are proved in §4.2. Here we make several comments on them. Theorem A implies Sanov's result (1.5) and Kostrikin's result (1.6) for degree $2p-1$. The condition $(C)_p$ in Theorem C arise in the author's paper [9] on Hughes' H_p problem; the construction given there succeeds in its aim precisely when $(C)_p$ holds. A proof that $(C)_5$ holds was outlined. Cannon [2] has subsequently verified by computer that $(C)_p$ holds for $p = 5, 7, 11$. This gives (1.7).

4.2. PROOFS

We take

$$A = A(n, 2p-1; \mathsf{F}_p),$$

so that $c = 2p-1$, $k = \mathsf{F}_p$.

PROOF OF THEOREM A. If $e_p(v), e_p(w) \in \hat{F}$, then

$$e_p(v)^p = e_p\left(v^p\right) = 1 + v^p,$$

$$e_p(v)^p e_p(w)^p = 1 + v^p + w^p.$$

It follows that $\hat{F}^p - 1$ is the subspace spanned by the elements v^p, $e_p(v) \in \hat{F}$.

In order to calculate v^p, we need only know the value of v modulo $\underline{\underline{a}}^{p+1}$. This information is provided by Lemma 9. Let $c_1, \ldots, c_t$ be a basis for $P^{(1)} + \ldots + P^{(p)}$, where $c_1, \ldots, c_n$ are $x_1, \ldots, x_n$ and $c_{n+1}, \ldots, c_t$ are in

$\underline{\underline{a}}^2$. Then

$$v = u + R(\lambda_1 x_1, \ldots, \lambda_n x_n)$$

where $u = \sum_1^t \lambda_i c_i$, $\lambda_i \in F_p$. Therefore

$$\begin{aligned} v^p &= u^p + \langle R(\lambda_1 x_1, \ldots, \lambda_n x_n), (p-1)u \rangle \\ &= u^p + \left[R(\lambda_1 x_1, \ldots, \lambda_n x_n) \mid \left(\sum_1^n \lambda_i x_i\right)^{p-1}\right] . \end{aligned}$$

Using the relation $\lambda_i^p = \lambda_i$, we express the right hand side in the form

$$\sum \lambda_1^{a_1} \ldots \lambda_t^{a_t} w_{a_1, \ldots, a_t} ,$$

where all $a_i < p$. Then the functions $\lambda_1^{a_1} \ldots \lambda_t^{a_t}$ are linearly independent and so $\hat{F}^p - 1$ is spanned by the coefficients $w_{a_1, \ldots, a_t}$.

The value of the index sum $a = \sum a_i$ may be $1, p$ or $2p - 1$. The coefficients corresponding to $a = 1$ are the x_i^p . Those corresponding to $a = p$ are either Kostrikin elements in $\underline{\underline{a}}^{p+1}$ or have the form

(1) $$\langle a_1 x_1, \ldots, a_n x_n \rangle + \ldots , \left(\sum_1^n a_i = p\right) ,$$

where the omitted terms are certain components of $\left[R(x_1, \ldots, x_n) \mid \left(\sum x_i\right)^{p-1}\right]$. Finally, the coefficients corresponding to $a = 2p - 1$ are

(2) $$\langle\langle a_1 x_1, \ldots, a_n x_n \rangle\rangle , \left(\sum_1^n a_i = 2p - 1 , \text{ all } a_i < p\right) .$$

It is evident from this list (and from the linear independence of the leading terms of the expressions (1)) that $\mathrm{gr}(\hat{F}^p - 1)$ is spanned by $P^{[p]}$ and the elements (2). This proves the Theorem.

PROOF OF THEOREM B. Let S, T be as in §3.5. We define a linear mapping

$$\theta : S^{(2p-1)} \to \sum(n, 2p-1)$$

by

$$\left(X_1^{m_1} \cdots X_n^{m_n}\right)\theta = \langle\langle \underbrace{\xi_1, \ldots, \xi_1}_{m_1 \text{ terms}}, \ldots, \underbrace{\xi_n, \ldots, \xi_n}_{m_n \text{ terms}} \rangle\rangle .$$

By Lemma 7 *(a)*, $\langle\langle \eta_1, \ldots, \eta_{2p-1} \rangle\rangle$ is a symmetric, multilinear function of the arguments η_i in $\Lambda(n, p)$. It follows that θ is an $SL(n, p)$-homomorphism. θ is clearly surjective and by Lemma 7 *(c)*, $T^{(2p-1)}\theta = 0$. Thus θ induces an $SL(n, p)$-epimorphism $\Delta^{(2p-1)} \to \Sigma(n, 2p-1)$. Since $\Delta^{(2p-1)}$ is irreducible, the Theorem follows.

PROOF OF THEOREM C. Let n be a fixed integer ≥ 3. By Theorem B, $\Sigma(n, 2p-1) = 0$ if, and only if, $\Sigma(3, 2p-1) = 0$. We may therefore take $n = 3$.

By Theorems A and B, $\Sigma(3, 2p-1) = 0$ if, and only if,

(1)
$$\langle\langle \xi_3, \underbrace{\xi_1, \ldots, \xi_1}_{p-1 \text{ terms}}, \underbrace{\xi_2, \ldots, \xi_2}_{p-1 \text{ terms}} \rangle\rangle = 0 .$$

By Lemma 8, Corollary, and since $P^{[p]} \cap L = E_{p-1}$, (1) is equivalent to

(2)
$$[x_3 \mid \Theta(x_1, x_2)] \in E_{p-1} .$$

We have to prove that this is equivalent to

(3)
$$\Theta(x_1, x_2) \in J ,$$

where J is the ideal of $A(2, 2p-2; F_p)$ defined in the Theorem.

Suppose (2) holds. Then

$$[x_3 \mid \Theta(x_1, x_2)] = \Sigma \langle u_1, \ldots, u_p \rangle$$

where each u_i is a basic Lie product in x_1, x_2, x_3. In each term $\langle u_1, \ldots, u_p \rangle$, we may assume that x_3 occurs in u_1 with degree 1 but does not occur in any of $u_2, \ldots, u_p$. Then u_1 has the form $[x_3 \mid \phi(x_1, x_2)]$ and so

$$\begin{aligned}\langle u_1, \ldots, u_p \rangle &= [u_1 \mid \langle u_2, \ldots, u_p \rangle] \quad \text{(by 3.10)} \\ &= [x_3 \mid \phi(x_1, x_2) \langle u_2, \ldots, u_p \rangle] .\end{aligned}$$

Thus

(4)
$$[x_3 \mid \Theta(x_1, x_2)] = [x_3 \mid \Phi(x_1, x_2)] ,$$

where $\Phi(x_1, x_2) = \Sigma \phi(x_1, x_2) \langle u_2, \ldots, u_p \rangle$. We may obviously assume that every term in $\Phi(x_1, x_2)$ has partial degrees $p-1$, $p-1$ in x_1, x_2. Then, looking at those terms in (4) which have left hand factor x_3, we conclude that

$\Theta(x_1, x_2) = \Phi(x_1, x_2)$. The form of $\Phi(x_1, x_2)$ now shows that (3) holds. The fact that (3) implies (2) is evident. This completes the proof.

References

[1] N. Bourbaki, *Éléments de mathématique.* XXVI. *Groupes et algèbres de Lie.* Chapitre 1: *Algèbres de Lie* (Actualités Scientifiques et Industrielles, 1285. Hermann, Paris, 1960). MR24#A2641.

[2] John J. Cannon, "Some combinatorial and symbol manipulation programs in group theory", *Computational problems in abstract algebra* (Oxford, 1967), pp. 199-203 (Pergamon, Oxford, 1970). MR41#8512.

[3] Nathan Jacobson, *Lie algebras* (Pure and Appl. Math., 10. Interscience [John Wiley & Sons], New York, London, 1962). MR26#1345.

[4] А.И. Кострикин [A.I. Kostrikin], "О связи между периодичускими группами и кольцами Ли" [On the connection between periodic groups and Lie rings], *Izv. Akad. Nauk SSSR Ser. Mat.* 21(1957), 289-310; *Amer. Math. Soc. Transl.* (2) 45 (1965), 165-189. MR20#898.

[5] M. Lazard, "Sur les groupes nilpotents et les anneaux de Lie", *Ann. Sci. École Norm. Sup.* (3) 71 (1954), 101-190. MR19,529.

[6] W. Magnus, "A connection between the Baker-Hausdorff formula and a problem of Burnside", *Ann. of Math.* 52 (1950), 111-126. Errata, *Ann. of Math.* 57 (1953), 606. MR12,476 (errata: MR14,723).

[7] Wilhelm Magnus, Abraham Karrass, Donald Solitar, *Combinatorial group theory* (Pure and Appl. Math. 13. Interscience [John Wiley & Sons], New York, London, Sydney, 1966). MR34#7617.

[8] И.Н. Санов [I.N. Sanov], "Установление связи между периодическими группами с периодом простым числом и кольцами Ли" [Establishment of a connection between periodic groups with period a prime number and Lie rings], *Izv. Adad. Nauk SSSR Ser. Mat.* 16 (1952), 23-58. MR13,721.

[9] G.E. Wall, "On Hughes' H_p problem", *Proc. Internat. Conf. Theory of Groups* (Canberra, 1965), pp. 357-362 (Gordon and Breach, New York, London, Paris, 1967). MR36#2686.

University of Sydney,
Sydney, NSW 2006.

PROC. SECOND INTERNAT. CONF. THEORY OF GROUPS, 20E15
CANBERRA, 1973, pp. 691-700.

COMPUTATION IN NILPOTENT GROUPS (THEORY)

J.W. Wamsley

1. Introduction

Classical methods of dealing with groups using a computer prove to be unsatisfactory in the case of nilpotent groups. Recent work by Macdonald shows that these are more efficiently handled by building the group via the lower central series using the Schur multiplicator.

In the paper we give the theory involved in such a method.

2. Preliminaries

Let G be a finitely generated nilpotent group with a generating set chosen from a cyclic central series for G. That is, G is generated by $\{b_1, \ldots, b_n\}$ where the subgroup N_i generated by $\{b_i, \ldots, b_n\}$ is normal in G and N_i/N_{i+1} is central in G/N_{i+1} for each i. Let F be the free group on the set $\{a_1, \ldots, a_n\}$ and R the kernel of the homomorphism, ϕ, from F onto G defined by $\phi(a_i) = b_i$ for each i.

If N_i/N_{i+1} is finite cyclic then we define ρ_i to be the order of N_i/N_{i+1} otherwise we leave ρ_i undefined. For $i = 1, \ldots, n$ let α_i be an integer in the range $0 \le \alpha_i < \rho_i$ if ρ_i is defined, and let α_i be any integer (positive or negative) if ρ_i is undefined. The set $\left\{a_1^{\alpha_1} \ldots a_n^{\alpha_n}\right\}$ is a set of coset representatives for R in F where the α run through the given range.

For $x \in F$ let (x) denote the coset representative for x. The set

$$S = \left\{\left(a_j a_1^{\alpha_1} \ldots a_n^{\alpha_n}\right)^{-1} a_j a_1^{\alpha_1} \ldots a_n^{\alpha_n} \neq 1\right\},$$

is a set of free generators for R.

Given j , it is easy to see that $\left(a_ja_1^{\alpha_1} \cdots a_n^{\alpha_n}\right)^{-1} a_ja_1^{\alpha_1} \cdots a_n^{\alpha_n} = 1$ if and only if $\left(a_ja_1^{\alpha_1} \cdots a_n^{\alpha_n}\right) = a_ja_1^{\alpha_1} \cdots a_n^{\alpha_n}$. This happens when the smallest i , such that $\alpha_i \neq 0$ is greater than or equal to j . Moreover

$$\left(a_ja_j^{\alpha_j} \cdots a_n^{\alpha_n}\right) = a_ja_j^{\alpha_j} \cdots a_n^{\alpha_n}$$

if and only if $\alpha_j < \rho_j - 1$. Hence,

$$S = \begin{cases} \left(a_ja_i^{\alpha_i} \cdots a_n^{\alpha_n}\right)^{-1} a_ja_i^{\alpha_i} \cdots a_n^{\alpha_n} , & \alpha_i \neq 0 , \ 1 \le i < j \le n , \\ \left(a_j^{\rho_j} a_{j+1}^{\alpha_{j+1}} \cdots a_n^{\alpha_n}\right)^{-1} a_j^{\rho_j} a_{j+1}^{\alpha_{j+1}} \cdots a_n^{\alpha_n} , & 1 \le j \le n , \end{cases}$$

where $\left(a_j^{\rho_j} a_{j+1}^{\alpha_{j+1}} \cdots a_n^{\alpha_n}\right)^{-1} a_j^{\rho_j} a_{j+1}^{\alpha_{j+1}} \cdots a_n^{\alpha_n}$ is deleted if N_j/N_{j+1} is infinite cyclic.

We order the integers in the following way: $0 < 1 < -1 < 2 < -2 < \ldots$. Note that this ordering satisfies the descending chain condition. We then order the set as follows. For $\alpha_s \neq 0$, $\beta_t \neq 0$, we define

$$\left(a_ja_s^{\alpha_s} \cdots a_n^{\alpha_n}\right)^{-1} a_ja_s^{\alpha_s} \cdots a_n^{\alpha_n} > \left(a_ia_t^{\beta_t} \cdots a_n^{\beta_n}\right)^{-1} a_ia_t^{\beta_t} \cdots a_n^{\beta_n}$$

if

$s < t$; or if

$s = t$, $j < i$; or if

$s = t$, $j = i$, $\alpha_s > \beta_s$; or finally if

$s = t$, $j = i$, $\alpha_s = \beta_s, \ldots, \alpha_{s+i-1} = \beta_{s+i-1}$, $\alpha_{s+i} > \beta_{s+i}$.

This ordering of S also satisfies the descending chain condition.

Now suppose $i < j$, $\alpha_i > 0$,

$$\left(a_ja_i\right) = a_ia_ja_{j+1}^{\beta_{j+1}} \cdots a_n^{\beta_n}$$

and

$$\left(a_ja_i^{\alpha_i} \cdots a_n^{\alpha_n}\right) = a_i^{\alpha_i} \cdots a_{j-1}^{\alpha_{j-1}} a_j^{1+\alpha_j} a_{j+1}^{\gamma_{j+1}} \cdots a_n^{\gamma_n} .$$

Let us conjugate $(a_j a_i)^{-1} a_j a_i$ with $a_i^{\alpha_i - 1} a_{i+1}^{\alpha_{i+1}} \dots a_n^{\alpha_n}$. We have

$$a_n^{-\alpha_n} \dots a_{i+1}^{-\alpha_{i+1}} a_i^{1-\alpha_i} (a_j a_i)^{-1} a_j a_i^{\alpha_i} a_{i+1}^{\alpha_{i+1}} \dots a_n^{\alpha_n}$$

$$= a_n^{-\alpha_n} \dots a_{i+1}^{-\alpha_{i+1}} a_i^{1-\alpha_i} (a_j a_i)^{-1} \left(a_j a_i^{\alpha_i} \dots a_n^{\alpha_n}\right) \left(a_j a_i^{\alpha_i} \dots a_n^{\alpha_n}\right)^{-1} a_j a_i^{\alpha_i} \dots a_n^{\alpha_n}$$

$$= \left[\left(a_j a_i^{\alpha_i} \dots a_n^{\alpha_n}\right)^{-1} (a_j a_i) a_i^{\alpha_i - 1} a_{i+1}^{\alpha_{i+1}} \dots a_n^{\alpha_n}\right]^{-1} \left(a_j a_i^{\alpha_i} \dots a_n^{\alpha_n}\right)^{-1} a_j a_i^{\alpha_i} \dots a_n^{\alpha_n}$$

and

$$\left(a_j a_i^{\alpha_i} \dots a_n^{\alpha_n}\right)^{-1} (a_j a_i) a_i^{\alpha_i - 1} a_{i+1}^{\alpha_{i+1}} \dots a_n^{\alpha_n}$$

$$= a_n^{-\gamma_n} \dots a_{j+1}^{-\gamma_{j+1}} a_j^{-1-\alpha_j} a_{j-1}^{-\alpha_{j-1}} \dots a_i^{-\alpha_i} a_i a_j a_{j+1}^{\beta_{j+1}} \dots a_n^{\beta_n} a_i^{\alpha_i - 1} a_{i+1}^{\alpha_{i+1}} \dots a_n^{\alpha_n}$$

$$= \left(a_j a_{j+1}^{\beta_{j+1}} \dots a_n^{\beta_n} a_i^{\alpha_i - 1} a_{i+1}^{\alpha_{i+1}} \dots a_n^{\alpha_n}\right)^{-1} a_j a_{j+1}^{\beta_{j+1}} \dots a_n^{\beta_n} a_i^{\alpha_i - 1} a_{i+1}^{\alpha_{i+1}} \dots a_n^{\alpha_n}$$

$$= \left(a_j a_{j+1}^{\beta_{j+1}} \dots a_n^{\beta_n} a_i^{\alpha_i - 1} a_{i+1}^{\alpha_{i+1}} \dots a_n^{\alpha_n}\right)^{-1} a_j \left(a_{j+1}^{\beta_{j+1}} \dots a_n^{\beta_n} a_i^{\alpha_i - 1} a_{i+1}^{\alpha_{i+1}} \dots a_n^{\alpha_n}\right) \times$$

$$\prod_{r,\gamma} \left(a_r^{\gamma+\varepsilon_r} a_{r+1}^{\beta_{r+1}} \dots a_n^{\beta_n} a_i^{\alpha_i - 1} a_{i+1}^{\alpha_{i+1}} \dots a_n^{\alpha_n}\right)^{-1} a_r^{\varepsilon_r} \left(a_r^{\gamma} a_{r+1}^{\beta_{r+1}} \dots a_n^{\beta_n} a_i^{\alpha_i - 1} a_{i+1}^{\alpha_{i+1}} \dots a_n^{\alpha_n}\right)$$

where $\varepsilon_r = 1$ if $\beta_r > 0$, $\varepsilon_r = -1$ if $\beta_r < 0$ and the product is taken over $r = j+1, \dots, n$ and $\gamma = \beta_r - \varepsilon_r, \dots, 0$. Whence

$$\left(a_j a_i^{\alpha_i} \dots a_n^{\alpha_n}\right)^{-1} a_j a_i^{\alpha_i} \dots a_n^{\alpha_n}$$

$$= \omega \times \left[a_i^{\alpha_i - 1} a_{i+1}^{\alpha_{i+1}} \dots a_n^{\alpha_n}\right]^{-1} (a_j a_i)^{-1} a_j a_i \left[a_i^{\alpha_i - 1} a_{i+1}^{\alpha_{i+1}} \dots a_n^{\alpha_n}\right],$$

where ω is the product of free generators which come earlier in the ordering of S.

Similarly, if we conjugate $(a_j a_i)^{-1} a_j a_i$ with $\left(a_i^{-1} a_j^{-1} a_i a_j a_i^{\alpha_i} \dots a_n^{\alpha_n}\right)$ for $i < j$, $\alpha_i < 0$, we get

$$\left(a_j a_i^{\alpha_i} \dots a_n^{\alpha_n}\right)^{-1} a_j a_i^{\alpha_i} \dots a_n^{\alpha_n}$$

$$= \omega_1 \times \left(a_i^{-1} a_j^{-1} a_i a_j a_i^{\alpha_i} \dots a_n^{\alpha_n}\right)^{-1} \left[(a_j a_i)^{-1} a_j a_i\right]^{-1} \left(a_i^{-1} a_j^{-1} a_i a_j a_i^{\alpha_i} \dots a_n^{\alpha_n}\right) \times \omega_2,$$

where ω_1 and ω_2 are the products of free generators which come earlier in the ordering.

Finally, for $i = j$, we have

$$\left(a_j^{\rho_j} a_{j+1}^{\alpha_{j+1}} \dots a_n^{\alpha_n}\right)^{-1} a_j^{\rho_j} a_{j+1}^{\alpha_{j+1}} \dots a_n^{\alpha_n}$$

$$= \omega \times \left[a_{j+1}^{\alpha_{j+1}} \dots a_n^{\alpha_n}\right]^{-1} \left(a_j^{\rho_j}\right)^{-1} a_j^{\rho_j} \left[a_{j+1}^{\alpha_{j+1}} \dots a_n^{\alpha_n}\right] .$$

Hence, inductively, R is generated by the set

$$T = \begin{cases} \left[a_i^{\beta_i} a_{i+1}^{\alpha_{i+1}} \dots a_n^{\alpha_n}\right]^{-1} (a_j a_i)^{-1} a_j a_i \left[a_i^{\beta_i} a_{i+1}^{\alpha_{i+1}} \dots a_n^{\alpha_n}\right] , & 1 \le i < j \le n , \\ \\ \left[a_{i+1}^{\alpha_{i+1}} \dots a_n^{\alpha_n}\right]^{-1} \left(a_i^{\rho_i}\right)^{-1} a_i^{\rho_i} \left[a_{i+1}^{\alpha_{i+1}} \dots a_n^{\alpha_n}\right] , & 1 \le i \le n , \end{cases}$$

where the α lie in the given range, $0 \le \beta_k < \rho_k - 1$ if ρ_k is defined and

$\left[a_{i+1}^{\alpha_{i+1}} \dots a_n^{\alpha_n}\right]^{-1} \left(a_i^{\rho_i}\right)^{-1} a_i^{\rho_i} \left[a_{i+1}^{\alpha_{i+1}} \dots a_n^{\alpha_n}\right]$ is deleted if ρ_i is undefined.

THEOREM 1. *T is a set of free generators for R.*

PROOF. Firstly we note that there is a natural correspondence between S and T which induces an order on T. If this correspondence is denoted by ϕ, then for $x \in S$ we have

$$\phi(x) = \omega_1(x_i) x \omega_2(x_i) ,$$

where each $x_i < x$, and similarly,

$$x = \omega_3(\phi(x'_j)) \phi(x) \omega_4(\phi(x'_j))$$

with $\phi(x'_j) < \phi(x)$. Now suppose we start with a finite set of elements of T with a general element $\phi(x)$. We construct two forests, X, $\phi(X)$, of trees as follows: in the forest $\phi(X)$ a tree starts at node $\phi(x)$ and branches go to $\{\phi(x_i) \cup \phi(x'_j)\}$, while in the forest X a tree starts at each node x and branches go to $\{x_i \cup x'_j\}$. We continue in the same way from each new node formed. Since the ordering satisfies the descending chain condition we cannot have a infinite path and so the process must end after a finite number of steps. When the process has ended then, since the nodes are in one-to-one correspondence, we have isomorphic forests.

Let T' denote the set of nodes of $\phi(X)$ and S' the set of nodes of X. Then T' and S' each have the same number of elements and each generate the same subgroup of R. Now, S is a set of free generators and therefore S' is a subset of free generators. It follows that T' is a subset of free generators. But the finite set of elements of T we started with is a subset of T' and hence a set of free generators. That is, every finite subset of T is a set of free generators, giving that T is a set of free generators for R.

We have also an algorithm for converting elements of S to elements of T and vice-versa.

Now suppose we have a group defined by the presentation

$$(1)\quad G = \left\{a_1, \ldots, a_n \mid a_i^{\rho_i} = a_{i+1}^{\alpha_{i+1}} \ldots a_n^{\alpha_n},\ 1 \le i \le n,\right.$$
$$\left. a_j^{-1}a_i^{-1}a_ja_i = a_{j+1}^{\beta_{j+1}} \ldots a_n^{\beta_n},\ 1 \le i < j \le n\right\} = F/R ,$$

where some of the relations $a_i^{\rho_i} = a_{i+1}^{\alpha_{i+1}} \ldots a_n^{\alpha_n}$ may not occur.

Then G is a finitely generated nilpotent group. We say this presentation is consistent if $|N_i/N_{i+1}| = \rho_i$ where $N_i = \{a_i, \ldots, a_n\}$.

Whether or not the presentation is consistent we may write down the set T, corresponding to the presentation. We now prove

THEOREM 2. *The presentation* (1) *for* G *is consistent if and only if* T *is a set of free generators for a normal subgroup of* F *in which case this subgroup is* R.

PROOF. We have shown that if the presentation is consistent then T is a set of free generators for R.

Conversely, suppose that T is a set of free generators for the normal subgroup R of F. We wish to show that if $a_i^{\tau_i}a_{i+1}^{\gamma_{i+1}} \ldots a_n^{\gamma_n} \in R$ then $\rho_i | \tau_i$. Suppose s is minimal such that $a_s^{\tau_s}a_{s+1}^{\gamma_{s+1}} \ldots a_n^{\gamma_n} \in R$, where $\rho_s \nmid \tau_s$, then, in the factor group formed by setting $a_{s+1}, \ldots, a_n$ equal to the identity, we shall have $a_s^{\tau_s} \in R$, where $\rho_s \nmid \tau_s$. That is, we may assume $s = n$ and that $a_n^{\tau} = 1$ where $\rho_n \nmid \tau$.

Let T^* be the set corresponding to a consistent presentation. Then $T^* = \left\{x_1, \ldots, x_m, a_n^{\tau}\right\}$, where $T = \left\{a_n^{-\gamma\tau}x_ia_n^{\gamma\tau}, a_n^{\rho_n}\right\}$, and the set T^* corresponds to T modulo a_n^{τ}.

Therefore, if $\omega(T) = a_n^{\tau}$ then, modulo a_n^{τ}, we have $\omega(T) \equiv 1$, whence the coefficient sums of the x_i are zero and hence the coefficient sum of a_n in $\omega(T)$ is a multiple of ρ_n. This yields a contradiction.

We now look more closely at what is involved in T generating a normal subgroup of F . Suppose r is a relator and $y^{-1}ry$ is an element of T . Then conjugation with x gives,

$$x^{-1}y^{-1}ryx = x^{-1}y^{-1}(yx) \cdot (yx)^{-1}r(yx) \cdot (yx)^{-1}yx$$
$$= \left[(yx)^{-1}yx\right]^{-1} \cdot (yx)^{-1}r(yx) \cdot (yx)^{-1}yx .$$

Hence, to show that T generates a normal subgroup, it is sufficient to show that the conjugate of each relator may be written as a word in T . That is, inductively, it is enough to show that

(1) $a_i^{-1}\left(a_k a_j\right)^{-1} a_k a_j a_i$, $1 \le i < j < k \le n$,

(2) $a_j^{1-\rho_j}\left(a_k a_j\right)^{-1} a_k a_j a_j^{\rho_j - 1}$, $1 \le j < k \le n$, and

(3) $a_j^{-1}\left\{a_k^{\rho_k}\right\} a_k^{\rho_k} a_j$, $1 \le j \le k \le n$,

may be written as words in T .

Let $\omega\left(a_i\right)$ be a word in F . Say, $\omega\left(a_i\right) = v a_k a_j a_i^{\alpha_i} \ldots a_n^{\alpha_n}$ where $j < i$, $j < k$. Then

$$\omega\left(a_i\right) = v a_j a_k \left(a_k, a_j\right) a_i^{\alpha_i} \ldots a_n^{\alpha_n} \times \left[a_i^{\alpha_i} \ldots a_n^{\alpha_n}\right]^{-1} \left(a_k a_j\right)^{-1} a_k a_j \left[a_i^{\alpha_i} \ldots a_n^{\alpha_n}\right] ,$$

where $\left(a_k, a_j\right)$ denotes the coset representative of the commutator $a_k^{-1} a_j^{-1} a_k a_j$.

Since we are only concerned with what is left over at each step, we need only know modulo a word in T that

$$v a_k a_j a_i^{\alpha_i} \ldots a_n^{\alpha_n} \equiv v a_j a_k \left(a_k, a_j\right) a_i^{\alpha_i} \ldots a_n^{\alpha_n} .$$

This process is called collection and the reader will note the similarity with other collection processes.

THEOREM 3. *Suppose we have a consistent presentation for* $G = F/R$. *If* x *and* y *are elements of* F *then* (xy) *is that word which is formed by carrying out the collection process on* $(x)(y)$.

PROOF. (xy) is a word of the form $a_1^{\alpha_1} \ldots a_n^{\alpha_n}$ and since the presentation is consistent the α_i are unique.

Given an equation $\omega_1\left(a_i\right) = \omega_2\left(a_i\right)$, we say that it is consistent relative to a presentation if collection of the left hand side yields the same as collection of the right hand side. Now, consider the words (1), (2) and (3).

Firstly, (1) gives

$$a_i^{-1}(a_k a_j)^{-1} a_k a_j a_i = \left[(a_k a_j a_i)^{-1}(a_k a_j)a_i\right]^{-1}(a_k a_j a_i)^{-1} a_k (a_j a_i)(a_j a_i)^{-1} a_j a_i ,$$

or the equation

$$(a_k a_j)a_i = a_k(a_j a_i) \quad \text{is consistent.}$$

Secondly, (2) gives

$$a_j^{1-\rho_j}(a_k a_j)^{-1} a_k a_j a_j^{\rho_j - 1} = \left[\left(a_k a_j^{\rho_j}\right)^{-1}(a_k a_j)a_j^{\rho_j - 1}\right]^{-1}\left(a_k a_j^{\rho_j}\right)^{-1} a_k \left(a_j^{\rho_j}\right)\left(a_j^{\rho_j}\right)^{-1} a_j^{\rho_j} ,$$

or the equation

$$(a_k a_j)a_j^{\rho_j - 1} = a_k\left(a_j^{\rho_j}\right) \quad \text{is consistent.}$$

Finally, (3) gives

$$a_j^{-1}\left(a_k^{\rho_k}\right)^{-1} a_k^{\rho_k} a_j = \left[\left(a_k^{\rho_k} a_j\right)^{-1}\left(a_k^{\rho_k}\right)a_j\right]^{-1}\left(a_k^{\rho_k} a_j\right)^{-1} a_k\left(a_k^{\rho_k - 1} a_j\right) \times$$
$$\left(a_k^{\rho_k - 1} a_j\right)^{-1} a_k\left(a_k^{\rho_k - 2} a_j\right) \ldots (a_k a_j)^{-1} a_k a_j ,$$

or the equation

$$\left(a_k^{\rho_k}\right)a_j = a_k\left(a_k^{\rho_k - 1} a_j\right) \quad \text{is consistent.}$$

Summing up, we have the following

THEOREM 4. *The presentation given for* G *is consistent if and only if the following equations are consistent:*

(1) $(a_k a_j)a_i = a_k(a_j a_i)$, $1 \le i < j < k \le n$,

(2) $(a_k a_j)a_j^{\rho_j - 1} = a_k\left(a_j^{\rho_j}\right)$, $1 \le j < k \le n$ *and*

(3) $\left(a_k^{\rho_k}\right)a_j = a_k^{\rho_k - 1}(a_k a_j)$, $1 \le j \le k \le n$.

The above theorem is equivalent to Lemma 2 of [1].

Note that in the theorems collection is performed in a specific manner. However it is easy to see inductively that the collection may be then performed in any convenient manner provided it is performed on the bracketted elements first. Of course, one must always substitute the right-hand side of the relations for the left and never vice-versa.

3. The multiplicator

Suppose now we are given a nilpotent group G with consistent presentation

$$G = \Big\{a_1, \ldots, a_n \mid a_i^{\rho_i} = a_{i+1}^{\alpha_{i+1}} \ldots a_n^{\alpha_n}, \ 1 \le i \le n,$$
$$\left(a_j, a_i\right) = a_{j+1}^{\beta_{j+1}} \ldots a_n^{\beta_n}, \ 1 \le i < j \le n\Big\} = F/R .$$

We give a method for obtaining the general extension of G by its multiplicator. We first partition the relations into three sets as follows.

Firstly, if $a_{i+1}, \ldots, a_n \in G'$ and G/G' is generated by $a_1, \ldots, a_i$ then the generators $a_{i+1}, \ldots, a_n$ are defined in terms of $a_1, \ldots, a_i$. That is, $n - i$ of the relators in the presentation define the elements $a_{i+1}, \ldots, a_n$. These $n - i$ relations are called definitions.

Secondly, there are at most i relations which give the orders of the generators $a_1, \ldots, a_i$ modulo G'. These are called relations of class 1.

Finally, the remaining relators are called consequences.

Now, suppose there are m consequences which we write $\{A_i = B_i, \ 1 \le i \le m\}$. Then we take m new generators $b_1, \ldots, b_m$ and H to be the group generated by $a_1, \ldots, a_n, b_1, \ldots, b_m$ with relations to be the definitions and relations of class 1 of the presentation for G together with the relations $\{A_i = B_i b_i, \ 1 \le i \le m\}$ and relations $\{(b_i, a_j) = 1, \ 1 \le j \le n, \ 1 \le i \le m\}$.

In effect we have added in every possible new generator of $(F' \cap R)/[F, R]$.

We now force the presentation to be consistent by solving the consistency equations of Theorem 4, which, since the presentation for G was consistent, results in a set of relations in $\{b_i\}$. If these relations are $R_1, \ldots, R_t$ then the multiplicator of G, denoted by $m(G)$, is the abelian group with presentation

$$m(G) = \{b_1, \ldots, b_m \mid R_1, \ldots, R_t\} .$$

The above method is basically that which was used to calculate the multiplicator of various groups given in [2] and [3].

The general extension of G by its multiplicator is then given by setting the relators of class 1 equal to arbitrary elements of $m(G)$.

4. Extending the class of a group

Suppose we are given now a group G with presentation

$G = \{x_1, \ldots, x_n \mid R_1, \ldots, R_m\}$. Let $\gamma_n(G)$ denote the n-th term in the lower central series of G . Suppose further, we have calculated a consistent presentation for $G/\gamma_n(G)$. We obtain a consistent presentation for $G/\gamma_{n+1}(G)$ as follows.

Firstly extend $G/\gamma_n(G)$ by its multiplicator and then insist that it satisfies the relations $R_1, \ldots, R_m$. This will give the elements in $m(G/\gamma_n(G))$ to which the class 1 relators are equal; and, in general, this will also give more relations in $m(G/\gamma_n(G))$.

We give an example.

We wish to construct the 2-generator free group of exponent 4 . We have

$$G/\gamma_3(G) = \{a_1, a_2, a_3 \mid a_1^4 = 1, a_2^4 = 1, a_3^2 = 1, [a_2, a_1] = a_3, [a_3, a_1] = 1, [a_3, a_2] = 1\} .$$

Set

$$K = \{a_1, a_2, a_3, b_1, b_2, b_3 \mid a_1^4 = 1, a_2^4 = 1, a_3^2 = b_1, [a_2, a_1] = a_3, [a_3, a_1] = b_2, [a_3, a_2] = b_3, b_i \text{ central}\} .$$

The consistency equations yield $b_1^2 = b_2^2 = b_3^2 = 1$. This means that the multiplicator of $G/\gamma_3(G)$ is $C_2 \times C_2 \times C_2$ and the general extension of $G/\gamma_3(G)$ by its multiplicator is

$$H = \{a_1, a_2, a_3, b_1, b_2, b_3 \mid a_1^4 = \omega_1, a_2^4 = \omega_2, a_3^2 = b_1, b_1^2 = 1, b_2^2 = 1, b_3^2 = 1, [a_2, a_1] = a_3, [a_3, a_1] = b_2, [a_3, a_2] = b_3, b_i \text{ central}\} ,$$

where ω_1 and ω_2 are elements of the subgroup generated by b_1, b_2 and b_3 .

Note that H is necessarily of class 4 . Therefore $G/\gamma_4(G)$ is the factor group of H gained by insisting that H be of exponent 4 . We have that $\gamma_4(H) = 1$ $\gamma_3^2(H) = 1$. Collection of $(xy)^4$ gives $(xy)^4 = x^4y^4(y, x)^6$, whence H is of exponent 4 if and only if $\gamma_2^2(H) = 1$ and the generators are of exponent 4 . Whence $b_1 = \omega_1 = \omega_2 = 1$ and

$$G/\gamma_4(G) = \{a_1, a_2, a_3, a_4, a_5 \mid [a_2, a_1] = a_3, [a_3, a_1] = a_4, [a_3, a_2] = a_5, a_1^4 = 1, a_2^4 = 1, \text{all other commutators and all other squares trivial}\} .$$

In actually calculating a group many shortcuts may be used. One may use the class condition to lessen the number of new generators added each step.

Expansion of the equation $(a_k a_j)a_i = a_k(a_j a_i)$ yields

$$(a_j, a_i, a_k)(a_j, a_i, a_k; a_k, a_j)(a_j, a_i, a_k; a_k, a_i) \times$$
$$(a_j, a_i, a_k; a_k, a_i; a_k, a_j)(a_k, a_j, a_i)$$
$$= (a_k, a_j; a_k, a_i)(a_k, a_j; a_j, a_i)(a_k, a_j; a_j, a_i; a_k, a_i)(a_k, a_i; a_j, a_i) \times$$
$$(a_k, a_i, a_j)(a_k, a_i, a_j; a_j, a_i) .$$

Hence if we assign a weight to each new generator as it occurs and as we then extend $G/\gamma_{n-1}(G)$ to $G/\gamma_n(G)$ we will get nothing from the equation

$$(a_k a_j)a_i = a_k(a_j a_i) \quad \text{if} \quad wt \cdot (a_i) + wt \cdot (a_j) + wt \cdot (a_k) \geq n .$$

Similarly, if $wt(a_i) + wt(a_j) \geq n$ then we get nothing from the equations

$$\left(a_j^{\rho_j}\right)a_i = a_j^{\rho_j-1}(a_j a_i)$$

and

$$a_j\left(a_i^{\rho_i}\right) = (a_j a_i)a_i^{\rho_i-1} .$$

If one calculates by hand then it seem easier to increase a full class at a time, in which case infinite groups can be handled. However if a computer is used it seems easier to go from one class to the next via bites of exponent p and to calculate the Sylow p-subgroups separately, in which case infinite groups become a problem.

Finally I would like to thank Dr I.D. Macdonald for making available to me a preprint of his paper [1] and for the many helpful discussions we had during my stay at the University of Stirling.

References

[1] I.D. Macdonald, "A computer application to finite p-groups", *J. Austral. Math. Soc.* 17 (1974), 102-112.

[2] T.W. Sag and J.W. Wamsley, "On computing the minimal number of defining relations for finite groups", *Math. Comp.* 27 (1973), 361-368.

[3] T.W. Sag and J.W. Wamsley, "Minimal presentations for groups of order 2^n, $n \geq 6$ ", *J. Austral. Math. Soc.* (to appear).

Flinders University of South Australia,
Bedford Park, South Australia 5042.

PROC. SECOND INTERNAT. CONF. THEORY OF GROUPS, 20D45
CANBERRA, 1973, 701-704.

ON GROUPS ADMITTING AN ELEMENTARY ABELIAN AUTOMORPHISM GROUP

J.N. Ward

Recently a number of theorems have been proved showing that if V is a fixed-point-free group of automorphisms of the finite group G then, with certain additional assumptions, G is soluble. These theorems may be found in [3], [4], [6] and [7].

In each of these theorems the assumption that V is fixed-point-free is used to deduce that for each prime divisor p of $|G|$ there exists a unique V-invariant Sylow p-subgroup of G. The collection of V-invariant Sylow subgroups of G is then shown to form a Sylow system. The theorem of P. Hall which characterises soluble groups by the existence of a Sylow system ([2], Theorem 6.4.3) is then applied to deduce that G is soluble.

There exist hypotheses different from the hypothesis that V is fixed-point-free which imply that there exists a unique V-invariant Sylow p-subgroup of G for each prime p. Thus it is natural to ask if any of these also imply that G is soluble. Apart from yielding more general theorems this gives us a greater insight into the proofs of Martineau's and Ralston's Theorems.

1. General results

LEMMA 1. *Assume that* V *is a group of automorphisms of the finite group* G. *Suppose that for each prime divisor* p *of* $|G|$ *there exists a unique* V*-invariant Sylow* p*-subgroup of* G. *Then* $C_G(V)$ *is nilpotent.*

This lemma, which is easy to prove, enables us to apply Glauberman's Factorization Theorem ([1], Corollary 1). We obtain

LEMMA 2. *Let* p *be a prime and* V *a group of automorphisms of the finite* p*-soluble group* G. *Suppose that for each prime divisor* q *of* $|G|$ *there exists a unique* V*-invariant Sylow* q*-subgroup of* G. *If* S *is a Sylow* p*-subgroup of* G *then*

$$G = O_{p'}(G)C_G\{Z(S)\}N_G\{J(S)\} .$$

Here $J(S)$ denotes the Thompson subgroup of S, as defined in [2], page 271.

For the remainder of this section we assume the following.

V is a group of automorphisms of the finite group G and $|V|$ is relatively prime to $|G|$. For each prime divisor p of $|G|$ there exists a unique V-invariant Sylow p-subgroup of G. The group G has no non-identity V-invariant proper normal subgroups. Any proper V-invariant subgroup of G is soluble.

Let p and q denote two distinct prime divisors of G. Denote by P and Q the V-invariant Sylow p- and q-subgroups of G. Define X (respectively Y) to be the largest V-invariant subgroup of P (respectively Q) which is permutable with Q (respectively P). It is easily seen that X and Y exist.

We are now in a position to repeat the arguments which are given by Martineau in [6], Section 2. We obtain the following results.

LEMMA 3. *Let H be a V-invariant $\{p, q\}$-subgroup of G containing $Z(Q)$. Then $H \cap P = O_p(H)(H \cap X)$.*

Let

$$\mathcal{H} = \{H \mid H \text{ is a maximal } V\text{-invariant } \{p, q\}\text{-subgroup of } G ,\ H \neq PY, H \neq XQ, O_p(H) \neq 1, O_q(H) \neq 1\} .$$

LEMMA 4. *Suppose that K is a maximal V-invariant $\{p, q\}$-subgroup of G distinct from PY and QX. Then $K \in \mathcal{H}$.*

LEMMA 5. *If $H \in \mathcal{H}$ and M is a V-invariant subgroup of $F(H)$ with $O_p(M) \neq 1$ and $O_q(M) \neq 1$, then H is the only maximal V-invariant $\{p, q\}$-subgroup of G containing M.*

LEMMA 6. *If $H \in \mathcal{H}$, then $Z(P) \leq H$ and $Z(Q) \leq H$.*

LEMMA 7. *If $H \in \mathcal{H}$, then $X \cap O_p(H) = 1$ and $Y \cap O_q(H) = 1$.*

LEMMA 8. *If $H \in \mathcal{H}$, M is a non-1 V-invariant subgroup of $F(H)$ and $M \leq K \in \mathcal{H}$, then $K = H$.*

2. Suitable hypotheses

We now seek conditions which will ensure that if V is a group of automorphisms of the finite group G then for each prime divisor p of $|G|$ there exists a unique V-invariant Sylow p-subgroup of G. By [2], Theorem 6.2.2 it follows that if $|G|$ and $|V|$ are relatively prime then any two V-invariant Sylow p-subgroups of G are conjugate by an element of $C_G(V)$. This yields

LEMMA 9. *Assume that* V *is a non-cyclic abelian group of automorphisms of the finite group* G . *Suppose that* $|V|$ *and* $|G|$ *are relatively prime and that for each element* v *of* $V^{\#}$, *elements of* $C_G(v)$ *commute with elements of relatively prime order in* $C_G(V)$. *Then for each prime divisor* p *of* $|G|$ *there exists a unique V-invariant Sylow p-subgroup of* G .

Perhaps the most striking special cases of this situation are when $C_G(V) = 1$, or when $C_G(v)$ is nilpotent for each $v \in V$.

3. The principal theorem

We are now in a position to state the more general theorems. Let r denote some fixed prime.

THEOREM 1. *Suppose that* V *is a non-cyclic abelian* r*-group of automorphisms of the finite group* G . *Suppose that* $|V|$ *and* $|G|$ *are relatively prime and that for each element* v *of* $V^{\#}$, *elements of* $C_G(v)$ *commute with elements of relatively prime order in* $C_G(V)$. *Then* G *is soluble.*

In order to prove this theorem we consider two cases separately. If $|V| > r^2$ then, in the notation of Section 2, we may prove, by the same argument as Martineau in [6], Section 3, that $|H| \leq 1$ for a minimum counterexample to the theorem. A contradiction is then reached exactly as in Martineau's paper.

If $|V| = r^2$ then the argument is different and resembles the argument given in [3].

Along the same lines as in [6] we can also show

THEOREM 2. *Let* V *be a group of automorphisms of the finite group* G *with* $|V|$ *and* $|G|$ *relatively prime. Suppose that* $Z(V)$ *contains a subgroup* A *which is elementary abelian or order* r^3 *for some prime* r . *Assume that for each* $a \in A^{\#}$, *the group* $C_G(a)$ *is soluble and elements of* $C_G(a)$ *centralise elements of relatively prime order in* $C_G(V)$. *Then* G *is soluble.*

At the time of writing, I have not tried to investigate what happens if we only assume that A has order r^2 .

The methods outlined here are also applicable to the study of soluble rank 3 signalizer functors on finite groups. A note of this fact has already been made by Martineau in [5]. We mention the following theorem.

THEOREM 3. *Let* r *denote a prime,* G *a finite group and* A *an elementary abelian* r*-subgroup of* G *with* $m(A) \geq 3$. *Let* θ *denote an A-signalizer functor* on G. *Assume that for each* $a \in A^{\#}$, *elements of* $\theta\left(C_G(a)\right)$ *centralize elements of* $\theta\left(C_G(A)\right)$ *which have relatively prime order. Then* θ *is complete.*

We are using the same notation here for signalizer functors as is used in [5]. Full proofs of these results appear in [8], [9].

References

[1] George Glauberman, "Failure of factorization in p-solvable groups", *Quart. J. Math. Oxford Ser.* (2) 24 (1973), 71-77.

[2] Daniel Gorenstein, *Finite groups* (Harper and Row, New York, Evanston, London, 1968). MR38#229.

[3] R. Patrick Martineau, "Solubility of groups admitting a fixed-point-free automorphism group of type (p, p) ", *Math. Z.* 124 (1972), 67-72. MR45#368.

[4] R. Patrick Martineau, "Elementary abelian fixed point free automorphism groups", *Quart. J. Math. Oxford Ser.* (2) 23 (1972), 205-212. Zbl.243.20031.

[5] R. Patrick Martineau, "Rank 3 signalizer functors on finite groups", *Bull. London Math. Soc.* 4 (1972), 161-162.

[6] R. Patrick Martineau, "Solubility of groups admitting certain fixed-point-free automorphism groups", *Math. Z.* 130 (1973), 143-147.

[7] Elizabeth Wall Ralston, "Solubility of finite groups admitting fixed-point-free automorphisms of order rs ", *J. Algebra* 23 (1972), 164-180.

[8] J.N. Ward, "On groups admitting a noncyclic abelian automorphism group", *Bull. Austral. Math. Soc.* 9 (1973), 363-366.

[9] J.N. Ward, "Nilpotent signalizer functors on finite groups", *Bull. Austral. Math. Soc.* 9 (1973), 367-377.

University of Sydney,
Sydney, NSW 2006.

PROC. SECOND INTERNAT. CONF. THEORY OF GROUPS, 20E10
CANBERRA, 1973, pp. 705-708.

VARIETIES DESCRIBED BY VERBAL SUBGROUPS

Paul M. Weichsel

Stewart [1] has given a complete classification of varieties of center-extended-by-metabelian groups of exponent p and class at most $p-1$. We will use this classification as a setting for some general results about join-irreducible varieties of p-groups and to motivate the title of this note. Full proofs will appear elsewhere.

Stewart shows that for each prime $p \geq 5$, the set of join-irreducible elements in the lattice of these varieties consists of a chain of metabelian varieties, one for each class, and one non-metabelian variety for each class from 4 to $p-1$. We illustrate the initial portion of such a lattice labelling each join-irreducible with its class, the metabelian chain appearing on the left.

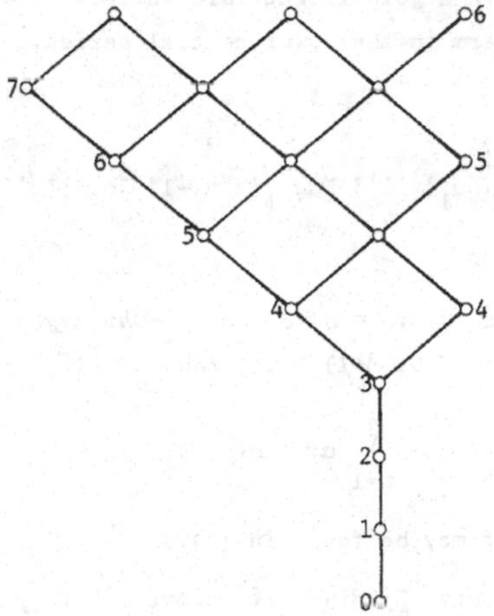

Each metabelian element in this lattice is characterized by its class and exponent. Now let $\underline{U}$ be a non-metabelian join-irreducible element of class d. Let F be a relatively free group of sufficient rank to generate $\underline{U}$. Then we claim, that

$$F_d = \bigcup_{r,s} (F_r, F_s) \ , \quad r+s=d \ , \quad r, s \geq 2 \ ;$$

and hence that this relation holds for all groups in $\underline{\underline{U}}$. We will verify this in two steps.

1. $\mathsf{U}(F_r, F_s) \neq 1$. This will follow from a more general result which will be Theorem 1 below.

2. If $\mathsf{U}(F_r, F_s) \lneq F_d$, then $F/\mathsf{U}(F_r, F_s)$ generates a proper subvariety of $\underline{\underline{U}}$ of class d . This is a contradiction since, as can be seen from the lattice diagram, every proper subvariety of $\underline{\underline{U}}$ has smaller class.

Now the class of verbal subgroups $\{F_\alpha\}$, the lower central series and the class $\left\{\left(F_{\alpha_1}, F_{\alpha_2}\right)\right\}$ with $\alpha_1, \alpha_2 \geq 2$ are examples of "hierarchies" of verbal subgroups which can be described in general as follows:

Let $f(x_1, \ldots, x_n)$ be a commutator (not necessarily left-normed) of weight n on the letters $x_1, \ldots, x_n$. Thus each letter appears exactly once in f . Let $\Lambda = \{\alpha_1, \ldots, \alpha_n\}$ be a set of positive integers. Now if G is a group we designate the verbal subgroup $f\left(G_{\alpha_1}, \ldots, G_{\alpha_n}\right)$ by $f(G, \Lambda)$. Thus for a given commutator f the sets of positive integers Λ generate the *hierarchy* of f . In order to describe the behavior of f in a join-irreducible variety we need to describe the analogue of the "successor" term in the lower central series. For a given f and Λ let

$$f(G, \Lambda+1) = \bigcup_{i=1}^{n} f\left(G_{\alpha_1}, \ldots, G_{\alpha_{i-1}}, G_{\alpha_i+1}, G_{\alpha_{i+1}}, \ldots, G_{\alpha_n}\right) .$$

We may now state Theorem 1.

THEOREM 1. *Let G be a p-group of class c which generates a join-irreducible variety. If $f(G, \Lambda) \neq 1$ but $f(G, \Lambda+1) = 1$, then*

$$\alpha = \sum_{i=1}^{n} \alpha_i \equiv c(p-1) .$$

(The proof of this result may be found in [2].)

We now return to the variety $\underline{\underline{U}}$ discussed above. Let $f = (x_1, x_2)$, and suppose that $\mathsf{U}(F_r, F_s) = 1$, $r+s=d$, $r, s \geq 2$. Then since $(F_2, F_2) \neq 1$, there must be a pair of integers $a, b \geq 2$ such that $(F_a, F_b) \neq 1$ but $\mathsf{U}(F_\alpha, F_\beta) = 1$ with $\alpha+\beta = a+b+1$, $\alpha, \beta \geq 2$. Now let $\Lambda = \{a, b\}$ in Theorem

1 and we conclude that $a + b = d$ since d is the class of F and $d \leq p - 1$. This proves the assertion that $U(F_r, F_s) \neq 1$, $r + s = d$, $r, s \geq 2$.

Theorem 1 allows us to obtain information about the "shape" of a hierarchy in case part of it is trivial. We give one example below whose proof is an amusing geometric exercise.

COROLLARY. *Let* G *be a* p*-group of class* $c < p$ *whose variety is join-irreducible. If* $(G_r, G_s) = 1$ *with* $2 \leq r \leq s \leq c/2$, *then* $(G_r, G_r) = 1$.

Consider the "lattice triangle" whose points are labelled by the pairs (i, j) representing the subgroups (G_i, G_j).

$$
\begin{array}{ccccc}
 & & (2, 2) & & \\
 & (2, 3) & & (3, 2) & \\
(2, 4) & & (3, 3) & & (4, 2) \\
 & & \vdots & & \\
(2, c-2) & & \ldots & & (c-2, 2)
\end{array}
$$

Now let $f = (x_1, x_2)$. It follows from Theorem 1 that whenever all pairs immediately below a given pair (i, j) are trivial, then either $(G_i, G_j) = 1$ or else $i + j = c$, the class of G. Hence the condition: $r \leq s \leq c/2$ simply guarantees the intersection of the two trivial triangles generated by (r, s) and (s, r).

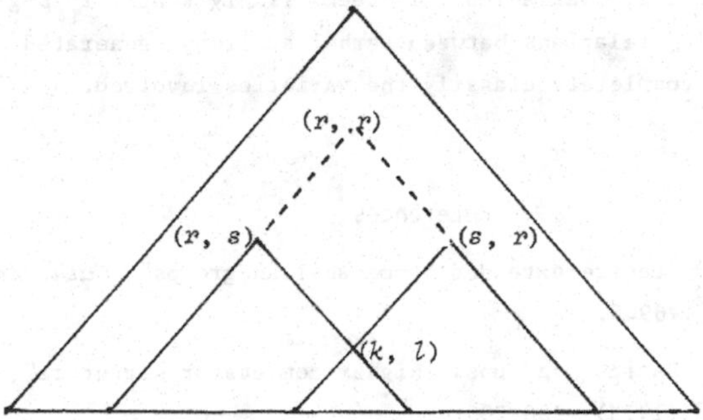

The pair immediately above the intersection, say (k, l) is either trivial or else $k + l = c$. The inequality $r \leq s \leq c/2$ rules out $k + l = c$ and so we eventually obtain the trivial triangle (r, r). In the case of triples (G_r, G_s, G_t) we can construct the analogous tetrahedron and argue similarly. The differences between these two cases are equivalent to the differences between a triangle and a tetrahedron and between the symmetry $(G_i, G_j) = (G_j, G_i)$ and that given by Hall's "3-subgroup" lemma.

The relation $F_d = \mho(F_r, F_s)$, $r + s = d$, $r, s \geq 2$ holds for all groups in $\underline{\underline{U}}$ and so must be equivalent to a law of $\underline{\underline{U}}$. The following theorem gives this law explicitly without any restriction on the derived structure of the group.

THEOREM 2. *Let* G *be a* p*-group of class* $c < p$ *and let* $w = (x, \underbrace{y, \ldots, y}_{c-1})$. G *satisfies the law* $w = 1$ *if and only if* $H_c = \mho(H_r, H_s)$, $r + s = c$, $r, s \geq 2$ *for all* $H \in \operatorname{var} G$.

(The proof of this result may be found in [3].)

All of the above suggests that certain varieties can be completely described by relations similar to $F_d = \mho(F_r, F_s)$, or, in general, of the form,

$$u(G) = v(G)$$

with u and v words or sets of words. It should be noted that such a relation on the group G is (in general) not varietal. Even if we insist that $u(H) = v(H)$ for all subgroups H of G as well, then it still may not be varietal. On the other hand many varieties, for example the varieties discussed by Stewart, can be completely described by such relations between verbal subgroups in a non-trivial way; that is, without either $u(G)$ or $v(G)$ always being trivial. The advantage of such a procedure is that only very restricted classes of words need be considered, a luxury not afforded by a description via laws. Thus the words used in Theorem 1 are single commutators, of various "shapes", which do not involve the troublesome property of multiple appearances of their variables. It seems likely that for p-groups of small class, for example, relations between verbal subgroups generated by such words might be sufficient to completely classify the varieties involved.

References

[1] A.G.R. Stewart, "On centre-extended-by-metabelian groups", *Math. Ann.* 185 (1970), 285-302. MR41#6945.

[2] Paul M. Weichsel, "Basic p-groups: higher commutator structure", *J. Austral. Math. Soc.* 16 (1973), 200-206.

[3] Paul M. Weichsel, "On Engel-like congruences", submitted.

University of Illinois,
Urbana, Illinois 61801, USA.

PROC. SECOND INTERNAT. CONF. THEORY OF GROUPS, CANBERRA, 1973, pp. 709-716.

20E30

A RELATION IN FREE PRODUCTS

M.J. Wicks

1. Introduction

A comparative study which contrasts the theory of free groups with the theory of free products has some general interest. One area where comparison can sometimes be made in an elementary way concerns relations, and their consequences: for example, the consequences of the relation $XY = YX$. It has been known for some years that, in a free group, this relation is a consequence of the relation of $X^2Y^2Z^2 = 1$ [1]. We now consider the analogous situation for free products.

Let G be a free product (without amalgamation) of a family of groups $\{G_p\}$. (Since the nature of the factorisation is not further specified we could assume that it is irreducible, and that each G_p is a free factor.) The elements of G are finite sequences of elements from the factors; such a sequence is a *word*. The number of terms in the sequence is the *length* of the word; for any word W, the length of W is denoted by $|W|$. A word of length 1 is a *letter*. The inverse of a word is obtained in the usual (combinatorial) way.

A word W is *fully reduced* if every part AB of W is such that A and B are non-trivial letters of distinct factors. (We shall use the same notation for a letter and the corresponding element in the factor.) A fully reduced word is *inert* if its initial and final letters are from distinct factors. Words V and W are *identical*, and we denote this fact by $V \equiv W$, if they have the same length and corresponding letters are equal (as elements of the factor to which they belong). The relation of equality in G, or in any one of the factors, is denoted in the usual way.

Let U be a word and U_0 be any one of its inert conjugates. An easy argument shows that if $|U_0| > 1$, then U equals a square if and only if the identity $X^2 \equiv U_0$ has a solution for the variable X. A necessary and sufficient condition for the latter is that $|U_0|$ be even and the initial and final halves of U_0 are identical.

We shall investigate the relation $X^2Y^2Z^2 = 1$ by considering the equation

(1) $$X^2Y^2 = U ,$$

where X, Y are variables and U is an element of a free product G such that U equals a square. Relations of G that derive from relations which hold in the factors are of no interest in the present context. The issue can be avoided in a simple way. In the first place, we assume that if A, B, C are elements of a factor such that $A^2B^2 = C^2$, then $AB = BA$. Secondly, it is clear that the relation of $X^2Y^2Z^2 = 1$ will hold in diverse ways if G has elements of order 2 . Hence, we further assume that there are no elements of order 2 . This can be assured if it is assumed that for any element A of a factor if $A^2 = 1$, then $A = 1$. This property has the additional consequence that the only fully reduced solution of the identity $X \equiv X^{-1}$ in G is the trivial one in which X is empty.

The full result may be stated, in a modified notation, as

THEOREM. *Let G be a free product such that each factor of the product satisfies the following conditions*

(i) for any X , if $X^2 = 1$, then $X = 1$;

(ii) for any X, Y, Z , if $X^2Y^2Z^2 = 1$, then $XY = YX$.

Then G itself satisfies (ii).

2. The equation $X^2Y^2 = U$

A general solution of equation (1), with U arbitrary, has been obtained in [2]. There is no loss of generality for present purposes in assuming that U is inert. Then the essential facts may be stated as follows: every solution of (1) is of the form $\left(VX_1V^{-1}, VY_1V^{-1}\right)$, where $\mathsf{X}_1 = \left(X_1, Y_1\right)$ is such that

(i) X_1 is certainly *abelian*, that is, it generates an abelian subgroup of G , or

(ii) there is a *primitive* solution X_0 of (1) such that the subgroups generated by X_0 and X_1 are conjugate.

Of course, it may still be the case that the subgroup generated by X_0 is abelian. Indeed, some of the X_0 are obtained by solving (1) in a factor, so in the present case these are again abelian. The remaining X_0 are obtained from solutions of identities in a way we describe below. We shall show that all these solutions are likewise abelian when U is a square, and hence, the theorem is proved. We proceed now to give the details in this last case.

Each of the identities from which X_0 may be obtained is of the form $W_0(X) \equiv U$, where X is an n-tuple of variables and W_0 is a word in these variables. The number of variables depends on the case and the variables are of various kinds. They are included among the following: X, Y, Z which denote fully reduced words of G, non-empty unless the contrary is stated; A, B, C, D which denote letters of a factor G_p; and L, M, N which denote letters of a factor G_q. It is not assumed that p and q are distinct, but this is sometimes implied by the fact that $W_0(X)$ is an inert word. With a single exception that is noted, the letters of G_p (which occur in W_0) are either all non-trivial, or all are trivial. This convention applies without exception to the letters of G_q.

Not every solution of an identity yields a primitive solution of (1). Necessary and sufficient conditions for this to be so are referred to as *consistency* conditions. The list which follows enumerates all the possibilities.

	W_0	Consistency Condition	Primitive Solution
I.	$AXBX$	$C^2 = AB^{-1}$	(C, BX)
II.	$AX^{-1}LX$	$L = M^2,\ A = B^2$	$(XBX^{-1}, M),\ X = 1$
		$L = BAB^{-1}$	(BX^{-1}, XAB^{-1}) or $(AX^{-1}B, B^{-1}X)$
III.	$AXBYCXDY^{-1}$	$BA = DC$	$(AC^{-1}Y^{-1}, YCXB),\ B, C = 1$ or $(XDY^{-1}A, A^{-1}YC)$
IV.	$AY^{-1}LXMXNY$	$A = B^2,\ M = NL$	$(YBY^{-1}, LXN),\ Y = 1$
V.	$AXBYCYDX$	$DA^{-1} = CB^{-1}$	$(AX, BYCB^{-1})$
VI.	$AXLYBZMYCXNZ^{-1}$	$C = BA,\ L = NM$	$(AXNZ^{-1}, ZMYB),\ Z = 1$ or $(CXLM^{-1}Z^{-1}B^{-1}, BZM)$ or $(CXLBZM, M^{-1}Z^{-1}B^{-1})$, $Y = 1$
VII.	$XAZ^{-1}LYMYNZBXC$	$C = AB,\ M = NL$	$(ZBXAZ^{-1}, LYN),\ Z = 1$

Some further explanation is needed. In the first place, the notation of [2] has been changed and some of the cases have been simplified. In some of the cases there are letters in the consistency condition which do not occur in W_0 ; for example, C in case I. The meaning of this is that the new letter satisfies the given condition; thus, in case I, $X = C$ is a solution of $X^2 = AB^{-1}$. The qualification which (sometimes) follows a primitive solution means that solutions satisfying the qualification are allowed; for example, in case II it is allowed that X may be empty. (Note then that either p and q are distinct, or all the letters are trivial.) The single exception to the convention for letter variables occurs in case III, the first alternative: either B or C , but not both, may be trivial.

We now consider the identities in greater detail for the case that U is a square, and dispose of the easier ones. The exception under case III cannot occur since it would imply that $|U|$ is odd. In all the other cases there are words W_1, W_2 such that $W_0 \equiv W_1W_2$ and $|W_1| = |W_2|$. Then, with U a square, there will be a solution of

$$W_1 \equiv W_2$$

if the original identity has a solution. There is an obvious converse.

The derived identity in case I is $AX \equiv BX$. It follows that $A = B$, and hence, that $C = 1$. The primitive solution is abelian.

In case II, we must have $X \equiv X^{-1}$, so X is empty. Thus AL is a (fully reduced) square only if A or L is trivial. Thus, in the first alternative under this case, either B or M is trivial, while in the second both A and L are trivial. The primitive solution is abelian in all these alternatives.

In case III, Y is empty and $A = C$, so it follows from the consistency condition that $B = D$. In any event, one component of X_0 is empty and the primitive solution is again abelian.

The remaining cases need a little more effort. The solutions can be made to depend on a few basic results. These are established in the lemmas of the next section.

3. Some simple identities

Since the results are of a general character we suspend the convention that the variables such as X denote non-empty, fully reduced words.

LEMMA 1. *If* A, B *are letters and* X, Y *are words such that* $XAY \equiv YBX$ *, then* $A = B$.

PROOF. It is clear, by considering lengths, that if one of A or B is trivial, then both are trivial. Also, if X and Y are of the same length, then they are identical and $A = B$.

There is an obvious symmetry which allows us to assume in the remaining cases that X is shorter than Y. If Y is a letter, then X is empty and we have $A = Y = B$.

We proceed in the general case by induction. It follows by comparing initial segments that there is a word Y' such that $Y \equiv XAY'$. Since A is non-trivial, we have $|Y'| < |Y|$. Moreover, substituting in the original identity and cancelling shows that $XAY' \equiv Y'BX$; the induction hypothesis applies to either X or Y' and the result follows.

LEMMA 2. *If* A *is a letter and* X, Y *are words such that* $XAY \equiv YAX$, *then there are integers* m, n *and a word* S *such that* $X \equiv (SA)^m S$ *and* $Y \equiv (SA)^n S$.

PROOF. The result is well known if A is trivial. Also, if X and Y have the same length, they are identical and we may take $m = n = 0$ and $S \equiv X$. The remaining cases follow by an induction similar to that in the previous lemma.

LEMMA 3. *If* X, Y *are words such that* $XY \equiv Y^{-1}X$, *then* Y *is empty.*

PROOF. If $|X| \le |Y|$, then, comparing initial and final segments, it follows that $Y \equiv Y_1X$ and $Y^{-1} \equiv XY_2^{-1}$. Hence, $Y_1X \equiv Y_2X^{-1}$ and X is empty. Then also, $Y \equiv Y^{-1}$ and Y is empty.

If the lemma is not true, there would be X of shortest length, which satisfies the identity, and which is such that $0 < |Y| < |X|$. It would follow that X is of the form X_1Y and $Y^{-1}X_2$, with X_1 shorter than X. Substituting and cancelling shows that X_1 and X_2 are identical, and thus, $X_1Y \equiv Y^{-1}X_1$. This is impossible.

A more general version of the preceding lemma is

LEMMA 4. *If* A, B *are non-trivial letters and* X, Y *are words such that* $XAY \equiv Y^{-1}BX$, *then there is an integer* m *such that either*

(i) $A = B$, $X \equiv A^m$, $Y \equiv 1$, *or*

(ii) $A = B^{-1}$, $X \equiv (AA^{-1})^m$, $Y \equiv A^{-1}$.

PROOF. If X is not longer than Y, then it follows as above that X is empty and we have $AY \equiv Y^{-1}B$. If Y is empty, we have alternative *(i)* with $m = 0$. If Y is non-empty, a simple argument shows that we have *(ii)* with $m = 0$.

The case $0 \le |Y| < |X|$ goes by induction. If Y is empty, which includes the case that X is a letter, we have $XA \equiv BX$, so Lemmas 1, 2 apply to show that *(i)* holds. In the general case we see that X is of the form X_1AY and $Y^{-1}BX_2$. It follows that X_1 and X_2 are identical and $|X_1| < |X|$. If X_1 is not longer than Y , we have the earlier case and *(i)* or *(ii)* holds with $m = 1$. Otherwise, the induction hypothesis applies and the result is easily verified.

LEMMA 5. *If A, B are letters and X, Y, Z are words such that $XAY \equiv ZBX$, then either*

(i) $Y \equiv Z$ *and Lemmas* 1, 2 *apply, or*

(ii) *there is an integer m and words S, T such that $X \equiv S(ATBS)^m$,*
$Y \equiv TBS$, $Z \equiv SAT$.

PROOF. If X and Y have the same length, then X, Y, Z are identical and *(i)* holds.

If $|X| < |Y|$, it follows that $Y \equiv Y'BX$ and $Z \equiv XAZ'$. Then Y' and Z' are identical and *(ii)* holds with $m = 0$, $S \equiv X$, $T \equiv Y'$.

Now suppose that $|X| > |Y|$. If Y is empty, so is Z and *(i)* holds. If not, we have X in the form X_1AY and ZBX_2 . It follows in the usual way that $X_1 \equiv X_2$. If X_1 is longer than Y , the induction hypothesis applies (even if A is trivial) and the result easily follows. Otherwise, the earlier argument applies to X_1 and the result is easily verified.

It may be noted that the lemmas have obvious converses, so that in each case the parametric solution is the general solution.

4. The analysis concluded

It remains now to consider the identities which result from cases IV - VII of Section 2.

In case IV we have $A = M$, so that $p = q$, and $XNY \equiv Y^{-1}LX$. If the letters are trivial, Lemma 3 applies to show that Y is empty. Since $B^2 = 1$, B also is trivial and X_0 is abelian. If the letters were not trivial, Lemma 4 would apply. Under alternative *(i)*, Y is empty and since $p = q$, W_0 would not be inert. In *(ii)*, we would have $L = N^{-1}$, and hence, $M = 1$ from the consistency condition. Neither of these cases can occur.

In case V, we have $A = C$ and $XBY \equiv YDX$. So $B = D$ by Lemma 1 and the consistency condition takes the form $BA^{-1} = AB^{-1}$. There are no elements of order 2 , so $A = B$ and all the letters are equal. This, and Lemma 2 which also applies, shows that X_0 is again abelian.

Case VI is quite simple. We have $A = M$, so that $p = q$; and $Z \equiv Z^{-1}$, so

Z is empty. It follows that all the letters are trivial, and hence, that $XY \equiv YX$. The primitive solutions are abelian.

The final case under VII is the most complicated. We have $C = M$, so that $p = q$, and if Z is empty, all the letters are trivial. We also have $XY \equiv YX$ so X_0 is abelian in this case. We shall show that Z must be empty.

The identity derived from VII is

$$XAZ^{-1}LY \equiv YNZBX . \tag{2}$$

If X and Y have the same length, they are identical and Z must be empty. Otherwise, symmetry allows us to assume that X is longer than Y and we have X in the form X_1LY and YNX_2. Hence, we may use Lemma 5 for the identity which follows from these two forms of X. For the alternative $X_1 \equiv X_2$, Lemma 1 gives that $L = N$, and then Lemma 2 shows that there are m, n and S such that $X_1 \equiv (SL)^mS$ and $Y \equiv (SL)^mS$. However, substituting directly in (2) gives $X_1AZ^{-1} \equiv ZBX_1$. If Lemma 3 or Lemma 4 *(i)* applies, then Z is empty. Lemma 4 *(ii)* implies that $A = B^{-1}$, and the consistency condition requires then that $C = 1$, hence all the letters are trivial. It follows again from Lemma 3 that Z is empty.

We consider now the consequences of Lemma 5 *(ii)*. There will be m, S and T such that

$$X_1 \equiv SNT , \quad X_2 \equiv TLS , \quad Y \equiv S(NTLS)^m .$$

Substituting in (2) and cancelling, we have

$$TLSAZ^{-1} \equiv ZBSNT . \tag{3}$$

If T and Z have the same length, it is immediate that Z is empty.

Now suppose that $|T| < |Z|$. It follows that Z is of the form TLZ_2, and Z^{-1} is $Z_1^{-1}NT$. Then T is empty, $L = N^{-1}$ and $Z_1 \equiv Z_2$. It follows from the consistency condition that $M = 1$, and hence, all the letters are trivial. Lemma 3 applies to (3) showing that Z is empty.

Finally, we have the case in which $|Z| < |T|$. Then T is of the form ZBT_2 and T_1AZ^{-1}. If T_1 and T_2 are identical, Lemmas 3 and 4 apply to show that Z is empty. Otherwise, substituting in (3) and cancelling, we find that $T_2LS \equiv SNT_1$. Lemma 5 *(ii)* applies and we have

$$T_1 \equiv T_0LS_0 \ , \quad T_2 \equiv S_0NT_0 \ , \quad S \equiv S_0(NT_0LS_0)^n \ .$$

Substituting in the expressions for T we have

$$T_0LS_0AZ^{-1} \equiv ZBS_0NT_0 \ .$$

This is (3) with S, T replaced by S_0, T_0. Since T is ZBS_0NT_0, it follows that T_0 is shorter than T if Z is not empty. Thus, a contradiction would follow from the last assumption if it is further assumed that T is the shortest word to satisfy (3). This completes the case.

We have shown that the primitive solution X_0 is always abelian and the Theorem is proved.

References

[1] R.C. Lyndon, "The equation $a^2b^2 = c^2$ in free groups", *Michigan Math. J.* 6 (1959), 89-95. MR21#1999.

[2] M.J. Wicks, "The equation $X^2Y^2 = g$ over free products", *Proc. Second Congress Singapore Nat. Acad. Science* (Singapore, 1971), to appear.

University of Singapore,
Singapore 10.

PROC. SECOND INTERNAT. CONF. THEORY OF GROUPS, 20E15
CANBERRA, 1973, pp. 717-718.

A NOTE ON SUBSOLUBLE GROUPS

John S. Wilson

A group is said to be hyperabelian if each of its non-trivial quotient groups has a non-trivial abelian normal subgroup, and subsoluble if each of its non-trivial quotient groups has a non-trivial abelian subnormal subgroup. In this note we settle a point raised by Robinson ([2], p. 87) by showing that subsoluble groups satisfying Min-n, the minimal condition for normal subgroups, need not be hyperabelian. More exactly, we construct a group G whose normal subgroups are well-ordered by inclusion, of order-type $\omega + 1$, having a perfect minimal normal subgroup N which is generated by its abelian normal subgroups, such that G/N is locally soluble and hyperabelian; G is obviously a group satisfying Min-n which is subsoluble but not hyperabelian. Our construction uses the notion of the treble product tower of a family of groups introduced in [1].

Let Λ be the field of real algebraic numbers, and let M be the McLain group defined with respect to Λ as linearly ordered set and (say) a finite prime field ([2], p. 14). The group X of order automorphisms of Λ,

$$x_{\alpha,\beta} ; \lambda \to \alpha\lambda + \beta ,$$

with $\alpha, \beta \in \Lambda$ and $\alpha > 0$, has the property that, for all λ_1, λ_2, μ_1 and μ_2 in Λ with $\lambda_1 < \lambda_2$ and $\mu_1 < \mu_2$, there is an $x \in X$ with $\lambda_1 x = \mu_1$ and $\lambda_2 x = \mu_2$. This property is sufficient to ensure that, if X operates on M in the natural way, the split extension MX of M by X has M as (unique) minimal normal subgroup (*cf.* the argument of Theorem 6.21 on p. 15 of [2]). Moreover X is itself a split extension of its minimal normal subgroup

$$B = \{x_{1,\beta} ; \beta \in \Lambda\}$$

by the subgroup

$$A = \{x_{\alpha,0} ; 0 < \alpha \in \Lambda\} .$$

We now consider a treble product tower

$$G = \mathrm{Trt}(MB, A_n;\ 0 < n < \omega) ,$$

with each group A_i isomorphic to A (and B) and each split extension $A_{n-1}A_n$ isomorphic to X. We take $A_1 = A$, so that $\langle MB, A_1 \rangle$ is just MX. By Lemma 7 of [1], $N = M^G$ is the unique minimal normal subgroup of G, and, being a direct power of M, N is perfect and is generated by abelian normal subgroups (Theorem 6.21 of [2]). Furthermore G/N is isomorphic to $\mathrm{Trt}(B, A_n;\ 0 < n < \omega)$, and this is a locally soluble hyperabelian group whose normal subgroups form an ascending tower of order-type $\omega + 1$ (*cf.* [1], Lemma 8). It follows that G has the properties claimed above.

References

[1] H. Heineken and J.S. Wilson, "Locally soluble groups with Min-n ", *J. Austral. Math. Soc.* (to appear).

[2] Derek J.S. Robinson, *Finiteness conditions and generalized soluble groups*, Part 2 (Ergebnisse der Mathematik und ihrer Grenzgebiete, Band 63. Springer-Verlag, Berlin, Heidelberg, New York, 1972). Zbl.243.20033.

University of Cambridge,
Cambridge CB2 1SB, England.

PROC. SECOND INTERNAT. CONF. THEORY OF GROUPS, 20E15
CANBERRA, 1973, pp. 719-729.

ON GENERALIZED SOLUBLE GROUPS

John S. Wilson

The work described here shows how some light is shed on two problems concerning $\overline{SI}$-groups and $\overline{SN}$-groups by consideration of analogous, but rather easier, problems in ring theory. In the case of the first problem, one can proceed quickly from a ring-theoretic counterexample to a group-theoretic one; in the case of the other, one gains enough insight to choose some very promising candidates for group-theoretic counterexamples and to see roughly what is entailed in establishing that they are actually counterexamples. More specific examples than those presented here, supported by detailed proofs, are to be found in [17] and [18]. It should perhaps be mentioned at this juncture that the second problem under discussion has also been solved by Noskov [14].

Before stating the problems, let me recall some definitions. A group is called an $\overline{SI}$-group if all of its chief factors are abelian and an $\overline{SN}$-group if all of its composition factors are abelian. Because the nomenclature is not universal, I would like to stress that I mean by a composition factor of a group an absolutely simple quotient group of a serial subgroup. H is a serial subgroup of G if there is a chain C of subgroups, closed with respect to unions and intersections, with $H \in C$ and $G \in C$, and having the property that every subgroup in C which has a successor in C is normal in its successor. An absolutely simple group is a group having no serial subgroups other than itself and the trivial subgroup.

For the purposes of this discussion, it will be enough to keep some special cases of serial subgroups in mind. First, subnormal subgroups are obviously serial subgroups. It is rather easy to show that arbitrary intersections of subnormal subgroups are serial subgroups; these, for reasons which will emerge later, I shall call closed subgroups. Finally, normal subgroups of closed subgroups are serial subgroups. Concerning absolute simplicity, we should remember that absolute simplicity implies simplicity, and that, while the converse implication does not hold, finite (and even finitely generated) simple groups are absolutely simple.

The questions that I want to discuss were raised in Problems VIII and X of the Kuroš and Černikov survey article [9] on generalized soluble and nilpotent groups, and

I will quote these problems in full.

PROBLEM VIII. *Does either of the properties* $\overline{SI}$ *or* $\overline{SN}$ *imply the other?*

PROBLEM X. *Is either of the properties* $\overline{SI}$ *or* $\overline{SN}$ *inherited by subgroups?*

There are now well established techniques for constructing $\overline{SN}$-groups which are not $\overline{SI}$-groups, using for example wreath powers or McLain groups (*cf.* Hall [3], Theorem 4), and I do not want to say any more about these, except to mention the fact, proved by Hall in [4], that there exist perfect simple $\overline{SN}$-groups. To answer Problem VIII negatively, it remains to show that *not every* $\overline{SI}$*-group is an* $\overline{SN}$*-group*.

Part of Problem X has been known for a long time to have a negative answer. Examples were given independently by Merzljakov [12] and Hall [5] of $\overline{SI}$-groups with non-abelian free subgroups. Such subgroups are obviously neither $\overline{SI}$-groups nor $\overline{SN}$-groups, because their finite perfect simple factor groups are both chief factors and composition factors. For each prime p we write Z_p for the ring of p-adic integers and A_p for the intersection of Z_p with the field of rationals; thus A_p is the ring of all rationals with denominators coprime to p. Merzljakov's example was the group of all elements of $GL_2(A_2)$ congruent to the identity matrix modulo the maximal ideal $2A_2$ of A_2; Hall's examples included the group of all elements of $SL_2(Z_p)$ congruent to the identity matrix modulo the maximal ideal pZ_p of Z_p, for p odd. Merzljakov and Hall showed that these groups are $\overline{SI}$-groups. But each of these groups has as a subgroup a group generated by a pair of matrices

$$\begin{pmatrix} 1 & m \\ 0 & 1 \end{pmatrix} \text{ and } \begin{pmatrix} 1 & 0 \\ m & 1 \end{pmatrix}$$

with m an integer greater than 1, and any such group is free, by a result of Brenner [2]. So subgroups of $\overline{SI}$-groups need not be $\overline{SI}$-groups, and to complete the negative solution of Problem X, it is left to show that *subgroups of* $\overline{SN}$*-groups are not necessarily* $\overline{SN}$*-groups*.

Merzljakov's and Hall's examples have much in common, both being congruence subgroups of classical groups over discrete valuation rings, and they seem to invite generalization. Of course generalizations exist, but before becoming involved with them it is worth asking why the examples should be so similar. It had to be shown somehow that these groups are $\overline{SI}$-groups. The only way I can think of for verifying that a group is an $\overline{SI}$-group, unless it is so as a result of a much stronger condition like local solubility (which would be of no use in this context) or by virtue of some general construction, is to locate all, or a large proportion, of its normal subgroups. This is just what Merzljakov and Hall did with their groups. They showed that the non-central normal subgroups all contain terms of the lower central series and that the groups are residually nilpotent. Their results are reminiscent of the classifications of normal subgroups of classical groups over rings obtained by a number of authors (for example, Klingenberg [7], [8], Mennicke [10], [11], Bass [1], Chapter 5, and Wilson [16]). The classifications that I have in mind, when valid, assert that if R is a ring and G a classical group over R, then each ideal I of R determines a largest normal subgroup L_I and a smallest normal subgroup S_I, with L_I and S_I not very far apart, such that every normal subgroup H of G satisfies $S_I \leq H \leq L_I$ for a uniquely determined I. Such a classification implies that the normal subgroup structure of the group G reflects quite accurately the ideal structure of the underlying ring R. This reflection of ideal structure in normal subgroup structure becomes rather blurred if one considers instead of G a normal subgroup K of G, because more subgroups of K will be normal in K than are normal in G. But, at least, in the cases considered by Merzljakov and Hall, non-nilpotent (central extensions of) residually nilpotent groups with all proper homomorphic images nilpotent were obtained from rings with the corresponding properties. I mean of course the rings $2A_2$ and pZ_p, and not A_2 and Z_p; it is really these with which Merzljakov's and Hall's results were concerned. If we choose our definitions correctly, these rings will be $\overline{SI}$-rings with subrings which are not $\overline{SI}$-rings.

We call a ring R an $\overline{SI}$-ring if $I^2 = 0$ whenever I is a minimal two-sided ideal of a homomorphic image of R. Serial subrings may be defined by letting the notion of being a two-sided ideal replace the notion of normality in the definition of serial subgroups; particular cases of serial subrings are subideals, and intersections of subideals, which we call closed subrings. We can continue to define absolutely simple rings, composition factors of rings and $\overline{SN}$-rings in an obvious fashion. With these definitions, one can verify that division rings are absolutely simple rings, and that, for commutative rings, the properties $\overline{SI}$, $\overline{SN}$ and the property of being a Jacobson radical ring all coincide. Thus the rings $2A_2$ and pZ_p, which are well known to be Jacobson radical rings, are $\overline{SI}$-rings; on the other hand, their intersections with the ring Z of rational integers have finite prime fields as homomorphic images, and therefore are not $\overline{SI}$-rings.

$\overline{SI}$-rings which are not $\overline{SN}$-rings will necessarily be non-commutative. Examples may be constructed as follows. Let F be any field of characteristic zero. To F we adjoin transcendentals x_i $(i \in Z)$, to obtain a field L, and we let s be the field automorphism of L fixing F and satisfying $x_i^s = x_{i+1}$ for all $i \in Z$. We let P be the set of all polynomials $\sum_{i=0}^{n} s^i k_i$ with coefficients k_i in L, and define addition in P in the obvious manner and multiplication using the relation $ks = sk^s$ for all $k \in L$. Then P becomes an associative ring of a type whose structure is well known (*cf.* Jacobson [6], Chapter 3). The non-trivial proper two-sided ideals of P are just the powers of the maximal ideal

(1) $$R = sP .$$

From this it follows that R has all proper homomorphic images nilpotent; further R is non-nilpotent and residually nilpotent. In particular, R is an $\overline{SI}$-ring.

But the set

(2) $$S = sF[s]$$

of all polynomials over F divisible by s is a subring of R, and moreover it is easily verified to be a closed subring of R. The multiplication in S is of course the usual multiplication of polynomials. Since $S/(s-1)S$ is isomorphic to F, S is not an $\overline{SI}$-ring and R is not an $\overline{SN}$-ring.

$S \le R$	$G_\Omega(S) \le G_\Omega(R)$	
$S \triangleleft R$	$G_\Omega(S) \triangleleft G_\Omega(R)$	
S cl R	$G_\Omega(S)$ cl $G_\Omega(R)$	
S ser R	$G_\Omega(S)$ ser $G_\Omega(R)$	
R an $\overline{SI}$-ring	$G_\Omega(R)$ an $\overline{SI}$-group	Ω infinite, *or* $\lvert\Omega\rvert \ge 3$ and R commutative
R has a division ring D as homomorphic image	$G_\Omega(R)'$ has an absolutely simple non-abelian homomorphic image	$\lvert\Omega\rvert \ge 3$, *or* D a field with char $D \ne 2, 3$
R nilpotent	$G_\Omega(R)$ nilpotent	
$\cap R^n = 0$	$G_\Omega(R)$ residually nilpotent	
R non-nilpotent; every ideal $\ne 0$ contains a power of R	$G = G_\Omega(R)$ non-nilpotent; H sn G and $[G, H] \ne 1$ iff $G/\cap\{H^g;\ g \in G\}$ nilpotent	Ω infinite, *or* $\lvert\Omega\rvert \ge 3$ and R commutative, *or* char $R \ne 2$ and R commutative

To pass back from rings to groups, we consider groups of matrices. Primarily for computational reasons, it is convenient to consider groups of arbitrary degree and not only groups of degree 2. Let Ω be any linearly ordered set with more than one element, and let R be any associative ring. Let R_1 be a ring obtained by adjoining an identity to R if R does not already have one, and let $R_1 = R$ otherwise. We write $G_\Omega(R)$ for the group of all invertible $\Omega \times \Omega$ matrices $p = (p_{uv})$ over R_1 such that $p_{uv} - \delta_{uv}$ is in R for all u and v and is non-trivial for only finitely many pairs (u, v). So Merzljakov's example mentioned above is $G_\Omega(2A_2)$ with $\Omega = \{1, 2\}$. The table above shows what happens to the properties of interest to us when the functor $G_\Omega(-)$ is applied. In each case, an assertion on the left implies an assertion in the middle column, subject possibly to a proviso on the right.

The notations X cl Y, X ser Y and X sn Y mean that X is a closed, serial or subnormal subgroup or subring of Y as the case may be. Some assertions in the table are completely trivial. Others require some work, and the provisos have been chosen more to give easier proofs than to give best possible results. The most difficult calculation is that involved in establishing the final implication in the case $|\Omega| = 2$; for this, the method of Lemma 1 of Hall [5] may be used (*cf.* Proposition 1 of [18]). It may be worth remarking that the assertion concerning division rings as homomorphic images amounts to an assertion that certain projective special linear groups over division rings are *absolutely* simple.

If we now appeal to the table, taking Ω infinite, and consider the rings R and S defined in (1) and (2), we see immediately that $G_\Omega(R)$ is an $\overline{SI}$-group which is not an $\overline{SN}$-group.

We can in fact say a little more than that $\overline{SI}$-groups are not necessarily $\overline{SN}$-groups. There are two reasons why the situation at the bottom of the table, in which R is a non-nilpotent, residually nilpotent ring all of whose proper homomorphic images are nilpotent, is a particular interesting one; on the one hand, we obtain groups which, in one sense, are as near as possible to being nilpotent, and are therefore very special $\overline{SI}$-groups. On the other hand, we can make R into a Hausdorff topological ring by taking the non-trivial ideals as a base of neighbourhoods of 0, and we can make $G_\Omega(R)$ into a Hausdorff topological group in either of two obvious ways - by taking the non-central subnormal subgroups as a base of neighbourhoods of 1 (the resulting topology we call the S-topology), or by using the topology on R to define a corresponding topology (the U-topology) of uniform convergence of matrices. The point is that these topologies are very closely related if any of the provisos in the bottom left hand corner of the table is satisfied. More precisely, U-closed subgroups are S-closed, and the two subspace topologies induced on the intersection of $G_\Omega(R)$ with the appropriate special linear group actually coincide; further the notions of closed subrings and closed subgroups introduced earlier are the same in this configuration as the topological notions of closed subrings and S-closed subgroups.

Using these ideas, I will sketch a proof of

THEOREM 1. *Let* α *be an infinite cardinal. There is a residually nilpotent group* G *of cardinality* α *, all of whose proper homomorphic images are nilpotent, which has a closed subgroup whose derived group is free of rank* α *.*

This theorem shows that the requirement that a group be an $\overline{SI}$-group does not place any restriction at all on its composition factors, and further that every group H is a homomorphic image of a serial subgroup of an $\overline{SI}$-group of cardinality $\max\{|H|, \aleph_0\}$. On the other hand, the only groups which can be homomorphic images of serial subgroups of $\overline{SN}$-groups are $\overline{SN}$-groups.

The group G of Theorem 1 is a group $G_\Omega(R)$, with Ω countably infinite, and with R as defined in (1), where the field F is chosen of cardinality α . From the results in the table, G is residually nilpotent and (since its centre is trivial) has all proper homomorphic images nilpotent; and naturally G has cardinality α . With $S = sF[s]$ as defined in (2) and with x and y distinct elements of Ω , we consider the group of all elements $p = (p_{uv})$ of G satisfying

$$p_{uv} - \delta_{uv} \in S \text{ for } \{u, v\} \subseteq \{x, y\}$$

and

$$p_{uv} - \delta_{uv} = 0 \text{ otherwise.}$$

This is a closed subgroup and is isomorphic to the subgroup K of $GL_2(F[s])$ of matrices congruent to the identity matrix modulo S . What we need to show is that K' is a free group of rank α . Let A and B be the subgroups of $GL_2(F[s])$ generated respectively by all elements of the form

$$\begin{pmatrix} 1 & f \\ 0 & 1 \end{pmatrix}$$

with $f \in S$, and all elements of the form

$$\begin{pmatrix} 1 & 0 \\ g & 1 \end{pmatrix}$$

with $g \in F[s]$. By Theorem 3 of Nagao [13], the subgroup E generated by A and B is isomorphic to the free product of A and B ; therefore E' is a free group. But an easy calculation shows that E is the group of matrices

$$\begin{pmatrix} a & b \\ c & d \end{pmatrix}$$

of $GL_2(F[s])$ with $a - 1, d - 1$ and b in S . Thus $E' \leq K \leq E$. So K' , having cardinality α , is a free group of rank α , as required.

We now leave $\overline{SI}$-groups which are not $\overline{SN}$-groups, and turn to the other problem - that of finding an $\overline{SN}$-group with subgroups which are not $\overline{SN}$-groups. To decide whether groups were $\overline{SI}$-groups or not, we obtained information about normal subgroups of groups $G_\Omega(R)$ using the fact that the functor $G_\Omega(-)$ takes some properties of the ideal structure of a ring into corresponding properties of the normal structure of a group. It is not clear what happens to the serial structure of a ring under the application of $G_\Omega(-)$; in particular, we do not know whether, under fairly general conditions, $\overline{SN}$-rings are taken to $\overline{SN}$-groups. We resort to the rather pedestrian approach of pinning down serial subgroups as closely as possible. We shall consider a group $G_\Omega(M)$ with M a residually nilpotent, non-nilpotent ring all of whose proper homomorphic images are nilpotent, because we know at least from the identification of the subnormal subgroups given in the table above that any composition factor H/K with H subnormal must be abelian. If we choose distinct elements x and y of Ω , then the subgroup of matrices of $G_\Omega(M)$ differing from the identity matrix only in the intersections of the rows and columns indexed by x and y is a closed subgroup H , and a classification of the serial subgroups of $G_\Omega(M)$ would require a classification of the serial subgroups of this subgroup. So we begin with $|\Omega| = 2$, and write $G_2(M)$ instead of $G_\Omega(M)$. The proviso in our table applicable to this configuration makes us choose M commutative. If S ser M , then $G_2(S)$ ser $G_2(M)$; we only want serial subgroups like $G_2(S)$ to arise when they must, that is, when $S = 0$ or $S \geq M^r$ for some integer r . Even for M a maximal ideal of a discrete valuation ring V it happens only rarely that the only non-trivial closed subrings of M are the subrings containing a power of M . However it does happen if V is the intersection of the ring Z_p of p-adic integers with a subfield of its field of fractions, for some prime p . So we are now back where we started, studying groups like those of Merzljakov and Hall.

Eighteen months ago, I proved that the non-subnormal serial subgroups of $G_2(pA_p) \cap \mathrm{SL}_2(A_p)$ are all metabelian, for each prime p . (From this and the fact that all soluble subgroups of $\mathrm{GL}_2(Q)$ are metabelian it follows easily that the non-subnormal serial subgroups of $G_2(pA_p)$ are metabelian, for each p .) Independently Noskov characterized the subnormal subgroups of $G_2(2A_2)$ and showed that this group has all non-subnormal serial subgroups soluble. So these groups are all $\overline{SN}$-groups with non-abelian free subgroups, and it follows from both my result and from Noskov's that *subgroups of* $\overline{SN}$*-groups are not necessarily* $\overline{SN}$*-groups*.

The proofs of these results fall into three parts. First, the subnormal subgroups are identified. We have seen from the table above that the characterization of the subnormal subgroups (as central or containing a term of the lower central series) holds in very much greater generality. Next, the closed subgroups are studied; the deduction that the non-subnormal ones are metabelian holds if pA_p is replaced by any ring $M = pV$ with V an intersection of Z_p with a subfield of its field of fractions. The final step from closed subgroups to serial subgroups is an extremely easy one, and it also generalizes without difficulty to the case when pA_p is replaced by M as above, provided that M consists entirely of algebraic numbers. There are $2^{\aleph_0}$ isomorphism types for such rings M. These generalizations lead to

THEOREM 2. *Let p be any prime. There is a set X of* $2^{\aleph_0}$ *pairwise non-isomorphic countable linear groups G with the following properties:*

(a) every proper homomorphic image of G is a finite p-group; further every non-trivial subnormal subgroup of G has p-power index in G,

(b) the set of isomorphism classes of proper homomorphic images of G is independent of the choice of $G \in X$,

(c) every non-subnormal serial subgroup of G is metabelian, and

(d) G has non-abelian free subgroups.

All of the groups of Theorem 2 are of course $\overline{SN}$-groups (and $\overline{SI}$-groups) with non-abelian free subgroups. Using our methods, we cannot exclude metabelian non-abelian serial subgroups; they arise, for instance, as subgroups of upper triangular matrices. I would like to mention in passing that, though the celebrated theorem of Tits [15] implies that all linear groups which are not soluble by locally finite have non-abelian free subgroups, the examples under discussion show that they need not have non-abelian free *serial* subgroups. Theorem 2 shows that there are (exactly) $2^{\aleph_0}$ isomorphism classes of countable $\overline{SN}$-groups with subgroups which are not $\overline{SN}$-groups. This can also be seen directly once one group with these properties has been found, because direct products of $\overline{SN}$-groups and abelian groups are $\overline{SN}$-groups.

The most significant difference in the methods used in my paper [18] and Noskov's paper [14] on $\overline{SN}$-groups is in the analysis of the closed subgroups. My treatment is purely group-theoretic, but is unintuitive and rather intricate. It can be made less intricate by requiring a weaker characterization of closed subgroups of $G_2(pA_p) \cap SL_2(A_p)$ than is actually given in [18]. Noskov's argument (which, given the characterization of subnormal subgroups as central or containing a term of the lower central series, is available for $G_2(pA_p)$ for all p and not just $p = 2$) is very intuitive, but uses quite heavy machinery from the theory of Lie groups. I will sketch a group-theoretic argument, similar in spirit to the Lie-theoretic one, which however breaks down seriously for $p = 2$.

Let M be the maximal ideal of an intersection V of Z_p with some subfield of its field of fractions. For any commutative ring N we write $S_2(N)$ for the subgroup of matrices of $G_2(N)$ with determinant 1 . For every $g \in S_2(M)$ with $g \neq 1$ we can write $g = 1 + p^r x$ with x a 2×2 matrix over V with $x \not\equiv 0$ modulo M , for some integer r . Write $f(g)$ for the image of x in the ring of 2×2 matrices over V/M and write $f(1) = 0$. We denote by $[a, b]$ the Lie product of two 2×2 matrices a and b over V/M . For $p \neq 2$, we have

(a) $f(g^p) = f(g)$ for all $g \in S_2(M)$,

(b) $f(g^r) = rf(g)$ for $0 \leq r < p$ and all g ,

(c) $\operatorname{tr} f(g) = 0$ for all g ,

(d) either $[f(g), f(h)] = 0$ or $f([g, h]) = [f(g), f(h)]$, and

(e) if $[f(g), f(h)] = 0$, with $g \neq 1$ and $h \neq 1$, then $f(g) = rf(h)$ for some integer r with $0 < r < p$.

Taken together, (a), (b), (c) and (d) show that the image $f(H)$ of any subgroup H of $S_2(M)$ is a Lie subalgebra of the Lie algebra of 2×2 matrices over V/M with trace zero. If $\dim f(H) = 3$, then, using (a), there is an n such that for all $m \geq n$, H contains matrices congruent to

$$\begin{pmatrix} 1 & p^m \\ 0 & 1 \end{pmatrix}, \quad \begin{pmatrix} 1 & 0 \\ p^m & 1 \end{pmatrix} \quad \text{and} \quad \begin{pmatrix} 1+p^m & 0 \\ 0 & (1+p^m)^{-1} \end{pmatrix}$$

modulo M^{m+1} . Thus $HS_2(M^m) = HS_2(M^{m+1})$, and $S_2(M^n) \leq HS_2(M^m)$ for all $m \geq n$. So, from either the topological definition or the definition in terms of subnormal subgroups, if H is closed it contains $S_2(M^n)$, and hence a term of the lower central series of $G_2(M)$. If on the other hand $\dim f(H) \leq 2$, then it is not quite immediate but is not hard to show that $f(H') = [f(H), f(H)]$, so that $\dim f(H') \leq 1$. From this it follows that $H'/\left(H' \cap S_2(M^n)\right)$ is cyclic for each n , so that H' is abelian and H is metabelian.

In conclusion, I would like to raise some further related problems. Obviously any $\overline{SI}$-group is either an $\overline{SN}$-group or is not an $\overline{SN}$-group. Most of our argument has been concerned with deciding which of these possibilities arose for two particular kinds of $\overline{SI}$-group. The gap between the manageable types remains wide. For example

QUESTION 1. *Is* $G_2(pZ_p)$ *an* $\overline{SN}$-*group?*

More generally, one might ask

QUESTION 2. *Is a linear* $\overline{SI}$*-group necessarily an* $\overline{SN}$*-group?*

Our work above shows that closed subgroups of $\overline{SI}$-groups are not necessarily $\overline{SI}$-groups. However, there remains

QUESTION 3. *Are normal subgroups of* $\overline{SI}$*-groups necessarily* $\overline{SI}$*-groups?*

References

[1] Hyman Bass, *Algebraic K-theory* (W.A. Benjamin, New York, Amsterdam, 1968). MR40#2736.

[2] Joël Lee Brenner, "Quelque groupes libres de matrices", *C.R. Acad. Sci. Paris* 241 (1955), 1689-1691. MR17,824.

[3] P. Hall, "The Frattini subgroups of finitely generated groups", *Proc. London Math. Soc.* (3) 11 (1961), 327-352. MR23#A1718.

[4] P. Hall, "On non-strictly simple groups", *Proc. Cambridge Philos. Soc.* 59 (1963), 531-553. MR28#129.

[5] P. Hall, "A note on $\overline{SI}$-groups", *J. London Math. Soc.* 39 (1964), 338-344. MR29#1262.

[6] Nathan Jacobson, *The theory of rings* (Mathematical Surveys, 2. Amer. Math. Soc. New York, 1943). MR5,31.

[7] Wilhelm Klingenberg, "Linear groups over local rings", *Bull. Amer. Math. Soc.* 66 (1960), 294-296. MR22#11047.

[8] Wilhelm Klingenberg, "Orthogonal groups over local rings", *Bull. Amer. Math. Soc.* 67 (1961), 291-297. MR23#A1725.

[9] А.Г. Курош и С.Н. Черников [A.G. Kuroš and S.N. Černikov], "Разрешимые и нильпотентные группы" [Soluble and nilpotent groups], *Uspehi Mat. Nauk (NS)* 2 (1947) no 3 (19), 18-59; *Amer. Math. Soc. Transl.* 80 (1953); reprinted *Amer. Math. Soc. Transl.* (1) 1 (1962), 283-338. MR10,677.

[10] Jens L. Mennicke, "Finite factor groups of the unimodular group", *Ann. of Math.* (2) 81 (1965), 31-37. MR30#2083.

[11] J. Mennicke, "Zur Theorie der Siegelschen Modulgruppe", *Math. Ann.* 159 (1965), 115-129. MR31#5903.

[12] Ю.И. Мерзляков [Ju.I. Merzljakov], " К теории обобщенных разрешимых и обобщеиных нильпотентных групп" [On the theory of generalized solvable and generalized nilpotent groups], *Algebra i Logika Sem.* 2 no. 5 (1963), 29-36. MR28#4030.

[13] Hirosi Nagao, "On $GL_2(K[x])$", *J. Inst. Polytech. Osaka City Univ. Ser. A* 10 (1959), 117-121. MR22#5684.

[14] Г.А. Носков [G.A. Noskov], "Субнормальное строение конгруэнц-группы Мерзлякова", [Subnormal structure of the congruence groups of Merzlyakov], *Sibirski Mat. Ž.* 14 (1973), 68 -683.

[15] J. Tits, "Free subgroups in linear groups", *J. Algebra* 20 (1972), 250-270. Zbl.236.20032.

[16] J.S. Wilson, "The normal and subnormal structure of general linear groups", *Proc. Cambridge Philos. Soc.* 71 (1972), 163-177. MR45#398.

[17] J.S. Wilson, "An $\overline{SI}$-group which is not an $\overline{SN}$-group", *Bull. London Math. Soc.* 5 (1973), 192-196.

[18] J.S. Wilson, "$\overline{SN}$-groups with non-abelian free subgroups", *J. London Math. Soc.* (to appear).

University of Cambridge,
Cambridge CB2 1SB, England.

PROC. SECOND INTERNAT. CONF. THEORY OF GROUPS, CANBERRA, 1973, pp. 730-732. 20D05

FINITE GROUPS WITH SYLOW 2-SUBGROUPS OF TYPE THE ALTERNATING GROUP OF DEGREE SIXTEEN

Hiroyoshi Yamaki

A 2-group is said to be of *type* X if it is isomorphic to a Sylow 2-subgroup of the group X. If G is a group with a Sylow 2-subgroup S of type X, we say that G has the *involution fusion pattern* of X if for some isomorphism θ of S onto a Sylow 2-subgroup of X, two involutions a, b of S are conjugate in G if and only if the involutions $\theta(a)$, $\theta(b)$ of $\theta(S)$ are conjugate in X. Also we say that a group G is *fusion-simple* if $G = O^2(G)$ and $O(G) = Z(G) = 1$.

The following results have been proved:

THEOREM A. *Let G be a fusion-simple finite group with Sylow 2-subgroups of type A_{16}. Then one of the following holds:*

(1) $G \cong A_{16}$ *or* A_{17},

(2) $G \cong A_9 \cdot E_{256}$, *the split extension of an elementary abelian group* E_{256} *of order* 256 *by* A_9 *with the action afforded by the* 8*-dimensional irreducible* GF(2)*-representation, or*

(3) G *has the involution fusion pattern of* $\Omega_9(3)$.

Here $\Omega_9(3)$ denotes the orthogonal commutator group of degree 9 over the field of 3-elements and A_m the alternating group on m letters.

In the process of proving Theorem A we obtain the following characterization.

THEOREM B. *Let G be a finite group with Sylow 2-subgroups of type A_{16}. If G has the involution fusion pattern of A_{16}, then $G/O(G) \cong A_{16}$ or A_{17}.*

The proofs of Theorems A and B are obtained in the following way which appears to be rapidly becoming standard (*cf*. Gorenstein and Harada [5, 6], Solomon [9]). Let S be a Sylow 2-subgroup of G and A be the unique elementary abelian subgroup of S

of order 2^8. At first we show that the fusion of elements of S is controlled by $N_G(A)$ and $C_G(Z_2(S))$ where $Z_2(S)$ is the second center of S, using results of Alperin [1] and Goldschmidt [2] on conjugation family. Since S/A is of type A_8 and $Z^*(N_G(A)) = O(N_G(A))$ as above, the structure of $N_G(A)/C_G(A)$ is determined by theorems of Harada [7] and Gorenstein and Harada [5, 6]. Here we can prove that if A is strongly closed in S with respect to G, then $G = N_G(A) \cong A_9 \cdot E_{256}$ by a recent result of Goldschmidt [4] on 2-fusion. Thus we may assume that A is not strongly closed in S with respect to G and so some involution of $S - A$ is fused into A by an element of $C_G(Z_2(S))$. Then the fusion possibilities of involutions follow immediately in the standard way. In particular, we get that G has the involution fusion pattern of A_{16} or $\Omega_9(3)$. Characterization theorems of Gorenstein and Harada [5, 6] and Solomon [9] permit the determination of $C_G(a)/O(C_G(a))$ for all involutions a in S. Now O is an A-signalizer functor and a signalizer functor theorem [3] implies that $W_A = \langle O(C_G(a)) ; a \in A^{\#} \rangle$ has odd order. Since we may assume $O(G) = 1$, $N_G(W_A)$ is strongly embedded in G provided $W_A \neq 1$. Since G has more than one conjugacy class of involutions, $W_A = 1$ and $O(C_G(a)) = 1$. Now Kondo's characterization theorem [8] implies that $G \cong A_{16}$ or A_{17}.

References

[1] J.L. Alperin, "Sylow intersections and fusion", *J. Algebra* 6 (1967), 222-241. MR35#6748.

[2] David M. Goldschmidt, "A conjugation family for finite groups", *J. Algebra* 16 (1970), 138-142. MR41#5489.

[3] David M. Goldschmidt, "2-signalizer functors on finite groups", *J. Algebra* 21 (1972), 321-340.

[4] David M. Goldschmidt, "2-fusion in finite groups", submitted.

[5] Daniel Gorenstein and Koichiro Harada, "On finite groups with Sylow 2-subgroups of type A_n, $n = 8, 9, 10, 11$ ", *Math. Z.* 117 (1970), 207-238. MR43#2095.

[6] Daniel Gorenstein and Koichiro Harada, "Finite groups with Sylow 2-subgroups of type $PSp(4, q)$, q odd", *J. Fac. Sci. Univ. Tokyo Sect. I A Math.* (to appear).

[7] Koichiro Harada, "Finite simple groups whose Sylow 2-subgroups are of order 2^7 ", *J. Algebra* 14 (1970), 386-404. MR41#8515.

[8] Takeshi Kondo, "On the alternating groups, III", *J. Algebra* 14 (1970), 35-69. MR41#5478.

[9] Ronald Solomon, "Finite groups with Sylow 2-subgroups of type $\underline{\underline{A}}_{12}$ ", *J. Algebra* 24 (1973), 346-378.

Osaka University,
Toyonaka, Osaka 560, Japan.

PROC. SECOND INTERNAT. CONF. THEORY OF GROUPS, CANBERRA, 1973, pp. 733-740.

PROBLEM SECTION

edited by John Cossey

A number of problems were presented, by conference participants and others: they are collected below. The problems are presented more or less as received: some minor tidying up has been done, but no attempt has been made to impose any uniformity of presentation. Many of the papers in the proceedings also contain problems: in particular, see Baumslag's paper, "Some problems on one relator groups", and the report "Hanna Neumann's problems on varieties of groups" by Kovács and Newman.

EDWARD T. ORDMAN

In his paper "A topological proof of Grusko's Theorem on free products", *Math. Zeit.* 90 (1965), 1-8, John Stallings has a construction on "binding ties" which may be expressed algebraically as follows: let G be a free product of two subgroups A and B with a subgroup $C = A \cap B$ amalgamated. Let s be an element of A not lying in C. Then G is the free product of its subgroups A and $\mathrm{gp}(B, s)$, with $\mathrm{gp}(C, s)$ amalgamated.

By repeating this, it is clear that G may have two factorizations $A \underset{C}{*} B$ and $A' \underset{C'}{*} B'$ with $A \subseteq A'$, $B \subseteq B'$ and $C \leq C'$. Call the first factorization *finer* than the second. Some groups have a minimal factorization: for example, if C is the center of G, there can be no finer factorization.

PROBLEM. *Can there be a minimal factorization in any other case (that is, for C greater than the center of G)? Given a factorization, can one find a finer minimal factorization?*

For further examples and a limited amount of machinery which may be applicable, see my paper "Factoring a group as an amalgamated free product", *J. Austral. Math. Soc.*, to appear.

J.-P. SERRE

PROBLEM 1. *Call a group* G coherent *if every finitely generated subgroup is finitely presented. A free group, a nilpotent group, is coherent. It has recently been proved (G.P. Scott, J. London Math. Soc. (2) 6 (1973), 437-440) that the fundamental group of a 3-dimensional manifold is coherent. This implies, in particular, that, if* A *is the ring of integers of an imaginary quadratic field, the group* $SL_2(A)$ *is coherent. Does this extend to other "arithemtic" (or "S-arithmetic") groups? For instance are* $SL_3(Z)$ *and* $SL_2(Z[1/p])$ *coherent? More optimistically: is* $GL_n(Q)$ *coherent?*

[EDITOR'S NOTES. J.C. Lennox and J. Wiegold point out that Example 4.22 of Wehrfritz's book *Linear Groups* gives a negative answer for $GL_n(Q)$, $n \geq 4$.]

Several people have pointed out that the answer is negative for $SL_n(Z)$ for $n \geq 4$, since $SL_4(Z)$ contains a subgroup isomorphic to $F_{x,y} \times F_{x,y}$, the direct product of two free groups of rank 2 , and the subgroup of this generated by (x, x), $(y, 1)$ and $(1, y)$ is not finitely presented.

G. Baumslag also observes $SL_2(Z[1/n])$ is not coherent if n is divisible by distinct primes p, q : for the subgroup generated by $\begin{pmatrix} 1 & 1 \\ 0 & 1 \end{pmatrix}$ and $\begin{pmatrix} p/q & 0 \\ 0 & q/p \end{pmatrix}$ is not finitely presented.]

PROBLEM 2. *Let* $k[t]$ *be the polynomial ring over a finite field* k . *The groups* $SL_n(k[t])$, $n \geq 3$, $SL_n(k[t, t^{-1}])$, $n \geq 2$, *are finitely generated. Are they* finitely presented?

(I would guess they are not.)

PROBLEM 3 (attributed to Laudenbach). *Say that a group* G *can be* killed by one relation *if there exists* $r \in G$ *such that the smallest normal subgroup of* G *containing* r *is* G *itself.*

Let $G = G_1 * G_2$ *be a free product of two groups. Assume* G_1 *is infinite cyclic, and* $G_2 \neq \{1\}$. *The problem is to prove (or disprove) that* G *cannot be killed by one relation.* (The difficult case is when G_2 is equal to its commutator subgroup; hence the first unsolved case is $G_2 = A_5$.)

If one is more optimistic, one can raise the following question: let $G = G_1 * G_2$ as above, with G_1 infinite cyclic. Let $r \in G$ be an element whose image by $G \to G_1$ is a generator of G_1 and let R be the smallest normal subgroup of G containing r . Prove (or disprove) that $R \cap G_2 = \{1\}$.

[EDITOR'S NOTE: P.M. Neumann points out that Problem 3, is attributed to M. Kewaire in Magnus, Karrass and Solitar, *Combinatorial Group Theory*, p. 403.]

GERHARD ROSENBERGER

In mehreren Arbeiten (z.b. Andreadakis, S.: *Proc. London Math. Soc.* (3) 15 (1965); Bachmuth, S.: *Trans. Amer. Math. Soc.* 118 (1965) sowie 122 (1966) sowie 127 (1967); Chein, O.: *Comm. Pure Appl. Math.* 21 (1968) und Rosenberger, G.: *Math. Z.* 129 (1972)) werden u.a. nichttriviale Beispiele von Gruppen G mit endlichem Rang $q \geq 1$ und folgenden Eigenschaften behandelt:

(i) Jeder Automorphismus von G wird von einem Automorphismus der freien Gruppe F_q vom Rang q induziert und

(ii) Jeder Automorphismus der freien Gruppe F_q vom Rang q induziert einen Automorphismus von G .

PROBLEM. *Gibt es dann zu einem beliebigen minimalen Erzeugenden system von G einen Automorphismus von G , der dieses System in ein gegebenes minimales Erzeugendensystem von G überführt?*

S.A. MESKIN

PROBLEM 1. *Is* $\langle a, b, c, d : a^2 = b^2, b^5 = c^5, c^3 = d^3 \rangle$ *a one relator group?*

This group satisfies the necessary condition, but not the sufficient condition described in Meskin, Pietrowski and Steinberg, "One relator groups with center", *J. Austral. Math. Soc.*, to appear.

PROBLEM 2. *Let G be a non-cyclic one relator group with non-trivial center, such that $G/G' \ncong Z \times Z$. Does G have a (one-relator) presentation $G = \langle a, b : R \rangle$ such that for suitable m, n , $a^m = b^n = z$, where z generates the center of G ?*

PROBLEM 3. *Is the fixed point subgroup of a periodic automorphism of a free group a ("usually" trivial) free factor?*

The case $n = 2$ has been settled by the author.

S. BACHMUTH

PROBLEM 1. *Is Tit's Theorem true for division rings (that is, is every subgroup of a matrix group over a division ring either soluble-by-finite or containing a non-cyclic free subgroup)?*

PROBLEM 2. *Does every insoluble automorphism group of a free soluble group contain a non-cyclic free subgroup?*

W.D. WALLIS

Perfect factorizations

For the background to this and the next two problems, see [2].

Let G be a group of order $2n + 1$; write H for $G\backslash\{1\}$. A *starter* S in G is a partition of H into unordered pairs $(x_1, y_1), (x_2, y_2), \ldots, (x_n, y_n)$ such that

$$H = \{x_1, x_2, \ldots, x_n, y_1, y_2, \ldots, y_n\}$$
$$= \left\{x_1y_1^{-1}, y_1x_1^{-1}, x_2y_2^{-1}, y_2x_2^{-1}, \ldots, y_nx_n^{-1}\right\} .$$

If K is a finite graph, a *one-factor* in K is a set of edges of K which between them contain each vertex of K precisely once; a *one-factorization* is a set of pairwise edge-disjoint one-factors whose union is K . A starter in G gives rise to a one-factorization of the complete graph whose vertices are the elements of G and a special element ∞ : the typical one-factor has edges

$$\{\infty, g\}, \{x_1g, y_1g\}, \{x_2g, y_2g\}, \ldots, \{x_ng, y_ng\}$$

where g is any element of G .

Anderson [1] calls a one-factorization of a complete graph *perfect* if the union of any two of the factors is a Hamiltonian cycle in the graph.

PROBLEM 1. *Is it true that (perhaps with finitely many exceptions) every group of odd order admits of a starter whose associated one-factorization is perfect?*

[1] B.A. Anderson, "Finite topologies and Hamiltonian circuits", *J. Combinatorial Theory*, 14B (1973), 87-93.

[2] W.D. Wallis, Anne Penfold Street, Jennifer Seberry Wallis, *Combinatorics: Room Squares, Sum-free Sets, Hadamard Matrices* (Lecture Notes in Mathematics, 292. Springer-Verlag, Berlin, Heidelberg, New York, 1972).

Room squares

Suppose S is a starter in G as in the preceding question. An *adder* for S is an ordered set

$$a_1, a_2, \ldots, a_n$$

of n elements of H such that

$$\{x_1a_1, x_2a_2, \ldots, x_na_n, y_1a_1, \ldots, y_na_n\} = H .$$

Given a starter and an adder, one can construct a Room Square of order $2n + 1$ as follows: label the rows and column of a square array of side $2n + 1$ with the elements of G ; given g and h , position (g, h) is left empty if hg^{-1} is not in the adder, but if $hg^{-1} = a_i$ then position (g, h) contains the pair $\{x_ih, y_ih\}$.

PROBLEM 2. *(i) Is there a starter with an adder in all but finitely many abelian groups?* (There are none in Z_3, Z_5 or $Z_3 \times Z_3$.)

(ii) Find starters and adders in non-abelian groups.

(iii) Find starters and adders which lead to special room squares (skew squares, squares with subsquares, Howell rotations) of new orders.

Strong starters

Continue the above notation. A starter in an abelian group is *strong* if the n objects x_iy_i are all different, and none equals 1 . A strong starter always has an adder, namely $a_i = (x_iy_i)^{-1}$.

PROBLEM 3. *Find reasonable analogues of strong starters, which "supply" their own adders, for non-abelian groups.*

E.E. SHULT.

PROBLEM 1. *Let* (G, Ω) *be a finite group of permutations on the set* Ω . *Let* K *be a union of classes of involutions in* G . *We say that* G has the unique transposition property with respect to K , *if for every pair of letters,* (α, β) , *there exists exactly one involution in* K *transposing* α *and* β . *If each involution in* K *fixes no letter, show that all members of* K *commute with one another.*

The conclusion is easily seen to hold if (G, Ω) is triply transitive. Although one might expect to obtain a proof for this when (G, Ω) is doubly transitive, induction can be used more often if the problem is proposed as above.

PROBLEM 2. *Let* (G, Ω) *be a doubly transitive group. If* x *and* x^g *both lie in* G_α *, it is easily seen that* $g = nh$ *, where* $h \in G_\alpha$ *and* $n \in N_G(G_{\alpha\beta})$ *for some* β *in* $\Omega\setminus\{\alpha\}$ *. Thus* G_α *controls its own fusion if* $N_G(G_{\alpha\beta})$ *is isomorphic to* $G_{\alpha\beta} \times Z_2$ *. In the case that* G *is triply transitive with this property, show that* G *is either the symmetric group, or* G *contains a normal regular 2-group.*

If G is $Sp(2n, 2)$ in either of its doubly transitive representations, or if G is the Higman-Sims group acting doubly transitively on 176 letters, then the normaliser of $G_{\alpha\beta}$ in G has the shape $G_{\alpha\beta} \times Z_2$. Indeed this is so for any unique transposition group (see Problem 1). Incidentally, since the Higman-Sims group is not a unique transposition group, the converse of the last remark fails.

PROBLEM 3. *Let* (P, B) *be a doubly transitive block design with* $\lambda = 1$ *. If there is an odd number of points in each block, it follows from results of O'Nan, Shult, and Aschbacher that if* $G = \mathrm{Aut}(P, B)$ *contains an involution fixing only one letter, then* G *contains a normal subgroup which is either an elementary abelian regular subgroup, or is* $PSU(3, 2^n)$ *for some* n *. What is* G *if* G *contains an involution whose fixed points are a block?*

If the block size is three, G is essentially known from results of M. Hall, Jr., and Shult on Steiner systems.

PROBLEM 4. *Classify all doubly transitive groups* G *in which* $N_G(G_{\alpha\beta})$ *is isomorphic to* $G_{\alpha\beta} \times Z_2$.

This generalises both Problem 1 and Problem 2.

PROBLEM 5. *Let* G *be doubly transitive, and suppose* $G_{\alpha\beta}$ *fixes exactly* $k > 2$ *letters. Show that one of the following holds:*

(i) $k = 3$, *G is a subgroup of* $GL(n, 2)$ *acting doubly transitively on the non-zero vectors, or* G *contains a regular normal 3-subgroup,*

(ii) $k = 4$, *G is* $S\Gamma L(2, 8)$ *on 28 letters, or* G *has a regular normal 2-subgroup,*

(iii) $k > 4$, *and* G *contains a regular normal elementary abelian subgroup.*

PROBLEM 6. *G is a doubly transitive group containing a normal subgroup* N *of index* p . *Then show either*

(i) N *is also doubly transitive,*

(ii) G *contains a regular normal subgroup (for example, many solvable groups appear here satisfying the hypotheses on* G *), or*

(iii) $G \cong S\Gamma L(2, 8)$ *and* $p = 3$.

PROBLEM 7. *Let* (G, Ω) *be triply transitive. Let* $L = \mathrm{Aut}(G)$ *, and* $H = N_L(G_\alpha) \cdot G$ *, so that* H *is that part of* L *which is still triply transitive. Show that* H/G *has* 2-*rank one.* (This result easily replaces the Schreier conjecture in showing any 2-fold transitive group is the alternating or symmetric group.) *Does this hold with* (G, Ω) *only doubly transitive?*

D. GORENSTEIN

There are two specific problem areas about simple groups that I can suggest for discussion. Both of these deal with simple groups of characteristic 2 type: that is, in which 2-local subgroups are 2-constrained and have trivial cores and in which $SCN_3(2)$ is nonempty. So all the discussion below pertains to such a simple group G .

A. In the N-group paper and again in his analysis of 3'-groups, Thompson had to study the following situation (actually slightly more general one, but this is good enough for the present purposes):

(a) Some 2-local subgroup of G has p-rank at least 3 for some odd prime.

(b) If q is an odd prime for which $SCN_3(q)$ is nonempty, then G possesses a strongly q-embedded subgroup - that is, a subgroup which contains a Sylow q-subgroup of G and contains $N_G(Q)$ for every q-subgroup $Q \neq 1$ of M .

Under these circumstances one must derive a contradiction - no simple groups should have such subgroups. The way Thompson does is to prove eventually that M is strongly embedded in the ordinary sense if M is determined from the prime p satisfying (a). The process by which he achieves this is very elaborate and has some subproblems as interesting special cases. Incidentally it is all right to assume that for the prime p , M is, in fact, a 2-local subgroup - one can add that to the assumptions if one wishes. Then one of the interesting special cases he has to consider is when $O_2(M)$ is of symplectic type. It looks as though many of Thompson's arguments carry through in part under much more general circumstances, but serious new problems arise in connection with composition factors of M which are of Lie type over fields of characteristic 2 . Progress on any facet of this problem area would be valuable.

B. Characterizations of groups of Lie type over fields of characteristic 2 in terms of the structure of the centralizers of elements of order three analogous to the characterizations that already exist for groups of odd characteristic in terms of the structure of the centralizers of involutions. If $H = C_G(z)$, z an element of order 3, an entirely new problem arises here that didn't occur before because we have no analogue of the Glauberman Z^*-theorem, so a priori there is no reason why z might not be an isolated element of order 3. So this is a subproblem. The real characterization begins when z is not isolated. We have in mind here "large" groups - that is, groups in which the Sylow 3-subgroup has large rank. In the contrary case, special difficulties will arise analogous to the situation for groups of low 2-rank. But in the large 3-rank cases, it looks as though much of the kind of arguments that worked for involutions will carry over for elements of order 3 and one will be able to construct (B, N)-pairs. This problem has a subproblem related to Problem A. At some point in the argument, one may manage to construct a strongly 3-embedded subgroup M in G and one would like to derive a contradiction from this. But as we have no analogue of Bender's Theorem for the prime 3, one will be forced to considerations of the type that arise in Problem A.

Vol. 215: P. Antonelli, D. Burghelea and P. J. Kahn, The Concordance-Homotopy Groups of Geometric Automorphism Groups. X, 140 pages. 1971. DM 16,-

Vol. 216: H. Maaß, Siegel's Modular Forms and Dirichlet Series. VII, 328 pages. 1971. DM 20,-

Vol. 217: T. J. Jech, Lectures in Set Theory with Particular Emphasis on the Method of Forcing. V, 137 pages. 1971. DM 16,-

Vol. 218: C. P. Schnorr, Zufälligkeit und Wahrscheinlichkeit. IV, 212 Seiten. 1971. DM 20,-

Vol. 219: N. L. Alling and N. Greenleaf, Foundations of the Theory of Klein Surfaces. IX, 117 pages. 1971. DM 16,-

Vol. 220: W. A. Coppel, Disconjugacy. V, 148 pages. 1971. DM 16,-

Vol. 221: P. Gabriel und F. Ulmer, Lokal präsentierbare Kategorien. V, 200 Seiten. 1971. DM 18,-

Vol. 222: C. Meghea, Compactification des Espaces Harmoniques. III, 108 pages. 1971. DM 16,-

Vol. 223: U. Felgner, Models of ZF-Set Theory. VI, 173 pages. 1971. DM 16,-

Vol. 224: Revètements Etales et Groupe Fondamental. (SGA 1). Dirigé par A. Grothendieck XXII, 447 pages. 1971. DM 30,-

Vol. 225: Théorie des Intersections et Théorème de Riemann-Roch. (SGA 6). Dirigé par P. Berthelot, A. Grothendieck et L. Illusie. XII, 700 pages. 1971. DM 40,-

Vol. 226: Seminar on Potential Theory, II. Edited by H. Bauer. IV, 170 pages. 1971. DM 18,-

Vol. 227: H. L. Montgomery, Topics in Multiplicative Number Theory. IX, 178 pages. 1971. DM 18,-

Vol. 228: Conference on Applications of Numerical Analysis. Edited by J. Ll. Morris. X, 358 pages. 1971. DM 26,-

Vol. 229: J. Väisälä, Lectures on n-Dimensional Quasiconformal Mappings. XIV, 144 pages. 1971. DM 16,-

Vol. 230: L. Waelbroeck, Topological Vector Spaces and Algebras. VII, 158 pages. 1971. DM 16,-

Vol. 231: H. Reiter, L^1-Algebras and Segal Algebras. XI, 113 pages. 1971. DM 16,-

Vol. 232: T. H. Ganelius, Tauberian Remainder Theorems. VI, 75 pages. 1971. DM 16,-

Vol. 233: C. P. Tsokos and W. J. Padgott. Random Integral Equations with Applications to stochastic Systems. VII, 174 pages. 1971. DM 18,-

Vol. 234: A. Andreotti and W. Stoll. Analytic and Algebraic Dependence of Meromorphic Functions. III, 390 pages. 1971. DM 26,-

Vol. 235: Global Differentiable Dynamics. Edited by O. Hájek, A. J. Lohwater, and R. McCann. X, 140 pages. 1971. DM 16,-

Vol. 236: M. Barr, P. A. Grillet, and D. H. van Osdol. Exact Categories and Categories of Sheaves. VII, 239 pages. 1971. DM 20,-

Vol. 237: B. Stenström, Rings and Modules of Quotients. VII, 136 pages. 1971. DM 16,-

Vol. 238: Der kanonische Modul eines Cohen-Macaulay-Rings. Herausgegeben von Jürgen Herzog und Ernst Kunz. VI, 103 Seiten. 1971. DM 16,-

Vol. 239: L. Illusie, Complexe Cotangent et Déformations I. XV, 355 pages. 1971. DM 26,-

Vol. 240: A. Kerber, Representations of Permutation Groups I. VII, 192 pages. 1971. DM 18,-

Vol. 241: S. Kaneyuki, Homogeneous Bounded Domains and Siegel Domains. V, 89 pages. 1971. DM 16,-

Vol. 242: R. R. Coifman et G. Weiss, Analyse Harmonique Non-Commutative sur Certains Espaces. V, 160 pages. 1971. DM 16,-

Vol. 243: Japan-United States Seminar on Ordinary Differential and Functional Equations. Edited by M. Urabe. VIII, 332 pages. 1971. DM 26,-

Vol. 244: Séminaire Bourbaki - vol. 1970/71. Exposés 382-399. IV, 356 pages. 1971. DM 26,-

Vol. 245: D. E. Cohen, Groups of Cohomological Dimension One. V, 99 pages. 1972. DM 16,-

Vol. 246: Lectures on Rings and Modules. Tulane University Ring and Operator Theory Year, 1970-1971. Volume I. X, 661 pages. 1972. DM 40,-

Vol. 247: Lectures on Operator Algebras. Tulane University Ring and Operator Theory Year, 1970-1971. Volume II. XI, 786 pages. 1972. DM 40,-

Vol. 248: Lectures on the Applications of Sheaves to Ring Theory. Tulane University Ring and Operator Theory Year, 1970-1971. Volume III. VIII, 315 pages. 1971. DM 26,-

Vol. 249: Symposium on Algebraic Topology. Edited by P. J. Hilton. VII, 111 pages. 1971. DM 16,-

Vol. 250: B. Jónsson, Topics in Universal Algebra. VI, 220 pages. 1972. DM 20,-

Vol. 251: The Theory of Arithmetic Functions. Edited by A. A. Gioia and D. L. Goldsmith VI, 287 pages. 1972. DM 24,-

Vol. 252: D. A. Stone, Stratified Polyhedra. IX, 193 pages. 1972. DM 18,-

Vol. 253: V. Komkov, Optimal Control Theory for the Damping of Vibrations of Simple Elastic Systems. V, 240 pages. 1972. DM 20,-

Vol. 254: C. U. Jensen, Les Foncteurs Dérivés de lim et leurs Applications en Théorie des Modules. V, 103 pages. 1972. DM 16,-

Vol. 255: Conference in Mathematical Logic - London '70. Edited by W. Hodges. VIII, 351 pages. 1972. DM 26,-

Vol. 256: C. A. Berenstein and M. A. Dostal, Analytically Uniform Spaces and their Applications to Convolution Equations. VII, 130 pages. 1972. DM 16,-

Vol. 257: R. B. Holmes, A Course on Optimization and Best Approximation. VIII, 233 pages. 1972. DM 20,-

Vol. 258: Séminaire de Probabilités VI. Edited by P. A. Meyer. VI, 253 pages. 1972. DM 22,-

Vol. 259: N. Moulis, Structures de Fredholm sur les Variétés Hilbertiennes. V, 123 pages. 1972. DM 16,-

Vol. 260: R. Godement and H. Jacquet, Zeta Functions of Simple Algebras. IX, 188 pages. 1972. DM 18,-

Vol. 261: A. Guichardet, Symmetric Hilbert Spaces and Related Topics. V, 197 pages. 1972. DM 18,-

Vol. 262: H. G. Zimmer, Computational Problems, Methods, and Results in Algebraic Number Theory. V, 103 pages. 1972. DM 16,-

Vol. 263: T. Parthasarathy, Selection Theorems and their Applications. VII, 101 pages. 1972. DM 16,-

Vol. 264: W. Messing, The Crystals Associated to Barsotti-Tate Groups: With Applications to Abelian Schemes. III, 190 pages. 1972. DM 18,-

Vol. 265: N. Saavedra Rivano, Catégories Tannakiennes. II, 418 pages. 1972. DM 26,-

Vol. 266: Conference on Harmonic Analysis. Edited by D. Gulick and R. L. Lipsman. VI, 323 pages. 1972. DM 24,-

Vol. 267: Numerische Lösung nichtlinearer partieller Differential- und Integro-Differentialgleichungen. Herausgegeben von R. Ansorge und W. Törnig, VI, 339 Seiten. 1972. DM 26,-

Vol. 268: C. G. Simader, On Dirichlet's Boundary Value Problem. IV, 238 pages. 1972. DM 20,-

Vol. 269: Théorie des Topos et Cohomologie Etale des Schémas. (SGA 4). Dirigé par M. Artin, A. Grothendieck et J. L. Verdier. XIX, 525 pages. 1972. DM 50,-

Vol. 270: Théorie des Topos et Cohomologie Etale des Schémas. Tome 2. (SGA 4). Dirigé par M. Artin, A. Grothendieck et J. L. Verdier. V, 418 pages. 1972. DM 50,-

Vol. 271: J. P. May, The Geometry of Iterated Loop Spaces. IX, 175 pages. 1972. DM 18,-

Vol. 272: K. R. Parthasarathy and K. Schmidt, Positive Definite Kernels, Continuous Tensor Products, and Central Limit Theorems of Probability Theory. VI, 107 pages. 1972. DM 16,-

Vol. 273: U. Seip, Kompakt erzeugte Vektorräume und Analysis. IX, 119 Seiten. 1972. DM 16,-

Vol. 274: Toposes, Algebraic Geometry and Logic. Edited by. F. W. Lawvere. VI, 189 pages. 1972. DM 18,-

Vol. 275: Séminaire Pierre Lelong (Analyse) Année 1970-1971. VI, 181 pages. 1972. DM 18,-

Vol. 276: A. Borel, Représentations de Groupes Localement Compacts. V, 98 pages. 1972. DM 16,-

Vol. 277: Séminaire Banach. Edité par C. Houzel. VII, 229 pages. 1972. DM 20,-

Vol. 278: H. Jacquet, Automorphic Forms on GL(2). Part II. XIII, 142 pages. 1972. DM 16,–

Vol. 279: R. Bott, S. Gitler and I. M. James, Lectures on Algebraic and Differential Topology. V, 174 pages. 1972. DM 18,–

Vol. 280: Conference on the Theory of Ordinary and Partial Differential Equations. Edited by W. N. Everitt and B. D. Sleeman. XV, 367 pages. 1972. DM 26,–

Vol. 281: Coherence in Categories. Edited by S. Mac Lane. VII, 235 pages. 1972. DM 20,–

Vol. 282: W. Klingenberg und P. Flaschel, Riemannsche Hilbertmannigfaltigkeiten. Periodische Geodätische. VII, 211 Seiten. 1972. DM 20,–

Vol. 283: L. Illusie, Complexe Cotangent et Déformations II. VII, 304 pages. 1972. DM 24,–

Vol. 284: P. A. Meyer, Martingales and Stochastic Integrals I. VI, 89 pages. 1972. DM 16,–

Vol. 285: P. de la Harpe, Classical Banach-Lie Algebras and Banach-Lie Groups of Operators in Hilbert Space. III, 160 pages. 1972. DM 16,–

Vol. 286: S. Murakami, On Automorphisms of Siegel Domains. V, 95 pages. 1972. DM 16,–

Vol. 287: Hyperfunctions and Pseudo-Differential Equations. Edited by H. Komatsu. VII, 529 pages. 1973. DM 36,–

Vol. 288: Groupes de Monodromie en Géométrie Algébrique. (SGA 7 I). Dirigé par A. Grothendieck. IX, 523 pages. 1972. DM 50,–

Vol. 289: B. Fuglede, Finely Harmonic Functions. III, 188. 1972. DM 18,–

Vol. 290: D. B. Zagier, Equivariant Pontrjagin Classes and Applications to Orbit Spaces. IX, 130 pages. 1972. DM 16,–

Vol. 291: P. Orlik, Seifert Manifolds. VIII, 155 pages. 1972. DM 16,–

Vol. 292: W. D. Wallis, A. P. Street and J. S. Wallis, Combinatorics: Room Squares, Sum-Free Sets, Hadamard Matrices. V, 508 pages. 1972. DM 50,–

Vol. 293: R. A. DeVore, The Approximation of Continuous Functions by Positive Linear Operators. VIII, 289 pages. 1972. DM 24,–

Vol. 294: Stability of Stochastic Dynamical Systems. Edited by R. F. Curtain. IX, 332 pages. 1972. DM 26,–

Vol. 295: C. Dellacherie, Ensembles Analytiques, Capacités, Mesures de Hausdorff. XII, 123 pages. 1972. DM 16,–

Vol. 296: Probability and Information Theory II. Edited by M. Behara, K. Krickeberg and J. Wolfowitz. V, 223 pages. 1973. DM 20,–

Vol. 297: J. Garnett, Analytic Capacity and Measure. IV, 138 pages. 1972. DM 16,–

Vol. 298: Proceedings of the Second Conference on Compact Transformation Groups. Part 1. XIII, 453 pages. 1972. DM 32,–

Vol. 299: Proceedings of the Second Conference on Compact Transformation Groups. Part 2. XIV, 327 pages. 1972. DM 26,–

Vol. 300: P. Eymard, Moyennes Invariantes et Représentations Unitaires. II. 113 pages. 1972. DM 16,–

Vol. 301: F. Pittnauer, Vorlesungen über asymptotische Reihen. VI, 186 Seiten. 1972. DM 18,–

Vol. 302: M. Demazure, Lectures on p-Divisible Groups. V, 98 pages. 1972. DM 16,–

Vol. 303: Graph Theory and Applications. Edited by Y. Alavi, D. R. Lick and A. T. White. IX, 329 pages. 1972. DM 26,–

Vol. 304: A. K. Bousfield and D. M. Kan, Homotopy Limits, Completions and Localizations. V, 348 pages. 1972. DM 26,–

Vol. 305: Théorie des Topos et Cohomologie Etale des Schémas. Tome 3. (SGA 4). Dirigé par M. Artin, A. Grothendieck et J. L. Verdier. VI, 640 pages. 1973. DM 50,–

Vol. 306: H. Luckhardt, Extensional Gödel Functional Interpretation. VI, 161 pages. 1973. DM 18,–

Vol. 307: J. L. Bretagnolle, S. D. Chatterji et P.-A. Meyer, Ecole d'été de Probabilités: Processus Stochastiques. VI, 198 pages. 1973. DM 20,–

Vol. 308: D. Knutson, λ-Rings and the Representation Theory of the Symmetric Group. IV, 203 pages. 1973. DM 20,–

Vol. 309: D. H. Sattinger, Topics in Stability and Bifurcation Theory. VI, 190 pages. 1973. DM 18,–

Vol. 310: B. Iversen, Generic Local Structure of the Morphisms in Commutative Algebra. IV, 108 pages. 1973. DM 16,–

Vol. 311: Conference on Commutative Algebra. Edited by J. W. Brewer and E. A. Rutter. VII, 251 pages. 1973. DM 22,–

Vol. 312: Symposium on Ordinary Differential Equations. Edited by W. A. Harris, Jr. and Y. Sibuya. VIII, 204 pages. 1973. DM 22,–

Vol. 313: K. Jörgens and J. Weidmann, Spectral Properties of Hamiltonian Operators. III, 140 pages. 1973. DM 16,–

Vol. 314: M. Deuring, Lectures on the Theory of Algebraic Functions of One Variable. VI, 151 pages. 1973. DM 16,–

Vol. 315: K. Bichteler, Integration Theory (with Special Attention to Vector Measures). VI, 357 pages. 1973. DM 26,–

Vol. 316: Symposium on Non-Well-Posed Problems and Logarithmic Convexity. Edited by R. J. Knops. V, 176 pages. 1973. DM 18,–

Vol. 317: Séminaire Bourbaki – vol. 1971/72. Exposés 400–417. IV, 361 pages. 1973. DM 26,–

Vol. 318: Recent Advances in Topological Dynamics. Edited by A. Beck, VIII, 285 pages. 1973. DM 24,–

Vol. 319: Conference on Group Theory. Edited by R. W. Gatterdam and K. W. Weston. V, 188 pages. 1973. DM 18,–

Vol. 320: Modular Functions of One Variable I. Edited by W. Kuyk. V, 195 pages. 1973. DM 18,–

Vol. 321: Séminaire de Probabilités VII. Edité par P. A. Meyer. VI, 322 pages. 1973. DM 26,–

Vol. 322: Nonlinear Problems in the Physical Sciences and Biology. Edited by I. Stakgold, D. D. Joseph and D. H. Sattinger. VIII, 357 pages. 1973. DM 26,–

Vol. 323: J. L. Lions, Perturbations Singulières dans les Problèmes aux Limites et en Contrôle Optimal. XII, 645 pages. 1973. DM 42,–

Vol. 324: K. Kreith, Oscillation Theory. VI, 109 pages. 1973. DM 16,–

Vol. 325: Ch.-Ch. Chou, La Transformation de Fourier Complexe et L'Equation de Convolution. IX, 137 pages. 1973. DM 16,–

Vol. 326: A. Robert, Elliptic Curves. VIII, 264 pages. 1973. DM 22,–

Vol. 327: E. Matlis, 1-Dimensional Cohen-Macaulay Rings. XII, 157 pages. 1973. DM 18,–

Vol. 328: J. R. Büchi and D. Siefkes, The Monadic Second Order Theory of All Countable Ordinals. VI, 217 pages. 1973. DM 20,–

Vol. 329: W. Trebels, Multipliers for (C, α)-Bounded Fourier Expansions in Banach Spaces and Approximation Theory. VII, 103 pages. 1973. DM 16,–

Vol. 330: Proceedings of the Second Japan-USSR Symposium on Probability Theory. Edited by G. Maruyama and Yu. V. Prokhorov. VI, 550 pages. 1973. DM 36,–

Vol. 331: Summer School on Topological Vector Spaces. Edited by L. Waelbroeck. VI, 226 pages. 1973. DM 20,–

Vol. 332: Séminaire Pierre Lelong (Analyse) Année 1971-1972. V, 131 pages. 1973. DM 16,–

Vol. 333: Numerische, insbesondere approximationstheoretische Behandlung von Funktionalgleichungen. Herausgegeben von R. Ansorge und W. Törnig. VI, 296 Seiten. 1973. DM 24,–

Vol. 334: F. Schweiger, The Metrical Theory of Jacobi-Perron Algorithm. V, 111 pages. 1973. DM 16,–

Vol. 335: H. Huck, R. Roitzsch, U. Simon, W. Vortisch, R. Walden, B. Wegner und W. Wendland, Beweismethoden der Differentialgeometrie im Großen. IX, 159 Seiten. 1973. DM 18,–

Vol. 336: L'Analyse Harmonique dans le Domaine Complexe. Edité par E. J. Akutowicz. VIII, 169 pages. 1973. DM 18,–

Vol. 337: Cambridge Summer School in Mathematical Logic. Edited by A. R. D. Mathias and H. Rogers. IX, 660 pages. 1973. DM 42,–

Vol: 338: J. Lindenstrauss and L. Tzafriri, Classical Banach Spaces. IX, 243 pages. 1973. DM 22,–

Vol. 339: G. Kempf, F. Knudsen, D. Mumford and B. Saint-Donat, Toroidal Embeddings I. VIII, 209 pages. 1973. DM 20,–

Vol. 340: Groupes de Monodromie en Géométrie Algébrique. (SGA 7 II). Par P. Deligne et N. Katz. X, 438 pages. 1973. DM 40,–

Vol. 341: Algebraic K-Theory I, Higher K-Theories. Edited by H. Bass. XV, 335 pages. 1973. DM 26,–

Vol. 342: Algebraic K-Theory II, "Classical" Algebraic K-Theory, and Connections with Arithmetic. Edited by H. Bass. XV, 527 pages. 1973. DM 36,–